Lecture Notes in Artificial Intelligence   4200
Edited by J. G. Carbonell and J. Siekmann

Subseries of Lecture Notes in Computer Science

T0180855

Ian F. C. Smith (Ed.)

Intelligent Computing in Engineering and Architecture

13th EG-ICE Workshop 2006
Ascona, Switzerland, June 25-30, 2006
Revised Selected Papers

 Springer

Series Editors

Jaime G. Carbonell, Carnegie Mellon University, Pittsburgh, PA, USA
Jörg Siekmann, University of Saarland, Saarbrücken, Germany

Volume Editor

Ian F. C. Smith
École Polytechnique Fédérale de Lausanne
Station 18, GC-G1-507, 1015 Lausanne, Switzerland
E-mail: ian.smith@epfl.ch

Library of Congress Control Number: 2006934069

CR Subject Classification (1998): I.2, D.2, J.2, J.6, F.1-2, I.4, H.3-5

LNCS Sublibrary: SL 7 – Artificial Intelligence

ISSN   0302-9743
ISBN-10   3-540-46246-5 Springer Berlin Heidelberg New York
ISBN-13   978-3-540-46246-0 Springer Berlin Heidelberg New York

Springer is a part of Springer Science+Business Media

springer.com

© Springer-Verlag Berlin Heidelberg 2006
Printed in Germany

Typesetting: Camera-ready by author, data conversion by Scientific Publishing Services, Chennai, India
Printed on acid-free paper   SPIN: 11888598   06/3142   5 4 3 2 1 0

Preface

Providing computer support for tasks in civil engineering and architecture is hard. Projects can be complex, long and costly. Firms that contribute to design, construction and maintenance are often worth less than the value of their projects. Everyone in the field is justifiably risk adverse. Contextual variables have a strong influence making generalization difficult. The product life cycle may exceed one hundred years and functional requirements may evolve during the service life. It is therefore no wonder that practitioners in this area have been so reluctant to adopt advanced computing systems.

After decades of research and industrial pilot projects, advanced computing systems are now being recognized by many leading practitioners to be strategically important for the future profitability of firms involved in engineering and architecture. Engineers and architects with advanced computing knowledge are hired quickly in the market place. Closer collaboration between research and practice is leading to more comprehensive validation processes for new research ideas. This is feeding development of more useful systems, thus accelerating progress. These are exciting times.

This volume contains papers that were presented at the 13th Workshop of the European Group for Intelligent Computing in Engineering. Over five days, 70 participants from around the world listened to 59 paper presentations in a single session format. Attendance included nearly everyone on the Scientific Advisory Committee, several dynamic young faculty members and approximately ten doctoral students. The first paper is a summary of a panel session on the Joint International Conference on Computing and Decision Making in Civil and Building Engineering that finished in Montreal nine days earlier. The remaining papers are listed in alphabetical order of their first author.

Organizational work began with requests for availability and funding in September 2004. This was followed by tens of personal invitations to experts from around the world during 2005. I would like to thank the Organizing Committee, and particularly from January 2006 its Secretary, Prakash Kripakaran, for assistance with the countless details that are associated with running meetings and preparing proceedings. The meeting was sponsored primarily by the Swiss National Science Foundation and the Centro Stefano Franscini. Additional support was gratefully received from the Ecole Polytechnique Fédérale de Lausanne (EPFL), the Technical Council on Computing and Information Technology of the American Society of Civil Engineers and the Ecole de Technologie Supérieure, Montréal.

July 2006   Ian F.C. Smith

Committees

Organizing Committee

Ian Smith, Acting Chair
Martina Schnellenbach-Held, Chair(resigned)
Gerhard Schmitt, Vice-Chair
Prakash Kripakaran, Secretary
Bernard Adam
Suraj Ravindran
Sandro Saitta

EG-ICE Committee

Ian Smith, Chair
John Miles, Vice-Chair
Chimay Anumba, Past Chair
Yaqub Rafiq, SecretaryTreasurer

International Scientific Advisory Committee

Chimay Anumba, Loughborough University, UK
Thomas Arciszewski, George Mason University, USA
Claude Bédard, ETS, Canada
Karl Beucke, Bauhaus-Universität, Weimar, Germany
Adam Borkowski, Academy of Sciences, Poland
Bo-Christer Björk, Hanken, Finland
Moe Cheung, Hong Kong University of Science and Technology, China
Chuck Eastman, Georgia Tech, USA
Gregory Fenves, UC Berkeley, USA
Steven Fenves, NIST, USA
Martin Fischer, Stanford University, USA
Ian Flood, University of Florida, USA
Thomas Froese, University of British Columbia, Canada
Renate Fruchter, Stanford University, USA
James Garrett, CMU, USA
John Gero, University of Sydney, Australia
Donald Grierson, University of Waterloo, Canada
Patrick Hsieh, National Taiwan University, Taiwan
Raymond Issa, University of Florida, USA

Table of Contents

Outcomes of the Joint International Conference on Computing and Decision Making in Civil and Building Engineering, Montreal 2006

Ian F.C. Smith

Ecole Polytechnique Fédérale de Lausanne
Station 18, 1015 Lausanne, Switzerland
`Ian.smith@epfl.ch`

The Joint International Conference on Computing and Decision Making in Civil and Building Engineering in Montreal finished nine days before the beginning of the 13th Workshop of the European Group for Intelligent Computing in Engineering in Ascona. This provided an opportunity at the workshop to organize a panel session in Ascona to discuss outcomes after a week of retrospection. The Joint International Conference in Montreal was the first time that five leading international computing organizations (ASCE, ICCCBE, DMUCE, CIB-W78 and CIB-W102) joined forces to organize a joint conference. The meeting was attended by nearly 500 people from 40 countries around the world. It was unique in that this was the first time so many branches of computing in civil engineering were brought together.

It was also unique because the majority of participants in Montreal were not active researchers in computational mechanics, the traditional domain of computing in civil engineering. While a movement away from numerical analysis has long been predicted by leaders in the field, this was the first clear confirmation of a general trend across civil engineering. Numerical analysis methods indeed remain important for engineers and they are often found to be embedded within larger systems. However, this area now seems mature for most new computing applications and many innovative contributions are found elsewhere.

The panel started with brief presentations by five leaders in the field who attended the conference in Montreal in various capacities. All panelists actively combine innovative research and teaching with strategic organizational activity within international and national associations. Their contributions are summarized below.

Hugues Rivard, Professor, Ecole de Technologie Supérieure, University of Quebec, Canada; Co-Chair, Joint International Conference on Computing and Decision Making in Civil and Building Engineering, Montreal 2006

Researchers in the field of computing in civil engineering generally present their research at two types of venues: conferences on sub-disciplines of civil engineering, such as structures, and conferences on computing in civil engineering. For the former, computing in engineering is often a single track lost in a large conference while for the latter, the numbers of attendees have decreased over the past several years. One reason for the drop in attendance is that there are too many conferences being proposed to researchers (EG-ICE, ASCE, ICCCBE, CIB-W78, CIB-W102, ECPPM,

I.F.C. Smith (Ed.): EG-ICE 2006, LNAI 4200, pp. 1–6, 2006.

DCC, DMUCE, ACADIA, CAADFutures, etc.) They all serve a purpose, some of them are regional and some are focused on particular topics. This results in a very fragmented research community. A researcher cannot attend all the events. Therefore, less people attend individual conferences.

At the Montreal conference, a survey was performed at the opening ceremony and it was observed that a majority of the attendees present had not previously attended any of the five conference streams. This indicated that bringing these groups together had created a synergy that attracted large numbers of new people.

Karl Beucke, Prof., University of Weimar, Germany; Secretary, International Society for Computing in Civil and Building Engineering

The focus of IT applications in civil and building engineering has greatly expanded over the recent years and this could be noticed explicitly in the Montreal conference. If previously the focus was predominantly on buildings and structures, now the areas of construction and infrastructure are growing in importance. This development also has a strong influence on proposed technologies and methodologies.

In building and structures the scientific focus has shifted from structural analysis to aspects of data and process integration and to distributed cooperation between engineering teams. In integration there is still a major effort noticeable towards establishing Industry Foundation Classes as an industry standard - even though some frustration was voiced over the rate of progress. In distributed cooperation, a strong group from Germany presented results from a German Priority Program with a focus on cooperative product and process models and on agent technologies.

Construction, infrastructure and transportation accounted for considerably more contributions than building and structures. Major aspects of scientific value included 4D-modeling techniques, cost performance issues and the utilization of sensor technologies. The influence of web-based and mobile technologies has had a major impact over recent years and much effort is now concentrated in these areas.

Overall, participants benefited from a broad range of aspects presented in the conference and from large numbers of experts from different fields. If integrated civil engineering remains an important field, conferences, such as Montreal, that bring together experts from many fields will retain their value - even if this is at the expense of many parallel sessions.

Renate Fruchter, Director Project Based Learning Laboratory, Stanford University; Chair, Executive Committee, Technical Council on Computing and Information Technology, American Society of Civil Engineers

Renate Fruchter provided perspectives in two capacities – as a citizen of the community of *Intelligent Computing in Engineering* and as the ASCE TCCIT EXCOM Chair. The speaker noted that a majority of participants at this workshop were also present at Montreal and that since there were parallel sessions and group meetings, her review covers only the sessions in which she could participate and her insights are thus related only to these sessions.

What is new?
Being a researcher associated with the field since the 90's, it is exciting to see a new generation of young assistant professors bringing in young students, sharing new ideas and strengthening as well as refreshing the computing community in CEE.

What has not changed?
We are overloaded with so many events that it is hard to obtain good attendance at all of them. There is a need to look at how to *maintain* and *strengthen* our community. Computing is evolving from being state-of-the-art to state-of-the-practice; this warrants activities that synchronize and synergize various computing communities. The interesting aspect of the Montreal conference was that the turnout was good. Hence, it was a success in that it cross-fertilized five communities (ICCCBE, ASCE TCCIT, DMUCE-5, CIB-W78, CIB-W102) under the computing umbrella.

What are the areas that are gaining momentum?
The areas that are gaining momentum are 3-D and 4-D modelling as well as infrastructure sensing and monitoring. There has also been a growing focus on globalization; in both industry and education.

What areas are losing momentum?
An area of research that had been a regular in earlier times but now seems to have lost steam is the finite element method. Areas such as environmental engineering and hydrology have seen lower numbers of papers while there are more on sensing and monitoring.

What areas need more science?
Educational efforts have often been presented. Such efforts have been ad-hoc documentation of, say, a new e-learning tool. However, these lack the same scientific rigour as research results in other fields. What we need to do is build on learning theory through a scientific blueprint and see assessment metrics and methods that highlight the students' learning growth.

What was missing?
There were two things conspicuous by their absence – panel sessions to discuss key aspects and industry track sessions. The first allows us to engage in larger discourses on important topics while the second leads to presenting innovation and advances in the field and helps to attract industry participation.

What was the best message?
The best message was the proof of industrial implementation of academic research, e.g. 4D CAD. It was exciting to see researchers demonstrate and validate theories and research models on real industry test-beds. At the same time, slow penetration of new technologies in the real world is a truth. Of particular note was the keynote talk by Arto Kiviniemi who reflected on IFC development. The gist of his talk can be generalized and summarized through one of his statements: "the creation, implementation and deployment of standards has progressed too slowly and we must work towards accelerating the process". Aptly, this calls for action on the part of our communities to work towards implementation.

James H. Garrett, Jr., Professor and Head of Department, Carnegie Mellon University; Vice Chair, Technical Committee, International Association for Bridge and Structural Engineering

Based on Arto Kiviniemi's presentation, IFCs do not seem to be finding sufficient resources required for their continued upkeep. Despite all discussions related to this subject and 10 years of research, there is an alarming erosion of financial support. However, it is obvious that the use of BIM in design and construction information technology is gaining momentum. In Montreal, there were a large number of talks that addressed building information modelling technology.

There is a noticeable change in the way one does research in the field of computing in civil engineering. While in the past, research was more technology driven, it is now more driven by engineering problems and industrial needs. Research focused solely on demonstrating the abilities of technologies, such as genetic algorithms (GA) or artificial neural networks (ANN), appears to be less prevalent while research driven to address the details and intricacies of actual problems considering a variety of different possible approaches is becoming more the norm. Nowadays research aims at best supporting people who are working in the field. Moreover, a much broader scope for BIM than just design and construction planning is being explored. For example, several studies have more recently focused on infrastructure lifecycle phases, such as construction monitoring, operation and maintenance.

Few areas outside of architecture, engineering and construction (AEC) were discussed at the Montreal conference, although information and communication technology (ICT) can support many domains other than AEC. For example, cyber-infrastructure researchers in the civil engineering community (computationally intensive activities within other research areas such as earthquake engineering) were completely missing from the meeting even though there is significant IT involved in this effort. Moreover, few people from the environmental engineering community attended the conference. Research driven by the initiative known as Fully Integrated and Automated Technology (FIATECH) was hardly discussed during the conference.

To conclude on a positive note, it was recognized and discussed in a variety of contexts that ICT can play a major role in delivering sustainable built infrastructure. It was also accepted that data exchange standards will always be in a state of flux and that we must develop tools to support this evolution.

Donald E. Grierson, Professor, University of Waterloo, Canada; Co-Chair 17th Analysis and Computation Conference, St. Louis, 2006

Montreal was a very good conference from the research point of view. Examples of interesting research areas are activities in 4D, IFC, etc.

A lot of "angst" was detected while speaking with people in Montreal. It seems that researchers were concerned that their research work was not used in industry, their question being "why is our work not used in practice?". One answer is that the process of going from research to industry is a slow one. This is especially true in the construction industry.

Such slow process between research and industry is normal. Indeed, it has always taken a long time to adopt new technology. As research brings new ideas, it needs time to get accepted and incorporated in industry. Therefore, research made twenty years ago is now coming into practice. This is the way it has always been.

For example finite element modelling and optimization have taken a long time to appear in industry, starting in the 60's with the program STRESS (developed by S. Fenves who was present at this workshop), then with the first microcomputers used in the 70's. He concludes that good software finally appeared in the 90's. This is an example of the slowness of research adoption by the industry. Furthermore, he observes that final adoption of research is usually imperceptible.

The big question seems to be "what to do then?" Grierson thinks that one answer could be "nothing or just wait". He makes a comparison with trying to sell bibles a long time ago, even though nearly no one was able to read. It is the same for new research. Only 5-20% of companies use new research technologies. In general, other companies do not use ideas and results directly from the research community.

Discussion

A lively discussion followed these remarks. It was agreed that the Montreal meeting was a huge success and that the organisers should be congratulated for their work and their innovative efforts to bring diverse groups together. This was appreciated by all people who attended.

Working in the spirit that even outstanding events can be improved upon, many people provided suggestions for subsequent meetings. Issues such as the scientific quality of papers, objectives of conferences, conference organization, media for proceedings, the emergence of synergies and the importance of reviewing were evoked.

Although a detailed discussion is out of the scope of this paper, the following is a non exhaustive list of suggestions that were provided by the audience. Classify papers into categories according to the degree of industrial validation so that well validated proposals are distinguished from papers that contain initial ideas. Maintain high quality reviewing. Ensure open access to all documentation. Encourage links to concurrent industrial events. Foster bridge building between research and practice. Do not sacrifice opportunities for synergy and transmission of new ideas for the sake of paper quality. Maintain scientific quality even when papers are short. Allow films and other media in electronic proceedings. Investigate the possibility to relax page limits to add more science. Encourage key references in abstracts to see foundations of ideas. Ensure that contributions recognize and build on previous work. Include review criteria for authors in the call for papers.

While some ideas were relatively new, many suggestions reflected well established challenges that have always been associated with all large meetings. There are multiple objectives that require tradeoffs. The best point on the "Meeting Pareto front" is difficult to identify especially since it depends on so many contextual parameters. Indeed, there is much similarity between the challenges of conference organization and the challenges we face when providing computing support for civil engineering tasks. The panel session at Ascona turned out to be a very useful validation exercise for conference designs!

I would like to thank the panellists whose insightful contributions did much to ensure that the subsequent discussion was of high quality. Finally, thanks are due to P. Kripakaran, S. Saitta, B. Adam, H. Pelletier and S. Ravindran who helped record the contributions of the panellists.

Self-aware and Learning Structure

Bernard Adam and Ian F.C. Smith

Ecole Polytechnique Fédérale de Lausanne (EPFL),
Applied Computing and Mechanics Laboratory, Station 18, 1015 Lausanne,
Switzerland
{bernard.adam, ian.smith}@epfl.ch

Abstract. This study focuses on learning of control commands identification and load identification for active shape control of a tensegrity structure in situations of unknown loading event. Control commands are defined as sequences of contractions and elongations of active struts. Case-based reasoning strategies support learning. Simple retrieval and adaptation functions are proposed. They are derived from experimental results of load identification studies. The proposed algorithm leads to two types of learning: reduction of command identification time and increase of command quality over time. In the event of no retrieved case, load identification is performed. This methodology is based on measuring the response of the structure to current load and inferring the cause. It provides information in order to identify control commands through multi-objective search. Results are validated through experimental testing.

1 Introduction

Tensegrities are spatial, reticulated and lightweight structures. They are composed of compressed bars and tensioned cables and stabilized by self-stress states. When equipped with an active control system, they are attractive for shape control. This topology could to be used for structures such as temporary roofs, footbridges and antennas. Over the past decade, it has been established that active shape control of cable-strut structures is a complex task. While studying shape and stress control of prestressed truss structures, difficulties were identified in validating numerical results through experimental testing [1]. Most studies addressing tensegrity structure control involve only numerical simulation and only simple structures [2 – 6]. One of the few experimental results of tensegrity active control on a full-scale structure is presented in [7]. Shape control involves maintaining the top surface slope through changes in active strut length. Displacements are measured for three nodes at the edge of the top surface: 37, 43, 48 (Figure 1), in order to calculate the top surface slope value (1). Since there is no closed-form solution for active-strut movements, control commands are identified using a generate-test algorithm together with stochastic search and case-based reasoning [8]. However, the learning algorithm did not use information provided by sensor devices of the structure and while command identification time decrease with time, no enhancement of control quality over time was demonstrated. Moreover, in these studies, it was assumed that both load position and magnitude were known.

I.F.C. Smith (Ed.): EG-ICE 2006, LNAI 4200, pp. 7 – 14, 2006.

Case-based reasoning systems build on findings that previous experience is useful. Humans resolve new challenges by first searching in their memory for similar tasks that they have successfully carried out in the past. Retrieved solutions are adapted [9 – 10]. The core component of the CBR system is the case-base, where past experience is stored. It was stated in [11] that solutions of prior tasks are useful starting points for solving similar current tasks, and that case bases should contain cases similar to anticipated new tasks. The most widely used procedure to compute similarity between cases is the K-nearest neighbor according to the Euclidian distances. Nevertheless, some workers propose other methods such as Kernel methods [12] to measure distances. Adaptation methods are summarized in [13]. These methods can be divided into three approaches: substitution, transformation and derivational analogy. The method of case-based reasoning is widely used [14]; and several applications have been proposed for structural engineering design, for example [15 – 16].

System identification supports load identification. It involves determining the state of a system and values of key parameters through comparisons of predicted and observed responses [17]. It involves solving an inverse problem to infer causes from effects. In structural engineering, system identification can be divided into three subareas: damage identification [18], load identification [19] and structural property identification [20]. However, most studies are validated on simple structures. Only [21] and [22] tested their methodologies on real civil structures. Errors due to measurement precision and modeling assumptions influence results. Solutions are usually a set of good candidate models rather then one single solution [23]. In most of these studies, measured effects are dynamic responses due to perturbations. Moreover, they focus on passive health monitoring measurements and rehabilitation of traditional civil structures. There is therefore little relation to control tasks.

This paper addresses learning and load identification for shape control of an active tensegrity structure in situations of unknown loading event. Learning allows decreasing command identification time and improving command quality. In the event of no retrieved case, load identification and multi-objective control command search are performed. Learning and load identification methodologies are reviewed in the next section. The results section provides an experimental validation of the proposed methodologies.

2 Methodologies

Studies on self diagnosis [24] observed that multiple loads at several locations can induce the same behavior. Multi-objective control commands are associated with sets of similar behavior, in order to create cases. Responses to applied loads define case attributes. Case retrieval involves comparing case attributes with responses of the structure to the current load.

In the event of no retrieved case, load identification is performed. This involves determination of load location and magnitude. Since precision errors of the active control system are taken into account, solutions are a set of good candidate pairs of locations and magnitudes. These candidate solutions exhibit a behavior that is similar to the behavior of the structure subjected to the current load. Once current loading is known, a control command is identified using a multi-objective search method and applied to the structure for shape control. Response of the structure, control command and experimentally observed slope compensation are then memorized in order to create a new case.

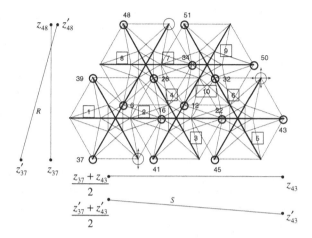

Fig. 1. View of the structure from the top with the 10 active struts numbered in squares and upper nodes indicated by a circle. Nodes 37, 43 and 48 are monitored for vertical displacement. Support conditions are indicated with large circles and arrows. Top surface slope S and transversal rotation R are indicated.

Response of the structure to unknown loading event is evaluated with respect to three indicators: top surface slope, transversal rotation and slope variation.

Top surface slope S is the first indicator. Since maintaining the top surface slope is the principal shape control objective, it is also used as the main indicator:

$$S = \left(z_{43} - \frac{z_{37} + z_{48}}{2} \right) \Big/ L \tag{1}$$

where z_i is the vertical coordinate of node i and L the horizontal length between node 43 and the middle of segment $37 - 48$ (Figure 1). For these calculations, slope unit is mm/100m.

Transversal rotation, R: This second indicator is the rotation direction of segment $37 - 43$ (Figure 1). It can be equal to 1 or -1:

$$R = \frac{\left(z'_{48} - z'_{37} \right) - \left(z_{48} - z_{37} \right)}{\text{abs}\left(\left(z'_{48} - z'_{37} \right) - \left(z_{48} - z_{37} \right) \right)} \tag{2}$$

where z'_i is the vertical coordinate of node i after load has been applied and z_i the vertical coordinate of node i before load has been applied.

Influence vector v is the third indicator. Slope variations due to 1mm elongation of each of the 10 active struts (Figure 1) are put together in order to create the influence vector v:

$$v = \left[\Delta S(1) \quad \cdots \quad \Delta S(10)\right]^T \qquad (3)$$

where $\Delta S(i)$ is the slope variation which results of 1mm elongation of active strut i.

2.1 Learning

Case-based reasoning supports incremental improvements of command identification. Both command identification time and command quality improve over time. It was observed in [24] that different loads of different magnitudes and in different locations cause a similar response of the structure. This property is the basis of retrieval. The response of the structure to current load is evaluated through measurement of the three indicators (1), (2) and (3). This response is compared with memorized case attributes.

If the response to the current load is similar enough to the attributes of a memorized case, the corresponding control command is retrieved. Similarity conditions are defined according to load identification experimental testing results [24]. They are described below:

$$\left|S'_c - S'_m\right| \leq 10 \ mm/100m \qquad (4)$$

where S'_c is the top surface slope of the structure subjected to the current load and S'_m is the top surface slope of the memorized case (1),

$$R_c = R_m \qquad (5)$$

where R_c is the transversal rotation direction of the structure subjected to the current load and R_m is the transversal rotation direction of the memorized case (2),

$$\left|v_c - v_m\right| = \sqrt{\sum_{j=1}^{10}\left(\Delta S_c(j) - \Delta S_m(j)\right)^2} \leq 0.15 \ mm/100m \qquad (6)$$

where v_c is the influence vector of the structure subjected to the current load and v_m is the influence vector of the memorized case (3). $\Delta S_m(j)$ is the measured slope variations of a case for 1mm elongation of active strut j, $\Delta S_c(j)$ the measured slope variations for 1mm elongation of active strut j on the laboratory structure subjected to the current load.

If conditions (4), (5) and (6) are true for a memorized case, the behavior of the structure subjected to the current load is similar enough to the memorized case for retrieving its control command. Once a case is retrieved, the control command is adapted for shape control of the structure subjected to current loading. For the purpose of this study, a simple adaptation function is proposed, based on a local elastic-linear assumption. The experimentally observed slope compensation of the case is used to adapt control command in order to fit to the top surface slope induced by current loading as follows:

$$CC_c = \frac{S'_c}{S'_m \cdot SC_m} CC_m \qquad (7)$$

where CC_c is the command for shape control of the structure subjected to current loading, S'_c is the top surface slope of the structure subjected to the current load, S'_m is the top surface slope of the case, SC_m is the experimentally observed slope compensation of the case and CC_m is the control command of the case.

Once the new control command is applied to the structure for shape control, experimentally observed slope compensation is measured. If experimentally observed slope compensation of the command is less than the precision of the active control system, adaptation function improves experimentally observed slope compensation quality and the memorized case is replaced by the current case.

In the event that no memorized case is close to the response of the structure subjected to the current load, the three indicators (1), (2) and (3) are used for load identification. A control command is then identified by a multi-objective search algorithm [25] and once applied successfully to the structure, a new case is created.

2.2 Load Identification

For the purposes of this paper, load identification task involves magnitude evaluation and load location. Since the structure is monitored with only three displacement sensors, system identification is used to identify load. The advantage of using system identification is that it requires neither intensive measurements nor the use of force sensors. The methodology is based on comparing measured and numerical responses with respect to the indicators (1), (2) and (3). In this study, the following assumptions are made regarding loading: loading events are single static vertical point loads. They are applied one at a time on one of the 15 top surface nodes (Figure 1).

Step 1: Top surface slope is the first indicator. When the laboratory structure is loaded, load magnitude evaluation involves numerically determining, for each of the 15 nodes, which load magnitude can induce the same top surface slope as the one measured in the laboratory structure. This evaluation is performed iteratively for each node. Load magnitude is gradually increased until the numerically calculated top surface slope is equal to the one measured on the laboratory structure. The load is incremented in steps of 50N.

Step 2: Transversal rotation is the second indicator. Candidate solutions exhibiting inverse transversal rotation with respect to laboratory structure measurements are rejected. Experimental measurements show that 0.1 is an upper bound for precision error for transversal rotation. In situations where transversal rotation is less than 0.1, no candidate solutions are rejected by this indicator.

Step 3: The influence vector is the third indicator. The influence vector is evaluated for the laboratory structure through measurements. For the candidate solutions, the influence vector is evaluated through numerical simulation. The candidate influence vector that exhibits the minimum Euclidian distance with the influence vector of the laboratory structure subjected to the current load indicates the candidate that is the closest to the laboratory structure according to (8). It is taken to be the reference candidate.

$$\min \left| \mathbf{v}_{can} - \mathbf{v} \right| \tag{8}$$

Practical applications of system identification must include consideration of errors. An upper bound for the error on slope variations for one single elongation of 1mm has been observed to be equal to $e_{ap} = 0.34$ mm/100m. This error is related to accuracy of the control system. Candidate solutions for which the Euclidian distance with the reference candidate is less than 10 times the error on active perturbation e_{ap} are also considered good load identification candidate solutions.

$$\left| \mathbf{v}_{ref} - \mathbf{v}_{can} \right| \leq 10 e_{ap} \tag{9}$$

This process results in a set of good candidate solutions. This information is used as input to identify a control command for shape control task [25].

3 Experimental Results

3.1 Learning

Experimental testing validates two types of learning: decreased command identification time and increased command quality over time. Since adaptation of a retrieved case is direct and the number of cases increases over time, command identification time decreases over time. Since the adaptation function involves using a fraction of slope-compensation commands of retrieved cases and since the number of cases increases over time, experimentally observed slope compensation quality increases over time.

3.2 Load Identification

The laboratory structure is loaded with 859 N at node 32 (Figure 1). The measured top surface slope is equal to 133.6 mm/100m when load is applied. Three candidate solutions exhibit a behavior that is close to the behavior of the laboratory structure: 770 N at node 32, 1000 N at node 51 and 490 N at node 48. For these three solutions, control commands are identified using a multi-objective search algorithm [25]. The three control commands have been applied to the laboratory structure. The evolution of top surface slope, where zero slope is the target, is shown in Figure 2 for these commands. Top surface slope is plotted versus steps of 1mm of active strut movement. Experimentally observed slope compensation ranges between 91 % and 95 %, even when the control command is associated with a load identification solution that does not exactly represent the real loading. The three solutions are thus considered to be equivalent. The best experimentally observed slope compensation of 95 % is the closest: 770 N at node 32.

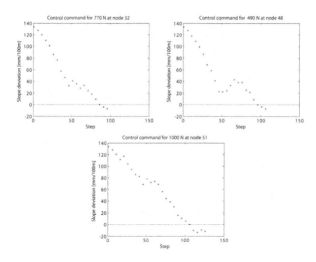

Fig. 2. Shape control for load case 5: 859 N at node 32, for the three load identification solutions: 770 N at node 32, 1000 N at node 51 and 490 N at node 48

4 Conclusions

The learning methodologies described in this paper allows for two types of learning: increased rapidity and increased quality over time. The success of both types is related to the formulation of retrieval and adaptation, as well as the number of cases. More generally, it is also demonstrated that interactivity between learning algorithms and sensor devices is attractive for control tasks.

System identification algorithms contribute to self-awareness in active structures and lead to successful load identification. Load identification solutions are used efficiently for shape control in situations of unknown loading event. Experimental testing supports the strategy involving initial generation of a set of good solutions rather than direct (and often erroneous) application of a single control command.

References

1. Kawaguchi, K., Pellegrino, S. and Furuya, H., (1996), "Shape and Stress Control Analysis of Prestressed Truss Structures", Journal of Reinforced Plastics and Composites, 15, 1226-1236.
2. Djouadi, S., Motro, R., Pons, J.C., and Crosnier, B., (1998), "Active Control of Tensegrity Systems", Journal of Aerospace Engineering, 11, 37-44.
3. Sultan, C., (1999) "Modeling, design and control of tensegrity structures with applications", PhD thesis, Purdue Univ., West Lafayette, Ind.
4. Skelton, R.E., Helton, J.W., Adhikari, R., Pinaud, J.P. and Chan, W., (2000), "An introduction to the mechanics of tensegrity structures", Handbook on mechanical systems design, CRC, Boca Raton, Fla
5. Kanchanasaratool, N., and Williamson, D., (2002), "Modelling and control of class NSP tensegrity structures", International Journal of Control, 75(2), 123-139

6. Van de Wijdeven, J. and de Jager, B., (2005), "Shape Change of Tensegrity Structures: Design and Control", Proceedings of the American Control Conference, Protland, OR, USA, 2522-2527

7. Fest, E., Shea, K., and Smith, I.F.C., (2004), "Active Tensegrity Structure", Journal of Structural Engineering, 130(10), 1454-1465

8. Domer, B., and Smith, I.F.C., (2005) "An Active Structure that learns", Journal of Computing in Civil Engineering, 19(1), 16-24

9. Kolodner, J.L. (1993). "Case-Based Reasoning", Morgan Kaufmann Publishers Inc., San Mateo, CA.

10. Leake, D.B. (1996), Case-based reasoning: Experiences, lessons, & future directions, D.B. Leake, ed., California Press, Menlo Park., Calif.

11. Leake, D. B., and Wilson, D. C. (1999)"When Experience is Wrong: Examining CBR for changing Tasks and Environments." *ICCBR 99, LNCS 1650, Springer Verlag.*

12. Müller, K. R., Mika, S., Rätsch, G., K., T., and Shölkopf, B. (2001). "An Introduction to Kernel-Based Learning Algorithms." *IEEE Transactions on Neural Networks*, 12(2), 181-201.

13. Purvis, L., and Pu, P. "Adaptation Using Constraint Satisfaction Techniques." *ICCBR-95, LNAI 1010, Springer Verlag*, Sesimbra, Portugal, 289-300

14. Marling C, Sqalli M, Rissland E, Munoz-Avila H, Aha D. "Case-Based Reasoning Integrations", AAAI, Spring 2002

15. Waheed, A. and Adeli, H., (2005), "Case-based reasoning in steel bridge engineering", Knowledge-Based Systems, 18, 37-46.

16. Bailey, S. and Smith, I.F.C., "Case-based preliminary building design", Journal of Computing in Civil Engineering, ASCE, 8, No. 4, pp 454-68,1994

17. Ljung, L., (1999), System identification-theory for the users, Prentice-Hall, Englewood Cliff, N.J.

18. Park, G., Rutherford, A.C., Sohn, H. and Farrar, C.R., (2005), "An outlier analysis framework impedance-based structural health monitoring", Journal of Sound and Vibration, 286, 229-250

19. Vanlanduit, S., Guillaume, P., Cauberghe, B., Parloo, E., De Sitter, G. and Verboven, P., (2005), "On-line identification of operational loads using exogenous inputs", Journal of Sound and Vibration, 285, 267-279

20. Haralampidis, Y., Papadimitriou, C. and Pavlidou, M., (2005), "Multi-objective framework for structural model identification", Earthquake Engineering and Structural Dynamics, 34, 665-685

21. Maeck. J. and De Roeck, G., (2003), "Damage assessment using vibration analysis on the z24-bridge", Mechanical Systems and Signal Processing, 17, 133-142

22. Lagomarsino, S. and Calderini, C., (2005), "The dynamical identification of the tensile force in ancient tie-rods", Engineering Structures, 27, 846-856

23. Robert-Nicoud, Y., Raphael, B. and Smith, I.F.C., (2005), "System Identification through Model Composition and Stochastic Search", Journal of Computing in Civil Engineering, 19(3), 239-247

24. Adam, B. and Smith, I.F.C, (in review), "Self Diagnosis and Self Repair of an Active Tensegrity Structure", Journal of Structural Engineering.

25. Adam, B. and Smith, I.F.C, (2006), "Tensegrity Active Control: a Multi-Objective Approach", Journal of Computing in Civil Engineering.

Capturing and Representing Construction Project Histories for Estimating and Defect Detection

Burcu Akinci, Semiha Kiziltas, and Anu Pradhan

Department of Civil and Environmental Engineering, Carnegie Mellon University,
Pittsburgh, PA 15213
{bakinci, semiha, pradhan, LNCS}@cmu.edu

Abstract. History of a construction project can have a multitude of uses in supporting decisions throughout the lifecycle of a facility and on new projects. Based on motivating case studies, this paper describes the need for and some issues associated with capturing and representing construction project histories. This research focuses on supporting defect detection and decision-making for estimating an upcoming activity's production rates, and it proposes an integrated approach to develop and represent construction project histories. The proposed approach starts with identifying the data needs of different stakeholders from job sites and leverages available automated data collection technologies with their specific performance characterizations to collect the data required. Once the data is captured from a variety of sensors, then the approach incorporates a data fusion formalism to create an integrated project history model that can be analyzed in a more comprehensive way.

1 Introduction

The history of a construction project can have a multitude of uses in supporting decisions throughout the lifecycle of a facility and on new projects. Capturing and modeling construction project history not only helps in active project monitoring and situation assessment, but also aids in learning from the trends observed so far in a project to make projections about project completion. After the completion of a project, a project history also provides information useful in estimations of upcoming projects.

Many challenges exist in capturing and modeling a project's history. Currently, types of data that should be collected on a job site are not clearly identified. Existing formalisms (e.g., time cards) only consider a single view, such as cost accounting view, on what data should be captured; resulting in sparse data collection that do not meet the requirements of other stake-holders, such as cost estimators and quality control engineers. Secondly, most of the data is captured manually resulting in missing information and errors. Thirdly, collected data are mostly stored in dispersed documents and databases, which do not facilitate integrated assessment of what happened on a job site. As a result, there are not many decision support systems available for engineers to fully leverage the data collected during construction.

This paper provides an overview of findings from various case studies, showing that: (1) current data collection and storage processes are not effective in gathering the

I.F.C. Smith (Ed.): EG-ICE 2006, LNAI 4200, pp. 15–22, 2006.

data needed for developing project histories that can be useful for defect detection and cost estimation of future projects; (2) sensing technologies enable robust data collection when used in conjunction with a formalized data collection procedure; however, the accuracy of the data collected using such technologies is not well-defined. Findings suggest a need for a formalized approach for capturing and modeling construction project histories. An example of such an approach, as described in this paper, starts with identifying different users' needs from project histories, provides guidance in collecting data, and incorporates a framework for fusing data from multiple sources. With such formalism, it would be possible to leverage project histories to support active decision-making during construction (e.g., active defect detection) and proactive decision-making for future projects (e.g., cost estimation of future projects).

2 Motivating Vignettes from Case Studies

Four case studies were conducted in commercial buildings with sizes ranging from 3,345 m^2 to 12,355 m^2, where laser scanners and temperature sensors were used in periodically collecting data from the job sites to actively identify defects [1]. In addition, we conducted a case study on a 19 km highway project, during which we identified a set of issues associated with data collection [2]. Currently, we are conducting another study on a 9 km of a roadway project, where we are trying to understand how to leverage and fuse the data collected by equipment on-board instrument (OBI) and other publicly available databases (e.g., weather database), to create a more comprehensive project history to support estimators in determining production rates in future projects [3]. Below summarizes some findings from these case studies.

The need for and issues associated with data collection to support multiple decisions: Current data collection at job sites seems to support mostly the needs of construction schedule and cost control, yet data from job sites are useful for other tasks, such as active defect detection and situation assessment in current projects and estimating the production rates of activities in future ones. Such varying uses place different requirements on data collection from sites. For example, highly accurate information on geometric features of components is necessary to identify defects [4], versus general information on processes and the conditions under which processes occur (i.e., contextual data) is helpful for estimating a production rate of a future activity [3].

The case study findings showed that most of the data collected on site focused on the resources used for a given task, but the contextual data was rarely collected [2]. Similarly, the available sources of data and related databases did not provide detailed information that can be used to assess why certain production rates were observed to be fluctuating when the same activity was performed in different zones and dates [3]. As a result, the collected data was not useful in helping estimators in picking a production rate among alternatives for a new project.

Issues with manual data collection and utilization of sensing technologies: Current manual data collection processes utilized at job sites do not enable collecting required data completely and accurately. One of the cases showed large percentages of missing data describing the daily productivities of activities and the conditions under which such daily productions are achieved [2]. Even when collected, the quantity information was described based on some indirect measures (e.g., number of truck-loads of dirt moved out); resulting in inaccuracies in the data collected and stored.

Utilization of sensing technologies (e.g., laser scanners, equipment OBIs), reduces the percentage of missing information. However, there can still be inaccuracy issues, if a sensor is not well calibrated and its accuracy under different conditions is not well defined. In certain cases, data collected from such sensors needs to be processed further and fused to be in a format useful for decision makers.

Issues with and the need for fusing data from multiple sources:
Currently, data collected on job sites is stored in multiple dispersed documents and databases. For example, daily crew and material data are kept on time cards, soil conditions are described on reports and production data are stored on databases associated with equipment OBIs. To get a more comprehensive understanding of how activities were performed, one needs to either fuse data stored in such various sources or rely on his/her tacit knowledge, which might not be accurate. In a case study, when an engineer was asked to identify reasons for explaining the fluctuations in the excavation work, he attributed it to fog in the mornings and the soil conditions. When the data collected from equipment OBIs merged along with the data collected in time-cards, the soil profiles defined by USGS and the weather data, it was observed that the factors identified by him did not vary on the days when there were large deviations on production [3]. While this showed the benefits of integrating such data to analyze a given situation, the research team observed that it was tedious and time-consuming to do the integration manually. For instance, it took us approximately forty hours to fuse daily production data of a single activity with the already collected crew, material, and daily contextual data for a typical month.

3 Vision and Overview of the Approach

We have started developing and implementing an approach that addresses the identified issues and needs based on the case studies. This approach consists of two parts focusing on: (1) formalizing a data collection plan prior to the execution of construction activities (Figure 1); and (2) fusing the data collected and creating integrated project histories to support decision-making (Figure 2). So far, our research has focused on using such formalism to support defect detection on construction sites and to support estimation of production rates of activities occurring in future projects. Hence, the corresponding figures and subsections below highlight those perspectives.

3.1 Formalization of a Data Collection Plan

The data requirements of decision makers from job sites need to be incorporated prior to data collection. The first step in doing that is to understand what these requirements are, and how they can be derived or specified (Figure 1). Since our research focus has been to support defect detection and cost estimation, we identified construction specifications and estimators' knowledge of factors impacting activity production rates as sources of information to generate a list of measurement goals.

The approach leverages an integrated product and process model, depicting the as-design and schedule information, and a timeframe for data collection, as input. Using this, it first identifies activities that will be executed during that timeframe and the corresponding measurement goals, derived based on specifications and factors affecting the production rates of those activities. Next, measurement goals identified for each activity are utilized to identify possible sources for data collection with the goal of reducing manual collection. Sources of data include sensors (e.g., laser scanners, equipment OBIs) and general public databases (e.g., USGS soil profiles, weather database). With this, the approach generates a data collection plan as an output.

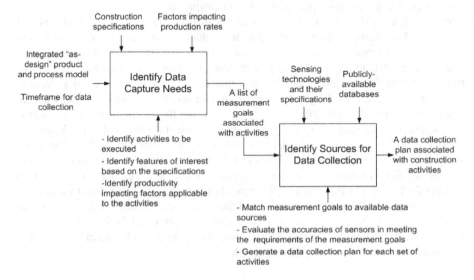

Fig. 1. An approach for generating a data collection plan

Construction specifications provide information on the expected quality of the components by specifying the features of a product to be inspected and the tolerances to deduce the required accuracy for measurements. Specifications can be represented in a computer-interpretable way and can be used to automate the generation of measurement goals for a given set of components [4].

Creating a project history model to be utilized by estimators requires not only product related data, but also contextual data representing the conditions under which the activities were executed, so that estimators can understand what happened in the past, compare it to the current project's conditions and make a decision accordingly. Therefore, within the scope of this research, project history can be defined as as-design project information augmented with activity-specific as-built project data. As built project data are enriched with contextual data, which are captured and stored on a daily basis. We built on the factors identified in the literature (e.g., [3]) and have further extended it based on findings from a set of interviews conducted with several senior estimators in heavy-civil and commercial construction companies. Table 1 provides examples from an initial list of factors identified for excavation, footing and wall construction activities. The factors identified can be grouped under the categories of *design-related factors, construction method-related factors, construction site-related factors* and *external factors*. Based on these factors, it would be possible to generate a list of measurement goals for each activity, map them to a set of sensors and publicly-available databases that would help in collecting some of the data needed. As a result it would be possible to generate a data collection plan associated with each group of activity to be executed.

Table 1. Initial findings on the factors affecting productivity of activities

Factor Groups	Specific examples of factors for excavation, foundation an wall construction activities
Design-Related Factors	Depth of cut, height, length and width of components
	Shape of cut/shape of component
	Total quantity of work for the entire project
	Number and sizes of openings in walls
	Existence of steps on walls and footings
	Rebar/Formwork to concrete ratio, rebar size
Construction Method-Related Factors	Type and capacity of equipment, number of equipments
	Crew size and composition
	Stockpile dirt vs. haul off
	Method of forming and type of bracing used, formwork size
	Material characteristics, such as concrete strength, soil type
Construction Site-Related Factors	Site access constraints and space availability
	Moisture content of soil
	Length, grade, direction, width of haul roads
External Factors	Time of year, weather, project location

3.2 Data Fusion and Analysis for Creating and Using Project Histories

Once a data collection plan is generated, it can be executed at the job site to collect the data needed. The next step is to process the data gathered from sensors and databases and fuse them to create an integrated project history model that can serve as a basis to perform analysis for defect detection and cost estimation (Figure 2). Different components of such an approach are described below.

3.2.1 Utilization of Sensors for Data Capture
Many research studies have explored utilization of sensors on construction sites for automated data collection (e.g., [1,5,6]). In our research, we have explored the utilization of laser scanners, Radio Frequency Identification (RFID), Global Positioning System (GPS), thermocouples, and equipment OBIs to capture data on job sites for supporting active defect detection and estimators' decision-making. The selections of these technologies are based on the lessons learned from past and ongoing research projects within our research group [1,6]. Our experiments demonstrated that such technologies can enable the capturing of some of the data needed for project history models in an automated way [6].

Our experiments also showed that the behaviors of such sensors vary greatly at job sites; hence manufacturers' specifications might not be reliable in varying situations. For example, we observed that the reading ranges of active RFID tags was reduced by about $1/4^{th}$ or $1/5^{th}$ of the specified ranges when they were used to track precast elements [6]. Similarly, laser scanner accuracy varies considerably based on its incidence angle and distance from the target object [7]. While in most cases, the accuracy and the reliability of the data were observed to be better than the manual approaches, it is still important to have a better characterization of accuracies of sensors under different conditions (e.g. incidence angle) when creating and analyzing project history models. Currently, we are conducting experiments for that purpose.

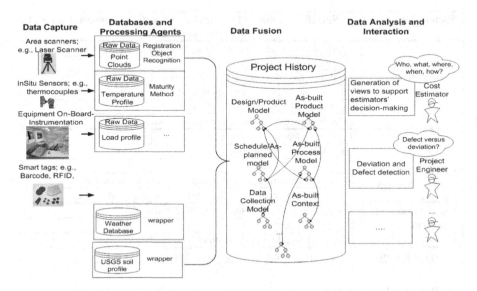

Fig. 2. An approach for data fusion and analysis for creating and using project histories

3.2.2 Formalization of Fusing Data from Multiple Sources

Data collected from multiple sources need to be fused to have a more comprehensive assessment of a project. We have started to develop and evaluate a system architecture for data fusion purposes, based on Dasarathy's fusion functional model [8], where the entire fusion processing is categorized into three general levels of abstraction as, the data level (*sensor fusion*), the feature level (*feature fusion*) and the decision level (*decision fusion*).

In *sensor fusion*, the raw data from multiple sensors, which are measuring the same physical phenomena, are directly combined. For example, the data collected from GPS and RFID readers can be directly combined after initial corrections to track the location and ID of components respectively [6]. However, some sensors, such as laser scanners, cannot measure a component and its geometric features directly and hence, the data collected needs to be processed further and fused with other data at a *feature and component level*. In one of our research projects, laser scanner is being used to detect geometric deviations, i.e. length, height and width of building components [1]. Since laser scanners provide point cloud data, the components and their features needed to be explicitly extracted from point clouds using 3D computer vision techniques [1]. The sensor and feature level fusions are done with appropriate processing agents (Fig 2).

The third level of fusion described in [8] is the *decision level fusion,* where the data fused at the sensor and feature levels are further integrated and analyzed to achieve a decision. We are leveraging different models such as-built product/process model and data collection model for decision level fusion (Fig 2). Decision level fusion is challenging compared to sensor and feature level fusions, since the formalisms used in sensor and feature level fusions are well defined and can be identical across multiple domains. However formalisms for decision-level fusion differ among domains since they need to support different decisions [9].. As discussed in Section 3.1., different tasks require different sets of data being collected and fused. Hence, the decision-level fusion requires customized formalisms to be developed to enable the integration and processing of the data to support specific decisions. In our approach, decision-level fusion formalisms are designed to generate the views (e.g. from the estimator's perspective) that are helpful in supporting decisions to select a proper production rate. These formalisms are not meant to perform any kind of predictions or support case-based reasoning.

3.2.3 Formalisms for Data Interaction and Analysis to Support Active Defect Detection and Cost Estimating

In this research, we have explored project history models to support defect detection during construction and in estimating production rates of future activities. An approach implemented for active defect detection leverages the information represented in as-design models, construction specifications, and the as-built models, generated by processing the data collected from laser scanners. It uses the information in specifications to identify the features of the components that are of interest for defect detection and compares the design and as-built models accordingly. When there is a deviation between an as-design and an as-built model, it refers to the specifications to assess whether the deviation detected exceeds the tolerances specified. If it exceeds the tolerances, then it flags the component as a defective component [4].

In supporting estimators' decision-making, we have been focusing on identifying and generating views from integrated project history models, so that estimators can navigate through the model and identify the information that they need to determine the production rates of activities in future bids. Initial interviews with several estimators from two companies showed that estimators would like to be able to navigate through production data in multiple levels (e.g., zone level, project level) and in multiple perspectives (e.g., based on a certain contextual data, such as depth of cut), and be able to compare alternatives (e.g., comparing productions on multiple zones) using such a model. These views will enable estimators to factually learn from what happened on a job site, and make the estimate for a similar upcoming activity based on this learning. We are currently implementing mechanisms to generate such views for estimators.

4 Conclusions

This paper describes the need for capturing and representing construction project histories and some issues associated with it for cost estimation and defect detection purposes. The approach described in the paper starts with identifying some data capture needs and creating data collection plan for each activity to satisfy those needs. Since several case studies demonstrated that manual data collection is inaccurate and unreliable, the envisioned approach focuses on leveraging the data already stored in publicly-available databases and data collection through a variety of sensors. Once the data is captured from a variety of sensors, they should be fused to create an integrated project model that can be analyzed in a comprehensive way. Such analyses include defect detection and situation assessment during the execution of a project, and generation of information needed for estimators in determining the production rates of future activities.

Acknowledgements

The projects described in this paper are funded by two grants from the National Science Foundation, CMS #0121549 and 0448170. NSF's support is gratefully acknowledged. Any opinions, findings, conclusions or recommendations presented in this paper are those of authors and do not necessarily reflect the views of the National Science Foundation.

References

[1] Akinci, B., Boukamp, F., Gordon, C., Huber, D., Lyons, C., Park, K. (2006) "A Formalism for Utilization of Sensor Systems and Integrated Project Models for Active Construction Quality Control." Automation in Construction, Volume 15, Issue 2, March 2006, Pages 124-138

[2] Kiziltas, S. and Akinci, B. (2005) "The Need for Prompt Schedule Update By Utilizing Reality Capture Technologies: A Case Study." Constr. Res. Cong., 04/2005, San Diego, CA.

[3] Kiziltas, S., Pradhan, A., and Akinci, B. (2006) "Developing Integrated Project Histories By Leveraging Multi-Sensor Data Fusion", ICCCBE,, June 14-16, Montreal, Canada.

[4] Frank, B., and Akinci, B. (2006) "Automated Reasoning about Construction Specifications to Support Inspection and Quality Control" , Automation in Construction, under review.

[5] Kiritsis, D., Bufardi, A. and Xirouchakis, P., "Research issues on product lifecycle management and information tracking using smart embedded systems", Advanced Engineering Informatics, Vol. 17, Numbers 3-4, 2003, pages 189-202.

[6] Ergen. E., Akinci, B. Sacks, R. "Tracking and Locating Components in a Precast Storage Yard Utilizing Radio Frequency Identification Technology and GPS." Automation in Construction, under review.

[7] Axelson, P. (1999). "Processing of Laser Scanner Data – Algorithms and Applications," ISPRS Journal of Photogrammetry & Remote Sensing, 54 (1999) 138-147.

[8] Dasarathy, B. (1997). "Sensor Fusion Potential Exploitation-Innovative Architectures and Illustrative Applications", IEEE Proceedings, 85(1).

[9] Hall, D. L. and Llinas, J. (2001). "Handbook of Mulitsensor Data Fusion," 1st Ed., CRC.

Case Studies of Intelligent Context-Aware Services Delivery in AEC/FM

Chimay Anumba and Zeeshan Aziz

Department of Civil & Building Engineering, Loughborough University, LE11 3TU, UK
{c.j.anumba, z.aziz}@lboro.ac.uk

Abstract. The importance of context-aware information and services delivery is becoming increasingly recognised. Delivering information and services to AEC/FM (architecture, engineering and construction/facilities management) personnel, based on their context (e.g. location, time, profile etc) has tremendous potential to improve working practices, particularly with respect to productivity and safety, by providing intelligent context-specific support to them. This paper discusses a vision of context-aware service delivery within the AEC/FM sector and presents three case studies to illustrate the concepts. It starts with a brief overview of context-aware computing and a system architecture which facilitates context capture, context brokerage and integration with legacy applications. This is followed by presentation of case-studies that relate to actual deployments on a simulated construction site, in a construction education environment and in a train station. The deployment process and findings from each of the case studies are summarised and the benefits highlighted. Conclusions are drawn about the possible future impact of context-aware applications in AEC/FM.

1 Introduction

The potential of mobile Information Technology (IT) applications to support the information needs of mobile AEC/FM workers has long been understood. To exploit the potential of emerging mobile communication technologies, many recent research projects have focused on the application of these technologies in the AEC/FM sector. However, from a methodological viewpoint, a key limitation of the existing mobile IT deployments in the construction sector is that they see support for mobile workers as a "simple" delivery of the information (such as project data, plans, technical drawings, audit-lists, etc.). Information delivery is mainly static and is not able to take into account the worker's changing context and the dynamic project conditions. Many existing mobile IT applications in use within the construction industry rely on asynchronous methods of communication (such as downloading field data from mobile devices onto desktop computers towards end of the shift and then transferring this information into an integrated project information repository) with no consideration of user-context . Even though in some projects real time connectivity needs of mobile workers are being addressed (using wireless technologies such as 3G, GPRS, WiFi), the focus is on delivering static information to users such as project plans and documents or access to project extranets. Similarly, most of the commercially available

I.F.C. Smith (Ed.): EG-ICE 2006, LNAI 4200, pp. 23–31, 2006.
© Springer-Verlag Berlin Heidelberg 2006

mobile applications for the construction industry are designed primarily to deliver pre-programmed functionality without any consideration of the user context. This often leads to a contrast between what an application can deliver and the actual data and information requirements of a mobile worker. In contrast to the existing static information delivery approaches, work in the AEC/FM sector, by its very nature, is dynamic. For instance, due to the unpredictable nature of the activities on construction projects, construction project plans, drawings, schedules, project plans, budgets, etc often have to be amended. Also, the context of the mobile workers operating on site is constantly changing (such as location, task they are currently involved in, construction site situations and resulting hazards, etc) and so do, their information requirements. Thus, mobile workers require that supporting systems rely on intelligent methods of human-computer interaction and deliver the right information at the right time on an as-needed basis. Such a capability is possible by a better understanding of the user-context.

The paper is organised as follows. Section 2 introduces the concept of context-aware computing and reviews the state of the art. Section 3 presents the service delivery architecture which facilitates context capture, context brokerage and integration with back-end systems using a Web Services model. Section 4 presents case-studies related to the deployment of context-aware applications. Conclusions are drawn about the possible future impact of context-aware service delivery technologies in the AEC/FM sector.

2 Context-Aware Computing – State of the Art

Context-aware computing is defined by Burrell et al [1] as the use of environmental characteristics such as the user's location, time, identity, profile and activity to inform the computing device so that it may provide information to the user that is relevant to the current context. Context-aware computing enables a mobile application to leverage knowledge about various context parameters such as who the user is, what the user is doing, where the user is and what mobile device the user is using. Pashtan [2] described four key partitions of context parameters, including user's static context (includes user profile, user interests, user preferences), user's dynamic context (includes user location, user's current task, vicinity to other people or objects), network connectivity (includes network characteristics, mobile terminal capabilities, available bandwidth and quality of service) and environmental context (include time of day, noise, weather, etc.).

Context-aware computing is an established area of research within computer science. The application of context-awareness for mobile users has been demonstrated in a large number of applications, including fieldwork [3], museums [4], route planning [5], libraries [6], meeting rooms [7], smart-houses [8] and tourism [9]. Location is a key context parameter and other projects that have specifically focused on location-based data delivery included Mobile Shadow Project (MSP) [10] and the GUIDE project [11]. The MSP approach was based on the use of agents to map the physical context to the virtual context while the GUIDE project focused on location-aware information provision to tourists. In the AmbieSense Project [12] a different approach was adopted by focusing on creating a tag-based digital environment that is aware of a person's presence, context, and sensitivity and responds accordingly. Lonsdale et al [13] implemented a prototype to facilitate mobile learning. In the implementation, mobile devices pass contextual information obtained from sensors, user input, and user profile to the context subsystem. The context sub-system then compared this metadata to the content metadata provided by the delivery sub-system and returned a set of content recommendations. In the Active Campus project [14], a prototype was developed, to demonstrate the potential of context-aware information delivery technology to support staff and students in an educational setting. In a similar piece of work [15], location-aware technologies were used in a laboratory environment to first collect and organise data where and when created and then make this information available where it is needed. Proximity to a particular object or location was sensed either via Radio Frequency Identification (RFID) badges or direct contact with a touch screen. Each researcher in the laboratory was given a RFID badge that uniquely identified him. This unique identifier provided authentication for access to laboratory applications as well as triggering the migration of the user interface from one display to another closer to the position of the researcher. Context-aware applications are also being investigated by other fields of research in computer science, including mobile, ubiquitous and wearable computing, augmented reality and human-computer interaction. However, the application of context-aware technology in the construction industry remains limited.

The awareness of user context can enhance mobile computing applications in the AEC/FM sector by providing a mechanism to determine information relevant to a particular context. In recent yeas, the emergence of powerful wireless Web technologies, coupled with the availability of improved bandwidth, has enabled mobile workers to access in real time different corporate back-end systems and multiple inter-enterprise data resources to enhance construction collaboration. Context-aware information delivery adds an additional layer on top of such real time wireless connectivity [16] offering the following benefits:

- Delivery of relevant data based on the worker's context thereby eliminating distractions related to the volume and level of information;
- Reduction in the user interaction with the system by using context as a filtering mechanism. This has the potential to increase usability by making mobile devices more responsive to user needs;
- Awareness of the mobile worker's context, through improved sensing and monitoring can also be used to improve security and health and safety practices on the construction site. At the same time, it is possible to use the knowledge of onsite activities to improve site-logistics, site-security, accountability and health and safety conditions on the site.

3   Context-Aware Service Delivery Architecture

Figure 1 presents a context-aware services delivery architecture that combines context-awareness and Web Services to create a pervasive, user-centred mobile work environment, which has the ability to deliver context-relevant information to the workers to support informed decision making. The key features of the architecture are discussed below:

3.1 Context-Capture

This tier helps in context capture and also provides users access to the system. The context is drawn from different sources, including:

- Current location, through a wireless local area network-based positioning system [17]. A client application running on a user's mobile device or a tag sends constant position updates to the positioning engine over a WLAN link. This allows real time position determination of users and equipment. It is also possible to determine a user's location via telecom network-based triangulation;
- User Device Type (e.g. PDA, TabletPC, PocketPC, SmartPhone, etc.), via W3C CC/PP standards [18]. These standards allow for the description of capabilities and preferences associated with mobile devices. This ensures that data is delivered according to the worker's device type;
- User identity (e.g. Foreman, Electrician, Site Supervisor, etc.), via the unique IP address of their mobile device. User profile is associated with user identity;
- User's current activity (e.g. inspecting work, picking up skips, roof wiring, etc.), via integration with project management/task allocation application;
- Visual context, via a CCTV-over-IP camera;
- Time via computer clock.

The use of IP-based technologies enables handover and seamless communication between different wireless communication networks such as wireless wide area networks, local area networks and personal area networks. Also, both push and pull modes of interaction are supported. Thus, information can be actively pushed to mobile workers (through user-configured triggers), or a worker can pull information through ad-hoc requests, on an as-needed basis. As application content, logic and data processing reside on the wired network, the mobile client is charged with minimal memory and processor consuming tasks.

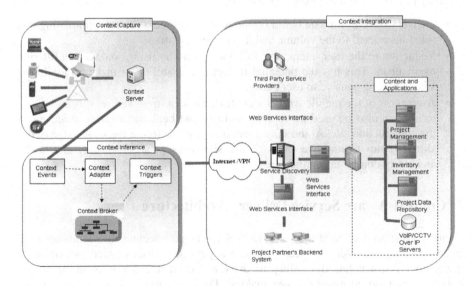

Fig. 1. The Deployment Architecture

3.2 Context Inference

This tier provides the ability to reason about the captured context using a Semantic-Web based model to describe a knowledge model for a corresponding context domain, thereby helping context description and knowledge access (by supporting information retrieval, extraction and processing) based on the inferred context. The understanding of semantics (i.e. meanings of data) enables the creation of a relationship between the context parameters and available data and services. Output from the context-inference tier is passed into AEC/FM applications to make them aware of events on the site. The context adapter converts the captured context (e.g. user id, user location, time, etc.) into semantic associations. RDF schema [19] is used to provide vocabulary and structure to express the gathered contextual information. Being XML-based, RDF also ensures the provision of context information in an application and platform-independent way. Also, using the RDF schema, the context broker maps the captured contextual information to available data and services. Mapping can include user profile to project data (mapping of information, based on the role of the user on site), location to project data (mapping user location to project data e.g. if electrician is on floor 3, he probably requires floor 3 drawings and services) and user task to project data (mapping information delivery to the task at hand). RDF was also used as a meta-language for annotating project resources and drawings. Such a semantic description provides a deeper understanding of the semantics of documents and an ability to flexibly discover required resources. A semantic view of construction project resources logically interconnects project resources, resulting in the better application of context information. At the same time, semantic description enables users to have different views of data, based on different criteria such as location and profile. As the user context changes (e.g. change of location, tasks), the context broker recalculates the available services to the users in real time.

3.3 Context Integration

Based on the captured context, this tier helped in service discovery and integration. Changes in the context prompt the context broker to trigger the pre-programmed events which may include pushing certain information to users or an exchange of information with other applications using Web Services, to make them aware of the events on the site. Web-services standards are used to allow applications to share context data and dynamically invoke the capabilities of other applications in a remote collaboration environment.

4 Case Studies of Context-Aware Information Delivery

This section presents three case studies which relate to the deployment of the context-aware services delivery architecture (Fig 1) in a simulated construction site, in a construction education environment, and in a train station. The choice of case studies was based on availability, and the need to explore a variety of deployment scenarios.

4.1 Construction Site Environment

This involved the deployment of the context-aware services delivery architecture to support construction site processes. As site workers arrived for work, the on-site wireless network detected the unique IP address of their mobile devices and prompted them to log-in. On a successful log-in, the site-server pushed the worker's task list and associated method statement (as assigned by the site supervisor using an administration application) based on the worker's profile (Fig 2 (a & b)). Completion of tasks were recorded in real-time and an audit trail was maintained. Also, application and service provisioning to site workers was linked to their context (i.e. location, profile and assigned task) e.g. based on changing location, relevant drawings and data was made available (Fig 3). The context broker played the key role of capturing the user context and mapping the user context to project data, at regular time intervals. Real-time location tracking of site workers and expensive equipment was also used to achieve health and safety and security objectives. Also, WLAN tags were used to store important information about a bulk delivery item. XML schema was used to describe the tag information structure. As the delivery arrives at the construction site, an on-site wireless network scans the tag attached to the bulk delivery and sends an instant message to the site manager's mobile device, prompting him/her to confirm the delivery receipt.

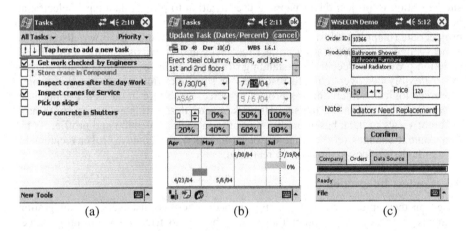

Fig. 2. Profile based task allocation (a & b) and inventory logistics support (c)

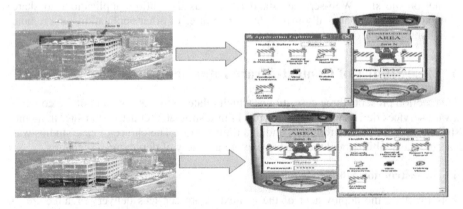

Fig. 3. Context-Aware Access to Project Data

The site manager browses through the delivery contents (Fig 2(c)) and records any discrepancies. Once the delivery receipt is confirmed, data is synchronized with the site server, resulting in a real-time update of the inventory database.

4.2 Construction Education Environment

This pilot was undertaken at Loughborough University to demonstrate the potential of context-aware information delivery in a construction education environment. The implementation addressed the key issues of using handheld devices for context-aware content delivery, access to online resources and real-time response for interactivity between lecturers and students. Different aspects of the implementation included:

- *Context-Aware Delivery of Learning Content:* An on-campus WLAN network was used to capture the students' context (i.e. identity, location, time, course enrolment, etc.). The captured contextual information was used as a filtering mechanism to query the students' virtual learning environment to determine relevant data for a particular class, which was subsequently pushed to the student's PDA or laptop.
- *Context-Aware Access to Online Resources:* Students were able to access various online resources (such as the virtual learning environment, library resources, etc.) based on their context thereby minimising the interaction required between the mobile device and the user. Also, access to some online resources (such as the Internet, chat applications, etc) was restricted during the lecture period.
- *Context-Aware Classroom Response:* Mobile devices were used to support the learning process by supporting interactivity between lecturers and students during tutorials. The lecturer could see students' responses to presented case-studies by accessing a back-end system (Fig 4). Such interactivity could be used to support class-room discussions, evaluate individual student performance or elicit feedback.

The feedback obtained from lecturers and students in this case study was positive and established the effectiveness of supporting learning in this way because of the system's portability, context-awareness and real-time communication features. However, it was shown that not all subjects/topics can effectively be supported in this way.

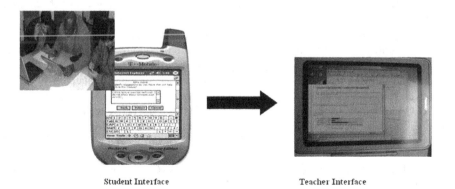

Student Interface   Teacher Interface

Fig. 4. Context-Aware Access to Learning Resources

4.3 Train Station

A proof-of-concept deployment was undertaken on a UK train station to provide an intelligent wireless support infrastructure for the station staff. The key objective of the deployment was to provide context-aware data support to the station staff based on their information needs (location, role), device processing capabilities (device type, bandwidth) and external events (train disruptions, security alerts). On account of a large number of user profiles (which included train managers, station managers, station supervisors, train dispatch staff, maintenance engineers), the interface was personalised based on the user log-in. Station staff were pushed information about disruptions to train services via integration with a customer information system using Web Services. After a successful log-in, the content was automatically updated with current information, personalised for the user's context. Two main applications were deployed:

- *Real-time Train Information:* Station staff were provided real-time access to train running information. Knowledge of the user context (e.g. station information, time of the day, date, etc.) was used to present the relevant information minimising the interaction required between the staff and the mobile device.
- *Security Alerts:* Using their handheld devices, station staff could generate and respond to security alerts. Also, based on their location, station staff could access video feeds of IP-based surveillance cameras. Once a security alert is generated, the closest station staff and security officer were immediately warned based on their proximity to the person or object generating the alert.

This case study is ongoing and a detailed evaluation is planned in the near future.

5 Conclusions

This paper has presented an architecture for context-aware services delivery and three implementation case-studies. Awareness of the user-context has the potential to cause a paradigm shift in AEC/FM sector, by allowing mobile workers access to context-specific information and services on an as-needed basis. Current approaches of supporting AEC/FM workers often involve the complexities of using a search engine, moving between files or executing complicated downloads. In comparison, context-awareness makes human-computer interaction more intuitive, thereby reducing the need for training. Also, new application scenarios are becoming viable by the ongoing miniaturisation, developments in sensor networking, the increase in computational power, and the fact that broadband is becoming technically and financially feasible. However, the case studies have demonstrated that context-aware services delivery in the AEC/FM sector needs to satisfy the constraints introduced by technological complexity, cost, user needs and interoperability. Also there is a need for more successful industrial case studies; these will be explored as part of further field trials.

Acknowledgements. This project was funded by EPSRC and Fanest Business Intelligence ltd.

References

1. Burrell, J. & Gay, K. (2001). "Collectively defining context in a mobile, networked computing environment," *CHI 2001 Extended abstracts,* May 2001.
2. Pashtan, A. (2005). *Mobile Web Services.* Cambridge University Press.
3. Kortuem, G., Bauer, M., Segall, Z (1999) "NETMAN: the design of a collaborative wearable computer system", MONET, 4(1), pp. 49-58
4. Fleck, M.F, Kindberg, T, Brien-Strain, E.O, Rajani, R and Spasojevic, M (2002) "From informing to remembering: Ubiquitous systems in interactive museums". IEEE Pervasive Computing 1: pp.13-21
5. Marmasse, N., Schmandt, C. (2002) "A User-Centered Location Model. Personal and Ubiquitous Computing", Vol:5, No:6, pp:318–321
6. Aittola, M., Ryhänen, T., Ojala, T. (2003), "SmartLibrary - Location-aware mobile library service", Proc. Fifth International Symposium on Human Computer Interaction with Mobile Devices and Services, Udine, Italy, pp.411-416
7. Chen, H., Finin,T & Joshi, A. (2004), "Semantic Web in the Context Broker Architecture", IEEE Conference on Pervasive Computing and Communications, Orlando, March 2004, IEEE Press,2004, pp. 277–286
8. Coen, M.H. (1999). "The Future Of Human-Computer Interaction or How I Learned to Stop Worrying and Love my Intelligent Room", IEEE Intelligent Systems 14(2): pp. 8–19
9. Laukkanen, M., Helin, H., Laamanen, H. (2002) "Tourists on the move", In Cooperative Information Agents VI, 6th Intl Workshop, CIA 2002, Madrid, Spain, Vol 2446 of Lecture Notes in Computer Science, pages 36–50. Springer
10. Fischmeister, S., Menkhaus, G., Pree, W. (2002), "MUSA-Shadows: Concepts, Implementation, and Sample Applications: A Location-Based Service Supporting Multiple Devices", In Proc. Fortieth International Conference on Technology of Object-Oriented Languages and Systems, Sydney, Australia. 10. Noble, J. and Potter, J., Eds., ACS. pp. 71-79
11. Davies,N., Cheverst, K., Mitchell, K & Friday, A. (1999), "Caches in the Air: Disseminating Information in the Guide System". Proc. of the 2nd IEEE Workshop on Mobile Computing Systems and Applications, Louisiana, USA, February 1999, IEEE Press, pp. 11-19
12. Goker, A., Cumming, H., & Myrhaug, H. I. (2004). *Content Retrieval and Mobile Users: An Outdoor Investigation of an Ambient Travel Guide.* Mobile HCI 2004 Conference, 2nd intl Workshop on Mobile and Ubiquitous Information Access , Glasgow, UK.
13. Lonsdale P., Barber C., Sharples M., Arvantis T. (2003) "A context-awareness architecture for facilitating mobile learning". In Proceedings of MLEARN 2003, London, UK
14. Griswold,W.G., Boyer,R., Brown,S.W., Truong, T.M., Bhasket, E., Jay,R., Shapiro, R.B (2002) "ActiveCampus: Sustaining Educational Communities through Mobile Technology", Uni. of California, Dept. of Computer Science and Engineering, Technical Report
15. Arnstein, L., Borriello,G., Consolvo,S., Hung, C & Su, J. (2002)."Labscape: A Smart Environment for the Laboratory", IEEE Pervasive Computing, Vol. 1, No. 3, pp. 13-21
16. Aziz, Z., Anumba, C.J., Ruikar, D., Carrillo., P.M., Bouchlaghem.,D.N. (2005), "Context-aware information delivery for on-Site construction operations," 22nd CIB-W78 Conf on ITin Construction, Germany, CBI Publication No:304
17. Ekahau (2006) Ekahau Positioning Engine [Online] http://www.ekahau.com
18. CC/PP (2003): http://www.w3.org/TR/2003/PR-CCPP-struct-vocab-20031015/
19. RDF (2005) [Online] *http://www.w3.org/RDF/*

Bio-inspiration: Learning Creative Design Principia

Tomasz Arciszewski and Joanna Cornell

George Mason University, University Drive 4400, Fairfax, VA 22030, USA
{tarcisze@gmu.edu, jcornell@gmu.edu}

Abstract. Reusing or modifying known design concepts cannot meet new challenges facing engineering systems. However, engineers can find inspiration outside their traditional domains in order to develop novel design concepts. The key to progress and knowledge acquisition is found in inspiration from diverse domains.

This paper explores abstract knowledge acquisition for use in conceptual design. This is accomplished by considering body armor in nature and that developed in Europe in the last Millennium. The research is conducted in the context of evolution patterns of the Directed Evolution Method, which is briefly described. The focus is on conceptual inspiration. Analysis results of historic and natural body armor evolution are described and two sets of acquired creative design principia from both domains are presented. These principia can be used to stimulate human development of novel design concepts. Creative design principia, combined with human creativity, may lead to *revolutionary changes*, rather than merely *evolutionary steps*, in the evolution of engineering systems.

1 Introduction

Intelligent computing is usually understood to utilize heuristics and stochastic algorithms in addition to knowledge in the form of deterministic rules and procedures. As a result, it may produce outcomes usually associated only with human/intelligent activities in terms of novelty and unpredictability. In particular, intelligent computing may lead to an emergence of unexpected patterns and design concepts, which are highly desirable, potentially patentable, and may drive progress in engineering. Unfortunately, novelty of results reflects the extent and nature of knowledge used. Therefore, if the goal is exploring novelties, the key issue in intelligent computing is not the computing algorithm but acquiring proper knowledge. In this context, our paper on bio-inspiration is directly related to intelligent computing. Bio-inspiration may be considered as a potentially attractive source of design-relevant knowledge.

Traditional conceptual design is typically deductive. In most cases, the approach is to select a design from a variety of known design concepts and, at most, slightly modify it. No unknown or new design hypotheses/concepts are generated and therefore no abduction takes place. In accordance to Altshuller, such design paradigms are called "selection" and "modification," respectively [1, 2, 3, 4]. Gero [5] calls such paradigms "exploitation," because he views the designer as probing a relatively small, static, well-known, and domain-specific design representation space. Exploitation is relatively well understood and design researchers work on various methods and exploitation tools, with recent efforts focusing on evolutionary design [6].

I.F.C. Smith (Ed.): EG-ICE 2006, LNAI 4200, pp. 32 – 53, 2006.

We live in a rapidly changing world – one that constantly generates new challenges and demands for engineering systems that cannot be easily met by reusing or modifying known design concepts. This creates a need for novel design concepts, which are unknown yet feasible and potentially patentable. Such new and unknown concepts can be generated only abductively using our domain-specific knowledge as well as knowledge from other domains. Altshuller [1] refers to such design processes as "innovation," "invention," and "discovery," depending on the source of the outside knowledge. Gero [5] refers to such design paradigm as "exploration," because knowledge from outside a given domain is utilized.

Exploration represents the frontier of design research. Little is known about how exploration might be achieved and how the entire process could be formalized and implemented in various computer tools. Existing computer tools for exploration, including, for example, IdeaFisher (IdeaFisher Systems, Inc.), MindLink (MindLink Software Corporation), and WorkBench (Ideation International) are based on Brainstorming [7] Synectics [8] and Theory of Solving Inventive Problems (TRIZ)) [1], respectively. They all provide high-level abstract knowledge for designers seeking inspiration from outside their own domains. Unfortunately, these tools require extensive training, are not user friendly, and, worst of all, their output requires difficult interpretation by domain experts.

Exploration can be interpreted in computational terms as an expansion of the design representation space by acquiring knowledge from outside the problem domain and conducting a search in this expanded space. The key to exploration is knowledge acquisition. It can be conducted automatically using machine learning or manually by knowledge engineers working with domain experts. Machine learning in design knowledge acquisition is promising, but, unfortunately, the last fifteen years of research have produced limited results. Research clearly demonstrates that the use of machine learning to acquire design rules is feasible in the case of specific, well-understood and relatively small design domains [9]. Unfortunately, the practicality of using machine learning to acquire more abstract design rules is still not known.

This paper takes a different approach to knowledge acquisition. Our focus is on human design knowledge acquisition. In particular, we are interested in acquiring abstract design rules from various domains. In contrast to TRIZ, briefly discussed in Section 2, we want to acquire knowledge from outside the field of engineering, specifically from the natural sciences fields and especially biology. Knowledge acquisition is understood here as the process of learning abstract design rules, or creative design principia, which can help designers develop novel design concepts. These rules are not deterministic and are not always right. They are heuristics, representing potentially useful knowledge, but without any guarantee of their actual usefulness, or even of their relevance.

Heuristics in conceptual design can be considered as a source of inspiration from outside a considered design domain. Therefore, inspiration can be described as knowledge from outside the problem domain, in the form of a collection of weak decision rules or heuristics. This knowledge is needed and potentially sufficient to produce novel designs. For example, inspiration from the domain of pre-stressed concrete arch bridges could be used in the development of novel design concepts for large span arch steel bridges. We are particularly interested in heuristics related to the evolution of both human and animal body armor. Such heuristics are called "patterns of evolution" in the Directed Evolution method [10, 11]. They provide superior understanding of the evolution of engineering systems over long time periods, and are most valuable to designers working to develop their products in a specific direction through the use of novel design concepts.

Bio-inspiration looks to natural environmental processes for inspiration in engineering design. Bio-inspiration in design can be used on several levels, including nano, micro, and macro levels. The nano-level deals with individual atoms in the system being designed, the micro-level deals with the individual system's components, and the macro-level deals with an entire engineering system. Pioneering research at MIT exploring bio-inspiration on the nano-level focuses on structural and functional design from mollusk shells, called mother-of-pearl, to potentially improve human body armor [12]. This research may have a potentially significant impact on on the development of new types of body armor. Evaluation of its results and implementation, however, may be many years ahead. Therefore, our research focuses on bio-inspiration on both the micro- and macro-levels. In this case, the results may be used within a much shorter time frame and may also provide additional inspiration for research on the nano-level.

2 TRIZ and Directed Evolution

Directed Evolution (DE) is a method for the development of a comprehensive set of lines of evolution for an engineering system (also called a "scenario") over a long time period [10, 11]. A line of evolution is understood as a sequence of design concepts for a given engineering system. Obviously, a system can evolve along several lines of evolution, as is often the case when it is developed by competing companies or in various countries. A comprehensive set of lines of evolution is supposed to cover the majority, if not all, of the feasible lines of evolution. Such a set is intended for planning and/or design purposes.

The method has been developed entirely in an engineering context without any inspiration from nature. Its development began in the early 1990's, pioneered by Clark [10]. The method is related to TRIZ Technological Forecasting, developed since the mid-1970's and based on the principia of TRIZ. Altshuller proposed the initial concept of TRIZ in the late 1940's [1] and gradually developed it. Zlotin and Zusman [3] developed TRIZ into a number of versions. The fundamental tenants of TRIZ are that the solving of inventive problems (whose solutions are unknown and patentable) requires elimination of technical contradictions (for example, between stiffness and weight) and that it may be done using abstract engineering knowledge acquired from existing patents. The most popular version, Ideation-TRIZ (I-TRIZ), was developed and commercialized by a group of TRIZ experts in the research division of Ideation International. Our description of DE is based primarily on publications related to I-TRIZ.

The basic premise of DE is that evolution of engineering systems is driven by paradigm changes leading to novel design concepts. These changes can be understood as objective patterns of evolution. An example of identification of evolution and domain-specific evolutionary patterns is provided in [13]. The subject of analysis is patented joints in steel structures.

Evolution studies of thousands of patents identified nine patterns related to various engineering systems developed in many countries over long periods of time. These patterns are listed below with short descriptions based on [10]. They can also be observed in nature, although they have not been formally identified and described in biology. The patterns reported here enable interpretation of human body armor evolution in their context and, more importantly, provide a conceptual link to understanding body armor evolution in nature as discussed in the following sections.

1. Stages of evolution
An engineering system evolves through periods of infancy, growth, maturity, and decline. Its major characteristic versus time can be plotted as an S-curve. *Example*: evolution of airplanes over the last hundred years versus their speed.

2. Resources utilization and increased ideality
An engineering system evolves in such a direction as to optimize its utilization of resources and increase its degree of ideality, which is understood [10] as the ratio of all useful effects to all harmful effects. *Example:* evolution of I-beams.

3. Uneven development of system elements
A system improves monotonically but the individual subsystems improve independently and individually. *Example:* evolution of ocean tankers in which evolution of the propulsion system is not matched by evolution of the braking system.

4. Increased system dynamics
As an engineering system evolves, it becomes more dynamic and parts originally fixed become moveable or adjustable. *Example:* evolution of landing gear in airplanes.

5. Increased system controllability
As an engineering system evolves, it becomes more controllable. *Example:* evolution of heating and cooling systems.

6. Increased complexity followed by simplification
As an engineering system evolves, periods of growing complexity are followed by periods of simplification. *Example:* electronic watches were becoming more and more complex with many functions before the process of simplification began, leading to mechanical-like watches with only one or two functions. These simpler watches are in turn becoming more complex with new functions being added continually.

7. Matching and mismatching of system elements
As an engineering system evolves, its individual elements are initially unmatched (randomly put together), then matched (coordinated), then mismatched (separate s-curves), and finally a dynamic process of matching-mismatching occurs. *Example:* evolution of car suspension from a rigid axis to dynamically adaptive pneumatic suspension.

8. Evolution to the micro-level and increased use of fields
An engineering system evolves from macro to micro level and expands to use more fields. *Example:* 1. Evolution to the micro-level: computer based on tubes evolves into one based on integrated circuits, 2. Increased use of fields: a mechanical system uses an electric controller, next an electromagnetic controller, etc.

9. Evolution toward decreased human involvement
With the evolution of an engineering system, required human involvement systematically decreases. *Example:* every year improvements are made to reduce human involvement in the operations of the car's engine, brakes, or steering.

The patterns reported here enable interpretation of human body armor evolution in their context and, more importantly, provide a conceptual link to understanding body armor evolution in nature as discussed in the following sections.

3   Bio-inspiration in Conceptual Design

Bio-inspiration is understood by the authors as the design use of knowledge from biology, based on observations from nature. It is different than bio-mimicking, which can be described as a mechanistic use in engineering design of observations from nature, particularly regarding the form of living organisms [14]. There is still an open research question if bio-inspiration exists. It is reasonable to claim that evolution of engineering systems occurs in a closed engineering world, which is not affected by knowledge and processes occurring outside it. In this case, any similarities between evolution of living and artificial systems is simply coincidental. However, we have assumed that bio-inspiration exists and in accordance to [15], inspiration in design has been classified as "visual," "conceptual," or "computational inspiration," considering its character. This section builds on that research and investigates bio- and historical inspiration. In the case of bio-inspiration, inspiration is in the form of biological knowledge acquired from observations of nature. Historical inspiration is knowledge in the form of heuristics acquired from evolution of historical designs, as discussed in Section 4.1.

"Visual inspiration" has been widely used by humans for centuries. In nature, an evolutionary strategy, known as mimicry, results in a plant or animal evolving to look or behave like another species in order to increase its chances of survival. Animals frequently advertise their unpalatability with a warning pattern or color [16]. Over time and with experience, predators learn to associate that signal with an unpleasant experience and seldom attack these prey. A classic example of visual inspiration in nature is the Batesian Mimicry system in which a palatable prey species protects itself from predation by masquerading as a toxic species. In such a system, protection is gained through visual mimicry. Another less known strategy is behavioral mimicry as exhibited by the mocking bird [17]. In nature, the situation wherein an organism has evolved to be, superficially, like another based upon presence or observation of that other organism has not been documented. But humans use visual inspiration from nature to integrate into their myths and customs; warriors wear skins of crocodiles or leopards or animal masks to instill fear. Such cases highlight that, with humans, it is not even necessary to mimic *real* animals, only that the concept instills fear. There are many forms and variations of mimicry, but for the sake of this article, the main point is that nature can inspire innovation in the human mind. We are not taking this a step further to evaluate what drives evolution, nor to get into the religious/scientific arguments, but we are simply saying that natural processes offer inspiration for engineering design.

In design, visual inspiration can be described as the use of pictures (visuals) of animals or their organs to develop similar-looking engineering systems or system components. For example, a beetle could have inspired this Japanese body armor as shown in Fig. 1[1].

Fig. 1.

Virtual inspiration has long been the subject of scholars, particularly in the context of utilization in design of forms found in nature. Many interesting examples of various natural forms, potentially design-inspiring, are provided in [14, 18]. Unfortunately, visual inspiration is only skin-deep. A human designer must fully understand the functions of the various animal organs and be able to copy these organs in a meaningful way. In other words, in such a way that their shape and essential function are preserved – for example the protection of internal organs -while other secondary features may be eliminated. There is always a danger that the final product will preserve primarily the secondary features of the original animal weakening the effectiveness of visual inspiration.

"Conceptual inspiration" can be described as the use of knowledge in the form of heuristics from outside the design domain in order to develop design concepts. It is potentially more applicable to engineering design than visual inspiration. Conceptual inspiration is more universal since it provides not visuals but knowledge representing our understanding of an outside domain applicable to the design domain. In this case, a designer uses principia found in nature for design purposes. Various examples of such principia are discussed in Section 4.2. Design principia can be also interpreted as design rules, or design patterns [19]. In this context, knowledge acquired from nature may be formally incorporated in model-based analogy design [20].

---

[1] Beetle photo: http://home.primus.com.au/kellykk/010jrbtl.JPG, Image of Japanese Samuri – Imperial Valley College Located at Pioneers Park Museum 373 East Aten Road (Exit I-8 at Hwy 111 North to Aten Road) Imperial, CA 92251 Phone: (760) 352-1165 http://www.imperial.cc.ca.us/Pioneers/SAMURAI.JPG

Using conceptual inspiration is challenging. To employ it, an engineer must be trained in abstract thinking in terms of design principia. As our experience in learning and teaching Synectics indicates, such training is time-consuming and difficult. Also, not all engineers are able to learn how to think in abstract terms. The limitations of some engineers can be at least partially mitigated through the use of computer tools like "Gymnasium" (MindLink software). Despite these difficulties, conceptual inspiration is the most promising kind of inspiration, since it stimulates and utilizes the "creative power" of the human mind. Therefore, it is the subject of our research reported here.

"Computational inspiration" occurs on the level of computational mechanisms and/or knowledge representations, which are inspired by nature. It is intriguing and poorly understood, but nature offers a promise to revolutionize conceptual design. The state of the art in this area, including the direction of current research, is discussed in [6]. The area can be roughly divided into evolutionary computation [20], including co-evolutionary computation, and cellular automata [21]. In the first area, there are many examples of various applications of evolutionary algorithms in design. For example, genetic algorithms were used in the design of gas pipelines [22, 23], evolutionary strategy algorithms in the design of steel skeleton structures in tall buildings [24, 25], and genetic programming algorithms were used in the design of computer programs, electric circuits, and mechanical systems [26, 27, 28]. In the area of cellular automata, initial applications to structural engineering design are provided in [29, 30].

4 Evolution of Body Armor

While there are countless styles of ancient post-neolithic armor, there are seven major types of body armor, as listed in Table 1. This table summarizes historical armor types and shows a natural analogue. The table reveals parallel spectra of historic and natural body armors. The left column refers to historic armor and the while the right one to natural armor. The table demonstrates that all types of historic armor have their analogues in nature. However, most likely an opposite statement is not correct and this may represent great promise for the development of a new generation of human body armor conceptually inspired by nature in which various evolution patterns from nature are utilized in engineering context. In fact, artificial and natural body armors are usually considered separately and there is very little commonality in our understanding of both domains. The development of a unifying understanding might significantly improve our ability to design modern novel body armor that satisfies ever-growing requirements.

4.1 Human Body Armor

This section focuses on European metal body armor, primarily plate body armor. Its evolution is compared with natural processes and discussed in the context of the tradeoff between protection and mobility. Plate armor consists of a solid metal plate, or several plates covering most of the body, with articulations only at joints. This armor is heavy, prohibits fast movement, and although it can provide extensive protection, it has many drawbacks. A possible line of evolution for plate armor is shown in Fig. 2[2].

Table 1. Historical armor types and their natural analogues

Historical Armor type	Natural analogues
Plate (solid metal plates covering most of body, articulated only at joints)	turtles, molluscs, glyptodonts, arthropods
Lorica segmenta (Bands of metal encircling body, each overlapping the next)	numerous arthropods, armadillos
Jazeraint (small pieces of metal or other rigid material, each overlapping others)	fish, reptiles, pangolins
Lamellar (similar in function and design to Jazeraint, more typical in Asia)	See above
Maille (rivetted metal rings linked together, typically in a 4-1 weave)	
Cuirbolli (leather boiled in oil or wax to harden it, usually molded to a torso shape)	countless animals with thick hides or thin, flexible exoskeletons
Brigandine, reinforced (leather or cloth backing with metal or other rigid studs or small plates)	nodosaurs, sturgeon, crocodiles
Shield (any object primarily held in front of body to block attack	box crabs

The examples provided are mostly of Austrian, German, and Polish ancestry. However, they are representative of European trends. In medieval Europe, the development and manufacturing of body armor was mostly concentrated at a limited number of centers in Germany (Cologne and Nurenberg), Austria (Innsbruck), Italy (Milan), and Poland (Cracow) [31]. There was continual exchange of information among "body armor builders" who traveled and worked in various countries.

---

[2] All nature photos taken by Joanna Cornell and all pictures of historic body armor taken by Tomasz Arcsizewski, Fig. 2.7 taken from Woosnam-Savaga, R.C. and Hill, A., "Body Armor," Merlin Publications, 2000, and Fig. 2.8 from http://us.st11.yimg.com/store1.yimg.com/I/security2020_1883_28278127

The development of body armor is driven by seven main objectives:

1. Maximize energy absorption
2. Maximize energy dissipation
3. Maximize mobility
4. Minimize deformations
5. Minimize penetration
6. Maximize visual impact
7. Maximize noise/sound impact

These objectives can be synthesized as a single ultimate objective: *maximize battlefield survivability*. The seven objects can also be formulated as a technical contradiction in accordance with TRIZ: [1] *maximize body protection and maximize mobility*. The evolution of body armor over centuries illustrates how designers dealt with this contradiction in various time periods. In general, there is almost always a natural balance in body armor between protection and mobility. Very heavy armors, such as those worn by wealthy knights were never the choice of the average warrior. Such heavy armor was too expensive, cumbersome and hot for the foot soldier. The finest armors of Italian manufacture of the 14th and 15th centuries fit well and while they were heavy, they articulated so smoothly that the wearer could move freely and fight with ease. They were undoubtedly superior in protection. On the other hand, they were so expensive, required such skill to produce, and were fitted so specifically to the wearer that they simply could not be worn by any but the most wealthy and powerful. Poorly fitted heavy armor, while it provided great protection, was ultimately a handicap - an unhorsed knight might find himself lying helplessly like a turtle on its back. In such a case, social structure may have been an added defense, with the fallen knight being protected by a retinue of personal guards due to his status.

Human armor generally developed toward lighter designs with change being prompted by new threats. Several times in history, a new technology made it possible to lighten armor - for example iron replaced bronze. It seems that soldiers wore a relatively constant amount of armor until the widespread adoption of guns, which rendered most armors of the time ineffective. At that point, added weight became a liability rather than a defense, and armor was reduced to progressively more minimal levels. Eventually, it was shed altogether or reduced to a steel helmet. Another force behind the reduction of armor, aside from weapon technology, was the environment. For example, Spanish conquistadors arrived in the New World equipped in the European style field plate armor. It did not take long for these men to realize that armors fit for Europe could be fatal to their wearer in a tropical climate. In addition, European armors were not practical in tropical climates due to humidity and salt interactions. Many of Cortez's men opted for the lighter, cotton batting vests of the natives, keeping only their helmets and perhaps their breastplates, swords, shields and guns. There were similar situations during the crusades when heavily armored crusaders succumbed to heat.

Fig. 2.1 provides an example of Polish plate armor from the 9th century. It is a single breastplate in the form of a nearly flat surface of uniform thickness, a plate from structural point of view. Next, the armor (Poland, 12th century) evolved into a strongly curved surface with non-uniform or differentiated thickness (Fig. 2.1.), largest in the central part, which can be considered a shell from structural point of view. It was produced by hammering, a type of forging. A curved shape (a shell

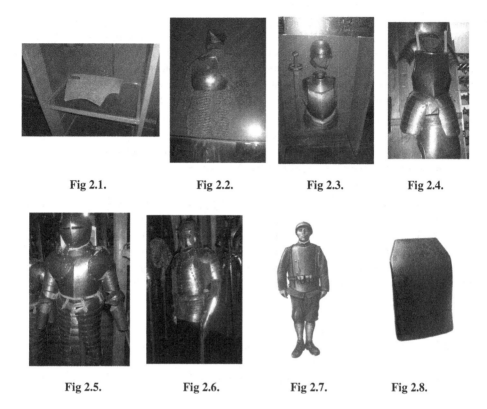

Fig 2.1.   Fig 2.2.   Fig 2.3.   Fig 2.4.

Fig 2.5.   Fig 2.6.   Fig 2.7.   Fig 2.8.

structure) is capable of better energy absorption capability than a flat surface (a plate structure). In nature, shells are usually curved. Again, we want to stress the point that those shells found in nature were not "inspired" by each other, but simply arrived at those designs because they worked better than any other. Natural selection is excellent at finding the "most perfect" (or "least imperfect") solution to any problem. Unfortunately, it is slow whereas our exponentially increasing computational power can evaluate all the dead ends and inefficient paths at a faster rate. The deformations of a shell structure are much smaller than those of a flat surface under the same impact force, significantly reducing internal injuries. In addition, forging hardens metal increasing its ability to withstand blows by sharp penetrating objects like spears or arrows. The evolution from a flat surface to a curved shell represents the use of the TRIZ's inventive principle of "spheroidaility." (This principle says, among others, "replacing flat surface with curved one.") [32]. There are also two specific armor design inventive principia (heuristics), which can be acquired from the described transition:

1. Differentiate thickness to reduce deformations
2. Use forging to reduce penetration

Both principia suggest ways to improve the behavioral characteristics of armor on a global level (the entire armor) or a local level (only the central part of the armor). One is related to the entire piece of armor and it provides a heuristic on a global level of the entire armor considered. The second one is related only to the central part of armor and provides a heuristic, which is valid only locally. Both, however, are complementary.

The third stage of evolution is illustrated in Fig. 2.3 with Polish 13th century one-piece body armor, which provides both front and back protection and even some side protection. The breastplate is so shaped that a rib is formed, which acts as a stiffener from the structural point of view. In this case, three inventive principia can be acquired:

3. Increase volume of a given piece of armor to absorb more energy
4. Increase the spatial nature of armor to improve its global stiffness
5. Introduce ribs to increase local stiffness wherever necessary

The next transformation resulted in a multi-piece armor in which front and back plates were separated (Fig. 2.4 - 14th century German armor). Also, additional multi-plate armor for upper arms and legs emerged, which allowed some degree of mobility. In this case, a well-known TRIZ inventive principle of segmentation is used ("divide an object into independent parts or increase the degree of object's segmentation"). This type of armor gradually evolved into full armor, shown in Fig. 2.5. (15th century Austrian armor). In this case, a simple inventive principle can be acquired:

6. To increase protection, expand armor

Fig. 2.6 shows Polish light cavalry ("husaria") armor from the 17th Century. The armor is significantly reduced in size and complexity and provides protection only for the vital parts of the body. This type of armor is considered a successful compromise between protection and mobility and was in use for several centuries. Its development was driven by a simple inventive principle:

7. Protect only battlefield vital body parts while providing maximum mobility

Finally, Fig. 2.6 and 2.7 provide examples of 20th century body armors. The first was developed in Italy during the 1st World War while the second is a modern ceramic breastplate, commercially available.

The entire identified line of evolution can be interpreted from a perspective of the Directed Evolution Method and its nine evolution patterns. In this way, better insight into the conceptual inspiration provided by the described evolution line can be acquired. The analysis provided below demonstrates also that engineering evolution patterns are valid for the considered line of evolution of body armor and provide its additional understanding.

1. S-curve Pattern
When battlefield survivability is considered, four periods of evolution can be distinguished. The first one, called "infancy," and represented by Fig. 2.1. and 2.2., ends approximately in the 10th century. During this period, first attempts were made in Europe to develop effective breastplates. Next, during the period of growth (approximately 10th – 14th century), rapid evolution of body armor can be observed, including the emergence of many new concepts and growing sophistication (see Fig. 2.3 and 2.4). The 16th century can be considered a maturity period, when progress stagnated and only various quantitative refinements occurred. Finally, after the 16th century the period of decline begins, when body armor was gradually reduced in size and complexity and its decorative function became progressively important. A new

S-curve begins in the 20th with the emergence of body armors during the 1st World War (Fig. 2.7) and of modern ceramic body armors (Fig. 2.8) developed only recently.

2. Resources Utilization and Increased Ideality

Evolution of body armor is driven both by rational and irrational factors. When relatively short time periods are considered (20-30 years), the resources are often utilized in a suboptimal way because of the tradition, autocratic inertia, culture, etc. However, when much longer time periods are considered (a century or two), the resources are usually utilized in an optimal way.

Ideality is understood as a ratio between the useful effects of body armor (survivability) and its harmful effects (immobility). New types of armor were developed, tested, and used over periods of time and survived the evolutionary process only because they had increased ideality with respect to their predecessors. Obviously, types of armor with decreased ideality were soon eliminated and we may not even know about them.

3. Uneven Development of System Elements

Unfortunately, this pattern is also valid in the case of body armor. The best example is a comparison between the development of corpus protection and eye protection. Breastplates rapidly evolved and improved in the early medieval centuries, providing excellent front protection. Evolution of helmets until recently has not provided sufficient eye protection.

4. Increased System Dynamics

In the case of body armor, increased system dynamics means increased mobility. This principle has been a driving force during the evolution of armor. A good specific example is the transition from a single breastplate (Fig. 4.5) to 14 interconnected narrow plates (Fig. 4.6). Such a configuration allows movement of individual plates and increases warrior mobility.

5. Increased System Controllability

Development of multi-piece body armor is a good example of this evolution pattern. Such armor allows precise positional control of the individual pieces with respect to adjacent pieces and allows their adjustment depending on battle conditions, mood of the knight, his changing weight as he ages, etc.

6. Increased Complexity followed by Simplifications

During stages 1 through 5 (Fig. 2.1 – 2.5) the complexity of armor grew, while stages 6 through 8 show subsequent simplifications (Fig. 2.6 – 2.8). The most recent ceramic body armor is multi-piece armor (not shown in Fig. 2), indicating that a new cycle has just begun. A single piece of ceramic body armor (Fig. 2.8) represents the beginning of this cycle.

7. Matching and Mismatching of System Elements

Body armor can be considered as a subsystem of a body protection system. At first, breastplates were simply put on the top of ordinary clothes (Fig. 2.1). Next, a breastplate was matched to the maille, but its evolution followed a separate S-curve and mismatching could be observed. Finally, the entire body protection system was considered as a single system and its subsystems, body armor and clothes, were being developed coordinating their development (Fig. 2.3 – Fig. 2.6).

8. Transition to the Micro-level and Increased use of Fields

The best example of transition to the micro level in the development of body armor is present research on armor materials in the context of nanotechnology. The gradually increased use of fields (temperature and stress fields) can be also observed. For example, in medieval times the initial cold hammering of armor plates evolved into various forging methods in which not only stress field (the result of hammering) but also heating and cooling (temperature field) were used. In modern days, production of ceramic plates requires sophisticated use of stress field during compression of ceramic materials and temperature field, which is applied to plates in an oven during the final baking process.

9. Transition to Decreased Human Involvement

One can interpret this to mean that body armor should gradually become easier to put on and adjust during use. This is most likely the case. During the last millennium, however, in Europe the labor costs did not affect the evolution of armor. On the contrary, during this period a culture of knighthood emerged in which complex armor-donning rituals held important social and psychological significance. A knight was constantly surrounded by many servants, whose official occupation was to take care of his armor. Their unofficial purpose was to serve as symbols of his social position and power. There was no strong push to reduce the number of servants or minimize the effort required to use body armor. The most recent experience with body armor in Iraq indicates that soldiers want effective, light, and easy-to-put-on armor, validating the decreased human involvement principle.

The above interpretation may also be applied to natural body armor, as discussed in the next section.

4.2 Animal Body Armor

The extent to which biological evolution (BE) and engineering evolution (EE) are directly comparable is still a research question. In BE, change proceeds as variations on a theme and must progress through incipient stages before a new functional structure is complete. Essentially, early mammals (or their genes) couldn't just *look* at a fly and then decide to grow wings and become bats. There had to be intermediate stages of design, which were inadequate for flight of any kind. EE can take leaps -- humans are capable of inventing armor on a different time scale than similar armor could evolve in a natural setting. It is the human brain -- human intelligence and creativity -- that distinguishes BE from EE. EE is not necessarily bound by the constraints of small baby steps. Also, living organisms evolve over log time periods in dynamic environments while a designer evolves his/her designs over a short time period operating in a closed world of his/her static body of knowledge representing the state of the art at the time of designing, as discussed in Section 3. For all these reasons, the chapter highlights the use of principia from natural evolution to stimulate and accelerate exploration of a design space. The ultimate goal is to find novel design concepts within a limited domain, which satisfy the requirements and constraints of a given engineering problem.

The parallels and principia that exist in the relationship between human armor and animal armor are particularly interesting because the evolution of both is driven by the same basic force: maximization of survivability. Poorly designed human armor results in higher mortality rates, and poorly evolved natural armor does the same and can even lead to extinction of a species. Although many other factors come into play when designing armor – for example, economics - increased survival is the basic driving force in the evolution of both human and animal armor.

In the natural world, a wide range of different species use plate armor for protection, with adaptations for movement. As discussed in Section 2, there are two contradictive requirements for armor: *maximize body protection and maximize mobility.* Interestingly, in nature separate lines of evolution can be distinguished in which the focus is only on a single requirement.

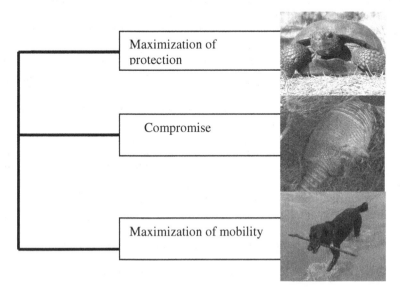

Fig. 3.

When considering the Gopher Turtle, Figure 3, it is evident that its body gains almost full protection from its plated shell. There is additional armoring along its legs. The fragile components of its body are protected underneath and on top with a thick plate. Humans have imitated this natural armor with increasing creativity. Early armor (see Fig. 2.4) weighed more than armor shown in Fig. 2.6 and impeded human movement. The Romans even imitated the function of the turtle shell with a military maneuver called the Tortoise or Turtle; in which soldiers marched in a rectangular formation with those at the head holding their shields in front, those on the side holding their shields to the side, and soldiers in the middle holding their shields over their heads. This created a box or turtle shell with all the men protected within. The analysis of body armor evolution in turtles results in several creative design principia/heuristics provided below:

  1.   Maximize size and volume of body armor
  2.   Create some smooth surfaces
  3.   Create multilayer body armor
  4.   Introduce shock absorbing layers

Over time, humans lightened armor and added more articulations. There are some evolutionary parallels that can be made and some heuristics learned. In this case, they can be formulated as:

5. *Minimize weight*
6. *Maximize articulation*

Although turtle shells have been comparatively stable morphologically for 200 million years, sustaining the popular conception of turtles as "living fossils," the turtle's skull, neck, and other structures have evolved diverse and complex specializations [33]. One interesting example is the Kayentachelys, the earliest known turtle to exhibit a shell that has all the features usually associated with an aquatic habitat. These include sharp, tapered edges along the low-domed shell, the absence of limb armor and coarse sculpturing on the shell [33]. Turtles have been around since the Mesozoic. Their basic body plan has served them well. However, not all turtles are heavily armored. Leatherbacks and softshells have, secondarily, lost their heavy shells. Apparently, these animals found such protection unnecessary. Rove beetles, too, are relatively lightly-armored against fast-moving predators, often ants. Apparently, speed is a better defense against outraged ants than all the armor in the world. Softshells and rove beetles are both examples of the flexibility of evolution, reverting back to speed instead of protection. Note that, in both cases, the original (pre-turtle or pre-beetle) condition was not the result, but a new version that achieved the same result.

Fig. 4[3].

Although a heavy-shelled turtle can successfully survive with its limitations, some organisms evolved by lightening their loads. An example is the three-spine stickleback, *Gasterosteus aculeatus*, Fig. 4. It is a widely studied fish featured in thousands of scientific papers [34]. It has three life-history modes: fully marine, resident freshwater, and anadromous (entering freshwater only to breed). Freshwater populations are theorized to have independently evolved from marine and anadromous ones [35, 36, 37]. Several of its marine characteristics changed repeatedly in the freshwater environment. For example, many aspects of extensive bony armor found in marine fish were reduced [37, 38, 39]. Marine sticklebacks are built for battle with prominent spines sticking out behind their lower fins and as many

---

[3] Photo of three-spined stickleback barrow, 2005. Aquarium Project. http://web.ukonline.co.uk/ aquarium/pages/threespinestickleback.html, due to copyright issues, we are unable to include photos that show the armored fish in salt water and the lack of armoring in freshwater, but the images are available online.

as 35 plates covering their bodies--presumably to fend off predators [34]. But once the sticklebacks evolved to live in freshwater habitats, their spines and plates were reduced or disappeared. There is an advantage to losing the armor [39]. Armoring and spines reduce speed. Fish living in lakes need to be faster because they typically have to hide from predators more often and there is an advantage in being able to dart into a hiding place quickly. Also, since fresh water lacks calcium reserves of salt water, bony armor could be too costly to make. Sticklebacks are unusual because a population can lose their armoring in just a few generations. It is this high rate of evolution that makes the Stickleback so popular to biologists, as it is rare that changes in nature follow such a quick time frame. The analysis yields a simple heuristic:

7. When operating in various environments, develop a flexible system, a body suite

Unlike turtles, armadillos can move quickly. Armadillos achieve a balance between armoring and mobility. The armadillo, considered to be an ancient and primitive species, is one of the only living remnants of the order Xenarthra. It is covered with an armor-like shell from head to toe, except for its underbelly, which is basically a thick skin covered with coarse hair [39] (Fig. 5).

Fig. 5.

The carapace (shell) is divided into three sections – a scapular shield, a pelvic shield, and a series of bands around the mid-section [40]. This structure consists of bony scutes covered with thin keratinous (horny) plates. The scutes cover most of the animal's dorsal surface. They are connected by bands of flexible skin behind the head, and, in most species, at intervals across the back as well. The belly is soft and unprotected by bone, although some species are able to curl into a ball. The limbs have irregular horny plates at least partially covering their surfaces, which may also be hairy. The top of the head is always covered by a shield of keratin-covered scutes, and the tail is covered by bony rings [41].

Limited information is available about the evolution of the armadillo. Its closest relatives are sloths and anteaters, which also belong to the order Xenarthra. Both relatives lack armoring. The order Xenarthra first arose around fifty million years ago [42]. Armadillos from 10,000 years ago were much bigger, so evolution decreased their size. One theory suggests that the giant armadillos became extinct due to an increase in large canine and feline predator migrations [43]. It could be theorized that a smaller size helped armadillo survival by lightening the weight of the body. In general, human-made armor has grown lighter in weight over time as long as protection was not significantly sacrificed. The advantage of the lorica segmenta design over traditional plate armor is that movement is much easier due to the increased number of moveable parts

Another unique example of balance between mobility and armoring is the pangolin [44]. A pangolin is covered with large, flat, imbricated, horny scales and it resembles the New World armadillo in terms of its feeding habits and use of a curled up, hedgehog-like defensive posture. Its body resembles a walking pinecone. The pangolin has what may be slightly less complete protection than that of the armadillo, but it has the benefit of greater flexibility. It is able to flex the entire length of its body in all directions and climb trees.

Regarding balance and tradeoffs between protection and mobility, it is important to consider a less obvious design constraint. Armoring does come at a cost to species -- otherwise animals of every lineage would be armored. As in the discussion of the stickleback, there may be metabolic costs associated with mineral-rich armor requiring specialized diets that are more laborious to procure. Also, while armadillos have the ability to move fast and even jump, their bodies are not flexible. This lack of flexibility may make them vulnerable to parasites, which could take up residence between their armor bands, transmitting diseases and remaining safe from the host's attempts to groom or dislodge them. These limitations apply to humans as well. Difficulties in removing armoring and maintaining proper hygiene may lead to consequences like illness or even death. Another major problem of the armadillo's design, as well as human armor throughout history, is one of thermoregulation. The armor surfaces do not insulate nearly as well as fur, nor can they sweat. This imposes limitations on the spatial and temporal range of the animal. Applied to human armor, thermo-regulation issues may result in illness or death, and this is clearly a concern to designers of modern body armor, intended for use in tropical climate zones. Both lack of flexibility and thermoregulation are issues in regards to natural and human body armor. These are just a few additional considerations to take into account when assessing mobility versus protection.

Modern armors exist which can protect a person against even the high-velocity rounds fired by assault, battle or sniper rifles. There are even complete body armors that, theoretically, can save a person from a direct blast by a modest sized bomb or Improvised Explosive Devices (IED). However, they are so bulky and restrictive that no army would field them in large numbers and no soldier would wear them for long periods. Throughout the centuries, there is a clear pattern found in both natural and human armor design: the simple principle to protect the head and vital organs first. Protection of everything else is perceived as a luxury. Limbs are usually left free to move, allowing troops to keep their best defense: mobility and the ability to fight back. The same is true in the animal world. Certain parts of the body are usually more heavily protected than others. Limbs are unnecessary for critical survival, but quite necessary for movement, and so are rarely armored. The heart, lungs, central nervous systems, etc. are usually well protected. Once again, constraints such as heat exhaustion and the need to remove the armor to perform basic body functions can limit the practicality of the heaviest armors.

The cost of armoring prevents all species from developing such protection. There is a range of other defenses, however. For example, canines and felines use speed, strength, and intelligence to survive without any armoring. It could be postulated that armor is exponentially more costly if the species is high on the food chain. There is an added cost to decreased mobility (speed and agility are critical to a predator), coupled with narrower energy margins limiting "disposable metabolic income" (i.e. an ecosystem can support far fewer lions than zebras, because of the inefficiency of predation and energy lost in digestion, respiration, defecation, and basic thermodynamics). A better comparison to the armadillo is the opossum, which is a similar-sized mammal, occurring in the same areas and feeding at a similarly mid-level trophic position, but without armor. The opossum lacks armor, but has other survival mechanisms, like a cryptic, nocturnal lifestyle and behavioral specializations.

Although our natural armoring discussion focused on several animals, armoring is evident in all living organisms. Fig. 5 shows a tropical palm tree armored to protect it from fire, and other environmental factors. Looking at the natural world through the lens of armoring results in fascinating observations. Humans are often able to think beyond mere survival and can apply creativity to abstract problem solving. In nature, the goal is more basic: to delay mortality while maximizing reproductive success.

Fig. 6.

In this paper we choose to focus on plate armor. However, a few points about maille armor should be made. Maille armor was the dominant form of armor for at least 2000 years, from before the Roman Empire until the 14th century. Many other types arose in the regions under discussion herein, but while several armors were used during that huge time span, none had the degree of flexibility, availability, ease of repair/replacement and general utility of maille. With proper padding, riveted maille was relatively lightweight and flexible and while it did not dissipate forces as well as the finest, fluted plate armors of the Renaissance, it was very effective at preventing the worst battlefield injuries. Perhaps maille's best attribute was that, while labor-intensive to produce, it required relatively little in the way of specialized skills, tools or facilities.

5 Summary

This chapter reports preliminary results of interdisciplinary research that addresses the challenging issue of conceptual bio-inspiration in design. This is accomplished in the context of evolution in nature and in engineering. In both cases, body armor is analyzed. The research proved to be more difficult than anticipated and can be considered only a first step in the direction of understanding conceptual bio-inspiration in design.

Evolution, as this chapter's central theme attests, is remarkably efficient at finding the optimum combination of traits for a given set of requirements and constraints. It is, both in nature and in the form of evolutionary computation, very slow because of its stochastic nature and depending on nothing more than guided trial and error and a lot of dead (literally in nature) ends. On the other hand, abstract design knowledge (design intelligence), driven by human creativity, can make *revolutionary* jumps, rather than merely *evolutionary* steps. Human creativity is the key to pushing the state of the art to new levels within directed evolution. By relocating the arena of the "design space" from the natural world to the human mind's theatre of conceptual abstraction and, now, to the digital theatre of computational modeling, many of those dead ends can be circumvented.

Directed Evolution Method, applicable only to engineering systems, can effectively circumvent many of those dead ends using abstract engineering knowledge in terms of evolution patterns. Whereas natural evolution relies on only what is given and on *random* operators as mechanisms producing raw variability, directed evolution can use "inspiration" from completely separate sources to push search for design concepts in very different directions. However, in the case of directed evolution still only engineering knowledge is used, although the entire engineering design space is searched, which is much larger than an engineering domain-specific design space. Such search is obviously an engineering exploitation with all consequences in the form of limited expectations to find truly novel design concepts.

Conducted analysis of evolution line of human plate body armor produced seven creative design principia/heuristics, which are abstract and may be used in the development of modern body armor. Similarly, the analysis of evolution of body armor in nature led to discovery of seven heuristics, which are also applicable to modern body armor. More importantly, the discovery of these heuristics means that the Directed Evolution Method can be expanded by incorporating knowledge from biology. Unfortunately, such expansion is still infeasible since much more heuristics must be discovered first. That will require extensive research involving both engineers and evolutionary biologists.

Evolution, both in nature and engineering, can be considered on various levels. The most fundamental level is that of basis operations (mutation and crossover) conducted on the genetic material in nature or on strings of allees describing design concepts. On this level, evolution has stochastic character and its results are often unpredictable. However, lines of evolution can be considered on the level of evolution patterns/heuristics driving evolution. Then, these principia can be discovered and compared for evolution in nature and in engineering. Such comparison can reveal missing "links" or principia for both types of evolution and may enable the creation of complete sets of heuristics. That raises an intriguing research question. If evolution principia in the cases of nature and engineering are comparable, is it possible to formulate a unified theory of evolution, which would be valid for both the biological and engineering evolution. If such theory is developed, it will have tremendous impact on our understanding of both nature and engineering. Even more importantly, such theory would change our understanding of design and would enable us to teach design in a truly holistic context. Also, such theory would help to design entirely different engineering systems inspired by nature.

Acknowledgement

The authors have the pleasure of acknowledging the help provided by Dr. Witold Glebowicz, the Curator, Department of Old History, Polish Army Museum, Warsaw, Poland. We would like to thank Andy May, a graduate student at University of South Florida, for his invaluable suggestions and technical feedback. Lastly, thanks to Paul Gebski, a Ph.D. student at George Mason University, for his technical assistance.

References

1. Altshuller, G.S.: Creativity as an Exact Science, Gordon & Breach, New York (1988).
2. Arciszewski, T., Zlotin, B.: Ideation/Triz: Innovation Key To Competitive Advantage and Growth. Internet http://www.ideationtriz.com/report.html (1998).
3. Terninko, J., Zusman, A., Zlotin, B.: Systematic Innovation, An Introduction to TRIZ, St. Lucie Press (1998).
4. Clarke, D.: TRIZ: Through the Eyes of an American TRIZ Specialist, Ideation (1997)
5. Gero, J.: Computational Models of Innovative and Creative Design Processes, special double issue. Innovation: the key to Progress in Technology and Society. Arciszewski, T., (Guest Editor), Journal of Technological Forecasting and Social Change, North-Holland, Vol. 64, No. 2&3, June/July, pp. 183-196, (2000)
6. Kicinger, R., Arciszewski, T., De Jong, K.A.: Evolutionary Computation and Structural Design: a Survey of the State of the Art. Int. J. Computers and Structures, Vol. 83, pp. 1943-1978, (2005).
7. Lumsdaine, E., Lumsdaine, M.: Creative Problem Solving: Thinking Skills for a Changing World, McGraw-Hill (1995).
8. Gordon, W.J.J.: Synectics, The Development of Creative Capacity (1961).
9. Arciszewski, T., Rossman, L.: (Editors) Knowledge Acquisition in Civil Engineering, The ASCE (1992).
10. Clarke, D.W.: Strategically Evolving the Future: Directed Evolution and Technological Systems Development. Int. J. Technological Forecasting and Social Change, Special Issue,: Innovation: The Key to Progress in Technology and Society, Arciszewski, T. (ed.), Vol. 62, No. 2&3, pp. 133-154, (2000).
11. Zlotin, B., Zusman, A.: Directed Evolution: Philosphy, Theory and Practice, Ideation International (2005).
12. Bruet, B.J.F., Oi, H., Panas, R., Tai, K., Frick, L., Boyce, M.C., Ortiz, C.: Nanoscale morphology and indentation of individual nacre tablets from the gastropod mollusc Trochus niloticus, J. Mater. Res. 20(9), (2005) 2400-2419 http://web.mit.edu/cortiz/www/Ben/BenPaperRevisedFinal.pdf Information about lab: http://web.mit.edu/cortiz/www/

13. Arciszewski, T., Uduma K.: Shaping of Spherical Joints in Space Structures, No.3, Vol. 3, Int. J. Space Structures, pp. 171-182 (1988).
14. Vogel, S.: Cats' Paws and Catapults, W. W. Norton & Company, New York and London, (1998).
15. Arciszewski, T., Kicinger, R.: Structural Design inspired by Nature. Innovation in Civil and Structural Engineering Computing, B. H. V. Topping, (ed.), Saxe-Coburg Publications, Stirling, Scotland, pp. 25-48 (2005).
16. Balgooyen, T.G.: Evasive mimicry involving a butterfly model and grasshopper mimic. The American Midland Naturalist Vol. 137 n1, Jan (1997) pp. 183 (5).
17. Wickler, W.: Mimicry in plants and animals. (Translated by R. D. Martin from the German edition), World Univ. Library, London, pp. 255, (1968).
18. D'Arcy, Thompson's, On Growth and Form: The Complete Revised Edition, Dover Publications, ISBN 0486671356, (1992)
19. Goel, A. K., Bhatta, S. R., "Use of design patterns in analogy based design", Advanced Engineering Informatics, Vol. 18, No 2, pp. 85-94, (2004).
20. De Jong, K.: Evolutionary computation: a unified approach. MIT Press, Cambridge, MA (2006).
21. Wolfram, S., New Kind of Science, Wolfram Media, Champaign, Il., (2002).
22. Goldberg, D.E., Computer-aided gas pipeline operation using genetic algorithms and rule learning, Part I: genetic algorithms in pipeline optimization, Engineering with Computers, pp. 47-58, (1987).
23. Goldberg, D.E., Genetic Algorithms in Search, Optimization, and Machine Learning, Addison-Wesley Pub. Co., Reading, MA, (1989).
24. Murawski, K., Arciszewski, T., De Jong, K.: Evolutionary Computation in Structural Design, Int. J. Engineering with Computers, Vol. 16, pp. 275-286, (2000).
25. Kicinger, R., Arciszewski, T., De Jong, K. A.: Evolutionary Designing of Steel Structures in Tall Buildings, ASCE J. Computing in Civil Engineering, Vol. 19, No. 3, July, pp. 223-238, (2005)
26. Koza, R. J., Genetic Programming II: Automatic Discovery of Reusable Programs, MIT Press, (1994).
27. Koza, John R., Bennett III, Forrest H, Andre, David, and Keane, Martin A.: Genetic Programming: Biologically Inspired Computation that Creatively Solves, MIT Press, (2001).
28. Ishino, Y and, Jin, Y., "Estimate design intent: a multiple genetic programming and multivariate analysis based approach", Advanced Engineering Informatics, Vol. 16, No 2, (2002), pp. 107-126.
29. Kicinger, R., Emergent Engineering Design: Design Creativity and Optimality Inspired by Nature, Ph.D. dissertation, Information Technology and Engineering School, George Mason University, (2004).
30. Kicinger, R., Arciszewski, T., and De Jong, K. A. "Generative Representations in Structural Engineering," Proceedings of the 2005 ASCE International Conference on Computing in Civil Engineering, Cancun, Mexico, July, (2005).
31. Arciszewski, T., DeJong K.: Evolutionary Computation in Civil Engineering: Research Frontiers. Topping, B.H.V., (Editor), Civil and Structural Engineering Computing pp. 161-185, (2001).
32. Zlotin, B. Zusman, A.: Tools of Classical TRIZ, Ideation International, pp. 266, (1999).
33. Eugene, S., Gaffney, J., Hutchison, H., Farish, A., Lorraine, J., Meeker, L.: Modern turtle origins: the oldest known cryptodire. Science, Vol. 237, pp. 289, (1987).

34. Pennisi, E.: Changing a fish's bony armor in the wink of a gene: genetic researchers have become fascinated by the threespine stickleback, a fish that has evolved rapidly along similar lines in distant lakes. Science, Vol. 304, No. 5678, pp. 1736, (2004).

35. McPhail, J. and Lindsey, C. Freshwater fishes of northwestern Canada and Alaska. Bulletin of the Fisheries Research Board of Canada (1970) 173:1-381.

36. Bell, M.: Evolution of phenotypic diversity in Gasterosteus aculeatus superspecies on the Pacific Coast of North America. Systematic Zoology 25, pp. 211-227, (1976).

37. Bell, M., Foster, S.: Introduction to the evolutionary biology of the threespring stickle-back. Editors: Bell, M.A., Foster, S.A., The evolutionary biology of the three spine stickleback, Oxford University Press, Oxford, pp. 1-27, (1993).

38. Bell, M., Orti, G., Walker, J., Koenings, J.: Evolution of pelvic reduction in threespine stickleback fish: a comparison of competing hypotheses. Evolution, No. 47, Vol. 3, pp. 906-914, (1993).

39. Storrs, E.: The Astonishing Armadillo. National Geographic. Vol. 161 No. 6, pp. 820-830, (1982).

40. Fox, D.L. 1996 January 18. Dasypus novemcinctus: Nine-Banded Armadillo. http://animaldiversity.ummz.umich.edu/acounts/dasypus/d._novemcinctus.html (November 3, 1999).

41. Myers, P.: Dasypodidae (On-line), Animal Diversity Web. Accessed February 08, 2006 at http://animaldiversity.ummz.umich.edu/site/accounts/information/Dasypodidae.html

42. Breece, G., Dusi, J.: Food habits and home range of the common long-nosed armadillo Dasypus novemcinctus in Alabama. In The evolution and ecology of armadillos, sloths and vermilinguas. G.G. Montgomery, ed. Smithsonian Institution Press, Washington and London, p. 419-427, (1985).

43. Stuart, A.: Who (or what) killed the giant armadillo? New Scientist. 17: 29 (1986)

44. Savage, R.J.G., Long, M.R.: Mammal Evolution, an Illustrated Guide. Facts of File Publications, New York, pp. 259, (1986).

Structural Topology Optimization of Braced Steel Frameworks Using Genetic Programming

Robert Baldock[1] and Kristina Shea[2]

[1] Engineering Design Centre, University of Cambridge, Cambridge, CB2 1PZ, UK
[2] Product Development, Technical University of Munich, Boltzmannstraße 15, D-85748
Garching, Germany
rdb27@cam.ac.uk, kristina.shea@pe.mw.tm.de

Abstract. This paper presents a genetic programming method for the topological optimization of bracing systems for steel frameworks. The method aims to create novel, but practical, optimally-directed design solutions, the derivation of which can be readily understood. Designs are represented as trees with one-bay, one-story cellular bracing units, operated on by design modification functions. Genetic operators (reproduction, crossover, mutation) are applied to trees in the development of subsequent populations. The bracing design for a three-bay, 12-story steel framework provides a preliminary test problem, giving promising initial results that reduce the structural mass of the bracing in comparison to previous published benchmarks for a displacement constraint based on design codes. Further method development and investigations are discussed.

1 Introduction

Design of bracing systems for steel frameworks in tall buildings has been a challenging issue in a number of high-profile building projects, including the Bank of China building in Hong Kong and the CCTV tower in Beijing, often due to unique geometry and architectural requirements. The complex subsystem interaction and design issues, coupled with the quantity of design constraints makes automated design and optimization of bracing systems difficult for practical use. These challenges are reflected in the volume of research within structural topology optimization that has addressed bracing system design, as discussed in the next section.

One difficulty with applying computational structural optimization in practice, especially to topological design, is that designers often find it difficult to interpret and trust the results generated, due to a lack of active involvement in design decisions during design evolution. Thus better means for following the derivation of and rationale behind optimized designs are required. In contrast to other evolutionary methods, Genetic Programming (GP) [1] evolves "programs" containing instructions for generating high-performance designs from a low-level starting point. This allows designers to examine the "blue-prints" of these by executing the branches of corresponding program trees. In common with other evolutionary methods, GP is population based and stochastic, facilitating the generation of a set of optimally-directed designs for further consideration according to criteria, such as aesthetic value, that are difficult to model computationally. Successive populations are developed through the genetic operations of reproduction, crossover and mutation. However, previous research using

I.F.C. Smith (Ed.): EG-ICE 2006, LNAI 4200, pp. 54–61, 2006.

genetic programming for structural optimization [2, 3] has been limited to evolving tree representations of designs, rather than programs for generating designs.

The next section discusses previous work in the field of bracing design for steel frameworks as well as relevant GP research. Section 3 introduces the proposed GP methodology. There follows a description and results of a test problem design task, taken from previous work by Liang et al [4]. Finally, results and conclusions are presented, noting potential extensions of the method for increased practicality and scale.

2 Related Work

Optimization of steel frame structures, including bracing systems, has been used as a demonstration problem for methods adopting discrete and continuum physical representations of bracing systems. Amongst discrete representations, Arciszewski et al [5] include general bracing system parameters in a demonstration of machine learning of design rules. Murawski et al [6] report a series of experiments in which evolutionary algorithms are used to seek optimal designs for a three-bay, 26-story tall building with type of bracing in each cell and connectivity of beams, columns and supports as variables. Section sizes are optimized using SODA [7]. Kicinger et al. [8] use an evolutionary strategy (ES), noting that this approach is more suited to small population sizes. This is relevant when objective function evaluation is computationally expensive, as is frequently the case in large-scale structural analysis. Kicinger [9] combines cellular automata and a genetic algorithm to generate and optimize designs, observing emergent behavior. Baldock et al. [10] previously applied a modified pattern search algorithm to optimize lengths of bracing spirals on a live tall building project. It is noteworthy that none of the above considers variation in size of basic bracing units, something that the current research aims to address.

Mijar et al. [11] and Liang et al. [4] use continuum structural topology optimization formulations to evolve bracing systems for simple two-dimensional multistory frames. A mesh of small 2D plane-stress finite elements is superimposed onto a vierendeel frame and elements are gradually removed by a deterministic process driven by minimizing the product of structural compliance and bracing tonnage. A three-bay, 12-story framework is adopted from Liang et al [4] as the test problem in this paper.

Genetic Programming (GP) is a class of evolutionary algorithm developed in the early 1990s [1], which manipulates tree representations containing instructions for solving a task, such as a design problem. Despite various attempts at using GP in civil engineering [12], in the field of structural topology optimization, to the authors' best knowledge, the full potential of GP has not been fully exploited. This is because functions have not taken the form of operations, but rather a component of the design itself [2] or an assembly of lower level components [3]. The current research aims to demonstrate how tree representations of the development of full bracing system designs from fundamental components can be manipulated by genetic operations to evolve optimally directed solutions.

3 Genetic Programming Method

Genetic Programming uses tree representations of solutions, with fundamental components as terminals or "leaves", operated on by internal function nodes. This has been applied literally in developing high performance computer programs as well as in electronic circuit design and other fields. Terminals take the form of constants or variables, while functions take a defined number of inputs and pass a result up the tree. Fig. 1 demonstrates a tree representation of the mathematical equation: $y = 5*(7-X) + 4/(X*X)$ evolved to fit a set of experimental data. Terminals may be constant integer values or the variable, X. Functions are chosen from the arithmetic operators: add '+', subtract '-', multiply '*', divide '/'.

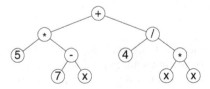

Fig. 1. The tree representation of a mathematical equation: $y = 5*(7-X) + 4/(X*X)$

The current implementation of the GP process is shown diagrammatically in Fig. 2. Applied to the problem of bracing system design, an orthogonal framework, with insufficient lateral rigidity, is taken as a starting point. An initial population of bracing system designs is created. Each individual is randomly generated by first seeding the framework with a number of bracing units (or *instances*) each occupying a single bay-storey cell, with column (x) and row (y) indices, as in Fig. 3a. A number of design modification operations (Fig. 4) are randomly selected and sequentially applied to single or united groups of instances to assemble a tree representation of a design (Fig. 3b). Each design modification operator has associated parameters describing the direction, frequency or magnitude of the operation relative to the instance on which it operates. The tree representation is the *genotype* of the design, on which the genetic operations are performed (eg. Fig. 3c,d) and which can be used to reassemble the physical representation of the design, its *phenotype*.

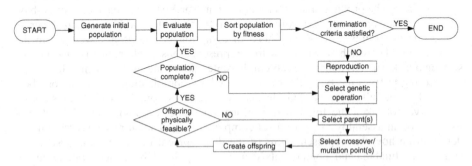

Fig. 2. Process flowchart for the genetic programming optimization process

Optimization parameters are population size, reproduction ratio (R), crossover probability (P_c), mutation probability ($P_m = 1-P_c$) and termination criterion. In creating a new generation, the best R% of individuals from the previous generation are *reproduced*, i.e. replicated exactly in the new population. Executing *crossover* or *mutation* operations, parent individuals are selected from the previous generation with linear weighting towards the fittest. Crossover is applied with probability P_c to two parent individuals, mutation is applied with probability P_m (=1-P_c) to a single parent. In both cases a branch (highlighted by dashed boxes in Fig. 3) is randomly selected from the parent and either replaced by a branch from another parent tree (crossover), or randomly regenerated (mutation). The optimization process terminates when neither the best-of-generation individual fitness nor the lowest average fitness of a generation has been improved for 10 generations. The method is implemented in Matlab.

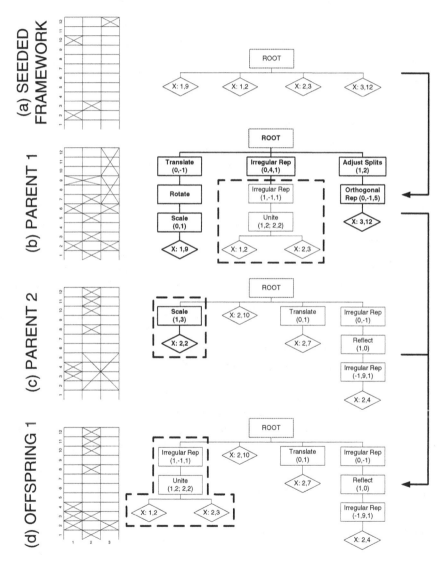

Fig. 3. Development of an individual in the initial population (a),(b) and the genetic crossover operation (c),(d)

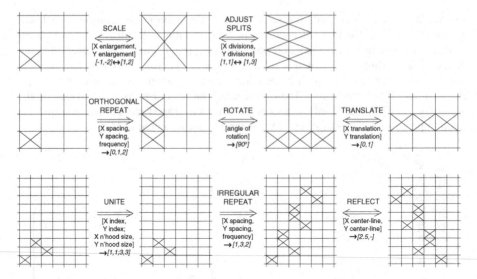

Fig. 4. Design modification operations, which are incorporated into GP trees as program functions. Beneath the arrows are detailed the function parameters appearing in Fig. 3.

4   Test Problem and Results

The test problem adopted in this paper was originally proposed by Liang et al [4] as a demonstration of compliance-driven Evolutionary Structural Optimisation. A planar 3-bay 12-story tall steel building framework is subjected to uniformly distributed loading on each side. The steel sections in the framework have been selected to meet strength requirements under gravity loading. Throughout the evolution of the bracing topology, the framework sections are fixed and gravity loading is neglected. The unbraced framework has a maximum lateral displacement of 0.660m[†] under the prescribed loading, well above the h/400 drift limit (h is total building height) of 0.110m. In the continuum solution published by Liang et al [4], using an element thickness of 25.4mm, the product of mean compliance and steel mass is minimized, yielding a total bracing volume of 4.82m^3 and a maximum lateral displacement 0.049m[†]. The ESO design is interpreted as a discrete bracing layout shown in Fig. 5, noting that further sizing optimization is required. The current research adopts a more practical objective of minimizing steel mass subject to a limit on maximum lateral displacement of 0.1m (just under h/400).

Applying the Genetic Programming method described to the above problem, the optimization model can be expressed as follows:

$$\text{Minimize:} \quad \overline{L} = \sum_{e=1}^{n} L_e + \max(0, p(\delta * - \delta_{\max})) \qquad (1)$$

[†] Reproduced and analysed in Oasys GSA 8.1 - also used for objective function evaluation.

where:

$\overline{L}$ = total length of bracing elements

L_e = length of bracing element e

n = total number of bracing elements

δ^* = limit on maximum lateral displacement

δ_{max} = maximum lateral displacement observed in structure

p = penalty factor imposed on designs violating constraint on maximum lateral displacement

The total number, length and location of bracing elements are variable in the evolutionary process. Fixed parameters in the structural model include framework geometry, applied loads and section size of bracing members (A_e), beams and columns. Issues of strength and buckling are recognized as important but not included at this stage for means of comparison.

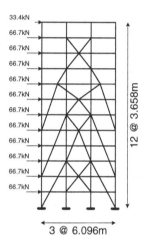

Fig. 5. Test problem geometry and loads, with discrete interpretation of optimal bracing layout from Liang et al [4]. Framework specifications can be found in [4].

The initial population of 30 individuals was developed by randomly applying between 2 and 12 design modification operations to between 1 and 5 seeded single-cell bracing units within the orthogonal framework. Bracing elements have a solid circular section of 190mm, 60mm or 30mm diameter, constant for a given run. Reproduction ratio, R=10%; crossover probability, P_c=0.8; mutation probability, P_m=0.2; penalty factor on infeasible designs, p=3000.

Fig. 6 shows a selection of randomly generated designs, illustrating the diversity obtained. Fig. 7 displays the best designs generated by each of four runs with different section sizes, also listing maximum lateral displacement, δ (with δ^* = 0.1m), total bracing length and total bracing steel volume.

A typical run of around 50 iterations required about 1500 function evaluations and took around 60 minutes to run on a PC with Pentium® 4 CPU 2.66 GHz and 512 MB RAM. An improvement in fitness of 20-30% was observed between the best initial randomly generated designs and the best evolved design in the final population.

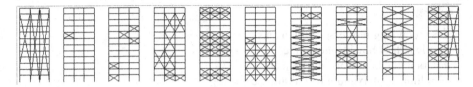

Fig. 6. Sample initial designs

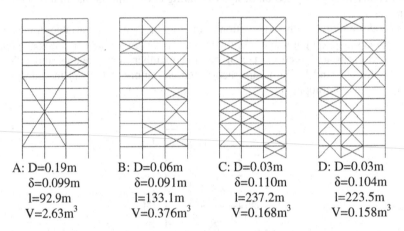

A: D=0.19m   B: D=0.06m   C: D=0.03m   D: D=0.03m
   δ=0.099m   δ=0.091m   δ=0.110m   δ=0.104m
   l=92.9m   l=133.1m   l=237.2m   l=223.5m
   V=2.63m³   V=0.376m³   V=0.168m³   V=0.158m³

Fig. 7. Sample best designs generated using different bracing section sizes

5 Discussion and Conclusions

This paper has demonstrated the potential for the application of genetic programming to the design of bracing systems in steel structures, through initial results on a small-scale example. The randomly generated initial designs of Fig. 6 demonstrate the increased diversity of design solutions offered by the current approach compared with previous research, e.g. [9], due to the capacity for bracing units to be larger than single bay-story cells. This is beneficial for creating novel solutions. For all section sizes used, the method has generated optimally directed solutions with considerably less steel volume than Liang et al's [4] continuum solution to meet the more practical h/400 displacement constraint. Although the solutions are not conventional in appearance, nor globally optimal, they efficiently provide required stiffness, with bracing in every story if required. There is substantial variation in the steel volume required for different section diameters. This indicates the necessity of simultaneous topology and section size optimization, which will be implemented in later stages of research. Previous research has included section size optimization using optimality criteria methods such as SODA [7], either for every design generated or as a selectively applied Lamarkian operator [3].

The method will now be extended to include local strength and buckling constraints, initially omitted for simplicity, possibly as further penalty terms in the objective function. Further extensions could be made to rationalize aesthetic requirements of size, pattern, and geometry.

Convergence studies, including parametric studies of optimization parameters will be carried out to fine-tune the performance and demonstrate robustness of the algorithm. In extending the complexity of the problem (larger or three-dimensional tubular framework), an increase in size of design space, and consequentially number of generations required for convergence, is expected, as well as time taken for each function evaluation.

Future research will endeavor to implement the additions described above and further demonstrate potential through application to a real-world building project of substantially greater scale, and in three-dimensions.

References

1. Koza, J.R.: Genetic programming: on the programming of computers by means of natural selection. Cambridge, MA: MIT Press. (1992)
2. Yang, Y., Soh, C.K. "Automated optimum design of structures using genetic programming" Computers and Structures (2002) 80: 1537-1546
3. Liu, P.: Optimal design of tall buildings: a grammar-based representation prototype and the implementation using genetic algorithms. PhD thesis. Tongji University, Shanghai.(2000)
4. Liang, Q.Q., Xie, Y.M., Steven, G.P.: Optimal topology design of bracing systems for multistory steel frames J. Struct. Engrg. (2000) 126(7) pp823-829.
5. Arciszewski, T., Bloedorn, E., Michalski, R.S., Mustafa, M., Wnek, J.: Machine learning of design rules: methodology and case study. ASCE J. Comp. Civ. Engrg. (1994) 8(2): 286-309.
6. Murawski, K., Arciszewski, T., De Jong, K.: Evolutionary computation in structural design. Engineering with Computers (2000) 16: 275-286.
7. Grierson, D.E., and Cameron, G.E.: Microcomputer-based optimization of steel structures in professional practice. Microcomput. Civ Eng. (1989) 4, 289-296.
8. Kicinger, R., Arciszewski, T., DeJong, K.: Evolutionary designing of steel structures in tall building. ASCE J. Comp. Civ. Engrg. (2005)
9. Kicinger, R.: Emergent engineering design: design creativity and optimality inspired by nature. PhD Thesis, George Mason University (2004)
10. Baldock, R., Shea, K., Eley, D.: Evolving optimized braced steel frameworks for tall buildings using modified pattern search. ASCE Conference on Computing in Civil Engineering, Cancun, Mexico.(2005)
11. Mijar, A.R., Swan, C.C., Arora, J.S., Kosaka, I.: Continuum topology optimization for concept design of frame bracing systems. J. Struct. Engrg. (1998) 5, p541-550
12. Shaw, D., Miles, J., Gray, A.: Genetic Programming within Civil Engineering. Organisation of the Adaptive Computing in Design and Manufacture Conference (2004) 20-22 April, Bristol, UK.

On the Adoption of Computing and IT by Industry: The Case for Integration in Early Building Design

Claude Bédard

École de technologie supérieure (ÉTS),
1100 Notre-Dame W., Montréal, Canada H3C 1K3
claude.bedard@etsmtl.ca

Abstract. Civil engineers were among the first professionals to embrace computerization more than 50 years ago. However computing applications in construction have been in general unevenly distributed across the industry. The significance of such a situation cannot be overstated, particularly in the North American context where fragmentation plagues the structure and the mode of operation of the industry. The paper attempts first to characterize the adoption of computing and IT tools by the industry, to describe the current status of this penetration as well as factors that prevent the practice from embracing the new technologies. Integrative approaches may hold the key to the development of a new generation of computing and IT tools that counteract effectively fragmentation in the industry. An on-going research project is briefly described to illustrate recent developments in the area of collaborative work and integration across disciplines for the conceptual design of building structures.

1 Introduction

Undoubtedly, professionals in the AEC industry (architecture, engineering, construction) are now routinely using computing and IT tools in many tasks. While this situation would indicate that the industry is keeping up with technological developments, a quick comparison with other industries such as automotive or aerospace reveals that computing applications in construction have been sporadic and unevenly distributed across the industry, with a major impact only on a few tasks/sectors. The significance of such a situation on the construction industry in North America cannot be overstated. It has resulted in a loss of opportunities, indeed competitiveness on domestic and foreign markets, and a level of productivity that lags behind that of other industries. Even among researchers and reflecting on the past 20 years of conferences about computing in construction, one can easily note a progressive "lack of enthusiasm" for computing research over the last few years, to the point where the frequency and size of annual events have been questioned (particularly true for ASCE-TCCIP, American Society of Civil Engineers – Technical Council on Computing and Information Technology).

This paper will first attempt to understand better the current status of IT use and developments in the AEC industry as well as the main roadblocks for widespread adoption of better tools and solutions by practitioners that should inform our collective R&D agenda. A research project will also be presented briefly to illustrate innovative ways of advancing integration in building design.

I.F.C. Smith (Ed.): EG-ICE 2006, LNAI 4200, pp. 62–73, 2006.

2 Computing and IT Use in AEC Industry

Civil engineers were among the first professionals to embrace computerization more than 50 years ago. Early prototype applications were rapidly developed for highly structured numerical tasks like bookkeeping, surveying and structural analysis. The adoption of computer-based solutions however entailed a significant level of investments in highly specialized resources – computer-literate technical staff, costly hardware, complex and unwieldy software – that only few organisations could afford like academia, some governmental services and large consulting firms. The availability of microcomputers in the early '80s signalled a turning point in the development, and subsequent adoption, of computer-based solutions by the majority of AEC firms. Twenty years later, one can argue that the majority of structured, single tasks have been successfully computerized and marketed to practitioners in the construction industry [1]. With the advent of new technologies like RFID, wireless, Bluetooth, GPS, internet-based services etc., computer-based solutions and tools appear to be accessible to all, mobile as well as ubiquitous.

Given the availability of such solutions, what can be said of their actual use and adoption by the AEC industry ? Three studies have been conducted in Canada in an attempt to answer these questions. On the current and planned use of IT and its impact on the industry, a survey by Rivard [2] in 2000 found that many business processes were almost completely computerized and the tendency was toward a greater computerization of the remaining processes. IT also raised productivity in most business processes and resulted in an increase in the quality of documents and in the speed of work, better communications, simpler and faster access to common data as well as a decrease in the number of mistakes in documentation. However, the benefits of IT came at a cost since the complexity of work, the administrative needs, the proportion of new operations and the costs of doing business all increased. Furthermore, although the Internet was adopted by most firms surveyed, design information was still exchanged in the traditional form. The two research topics that were clearly identified as the most important by industry were computer-integrated construction and better support for concurrent and conceptual design.

A second and related study in 2004 reported on eleven case studies from across Canada to define an initial compendium of Best Practice in the use of IT in construction [3]. The professionals interviewed included architects, engineers, general contactors and owners at the cutting edge in the use of IT. The documentation of their pioneering use of IT demonstrated how useful these technologies can be and what potential pitfalls are of concern. The following technologies were demonstrated : 3D CAD, commercial Web portals, and in-house software development. However, such a select group of professionals also pointed to a number of pragmatic issues that can impede significantly the use of IT in construction : a) the speed at which projects progress, b) money (always !), c) the difficulty of introducing a new CAD system, d) the cost to maintain trained personnel, e) the difficulty to champion IT when collaborators lag behind (e.g. small contactors), f) the necessity to maintain some paper work, and finally g) the implementation of an information system which has to focus on the construction process, i.e. on the work culture rather than on the technology.

Computing and information technologies can affect profoundly how information is generated and exchanged among collaborators in an industry that is highly fragmented as the AEC in North America. A third study was carried out recently among various stakeholders in construction projects to better understand the impact of information exchange and management [4]. The preliminary results indicate that people in construction prefer traditional, low-tech communication modalities. Table 1 shows to what extent each technology or communication mode is used by participants and how they perceive that such technology makes them more efficient. E-mails, with or without attached documents, is the most frequently used method of communication, followed by phone calls and face-to-face meetings. Similarly these methods of communications are perceived to contribute to personal efficiency. At the other end of the spectrum, groupware, planners with cell phone capacity, walkie-talkie type cell phones and chat appear to not be used frequently. Research participants also do not perceive these IT to contribute to their efficiency. Hence, there is consistency between IT usage and perceived contribution to personal efficiency for high and low frequency of IT usage. Documents obtained on FTP sites and regular cell phones are not contributing either to higher efficiency. In terms of which technology or communication mode was considered the most (or the second most) efficient as a

Table 1. Technology or communication mode. Frequency of usage and perceived efficiency (M: mean, SD: standard deviation).

a)   Frequency of usage

Technology or communication mode	IT usage	
	M	SD
Email without attached document	4.58	0.64
Email with attached document	4.54	0.58
Phone with one colleague	4.50	0.65
Face-to-face meetings	4.35	0.63
Fax	4.12	0.86
Regular cell phone	3.58	1.27
Private courier	3.42	0.90
Electronic planner without cell phone capacity	2.85	1.29
Phone or video conferencing	2.75	0.53
Document obtained from an FTP site	2.72	0.89
Portable computer on construction site	2.58	1.10
Pager	2.31	0.84
Chat	2.29	1.04
Walkie-talkie type cell phone	2.28	1.10
Electronic planner with cell phone capacity	2.24	1.09
Document obtained from web portal	2.17	0.95
Groupware	2.00	1.08
Note: Scale for frequency: 1=unknown technology, 2=never, 3=sometimes, 4=often, 5=very often.		

b)   Perceived efficiency

Technology or communication mode	Perceived efficiency because of IT usage	
	M	SD
Email with attached document	4.80	0.58
Face-to-face meetings	4.76	0.52
Email without attached document	4.76	0.52
Phone with one colleague	4.72	0.61
Fax	4.44	0.65
Private courier	4.12	1.01
Document obtained from an FTP site	3.78	1.54
Regular cell phone	3.64	1.66
Phone or video conferencing	3.33	1.55
Electronic planner without cell phone capacity	2.61	1.83
Document obtained from web portal	2.42	1.77
Portable computer on construction site	2.39	1.67
Chat	1.75	1.26
Walkie-talkie type cell phone	1.70	1.40
Groupware	1.68	1.29
Electronic planner with cell phone capacity	1.67	1.34
Pager	1.65	1.19

Note: Scale for efficiency: 1=does not apply, 2=strongly disagree, 3=somewhat disagree, 4=somewhat agree, 5=strongly agree.

function of key stakeholder, results clearly show that the telephone is the method of choice. Overall, participants favored using the phone individually to communicate with internal team members (69 %), with internal stakeholders (73 %), with clients (54 %), with professionals (62 %), with general contactors (50 %), and with higher management (58 %). With respect to which technology or communication mode was considered the most (or the second most) efficient as a function of project phase, results are also quite clear. Participants favored face-to-face meetings to communicate during the feasibility study (50 %), during construction design (46 %), during construction to coordinate clients, professionals and contractors (50 %), during construction to manage contactors and suppliers (54 %), commissioning (46 %), and during project close-out (39 %). Hence, participants clearly favoured traditional communication modalities such as the phone or face-to-face meetings, irrespective of project phase and internal or external stakeholders.

3   Impediments to Wider Use of Computing and IT

It is well known that the AEC industry represents a major segment of national economy, accounts for a significant proportion of the gross domestic product and the total workforce, yet lags behind other industrial sectors in terms of productivity, innovation and competitiveness, especially in the North American context. The deeply fragmented structure and mode of operation of the construction industry are to be blamed for such a situation. The implementation of integrative solutions throughout the entire building delivery process, i.e. among various people and products involved from project inception until demolition, would appear as key to counteract such fragmentation, with the adoption of computing and IT by the industry playing a capital role in facilitating the development of such integrated solutions. The aforementioned studies reveal a contradiction in the adoption of new technologies: on the one hand, computerization and IT can now be relied upon in many tasks performed by the majority of stakeholders in the AEC industry, yet on the other hand, promises brought by the new technologies remain unfulfilled, thus leaving practitioners to contend with new complexities, constraints and costs that make them stick with traditional approaches, with the ensuing poor performance. Many factors were pointed out in the above studies as impeding the adoption of computing and IT, and these corroborate the findings of other researchers.

At the 2003 conference of CIB W78 on Information Technology for Construction, Howard identified patterns in the evolution of IT developments over a 20 year-period in six areas as hardware, software, communication, data, process and human change. While he qualified progress in the first three as having surpassed initial expectations, he deplored only slow progress in the remaining areas – the lack of well organized, high quality building data and our inability to change either processes or peoples' attitudes [5]. Whereas CIB reports on the conditions of the construction industry world-wide, the above comments would only be more relevant to the North American context with a profoundly fragmented industry that is incapable of developing a long-term coherent vision of its own development nor to invest modest amounts to fund its own R&D. The few notable exceptions only cater to the R&D needs of their own members, such as FIATECH which groups a number of large capital projects construction/consulting companies in the US. Similarly with reference to computing support in the field of structural engineering, Fenves and Rivard commented on the drastic disparity between two categories of environments, generative (design) systems vs analysis tools, in terms on their impact on the profession. Generative systems produced by academic research have had negligible impact on the profession, unlike analysis tools, possibly because of a lack of stable and robust industrial-strength support environment [6]. One can argue also that engineers worldwide are still educated to view design as a predominantly number-crunching activity, like analysis for which computers represent formidable tools, rather than a judgment-intensive activity relying on qualitative (as well as quantitative) decisions.

In short, computing and IT advances have been numerous and significant in the AEC industry in terms of hardware, software and communications. However the industry remains profoundly divided and under-performing compared to its peers because these technologies are still incapable of accounting properly for human factors like :

- the working culture, style and habits, which ultimately determine the level of acceptation or resistance to change toward new environments ;
- the training needs of individuals who have to feel "at ease" with new technology in order to maintain interest and adopt it on a daily basis ;

- the interdisciplinary nature of communications, decision-making and projects which is poorly captured in automated support environments ;
- the intrinsic complexity and uncertainty of information used at the early stages of project development i.e. at the time when decisions have the greatest impact on the final product performance.

Research agendas for the development of computing and IT tools in construction must address the above human factors in priority if wider acceptance by the practice is pursued. Examples of promising avenues are given elsewhere [7]. As mentioned above, one of the most effective ways to counteract fragmentation in the industry is to promote the development of integrative solutions. In the long-term, integration should be as broad as possible and enable decision-making as early as possible in the process, at a time when decisions have the greatest impact on the overall facility life-cycle performance. This ambitious goal may not be reached for quite a long time yet although numerous IT developments to date have addressed some aspects of integration, like improved communications by means of exchange protocols. This low level of integration was made possible more than 20 years ago by industry-driven exchange protocols like IGES and DXF files for drawings, lately followed by the more general IFC's [8] which are progressively making possible effective communications across firms that are geographically dispersed, even among different disciplines and distinct project phases. However too many tasks in the building delivery process still lack the ability to communicate effectively with each other, by means of IFC's or otherwise i.e. to "interoperate". A recent survey about the situation in the US alone for capital projects evaluates the annual cost of such a lack of interoperability at 15.8 G $ [9].

There are many other characteristics of the construction industry that contribute also to slowing down, even hindering, the penetration of IT and computing in practice. For example, the fact that building projects produce a single unique product, erected once in an unprotected natural environment — unlike mass production in a manufacturing environment — has been discussed and documented for a long time [10], thus does not need repeating here. However what may be useful at this point is the presentation of a research project that attempts to achieve an integrated solution while accounting for some of the aforementioned characteristics. In the next section, the development of an innovative approach that endeavours to advance integration at the early stages of building design is described briefly.

4 Enabling Interactivity in the Conceptual Design of Building Structures

Nowadays, advanced computer modeling tools are available to support structural system generation, analysis, and the integration to the architecture [11]. This kind of support is model-based since it relies on the geometric and data modeling capabilities of a building information model (BIM) that combines the building architecture with other disciplines. Explicit knowledge can be used in conjunction with BIM's in the form of requirements. These requirements constrain the model and maintain its consistency when changes take place. This type of knowledge support could be called passive since it validates or confirms design decisions that have already been made. However, these tools lack the knowledge required to assist the engineer to explore design alternatives and make decisions actively. A knowledge-based approach is proposed that aims at providing interactive support for decision-making to help the engineer in the exploration of design alternatives and efficient generation of structural solutions. With this approach a structural solution is developed by the engineer from an abstract description to a specific one, through the progressive application of knowledge interactively.

Researchers have applied artificial intelligence (AI) techniques to assist engineers in exploring design alternatives over a vast array of possible solutions under constraints. Relevant techniques that have been explored over the last 30 years are: expert systems, formal logic, grammars, case-based reasoning (CBR) systems, evolutionary algorithms and hybrid systems that combine AI techniques such as a CBR system with a genetic algorithm. The impact of AI-based methods in design practice however has been negligible mainly because the proposed systems were standalone with no interactions with design representations currently employed in practice, such as BIM's. In fact, only few of the research projects [12] used architectural models with 3D geometry as input for structural synthesis. In the absence of such models, only global gravity and lateral load transfer solutions could be explored to satisfy overall building characteristics and requirements. These solutions needed actual architectural models to be substantiated and validated. Another disadvantage of the above research systems that hindered their practical use was that the support provided was mainly automatic and the reasoning monotonic (i.e. based on some input, these systems produced output that met specified requirements).

By contrast, a hierarchical decomposition/refinement approach to conceptual design is adopted in this research [13] where different abstraction levels provide the main guidance for knowledge modeling. This approach is based on a top-down process model proposed by Rivard and Fenves [14]. To implement this approach the structural system is described as a hierarchy of entities where abstract functional entities, which are defined first, facilitate the definition of their constituent ones.

Figure 1 illustrates the conceptual structural design process. In Figure 1, activities are shown in rectangles, bold arrows pointing downwards indicate a sequence between activities, arrows pointing upwards indicate backtracking, and two horizontal parallel lines linking two activities indicate that these can be carried out in parallel. For clarity, in Figure 1 courier bold 10 point typeface is used to identify structural entities. As shown in Figure 1, the structural engineer first defines independent structural volumes holding self-contained structural skeletons that are assumed to behave as structural wholes. These volumes are in turn subdivided into smaller sub-volumes called structural zones that are introduced in order to allow definition of structural requirements that correspond to architectural functions (i.e. applied loads, allowed vertical supports and floor spans). Independent structural volumes are also decomposed into three structural subsystems, namely the horizontal, the vertical gravity, and the vertical lateral subsystems (the foundation subsystem is not considered in this research project). Each of these structural subsystems is further refined into structural assemblies (e.g. frame and floor assemblies), which are made out of structural elements and structural connections. The arrangement of structural elements and structural connections makes up the "physical structural system". During activity number 2 in Figure 1 (i.e. Select Structural Subsystems), the engineer

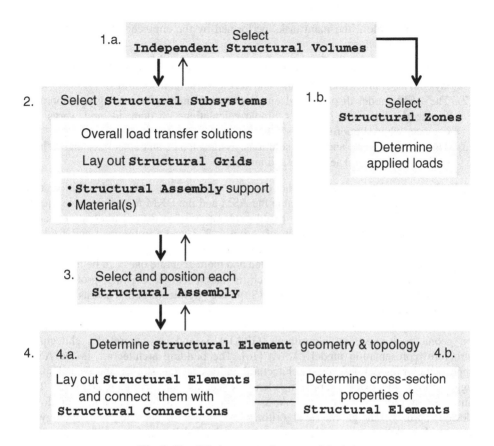

Fig. 1. Simplified conceptual structural design

defines overall load transfer solutions described in terms of supporting structural assemblies and corresponding material(s). Structural grids are also laid out during activity number 2 to assist in the validation of subsystem choices. These grids determine tentative vertical supports (at gridline intersections), structural bays, likely floor framing directions, and floor spans.

Interactivity is intended between a structural engineer, a simplified model of the building architecture and the structural system, Architecture-Structure Model (ASM) simplified for conceptual design, and a structural design knowledge manager (DKM). During the synthesis process, an architectural model is made available first to the engineer. Then, with the progressive use of knowledge from the DKM the structural system is integrated to the architecture and the result is an integrated architecture-structure model (ASM). Table 2 summarizes the types of interactions that take place at each step of the process between the engineer, the ASM and the DKM. In Table 2 a pre-processing and a post-processing activity in the process are included (unlike Figure 1). The pre-processing activity is an inspection of the architectural model, whereas the post-processing activity is the verification of the structural model.

As seen in Table 2 the main tasks performed by the engineer, the ASM and the DKM are the following:

(1) The engineer queries the ASM model, selects entities, specifies, positions and lays out assemblies and elements, and verifies structural solutions.
(2) The ASM model displays and emphasizes information accordingly, elaborates engineer's decisions, performs simple calculations on demand, and warns the engineer when supports are missing.
(3) The DKM suggests and ranks solutions, assigns loads, and elaborates and refines engineer's structural selections and layouts.

Each activity performed by the engineer advances a structural solution and provides the course of action to enable the ASM and the DKM to perform subsequent tasks accordingly. The knowledge-based exploration of structural alternatives takes place mostly at the abstraction levels of activities 2, 3, and 4 in Figure 1 and Table 2. At each subsequent level more information and knowledge is made available so that previously made decisions can be validated and more accurate ones can be made.

The implementation of the approach is based on an existing prototype for conceptual structural design called StAr (Structure-Architecture) that assists engineers in the inspection of a 3D architectural model (e.g. while searching for continuous load paths to the ground) and the configuration of structural solutions. Assistance is based on geometrical reasoning algorithms (GRA) [15] and an integrated architecture-structure representation model (ASM) [16]. The building architecture in the ASM representation model describes architectural entities such as stories, spaces and space aggregations, and space establishing elements such as walls, columns and slabs. The structural system is described in StAr as a hierarchy of entities to enable a top-down design approach. The geometric algorithms in StAr use the geometry and topology of the ASM model to construct new geometry and topology, and to verify the model. The algorithms are enhanced with embedded structural knowledge regarding layout and dimensional thresholds of applicability for structural assemblies made out of cast-in-place concrete. However, this knowledge is not sufficient for assisting engineers during conceptual design. StAr provides the kind of support described in the second column of Table 2, plus limited knowledge-based support (column 3) at levels 1.b and 4. Therefore, StAr is able to generate and verify a physical structure based on information obtained from precedent levels. However, no knowledge-based support is provided by StAr for exploration at levels 2, 3 and 4.

A structural design knowledge manager (DKM) is therefore developed that gets architectural and/or partial structural information from the ASM directly or via GRA to assist the engineer to conceive, elaborate and refine structural solutions interactively. Once the engineer accepts a solution suggested by the DKM, it automatically updates (i.e. elaborates or refines) the partial ASM. Architectural requirements in the form of model constraints (e.g. floor depths, column-free spaces, etc.) from the ASM model are also considered by the DKM for decision-making. The DKM encapsulates structural design knowledge by means of a set of technology nodes [17]. The type of knowledge incorporated in the nodes is heuristic and considers available materials, construction technologies, constructability, cost and

Table 2. Interactivity table between the engineer, the ASM and the DKM

Engineer	ASM	DKM
Architectural Model Inspection		
Query – Look for potential structural problems, continuous load paths to the ground and constraints. Select - Select elements that may become structural	Display the architectural model Emphasize continuous physical elements from this model Highlight architectural grids (i.e. main functional dimensions) Display global dimensional/layout constraints	N/A
1.a. Select **Independent Structural Volumes (ISV)**		
Query - Verify building shape, occupancies, lengths and proportions. Select - Select ISV by grouping spaces.	Emphasize spaces Compute overall building dimensions and aspect ratios	Suggest seismic/expansion joints if applicable
1.b. Select **Structural Zones**		
Query - Check types of spaces and associated constraints Select - Select structural zones by grouping spaces	Emphasize spaces Show space occupancies Display space layout/dimensional constraints	Assign loads to each zone based on its occupancy
2. Select **Structural Subsystems**		
Query - Inspect the model globally Select - Select structural subsystems and materials • Structural assembly support • Material(s) • Lay out structural grids	Display overall building characteristics Display global architectural layout/dimensional constraints Emphasize architectural elements selected to become structural	Suggest structural subsystems and materials Rank overall structural solutions
3. Select and position **Structural Assemblies**		
Select - Select each structural assembly Verify – Validate the initial description from level 2 Specify - Position each assembly Lay out - May determine preferred floor framing directions	Display structural grids Display applied loads Display local architectural layout/dimensional constraints Emphasize architectural elements selected to become structural	Suggest feasible structural assemblies Rank structural assemblies
4. Determine **Structural Element** geometry and topology		
Verify- Anticipate problematic supporting conditions locally Lay out - May position special structural elements and supports locally	Emphasize openings and irregularities in assemblies Elaborate - Make selected architectural elements structural Compute element loads based on tributary areas	Elaborate - Lay out and connect primary structural elements (within gridlines) Elaborate – Lay out and connect secondary structural elements Refine – Select preliminary cross-section shape and size of structural members
Structural system verification		
Verification - Verify and support still unsupported members Verification - Verify critical members	Warn about lack of supports and show unsupported elements	N/A

weight. A technology node represents the knowledge required to implement one design step (in the top-down hierarchy) utilizing a specific construction system or component. Nodes are organized into a hierarchy ranging from nodes dealing with abstract concepts (e.g. a structural subsystem) to those dealing with specific building entities (e.g. a reinforced concrete beam). The application of a technology node to a building entity from the ASM can be interpreted as making one decision about a design solution. Technology nodes support non-monotonic reasoning since they let the engineer retract any decision node and select another path in the technology tree.

72   C. Bédard

A fundamental difference between this approach and the AI-based techniques discussed above is that here the architectural model is created by an architect and not by an architecturally constrained AI system, and alternative structural subsystems and layouts are proposed by the engineer and not by the computer. The computer only evaluates alternatives and suggests solutions on demand. Following this approach, significant advantages accrue over commercial applications for structural model generation: (1) it facilitates design exploration by proposing feasible design alternatives and enabling non-monotonic reasoning, (2) it constitutes a more efficient method for conceptual structural design because it simplifies the design problem by decomposition/refinement, (3) it enables more integrated design solutions because it uses structural design knowledge to evolve an architecturally constrained building information model, and (4) it facilitates decision-making and early architect-engineer negotiations by providing quantitative evaluation results. This research work is in progress. A more detailed description is given elsewhere [13].

5 Conclusions

Practitioners in the AEC industry have benefited from computing and IT tools for a long time, yet the industry is still profoundly fragmented in North America, which translates into poor productivity and a lack of innovation compared to other industrial sectors. Recent surveys reveal a contradiction in the adoption of new technologies: on the one hand, they appear to be used in many tasks performed by the majority of stakeholders in the industry, yet on the other hand, they fall short of delivering as promised, thus leaving practitioners to contend with new complexities, constraints and costs that make them stick with traditional approaches, with the attending poor performance. The fact that critical human factors are not given due consideration in the development of new computing and IT tools can explain in part why such technologies are often not adopted by the practice as readily as expected. In this context, the development of integrated approaches would appear highly effective in counteracting the currently fragmented approaches to multidisciplinary building design. An on-going research project is presented briefly to illustrate innovative ways of advancing integration in the conceptual design of building structures.

References

1. Bédard, C. and Rivard, H.: Two Decades of Research Developments in Building Design. Proc. of CIB W78 20th Int'l Conf. on IT for Construction, CIB Report: Publication 284, Waiheke Island, New Zealand, April 23-25, (2003) 23-30
2. Rivard, H.: A Survey on the Impact of Information Technology on the Canadian Architecture, Engineering and Construction Industry. Electronic J. of Information Technology in Construction, 5(http://itcon.org/2000/3/). (2000) 37-56
3. Rivard, H., Froese, T., Waugh, L. M., El-Diraby, T., Mora, R., Torres, H., Gill, S. M., & O'Reilly, T.: Case Studies on the Use of Information Technology in the Canadian Construction Industry. Electronic J. of Information Technology in Construction, 9(http://itcon.org/2004/2/). (2004) 19-34

4. Chiocchio, F., Lacasse, C. Rivard, H., Forgues, D. and Bédard, C.: Information Technology and Collaboration in the Canadian Construction Industry. Proc. of Int'l Conf. on Computing and Decision Making in Civil and Building Engineering, ICCCBE-XI, Montréal, Canada, June 14-16, (2006) 11 p.

5. Howard, R.: IT Directions – 20 Years' Experience and Future Activities for CIB W78. Proc. of CIB W78 20th Int'l Conf. on IT for Construction, CIB Report: Publication 284, Waiheke Island, New Zealand, April 23-25, (2003) 23-30

6. Fenves, S.J. and Rivard, H.: Generative Systems in Structural Engineering Design. Proc. of Generative CAD Systems Symposium, Carnegie-Mellon University, Pittsburgh, USA (2004) 17 p.

7. Bédard, C.: Changes and the Unchangeable : Computers in Construction. Proc. of 4th Joint Int'l Symposium on IT in Civil Engineering, ASCE, Nashville, USA (2003) 7 p.

8. IAI (International Alliance for Interoperability) www.iai-international.org (2006)

9. NIST (National Institute of Standards and Technology): Cost Analysis of Inadequate Interoperability in the US Capital Facilities Industry. NIST GCR 04-867 (2004)

10. Bédard, C. and Gowri, K.: KBS Contributions and Tools in CABD. Int'l J. of Applied Engineering Education, 6(2), (1990) 155-163

11. Khemlani L.: AECbytes product review: Autodesk Revit Structure, Internet URL: http://www.aecbytes.com/review/RevitStructure.htm (2005)

12. Bailey S. and Smith I.: Case-based preliminary building design, ASCE J. of Computing in Civil Engineering, 8(4), (1994) 454-467

13. Mora, R., Rivard, H., Parent, S. and Bédard, C.: Interactive Knowledge-Based Assistance for Conceptual Design of Building Structures. Proc. of the Conf. on Advances in Engineering, Structures, Mechanics and Construction. University of Waterloo, Canada (2006) 12 p.

14. Rivard H. and Fenves S.J.: A representation for conceptual design of buildings, ASCE J. of Computing in Civil Engineering, 14(3), (2000) 151-159

15. Mora R., Bédard C. and Rivard H.: Geometric modeling and reasoning for the conceptual design of building structures. Submitted for publication to the J. of Advanced Engineering Informatics, Elsevier (2006)

16. Mora R., Rivard H. and Bédard C.: A computer representation to support conceptual structural design within a building architectural context. ASCE J. of Computing in Civil Engineering, 20(2), (2006) 76-87

17. Fenves S.J., Rivard H. and Gomez N.: SEED-Config: a tool for conceptual structural design in a collaborative building design environment. AI in Engineering, 14(1), Elsevier, (2000) 233-247

Versioned Objects as a Basis for Engineering Cooperation

Karl E. Beucke

Informatik im Bauwesen, Fakultät Bauingenieurwesen, Bauhaus-Universität Weimar
99423 Weimar, Coudraystr. 7, Germany
`karl.beucke@informatik.uni-weimar.de`

Abstract. Projects in civil and building engineering are to a large degree dependent upon an effective communication and cooperation between separate engineering teams. Traditionally, this is managed on the basis of Technical Documents. Advances in hardware and software technologies have made it possible to reconsider this approach towards a digital model-based environment in computer networks.

So far, concepts developed for model-based approaches in construction projects have had very limited success in construction industry. This is believed to be due to a missing focus on the specific needs and requirements of the construction industry. It is not a software problem but rather a problem of process orientation in construction projects. Therefore, specific care was taken to take into account process requirements in the construction industry.

Considering technological advances and new developments in software and hardware, a proposal is made for a versioned object model for engineering cooperation. The approach is based upon persistent identification of information in the scope of a project, managed interdependencies between information, versioning of information and a central repository interconnected via a network with local workspaces.

An implementation concept for the solution proposed is developed and verified for a specific engineering application. The open source project CADEMIA serves as an ideal basis for these purposes.

1 State of the Art in Engineering Cooperation

Projects in civil and building engineering generally require a close cooperation between separate engineering teams from various disciplines towards a common goal. Engineering cooperation is based upon communication processes that need to support the specific needs of engineering projects. Currently, by far most projects are still relying for these purposes on a set of Technical Documents that is exchanged between engineering teams. This is often referred to as the document-oriented approach.

The problems associated with this approach are well documented and quite apparent. Information is replicated in multiple ways and the consistency of information over the complete set of documents is difficult to ensure – if not impossible. It is safe to state that there is no major construction project with a complete and consistent set of Technical Documents.

I.F.C. Smith (Ed.): EG-ICE 2006, LNAI 4200, pp. 74–82, 2006.

Digital technologies have not changed this approach fundamentally yet. Most often application software is used to produce the same documents as before, digital exchange formats are used to interchange information between separate software applications and computer networks are used to speed up the process of information interchange. This optimizes individual steps in the process but does not change the fundamental approach with all its inherent problems.

For several years now, it was proposed to change this fundamental paradigm of engineering cooperation to a new approach based on a single consistent model as a basis for engineering cooperation. This is often referred to as the model-oriented approach. So far, success has been very limited. A lack of acceptance in the construction industry has even led to major uncertainties regarding commercial products developed for these purposes.

2   Specific Requirements for Engineering Cooperation

Engineering cooperation in major construction projects can never be regarded as a linear process of defining and refining information in a continuous manner with new solutions building upon reliable and final results of previous steps. Rather, because of the complexity of the problem separate engineering teams must be able to work synchronously in parallel (synchronous cooperation) and they must be able to synchronize the results of their work with results produced by the other teams at specified times of their choice.

Engineering solution finding does not take place as a steady process where each intermediate step of solution finding is immediately propagated to all others involved in the project. Rather, a design process generally requires longer periods of time working in isolation while trying and searching for an acceptable solution. Only the *result* of an elaborate design process is then communicated to the others but not each intermediate step of iteration towards a solution found acceptable by the engineer.

Interdependencies between information generated or modified simultaneously by other engineering teams are also important to handle. Currently, such interdependencies are mainly identified via the use of Technical Documents distributed between engineering teams and by personal communication between engineers. Application software today offers little functionality for monitoring such interdependencies and for helping the engineer to judge the consequences of modifications with respect to other information dependent upon that modification.

Finally, the current state of progress in engineering cooperation needs to be identified and communicated. This is commonly done with specified revisions of Technical Documents or on a digital basis with specified versions of computer files. Document Management Systems (DMS) have been developed in order to help the engineers for these purposes. Technologies like redlining were developed as an aid in that process.

Based upon these requirements, the following three-phase model is proposed as an appropriate model for supporting engineering cooperation in large construction projects. It is a model commonly used in process industry and it is also used in engineering teams working on the basis of Technical Documents.

76   K.E. Beucke

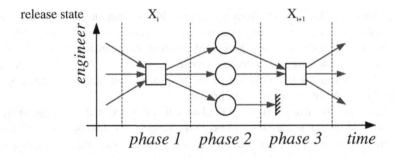

Fig. 1. Three phase model for engineering cooperation

In phase 1 all relevant information in a project defined up to that point is combined into a complete consistent set of information. This is commonly achieved in joint project meetings defining a new set of Technical Documents and associating with it a uniquely defined state of information (release state). Consistency of information is a prerequisite before a subsequent distribution.

Next, the consistent set of information is distributed to separate engineering teams. These are then allowed to work synchronously in parallel, accepting temporarily inconsistent states to develop in separate solutions. Each individual solution is consistent in its own, limited context but not necessarily consistent in the context of the complete project.

In phase 3 solutions developed separately in parallel need either be combined and synchronized again into a new consistent state of information in the context of the complete project or to be discarded. The new release state is again uniquely identified.

This process takes place iteratively a number of times until the required level and refinement of project information is achieved. In process industry, separate release states (X_i) are identified via a standardized naming system, thus enabling engineers to communicate progress of project information via the use of such standardized names, i.e. release state X_i is used to identify a specific level of detailing in the progress of a project.

3   Technical Advances and New Developments

Traditionally, engineering software had to cope with severe limitations on resources regarding storage capacities, speed of information processing and network capabilities. This has led to application software design that was highly optimized regarding an efficient use of these resources. Only data that was absolutely essential for an application was stored persistently. Other information was regenerated when needed or discarded. The complexity of algorithms for an evaluation of data was closely observed and optimized in order to ensure reasonable processing times. Information to be transferred via networks was condensed to an absolute minimum.

This situation has changed dramatically over the recent years. The capacity of permanent storage devices has increased to an extent believed to be impossible just a few years ago. There does not seem to be the need anymore for highly optimized and condensed data formats. The processing power of small computers is at a state that makes large High-Performance-Computers dispensable for most but the very largest problems in civil and building engineering. The bandwidth of computer networks has increased to such an extent that we are now able to share large sets of data between locally dispersed engineering teams.

Software technology has advanced from former procedural methods to object-oriented methods for design and implementation of application systems. Modern application software is built around a set of objects with attributes and methods that are strongly interrelated and connected amongst each other. Links between objects are modeled in different ways. The most common concept are internal references between objects. However, most applications still work with a proprietary object model that is not available to others. Some will allow application programmers to extend the object model with own, individual object definitions that are transparently embedded within the application, but still the core of the software is not transparent.

Object-oriented software systems conceptually would be able to support a persistent identification of information (objects) in the scope not only of its own system but even in a global sense in form of, for example, Globally Unique Identifiers (GUID) as required by systems that are distributed over the Internet. Persistent identification of information is believed to be of crucial importance for engineering cooperation via computer networks. Many "old" application systems will still not support such a concept. They will not store identifiers of objects permanently but rather generate them at runtime when an application is started. Data serialization into files in Java will also generate identifiers but these will be totally independent from internal identifiers in an application. Therefore, the identifiers generally will change from session to session and uniqueness of identification can not be ensured. Uniqueness of links between objects in separate applications would also require unique identifiers in the scope of the complete project. This must also be ensured when corresponding solutions are defined.

4   A Versioned Object Model for Engineering Cooperation

The first main aspect for application systems supporting engineering cooperation in networks is based upon establishing and maintaining permanent links between objects in application systems beyond the scope of the application in a namespace of a complete project. This idea was formulated in [1]. These links were called bindings in that publication in order to differentiate them from internal references in the object model of an application. Bindings can be generally formulated with the mathematical concepts of graphs and relations. The state of bindings between information can be recorded, organized and stored via binding relations in a binding graph.

The second main aspect is based upon versioning of objects. Versioning has long been in the focus of scientific research for different purposes. The early state of theoretical work was summarized in [2]. Specifications and requirements were formulated many of which are now generally available via the concepts of object orientation. Object versioning with a focus on applications in civil and building engineering was discussed in [3]. This work goes beyond theoretical concepts towards a system specification and implementation for the specific requirements in civil and building engineering. Information about objects in this context - when created, modified or deleted - is not lost but rather the "old" state is preserved and relevant "new" states will be generated in addition as new versions. In addition to the history of versions of objects, the evolution of an object is recorded and stored via version relations in a version graph. This was not considered to be reasonable before because of its requirements on resources. The concept of versioned objects provides a much better basis for flexible configurations of information in construction projects as opposed to a single rigidly defined configuration.

One major advantage of this approach is the opportunity to preserve the validity of bindings between objects (referential integrity). If, in the context of a project, bindings between objects are established within an application by different users or between separate applications, these bindings will possibly be invalid or wrong when the original object referred to was changed or deleted. If, however, the "old" object is preserved as a specific version and any modifications are reflected in a "new" version of that object, all bindings to the "old" object will still remain valid thus preserving referential integrity in the context of a complete engineering project. This is much like a new edition of a book in a library, where previous editions are not removed but rather kept in the library for any references to that book in order to remain valid and accessible.

The third main aspect is based upon the idea of private workspaces for supporting cooperation in phase 2 of the concept above. In phase 2 separate, isolated states of information are accepted to develop which are not necessarily consistent between each other. Each separate state is developed using specific application software for a period of days and maybe even weeks. This phase is regarded as a single long transaction. At specified points in time an engineer can decide or project guidelines may require to synchronize separate, individual results of such long transactions with a central data store called the repository. Any information produced in phase 2 is not immediately propagated into the repository but rather maintained in the private workspace and regarded as a deferred transaction.

The fourth aspect is based upon a distinction between application specific information kept in the object model of an application in form of attributes to objects and application independent information kept and maintained in the project and private workspaces in form of elements with specific features. This is necessary since object models of different applications must be supported which may even in some cases not be transparent to the users. Also, the process of selection of information from the repository must be supported independent from the functionality and models of individual applications. The engineers working in phase 2 must be able to query the repository or the private workspace with functionality that reflects their specific needs independent from specific applications and across the functionality of different applications. Such features of elements are modeled via an approach developed originally in [4]. The original approach was adopted in [5] for problems related to Software Configuration Management (SCM). In the context of this work it is called Feature Logic and serves as the basis for a corresponding query language.

Unique links will be established between the elements and application specific objects. Not considered in this contribution but a matter of further research would be the implementation of the elements proposed in this context as Industry Foundation Classes (IFC) developed under the guidance of the International Alliance for Interoperability (IAI) Modeling Support Group [6].

Finally, much work has been done on the topics of Change Management and Revision Management. Most of this work has been done in collaborative software development under the term Software Configuration Management - SCM (e.g. [7]). Major projects with hundreds of contributors, thousands of files and millions of lines of code are managed with Version Control Systems (VCS). Several of these systems were investigated for its suitability in engineering cooperation. An initial approach was based upon the software objectVCS [8]. Eventually, it was concluded that much of the functionality offered by these systems can be used very well in the context of engineering cooperation. The software Subversion [9] was consequently selected for the purposes of this work [10]. It is based upon the concept of a central repository and an additional set of several local environments – called Sandboxes. A Sandbox consists of project data and additional versioning information required for the synchronization with the central repository. It is stored in a specific hierarchy in the file system. Operations required for that approach are given in Fig. 2.

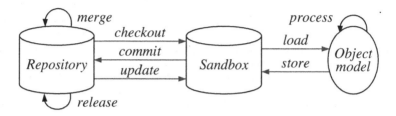

Fig. 2. Operations required for the Application with a specific Object model, for the Sandbox and Repository

An application with a corresponding object model can be processed with the functionality provided by the application which is further enhanced to load objects from and store objects into a local Sandbox. The Sandbox may be connected to a centrally organized Repository with functionality for checking out objects, for committing objects into it and for updating objects in the Sandbox. Specific release states may be defined for the Repository and it may be merged with another Repository.

5 Implementation Concept

Based on the concepts outlined above the following proposal for an implementation was developed for the support of synchronous engineering cooperation in civil and building engineering projects (Fig. 3):

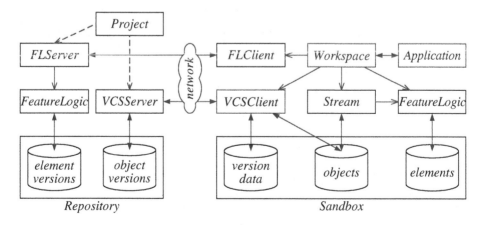

Fig. 3. System Architecture: Components of the Sandbox and Repository

The complete project data are stored and maintained in a Repository on a central server accessible via Internet technology. Object versions are maintained by a version control system (VCS). Element information needed for a specific purpose can be selected with extended functionality of the Feature Logic.

A number of private workspaces are connected to an application and are able to work independently from a connection to the central project data within a Sandbox environment which contains objects, elements and version data.

The system architecture and implementation concept are explained in detail in [10]. The key elements of this concept are the ideas that an application should not just store the latest version of an object but rather that it should store the version *history* of the objects involved in a project. Also, in order to support independent engineering work, a Sandbox environment allows to work independent from network restrictions. Finally, local Sandbox information may be synchronized with central project information.

Applications operating on workspace information can either be specific implementations designed for such an environment or also commercially available products if they satisfy certain requirements. The implications in utilizing the concept proposed in conjunction with existing commercial applications are discussed in [11]. Such systems must be built upon object-oriented technology with an individual object model and they must provide an Application Programming Interface (API) in order to extend the standard product by the functionality and commands required for the workspace connections. An example would be the software AutoCAD with its API called AutoCAD Runtime Extension (ARX).

6 Engineering Applications

The Open Source project CADEMIA [12] is an engineering platform that serves as an ideal basis for a verification of the concepts outlined above.

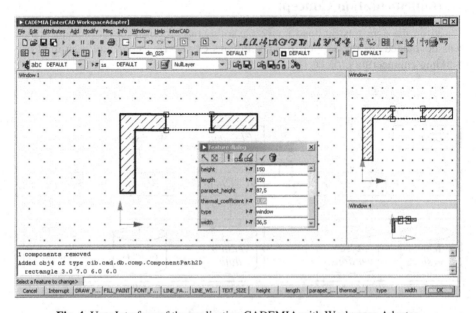

Fig. 4. User Interface of the application CADEMIA with Workspace Adapter

The project was implemented in the object-oriented programming environment Java [13]. The source code is fully accessible via the Internet and can easily be extended by individual commands that are fully embedded into the software and by individual object definitions.

The existing software was enhanced by the workspace functionality and features required and verified for the concepts developed. A detailed discussion of the system design is beyond the scope of this paper and can be found under [12].

The engineering application CADEMIA in conjunction with the proposed concept for engineering cooperation so far was only tested under research conditions. The next step will be to verify the concept in a practical environment. For these purposes, a proposal for a so-called "transfer project" with the German Research Foundation (DFG) is under preparation. A transfer project requires the involvement and active contribution of an industrial partner. The largest German construction company has committed to act as a cooperation partner in the proposed transfer project.

A second application is currently being developed for structural analysis. A new object-oriented FEM code is developed [14] in accordance with the proposed concept for engineering cooperation. Specific requirements for engineering cooperation in structural analysis are under investigation.

Acknowledgements

The author gratefully acknowledges the financial support by the German Research Foundation (Deutsche Forschungsgemeinschaft DFG) within the scope of the priority program 'Network-based Co-operative Planning Processes in Structural Engineering'.

References

1. Pahl, P.J., and Beucke, K., "Neuere Konzepte des CAD im Bauwesen: Stand und Entwicklungen". Digital Proceedings des Internationalen Kolloquiums über Anwendungen der Informatik und Mathematik in Architektur und Bauwesen (IKM) 2000, Bauhaus-Universität Weimar.
2. Katz, Randy H., "Towards a Unified Framework for Version Modeling in Engineering Databases", ACM Computing Surveys, Vol. 22, No. 4, December 1990.
3. Firmenich, B., „CAD im Bauplanungsprozess: Verteilte Bearbeitung einer strukturierten Menge von Objektversionen", PhD thesis (2001), Civil Engineering, Bauhaus-Universität Weimar.
4. Smolka, G., "Feature Constraints Logics for Unification Grammars", The Journal of Logic Programming (1992), New York.
5. Zeller, A., "Configuration Management with Version Sets", PhD thesis (1997), Fachbereich Mathematik und Informatik der Technischen Universität Braunschweig.
6. Liebich, T., "IFC 2x, Edition 2, Model Implementation Guide", Version 1.7, Copyright 1996-2004, (2004), International Alliance for Interoperability.
7. Hass, A.M.J., "Configuration Management Principles and Practice", The Agile software development series. Boston[u.a.], (2003), Addison-Wesley.

8. Firmenich, B., Koch, C., Richter, T., Beer, D., "Versioning structured object sets using text based Version Control Systems", in Scherer, R.J. (Hrsg); Katranuschkov, P. (Hrsg), Schapke, S.-E. (Hrsg.): CIB-W78 – 22nd Conference on Information Technology in Construction: Institute for Construction Informatics, TU Dresden, Juli 2005.
9. Collins-Sussman, B., Fitzpatrick, B. W., Pilato, C.M, "Version Control with Subversion", Copyright 2002-2004, http://svnbook.red-bean.com/en/1.1/index.html (2004).
10. Beer, D. G., „Systementwurf für verteilte Applikationen und Modelle im Bauplanungsprozess", PhD thesis (2006), Civil Engineering, Bauhaus-Universität Weimar.
11. Beucke, K.; Beer, D. G., Net Distributed Applications in Civil Engineering: Approach and Transition Concept for CAD Systems. In: Soibelman, L.; Pena-Mora, F. (Hrsg.): Digital Proceedings of the International Conference on Computing in Civil Engineering (ICCC2005)American Society of Civil Engineers (ASCE), July 2005, ISBN 0-7844-0794-0
12. Firmenich, B., http://www.cademia.org, (2006).
13. SUN, JavaTM 2 Platform, Standard Edition, v 1.5, API Specification, Copyright 2004 Sun Microsystems, Inc.
14. Olivier, A.H., „Consistent CAD-FEM Models on the Basis of Object Versions and Bindings", International Conference on Computing in Civil and Building Engineering XI, Montreal 2006.

The Effects of the Internet on Scientific Publishing – The Case of Construction IT Research

Bo-Christer Björk

Department for Management and Organisation,
Swedish School of Economics and Business Administration,
Helsinki, Finland
Bo-Christer.Bjork@hanken.fi

Abstract. Open access is a new Internet-enabled model for the publishing of scientific journals, in which the published articles are freely available for anyone to read. During the 1990's hundreds of individual open access journals were founded by groups of academics, supported by grants and unpaid voluntary work. During the last five years other types of open access journals, funded by author charges, have started to emerge. In addition a secondary route for better dissemination of research results has been established in the form of either subject based or institutional repositories, in which researchers deposit free copies of material, which has been published elsewhere. This paper reports on the experiences of the ITcon journal, which also has been benchmarked against a number of traditional journals in the same field. The analysis shows that it is equal to its competitors in most respects, and publishes about one year quicker.

1 Introduction

Scientific communication has gone through a number of technology changes, which fundamentally have changed the border conditions for how the whole system works. The invention of the printing press was of course the first, and IT and in particular the Internet the second. Currently we are witnessing a very fast change to predominantly electronic distribution of scientific journal articles. Yet the full potential of this change has not been fully utilised, due to the lack of competition in the area of journal publishing, and the unwillingness of the major publishers to change their currently rather profitable subscription-based business models. In the early 1990's scattered groups of scientists started to experiment with a radical new model, nowadays called Open Access, which means that the papers are available for free on the Internet, and that the funding of the publishing operations are either done using voluntary work, as in Open Source development or Wikipedia, or lately using author charges.

Originally scientific journals were published by scientific societies as a service to their members. Due to the rapid growth in the number of journals and papers during the latter half of the 20th century, the publication process was largely taken over by commercial publishers. Due to the enormous growth in scientific literature a network of scientific libraries evolved to help academics find and retrieve interesting items, supported by indexing services and inter-library loan procedures.

I.F.C. Smith (Ed.): EG-ICE 2006, LNAI 4200, pp. 83–91, 2006.
© Springer-Verlag Berlin Heidelberg 2006

This process worked well until the mid nineteen nineties. The mix between publicly funded libraries on one hand and commercial publishers and indexing services on the other was optimal, given the technological border condition. The quick emergence of the Internet, where academics were actually forerunners as users, radically changed the situation. At the same time there has been a trend of steadily rising subscription prices ("the serials crisis") and mergers of publishers. Today one publisher controls 20 % of the global market. In reaction, a new breed of publications emerged, published by scientists motivated not by commercial interests, but by a wish to fulfil the original aims of the free scientific publishing model, now using the Internet to achieve instant, free and global access. The author of this paper belongs to this category of idealists, and has since 1995 acted as the editor-in-chief of one such publication (the Electronic Journal of Information Technology in Construction). He was participated in the EU funded SciX project, which aimed at studying the overall process and at establishing a subject-based repository for construction IT papers.

There is widespread consensus that the free availability of scientific publications in full text on the web would be ideal for science. Results from a study carried out by this author and his colleague [1] clearly indicated that scientists prefer downloading papers from the web to walking over to a library. Also, web material that is readily available, free-of-charge is preferred to that which is paid for or subscription based (Figure 1).

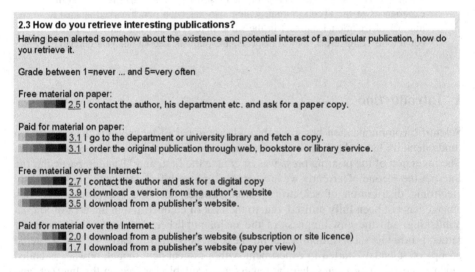

Fig. 1. Some results from a web-survey of the reading and authoring habits of researchers in construction management and construction IT [2]. Material, which is available free-of-charge on the web, is the most popular means for accessing scientific publications.

There is, however, a serious debate about the cost of scientific publishing on the web [3]. It is clearly in the interest of commercial publishers to claim that web publishing is almost as expensive as ordinary paper based publishing, in order to justify the increasingly expensive subscriptions. Advocates of free publishing cite case examples of successful endeavours where costs have been markedly lower [4]. While commercial publishers state that the publishing costs per article are from 4000 USD upwards, it is interesting to note that the two major open access publishers, commercially operating BioMedCentral and the non-profit Public Library of Science, charge authors a price of between 1000 and 1500 USD for publishing an article. Recently a number of major publishers (Springer, Blackwell, Oxford University Press) have announced possibilities for authors to open up individual articles at prices in the range of 2500-3000 USD.

It is not only the publishing itself, which is becoming a battleground between commercial interests and idealistic scientists. Since the emergence of data base technology in the 1960's a number of commercial indexing services have emerged, which libraries subscribe to. Traditionally these have relied on manual and or highly structured input of items to be included, a costly and also selective (and thus discriminatory) process. Now scientists are building automated web search engines which use the same web crawler techniques as used by popular tools, such as Google, and which apply them to scientific publications published in formats such as PDF or postscript. These are called harvesters and rely of the tagging of Open Access content using a particular standard (OAI) If technically successful, such engines can be run at very low cost and thus be made available at no cost. The combination of free search engines and eprint repositories is providing what is called the green route to Open Access. Currently around 15 % of journal papers are estimated to be available via this route.

A repository for construction IT papers was set up as part of the EU-funded SciX project (http://itc.scix.net/). Currently the repository houses some 1000+ papers, with the bulk consisting of the proceedings of the CIB W78 conference series going as far back as 1988. This was achieved via digitising the older proceedings. The experiences concerning the setting up of the repository are described more in detail elsewhere [5].

The overall experience with the ITC repository is mixed. Ideally agreements should have been made with all major conference organisers in our domain for uploading their material, at least in retrospect. This was, however, not possible due to copyright restriction, the ties between conference organisers and commercial publishers, fears of losing conference attendees or society members if papers were made freely available etc. As a concrete example take this conference. After this author has signed the copyright agreement with Springer he is still allowed to post a copy of the paper on his personal web pages (the publisher recommends waiting 12 months) but it would be illegal to post a copy of the paper to the ITC repository.

Due to problems like this the repository has not reached the hoped for critical mass. On the other hand the technical platform built for the repository has successfully been used for running a number of repositories in other research areas. The papers in the repository are also easy to find via general search engines. For instance a Google search with the following search terms: "Gielingh AEC reference model" will show a link to the paper shown in figure 2.

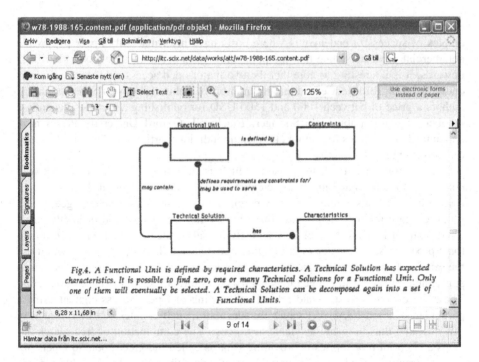

Fig.4. A Functional Unit is defined by required characteristics. A Technical Solution has expected characteristics. It is possible to find zero, one or many Technical Solutions for a Functional Unit. Only one of them will eventually be selected. A Technical Solution can be decomposed again into a set of Functional Units.

Fig. 2. In the setting up of the ITC Open Access repository the conference series from CIB W78 was scanned as far back as 1988. Some of the papers are important contributions to our discipline which otherwise would be very difficult to get hold of.

2 Experiences with ITcon

The Electronic Journal of information technology was founded in 1995, at a time when publishers were still only delivering on paper. It was the first Open Access journal in civil engineering and has since been followed by the International Journal of Design Computing and the Lean Construction Journal. After an initial struggle to get acceptance and a sufficient number of submissions, ITcon is now well established and publishes around 25 papers per year, on a par with a traditional quarterly journal. ITcon uses the normal peer review procedure and the papers have a traditional layout. Other Open Access journals have experimented with alternative forms of peer review and more hyper-media like user interfaces, but our experience is that what is most important to authors and readers is rapid publication and easy access.

The central problem in getting ITcon launched has been to overcome the "low quality" label that all only electronically published journals had, particularly in the early days. Researchers in our domain eagerly embraced open access journals as readers, but as authors they mostly chose established journals for their submissions, often more or less forced to by the "academic rules of the game" of their countries of universities.

3   Benchmarking ITcon

ITcon has recently been benchmarked against a group of journals in the field of construction information technology [6], [7]. This sub-discipline numbers a few hundred academics worldwide, mostly active in the architectural and civil engineering departments of universities as well as in a few government research institutes. It is a relatively young field where speed of publication should be a very important factor, due to the fast developments in the technology. Despite the fact that the field is relatively small there are half-a-dozen peer reviewed journals specialised in the topic, most with circulations in the hundreds rather than exceeding one thousand copies. In 2004 these journals published 235 peer-reviewed articles. The benchmarking at this stage concentrated on factors which were readily available or could be calculated from journal issues. For some of the factors, journals in the related field of construction management, which often publish papers on construction IT, were also studied to get a wider perspective. The following factors were studied:

- Journal subscription price
- Web downloads
- Impact factors
- Spread of authorship
- Publication delay
- Acceptance rate

The *subscription prices* are easily available from the journal web sites. The institutional subscriptions to electronic versions are by far the most important and were used. In order to make the results comparable the yearly subscription rates were divided by the number of scientific articles. The price per article ranged from 7.1 to 33.3 euro (Figure 3). Two of the journals compared were open access journals.

Readership is one factor for which it very difficult to obtain data. First the number of subscribers, in particular institutional subscribers, does not equate to the number of readers. Second most journals tend to keep information about the number of subscribers as trade secrets, since low numbers of subscribers might scare off potential submitting authors.

Society published journals tend on the average to have much lower prices. In economics the price ratio, per article, between society journals and purely commercial journals, is 1 to 4 [8]. In practice this means that commercial journals have often opted for much smaller subscription bases where their overall profits are maximised.

Society journals often offer very advantageous individual subscription to members, which tends to increase the readership. Consider for instance the above figure, in which ECAM, CACIE and CME are published by big commercial publishers, and JCCE by a Society. Also IJAC is essentially a journal published by the eCAADe society, with its sister organisation on other continents.

Data on the *downloads of published papers* by readers could only be studied for one of the journals. It would be a very useful yardstick to compare journals. For ITcon the web download figures from the past three years were used. In order to make the

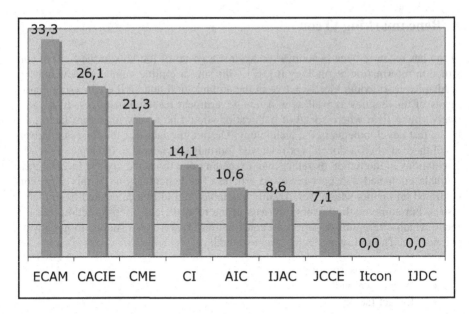

Fig. 3. Price per article (Euros per article)[1]

data usable, downloads by web search engines and other non-human users were as far as possible excluded (which resulted in a reduction of the figures by 74%). The downloads of the full text PDFs were counted, since this would come closest to actual readings. Over the three-year period each of the 120 published papers was on the average downloaded 21.2 times per month (with a spread of 4.7 –47.3). In addition to the number of average downloads per month, the total readership for each article over a longer period, as well as differences in level of readership between articles, are interesting (Figure 4).

Three of the journals are indexed in the Science Citation Index but with rather low *impact factors* (0.219 – 0.678) and none of the journals in the whole sample is clearly superior to the others in prestige or scientific quality. This is in contrast to many other scientific areas, where there often is one journal with a very rigorous peer review and low acceptance rate which is clearly superior in quality.

It is relatively straightforward to calculate the *geographic spread of journal authors* since the affiliations of the authors are published with the articles. The actual analysis was done on a country-by-country basis from articles published in 2001-2005. Thus European authors had 28 % of authorships, North American 37 % and Asian 30 %., from where 28 % of the articles stem, has been divided into four regions (UK, Central

---

[1] ECAM = Engineering, Construction and Architectural Management, CACIE = Computer Aided Civil and Infrastructure Engineering, CME = Construction Management and Economics, CI = Construction Innovation, AIC = Automation in Construction, IJAC = International Journal of Architectural Computing, JCCE = Journal of Computing in Civil Engineering, Itcon = Electronic Journal of Information Technology in Construction, IJDC=International Journal of Design Computing.

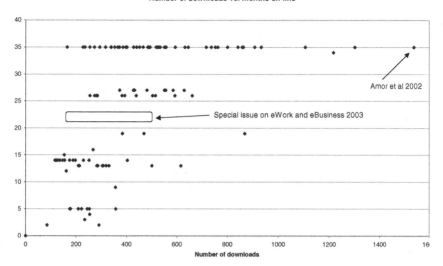

Fig. 4. Total number of downloads per IT-con paper over a three-year period as a function of the number of months on the web

Europe, Scandinavia, Eastern and Southern Europe). The more precise figures for the individual journals show a wide variation [6]. For instance JCCE had 67 % North American authors and AIC 49 % Asian authors. ITcon and the Open Access mode of publishing has been embraced by in particular European authors (69%), less so by North Americans (20 %) and significantly little by Asian authors (8 %).

The *speed of publication* (from submission to final publication of accepted papers) is an important factor for submitting authors. For ITcon the full publication delays where calculated from available databases. For some journals complete or incomplete information could be gathered from the submission and acceptance dates posted with the articles and the publication delay ranged from 7.6 to 21.8 months. For other journals this calculation was not possible to do. The figure for the IEEE journal Transactions on Geoscience and Remote Sensing has been reported by Raney [9].

The recent study on open access publishing performed by the Kaufman-Wills Group [10] provides statistics on *acceptance rates* for around 500 journals from different types of publishers, covering both subscription based and open access journals. Thus the average acceptance rate for the subscription-based journals published by the Association of Learned and Professional Society Publishers was 42%. The average for open access journals indexed by the Directory of Open Access Journals (DOAJ) was 64%, but if one excludes two large biomedical open access publishers (ISP and BioMedCentral) the average was 55%. Construction Management and Economics has made quite detailed statistics on submissions and acceptance rates available on its web site [11]. Over the period 1992-2004 the acceptancy rate was 51%. Also the ASCE journal for Computing in Civil Engineering has recently reported its

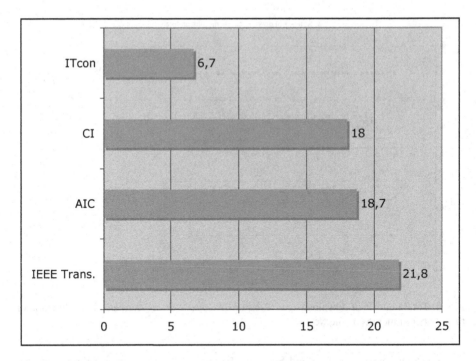

Fig. 5. The speed of publication (from submission to final publication of accepted papers)

acceptance rate to be 47 % [12]. The overall acceptance rate for ITcon was calculated from the records and proved to be 55%, which is to very close to the DOAJ average excluding the two biomedical publishers.

4 Conclusions

All in all the experience with ITcon has shown that it is possible to publish a peer-reviewed journal which is on par with the other journals in its field in terms of scientific quality, using an Open Source like operating model, which requires neither subscriptions nor author charges. As the authorship study shows ITcon has a very globally balance range of authors. ITcon outperforms its competitors in terms of speed of publication. Concerning the total amount of readership it is impossible to obtain comparable figures for other journals. The analysis of journal pricing does, however, indicate that the pricing of some journals is so high that the number of subscribers is likely to be low.

The one parameter where ITcon still lags many of its competitors is "prestige", in terms acceptance of an ITcon article as an equally valuable item when comparing CVs for tenure purposes, research assessment exercises etc. Here inclusion in SCI, or having a well known society or commercial publisher, still makes a difference. Only time and more citations can in the long run remedy this situation.

References

1. Björk, B.-C., Turk, Z.: How Scientists Retrieve Publications: An Empirical Study of How the Internet Is Overtaking Paper Media. *Journal of Electronic Publishing*: 6(2),(2000).
2. Björk, B-C., Turk, Z.: A Survey on the Impact of the Internet on Scientific Publishing in Construction IT and Construction Management, ITcon Vol. 5, pp. 73–88 (2000). http://www.itcon.org/
3. Tenopir, C., King, D.: Towards Electronic Journals, Realities for Scientists, librarians, and Publishers, Special Libraries Association, Washington D.C. 2000.
4. Walker, T. J.: Free Internet Access to Traditional Journals, American Scientist, 86(5) Sept-Oct 1998, pp. 463–471. http://www.sigmaxi.org/amsci/articles/98articles/walker.html
5. Martens, B., Turk, Z., Björk, B.-C.: The SciX Platform - Reaffirming the Role of Professional Societies in Scientific Information Exchange, EuropIA 2003 Conference, Istanbul, Turkey.
6. Björk, B.-C., Turk, Z., Holmström, J.: ITcon - A longitudinal case study of an open access scholarly journal. *Electronic Journal of Information Technology in Construction*, Vol 10. pp. 349–371 (2005). http://www.itcon.org/
7. Björk, B.-C., Holmström, J (2006). Benchmarking scientific journals from the submitting author's viewpoint. Learned Publishing. Vol 19 No. 2, pp. 147–155.
8. Bergstrom, C. T., Bergstrom, T. C (2001). The economics of scholarly journal publishing. http://octavia.zoology.washington.edu/publishing/
9. Raney, K.: Into the Glass Darkly. *Journal of Electronic Publishing*, 4(2) (1998). http://www.press.umich.edu/jep/04-02/raney.html
10. Kaufman-Wills Group, *The facts about Open Access*, ALPSP, London, 2005.
11. Abudayyeh, O., DeYoung, A., Rasdorf, W., Melhem, H (2006).:Research Publication Trends and Topics in Computing in Civil Engineering. *Journal of Computing in Civil Engineering,* 20(1) 2–12
12. CME. Home pages of the journal Construction Management and Economics. http://www.tandf.co.uk/journals/pdf/rcme_stats.pdf).

Automated On-site Retrieval of Project Information

Ioannis K. Brilakis

Dept. of Civil and Environmental Engineering, University of Michigan,
Ann Arbor, MI 48109, USA
brilakis@umich.edu

Abstract. Among several others, the on-site inspection process is mainly concerned with finding the right design and specifications information needed to inspect each newly constructed segment or element. While inspecting steel erection, for example, inspectors need to locate the right drawings for each member and the corresponding specifications sections that describe the allowable deviations in placement among others. These information seeking tasks are highly monotonous, time consuming and often erroneous, due to the high similarity of drawings and constructed elements and the abundance of information involved which can confuse the inspector. To address this problem, this paper presents the first steps of research that is investigating the requirements of an automated computer vision-based approach to automatically identify "as-built" information and use it to retrieve "as-designed" project information for field construction, inspection, and maintenance tasks. Under this approach, a visual pattern recognition model was developed that aims to allow automatic identification of construction entities and materials visible in the camera's field of view at a given time and location, and automatic retrieval of relevant design and specifications information.

1 Introduction

Field construction tasks like inspection, progress monitoring and others require access to a wealth of project information (visual and textual). Currently, site engineers, inspectors and other site personnel, while working on construction sites, have to spend a lot of time in manually searching piles of papers, documents and drawings to access the information needed for important decision-making tasks. For example, when a site engineer tries to determine the sequence and method of assembling a steel structure, information on the location of each steel member in the drawings must be collected, as well as the nuts and bolts needed for each placement. The tolerances must be reviewed to determine whether special instructions and techniques must be used (i.e. for strict tolerance limits) and the schedule must be consulted to determine the expected productivity and potential conflicts with other activities (e.g. for crane usage).

All this information is usually scattered in different sources and often conflicts with expectations or other information, which makes the urgency and competency of retrieving all the relevant textual, visual or database-structured data even more important. However, manual searches for relevant information is a monotonous, time-consuming process, while manual classification [1] that really helps speed up the

I.F.C. Smith (Ed.): EG-ICE 2006, LNAI 4200, pp. 92–100, 2006.

search process only transfers that problem to the earlier stage. As a possible alternative to user-based retrieval, this paper builds on previous modeling, virtual design and collaboration research efforts (i.e. [2]) and presents a computer vision type approach that, instead of requiring browsing through detailed drawings and other paper based media, it can automatically retrieve design, specifications and schedule information based on the camera's field of view and allow engineers to directly interact with it in digital format.

The computer vision perspective of this approach is based on a multi-feature retrieval framework that the author has previously developed [3]. This framework consists of complementary techniques that can recognize construction materials [4; 5] and shapes [6] that, when augmented with temporal and/or spatial information, can provide a robust recognition mechanism for construction-related objects on-site. For example, automatically detecting at a certain date and time (temporal) that a linear horizontal element (shape) made out of red-painted steel (material) is located on the south east section of the site (location) is in most cases sufficient information to narrow down the possible objects matching such description to a small and easily manageable number.

This paper initially presents previous work of the author that serves as the base for the computer vision perspective of this research and continues with the overall approach that was designed and the relationship between its various components. Conclusions and future work are then presented. At this stage, it is important to note that this work is a collaboration effort with the National Institute of Standards and Technology (NIST) in steel structure inspection, and therefore, all case studies and examples are focused on steel erection.

2 Previous Work

The following two sub-sections present the findings of recent research efforts of the author in construction site image classification based on the automatic recognition of materials and shapes within the image content [3; 4; 5; 6], which is the basis for the proposed on-site project information retrieval approach that will be presented in the following sections. The purpose is to familiarize the reader with some of the main concepts used in the mechanics of this research.

2.1 Recognition of Construction Materials

The objective of this research [4; 6] was to devise methods for automating the search and retrieval of construction site related images. Traditional approaches were based on manual classification of images which, considering the increasing volume of pictures in construction and the usually large number of objects within the image content, is a time-consuming and tedious task, frequently avoided by site engineers. To solve this problem, the author investigated [7] using Content Based Image Retrieval (CBIR) tools [8; 9; 10] from the fields of Image and Video Processing [11] and Computer Vision [12]. The main concept of these tools is that entire images are matched with other images based on their features (i.e. color, texture, structure, etc). This investigation revealed that CBIR was not directly applicable to this problem and had to be redesigned and modified in order to take advantage of the construction domain characteristics. These modifications were based on the need for:

1) Matching parts of each image instead of the entire content. In most construction site images, only a part of each picture is related to the domain while the remaining parts are redundant, misleading and can possibly reduce the quality of the results. For this purpose, it was necessary to effectively crop the picture in order to isolate construction-related items (pavement, concrete, steel, etc.) from picture background (sky, clouds, sun, etc.) or foreground (trees, birds, butterflies, cars, etc.).

2) Comparing images based on construction-related content. Each relevant part of the picture needs to be identified with construction-related terms. The comparison of images with other images or with objects in a model based system should not be performed at a low level (using color, texture, etc.). Instead, the comparison could be based on features such as construction materials, objects and other attributes that site engineers are more familiar with.

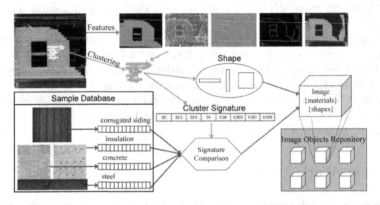

Fig. 1. Construction Materials and Shapes Recognition [6]

Overall, this material-based classification method is comprised of 4 steps (Fig. 1). In the first step, each image is decomposed into its basic features (color, texture, structure, etc.) by applying a series of filters through averaging, convolution and other techniques. The image is then cropped into regions using clustering and the feature signatures of each cluster are computed. During the fourth step, the meaningful image clusters are identified and isolated by comparing each cluster signature with the feature signatures of materials in a database of material image samples called "knowledge base". The extracted information (construction materials found) are then used to classify each image accordingly. This method was tested on a collection of more than a thousand images from several projects. The results showed that images can be successfully classified according to the construction materials visible within the image content.

2.2 Recognition of Construction Shapes

The objective of this research [6] was to enhance the performance of the previously presented material-based image classification approach by adding the capability of recognizing construction shapes and, by cross-referencing shape and material information, detect construction objects, such as steel columns and beams. This information (materials and shapes) was then integrated with temporal and spatial information in a flexible, multi-feature classification and retrieval framework for construction site images [3].

The motivation behind the need for more flexibility was that several materials are frequently encountered in construction site images (e.g. concrete, steel, etc.) and, unless accurate spatial and temporal information are also available, image retrieval based on such materials could retrieve an overwhelming amount of pictures. In such circumstances, it is necessary to classify images in even smaller, more detailed groups based on additional characteristics that can be automatically recognized from the image content. Earth, for example, can be classified into the several different types of soil [13] while concrete and steel objects can be classified according to their shape (columns, beams, walls, etc.). The latter is what this shape recognition approach can successfully recognize. In this work, shape is represented as the dimensions of each material region and is stored as an additional feature in the multi-feature vector used to mathematically describe each material.

This approach operates by skeletonizing construction objects if such a skeleton exists. Objects in this case are presumed to be image areas of similar characteristics (e.g. similar color distribution, similar texture, or similar structure) with a certain degree of uniformity (since construction materials are often characterized by consistent colors, textures and structures). These image areas are selected using a flooding-based clustering algorithm [5] with high accuracy, and the materials that comprise each cluster (group of pixels) are identified.

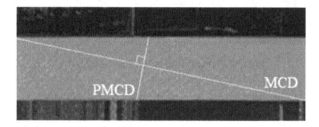

Fig. 2. Steel cluster and measurements [6]

The linearity and (if linear) orientation of the "object's spine" of each cluster is evaluated. Both are determined by computing the maximum cluster dimension (MCD) and the maximum dimension along the perpendicular axis of MCD (PMCD) (Fig. 2). These dimensions are then used to determine the linearity and orientation under three assumptions (i) If MCD is significantly larger than PMCD, then the object is linear, (ii) If the object is linear, then the tangent of the MCD edge points represents its direction on the image plane; the object's "spine", (iii) If the computed direction is within 45 degrees from the vertical/horizontal image axis then the linear object is a column/beam, respectively. This method was tested on the same collection of more than a thousand images from several projects. The results showed that images can be successfully classified according to the construction shapes visible within the image content.

3 On-site, Vision-Based Information Retrieval Model

The primary goal of this research is to minimize the time needed for on-site search and retrieval of project information, and by consequence, to reduce cost and effort needed for this process. In order to achieve this objective, the author investigated the requirements of a vision-based approach that focuses on automatically retrieving relevant project information to the user for on-site decision-making in construction, inspection, and maintenance tasks. Figure 3 summarizes the mechanics of the novel information retrieval model:

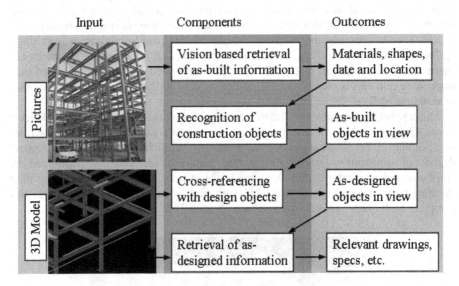

Fig. 3. On-site Information Retrieval Model

3.1 Retrieval of as-built Information and Objects Recognition

This component aims to detect all possible construction-related visual characteristics within the image content, such as surface materials and object shapes, and the relative position of each in reference to the image plane. This information is extracted using the materials and shapes detection tools described above [4, 6]. The input in this case is construction site time/location/orientation-stamped photographs (using a GPS digital camera), and a set of user-pre-selected material image samples needed for the vision algorithms involved to understand what each material looks like [4]. The image components (red, green, blue and alpha [transparency info]) are initially separated for further analysis. Image and video processing filters then extract the normalized color distribution and color histograms, the texture response of each frame to sets of texture kernels, the wavelet coefficients (when wavelets are used) and other mathematical image representations. The values of these representations are then grouped by image areas (clusters) that contain the same materials, and compacted into cluster signatures using statistical measures such as mean, mode, variance, etc. The signature of each cluster is then compared with those of the pre-classified material samples so as to detect their existence within the image content. The outcome is material and shape information grouped in vectors (signatures, Fig. 4) by relative position in reference to the image plane, along with the hardware provided time, location and orientation data.

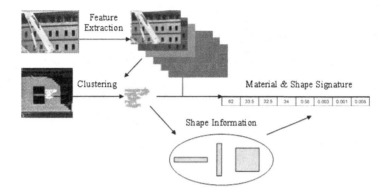

Fig. 4. Materials/Shapes Recognition – Representation with Image Signatures

The as-built objects are then recognized based on Euclidian distance matching. Each attribute (texture response, shape directionality, etc) in the multi-dimensional material and shape signature represents a different dimension of comparison. By comparing the distance of each attribute of the extracted signatures with the corresponding attributes of the object types in the 3D CAD model, the similarity of each signature with each object type in the model can be represented mathematically. The design object type with the highest similarity (least distance) is then selected to represent the recognized object.

3.2   Cross-Referencing Detected Objects with Design Objects and Retrieving Design Information

The position where the camera was located on the site and the direction in which it was facing are useful in narrowing down the possible construction objects that might match the detected object and its material and shape information [3]. This is where off-the-shelf GPS cameras can be really useful since the location and orientation information that they provide is enough to determine the camera's line-of-sight and the corresponding viewing frustum. This information, along with a camera coordinate system that is calibrated with the coordinate system used in creating the design of the constructed facility, can then assist in more accurately matching with the design objects that are expected to be in the camera's view. Calibration is essential in this case, since CAD designs typically use a local coordinate system.

In this approach, the object attributes are enhanced with camera position and orientation information and a Euclidian distance matching is repeated. The difference in this step is that specific objects are sought instead of generic object types. For example, while any steel beam is sufficient to determine the type of an as-built steel beam object in the previous step, the specific steel beam that it corresponds to in the design model is needed in this case. The CAD models used for these comparisons were based on the CIS/2 standard that provides data structures for multiple levels of detail ranging from frames and assemblies to nuts and bolts in the structural steel domain (Fig. 5). The CIS/2 standard is a very effective modeling standard and was successfully deployed on a mobile computing system at NIST [14].

Fig. 5. CIS/2 product models: (left) Structural frame of large, multistory building and (right) Connection details with bolts [14]

After minimizing the number of possible matches, the next step is to provide the user with the relevant design information needed. In this case, the information related to each possible match is acquired from the model objects and isolated. This way, the user need only browse through small subsets of information (i.e. a few drawings, a few specification entries, a segment of the schedule, etc.).

4 Conclusions

Designing and implementing a pattern recognition model that allows the identification of construction entities and materials visible in a camera's field of view at a given time was the base for this ongoing research work. The long-term goal is to reduce the cost and effort currently needed for search and retrieval of project information by using the automatically detected visual characteristics of project-related items to determine the possibly relevant information that the user needs. Thus, the innovative aspects of this research lie in the ability to automatically identify and retrieve project information that is of importance for decision-making in inspection and other on-site tasks and, to achieve this, a new methodology that can allow rapid identification of construction objects and subsequent retrieval of relevant project information for field construction, inspection, and maintenance was developed. The merit of its technical approach lies in taking advantage of the latest developments in construction material and object recognition to provide site personnel with automated access to both as-built and as-designed project information. Automated retrieval of information can also, for example, serve as an alerting mechanism that can compare the as-built and as-designed information and notify the site (or office) personnel of any significant deviations, like activities behind schedule and materials not meeting the specifications. Reducing the human-intervention from this tedious and time-consuming process is also expected to reduce man-made mistakes. Eventually, it is anticipated that the designed model will allow construction personnel to increase their productivity in field tasks such as inspection and maintenance, thereby achieving cost and time savings and lesser life cycle costs in constructed facilities.

5 Ongoing and Future Work

The presented model will be validated in 2 stages. The purpose behind the first stage of testing is to explore the limits of the materials and shape recognition algorithms in detecting installed members and cross-referencing them with their design information. A case study is planned to be conducted at the NIST structural steelwork test bed in Gaithersburg, MD. Based on a CIS/2 model for a multistory steel frame that was erected with many errors, several experiments will be designed to evaluate the ability of the designed prototype to identify, extract, and present relevant information to an inspector attempting to detect errors and irregularities in the structure. Based on the observed performance, one can check whether this situation is better than situations wherein the relevant information was manually identified and recovered. The second stage of testing will be done in collaboration with industrial partners on real projects. This includes retrieval of design information and instructions to serve as an assembly aid as well as design retrieval for evaluating compliance with specifications.

References

1. Abudayyeh, O.Y, (1997) "Audio/Visual Information in Construction Project Control," Journal of Advances in Engineering Software, Volume 28, Number 2, March, 1997
2. Garcia, A. C. B., Kunz, J., Ekstrom, M. and Kiviniemi, A., "Building a project ontology with extreme collaboration and virtual design and construction", Advanced Engineering Informatics, Vol. 18, No 2, 2004, pages 71-85.
3. Brilakis, I. and Soibelman, L. (2006) "Multi-Modal Image Retrieval from Construction Databases and Model-Based Systems", Journal of Construction Engineering and Management, American Society of Civil Engineers, in print
4. Brilakis, I., Soibelman, L. and Shinagawa, Y. (2005) "Material-Based Construction Site Image Retrieval" Journal of Computing in Civil Engineering, American Society of Civil Engineers, Volume 19, Issue 4, October 2005
5. Brilakis, I., Soibelman, L., and Shinagawa, Y. (2006) "Construction Site Image Retrieval Based on Material Cluster Recognition", Journal of Advanced Engineering Informatics, Elsevier Science, in print
6. Brilakis, I., Soibelman, L. (2006) "Shape-Based Retrieval of Construction Site Photographs", Journal of Computing in Civil Engineering, in review
7. Brilakis, I. and Soibelman, L. (2005) "Content-Based Search Engines for Construction Image Databases" Journal of Automation in Construction, Elsevier Science, Volume 14, Issue 4, August 2005, Pages 537-550
8. Rui, Y., Huang, T.S., Ortega, M. and Mehrotra, S. (1998) "Relevance Feedback: A Power Tool in Interactive Content-Based Image Retrieval", IEEE Tran on Circuits and Systems for Video Technology, Vol. 8, No. 5: 644-655

9. Natsev, A., Rastogi, R. and Shim, K. (1999) "Walrus: A Similarity Retrieval Algorithm for Image Databases", In Proc. ACM-SIGMOD Conf. On Management of Data (SIGMOD '99), pages 395-406, Philadelphia, PA
10. Zhou, X.S. and Huang, T.S. (2001) "Comparing Discriminating Transformations and SVM for learning during Multimedia Retrieval", ACM Multimedia, Ottawa, Canada
11. Bovik, A. (2000) "Handbook of Image and Video Processing". Academic Press, 1st edition (2000) ISBN:0-12-119790-5
12. Forsyth, D., and Ponce, J. (2002) "Computer Vision - A modern approach", Prentice Hall, 1st edition (August 14, 2002) ISBN: 0130851981
13. Shin, S. and Hryciw, R.D. (1999) "Wavelet Analysis of Soil Mass Images for Particle Size Determination" Journal of Computing in Civil Engineering, Vol. 18, No. 1, January 2004, pp. 19-27
14. Lipman R (2002). "Mobile 3D Visualization for Construction", Proceedings of the 19th International Symposium on Automation and Robotics in Construction, 23-25 September 2002, Gaithersburg, MD

Intelligent Computing and Sensing for Active Safety on Construction Sites

Carlos H. Caldas, Seokho Chi, Jochen Teizer, and Jie Gong

Dept. of Civil, Architectural and Environmental Engineering,
The University of Texas, Austin, TX USA 78712
caldas@mail.utexas.edu

Abstract. On obstacle-cluttered construction sites where heavy equipment is in use, safety issues are of major concern. The main objective of this paper is to develop a framework with algorithms for obstacle avoidance and path planning based on real-time three-dimensional job site models to improve safety during equipment operation. These algorithms have the potential to prevent collisions between heavy equipment vehicles and other on-site objects. In this study, algorithms were developed for image data acquisition, real-time 3D spatial modeling, obstacle avoidance, and shortest path finding and were all integrated to construct a comprehensive collision-free path. Preliminary research results show that the proposed approach is feasible and has the potential to be used as an active safety feature for heavy equipment.

1 Introduction

According to the United States Bureau of Labor Statistics' 2004 Census of Fatal Occupational Injuries (CFOI) study, out of a total of 1,224 on-the-job fatalities that occurred in the construction industry, accidents involving from heavy equipment operation (e.g.: transportation accidents and contact incidents with objects and equipment) represented about 45% [1]. Clearly, attention to the safety issues surrounding heavy equipment operation plays an important role in reducing fatalities. However, since most construction sites are cluttered with obstacles, and heavy equipment operation is based on human operators, it is virtually impossible to avoid the general lack of awareness of work-site-related hazards and the relative unpredictability of work-site environments [2]. With the growing awareness of the risks construction workers face, the demand for automated safety features for heavy equipment operators has increased.

The main objective of the research presented here is to develop a framework and efficient algorithms for obstacle avoidance and path planning which have the potential not only to prevent collisions between heavy equipment vehicles and other on-site objects, but also to allow autonomous heavy equipment to move to target positions quickly without incident. A research prototype laser sensor mounted on heavy equipment can monitor both moving and static objects in an obstacle-cluttered environment [3] [4]. From such a sensor's field of view, a real-time three-dimensional modeling method can quickly extract the most pertinent spatial information of a job site, enabling path planning and obstacle avoidance. Beyond generating an efficient and effective real-time 3D modeling approach, the proposed framework is expected to contribute to the development of active safety features for construction job sites.

I.F.C. Smith (Ed.): EG-ICE 2006, LNAI 4200, pp. 101–108, 2006.
© Springer-Verlag Berlin Heidelberg 2006

2 Framework for Path Planning

This section presents an overview of the framework for real-time path planning which includes real-time job site modeling. The proposed path planning framework can be described in two parts: the static search and the dynamic search. The static search is based on the entire static world (i.e. work space of a construction operation). The dynamic search is based on the dynamic local environment derived and limited to the field of view of the actual sensors mounted on mobile equipment.

In the static search, all information about static objects is used for constructing a world map of the job site. First, a priori knowledge about the entire environment such as heavy equipment fleet information or CAD data is considered and converted into the world model. Based on this prior knowledge of the environment, the site's world map is constructed. From this map, basic paths for mobile equipment are initiated from starting positions to goal positions. In the dynamic search, however, all dynamic and static objects in the local area –the area determined by the sensor's field of view – are registered and tracked. The real-time 3D model is derived from an occupancy grid approach, and all dynamic and static objects are built into the local map. This 3D image represents the local environment. Once it is created, the local map is systematically superimposed onto the baseline global map, and any differences between the two maps play a key role in defining unknown objects and moving objects. The unknown objects should not appear on the global map, but do appear on the local map. The moving objects could be in the both maps, but the positions of these objects will vary. From this regularly updated local information, the initial path of the equipment is periodically revised as the dynamic environment evolves. To build optimized paths, nodes (points) that are designated as the algorithms perform dynamic searching, sensing, and reasoning functions in the environment.

3 Real-Time 3D Job Site Modeling

In the path planning algorithm, real-time 3D job site modeling is the first step. This modeling algorithm is based on an occupancy grid 3D modeling algorithm already developed by the Field Systems and Construction Automation Laboratory (FSCAL) at the University of Texas at Austin [5]. The occupancy grid method, first pioneered by H. Moravec and A. Elfes in 1985 [6], is one of the most popular and successful methods of accounting for uncertainty. Occupancy grids divide space into a grid of regular 3D cells, which are scanned by a sensor that registers any surfaces filling them. From these readings, the modeling algorithm can estimate the probability of any one cell being occupied by any of these surfaces. The data points of a surface coalesce as real images that can then be plotted into one of the predefined virtual cells. If enough data points are plotted in a cell, that cell is considered occupied. If the cell is occupied, the cell has a value 1 and if not, the cell has an initial value 0. After conducting the proper noise removal process, a set of occupied cells builds an image of one object which can be represented as either static or dynamic by means of a certain clustering method. Since occupied cells represent real objects, it is easy to cluster an object and track its moving conditions. Also, the processing time is fast enough to attain an effective real-time application because only occupied cells are concerned in the process.

3.1 Occupancy Grid Modeling Processing Techniques

First, the world model and the sensor's field of view need to be divided into a 3D grid system. When dividing the entire space into a 3D grid system, it is important to first determine the grid size. If the tracking of small objects is required, a high resolution grid map should be used. However, when higher resolution grids are used, the processing time is greater, and because noise becomes more prevalent, results are generally poorer. Therefore, the effective cell size should be chosen in accordance with the local environment's type and size, keeping in mind the incoming data processing capability of the hardware.

After the grid map is built, the sensor range data can be plotted into the cells of the grid. Each cell of the grid can have zero, single, or multiple range points. Once each cell meets a certain threshold count of range points, its center is filled with the value 1 and can be called occupied. All other cells which have fewer range points than the threshold count, such as zero-occupied or only one-occupied, are considered to hold an extreme range value and fall in the category of noise. For example, if the threshold value for counting cells as occupied is three, only cells which have more than three range points are considered occupied cells. Cells which have less than three range points are considered noise. In this case, occupied cells have the value 1, and noise cells have the value zero.

A reliable way of reducing the number of points in cells without losing valuable data is important for noise treatment and makes for faster image processing speed. If the above-mentioned noise removal, which is based on counting range points, is considered the first level of noise removal, the second level of noise removal only deals with occupied cells which have a value of 1. Second-level noise can happen when occupied cells exist alone in the 3D space – cells that are actually empty but that project a virtual image. To safely eliminate single-occupied noise cells, their surrounding neighbor cells should be investigated. If a certain number of neighbor cells around these cells are also occupied, the original value is kept as a value 1. If not, the original value is rejected as noise. All the above parameters (grid size, range point threshold value, and neighbor threshold value) are user input data. Many different sets of grid mapping parameters are available; their variety helps users find modeling conditions best-suited for the real environment.

A set of occupied cells can be made to represent one object by applying a cell clustering method. In this research, a nearest neighbor clustering algorithm [7] was adapted by following several steps. First, positions of every occupied cell were compared with each other, and if a distance between two cells was less than a given threshold value, it meant that the two cells belonged in the same cluster. Conversely, if a distance was larger than the threshold value, it meant that the two cells belonged in different clusters. This cell-to-cell comparison was iterated between whole occupied cells and as a result, all cells were modeled into the correct clusters.

After grouping occupied cells, cluster information such as the center of gravity value of each cluster and the cluster size was determined. This cluster information is valuable for tracking objects and for the path planning of objects because knowing the center of gravity value and the cluster size is basic to calculating the conditions of moving objects like velocity and acceleration vectors.

4 Path Planning

The ultimate purpose of real-time obstacle detection and environmental modeling is to plan a collision-free path under the real-world constraints of a job site, and the planned path represents the shortest, safest, and most visible path [8] [9] [10]. The first step of the proposed path planning algorithm is to determine a starting position, an ending position, and interim path nodes within the static environment, with safety margins established around static objects. Then the interim nodes are revised as the dynamic object's moving conditions are tracked according to a dynamic path tree algorithm. This dynamic path tree algorithm uses dynamically allocated points through real-time searching, sensing, and reasoning in the environment. This algorithm is able to find the visible points of any local position in the environment and; from that data, can plan a collision-free path and motion trajectory by projecting angles to partition the obstacle-space. By making the node-with-no-obstacles state a higher priority, this algorithm chooses the shortest cost state as the discrete goal and keeps iterating this goal-oriented action until the mobile vehicle reaches its planned destination.

4.1 Path Planning Processing Techniques

The real-time object detection and environmental modeling approach is based on a 3D environment because the 3D modeling approach can represent a real environment more accurately and more effectively than a 2D approach. However, at this stage of the research, the path planning algorithm is based on a 2D environment without elevation information. Current research is being conducted to incorporate 3D path planning algorithms into the proposed framework.

The first step of the path planning algorithm is to set the task and interim nodes on the map. Task nodes contain starting and target positions of the autonomous heavy equipment vehicle. After setting these two task nodes, interim nodes around static objects begin to be set. A safety zone is created around each static object to prevent collisions between objects and the autonomous vehicle, and four interim nodes are set at every vertex of the end edge of each object's safety zone (Figure 1).

The second step of the algorithm is creating possible discrete paths. A discrete path is any possible path between any of the nodes and consists of a beginning position, a stopping position, and a distance between both positions. In Figure 1, a path between a starting position and interim node 1 is a discrete path which has a distance d. The notation for a path is: *Path name(Beginning node, Stopping node, Distance)*. To determine the possibility of a collision-free path, a distance between a path and every vertex of a static object need to be investigated. This calculated distance should be compared with a safety threshold value, and if the compared distance is larger than the threshold, the path is possible to track. Figure 2 shows several possible paths, such as *Path 1(S, 1, Dist 1)* or *Path 2(S, 2, Dist 2)*.

The next step of the algorithm is to consider the dynamic object's moving positions. After generating all possible discrete paths from the nodes that are set around static objects, dynamic objects should be incorporated into discrete paths. First, a certain discrete path is selected and compared to a moving object to determine whether a moving object intersects the autonomous vehicle's possible path. Both the

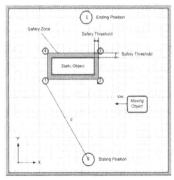

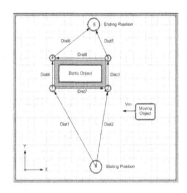

Fig. 1. Task and interim node settings   **Fig. 2.** Created discrete paths

autonomous vehicle's position and the moving object's position at a certain time *t* are determined. Then, the distance between the two positions are calculated to figure out whether the autonomous vehicle's path is influenced by the moving object. If the distance is larger than a safety threshold, there is no danger of the vehicle colliding with the moving object. However, if the distance is smaller than the threshold, one more node should be added onto the map to avoid collision (Figure 3). After adding a new node, a new path is created and designated as *New_Path(Starting, New, Dist_New)*. The previous path, *Path 1(S, 1, Dist 1)*, is deleted from the set of discrete paths and *New_Path* replaces *Path 1*. Once a new node is added, it is necessary to repeat the entire path creation process, incorporating the new set of nodes. The process of creating discrete paths and considering dynamic objects should be repeated until all discrete paths become collision-free paths.

The final step of the algorithm is to calculate a shortest path. All possible paths from the starting position to the goal position are considered and their total travel distances are stored for comparison with each other. Finally, the shortest path for the autonomous vehicle is determined from among all possible trajectories. This selected path allows the autonomous vehicle to reach the target position in the least amount of time without any collision, even within an obstacle-cluttered environment (Figure 4).

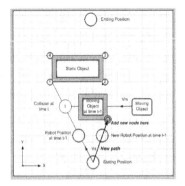

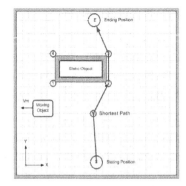

Fig. 3. Dynamic object consideration   **Fig. 4.** Calculated shortest path

5 Simulation Results

With the proposed real-time 3D modeling path planning algorithms providing the virtual model environment, a computer simulation was generated using the C++ programming language in Microsoft Visual Studio .NET 2003. This experimental environment was constructed in the FSCAL. The experimental environment consisted of a 3D video range camera sensor (FlashLADAR), a static box, a moving wire-controlled cart transporting a vertically mounted pipe, and a background wall.

The simulation held five basic assumptions: (1) Path planning is based on the two-dimensional approach. (2) An autonomous mobile vehicle first plans its collision-free paths based on a path planning algorithm in a static position, and then starts moving with a constant forwarding velocity (cm/sec). (3) A moving cart transporting a pipe moves with a constant forwarding velocity (cm/sec), and no acceleration. For the path planning simulation, only four frames captured within 0.14 seconds are used to calculate the moving object's velocity. The autonomous vehicle waits until four frames are captured to avoid measuring the initial acceleration of the moving object. 0.14 seconds is also a short enough time span not to cause meaningless idling time of the autonomous vehicle. (4) The 0.14-second time span is also enough time for the vehicle to update image frames while it is moving to its target position. While the proposed path planning algorithm allows the vehicle to update local image frames every 0.14 seconds, in the current simulation, it trusts the planned path without any new image update while it is moving. (5) All frames are captured from a static sensor.

5.1 Occupancy Grid

There were four major saved data derived from the 3D modeling process: occupancy grid data, cluster information, a sensor position, and velocity vectors. These simulation results were exported into Matlab software to show how well the saved data represented the local environment as a 3D image. All results were based on a 10cm occupancy grid size, on a three-point threshold for determining occupied cells, and on occupied cells having four occupied neighbors to establish their validity.

5.2 Path Planning Results

First, velocity vectors of moving objects were calculated from frame captures using the Flash LADAR. Next, an initial sensor position was set as a starting position for the autonomous equipment. These data came from the results of the occupancy grid processing. After the local conditions were considered, the goal position and the safety threshold value were set. The simulation results showed that, when the autonomous vehicle's speed was low, the shortest path was influenced by the moving object's position, and as a result, a new node was incorporated into the shortest path (Figure 5). However, when the autonomous vehicle moved at a high speed, the position of the moving object did not intersect the shortest path; therefore, the shortest path did not incorporate any new node (Figure 6).

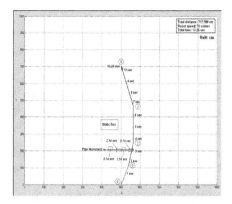 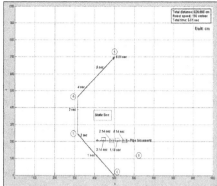

Fig. 5. Results with 70cm/sec – shortest path   **Fig. 6.** Results with 150cm/sec - shortest path

6   Conclusions and Discussion

A preliminary study of an obstacle avoidance and path planning method based on a real-time 3D modeling approach was described in this paper. The preliminary results suggest that the proposed framework and algorithms work well in a dynamic environment, cluttered with both static and moving objects. The occupancy grid algorithms successfully build a suitable 3D local model in real-time, and the path planning algorithms are able to produce a collision-free motion trajectory. Such situation-specific trajectories can then assist heavy equipment operators plan safer, more efficient, and no longer arbitrary routes. Using this technology, operators can guard against striking unexpected site objects, especially personnel moving outside operators' range of visibility. As a result, the proposed approach has the potential to improve safety in situations where heavy equipment is in use. The collision-free path can be determined even under low-visibility job site conditions. Used as an active safety feature, it has the potential to reduce accidents caused by operators' inattention, to detect unknown dynamic obstacles, and eventually to minimize fatalities and property damage resulting from unexpected situations.

The proposed framework uses a research prototype laser scanning sensor to acquire spatial information, technology which costs approximately $7000. For a large scale construction site, the sensors could be either placed at strategic positions on the site or installed on selected heavy equipment, depending on what type of sensor coverage is needed. In this paper, the cost-benefit ratio of applying such technology has not been fully investigated and leaves room for future research. Such future research would also extend the proposed preliminary path planning algorithm into 3D-based approaches. In addition, additional experiments on actual construction sites should be conducted to further validate the feasibility of the proposed framework.

Acknowledgements

This material is based in part upon work supported by the National Science Foundation under Grant Number CMS 0409326. Any opinions, findings, and conclusions or recommendations expressed in this material are those of the authors and do not necessarily reflect the views of the National Science Foundation.

References

1. BLS (Bureau of Labor Statistics). U.S. Department of Labor, Washington D.C., http://stats.bls.gov/iff/home.htm, Accessed on November 21, 2005.
2. Kim, C. Spatial Information Acquisition and Its Use for Infrastructure Operation and Maintenance. Ph.D. Diss., Dept. of Civil Eng., The University of Texas at Austin (2004).
3. Gonzalez-Banos, H.H., Gordillo, J.L., Lin, D., Latombe, J.C., Sarmiento, A., and Tomasi, C.: The Autonomous Observer: A Tool for Remote Experimentation in Robotics. Telemanipulator and Telepresence Technologies VI, November 1999, vol. 3840.
4. Gonzalez-Banos, H.H., Lee, C.Y., and Latombe, J.C.: Real-Time Combinatorial Tracking of a Target Moving Unpredictably Among Obstacles. IEEE International Conference on Robotics and Automation, Washington, DC (2002)
5. Teizer, J., Bosche, F., Caldas, C.H., Haas, C.T., and Liapi, K.A.: Real-Time, Three-Dimensional Object Detection and Modeling in Construction. Proceedings of the 22nd Internat. Symp. on Automation and Robotics in Construction (ISARC), Ferrara, Italy (2005)
6. Moravec, H. and Elfes, A.: High-resolution Maps from Wide-angle Sonar. Proc. of IEEE Int. Conf. on Autonomous Equipments and Automation, 116-121, Washington, DC (1985)
7. Ertoz, L., Steinbach, M. and Kumar, V.: A New Shared Nearest Neighbor Clustering Algorithm and its Applications. Workshop on Clustering High Dimensional Data and its Applications at 2nd SIAM International Conference on Data Mining (2002)
8. Soltani, A.R., Tawfik, H., Goulermas, J.Y., and Fernando, T.: Path Planning in Construction Sites: Performance Evaluation of the Dijstra, A*, and GA Search Algorithms. Advanced Engineering Informatics, 16(4), 291-303 (2002)
9. Wan, T.R., Chen, H., and Earnshaw, R.A.: A Motion Constrained Dynamic Path Planning Algorithm for Multi-Agent Simulations. Proc. Of the 13-th International Conference in Central Europe on Computer Graphics, Plzen, Czech Republic (2005)
10. Law, K., Han, C., and Kunz, C.: A Distributed Object Component-based Approach to Large-scale Engineering Systems and an Example Component Using Motion Planning Techniques for Disabled Access Usability Analysis. Proc. of the 8th International Conference on Computing in Civil and Building Engineering. ASCE, Stanford, CA (2000)

GENE_ARCH: An Evolution-Based Generative Design System for Sustainable Architecture

Luisa Caldas

Instituto Superior Técnico, Technical University of Lisbon, Portugal
luisa@civil.ist.utl.pt

Abstract. GENE_ARCH is an evolution-based Generative Design System that uses adaptation to shape energy-efficient and sustainable architectural solutions. The system applies goal-oriented design, combining a Genetic Algorithm (GA) as the search engine, with DOE2.1E building simulation software as the evaluation module. The GA can work either as a standard GA or as a Pareto GA, for multicriteria optimization. In order to provide a full view of the capacities of the software, different applications are discussed: 1) Standard GA: testing of the software; 2) Standard GA: incorporation of architecture design intentions, using a building by architect Alvaro Siza; 3) Pareto GA: choice of construction materials, considering cost, building energy use, and embodied energy; 4) Pareto GA: application to Siza's building; 5) Standard GA: Shape generation with single objective function; 6) Pareto GA: shape generation with multicriteria; 7) Pareto GA: application to an urban and housing context. Overall conclusions from the different applications are discussed.

1 Introduction

GENE_ARCH is an evolution-based Generative Design System (GDS) that uses adaptation to shape architectural form [1]. It was developed to help architects in the creation of energy-efficient and sustainable architectural solutions, by using goal-oriented design, a method that allows to set goals for a building's performance, and have the computer search a given design space for architectural solutions that respond to those requirements. The system uses a Pareto Genetic Algorithm as a search engine, and the DOE2.1E building simulation software as the evaluation module (Fig. 1). Other existing GDS related to architecture include those by Shea [2] and Monks [3].

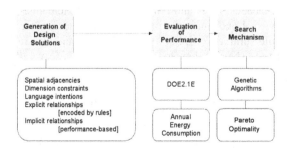

Fig. 1. GENE_ARCH's components

I.F.C. Smith (Ed.): EG-ICE 2006, LNAI 4200, pp. 109–118, 2006.

DOE2.1E is one of the most sophisticated building energy simulation packages in the market, what provides significant confidence in the results obtained by GENE_ARCH. For each of the thousands of alternative solutions it creates in a typical run, a full DOE2.1E hourly simulation is done, performed for the whole year and based on actual climatic data of the building's location.

2 Applications and Case Studies

A number of applications of GENE_ARCH, most of them previously published, are discussed in order to provide an overall view of the software and its capabilities.

2.1 Initial Testing in a Simplified Test Building

The software was initially tested within a test building with a simple geometry, with similar box-like offices facing the four cardinal directions [4]. GENE_ARCH's task was to locate the best window dimension for each space and orientation. The problem was set up in such way that the optimal solutions were known, despite the considerable size of the solution space, over 16 million. The testing was performed with a Micro Genetic Algorithm [5] and did not apply Pareto optimization, but a single fitness value. The objective function used was annual energy consumption of the building, which combined, even if as a simple average, both energy spent for space conditioning (heating, cooling and ventilating the building) and for illumination. Those are the two main final energy uses in buildings, and are usually in conflict with each other, as solutions that are more robust in terms of thermal performance - typically by reducing the number and size of openings in the building envelope - tend to score not so well in terms of capturing daylight, and vice-versa. The effectiveness of the building in capturing daylight was measured by placing virtual photocells at two reference points in each room of the building, and simulating a dimming artificial lighting system, which, at any point in time, would provide just enough artificial light to make up for the difference between the available daylight in the space (in lux), and the desirable lighting levels - determined by the architect, and in this case set to 500 lux, the typical illumination level recommended for office buildings. The simulations were done using real weather date for selected sites, in TMY format, which represents a Typical Meteorological Year based on statistical analysis of 30 years of actual measurements. Simulations were carried out hourly, in a total of 8760 hours per simulation, performed over a complete three-dimensional model of the building. This included a detailed geometrical description of spaces, facades, roofs and other construction elements, building materials, including their thermal and luminous properties, and much other information regarding not only architectural aspects, but also mechanical and electrical installations. The results from the tests showed that, for a solution space of over 16 million, solutions found by GENE-ARCH were within around 0.01% of the optimal.

2.2 Application to Álvaro Siza's School of Architecture at Oporto

Given the confidence gained in the quality of results, GENE-ARCH was applied to an actual architectural context, in the study of an existing building by Portuguese architect Álvaro Siza, the School of Architecture at Oporto [6]. The objectives were to test the software in a complex design context, assess methods for encoding design intentions, and analyse the trade-offs reached by the system when dealing with conflicting requirements. In this study, the overall building geometry and space layout were left unchanged, and the system was applied solely to the generation of alternative façade solutions. GENE-ARCH worked over a detailed three-dimensional description of the building and used natural lighting and year-round energy performance as objective functions to guide the generation of solutions. The experiments also research the encoding of architectural design intentions into the system, using constraints derived from Siza's original design, that we considered able to capture some of the original architectural intentions. Experiments using this generative system were performed on three different geographical locations to test the algorithm's capability to adapt solutions to different climatic characteristics within the same language constraints, but only results for Oporto are presented here.

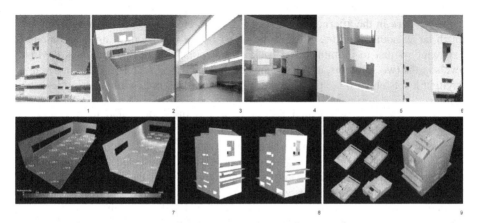

Fig. 2. Application of GENE_ARCH to the study of a building by Álvaro Siza: 1. Existing building; 2. Model of 6th floor roof: GENE_ARCH suggested a significant reduction of the south clerestory window, since it coincides with the area of space already lit by the south-facing *loggia*, and represents an important source of heat loss; 3. Photograph of northern end of the 6th floor: GENE-ARCH suggested an increase of the north-facing clerestory, as it is the only light source of that area, currently with very reduced dimensions; 4. 6th floor: Interior view of south-facing window of existing *loggia*; 5. *Loggia* design proposed by GENE-ARCH: simultaneously with the reduction of south clerestory, the software proposes a considerable increase in loggia fenestration, since it has a better solar orientation and is adequately shaded; 6. Existing *Loggia* seen from the outside: the small windows, inside the *loggia* recess, cause deficient daylighting levels; 7. 4th floor: Comparison of daylight levels in an east-facing Studio teaching room, at 3pm: the existing solution (left) has only about 1/3 of the lighting levels of GENE_ARCH's solution (right); 8. 3D models of GENE_ARCH's solution (left) and existing solution (right); 9. Rapid Prototyping of solutions: 3D-Printing (FDM – Fuse Deposition Machine) of GENE_ARCH's solution: in the left, separate floors allow a detailed observation of space. Daylight Factor measurements using the model are also possible.

Some of the more interesting results generated related to implicit trade-offs between façade elements. These are relations that were not explicitly incorporated in the constraints (as were the compositional axes, for example), but, being performance-based, emerged during the evolutionary process. An example is the trade-off between shading and fenestration elements in the south-facing studio teaching rooms, where the existing deep overhangs (2 meters depth) forced the system to propose window sizes as large as permitted by the constraints, in order to allow some daylight into the space (see figure 2.8). In a subsequent experiment, when the system was able to change overhang depth too, it did propose much shallower elements (60 cm), which could still shade the high-level south sun, while simultaneously allowing into the space both natural light and useful winter solar gains.

The east-facing 4th floor studio was another example of emerging implicit relations. In Siza's design, a large east-facing strip window illuminates most of the room, while a much smaller south-facing window occupies the end wall (figure 2.7, left). The morning sun makes the room overheat and is a cause of glare, as could be observed during a visit to the building, where students glued large sheets of drawing paper to the windows in order to have some comfort. Simultaneously, the room tends to become too dark in the afternoon. GENE_ARCH detected this problem and proposed a much larger south window, close to the upper bound of the constraints, and a small east window just to illuminate the back of the room. Figure 2.7 compares daylight levels in the two solutions, at 3pm in the afternoon. GENE_ARCH's solution displays daylight levels about three times higher than the existing one, while causing less discomfort.

Finally, the single space that occupies the top floor, dedicated to life drawing classes, provided another interesting case study. The system implicitly related the design of the north-facing strip clerestory windows to that of the south-facing loggia. The large clerestory window was reduced because it represented a significant heat loss source, and it coincided, it terms of daylighting, with the area covered by the loggia. The system simultaneously proposed a significant increase in the fenestrations inside the loggia, since they were already shaded and had a convenient solar orientation. As for the northern clerestory, the system proposed it should be increased, as it is the only light source of that side of the room (figure 2.3).

The results generated by this experiment were extremely interesting, as they related to an actual building and proved that GENE_ARCH could indeed deal with the complexity of a real case. Capturing the architect's architectural intention into the generative design system becomes a major challenge. It was also interesting to notice that, for the north façade, the system generated a solution that almost exactly resembled that of Siza, apart from some 'melodic' variations in the original design.

2.3 Pareto Genetic Algorithms Applied to the Choice of Building Materials and Respective Environmental Impact

The next experiment focused on the application of GENE-ARCH to the choice of building materials [7]. The question addressed was the conflict existent between the initial cost of materials, and the energy performance of the building. Typically, the more is spent on higher quality materials, the more will be saved in the building's life-cycle, in terms of energy expenditure. GENE-ARCH was used to find frontiers of trade-offs between these two situations. In another experiment, the trade-offs analyzed were considered only in terms of environmental impacts: the system considered both the energy saved locally in the building, by using better construction terms, and the embodied energy of materials, that is, the energy spent to manufacture them. The reasoning was that, in global greenhouse gas emissions terms, it might not make sense to save energy at the building level if more energy is being spent, even if at a remote location, to build those materials.

In these experiments, Pareto Genetic Algorithms were used as the optimization technique, as the problem involved conflicting design criteria. Pareto Genetic Algorithms provide a frontier of solutions representing the best trade-offs for a given problem [8], instead of single, optimal solutions as more traditional methods do, often based on sometimes arbitrary weighting factors assigned to each objective. GENE_ARCH was given a test building and a library of building materials, including thermal and luminous properties, typical costs/m^2, and Global Warming Potential [GWP] expressed in $KgCO_2/Kg$. The three objective functions used were annual energy consumption of the building, initial cost of materials, and GWP. Experiments generated well-defined, uniformly sampled Pareto fronts between the criteria considered, by using appropriate ranking and niching strategies [9]. Figure 3 shows the evolution of an experiment along 200 generations, from an initial series of scattered point, to a well-defined frontier of trade-offs, where each point represents a different wall configuration. Results suggest this is a practical way for choosing construction materials, and that new solutions emerge that may represent viable and energy-efficient alternatives to those commonly used in construction.

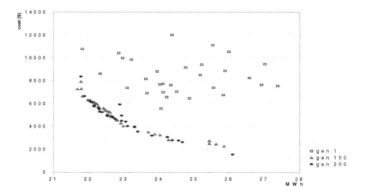

Fig. 3. Pareto front results for Phoenix climate

2.4 Pareto Genetic Algorithms Applied to Siza's School of Architecture

In the forth application, Pareto optimality was applied to Siza's building, using two conflicting objectives: daylight use and thermal performance. Figure 4 shows the five points (from a Pareto Micro Genetic Algorithm) that formed the final frontier. The top image shows the best solution in terms of heating energy, which simultaneously achieves the best possible daylight performance without degrading the thermal – a characteristic of Pareto optimality. The bottom image shows the best performance for daylight, with the best possible trade-off with heating. The other images show the remaining points of the Pareto frontier, representing other possible trade-offs.

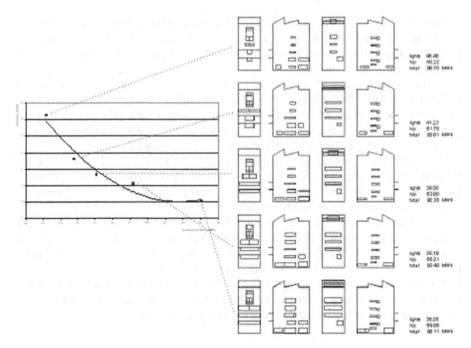

Fig. 4. Pareto optimal solutions for Siza's building

2.5 Application to Shape Generation

In the fifth application, GENE_ARCH was used to evolve three-dimensional architectural forms that were energy-efficient, while complying to architectural design intentions expressed by the architect [10]. Experiments were carried out for different climates, namely Chicago, Phoenix and Oporto. GENE_ARCH adaptively generated populations of alternative solutions, from an initial schematic layout and a set of rules and constraints designed by the architect to encode design intentions. The problem set was quite simple, consisting of a building with 8 spaces, and known adjacencies. There were four rooms in the 1st floor, all with the same height, and another four rooms in the 2nd floor, which could have different heights, different roof tilts and directions, and a clerestory window under each roof that was formed. Each room had two windows, which could be driven to zero-dimension if the system determined they were unnecessary. Figure 5, on the left end, displays the basic problem set. The right side shows a number of random configurations initially generated, illustrating how a simple problem set like this could still give rise to very different geometries.

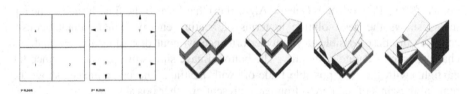

Fig. 5. Left: Problem schematics; Right: Some random geometries generated by GENE_ARCH

A problem faced was that the most immediate way GENE_ARCH found to reduce energy consumption was to decrease the overall building size, to the minimum allowed by the dimensional constraints of each room. In order to force solutions to be within certain areas set by the architect, a system of penalty functions was introduced. Penalties degraded the energy-based fitness function to an extent dependent on area violation. However, this tended to confound the algorithm, since it was giving mixed information to the GA, combining both energy performance and floor area in a single fitness-function. For that reason, another outcome measure was applied: Energy Intensity Use, which represented the amount of energy used per unit area. This approach also had its limitations, as discussed in reference [11], but was the basis for the results shown in figure 6. It is also interesting to notice how more extreme geometries initially generated, later stabilized in rather more compact and discrete ones.

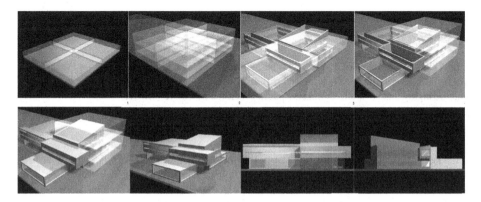

Fig. 6. GENE-ARCH's generation of 3D architectural solutions for Oporto, using Energy Use Intensity as fitness function: 1. 1st floor constraints; 2. Overall constraints, including roofs; 3, 4, 5. Partial views of generated solution within constraints. 6. SE view of final solution; 7. South elevation of final solution; 9. West elevation.

2.6 Pareto Genetic Algorithms: Application to Shape Generation

The sixth application concerns again 3D shape generation, but this time responding to conflicting objectives. To achieve this, Pareto Genetic Algorithms were applied once more [11]. The two conflicting objective functions were maximizing daylighting, and minimizing energy for heating the building, in the cold Chicago climate. GENE_ARCH generated a uniformly sampled, continuous Pareto front, from which seven points were visualized in terms of the proposed architectural solutions and environmental performance (Fig. 7). It can be seen that the best solution in terms of heating bases its strategy on creating a deep, compact volume which is surrounded, to the south and partially to the west, by narrow, highly glazed spaces that act as greenhouses, collecting solar gains to heat up the main space. Although it is hard to daylit those deep areas, savings in heating energy compensate for the spending in artificial lighting. On the contrary, the best solution for lighting generates narrow spaces that are easy to lit from the periphery, and tend to face the sun's predominant direction, in a flower-like configuration. However, it is also possible to notice the use of south-facing sunspaces, highly glazed, like in the previous solution. The intermediate points in the frontier represent other good trade-offs, and it is rather interesting to notice how solutions tend to gradually 'morph' from solution 1 to 7.

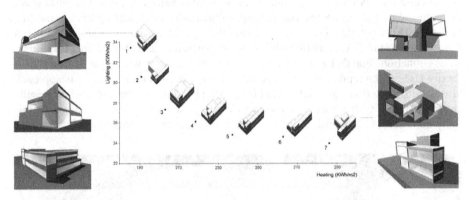

Fig. 7. Pareto frontier for Chicago, with best solution for heating (1) visualized on the left, and best solution for lighting (7) on the right; Both bottom images illustrate southeast views, showing the highly glazed south-facing sunspaces generated by GENE_ARCH

2.7 Application to an Urban and Housing Islamic Context

The last, ongoing application concerns the incorporation in GENE_ARCH of an urban shape grammar for an area of the Medina of Marrakesh, in Morocco [12]. The goal is to create the basis for a system that can capture some of the characteristics of the rich existing urban fabric, based on patio-houses and often-covered narrow streets, and apply them in contemporary urban planning. From the historical analysis and fieldwork in Marrakech, it was possible to identify three sub-grammars necessary to encode the complexity of the urban pre-existences: the urban grammar, negotiation grammar (shifting spaces between adjacent lots), and patio house grammar.

The patio-house shape grammar, developed by Duarte [13], is being combined with GENE_ARCH to generate new housing configurations based on the existing rules, while providing modern living standards in terms of daylighting, ventilation, thermal performance, and other environmental parameters. A novel approach is being introduced, departing from the standard shape grammar described in [13], and transforming it into a "subtractive shape grammar". This method departs from the most complex design achievable within the grammar, and analyzes the possible exclusion of each element, and the impact that would have on all the other elements of the design. The coding of these possible sequences will then be used to communicate with GENE_ARCH.

3 Discussion and Conclusions

This paper consists mainly of an overview of the capabilities and applications of GENE_ARCH up to this date. The software has proven to be robust and applicable in actual buildings of considerable complexity, and to help finding architectural solutions that are more sustainable and consume less energy. Part of the robustness of the program comes from applying as the calculation engine, for energy simulation, the software DOE2.1E, that is well respected in the field and is able to consider a very wide range of building variables in its calculations. In terms of the search engine, the standard GA has proven to be able to locate high quality designs in large solution spaces. Since GA's are heuristic procedures, and it is usually not possible to know the optimal solution for the type of problems faced in architecture, it is difficult at this stage to know what are the limits for the size of problems to be approached by GENE_ARCH. Given reasonable solution spaces, the system seems to have facility in solving problems like façade design, including openings geometry, materials and shading elements, given relatively stable geometries for the building.

The generation of complete 3D architectural solutions, that is, of a complete building description, poses much more complex questions. First of all, there are issues of representation of the architectural problem, in such way that both expresses the architect's design intentions, and allows the system to manipulate them and generate new solutions. Secondly, there are complex questions in terms of the method to evaluate solutions, since the issues involved are not only energy-related, but include functional and spatial characteristics and compliance with given requirements and intentions. The on-going experiments with shape grammars suggest that the method may be too limited to provide the necessary handles on complex three-dimensional problems, suggesting the need for other paradigms.

Acknowledgements

This paper was developed with the support from project POCTI/AUR/42147/2001, from Fundação para a Ciência e a Tecnologia, Portugal. Some of the graphical images in figure 2 were developed with the collaboration of João Rocha.

References

1. Caldas, L.G.: An Evolution-Based Generative Design System: Using Adaptation to Shape Architectural Form, Ph.D. Dissertation in Architecture: Building Technology, MIT(2001)
2. Shea, K. and Cagan J.: Generating Structural Essays from Languages of Discrete Structures, in: Gero, J. and Sudweeks, F., eds., Artificial Intelligence in Design 1998, Kluwer Academic Publishers, London (1998) 365-404
3. Monks, M., Oh, B. and Dorsey, J.: Audioptimization: Goal based acoustic design, IEEE Computer Graphics and Applications, Vol. 20 (3), (1998) 76-91
4. Caldas, L and Norford, L: Energy design optimization using a genetic algorithm. Automation in Construction, Vol. 11(2). Elsevier (2002) 173-184
5. Krishnakumar, K.: Micro-genetic algorithms for stationary and non-stationary function optimization, in Rodriguez, G. (ed.), Intelligent Control and Adaptive Systems, 7-8 Nov., Philadelphia. SPIE – The International Society for Optical Engineering (1989) 289-296
6. Caldas, L., Norford, L., and Rocha, J.:An Evolutionary Model for Sustainable Design, Management of Environmental Quality: An Int. Journal, Vol. 14 (3), Emerald (2003) 383-397

7. Caldas, L.: Pareto Genetic Algorithms in Architecture Design: An Application to Multicriteria Optimization Problems. Proceedings of *PLEA'02*, Toulouse, France, July 2002, 37-45
8. Fonseca, C. and Fleming, P.: 1993, Genetic Algorithms for Multiobjective Optimization: formulation, discussion and generalization, *Evolutionary Computation,* 3(1), pp. 1-16.
9. Horn, J., Nafpliotis, N., and Goldberg, D.: 1994, Niched Pareto Genetic Algorithm for Multiobjective Optimization. *Proceedings of the 1st IEEE Conference on Evolutionary Computation*, Part 1, Jun 27-29, Orlando, FL: 82-87
10. Caldas, L.:Evolving Three-Dimensional Architecture Form: An Application to Low-Energy Design, in: Artificial Intelligence in Design 2002, ed. by Gero, J., Kluwer Publishers, The Netherlands (2002) 351-370
11. Caldas, L.:Three-Dimensional Shape Generation of Low-Energy Architecture Solutions using Pareto GA's, Proceedings of ECAADE'05, Sep. 21-24, Lisbon (2005) 647-654
12. Duarte, J., Rocha, J., Ducla-Soares, G., Caldas, L.: An Urban Grammar for the Medina of Marrakech: A Tool for the Design of Cities in Developing Countries. Accepted for publications in Proceedings of Design Computing and Cognition 2006
13. Duarte, J., Rocha, J., Ducla-Soares, G.: A Patio-house Shape Grammar for the Medina of Marrakech. Accepted for publications in Proceedings of ECAADE'06

Mission Unaccomplished:
Form and Behavior But No Function

Mark J. Clayton

Texas A&M University, College of Architecture, College Station, TX 77845 USA
mark-clayton@tamu.edu

Abstract. Tools for modeling function may be an important step in achieving computer-aided design software that can genuinely improve the quality of design. Although researchers have included function in product models for many years, current commercial Building Information Models are focused upon representations of form that can drive models of behavior but lack models of function. If a model of function is added to a BIM, then the building model will be much more capable of representing the cognitive process of design and of supporting design reasoning. The paradigm of a form model, a function model, and a behavior model may suggest ways to reorganize architectural and engineering practice. Design teams could also be organized into roles of form modelers, function modelers, and behavior modelers. Although this would be a radical and novel definition of roles in a team, it parallels principles that have arisen naturally in contemporary practice.

1 Introduction

A review of architectural CAD research of the last twenty years reveals a quandary. Although many of the ambitions and expectations of researchers have been achieved, the expected promised land of high quality design has not been reached. The quandary is revealed by comparing two papers written in the mid 1980's by influential researchers in architectural computing. In a paper by Don Greenberg, he expressed optimism that newly invented radiosity and ray tracing methods could lead to great steps forward in design quality [1]. The computer would then enhance the essential visual and graphic processes of design. Yessios rebutted the assertion in a subsequent paper, warning that non-graphic information is critically necessary to support engineering analysis [2]. He suggested that *"... architectural modeling should be a body of theory, methods, and operations which (a) facilitate the generation of informationally complete architectural models and (b) allows them to behave according to their distinct architectural properties and attributes when they are operated upon."*

Interestingly, these two papers foreshadowed the themes of commercial CAD development over the next two decades. Ray tracing and radiosity rendering are the culmination of Greenberg's dream of complete and accurate visual representations of designs. In the late 1980's and early 1990's, software such as AutoCAD, 3D Studio, and Microstation achieved great strides forward in rendering ability, steadily bringing photorealism to the typical architectural firm. Answering Yessios' critique, the late

I.F.C. Smith (Ed.): EG-ICE 2006, LNAI 4200, pp. 119–126, 2006.

1990s and early 2000's brought Building Information Modeling (BIM) as a commercially viable way to integrate non-graphic information with the geometric models that have been the core of CAD systems. Energy modeling, finite element analysis, and computational fluid dynamics enabled much of the dream expressed by Yessios of comprehensive predictions of performance to come to fruition. BIM tools combine the two trends, offering powerful geometric modeling, and high quality rendering coupled to an extensive capability to store non-graphic information, building semantics, and quantity surveying.[1]

Although BIM is a major advance in representing buildings to support design, much research theory would declare it inadequate to support the cognitive processes of design. The design process remains complex, unintegrated, and awkward, and often produces products with obvious and egregious flaws. Ad hoc, informal, uncomputerized methods must still be employed by architects and engineers in fundamental design activities. I suggest that BIM formalizes and integrates the modeling of form and facilitates the modeling of behavior, but as yet fails to explicitly model the function of buildings. I offer a definition of design as the process of representing and balancing the form, function, and behavior of a future artifact. This definition implies a new and novel way of organizing a design team.

2 Form, Structure, Function, Requirements, Performance, Behavior

The research community has long postulated more complex and sophisticated design representations than those offered by the popular software vendors. As BIM penetrates the markets, achieves dominance, and is further developed, the research models are worth revisiting. I will explain three notable contributions: the development of the General AEC Reference Model (GARM), the logical arguments and empirical investigations of John Gero into structure, function and behavior, and my own largely overlooked working prototype software model of form, function, and behavior, the Virtual Product Model (VPM).

2.1 The General AEC Reference Model

An influential exploration of comprehensive digital building models was the RATAS, developed in Finland [3]. This effort received relatively large backing from industry and applied object-oriented programming methods to the development of product models for the building industry. It initiated a goal of representing all concepts and objects that constitute a building, both the physical aspects and ephemeral mental constructs. Explicitly included was a wish to represent functions and requirements, although the literature records little progress toward this goal.

The General AEC Reference Model employed an approach aligned with the Standards for Exchange of Product Information (STEP) of the International Standards Organization (ISO) to apply object-oriented product modeling to the building domain [4]. An application of the principles of the GARM has been well described using an example of life cycle building costs and renovation scheduling [5]. As implemented

---

[1] Communications tools have also emerged to enable designers to share information with alacrity. They represent a third major theme in architectural CAD development.

by Bedell and Kohler, the GARM defined a dynamic, user-configurable relation between a Functional Unit (FU) and a Technical Solution (TS). An FU defines the requirements for the design, while a TS defines a particular material, geometry and production method for satisfying the requirements. FU's and TS's may be composed hierarchically or decomposed into less complex units. An FU may be defined independently of the TS. To explore alternative designs, a particular TS may be swapped with a different TS. A TS may be reused from one project to another by combining it with a different FU.

The clear distinction of function representations from the solution representation was an important contribution that deserves revisiting.

2.2 Structure, Function and Behavior

A long thread of research has investigated a notion that there are three fundamental kinds of representations that support design: those defining the intent for the artifact, those defining the artifact itself, and those defining how the artifact performs. The idea already appears in investigations that helped formulate artificial intelligence as a field [6]. Simon stated that "Fulfillment of purpose or adaptation to a goal involves a relation among three terms: the purpose or goal, the character of the artifact, and the environment in which the artifact performs." (p. 6).

Gero has elaborated and extended a similar theme in his notion of Function, Behavior and Structure (FBS) [7, 8]. A design object is described by three kinds of variables. Function variables describe what the object is for. Behavior variables describe what the object does. Form variables describe what the object is, in terms of components and their relationships. A designer derives behavior from structure and ascribes function to behavior.

In the FBS theory, there are eight processes (or operations) in design. These processes include the definition of function in terms of expected behavior, the synthesis of structure, analysis that derives predicted behavior, evaluation that checks whether the predicted behavior is as expected, and documentation. The final three processes are reformulation of structure, behavior and function that constitute various iterative loops in the design process.

This cognitive model of design has been examined in several empirical observations of designers at work [9, 10].

3 Combining Form, Function, and Behavior in a Software Prototype

The Virtual Product Model (VPM) was my attempt to test the idea that design process can be formalized into computable steps for representing the function of a design, producing a form as a hypothetical solution for those functions, and deriving behaviors that enable a test of that hypothesis [11, 12]. This formulation of function, behavior, and form corresponds closely to Gero's function, behavior and structure.[2]

The VPM was a working software system that ran on a network of Sun computers. The user could draw a building in 3D using AutoCAD, assign functions to the

---

[2] My use of *form* rather than *structure* is meant as an homage to the great architect Louis Sullivan's aphorism "Form ever follows function."

drawing in a process of mapping the drawing elements to models of performance, and then run engineering software that derived the performance and enabled assessment of whether the functions were satisfied. The software used Interprocess Communication and Internet Protocol to launch processes on various networked computers, thus also demonstrating the feasibility of an Internet-based, Web-enabled distributed CAD system. In 1995, the VPM was unusual in product modeling research and cognitive design research because it was implemented as working software rather than merely a theoretical construct or software design.

3.1 Interpret, Predict, and Assess Methods in the VPM

The VPM was derived from a cognitive model of design rather than a product model. While most product modeling research has attempted to describe exhaustively the physical, economic, and production characteristics of design products, the VPM derived from a theory of the patterns of thought that produce a design. It provided an object-oriented representation of the conceptual objects supporting design cognition. The objects for representing a design in the VPM were derived from three related but independent hierarchies: form objects, function objects, and behavior objects. In the VPM, the form was defined as the geometry and materials. AutoCAD was used as the form modeler. Function was defined as the requirements, intents, and purpose of the design. Behavior was defined to be the performance of the design artifact. These root objects were related through fundamental operations (methods) of design cognition: *interpret*, which mapped function objects to the design form; *predict*, which mapped form objects to behavior objects; and *assess* which compared behaviors to the functions to determine whether the design was satisfactory.

The software adopted a hermeneutic philosophical stance that declares that the meaning in designed objects is ascribed by people rather than being inherent in the object [13]. The interpret method in the software implemented a key capability. It allowed one or more designers to apply engineering expertise from a specific domain to the pure form model by identifying and classifying the features of the design form that affect performance in that domain. Thus, in theory an architect could work freely in a graphic environment and then pass the model to an engineer who would interpret the CAD graphic model into an active engineering model that could derive performance. To test the generality of the ideas, four example performance models and supporting "interpretation" interfaces were constructed: energy analysis, cost analysis, building code analysis, and spatial function analysis. Each interpretation defined subclasses of function and behavior that could plug into the VPM framework and respond appropriately to method calls. The process of developing the various interpretations was the iterative creation of object hierarchies of function and behavior and careful consideration of where in the hierarchies a particular concept should be located. The development process was exploratory and heuristic; more recent research has formulated principles for the development of an object hierarchy for function [14].

The VPM produced model-based evidence in support of the form, function, and behavior (or function, behavior, structure) theory of design. It showed that the theory was sufficiently complete to permit working, usable software that appeared to closely conform to natural design cognitive processes. A handful of testers used the software to analyze three designs for a small medical facility and produced results that were in some ways better than results done with manual methods. Although the evidence did not prove that the theory is accurate in describing natural processes, the evidence did not disprove that the theory is accurate. In the field of artificial intelligence, such a level of proof is generally accepted as significant.

3.2 Hierarchies of Form, Function, and Behavior

In conjunction with the normal CAD modeling operations for defining the geometry of the design, *interpret*, *predict* and *assess* produced extensive object hierarchies of form, function and behavior. The user interface allowed one to follow the relations of a form to a function and to a behavior as a triad of objects addressing an engineering domain. A successful design resulted in the resolution of form, function, and behavior objects into stable, balanced triads in which the behavior, computed automatically from the form, fell into acceptable ranges defined by function objects.

An unexpected side effect of the software was that, as a user manipulated the CAD model, interpreted it into various engineering models, predicted the performance, and assessed the behavior against functions, the software constructed elaborate hierarchies of forms, functions, and behaviors. These constituted the "virtual" product model; rather than a predetermined product model schema established as a rigid class hierarchy of building components that the user explicitly instantiated, the VPM invented new combinations of forms, functions, and behaviors on the fly in response to the designers' actions. There were no components in the usual sense of walls, columns, doors, and windows, but merely geometric forms that had a declared function of being a wall, column, door, or window from whatever conventional or conflicting sets of definitions necessary to support the engineering interpretation. A Virtual Component object aggregated and managed the relations among a form object and various functions and behaviors.

Although the possible list of functions and behaviors was necessarily finite, the software achieved extensibility by providing a mechanism for very late binding and polymorphism that allowed the addition of new functions and behaviors by downloading applications over the Internet. Even if no prewritten software existed, a development interface allowed a software developer to add rudimentary functions or behaviors that required human editing and inspection to determine whether the design satisfied the functions. However, the software lacked a user interface for defining new functions.

A concrete example can explain the utility of the interpretation capability. Using the VPM, one could use AutoCAD to draw a cylindrical tube. This tube could be interpreted as a structural column, or it could be interpreted as pipe for conveyance of fluids, or it could be interpreted as a handrail. Actually, it could be simultaneously interpreted to be all three. One could easily draw non-vertical walls, or walls with sloped sides, or columns with a star-shaped cross section, or any other clever or novel architectural form and then declare the intent for that form in terms related to performance. The software did not constrain inventiveness with respect to the building form as do CAD systems (or BIM systems) that rely upon predefined components that unify form and function. With the VPM, there was no need for a stair tool, or a roof tool, or a wall tool that confounded form and function and constrained the designer to conventional shapes, materials, parameters and modeling sequences.

4 Problem Seeking and the Distinction of Roles in Design

The research stopped before exploring another way of looking at the modeling process. The user interface of the software prototype was designed to mimic a relatively conventional design process by which a design architect developed a description of form and engineers (or architects) then analyzed that form. Largely, the users focused on the triplets consisting of a form object, a function object, and a behavior object. However, the elaborate hierarchies of form, function and behavior could also be manipulated independently using developer interfaces. Never built was an alternative user interface that would allow a user to construct a relatively complete representation of function that could be passed to a design architect, who could then develop a relatively complete representation of form that could be passed to the engineers, who would develop a relatively complete representation of behavior.

4.1 The Architectural Programming Profession

Some architectural practice theory endorses this alternative way of conceiving the building design process and the roles of designers. A method called "problem seeking" distinguishes between the architectural programmers who are responsible for developing the description of needs and requirements (called a "brief" in the United Kingdom) and design architects, who synthesize the proposed solution and represent it so that it can be evaluated [15]. The argument is that the design architects are overly permissive towards changing the program to match the beautiful forms that they invent. The programmers should have sufficient authority to establish a complete, thoughtful, and relatively fixed program to which the design architects must comply. Through a process akin to knowledge engineering, specialists in programming, if given independence and authority, can work closely with the architect's client to establish a clear and complete statement of the requirements.

Similarly, design architects have become increasingly reliant upon consultants and engineers to analyze and predict the expected performance of the building design. In the United States, it is common to have a design team that includes forty or fifty consultants in addition to the design architects.

4.2 Three Branches of the Design Professions

Perhaps the design professions are trifurcating into function experts, form experts, and behavior experts. The concept of integrated but distinguished function models, form models, and behavior models neatly mirrors the kinds of talents among designers and the division of labor on design projects. In the future, the industry could reorganize its design professions into specialists in each of these three areas. A function architect would work with the client to define explicitly the requirements for the project. A form architect would prepare a CAD model to depict a solution alternative. The behavior architect would use software tools to predict the performance of the solution. The function architect would then reenter the process to verify whether functions have been satisfied and if not, to consider altering the function definitions and reinitiate the process.

Clearly, all three of these kinds of architects must document their work. Thus, documentation is a simultaneous, ancillary activity that parallels the design process.

As described, this new process appears to be a waterfall process that is highly sequential and departmentalized. Probably an iterative process would be better and would allow the emergence of functions during design and the emergence of forms during analysis. A team of function architect, form architect, and behavior architect working synchronistically may be better than individuals working sequentially.

This argument suggests that future software must integrate tools for documenting function. Once the function representations are integrated, the software can truly support the cognitive processes of design. As BIM thoroughly represents form and provides strong connections to simulation software that can predict behavior, the addition of function would create a complete tool for supporting design cognition.

Building Information Modeling plus function (BIM+Fun) could provide support for a cognitive model of design rather than merely the outward product of design. Such software might enable profound transformation of the design industries and even contribute to better design.

References

1. Greenberg, D. P.: Computer Graphics and Visualization. In: Pipes, A. (eds.): Computer-Aided Architectural Design Futures. Butterworth Scientific, Ltd., London (1986) 63-67.
2. Yessios, C.: What has yet to be CAD. In: Turner, J. (ed.) Architectural Education, Research and Practice in the Next Decade. Association for Computer Aided Design in Architecture (1986) 29-36
3. Bjork, B-C.: Basic structure of a proposed building product model, Computer-Aided Design, Vol. 21, No. 2, March. Butterworth Scientific, Ltd., London (1989) 71-78.
4. Willems, P.H.: A Meta-Topology for Product Modeling. In Proceedings: Computers in Building W74 + W78 Seminar: Conceptual Modelling of Buildings. Lund, Sweden, (1988) 213-221.
5. Bedell, J. R., Kohler, N.: A Hierarchical Model for Building Applications. In: Flemming, U., Van Wyk, S. (eds.): CAAD Futures '93. Elsevier Science Publishers B.V., North Holland (1993). 423-435.
6. Simon, H. A.: The Sciences of the Artificial. The M.I.T. Press, Cambridge, MA. (1969).
7. Gero, J. S.: Design prototypes: a knowledge representation schema for design, AI Magazine, Vol. 11, No. 4. (1990) 26-36.
8. Gero, J. S.: The role of function-behavior-structure models in design. In Computing in civil engineering, vol. 1. American Society of Civil Engineers, New York. (1995) 294 - 301.
9. Gero, J. S., Kannengiesser, U.: The situated function–behaviour–structure framework , Design Studies Vol. 25, No. 4. (2004) 373-391.
10. McNeill, T., Gero, J. S., Warren, J. Understanding conceptual electronic design using protocol analysis, Research in Engineering Design Vol. 10. (1998) 129-140..
11. Clayton, M. J., Kunz, J. C., Fischer, M. A.: Rapid conceptual design evaluation using a virtual product model, Engineering Applications of Artificial Intelligence, Vol. 9, No. 4 Elsevier Science Publishers B.V. North Holland. (1996) 439-451.
12. Clayton, M. J., Teicholz, P., Fischer, M., Kunz, J.: Virtual components consisting of form, function and behavior. Automation in construction, Vol. 8. Elsevier Science Publishers B.V. North Holland. (1999) 351-367.

13. Winograd, T. and Flores, F. Understanding computers and Cognition, A New Foundation for Design. Reading, Massachusetts: Addison-Wesley Publishing Company, (1986).
14. Kitamura, Y., Kashiwase, M., Fuse, M: and Mizoguichi, R., Deployment of an ontological framework of functional design knowledge. Advanced Engineering Informatics, Vol. 18, No 2. (2004) 115-127.
15. Peña, W., Parshall, S., Kelly, K.: Problem Seeking -- An Architectural Programming Primer, 3rd edn. AIA Press, Washington (1987).

The Value of Visual 4D Planning in the UK Construction Industry

Nashwan Dawood and Sushant Sikka

Centre for Construction Innovation Research, School of Science and Technology,
University of Teesside, Middlesbrough, TS1 3BA, UK
Phone: +1642/ 342405
n.n.dawood@tees.ac.uk, E5096253@tees.ac.uk

Abstract. Performance measurement in the construction industry has received considerable attention by both academic researchers and the industry itself over a past number of years. Researchers have considered time, cost and quality as the predominant criteria for measuring the project performances. In response to Latham and Egan reports to improve the performance of construction processes, the UK construction industry has identified a set of non-financial Key Performance Indicators (KPIs). There is an increased utilisation of IT based technologies in the construction industry and in particular 4D (3D+time). Literature reviews have revealed that there is an inadequacy of a systematic measurement of the value of such systems at both quantitative and qualitative levels. The aim of this ongoing research is to develop a systematic measurement framework to identify and analyse key performance indicators for 4D planning. Two major issues have been addressed in the research: the absence of a standardised set of 4D based KPIs and the lack of existing data for performance evaluation. In this context, the objective of this paper is to establish the benefits of 4D planning through identifying and ranking a set of KPIs on the basis of semi-structured interviews conducted with UK based project managers. The ultimate objective of this research is to deliver a set of industry based 4D performance measures and to identify how project performance can be improved by the utilisation of 4D planning.

1 Introduction

This study is a collaborative research project between the Centre for Construction Innovation & Research at the University of Teesside and Architectural3D. The aim of this study is to deliver a set of industry based 4D performance measures and to identify how project performance can be improved by the utilisation of 4D planning. Visual 4D planning is a technique that combines 3D CAD models with construction activities (time) which has proven to be more beneficial than traditional tools. In 4D models, project participants can effectively visualise, analyse, and communicate problems regarding sequential, spatial, and temporal aspects of construction schedules. As a consequence, more robust schedules can be generated and hence reduce reworks and improve productivity. Currently, there are several research prototypes and commercial software packages that have the ability to generate 4D models as a tool for analysing, visualising, and communicating project schedule.

I.F.C. Smith (Ed.): EG-ICE 2006, LNAI 4200, pp. 127 – 135, 2006.
© Springer-Verlag Berlin Heidelberg 2006

However, the potential value and benefits of such systems have not been identified. This has contributed to a slow intake of such technologies in the industry.

The industry based Key Performance Indicators (KPIs) that were developed by the Department of Trade and Industry (DTI) sponsored construction best practice program are too generic and do not reflect the value of deploying IT system for construction planning and in particularly 4D planning. The objective of this research study is to overcome the presence of a generalised set of KPIs by developing a set of 4D based KPIs at project level for the industry. Information Technology applications are progressing at a pace and their influence on working practice can be noticed in almost every aspect of the industry. The potential of IT applications is significant in terms of improving organisation performance, management practices, communication, and overall productivity. 4D planning allows project planners to visualise and rehearse construction progress in 3D at any time during the construction process. According to *Dawood et al. (2002,* using 4D planning, participants in the project can effectively visualise and analyse the problems considering the sequencing of spacial and temporal aspects of the construction time schedule. The thrust for improved planning efficiency and visualisation methodology has resulted into the development of 4D planning.

The Construction Industry Institute (CII) conducted research in the use of three-dimensional computer models on the industrial process and commercial power sector of AEC (architectural, engineering and construction) from 1993 to 1995 (*Griffis et al., 1995*). Major conclusions of the CII research are that the benefits of using a 3D technology include reduction in interference problems; improved visualisation; reduction in rework; enhancement in engineering accuracy and improved jobsite communications. *Songer (1998)* carried out a study to establish the benefits and appropriate application of 3D CAD for scheduling construction projects. *Songer (1998)* has demonstrated that the use of 3D-CAD and walk-thru technologies during planning stage can assist in enhancing the scheduling process by reducing the number of missing activities, missing relationships between various activities, invalid relationships in the schedule and resource fluctuation for complex construction processes. Center for Integrated Facility Engineering (CIFE) research group at Stanford University has documented the applications and benefits of 3D and 4D modelling in their CIFE technical reports (*Koo & Fischer-1998, Haymaker & Fischer-2001 and Staub-French & Fischer-2001*). These reports discuss the benefits of 3D and 4D modelling by considering individual project separately. The application of Product Model and Fourth Dimension (PM4D) approach at Helsinki University of Technology Auditorium Hall 600 (HUT-600) project in Finland has demonstrated the benefits of 4D modelling approach in achieving higher efficiency; better design and quality, and early generation of reliable budget for the project *(Kam et al. 2003)*.

The above studies lack a well-established metrics that would allow the quantification of 4D planning at project level. In the absence of well-defined measures at the project level, the priority of this research project is to establish a set of key performance indicators that will reflect the influence of 4D applications on construction projects. This will assist in justification of investments in advanced technologies in the industry. The remainder of the paper will discuss the research methodology adopted and research findings.

2 Research Methodology

The methodology compromises of three interrelated phases:

- Identification of performance measures through literature review and authors experience in the application of 4D.
- Conducting semi-structured interviews with project managers/planners to establish and prioritise the performance measures.
- Data collection to quantify the identified performance measures.

Three major construction projects in London (currently under construction and combined value of project is £230 million) were selected for study and for data collection. Project managers and construction planners from the three projects were selected for interviews to identify and prioritise the 4D KPIs. A semi-structured interview technique was used to elicit information from the project managers, and construction planners' viewpoint about the key performance indicators at project level. The semi-structured interviews used a methodological procedure known as the Delphi technique for data collection. This technique is ideal for modelling real world phenomena that involve a range of viewpoints and for which there is little established quantitative evidence *(Hinks & McNay 1999)*. The subsequent sections of the paper describe the process of identifying KPIs on the basis of semi-structured interviews conducted with project managers, ranking of 4D KPIs and research findings.

3 Identification and Selection of Key Performance Indicators (KPIs)

The development of the performance measures list has taken account of the performance measurement characterised by Rethinking Construction, the construction best practice program has launched the industry wide KPI for measuring the performance of construction companies *(CBPP-KPI-2004)*. The Construction Best Practice Program has identified a framework for establishing a comprehensive measurement system within both organisation and project level. Other literature includes *Kaplan & Norton (1992); Li. et.al (2000); Chan et.al (2002); Cox et.al (2003); Robert. et.al (2003); Albert & Ada (2004); Bassioni et.al (2004)* and the knowledge of authors in the area of performance measurement in the construction

Table 1. Definition of the identified measures

Measure	Definition
Time	It can be defined as percentage number of times projects is delivered on / ahead of schedule. The timely completion of project measures performance according to schedule duration and is often incorporated to better understand the current construction performance. Schedule performance index (Earned value Approach) has been identified to monitor the performance of schedule variance.

Table 1. (*continued*)

Safety	It can be defined as a measure of the effectiveness of safety policy and training of the personnel engaged in activities carried out on site. Safety is a major concern for every construction company, regardless of the type of work performed. Safety is measured quantitatively through time lost as a result of accidents per 1000 man hrs worked, Number of accidents per 1000 man hrs worked.
Communication	Information exchange between members using the prescribed manner and terminology. The use of a 4D interface allows the project team to explore the schedule alternatives easily and assist in deploying 4D approach. Communication can be quantified in terms of number of meetings per week and time spent on meetings (Hrs) per week.
Cost	Percentage number of times projects is delivered on/under budget. Cost performance index (Earned value Approach) has been identified to monitor the performance of cost variance.
Planning Efficiency	It represents the percentage progress of construction activities scheduled to be performed on a weekly basis. It is measured as number of planned activities completed divided by the total activities planned to determine the project progress on a weekly basis.
Client satisfaction	Client satisfaction can be defined as how satisfied the client was with the finished product/facility. Usually measured weekly/monthly or shortly after completion and handover.
Team Performance	Ability to direct and co-ordinate the activities of other team members in terms of their performance, tasks, motivation and the creation of a positive environment.
Productivity Performance	This method measures the number of completed units put in place per individual man-hour of work. Some of the identified productivity performance measures are; number of piles driven/day, number of piles caps fixed / day, tonnes of concrete used / day/m^3 and pieces of steel used per day or week.
Quality	Quality has been considered here in terms of rework. Rework can be defined as the activities that have to be done more than once in the project or activities which remove work previously done as a part of the project. By reducing the amount of rework in the pre-construction and construction stages, the profits associated with the specific task can be increased. Rework can be represented in terms of number of changes, number of error, number of corrections number of request for information to be generated, non-programme activities, number of claims and number of process clashes spotted due to sequencing of activities.

industry were also used for the identification of performance measures. Project managers and construction planners from three construction projects were invited for interviews. Table 1 shows a brief definition of the identified KPIs.

The first task for the interviewees were to identify and rank the performance measures using a four (4) point Likert Scale. The second task was to identify the information required to quantify each measure. Their input was considered to be critical in the success of this research. The concept behind conducting semi-structured interviews was to evaluate how mangers and planners perceive the importance of performance measures. This will assist in the identification of industry based performance measure that can be used to quantify the value of 4D planning. The interview included both open and closed questions to gain a broad perspective on actual and perceived benefits of 4D planning. Due consideration has been given to the sources from where data has to be collected in a quantitative or qualitative way. So far ten semi-structured interviews have been conducted with senior construction planners. The research team intends to continue the interviewing process for ten more senior construction planners. This will assist in gathering more substantial evidence about KPIs.

4 Findings and Ranking of KPIs

Interviewees were asked to rank the identified KPIs. The ranking of the KPIs was done by using a four (4) point Likert Scale. For the prioritisation process, each KPI can be graded on a Likert scale of 1 to 4 (where 1= Not important, 2 = fairly important, 3 = Important and 4 = Very important) to measure the importance of each performance measure. The benefits of 4D planning will be quantified on the basis of prioritised KPIs. The performance measures will be further classified in qualitative terms (rating on a scale) and quantitative terms (measurement units).

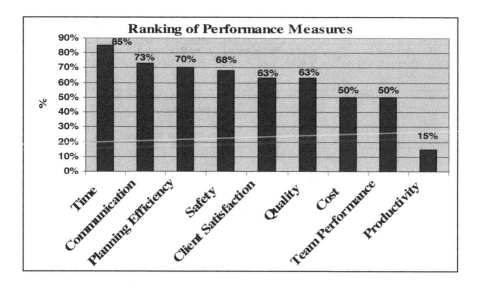

Fig. 1. Ranking of Performance Measures

Table 2. 4D based KPIs in order of priority

Ranking	KPIs	Indices	Performance Measures	Stages of Construction
1	Time	Schedule Performance Index	(i) Schedule Performance	Pre-construction and Construction
2	Communication	Communication Index	(i) Number of meeting per week (ii) Time spent on meetings per week	Pre-construction & Construction
3	Planning Efficiency	Hit Rate Index	(i) Percentage of activities completed per week (ii) Number of milestones delivered	Construction
4	Safety	Safety Index	(i) Number of accidents per 1000 man hrs worked (ii) Time lost in accidents per 1000 man hrs worked	Construction
5	Client Satisfaction	Satisfaction Index	(i) Number of client queries (ii) Satisfaction questionnaire (iii) Number of claims (Completion time/Cost etc.)	Pre-construction & Construction
6	Quality	Rework Index	(i) Number of changes (ii) Number of error (Drawing/Design) (iii) Number of corrections (Drawing/Design) (iv) Number of request for information generated (v) Non-Programme activities (vi) Number of claims (Quality) (vii) Number of process clashes spotted due to sequencing of activities.	Pre-construction & Construction
7	Cost	Cost Performance Index	(i) Cost Performance	Pre-construction and Construction
8	Team Performance	Team Performance Index	(i) Personnel turnover & productivity (ii) Timeliness of information from team	Pre-construction & Construction
9	Productivity	Productivity Index	(i) Tonnes of concrete used per day / m^3 (ii) Pieces of steel used /day or week (iii) Number of piles driven / day (iv) Number of pile caps fixed / day	Construction

Using responses from a four (4) point Likert Scale, the average percentage value for each of the performance measures was calculated. Figure 1 represents the ranking of the performance measures in descending order on the basis of the views of the respondents. The performance measures perceived as being highly important by the respondents are: *time, communication, planning efficiency, safety and client satisfaction.* As shown in figure 1, time and communication has scored the top ranking as compared to other performance measures.

Table 2 represents the ways to quantify the priorities of 4D KPIs at the different stages of a construction project. For example, 'Time' has been ranked (85%) as top KPI by the respondents and we propose to use 'Schedule Performance Index' to measure it. Schedule performance index (Schedule efficiency) can be defined as the ratio of the earned value created to the amount of value planned to be created at a point in time on the project. Similarly, we propose to measure 'Safety' in terms of 'Safety Index' i.e.: Number of accidents per 1000 man hrs worked and time lost in accidents per 1000 man hrs worked. Further, identified KPIs have been represented in their respective indices form to indicate the effect of any given change in the construction process.

5   Future Research Activities

This paper reports on the first stage of the research project. The current and future research activities will include:

- Continuing the interview process to further confirm the 4D KPIs and method of data collection.
- Establish a methodology for data collection and quantification of the KPI indices for the three identified construction projects.
- Benchmarking the KPIs indices with industry norms and identifying the improvements in construction processes resulted due to the application of 4D planning.
- Identifying the role of supply chain management in the development and updating of construction schedule for the 4D planning. As per the main contractor's viewpoint 4D is unable to bring any confirmed value as compared to their own planning system. Interviews with project managers have revealed that there are varying views between the main contractors and trade contractors on the usage of 4D planning on a construction project. The concern at the moment is the availability of the information, time used in the collection of information and cost factor attached in the implementation of the 4D technology. All the stakeholders were agreed that an early deployment of 4D brings about lot of transparency to resolve the conflicts among the various trades during the preconstruction phase.

6   Conclusions

Research studies and industrial applications have highlighted the benefits of 4D in a subjective manner and it has been stipulated that 4D can improve the overall project performance by identifying clashes, better communication and improved coordination. The evaluation of 4D planning in the construction management literature has not been addressed seriously from performance measurement viewpoint. The evaluation and justification of 4D planning is crucial to promote the value embedded in it. The study has developed five key performance indicators consistently perceived as being highly significant at project level are: time, communication, planning efficiency, safety and client satisfaction.

References

1. Al-Meshekeh, H.S., Langford, A.: Conflict management and construction project effectiveness: A review of the literature and development of a theoretical framework. J. Construction. Procurement. 5(1) (1999) 58-75
2. Albert, C., Ada C.: Key Performance Indicators for Measuring Construction Success. Benchmarking: An International Journal. Vol.11. (2004) 203-221
3. Bassioni, A.H., Price, A.D., Hassan, T.M.: Performance Measurement in Construction. Journal of Management in Engineering. Vol. 20. (2004) 42-50
4. Chan, A.P.C. Determining of project success in the construction industry of Hong Kong. PhD thesis. University of South Australia, Australia (1996)
5. Chan, A.P.C., David, S., Edmond, W.M.L.: Framework of Success Criteria for Design/Build Projects Journal of Management in Engineering. Vol. 18. No 3. (2002) 120-128
6. Construction Best Practice Program- Key Performance Indicators (CBPP-KPI-2004), (available at http://www.dti.gov.uk/construction/kpi/index.htm
7. Dawood, N., Eknarin, S., Zaki, M., Hobbs, B.: 4D Visualisation Development: Real Life Case Studies. Proceedings of CIB w78 Conference, Aarhus, Denmark, 53-60.
8. Egan, J. Sir.: Rethinking Construction: The Report of the Construction Task Force to the Deputy Prime Minister. Department of the Environment, Transport and the Regions, Norwich (1998)
9. Griffs, Hogan., Lee.: An Analysis of the Impacts of Using Three-Dimensional Computer Models in the Management of Construction. Construction Industry Institute. Research Report (1995) 106-11
10. Haymaker, J., Fischer, M.: Challenges and Benefits of 4D Modelling on the Walt Disney Concert Hall Project. *Working Paper 64,* CIFE, Stanford University, Stanford, CA (2001)
11. Hinks, J., McNay.: The creation of a management-by-variance tool for facilities management performance assessment. Management Facilities, Vol.17, No. 1-2, (1999) 31-53
12. Kam, C., Fischer, M., Hanninen, R., Karjalainen, A., Laitinen, J.: The Product Model And Fourth Dimension Project. IT Con Vol. 8. (2003) 137-165
13. Kaplan, R.S., Norton, P.: The Balanced Scorecard -- Measures that Drive Performance. Harvard Business Review, Vol. 70, No. 1 (1992) 47-54
14. Koo, B., Fischer, M.: Feasibility Study of 4D CAD in Commercial Construction. CIFE Technical Report 118. (1998)
15. Latham, M. Sir.: Constructing the Team: Final Report of the Government/Industry Review of Procurement and Contractual Arrangements in the UK Construction Industry. HMSO, London (1994)
16. Li, H., Irani, Z., Love, P.: The IT Performance Evaluation in the Construction Industry. Proceedings of the 33rd Hawaii International Conference on System Science (2000)

17. Naoum, S. G.: Critical analysis of time and cost of management and traditional contracts. J. Construction Management, 120(4), (1994) 687-705
18. Robert, F.C., Raja, R.A., Dar, A.: Management's Perception of Key Performance Indicators for Construction. Journal of Construction Engineering & Management, Vol. 129, No.2, (2002) 142-151.
19. Songer, A.: Emerging Technologies in Construction: Integrated Information Processes for the 21st Century. Technical Report, Colorado Advanced Software Institute, Colorado State University, Fort Collins, CO 80523-1 873 (1998)
20. Staub-French, S., Fischer, M.: Industrial Case Study of Electronic Design, Cost and Schedule Integration. CIFE Technical Report 122. (2001)
21. Yin, R. K.: Case study research design and method. 2nd edition. Sage publication Inc, CA (1994)

Approximating Phenomenological Space

Christian Derix

University of East London
School of Architecture & Visual Arts
4-6 University Way
London E 16 2RP
c.derix@uel.ac.uk

Abstract. Architectural design requires a variety of representations to describe the many expressions a building can be observed through. Commonly, the form and space of a building are represented through the visual abstraction of projective geometry. The medium of geometric representation has become synonymous with architectural space.

The introduction of computational design in architecture has not changed our understanding or representation of architectural space, only of its geometric description and production processes. addition of the computer as a medium should allow us to open new 'ways of seeing' since the medium allows for novel descriptions and expressions via data processing hitherto impossible. This paper would like to propose some computational methods that could potentially describe and generate non-geometric but rather phenomenal expressions of architectural space.

1 Introduction

Bill Mitchell published the Logic of Architecture in 1990 [1] and thus laid some of the most influential foundations for computational design in architecture. He did make it clear that he would 'treat design primarily as a matter of formal composition [...] in order to produce beauty'. Beauty for Mitchell was an expression of the functional fit of (visual and geometric[1]) elements and subsystems to the program of the building.

The computer as design tool has since been perceived as either an optimization tool for functional aspects of the building or as graphic pattern generator where the pattern represents architecture or a sub-set of it (façade), the algorithm the design process.

This representation of architecture in computational design is no surprise if one regards the history of architectural representation. With few exceptions, architecture has been represented as geometric projections. Metric descriptions of buildings and spaces make sense for visual impressions and construction information, and thus enhance imagination of space. But as Robin Evans would argue, it only reflects one dimension or expression of space and it rarely translates into the built object [2].

---

[1] It is extraordinary to find that in his book about the Logic of Architecture, only drawings are found, which the rules for the grammars and vocabulary are extracted from. No photographs, diagrams or other representations.

I.F.C. Smith (Ed.): EG-ICE 2006, LNAI 4200, pp. 136–146, 2006.

Often, architects tend to design on the basis of their tools or media, not on the basis of actual space or the construction of it. The tools used in the medium of projective representation are non-dynamic inert objects like ruler and pen and especially the drawing board, forming part of a dynamic system with the designer. According to Marshall McLuhan, every medium contains another medium [3]. If the drawing board, ruler and pen contained the medium of the projective geometry, then what medium does the computer contain in the case of architectural design? Why do computational designers emulate the same medium of projective representation like the first cast iron bridges mimicked wooden constructions? Shouldn't the computer with its capacity to represent data dynamically via any interpretation and its immense processing capacity be used to describe dynamic relations of data? The computer as a medium should therefore contain non-static representations of space – phenomena of spatial and social nature, Gestalten.

2 Phenomenal Space

'When, in a given bedroom, you change the position of the bed, can you say you are changing rooms, or else what?' -Georges Perec [4]

While Mitchell follows Chomsky's structural linguistic model, where meaning can be produced given a rigid syntax and a well-considered vocabulary, Peter Eisenman and other architectural theorists like Jenks advocated Derrida's linguistics of deferred meaning, where syntax and vocabulary are ephemeral, representing a function of meaning [5].

That embodies a direct echo of the Gestalt theory and of course initiated the call for complexity theory as a role model for architectural design. The medium for designing such complex architecture is supposed to be the computer. One should be able to simulate any kind of system through programmed algorithms.

Gestalt theory as much as Derrida's language model or complexity theory all try to describe the idea of the whole being greater than the sum of its parts [6]. This whole as an emergent phenomenon organizes the perception of its parts and their construction rules. By perception, the observer's perception is intended.

If one stands outside a house, a wall forms part of the delimiting shell, standing at the inside, say within a room, the same wall becomes commodity to hang things off. The perceived phenomenon of 'house' or 'room' makes the same wall appear to us as different elements within its context.

Maturana [7] and Luhmann [8] argued that through structural coupling of systems communicative and perceptual domains would emerge that have their lead-distinctions which are defining those domains and organizing all possible instances that can occur (Leitdifferenz). Those distinctions can only be binary when seen from within one of those domains (i.e., lawfulness is the Leitdifferenz for the domain of the judiciary). The gestalt psychologist' equivalent to domains were ordering 'schemata', or the learnt context within which one perceives the role of a part of a system [9].

In Wittgenstein's words: 'The form of the object consists through the possibilities of its appearances (seines Vorkommens) via relationships of instances (Sachverhalte). [...] There is no "thing in itself" (Ding an sich)' [10].

The Gestalt psychologist most striking example was the Phi-phenomenon to explain the idea of a phenomenon: taken two light sources in a dark room, switching them on and off in sequence with a precise distance between them doesn't create the sensation of seeing two light bulbs being switched on and off in a row, but to perceive 'movement' or a line [9]. Enter the cinema.

Properties of objects, be they architectural or other, don't rest within the object itself but form part of phenomenal whole that we observe and interpret. Thus parts and their expression through an ordering phenomenon are dependent on the context they occur in and the intention of the observer.

Additionally, the configuration of the parts of the occurrence is important to interpretation and perception. The Phi-phenomenon would not work if the light bulbs were too distant or the sequential switching too slow or too fast. They would appear as either two points or a line with direction. Meaning or quality doesn't reside in the objects but in their relationships and the observer. Relationships and ordering principle of phenomena are equivalent to topologies. If the topology of an organizing system changes, so does its gestalt. Perec wanted to question that condition.

Heinz von Foerster states that sensual stimuli don't convey qualities but changing quantities [11]. Quantities of sensual stimuli are computed on via neuro-physiologcal configurations – neuronal patterns. Lots of simple gates (neurons) computing differences to previous or later patterns. Interpretation or meaning is distributed over the network of neurons and occurs when a change of pattern is generated via stimuli.

Arnheim would concur with von Foerster's description when he said that 'space between things turns out not to look simply empty' [12].

Distributed representation should therefore afford the representation of phenomena. The architecture of the computer does just that: patterns of differences of simple logical gates generating via organizational rules various types of representation. The computing of architecture should afford the designer to generate phenomena of architecture – distributed representation of space, rather than just projective and geometric patterns[2].

3 Precedents of Phenomenal Representations in Architecture

There are few proponents in the design world who would argue that one can either identify drivers for phenomena of space or even incorporate catalysts for such qualities into generative computer code.

The paradigm of systems and complexity theory has been adapted in concept thinking by some practising architects in the past but it has hardly ever translated into their design process (with rare exceptions like Tschumi's Park della Villette), let alone been implemented via computational design.

---

[2] Miranda puts it as follows [13]: 'Electronic digital computation is built upon processes of accumulations of electric charges and their transformation, on to which, as Claude Shannon found in the late 1930s, it is possible to map the rules and logic syntax of Boolean algebra. From here, […], it is possible to transfer into Boolean algebras any phenomenon describable in terms of logic, and that, according to the project of natural sciences, accounts for about everything.'

In this section, I would like to look only at some examples of computational designer who have attempted to generate emergent phenomena in architecture.

Within such a discussion, one cannot omit the two towering inventors of analogue computing in architecture, Gaudi and Frei Otto. Both introduced the architectural design world to the notion of representing form not through projective geometry but through natural parallel computation. The drawings were strictly necessary for the builders and visualization not for finding from and space. All the hallmarks of gestalt theory were present in their work since the elements that compute the edges, surfaces and spaces produced an emergent whole that instilled the observer to think of it as architectural expression – a phenomenon nowhere to be found in the description of the system that produced it.

Although Christopher Alexander [14], Christian Norberg-Schulz [9] and Rudolf Arnheim [12] managed to *decode* some architectural phenomena it took another 20 years before Bill Hillier showed through computational simulation how an architectural phenomenon could be quantified and expressed through simple algorithms[3]. In his seminal book 'The Social Logic of Space', Hillier attempts to disclose correlations between configurations of shapes on plan and how those configurations influence the occupants' perception of space and subsequently, how those pattern influence actions [15].

One of his key concepts is the justified graph or depth map. He used graphs representations to show that a building would be perceived differently from one space to another. Therefore, adjacency understanding would change and perception of distances throughout a building. His graphs represent graphically this change of perception or the phenomenon of perceived distance, not in metric but through topology.

Steadman and March had already used graph visualization to generate topological representations of buildings but limited themselves to objective space configurations where a graph of a building would remain homogeneous from any location – the external observer [16]. Hillier opened the way to heterogeneous graph representations dependent on location in buildings – the internal or embedded observer.

Via axial lines analysis and convex shapes in the plan, he could demonstrate perceived hierarchies of urban tissue. This type of analysis (later done via Benedikt's concept of the isovist [17]) expressed the integration of spatial locations via visual connectivity. If a location was better connected than another, pedestrians tend to probabilistically end up more likely in the well connected location. No further assumptions were needed to understand space as configurational or topological mechanisms ('Space is the Machine') [18]. The elements of the system – locations – and the mechanism of analysis – isovists – are distributed and don't describe the observed phenomenon.

While Hillier managed to give architecture an insight into distributed representation to understand second-order phenomena, he never tried to understand if his methods could be used to generate gestalt.[4]

In computational design it was notably John Frazer who advocated a post-structuralist distributed and computed representation of architectural space [20].

---

[3] Robin Evans shortly after also shows the relationship between architectural lay-outs and social hierarchies and relationships [2].

[4] Lately, a few generative applications based on Visual Graph Analysis (VGA) are being developed, i.e. Kraemer & Kunze 'Design Code' [19].

Related to Nicholas Negroponte's Architectural Machine Group [21], Frazer essentially changed the set-up of Negroponte's system by eliminating teleology. That is to say, that Frazer used discrete cellular systems to generate emergent architectures, whereas Negroponte was interested in generating specific conditions (hoping he could teach the computer how to assume semantic descriptions from geometric input). Both had in common that they attempted to bridge the reduction in complexity of the input by building physical computers.

Frazer (and Coates [22]) perceived architectural space as distributed representations and were the first to apply computational methods to try and design architectural gestalt. De-centralized non-linear media such as cellular automata and evolutionary techniques were used to generate descriptions of space.

Frazer collaborated with Cedric Price on the Generator project for the Gilman Paper Corporation in 1978 [20]. It was also due to Price's understanding of system's theory (they were supporters of Gordon Pask's cybernetic interpretation of architecture) that Frazer could experiment on a 'live' architectural brief. In the spirit of Archigram-like mobile architectures, Price and Frazer envisaged a kind of intelligent structure in cubic volumes (akin to sentient cellular rooms) which would be able to communicate with their neighbouring cellular structures and reconfigure according to external and internal conditions. Thus, the architect became a system designer who specifies the relationships between elements but doesn't order the gestalt. This is done by the system itself. Frazer intended a truly self-organizing architecture that would be able to learn from the occupation of itself and the changes it made – building as self-observing organism. When Frazer states that 'the building could become "bored" and proposed alternative arrangements for evaluation' then he hints at observed phenomena that could emerge from the system's interaction with its context and its parts.

Frazer thought that each structural cell would be equipped with processors as discrete nodes that are trained first by an external computer and later take over the organization itself. This essentially introduced the idea of neural network architectures.[5]

4 Mapping Phenomenal Space

Neural networks as analogous mechanisms for the translation of gestalt and complexity theory are scrutinized and eventually proposed by Paul Cilliers [6]. Generally neural networks are not ideal for distributed representations of phenomena since they mostly solve goal directed problems. Teuvo Kohonen's self-organizing feature map – or SOM - on the other hand offers the chance to explore relationships between properties of objects that result in general categories of distinctions [23].

SOMs share many properties with complex systems and the notion of gestalt. Their learning mechanism reflects closely the recursive and self-referential differentiation process identified by Heinz von Foerster for cognition. Generating classes of feature

---

[5] Apart from those early proponents of distributed representation of phenomena in architecture, there have been a few more individuals who attempt to design spatial phenomena through computation (and I don't mean the hordes of flashy algorithmic patterns). A few notable ones are Pablo Miranda's swarm architectures at the Interactive Institute Sweden, Lars Spuybroek of Nox Architects, individuals at the CAAD chair of ETH Zurich and Kristi Shea at TU Munich to name but a few.

sets on the basis of their differences is the key axiom of gestalt theory. In complexity theory this represents the occurrence of emergent phenomena based on the interaction of simple elements.

Thus, three steps in the application and development of SOMs will be described below. At the end, I will attempt a forecast of the next development on this approach.

4.1 SOM as Spatial Gestalten

For the first application of the SOM, a representation of space was to be found that would be rooted in Euclidean space but otherwise as free of assumptions as possible. This tool was intended to help understand alternative descriptions of spatial qualities [24].

The sets of signals to train the network with were composed purely of three dimensional vertices. Those vertices were taken from a CAD model describing a site in London and stored in an array. No inferences towards higher geometric entities like line, face or volume were given. The topology of the network consisted of a general three dimensional orthogonal neighbourhood. The learning function a general Hebb rule:

$$w_{ij}(t+1) = w_{ij}(t) + k_{id}(t)[x - w_{ij}(t)] . \tag{23}$$

where w_{ij} represents the weight of a node at topological position I,J in the network, x the input signal and k the learning rate dependent on the position within the neighbourhood of the winning node [23].

The network 'learned' by adjusting the volume it occupied at any point to a new input volume. To do so, it had to distribute the difference from present shape to input space to all its nodes. As the nodes are constrained by a topology, the nodes would distribute within the input volume and thus describe potential sub-volumina of the input space.

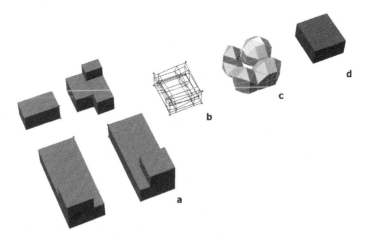

Fig. 1. a – vertices of site model; **b** – network adaptation; **c** – description of network space via an implicit surface and **d** – the same input space as generally described

The resulting network structure, consisting of point nodes and lines for synaptic connections, were visualized through an implicit surface (marching cube) algorithm. This seemed to be a good choice since implicit surfaces indicate the boundary between densities of elements. Therefore, the surface outlines the perceptive fields of the network's clusters and distinguishes the outer boundary towards the environment. That boundary coincides with what is called the probability density, which describes the probability with which a signal can occur within an area of the total input space.

Since the intention was to allow the network to select its own input from the site, each node's search radius was limited by its relationship to its topological neighbours. This bias ensured that the network would be able to collect new points from the site according to its previous learning performances.

This SOM experiment was successful in the sense that it not just implemented all salient properties of complex systems and their underlying mechanism of distributed representation of contextual stimuli but it also led to some unexpected outcomes:

Firstly, although it established an isomorphic relationship between the signal space and its own structure, it also highlighted the differences between objects through its shape and location on site. It mapped out spaces of differences between generally perceived geometric objects, pointing towards Arnheim's statement that 'space between things turns out not to look simply empty' [12].

Secondly, an unpredictable directionality of movement could be observed. I suppose it has something to do with the bias on the search space of the nodes but that doesn't necessarily explain why it would tend towards one direction rather than another if the general density of new signals surrounding the network is approximately even. Thus, locations with 'richer' features would be discerned from 'less interesting' locations.

Further, since signal spaces depend on the structural make-up of the network after a previous learning phase, it could be argued that the learning or perception capacity of this SOM is body or structure dependent. Heinz von Foerster said that 'change of shape' causes a 'change of perception' and vice versa, generating ever new descriptions of reality [11].

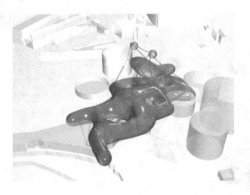

Fig. 2. An expression of the SOM in the site model at the end of a training period

4.2 Further Implementations of Spatial SOMs

One of the key problems for a good match between the dimension of the network and the dimension of the signal space in many applications is that the size of the signal space is not known yet; especially if the signal space is dynamic.

Hence, a growing neural gas algorithm was implemented as a variation of the standard SOM to account for fluctuations of density in the signal space [25]. The topology of the network would grow according to occurrences of signals in space.

This approach has just been started and is promising to help understand the relationship between events and spatial locations. It could be envisaged that cellular automata states or agent models serve as dynamic signals.

Another new problem posed itself from the first SOM implementation. The isomorphic representations of space produce interesting alternative descriptions of the qualities of locations on site. The differences could be observed over generations. However, it would seem helpful to understand if recurring or general patterns of descriptions occur. This would indicate not just differences of feature configurations but also point towards spaces that share non-explicit qualities.

To this end, a dot product SOM was trained reminiscent of Kohonen's examples of taxonomies of animals [23].

The input consisted as previously of three dimensional points for randomly configured Euclidean spaces. On the basis of the differences between the total coordinates, the map organized the input spaces into categories [26].

Although the input was highly reduced in complexity, the results were very successful, since the observer could understand the perceived commonalities between the inputs in each category and the differences to other emerging categories according to semantic and spatial phenomena. The SOM had mapped the input spaces inadvertently into binary distinctions of 'narrow' and 'wide', 'long' and 'short', 'high' and 'low'.

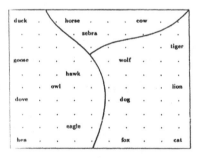 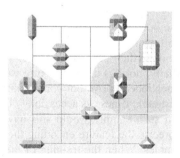

Fig. 3. Left: Kohonen's animal categories; **Right:** emergent categories of phenomenal decriptions of space

5 Designing with Learned Phenomena

The first application of the SOM to generate spatial Gestalten was not strictly meant for designing new architectures but to understand implicit qualities of space.

On one occasion, a model of several spatial SOMs based on the first applications was built that would generate design proposals for volumetric building layouts. The building was not conceived of as rooms but rather activities and the commonalities between the described activities would generate spatial diagrams as suggestive building layouts where each activity was represented by a SOM [27].

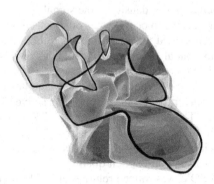

Fig. 4. Multiple SOMs reading each other as input to generate spatial categories of activities

While this type of generative use of self-organizing maps was interesting since the networks would try to group themselves into volumes according to given features, it does not use the SOMs to 'find' implicit relationships between features.

The next steps to be taken to test the capacity of SOMs or other neural networks to disclose configurations of phenomena should be twofold:

5.1 Indicative Mappings

Training neural networks to distinguish between samples of architectures or spaces that can be clearly labelled with 'good'/ 'bad', 'interesting'/ 'boring', 'scary'/ 'safe', 'comfortable'/ 'uncanny', etc, but where we fail to understand which circumstances contribute to their perceived states.

Initially, notions of architecture wouldn't necessarily have to be as abstract as the ones mentioned above but could comprise less complex expressions with some kind of measurable fitness as accessible/ obstructed, view-facilitation or others.

Each such sample would still have to be coded features. It seems clear that the more precise the feature encoding the larger the set of possible relationships between the samples. When the training of the feature map has been completed, new designs of which no general opinion or performance judgement exist can be read into the map. The location the new design would take within the categories of the trained feature map would give the designer an indication of what kind of qualities the new design produces.

More importantly though, the mapped categories also reveal the combinations of feature magnitudes (parameters) that produce a desired/ unwanted spatial quality. Hence, the designer might be able to find out what features and their blends are necessary to achieve certain phenomena.[6] Such an approach could for example map out the configuration necessary between light sources and time delay to produce the gestaltist Phi-phenomenon.

5.2 Cognition-Aided-Design (CAD)

The more advanced and ambition application of feature mapping in architectural design, would entail artificial software designers which feed of databases of learned input mappings.

The above mentioned project of multiple-SOMs generating volumetric diagrams according to activity features, should first be trained with a large cross-section of spatial configurations that contain given activities[7]. This would then lead to a cognitive artificial designer who aids the architect who trained the network. The samples chosen to train the networks would ensure that designs stay personalized but can also lead to common public databases (standards) that produce a vast amount of similar designs.

6 Conclusion

The SOMs discussed briefly above all produced a step towards the understanding of architectural space and phenomena as configurations of features, some dynamic some stable. The qualities mapped so far are of low complexity whereas the generative approaches have been a little more ambitious.

I hope to build (with the help of my students at the University of East London) a model within the near future that would help the designer to find indications as what features and their magnitudes might give rise to certain phenomena.

In the meantime also other methods for distributed representation of space should be continued to be explored, as the SOM represents by its very nature a good method but by no means the only one.

References

1. Mitchell, W. J.: The Logic of Architecture. Design, Computation, and Cognition. MIT Press, Cambridge, Massachusetts (1990)
2. Evans, R.: Translations from Drawing to Building and Other Essarys. Architectural Association Publications, London (1997)
3. McLuhan, M.: Understanding Media. The Extensions of Man. Routledge, London and New York (1964)
4. Perec, G.: Species of Spaces and Other Pieces. Editions Galilee, Paris (1974)

---

[6] Mixing immaterial features to create dynamics within space is reminiscent of Yves Klein's 'air-architectures'. Klein used elements like heat, light and sound in order to influence how people would occupy spaces [28].

[7] Professor Lidia Diappi at the Politecnico di Milano has trained SOMs with changes in urban land-use patterns. The resulting differences were used as a basis for transition functions of cellular automata to predict future changes [29].

5. Jenks, C.: The Architecture of the Jumping Universe, Academy, London & NY (1995). Second Edition Wiley (1997)
6. Cilliers, P.: Complexity and Postmodernism. Routledge, London (1998)
7. Maturana, H. and Varela, F.:The Tree of Knowledge: Goldmann, Munich (1987)
8. Kneer and Nassehi: Niklas Luhmanns Theorie sozialer Systeme. Wilhelm Finkel Verlag, Munich (1993)
9. Norberg-Schulz, C.: Intentions in Architecture. MIT Press, Cambridge, (1965)
10. Wittgenstein, L.: Tractatus Logico-Philosophicus. Routledge, London (1922)
11. von Foerster, H.: Understanding Understanding. Essays on Cybernetics and Cognition. Springer, New York (2003)
12. Arnheim, R.: Dynamics of Architectural Form. University of California Press, Berkley and Los Angeles (1977)
13. Miranda Carranza, P.: Out of Control: The media of architecture, cybernetics and design. In: Lloyd Thomas, K.: Material Matters. Routledge, London (forthcoming 2006)
14. Alexander, C.: Notes on the Synthesis of Form. Harvard University Press, Cambridge, Mass (1967)
15. Hillier, B. and Hanson, j.: The Social Logic of Space. Cambridge University Press, Avon (1984)
16. March, L. and Steadman, P.: The Geometry of Environment. MIT Press, Cambridge, Mass (1974)
17. Benedikt, M.L.: To take hold of Space: Isovists and Isovist Fields. In: Environment and Planning B 6 (1979)
18. Hillier, B.: Space is the Machine. Cambridge University Press, Cambridge (1996)
19. Kraemer, J. and Kunze, J. O.: Design Code. Diploma Thesis, TU Berlin (2005)
20. Frazer, J.: An Evolutionary Architecture. Architectural Association, London (1995)
21. Negroponte, J.: The Architecture Machine. MIT Press, Cambridge Mass (1970)
22. Coates, P. S.: New Modelling for Design: The Growth Machine. In: Architects Journal (AJ) Supplement, London, (28 June 1989) 50-57
23. Kohonen, T.: Self-Organizing Maps, Springer, Heidelberg (1995)
24. Derix, C.: Self-Organizing Space. Master of Science Thesis, University of East London (2001)
25. Coates, P., Derix, C., Lau, T., Parvin, T. And Pussepp, R.: Topological Approximations for Spatial Representations. In proceedings: Generative Arts Conference, Milan (2005)
26. Coates, P., Derix, C. And Benoudjit, A.: Human Perception and Space Classification. In proceedings: Generative Arts Conference, Milan (2004)
27. Derix, C. And Ireland, T.: An Analysis of the Poly-Dimensionlity of Living. In proceedings: eCAADe Conference, Graz (2003)
28. Noever, P.: Yves Klein: Air Architecture. Hatje Cantz Publishers (2004)
29. Diappi, L., Bolchi, P. and Franzini, L.: The Urban Sprawl Dynamics: does a Neural Network understand the spatial logic better than a Cellular Automaton?. In proceedings: 42nd ERSA Congress, Dortmund (2002)

KnowPrice2: Intelligent Cost Estimation for Construction Projects

Bernd Domer[1], Benny Raphael[2], and Sandro Saitta[3]

[1] Tekhne management SA, Avenue de la Gare 33, 1003 Lausanne, Switzerland
[2] National University of Singapore, 4 Architecture Drive, Singapore 117566, Singapore
[3] Ecole Polytechnique Fédérale de Lausanne, IMAC, 1015 Lausanne, Switzerland
domer@tekhne.ch, bdgbr@nus.edu.sg, sandro.saitta@epfl.ch

Abstract. Correct estimation of costs of construction projects is the key to project success. Although mostly established in early project phases with a rather limited set of project data, estimates have to be precise. In this paper, a methodology for improving the quality of estimates is presented in which data from past projects along with other knowledge sources are used. Advantages of this approach as well as challenges are discussed.

1 Introduction

Correct estimation of construction costs is one of the most important factors for project success. Risks vary with the contractual context in which estimations or price biddings are presented. Table 1 discusses three standard configurations.

Table 1. Possible contractual configurations for construction projects

	General planner	**General contractor**	**Total services contractor**
Design contract	One contract with general planner	One contract with general planner	One contract with total services contractor for design and construction
Contract for-construction works	Multiple contracts with each sub-contractor	One contract with general contractor	
Type of cost estimation	Cost estimation	Price bidding	Price bidding
Risk, when estimate is too high	Project will be stopped	Project might be executed by another contractor with a lower bid	Project might be executed by another contractor with a lower bid
Risk, when estimate is too low	Project will suffer quality problems due to tight budget	Bankruptcy of sub contractors	Bankruptcy of sub contractors or total services contractor

Decisions that have the biggest impact on project success (that is, related to cost) are taken in early project phases. Cost control over project duration can only prevent

I.F.C. Smith (Ed.): EG-ICE 2006, LNAI 4200, pp. 147–152, 2006.

an established budget from getting out of control. It cannot correct fundamental errors made in the beginning.

At the start of a project, many details are unknown and cost estimators have to make several assumptions based on their experience with similar projects in the past. Construction managers are not only interested in accurate estimates, but also in the level of risk associated with estimates. A systematic methodology for performing this task is desirable. This paper presents a methodology for cost estimation that has been developed and implemented in a software package called KnowPrice2. The primary objective of this paper is to compare the methodology with current practices in the industry, rather than to provide a detailed description of the approach, which is given in [10].

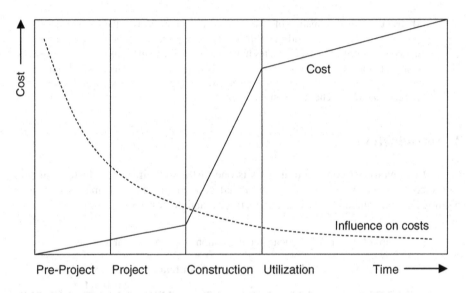

Fig. 1. Relationship between cost and the possibility to influence them in different project phases [1]

2 Estimating Construction Costs – Existing Methodologies and Data

Since the preparation of cost estimates is not a new challenge, several proposals on how to establish and structure construction budgets have been made. Databases with cost data of past projects have been prepared as well.

In most European countries, building codes define how to measure building surfaces and volumes [eg. 2, 3] and to structure building costs in cost groups [eg. 4, 5]

Although building codes for structures have been unified at European level, this has yet to be done for the above mentioned guidelines. A proposal has been made [6].

This means that even today the way building surfaces are measured as well as the structure of cost groups differ from country to country. As a result, the comparison of cost data from different countries is difficult.

The Swiss Research centre for Rationalization in Building and Civil Engineering (CRB) proposes methodologies such as the costs by element method [7]. Cost data of construction elements is provided and updated each year, but it does not account for regional variation of prices or building types. This approach needs a very detailed breakdown of all building elements and is therefore not the method of choice for quick estimates.

Data of existing buildings have been collected as well. Germany's BKI, the community of architectural chambers publishes each year a well structured and exhaustive catalogue of project costs [8]. The structure divides buildings into different types, gives examples for each type and provides standard deviation for price data. Regional data is included as well, since prices vary locally.

Although it is very tempting to use this data collection for swiss projects, construction is not yet as "globalized" as other industries. As discussed previously, data collections of other countries cannot be re-used without serious cleaning and adaptation.

3 KnowPrice2 Strategy

3.1 Background

KnowPrice2 is a software package for cost estimation that was developed by the Swiss Federal Institute of Technology (EPFL) in collaboration with an industrial partner, Tekhne management SA. It links case data with a unique approach for establishing construction project budgets. The employed methodology differs significantly from other database approaches such as [9], since case-based reasoning strategies are combined with relationships between variables that are discovered by data mining.

3.2 Methodology

The total project cost is estimated using knowledge from different sources. Knowlege sources include generic domain knowledge in the form of rules, cases consisting of data related to past projects and relationships that are discovered by mining past project data. Depending on what data is known about the current project, appropriate type of knowledge is used. For example, if all the variables that are needed for applying the rules are available, generic domain knowledge is used. Otherwise, relationships that are discovered by data mining are used. Only when relevant relationships are not available, case data is used.

Generic domain knowledge contains equations for computing costs of building elements by summing up costs of components or using unit costs. Each case contains characteristics of a building which includes information such as types, quantities and costs of elements. Data mining aims to discover relationships between building characteristics and costs. Rules of the following form are discovered through this process:

Where q_1, q_2, etc. are quantities and c_1, c_2, .. b are coefficients that are determined by regression using relevant case data. The data mining process is guided by the knowledge of dependencies which is provided by domain experts. A dependency relationship indicates that certain symbolic and quantitative variables might influence a cost variable. The exact relationship between variables is determined by the data mining module through analysing case data. More details of the data mining technique are presented elsewhere [10].

The CBR module selects relevant cases to a new project using a similarity metric. By default, a similarity metric that gives equal importance to all the input variables is used. In addition, users can define their own similarity metrics by specifying relevant variables and weights. Selected cases are used to estimate the variations in the values of independent variables.

$$\text{IF symbolic\_variable EQUALS value THEN}$$
$$\text{cost\_variable} = c_1 * q_1 + c_2 * q_2 + c_3 * q_3 + c_n * q_n + b. \tag{1}$$

Steps involved in the application of the cost estimation methodology are the following:

1. Users input known data related to a new project
2. The system creates a method for computing the total cost using generic rules and relationships that are discovered by data mining
3. Variations in the values of independent variables are determined from similar past cases and are represented as probability density functions (PDF).
4. Monte-carlo simulation is carried out for obtaining the probability distribution of total cost using PDFs of independent variables.

Since the methodology computes the PDF of the total cost instead of a deterministic value, information such as the likelihood of an estimate exceeding a certain value is also available. This permits choosing the bid price at an acceptable level of risk. This is not possible using conventional deterministic cost estimation.

3.3 Program Structure

The knowledge base is structured into five modules, namely, Generic domain knowledge, Dependencies, Cases, Ontology and Discovered knowledge.

The first module "Generic domain knowledge" implements rules to achieve the objective. Here, the objective is to compute the overall building cost, obtained by the summation of building cost classifications (BCC).

The second module "Dependencies" describes dependencies between a) the dependent variable and b) the influencing variable. This module is, from the industrial partner's point of view, one of the major achievements of KnowPrice2. Whereas in other programs the user has to describe relationships in a deterministic way, KnowPrice2 employs a different approach. In most cases, cost estimators cannot provide the deterministic expressions to relate building characteristics with BCCs directly. It is much easier to indicate that there is a relationship without giving precise values. The major achievement of Knowprice2 is that it evaluates the relationship between variables using existing data and data mining techniques [10].

Cases are input in a semi-automatic procedure using an Excel spreadsheet in which data is grouped into building characteristics such as surfaces, volumes, etc. and costs, structured according to the Swiss cost management system [5]. The same interface allows to management of cases.

The ontology module contains the decomposition hierarchy for organising variables, the data type of each variable, default values of variables and possible values of symbolic variables.

The module "discovered knowledge", contains a decision tree that organises all the relationships that are discovered by data mining.

3.4 Challenges

Collaboration between EPFL and Tekhne has been very close which means that the project did not suffer from major problems. One main challenge was to adapt the Dependencies module such that it can treat non-deterministic relations.

The second challenge was (and still is) to provide data for testing. In a first effort, a database with past projects of Tekhne has been created. This database has been used to test the basic functionality of KnowPrice2. It is not sufficient to do the fine tuning of the software.

4 Advantages of KnowPrice2

From the industrial partner's point of view, KnowPrice offers several advantages:

The approach used for the cost estimation depends no longer on personal experience only but links it with intelligent computational techniques to support the user. Employed methods have a sound scientific basis and increase the client's confidence in budgets thereby.

Costs can be estimated with incomplete project data (this means, when all project details are not known). It is up to the user to increase the degree of precision by increasing the number of case data and case variety.

Cases can be entered via Excel spreadsheets. This is very convenient for the user. Values for descriptive variables are proposed to guide user input. The amount of data entered depends on information present and not pre-fixed by the software.

The final cost is presented as a price range with associated probabilities. The client can choose the degree of risk he would like to take: either going for a low budget with a rather high risk of exceeding the estimate or to announce higher budgets with lower risks.

Costs can be related to similar previous projects and can thus be justified.

5 Future Work

The quality of results highly depends on the number and variety of cases entered. So far, only cases of the industrial partner (Tekhne) have been entered. KnowPrice2 works correctly so far, but needs definitely more data for proper testing.

Even though building cost classifications are defined in codes, they are not unambiguous. They leave space for interpretation and each user might apply them differently. The effects of this have not yet been examined.

KnowPrice2 has to be tested under real project conditions. This test might reveal necessities for changes in the program itself and necessary user interface adaptation.

Acknowledgements

This research is funded by the Swiss Commission for Technology and Innovation (CTI) and Tekhne management SA, Lausanne.

References

1. Büttner, Otto: Kostenplanung von Gebäuden. Phd Thesis, University of Stuttgart (1972)
2. DIN 277: Grundflächen und Rauminhalte von Bauwerken im Hochbau. Beuth Verlag (2005)
3. SIA 416: Flächen und Volumen von Gebäuden. SIA Verlag, Zürich (2003)
4. DIN 276: Kosten im Hochbau. Beuth Verlag (1993)
5. CRB (Ed.): Building cost classification. CRB, Zürich (2001)
6. CEEC (Ed.): Le code européen pour la planification des coûts, CEEC (2004)
7. CRB (Ed.): Die Elementmethode – Informationen für den Anwender. CRB, Zürich (1995)
8. BKI (Ed.): Baukosten Teil 1: Statistische Kostenkennwerte für Gebäude. BKI, Stuttgart (2005)
9. Schafer, M., Wicki, P.: Baukosten Datenbank "BK-tool 2.0", Diploma thesis, HTL Luzern (2003)
10. Raphael, B., et. Al.: Incremental development of CBR strategies for computing project cost probabilities, submitted to Advanced Engineering Informatics, 2006.
11. Campi, A., von Büren, C.: Bauen in der Schweiz – Handbuch für Architekten und Ingenieure, Birkhäuser Verlag (2005)

RFID in the Built Environment: Buried Asset Locating Systems

Krystyna Dziadak, Bimal Kumar, and James Sommerville

School of the Built and Natural Environment, Glasgow Caledonian University, Glasgow,
G4 OBA,
Scotland

Abstract. The built environment encompasses all buildings, spaces and products that are created or modified by people. This includes homes, schools, workplaces, recreation areas, greenways, business areas and transportation systems. The built environment not only includes construction above the ground but also the infrastructure hidden under the ground. This includes all buried services such as water, gas, electricity and communication services. These buried services are required to make the buildings functional, useful and fully operational: an efficient and well maintained underground infrastructure is required.

RFID tags (radio frequency identification devices) are in essence transceivers consisting of three components that make up a sophisticated transponder. Once activated, the tag transmits data back to a receiving antenna: the technology does not require human intervention and further benefits from the fact that no line of sight is needed to control/operate the system. The tags can have both read and write abilities and their performance characteristics can be tailored/changed to accommodate a range of situations.

Within this paper we argue that utility provision (the hidden services) is an area where RFID technology may be able to identify location of buried pipes and others underground equipments. Early results from field trials carried out so far will be presented. The issues and concerns relating to developing such an application using RFID technology will also be highlighted.

Keywords: Buried Assets, Built Environment, RFID Technology, Tracking.

1 Introduction

Building services and hidden infrastructure i.e. buried pipes and supply lines carry vital services such as water, gas, electricity and communications. In doing so, they create what may be perceived as a hidden map of underground infrastructure.

In the all too common event of damage being occasioned to these services, the rupture brings about widespread disruption and significant 'upstream' and 'downstream' losses. Digging in the ground without knowledge of where the buried assets lie could isolate a whole community from emergency services such as fire, police and ambulance, as well as from water, gas and electricity services. It is not only dangerous for people who are directly affected by the damage but also for workers who are digging, for example, near the gas pipes without knowing their specific location (Dial-Before-You-Dig, 2005).

I.F.C. Smith (Ed.): EG-ICE 2006, LNAI 4200, pp. 153–162, 2006.
© Springer-Verlag Berlin Heidelberg 2006

Various methods are used to pinpoint the location of buried assets. Some of these approaches utilise destructive methods, such as soil borings, test pits, hand excavation, and vacuum excavation. There are also geophysical methods, which are non-destructive: these involve the use of waves or fields, such as seismic waves, magnetic fields, electric fields, temperature fields, nuclear methods and gas detection, to locate underground assets (Statement of need, 1999).

The most effective geophysical method is Ground Penetrating Radar (GPR). This technique has the capability to identify metal assets but is not able to give accurate data about the depth of the object, which is important information for utility companies (Olheoft, 2004). GPR has been used for pipe location with varying success, partly because radar requires a high-frequency carrier to be injected into the soil. The higher the frequency is, the greater the resolution of the image. However, high-frequency radio waves are more readily absorbed by soil. Also, high-frequency operation raises the cost of the associated electronics (GTI, 2005). This system is also likely to be affected by other metallic objects in close proximity to the asset being sought.

Another widely used method of locating underground infrastructure is Radio-detection, which is based on the principle of low frequency electromagnetic radiation which reduces the cost of electronics and improves depth of penetration. This technique is unable to detect non-metallic buried plastic, water, gas and clay drainage pipes (Radio-detection, 2003). Combining Radio-detection with GPR opens up the possibility of locating non-metallic pipes (Stratascan, 2005). However, the technique becomes complicated and expensive.

All of the above methods are useful in varying degrees and each of them has its benefits but none gives the degree of accuracy required by SUSIEPHONE and UK legislation e.g. the New Roads and Street Works Act 1991, the Traffic Management Act 2004 and Codes of Practice. Unfortunately, thus far none of these methods is able to provide accurate and comprehensive data on the location of non-metallic buried pipes (ITRC, 2003). The shortcomings of the above methods are summarized below:

- They cannot locate non-metallic utilities.
- They cannot be used in all types of soils.
- They cannot penetrate to required depths.
- They use perilous/dangerous/complex equipment that increases risks and costs of operation.

The problems associated with inaccurate location of underground infrastructure have been a serious issue for many years and will become even worse because of lack of precise location system which will facilitate identification of these services. At the moment all the existing data on buried assets is usually inaccurate or incomplete.

By applying RFID technology within the provision and management of utilities, it may be possible to identify the location of non-metallic buried pipes and other underground equipment with a greater degree of accuracy that is currently possible.

Use of an RFID based system may bring about significant benefits for those locating buried assets and provide a more accurate underground mapping system.

2 The Potential of RFID

A contactless identification system called Radio Frequency Identification (RFID) is broadly implemented into a large number of business areas/fields. This indicates that the technology is worth/merits close examination and should be consider seriously.

Generally RFID application can be divided into two main categories which include: short-range (SR) applications and long-range (LR) applications. The feature that distinguishes short- and long- range systems is that in SR applications the transponder and readers have to be in close proximity to one another whereas in LR systems the distance can be much greater. That/it is usually caused by the use of active tags, which are powered internally by a battery (Shepard, 2004). Within short-range there are mainly applications such as access control, mass transit ticketing, personnel identification, organ identification, vehicle identification and pigeon racing. Long-range applications include: supply chain management, parcel and mail management, garment tags, library sector, rental sectors and baggage tagging (UPM Rafsec, 2004).

This technology can be implemented to monitor use and maintenance of construction equipment. Hours of operation, critical operating data (such as temperature or oil pressure), maintenance schedule, maintenance history and other relevant data can be gathered and stored on the tag for use by safety and maintenance personnel. RFID can also increase the service and performance of the construction industry with applications in materials management, tracking of tools and equipment, automated equipment control, jobsite security, maintenance and service, document control, failure prevention, quality control, and field operations

Table 1 Highlights a number of application areas where RFID can improve the overall efficiency of Facilities Management (FM) systems.

Table 1. RFID applications

Application	Target activity	Tag type
Access Control of the overall facility.	Doorway entry at various points on a building	Passive/ Active
Asset Tracking	Locating vehicles within a freight yard	Active
Asset Tagging	Tracking corporate computing hardware	Passive
Baggage/Mail Tracking	Positive bag/envelope matching	Passive
Supply Chain Management (SCM) (Container Level)	Tracking containers at distribution terminals	Active
SCM (Pallet Level)	Tracking each pallet in yard/store	Active/ Passive
SCM (Item Level)	Identifying each individual item/package	Passive

3   System Design Configuration

The project was bifurcated into two phases:

3.1   Phase 1

This phase determined an appropriate RFID tag, antennae and reader configuration which would give accurate depth and location indications at up to, and including, 2.0m below surface level. It will result in indications as to the size and shape of antenna which can achieve the required depth and accuracy.

Depth of 2m was set as a target in phase 1. Most of the existing pipes are located at depth between 0.5-3m below the ground. Second reason behind it is RFID specific devices and operating frequency that we are allowed to work on.

3.1.1   Laboratory Tests
Initial air tests were carried out at a construction industry training facility near Glasgow.

A series of air tests were run with the aim of ascertaining the connectivity between each of the three tags (transponders) with each of the four antennae. The data generated from these test is presented below:

Table 2. Tag's specification

SYMBOL	TRANSPONDER
T1	LTag
T2	MTag
T3	STag

Table 3. Antennae's specification

SYMBOL	ANTENNAE
AI	L1
AII	L2
AIII	M1
AIV	S1

These tests were run to determine the greatest signal reception range between the antennae and the tags. The best results are summarized in the *Table 5* below.

Table 4. Results

	L tag	M tag	S tag
	metres	metres	Metres
AI	2.7	2.4	1.75
AII	0.664	0.485	0.455
AIII	0.895	0.69	0.53
AIV	1.185	0.885	0.805

Table 5. Results

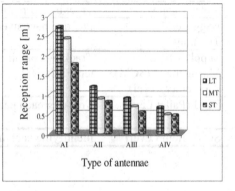

3.1.2 Data Analysis

To make sure that the measurements are accurate the distance presented in *Table 4* was measured when the signal sent from the antennae to the tag was continuous, without any interference.

These results show that the longest acceptable signal reception ranges can be achieved when antenna AI is connected with T1 or with T2. Air tests also show that the worst performances are between antennae AII when tested in conjunction with all tag types. Hence, AII was eliminated from further examination. Antennae AI, AIII and AIV were then tested with an underground signal.

Air tests allow testing effective performance of each tag and reader combination and create zones of magnetic field between each of the tags with each of the antennae. This information shows the range of magnetic field within which the technology can operate. With the aid of AutoCAD (design program) and data from the air tests, we created the range of the signal patterns between all the antennae and tags.

Figures: 1, 2 and 3 present a range of signal patterns created between antenna AI and tag T2 depending on the antenna position.

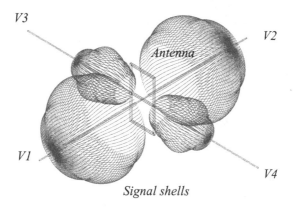

Fig. 1. Antenna positioned vertically

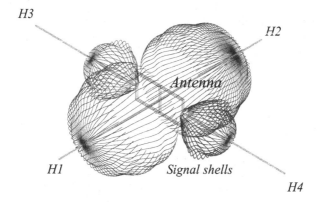

Fig. 2. Antenna positioned horizontally

In *Figure 1* the antenna was positioned vertically. There are two sizes of shells; bigger shells lie on axes *V1* and *V2* and smaller on *V3* and *V4*. The reason for this is the size of the antenna: the larger the antenna, the greater the capture of the magnetic field/signal generated by the tag.

Figure 2 shows the antenna in horizontal orientation. The description is similar to the one given in *Figure 1*. Again we can observe two sizes of the shells which show the reception range of the signal in this orientation.

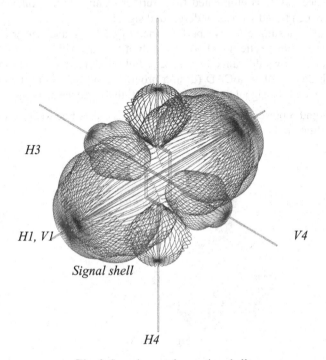

Fig. 3. Superimposed reception shells

Figure 3 indicates the combined reception shells for both orientations. It is clear that the antenna is capable of directionally locating the tag. This directional capability allows us to eliminate spurious signals and so concentrate on the desired signal from the tag i.e. the larger signals can be attenuated.

3.1.3 Data from Real Implementation

In this part of the first phase a range of passive tags were fixed to a small wheeled 'chariot', which was lowered into the pipe using a tape measure. The tag's return signal was received using a LF antenna and reader on the surface. The chariot was lowered until it reached the point of signal loss and from that maximum read depth was determined. Afterwards the chariot was located at pre-determined depths and the surface antenna was raised until the point of signal loss. The distance between the surface and the antenna was noted and this enabled the ground depth of a tag to be determined.

At this stage of the field trials each of the antennae and each of the tag were successfully tested. Tests were carried out at increasingly different depths until the required 2m depth was achieved.

An implicit part of the investigation is aimed at ascertaining the extent to which soil conditions that could affect the reception of the reading signal.

For completeness we carried out and compared tests when:

- the separation between the tag and antenna was only soil (*Figure 4*)
- half of the distance was in soil and the other half was air (*Figure 5*)

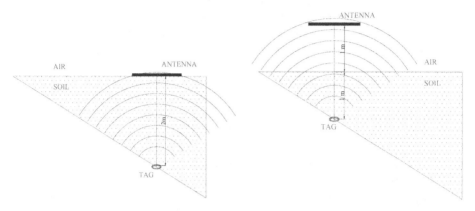

Fig. 4. Only soil   **Fig. 5.** Mixed

These tests showed that the results in the presence of soil lose only 3% of the reading distance in comparison with the results achieved in ideal condition (*Table 4*). However, in the United Kingdom there are six general types of soil: clay, sand, silt, peat, chalk, and loam, all of which have their own characteristics. The most important properties of soil are hydraulic conductivity, soil moisture retention and pathways of water movement (Jarvis, 2004) and it is possible that different soil condition/types can affect the performance and its accuracy.

Parameters such as the operating frequency, tag size and type (active or passive) and antenna size and shape can affect the performance characteristics of the system and therefore the maximum depth that the tag can read. This is why during this phase our target was to modify tag's and antennae's specifications in order to find out the best correlation between them.

In the first phase the efficacy of the RFID location system was proven, enabling us to move to the second phase.

4   Future Work

Future work will focus on the *Phase 2* of the research, which is presented below.

4.1   Phase 2

After the principles of the location system have been proven in Phase 1, Phase 2 focus on the following steps:

- Improving the tag reading performance to 3m below ground.
- Improving depth and positional accuracy to 5cm.
- Making the locating system mobile by providing a Global Positioning System (GPS) fix for the asset.
- Providing more accurate data on performance through differing types of ground/soil material.
- Storing the depth, latitude and longitude in a format compatible with the Digital National Framework (DNF)
- Applying the DNF information to topographical mapping tools to enable visualisation of underground infrastructure.

4.2 General Plan of Work

The Location Operating System (LOS) was created to facilitate the connection between the data captured during the field work and its later processing/configuration. A general operating of the system and its components is presented in *Figure 6* below.

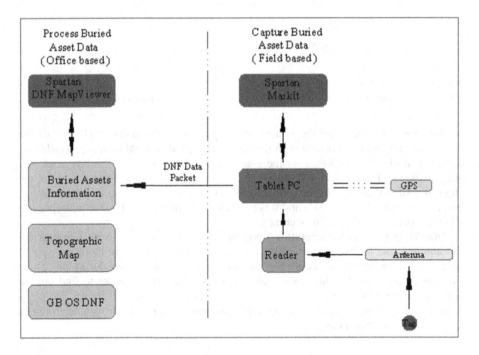

Fig. 6. The location operating system

The LOS scheme is divided into two parts: components which are geared towards Capturing Buried Asset Data (CBAD) and a system for Processing Buried Asset Data (PBAD).

The first part contains components that will help users to capture the data from the field. The latitude and longitude data will be captured using a Global Positioning System Device (GPSD). However, the depth of the buried asset will be ascertained using RF tags, antennae and reader. All this information will be captured by a waterproof and portable computer – Tablet/PC.

In the second part the data from the Tablet/PC will be sent and stored in the Buried Asset Information (BAI) system: the data will be processed to allow user visualization of buried assets using the Digital National Framework (DNF) compliant Topographic Map overlay. When processed, the necessary/required information about the underground services will be stored in the Ordnance Survey (OS) DNF format.

5 Conclusions

From what was achieved at this stage of research project the most significant results are, that:

1.) Air tests allowed to identify the ideal combination of antennae and tags. These tests also allowed to establish reception shells and expected reception ranges. These ranges facilitated expansion of the testing into appropriate site conditions.

2.) Underground tests enabled to establish reception at a range of depths through one soil type. As the tests progressed we were able to receive a signal at the target depth outlined in *Phase 1* (2m). We also discovered that soil characteristic i.e. saturation, soil type, etc. may not have an adverse effect on the signal reception.

These early results are encouraging and they seem to indicate that an answer to identifying non-metallic buried assets does lie in the use of RFID technology. Although there is not single solution to the problem concerning utility services, it may be that RFID will be able to contribute to a part of the problem related to locating buried assets.

As stated earlier, a considerable amount of development work is still to be done to arrive at a fully operational system. A successful beginning has at least been made. The next step will focus on improving the accuracy of reception range. Also more tests will be provided changing the condition of the soil, types of the pipes and different surfaces layers respectively.

RFID technology is becoming ubiquitous: as the RFID systems become more widespread, the technology itself becomes smaller and cheaper. The proliferation of RFID systems suggests that it will be all pervasive, and there is no doubt that RFID is set to have a tremendous impact on all major industries.

References

1. *Business Benefits from Radio Frequency Identification (RFID)*, SYMBOL, 2004 http://www.symbol.com/products/whitepapers/rfid_business_benefits.html
2. *Capacitive Tomography for Locating Buried Plastic Pipe,* Gas Technology Institute (GTI), Feb. 2005. http://www.gastechnology.org/webroot/app/xn/xd.aspx?it=enweb&xd= 4reportspubs%5C4_8focus%5Ccapacitivetomography.xml

3. *Consideration for Successful RFID Implementations*, MICROLISE, July, 2002. http://whitepapers.zdnet.co.uk/0,39025945,60089849p-39000532q,00.htm
4. *Practical research: planning and design*, Paul D. Leedy, Jeanne Ellis Ormrod, 2001.
5. *RADIODETECTION Application Note. Planning and Maintaining Locate Tracer Wire for non-conductive pipeline systems*, August, 2003. http://www.radiodetection.ca/docs/tracerw.pdf
6. *RFID applied to the built environment: Buried Asset Tagging and Tracking System*, K.Dziadak, J.Sommerville, B.Kumar, CIBW78 Conference Paper, Dresden, 2005.
7. *RFID*, Steven Shepard, ISBN 0-07-144299-5, 2005.
8. *RFID Handbook: Fundamentals and Applications in Contactless Smart Cards and Identification*, Klaus Finkenzeller, 2004.
9. *RFID Tagging Technology*, MICROLISE, 3rd January 2003 http://www.microlise.com/Downloads/MEDIA/white_papers/WP_1.0_RFID_Tagging_Tech.pdf
10. *Soil Information and its Application in the United Kingdom: An Update* Michael G. Jarvis Soil Survey and Land Research Centre, Cranfield University, Silsoe Bedfordshire, MK45 4DT, United Kingdom.
11. *The Cutting Edge of RFID Technology and Applications for Manufacturing and Distribution*, Susy d'Hont, Texas Instrument TIRIS http://www.ti.com/rfid/docs/manuals/whtPapers/manuf_dist.pdf
12. *Use of a common framework for positional referencing of buried assets*, Martin Cullen, 2005. http://www.ordnancesurvey.co.uk/oswebsite/business/sectors/utilities/docs/BSWG_report2%20KHB_final.pdf
13. *Statement of need: Utility Locating Technologies*, 1999, http://www.nal.usda.gov/ttic/utilfnl.htm
14. *New Techniques for Precisely Locating Buried Infrastructure*, 2001 http://www.awwarf.org/research/topicsandprojects/execSum/2524.aspx
15. *Underground Utility mapping*, Stratascan http://www.stratascan.co.uk/eng-utility.html
16. *Irrigation Training and Research Centre*, Report No.R03-010, 2003. http://www.itrc.org/reports/aewsd/undergroundpipe.pdf
17. UPM Rafsec, 2004. http://www.rafsec.com
18. *Georadar^{TM}* by Gary R.Olheoft, PhD, 2004 http://www.g-p-r.com/

New Opportunities for IT Research in Construction

Chuck Eastman

Professor in the Colleges of Architecture and Computing, Georgia Institute of Technology,
Atlanta, GA 30332-0155
chuck.eastman@coa.gatech.edu

Abstract. The transition to parametric 3D Building information Modeling (BIM) is predicated on the many uses of a fully digital building representation. This paper explores two important uses: (1) the embedding of design and construction expertise into modeling tools, and (2) the movement of building type design guides to new more integrated digital formats.

1 Introduction

3D parametric modeling and rich attribute handling are making increasing inroads into standard construction practice, worldwide. This fundamental change of building representation from one that relies on human readability to a machine readable building representation, opens broad new opportunities for enhancing design and construction in ways that have been dreamed about over the last two decades. The new technology and its facilitated processes are called Building Information Modeling, or BIM. BIM design tools provide a few direct and obvious benefits. These are based on a single integrated representation from which all drawings and reports are guaranteed to be consistent, and the easy catching of spatial conflicts and other forms of geometrical errors. Even the first step of realizing these basic BIM capabilities requires new practices regarding design development and coordination between design teams. Many other benefits are available, such as integrated feedback from analysis/simulation and production planning tools. BIM allows tools for structural, energy, costing, lighting, acoustic, airflow, pedestrian movement and other analyses to be more tightly integrated with design activities, moving these tools from a long loop iteration to one that can be used repetitively to fine tune architectural design to better achieve complex mixes of intentions. Many of these capabilities have been outlined in the research literature for decades and they now have the potential to be realized [1].

It is unclear whether there is a single integrated model from which all abstractions are derived, or more likely whether there is a federation of associated representations that are consistent internally and at their points of interaction. Issues of representation and model abstraction have not been resolved in manufacturing research [2] and will continue in construction. Issues of the specification of abstraction for new modeling tools, such as CFD fluid flows, and automating them, are open research issues that may be discussed during this workshop.

The definition of a construction-level building model is a complicated undertaking, requiring the definition and management of millions of component objects. Most

I.F.C. Smith (Ed.): EG-ICE 2006, LNAI 4200, pp. 163–174, 2006.

reputed BIM efforts for architectural use actually only partially define the building in machine readable form, carrying out most architectural detailing using drawn sections based on the older drafting technology. This mixed approach facilitates the transition from drawing to modeling and reduces the functional and performance requirements of the software. We should all be aware however of this limitation and the hugely varying scale of knowledge embedded in various so-called building models. Are the following building elements defined as fully machine readable information: fresh water and waste water piping? electrical network layouts? reinforcing bars and meshes? window detailing? ceiling systems? interior trimwork? A fully compliant building model allows detailed bills of material for all these components and provides the option for their offsite fabrication. Current architectural building models promoted as examples of BIM generally fail at the detail level.

Other systems have been developed to begin addressing the fabrication-level detailing of building systems, for steel, concrete, precast, electrical, piping and other systems. These systems embed different knowledge than that carried in architectural systems. The two different types of system have articulated the different level of information traditional carried in design documents and shop-level documents.

Resolution of the level-of-detail problem requires the extension of current BIM tools to support the definition and parametric layout of the components making up all building subsystems and assemblies. This is being undertaken incrementally, with all the current BIM design tools providing parametric objects to different levels of detail.

2 Design Expertise

Parametric building objects in BIM tools encapsulate design knowledge and expertise. The embedded knowledge facilitates definition and automatic editing. It distinguishes a BIM design tool form amore general parametric modeling tool, such as Solidworks®, CATIA® or Autodesk Inventor®. The most basic parametric capability of an architectural BIM tool is the layout of walls, doors and windows. All BIM tools allow easy placement and editing of wall segments and insertions of doors, windows and other openings. Some systems maintain the topological relations between walls and respond to changes to the walls they butt into. Most of the wall objects used in BIM design tools also incorporate *layers* of construction as a set of ordered parameters of the wall defined in a vertical section, with a structural core, optional insulation, and layers of finishing on both sides, to obtain a built-up wall and implicit material quantities, based on the wall area. ArchiCad® and Revit® support varying section properties along the vertical section, allowing horizontal variation of construction and finishes. Changes in the horizontal direction require a change in the wall element. We can say that the wall model incorporates this level of design knowledge.

BIM tool users will encounter the limitations of this level of definition of a wall if they try to design outside this limited conceptualization. Often building walls have small regions with different finishes or internal composition. External walls are commonly made up of such regions, as shown in Figure 1. Only one BIM design tool supports segments, to my knowledge, each geometrically defined region consisting of its set of layers and finishes. Built-in wall regions allows walls to have mixed composition and defines another level of wall architectural design expertise.

Fig. 1. Example walls that are not easily represented in current parametric models. They each show mixes of materials and properties that will not be reflected in costs or energy assessments.

Other potential limitations include the definition of the connection conditions between heterogeneous walls. Do they incorporate a connection detail, beyond a drawn section, that potentially has cost, energy, acoustic and other properties? See the section example in Figure 2. Can I define the detail for the wall-joint in such a way that thermal or acoustic properties can be assigned to the connection? Can I control that the interior wall goes through the insulation? Other cases include the automatic management of walls on sloped surfaces, the updating of non-vertical sloped walls, and other uncommon cases. I emphasize these examples because walls are universal components of BIM design tools, but the various levels of design knowledge they are based on varies widely. If the capability is not included, a designer must resort to manual definition of details and components, without parametric modeling updates, eliminating easy editing.

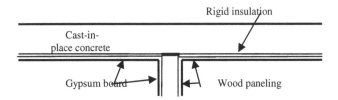

Fig. 2. Three walls coming together and the detail that effects energy, sound and other behaviors

Software programmers with a few architectural advisors have determined the vocabulary of shapes and behavior that are supported by the common BIM tools, determining to a significant degree the working vocabulary readily available to architectural designers. There is no deep study of cases, definition of best practices, direct industry input or careful review of various codes. I summarize the main issues:

1. software companies develop products with only limited involvement of end-users, relying on existing platforms and internal expertise for embedding expertise in their systems. **They do not clearly distinguish construction domain expertise from software development expertise;**

2. **products initially often only poorly meet the requirements of the end users**, and require iterative extension and modification. While this provides a context that facilitates feedback, it is inefficient for both software developers and end users;
3. on the other hand, if an advanced-level product is introduced, **end users are typically naive and attempt to use the product in an evolutionary way, trying to make it fit older practices**. A gap exists between where users are and where they will be, say, three years in the future.

There are many approaches that can address these issues, including, the development of standards regarding object behavior, the organization of panels or consortia to define the needed behavior in BIM tools, and more generally, the definition of procedures for specifying the knowledge to be embedded in AEC design and engineering tools.

The author has been responsible for an industry-wide endeavor for the maintenance and refinement of the CIS/2 product model for steel fabrication [3], for an industry consortium-led project that specified fabrication-level design software for precast concrete fabrication, and now another consortium for the specification of a engineering product for reinforced (cast-in-place) concrete. The software for precast concrete was developed out of an open bid for proposals to software companies, with 12 submissions and 6 months of evaluation prior to selection. This process and the resulting product have been widely reported [4],[5],[6]. The current work to develop an advanced system for reinforced concrete design and construction involves a consortium of interested engineering and construction companies and a pre-selected software developer.

The challenge in these activities has been to form a broad-based industry consortium, capture the knowledge and expertise of experienced designers and engineers, resolve stylistic differences, and to specify in an implementable format the functionality and behavior the software is to encapsulate. Based on our experiences, my associates and I have evolved the following general guidelines:

1. *jointly define the range of product types and corresponding companies with in-house expertise in the design and fabrication of those product types; this group becomes the technical team;*
2. *model the processes used in the design/fabrication of each product type, capturing the process and information needed/used/generated in each, along with required external resources involving information exchange;*
3. *for each product type, define the functionality of an idealized system, characterizing in broad terms how the system would operate;*
4. *select (if necessary) and learn the existing functionality of the platform system, so that gaps in functionality from the system capabilities upward can be defined;*
5. *identify the needed system objects -- beams, walls, columns, connections, etc. that are needed to define the components of the system in each type of product; some of the objects may be abstract and transfer their components to other objects, in part or whole;*
6. *resolve overlaps among different product types and define the combined behavior for the set of target objects;*
7. *for each object, identify (and gain agreement) on the object's initial definition, including all control parameters;*

8. *for each object, identify in detail the possible changes to its context; for each of the identified changes, define the desired update behavior;*
9. *identify basic types of drawings and other reports needed; for each, identify desired automatic layout;*
10. *as the system is implemented, undertake testing to determine if the functionality is correct and that the various embedded cases have been properly identified.*

In each case, it is necessary to resolve conflicting requirements among the team members. It is also necessary to translate the system requirements into an incremental sequence of discrete functional capabilities that can serve as software development steps and be implemented. For the precast concrete specification effort, the first four steps took 18 months. The last six steps took 30 months, with about a year of that time being software development time. The specification contained 626 development items in 31 distinct areas. The precast concrete process is now completed, with an effective commercial product.

Universities should be undertaking industrial development for multiple reasons, among them being:

-   to develop operational knowledge about problem domains as they are in reality, rather than textbooks;
-   to develop strong working relationships with innovative organizations and help them adopt innovative technologies or concepts;
-   to use the application of research as a springboard to undertake supporting research in related areas.

Our work with the North American precast concrete industry has led to the development of a set of new technologies to support expert knowledge capture and utilization. For step #2 above, we developed process modeling tools that allow full capture of different corporate processes, them merging into a single integrated data model supporting implementation [7],[8]. This work led to new ways to define and extend the integration of process models with product models. For steps #7 and #8, we developed a notation for representing parametric behavior [9], so that the technical team could effectively communicate complex behaviors.

An example of the behaviors we were dealing with is shown in Figures 3 and 4 showing the definition of pocket connections for double tee members. The process

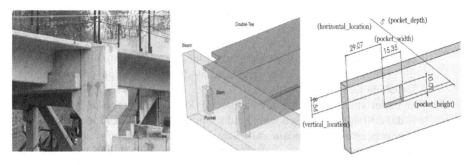

Fig. 3. The precast elements to be modeled, example of the pocket connections, a portion of the parameters involved

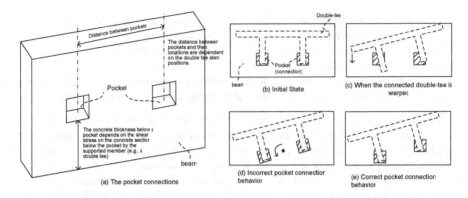

Fig. 4. Examples of the design rules developed for the precast concrete design tool

described moves from a general condition regarding one or more related system objects and definition of their parameters. Then their behavior in response to different external conditions is identified.

Those who participated in the consortium-led product development process have been very satisfied with the result. End-users participating in the specification have developed an early but sophisticated view of what the system should do. While they are not always satisfied with a 90 percent solution, they are well down the path of re-organizing company processes to take advantage of the new technology by the time it is released. The software company involved has developed products faster and with less expense than using traditional practices. Note: This work was undertaken collaboratively with Rafael Sacks and Ghang Lee, who deserve much of the credit for its success.

I offer several points growing out of these efforts:

1. while the current effort to capture domain expertise and transfer it to computer systems for production use is very visible at this time in history, the definition and translation of this knowledge will be on-going and evolutionary; people will continue to learn by doing, requiring later translation to knowledge embedded software; further development of methods for making this transfer is needed;

2. the tools for representing and communicating desired parametric object and assembly behavior is weak and not well developed.. We relied on a specialized form of story-boarding. I could imagine animation tools with reverse code generation (examples exist in the robotics programming [10]) allowing the desired behavior to be programmed by direction manipulation of objects, for example;

3. an underlying issue in the development of knowledge embedded design tools is the definition, representation, and refining of processes. The design of design processes is a fundamental issue in the development of advanced software for engineering and design; the old process modeling tools, such as IDEF0, (which grew out of SADT in the 1960s,) are terribly outmoded and should be replaced by methods that are more structurally integrated in

machine processable methods; needed are process modeling methods that capture actions, the data the actions use and also constraints applied by the actions [11];

4. developing parametric modeling of assemblies has gotten much easier over the last decade. When I began, it required programming in C or FORTRAN. Today, it requires only annotated sketching tools and scripting. However, there is still further possible steps, drawing from the example of robot programming [10] and the further application of graphical programming tools, as exemplified by some tools supporting UML [12].

3   New Technologies for Design Guides

The development of machine readable architectural and engineering design information changes the way that we can think about design information and processes. Previously, it was reasonable for all information to be in a format only readable by a person, as a person had to interpret conditions for the design, carry out all operations, and to derive all implications. With BIM, the opportunity to share these activities much more with the computer becomes practical. This opens opportunities for re-structuring how design knowledge can be better represented for easy integration with the process of designing. Here, I focus on design support literature, ranging from Ramsey and Sleeper's *Architectural Graphic Standards* [13] and Neufert's *Architect's Data* [14] to special topics such as sustainable design [15] and types of structural systems [16], to studies of particular building types [16],[17]. These are currently in book form (a few have CDs for searching on-line) and currently fill libraries and office shelves as information sources (collected best practices) for the field.

We envision the day when the knowledge these sources carry are integrated with design tools. For example, the team that developed the parametric design, engineering and production modeling tool for precast concrete included some of the authors of sections of the PCI Design Guide [18]. It was one of the important sources for the effort and we implemented some sections of the Guide. In others cases, the information is less structured and can be represented as case studies. An example here is the CourtsWeb project [19], funded by the US General Services Administration and the US Courts. It consists of an on-line database of plans, sections, 3D models, and issues that provide background for architects, courthouse clients and administrators. There are a wide variety of other examples.

3.1   Representation of Design Knowledge

How might design information be structured for use in advanced design/engineering tools? In developing an answer, I return to the early days of artificial intelligence. At that time, there was great interest in how information can be used in the context of problems [20]. These early efforts set out to identify the intuitive problem solving strategies that made people intelligent. They studied what distinguished novices from experts. This line of research involved development of problem solving heuristics and tested in such programs as GPS (General Problem Solver) [21]. As part of some early work at the beginning of my career, I undertook an analysis of these concepts [22].

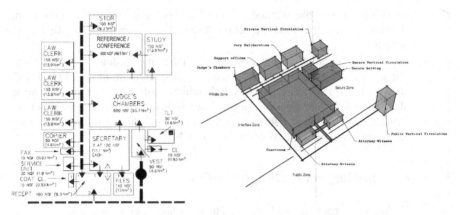

Fig. 5. Content base for a courthouse design guide

Any type of effective problemsolving process requires some level of bookkeeping; requiring, for example, one to keep track of what solutions have been tried; so they are not tried again. Also, changing a design aspect that was the basis for later decisions requires that the later ones be checked and potentially re-solved. More generally, information can be classified as to its problemsolving *power*:

(1)   Much design information is in the form of examples and considerations to be applied during design. These are observations – examples that are good or bad. The information is relatively unstructured. This is qualitative and unstructured data that requires interpretation and application by humans. It is the weakest form of problemsolving information. Such information can be organized as case information and structured as a *case-based reasoning* system [23]. Given some case or condition, here is what is known about it? Given the case of 'circulation', the help system might identify alternative building configurations; for the case of 'courtroom' it would provide cases and design guidelines for that space type. Case information can be structured along the lines of an augmented help system. For example, it could be organized in a help-type database and accessed through the Microsoft Helpdesk tools, using keywords, section organization, and other structured information mechanisms. It could be augmented, because some carried geometrical information could be structured in a portable graphic format (such as DXF), allowing it to be dragged-and-dropped into the design tool being used. This is the weakest structure of digital design information.

(2)   A more powerful kind of knowledge aggregates the information in cases into some metric that can score a design – either pass/fail, (building net area/gross area must be less than 0.67) defined in problemsolving as *"generate-and-test"*. In other situations, the design cases may have been analyzed to derive metrics. or a numerical score without knowledge to improve the design, (defined in problemsolving as *"hill-climbing"*) In the latter cases, a numerical score has been developed that indicates how good is the current design state. Examples are estimated building cost, energy usage, flow rates derived from Monte Carlo simulation of pedestrians. In these cases, we have a numerical score that tells how good the current solution is. In the worst case, however, we do not know

how local actions lead to improvements in the overall score. These latter cases often are the result of analysis/simulation applications, that apply models of behavior to determine the metric. The metric alone, however, is of only limited benefit without additional knowledge that gives insight what how operations change the metric.

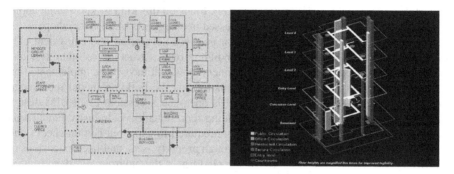

Fig. 6. Checking circulation path interferences

(3)   Other design information can be defined as rules or metrics, for example for security issues or for costs. The rules and metric can provide sub-goals for the design. In these cases a design can be assessed whether it satisfies the goal or not. These may be simply local tests that have no way to summarize (such as safety), while others may have a relation to an overall score, (elimination of blind corners in heavy pedestrian circulation routes affects circulation efficiency). The check may be quantitative, for example an area requirement. Alternatively, it may be qualitative, dealing with multiple-dimensional issues, such as the acoustics of a space.   These rules were called in problemsolving *"means-ends-analysis"*.

(4)   A stronger level of design knowledge are rules that can be embedded into generative procedures. In these cases, testing of the design is not necessary, it is guaranteed by the generation process. That is, a set of operators exist that embed the goal within them. Newell calls this method *"induction"*. This is the manner of implementation of design knowledge within a parametric modeling tool, that relies on parametric objects that are self-adjusting to their context yet are guaranteed to update in a manner that maintains the desired design rules. When an external input requires that the layout change, the changes are made automatically, or the system reports that the change led to conditions where the embedded rules cannot be satisfied. An example is the automated connection details found in some structural detailing applications such as Tekla Structures® and Design Data's SDS/2®.

Each of these levels of knowledge suggest different methods of information delivery to designers or engineers. The methods for making knowledge available have different implementations. For the first method, help systems are mostly easily implemented using a Help toolkit, such as Microsoft's, which is compatible with almost all CAD system environments. However, richer toolkits are possible. How can one build a context sensitive knowledge base that can work with different design tools? How can a program decipher the design intention within a design context, in order to provide desired information? Development of a case-based design information system platform is an important research (and possibly business) endeavor.

The second kind of knowledge is typically embedded in analysis and simulation tools. Support for iterative use of such tools and keeping track of multiple runs is important in practical use. However, I will not focus on this kind of knowledge application; I assume it will be extensively addressed in other parts of this workshop.

The third type of knowledge application involves developing the equivalent of a spell and grammar checker for particular building types, structural or other systems, or even design styles. Such building assessment tools will have to be implemented on some software platform. One part of the platform is the building model representation. Here, we rely on a public format that is open and accessible to all BIM design tools. In this case, the public standard building representation is Industry Foundation Class (IFC) [24]. IFC may not have the data required to carry out certain checks and these may require temporary extensions through property -sets, and later extensions to the IFC schema. The second aspect of the platform is the environment that reads in the building model data and provides the software environment to support calculating properties not directly stored and developing tests to assess the base or derived properties. Several rule-checking platforms exist, such as Solibri (see: http://www.solibri.com/services/public/main/main.php and EDM Model Server (see: http://www.epmtech.jotne.com/products/index.html).

For the fourth method of information usage, parametric models of building elements and systems provide a rich toolkit for defining generative design tools that can respond to their context. Currently parametric models are not portable, but can only be implemented for a particular parametric design tool. Today, the technology does not exist to define cross-platform parametric models. Further research is needed before such a production undertaking is warranted.

All four methods of information-capture can be applied to BIM design environments. They augment the notion of a *digital design workbench*. We expect that all designers and engineers will increasingly work at such workbenches from now on.

Two lines of study are embedded in this discussion. One deals with the abstract study of the power of different kinds of information in solving problems. What are the abstract classifications and what is their essential structure? I have proposed a classification based on problemsolving theory. The second line of study is to identify effective ways to delivery particular classes of design knowledge I have outlined methods for the four types of information. This suggests that a science of information delivery in design is possible, built upon the classic knowledge of problem solving. Last, we have the exercise of packaging and delivering the design information to end-users. This will become a major enterprise, a replacement for the current generation of material embedded in paper-based and electronic books.

4 Conclusion

The development of machine readable building models, first at the design, then at the construction stages, is leading to major changes in how we design and fabricate buildings. These transitions will impact all parts of the construction industry and lead to major restructuring of the industry, I believe. The transition provides a rare opportunity for strong collaboration between schools and practitioners. We are at the start of an exciting era of building IT research.

References

1. Eastman C, Building Product Models: Computer Environments Supporting Design and Construction, CRC Press, Boca Raton, FL 1999; Chapter 2.
2. Li WD, Lu WF, Fuh JYH, Wong YS, Collaborative Computer-aided design – research and development status, Computer-Aided Design, 37:9, (August, 2005), 931-940.
3. Eastman C., F. Wang, S-J You, D. Yang Deployment of An AEC Industry Sector Product Model, Computer-Aided Design 37:11(2005), pp. 1214–1228 .
4. Eastman C, Lee G, Sacks R, Development of a knowledge-rich CAD system for the North American precast concrete industry, in: K. Klinger (Ed.), ACADIA 22 (Indianapolis, IN, 2003) 208-215.
5. Lee G, Eastman C, Sacks R, Wessman R, Development of an intelligent 3D parametric modeling system for the North American precast concrete industry: Phase II, in: ISARC - 21st International Symposium on Automation and Robotics in Construction (NIST, Jeju, Korea, 2004) 700-705.
6. Eastman, C. M., R. Sacks, and G. Lee (2003). The development and implementation of an advanced IT strategy for the North American Precast Concrete Industry. ITcon International Journal of IT in Construction, 8, 247-262. http://www.itcon.org/
7. Eastman C, Lee G, and Sacks R, (2002) A new formal and analytical approach to modeling engineering project information processes, in: CIB W78 Aarhus, Denmark, 125-132.
8. Sacks R, Eastman C, and Lee G, Process model perspectives on management and engineering procedures in the North American Precast/Prestressed Concrete Industry, the ASCE Journal of Construction Engineering and Management, 130 (2004) pp. 206-215.
9. Lee G, Rafael Sacks , Eastman C, Specifying Parametric Building Object Behavior (BOB) for a Building Information Modeling System, Automation in Construction (in press)
10. Bolmsjö, G, Programming robot systems for arc welding in small series production, Robotics & Computer-Integrated Manufacturing. 5(2/3):199-205, 1989.
11. Eastman CM, Parker DS, Jeng TS, Managing the Integrity of Design Data Generated by Multiple Applications: The Theory and Practice of Patching, Research in Engineering Design, (9:1997) pp. 125-145.
12. Schmidt C , Kastens U, Implementation of visual languages using pattern-based specifications, Software—Practice & Experience, 33:15 (December 2003): 1471-1505
13. Ramsey CG , Sleeper HR, and Hoke JR Architectural Graphic Standards, Tenth Edition Wiley, NY (2000).
14. Neufert E & G, Architects' Data, Blackwell Publishing, 2002.
15. Kilbert CJ, Sustainable Construction: Green Building Design and Delivery, John Wiley & Sons, NY (2005).
16. Butler RB, Architectural Engineering Design -- Structural Systems McGraw-Hill, NY (2002)
17. Kobus RL, Skaggs RL, Bobrow M, Building Type Basics for Healthcare Facilities, John Wiley & Sons, Inc.NY (2000).
18. College of Architecture, Georgia Institute of Technology, Conducting Effective Courthouse Visits, General Services Administration, Public Building Service and Administrative Office of the U.S. Courts, 2003.
19. PCI 2004 PCI Design Handbook : Precast And Prestressed Concrete 6th Edition, Precast/Prestressed Concrete Institute, Chicago

20. See: http://www.publicarchitecture.gatech.edu/Research/reports/techreports.htm
21. Newell A, . Heuristic programming: Ill-structured problems. In: Arofonsky, J. (Ed.). Progress in Operations Research, Vol III, (New York,1968) pp. 360-414.
22. Newell, A. Shaw, JC.; Simon, HA. (1959). Report on a General problem-solving program. Proceedings of the International Conference on Information Processing, pp. 256-264.
23. Eastman C, "Problem solving strategies in design", EDRA I: Proceedings of the Environmental Design Research Association Conference, H. Sanoff and S. Cohn (eds.), North Carolina State University, (1970).
24. Kolodner J, Case-based Reasoning, Lawrence Earlbaum and Assoc., Hillsdale, NJ. (1993).
25. Liebich T,. (Ed.) Industry Foundation Classes, IFC2x Edition 2, Model Implementation Guide Version 1.6, (2003).

Infrastructure Development in the Knowledge City

Tamer E. El-Diraby

Dept. Of Civil Engineering, University of Toronto,
35 St. George St., Toronto M5S1A4, Canada
tamer@ecf.utoronto.ca

Abstract. This paper presents a roadmap for establishing a semantic Web-based environment for coordinating infrastructure project development. The proposed roadmap uses semantic knowledge management and web service concepts to integrate all aspects of infrastructure project development. The roadmap focuses on integrating business, safety and sustainability dimensions in addition to traditional engineering aspects. The roadmap also emphasizes a process-oriented approach to the development of e-infrastructure.

1 Introduction

Two features are believed to dominate the design of civil infrastructure in the 21st century: consideration for impacts on sustainable development and analysis of infrastructure interdependency. Parallel to that, computer based systems are evolving from focusing on data interoperability and information sharing into knowledge management. In fact, the city of tomorrow is shaped as a knowledge city that promotes progressive and integrated knowledge culture. The main characteristics of a modern knowledge city include [1]:

- Knowledge-based goods and services;
- Provision of instruments to make knowledge accessible to citizens;
- Provision of dependable and cost competitive access to infrastructure to support economic activity;
- An urban design and architecture that incorporates new technologies; and
- Responsive and creative public services.

It can be argued that the design of infrastructure in the 21st century requires the deployment of an effective knowledge management system with two core components:

1. Theoretical Components: a shared knowledge model (*ontology*) of interdependency and sustainability knowledge.
2. Implementation Components: the *computer systems*, *inter-organizational protocols* and *government polices* that use the knowledge model.

The promise of implementing a common knowledge management system is envisioned to allow stakeholders to work so closely that there are interoperable computer systems that allow partner A to seamlessly access the corporate data of partner B, manipulate certain aspects of their designs, and send a message: "we have changed the schedule of your activity K or the design of your product M to achieve more

I.F.C. Smith (Ed.): EG-ICE 2006, LNAI 4200, pp. 175 – 185, 2006.

optimized design/operation of both our infrastructure systems". This is in compliance with pre-agreed *profit sharing/collaboration protocols* that are encouraged by incentives from *government policies[1]*.

This paper presents a roadmap to support the establishment of collaborative design of infrastructure systems. The roadmap aims at integrating the business and engineering processes between different stakeholders and across various projects to assure consistency in design, coordinated construction, consideration of sustainability, and better handling of infrastructure interdependency.

2 Outline of the Proposed Roadmap

Figure 1 shows a general view of the proposed architecture. An interoperable knowledge model is at the core of the proposed architecture. This includes a set of ontologies that encapsulate knowledge from different domains:

Product Ontology: representation of the knowledge about infrastructure products in different sectors (electrical, telecommunication, water systems, etc.).

Process Ontology: a model for the processes of design and construction of infrastructure systems in various sectors.

Actor (Stakeholder) ontology: semantic profile of stakeholders in infrastructure development, for example government, owners/operators, local community. The ontology models responsibilities, and authority of various actors and their needs for information.

Ontology for sustainability in civil infrastructure: this ontology represents knowledge about the sustainability features and impacts of civil infrastructure systems.

Ontology for legal constraints: This ontology models the objectives and role of Canadian regulations, and how they support collaboration.

The composition of project stakeholders varies from one project to another, as the roles and requirements of the same stakeholder can change between projects. Consequently, the proposed architecture is built to provide stakeholders freedom in expressing and using their own ontologies. i.e. each stakeholder can either use the proposed ontologies to represent their knowledge (regarding products, processes, and actors) or provide their own ontology. A semi-automated ontology merger system is proposed to establish semantic interoperability in the portal.

Built on top of this layer, is a layer of web services that support the addition, access, retrieval and modification of information. Through using the ontology a set of interoperable web services, agents and other software tools will be built to provide services to stockholders, analyze information exchange patterns and wrap existing software systems.

---

[1] Academic models of such open cooperation started about 5-8 years ago (Hammer, 2001). Lately, some industries have started implementing these models (see for example: Fischer and Rehm, 2004).

The architecture provides access to information at three levels: corporate level (full access to corporate information), partner level (access to certain information in partner organizations) and public level (free access to a limited set of information). Orthogonal to these levels, information can be viewed in portals dedicated to products, processes and actors. All technical details of the infrastructure system being designed or managed will show up on the product view. Workflow and business transactions will show on the process view. Status, interest, and tasks of stakeholders will be shown in the actor view. Proper links will be made to relevant policies, codes and regulations that have bearing on any of these views.

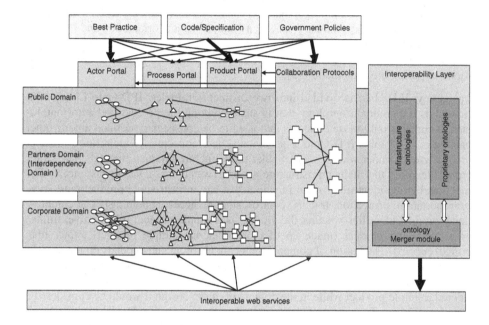

Fig. 1. Proposed Architecture

3 Ontology Development

This section summarizes the progress made so far in ontology development. The ontologies were developed in OWL (using Protégé) with the axioms molded in SWRL (Semantic Web Rule Language).

3.1 Infrastructure Product Ontology

This ontology encompasses all facets of infrastructure products (mainly physical products). An *infrastructure product* (IP) is produced through a set of *processes,* where *actors* are involved. Each process and/or actor uses a set of mechanisms to support their work (software, theories, best practices, rules of thumb). Each product has a set of *product attributes* and is constrained by 1 or more *constraints* and is related to a set of sustainability indicators/features. The IP attributes could include cost, material, security, performance index that relates to the IP performance and surrounding conditions attributes (see Figure 2).

The IPD-Ontology was created based on review of existing information modeling efforts in the various infrastructure domains (water, wastewater, electricity, telecommunication, and gas). IPD-Onto reused existing taxonomies whenever possible and created an upper level classification common to all products. IPD-Onto is currently implemented in OWL and contains around 1,200 concepts and relationships.

In this regard, it is worth noting that several initiatives for interoperability in the infrastructure product realm have been attempted (e.g. LandXML [2], SDSFIE [3], MultiSpeak [4], etc...). Nevertheless, these models lacked: 1) The ability to represent knowledge rather than data in a domain, 2) Interoperability among various infrastructure domains due to their industry-specificity and, 3) Object orientation and its associated benefits in information modeling. Other more application-oriented initiatives focused on the data interoperability between CAD and GIS for specific use case scenarios requiring their interaction [5].

Taxonomy: IPD-Onto is divided into two distinct ontologies. IPD-Onto Lite is considered as the common ontology that is shared among the process and actor ontologies. It contains only those concepts that need to be consistently defined among other ontologies. Currently IPD-Onto Lite contains 132 concepts. It identifies 3 distinct product groups under which any particular infrastructure product must fall. The sector group identifies the main infrastructure sectors (water, wastewater, gas, etc...) The functional group identifies 7 main functions that any infrastructure product must serve (transportation, protection, tracing, control, storage, access, pumping). The compositional product identifies whether the product is a simple product (pipe, valve, fitting) or a compound product (made up of more than one simple product) (water line, bridge, culvert). The notion of composition is not absolute and depends on the domain and setting considered (hence the need for categorization concepts at the root level). For example, in the infrastructure asset management domain a pump would be considered a simple product while in the domain of pump design it would be considered a complex product. Two concepts were central to the ontological model in this regard: attributes (as they present characteristics that fully describe any product) and constraints (as they present concepts that impact all aspects relating to a product). Other concepts like techniques and measures are also extensively utilized in the model.

Relationships: Taxonomical relationships are in the form of is-a relations (e.g. ElectricSwitch is-a ControlProduct). Non taxonomical relations relate different concepts together through a semantic construct for the relation. Some of the upper-level relations in IPD-Onto Lite include:

- InfrastructureProduct has_attribute InfrProductAttribute
- InfrastructureProduct has_technique InfrProductTechnique
- InfrastructureProduct has_constraint Constraint
- InfrProductAttribute has_domain Domain

Ontological modeling allows for creating taxonomies of relationships. As such, the following 4 relationships are considered to fall under a class hierarchy of descending abstraction (has, has_technique, has_method, has_repairmethod).

Axioms: Axioms serve to model sentences that are always true in a domain. They are used to model knowledge that cannot be represented by concepts and relationships. Axioms can be very useful in inferring new knowledge. Examples of some axioms (and their equivalent in first order logic) defined in IPD-Onto Full include:

- PVC pipes have an attribute of high resistance to aggressive soils:
  $\forall$ x (Pipe(x) $\wedge$ has_MaterialType(x, PVC)) $\supset$ has_SoilResistance(x, High)
- Steel pipes has an attribute of high strength:
  $\forall$ x (Pipe(x) $\wedge$ has_MaterialType(x, Steel)) $\supset$ has_Strength(x, High)
- Fiber optic cables that do not have a casing are likely to be damaged during construction:
  $\forall$ x,σ,t (FiberOpticCable(x) $\wedge$ hasCasing(x, None) $\wedge$ ExcavationProcess(σ) $\wedge$ Occurs(σ, t)) $\supset$ holds(has_attribute(x, damaged), t))

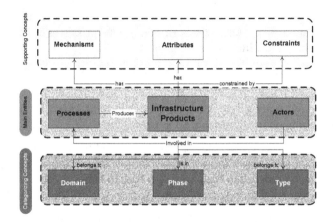

Fig. 2. Ontological Model for the Product Ontology

3.2 Infrastructure Process Ontology

This ontology captures process knowledge. A process has life cycle (expressed in a set of phases) including conceptualization (capturing the requirements, identifying the constraints), planning (who will do what at which time)[2] development (alternative development and evaluation), implementation (development of the final output). In addition to phasing, a full description of each process will require linkage to other concepts, such as actors (the people and organizations) involved in the process, roles (responsibilities of each actor), constraints (rules, codes, environmental conditions) and the supporting mechanisms (theories, best practice, technologies) that support the execution of the process. The ontological model of processes is perceived to be an extension of the basic IDEF0 model (input, output, constraints and mechanisms).

---

[2] Please notice that even the Planning Process has a planning phase of who will do what to develop the plan.

Processes are defined using the following major dimensions:

1. Project phases: this is the main dimension for categorizing. Processes are categorized per their position in the project life cycle. The main phases include: business planning, pre-project planning, execution and operation. These phases were adopted from CII publications [6] literature as an effort to comply with existing industry standards.
2. Domain: processes can belong to a set of domains, such as business, administrative, engineering.
3. Sector: a process can belong to one sector of infrastructure, such as transportation, water utilities, etc.
4. Level: some processes are enacted at the corporate level, others only exist at the project level

Two distinct types of processes are defined in this ontology, continuous processes and discrete processes. Continuous processes are those processes that continue to exist throughout the project life cycle, such as communication management processes and project co-ordination processes. This is in contrast to processes that have specific duration within the project life cycle. These include:

Design Process: includes pure technical design processes such as 'alignment design process', 'geometric design process', 'structural design process', etc.

Field / Construction Process: includes pure technical field processes such as 'survey process', 'concerting process', 'earth work process', etc.

Scope management Process: includes processes required to ensure that project scope is properly defined and maintained to reduce possible scope risks such as poor scope definition and scope creep; for example, 'scope verification process', 'scope definition process', 'scope monitoring process', etc..

Risk Management Process: includes processes required to properly allocate and manage project potential risks, such as 'risk assessment process', 'risk allocation process', 'risk handling process'.

Stakeholder Management Process: includes processes performed to capture and incorporate stakeholder input in the city / project development, such as 'stakeholder involvement program design process', 'stakeholder participation process, and 'stakeholder input classification and analysis process'.

Procurement Management Process: includes processes required to obtain necessary resources from external sources, such as 'tendering', 'sub-contracting process', etc..

Money Management Process: includes 'estimating process', 'budgeting process', 'accounting process', 'financial management process', and 'cost management process'.

3.3 An Ontology for Sustainability in Infrastructure

The proposed ontology has the following main concepts/domains (each is the root of a taxonomy): **Entity** (including *Project, Process, Product, Actor,* and *Resource*), **Mechanism** and **Constraint**. Any *project* (e.g. renovation of a street, construction of a new street, a new transit system) produces a set of *products* (e.g. new lanes, new bridge, dedicated lanes, transit tracks, new traffic patterns, and signals). Each of these *products* has a set of possible *design options*. The options are developed through a set of interlocked *processes*, where *actors* (e.g. design firms, Dept. of Transportation) make *decisions* (e.g. set project objectives, develop options, configure options, and approve an option). Each option has a set of *impacts* on various *sustainability elements*, such as health hazards, increased user cost, negative impacts on local business, and enhancement to traffic flow. These elements include *stakeholders* (Actor), such as a business or community group, or basic environmental elements, such as air, water, and soil. For each of these impacts, a set of *strategies* could be used to reduce any negative *consequences* on the *impacted elements*.

The ontological representation of highway sustainability management process is at the intersection of this ontology and the aforementioned process ontology. Each *Sustainability process* consists of two major phases: planning and management. Each phase is subdivided into sub-processes.

For example, the Sustainability planning process encompasses five major sub processes: Analysis of existing elements process, Impact & risk identification process, Impact & risk assessment process, Impact & risk mitigation process, and Code/policy enforcement process. On the other hand, three themes of sustainability: Natural environment, Society and Economy, have to be taken into account during any *Sustainability process*. Therefore, a matrix is formed with the columns representing the three themes and the rows representing the two phases. The first level sub processes of the highway sustainability optimization process is shown in the matrix in Figure 4. Each Planning process includes the following sub-processes: analysis of existing systems, identification of risks, risk assessment, development of risk mitigation tools, and code compliance check. Each management process includes two sub processes: development of risk/impact controls and evaluation process. For instance, the *Analysis of existing natural environment elements process* is at the intersection of the *analysis of existing elements process* and *natural environment sustainability process*. This is because it covers both domains of knowledge: looking at existing conditions (in contrast to future/suggested conditions) and only considering the environmental aspects of these conditions (in contrast to social and economic aspects).

4 Implementation

4.1 Prototype GIS System

A prototype GIS system (*StreetManager*) was developed to test how multi-organization constraint satisfaction can be accomplished to support micro-level utility routing. Primary users of the portal include local municipalities and utility companies who own or mange infrastructure within a ROW. The system relies on three main components: (1) An object oriented geo data model that is built on an Infrastructure Product Ontology developed, (2) an XML-spatial constraint model, and (3) A dynamic spatial constraint knowledge base which is built according to the XML-schema.

The constraint model acts as a generic schema for representing constraints. An XML-Constraint schema built upon the constraint model is used as the common structure for exchanging constraints among stakeholders. Any number of constraints can be represented in XML that will abide by the XML-Constraint schema. The constraint file is then used to generate the necessary constraint checking code using the ArcObjects programming language. The designer of a new utility system can consistently check the proposed route of his/her utility throughout the design process against any number of constraints that are shared and made explicit by other utility companies or regulating bodies.

The primary use-case of the system assumes the following process flow (see Figure 3). The designer of a new utility system uploads a new design to the system in either CAD or GIS format. The system will start resolving semantic differences between the uploaded data and that of existing utilities in the street. Examples of semantic inconsistencies include layer, attribute and value naming (e.g. the uploaded data might refer to a 'Gas_Pipe' whereas the OO geo-data model uses 'GasLine'). The semantic matching is made possible by the Infrastructure Product Ontology running at the back-end, but nonetheless the user is prompted to confirm semantic matching. This semantic conflict resolution is similar to that performed by [7] in the context of collaborative editing of design documents.

After all semantic differences are resolved, the existing geospatial utility data is appended with the new design. The user selects which subset of constraints to check for, based on the spatial constraint model. For example, the user may want to check the design against 'hard' constraints first to ensure that all minimum clearance requirements are satisfied and then check 'advisory' constraints to know how the design may be improved. Alternatively the user may want to select those constraints that have to do with Telecom infrastructure or those that are related to maintenance issues, etc. Based on the selected constraint subset, the GIS system invokes a series of spatial queries that are stored in the spatial constraint knowledgebase in XML format. The output of this process is a violated constraint list that registers all constraints that were violated by the proposed design.

The user can amend the design accordingly until it is ready for final submission after which other affected parties (agencies that have utilities within the ROW) are notified. These agencies can then view the proposed new design using the system and

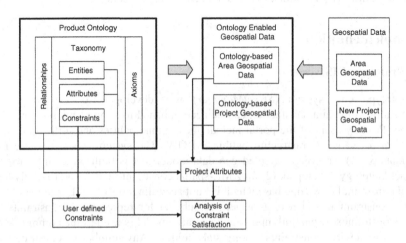

Fig. 3. Architecture of The Product Portal

invoke any subset of constraints to check the quality of the design against the knowledge base. The system allows for approvals and comments to be communicated among the collaborators to expedite the design coordination process. The collaborative web portal eliminates current practices of drawing exchange and review cycles that create bottlenecks in the design process. The designer of a new utility system can consistently check the proposed route of his/her utility throughout the design process against any number of constraints that are shared and made explicit by other utility companies or regulating bodies

4.2 Integrated Process Portal and Ontology Merger

A prototype portal for integrating work processes across different organizations has been implemented. The portal aims at integrating these processes based on knowledge flow. i.e. a consolidated process structure is created by matching (in a semi-automated fashion) closely aligned activities of the collaborating organizations. The following main steps are included in the implementation (see Figure 4):

1. present processes: the user of the portal can use the proposed process ontology (in a drag-and-drop fashion) to build the structure of their processes. If the user prefers not to use the proposed ontology, they can upload and use their own ontology to represent their processes. If the user does not want to use an ontology to present their processes, they are requested to fill out a simple table of the main tasks and their related actors and products before they document their process. The table is then transformed into a small ontological model using Formal concept analysis.

2. ontology merger: a separate module is then invoked to provide interoperability between the different ontologies of all collaborating organizations (see next section).

3. Establish collaborative process: the portal sorts out similarities in the different organization's processes. A user (called the coordinating officer) can use these similarities in developing a common process. Basically, the coordinating officer can access all the processes and drag-and-drop any activity from any organization into the combined process. The combined process can show the flow of information between different stakeholders. It can also show: who is involved in the project at which time, what products (or parts of products) are being designed at which time and by whom, and what attributes (of products) are being considered at which time?

4.2.1 Ontology Merger
The proposed merging methodology consists of three main steps: 1) encoding, 2) mapping of concepts and relations, 3) merging of concepts and relations using formal concept analysis and lattice algebra.

Encoding: The encoding stage aims at transferring the ontology / process model into a set of formal concepts for subsequent lattice construction. Ontology concepts (including attributes of concepts) are presented in context K_1 and ontology relations (including attributes of relations) are reflected in context K_2. The concept-relation link is reflected in both contexts as relational attributes, out-relation vs. in-relation for the 'ontology concepts context' and source-class vs. target-class for the 'relation context'.

Taxonomy & Relationship Mapping: Concepts will be matched using four main types of heuristics: 1) name-similarity heuristic, where mappings will be suggested based on similarity between concept names from a linguistic perspective, 2) definition-similarity heuristic, where mappings will be suggested based on similarity of natural language definitions from a linguistic point of view, 3) hierarchical-similarity matches, where mappings are performed based on similarity between taxonomical hierarchies and is-a relationships, i.e. compare the closeness of two concepts in the taxonomical concept hierarchy, and 4) relational-similarity matches, where mappings are suggested based on similarity of ontological relations between concepts.

Taxonomy & Relationship Merging: This work extends the work done by Rouane *et al* [8] to relationship mapping. An initial lattice is constructed from concepts and non-relational attributes. A lattice $L^{0}_{concept}$ is built from context $K^{0}_{concept}$. This lattice constitutes the first iteration of the construction process. Once each ontology is translated into a lattice. The rules of lattice algebra are applied to merge (add) the two lattices. The second iteration starts by relational scaling based on lattice $L^{0}_{concept}$, resulting in relational attributes scaled along the lattice. Thus, the attribute name in the scaled context will have reference to both the relation type and the formal context of the preceding lattice. A process of mutual enrichment continues until isomorphism between two consecutive lattices is achieved.

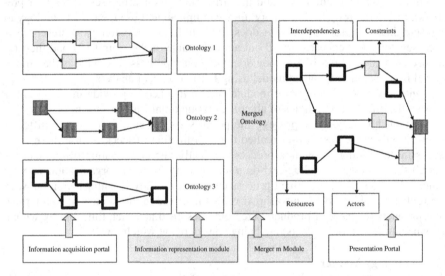

Fig. 4. Process Portal

5   Ongoing/Future Work

Actor ontology: given that many actors are going to be involved in the exchange of knowledge during the collaborative processing of infrastructure design, a substantial flow of information is expected. Furthermore, the consideration of sustainability adds a substantially new domain of knowledge, with very subjective and conflicting contents. This ontology will attempt to link the roles and responsibilities of various actors (including the general public) to their information needs. An agent-based system will then be implemented to filter relevant information to interested actors based on their profile (role/responsibilities, skills, interest), the current status of product development and the current stage of the collaborative process.

Grid-enabled Information Exchange: The Grid can be viewed as an application framework that defines standard mechanisms for creating, managing and exchanging information Grid services. A Grid service is a software system that represents a physical or logical resource. A resource is a database module, device, or even application logic. A Grid service is designed to support interoperable interaction with other Grid services. The Grid provides a standard means of this interoperation.

The use of ontologies along with the structured information exchange work will result in *semantic grids*, rather than just computational and data grids. The grid architecture will not only be technically robust, but will also understand the processes that take place within the infrastructure environment ....

Best Practice Map: The research will monitor actual use of the portal by industry practitioners. Data mining tools will be used to identify industry needs, problems, risks, best practice and the impacts of different code and policies on the ability of organizations to manage infrastructure risks. The research will study the interaction between risks (during different scenarios), regulations that could help manage these risks, proper business protocols (between partners) to enhance risk identification and control and government policies that provide incentives and tools to support a collaborative means to address these risks. Government and other stakeholders can then use this map for developing/ enhancing code, regulations and public policies.

References

1. Ergazakis, K., Metaxiotis, K., and Psarras, J. (2004). "Towards knowledge cities: conceptual analysis and success stories", J. of Knowledge Management, Vol. 8, No.5.
2. LandXML. http://www.landxml.org Accessed July 2005
3. SDSFIE. Spatial Data Standard for facilities, infrastructure, & environment – Data Model & Structure, U.S. Army CADD/GIS Technology Center, 2002
4. MultiSpeak. http://www.multispeak.org/whatisit.php Accessed July 2005
5. Peachavanish, R., Karimi, H. A., Akinci, B. and Boukamp, F. "An ontological engineering approach for integrating CAD and GIS in support of infrastructure management", Advanced Engineering Informatics, Vol. 20, No 1, 2006, pages 71-88.
6. CII-Construction Industry Institute. (1997). "Pre-Project Planning Handbook," University of Texas at Austin.
7. Gu, N., Xu, J., Wu, X., Yang J. and Ye, W., "Ontology based semantic conflicts resolution in collaborative editing of design documents", Advanced Engineering Informatics, Vol. 19, No 2, 2005, pages 103-112.
8. Rouane, M., Petko V., Houari S., and Marianne H. Merging Conceptual Hierarchies Using Concept Lattices. MechAnisms for SPEcialization, Generalization and inHerItance Workshop (MASPEGHI) at ECOOP 2004, Oslo, Norway, June 15, 2004.

Formalizing Construction Knowledge for Concurrent Performance-Based Design

Martin Fischer

Department of Civil and Environmental Engineering, Terman Engineering Center,
380 Panama Mall,Stanford, CA 94305-4020, USA
fischer@stanford.edu

Abstract. The capability to represent design solutions with product models has increased significantly in recent years. Correspondingly the formalization of design methods has progressed for several traditional design disciplines, making the multi-disciplinary design process increasingly performance and computer-based. A similar formalization of construction concepts is needed so that construction professionals can participate as a discipline contributing to the model-based design of a facility and its development processes and organization. This paper presents research that aims at formalizing construction concepts to make them self-aware in the context of virtual computer models of facilities and their construction schedules and organizations. It also describes a research method that has been developed at the Center for Integrated Facility Engineering at Stanford University to address the challenge of carrying out scientifically sound research in a project-based industry like construction.

1 Introduction

Virtual Design and Construction (VDC) methods are enabling project teams to consider more design versions from more perspectives than possible with purely human and process-based integration methods [1]. Advancements in product modeling (or building information modeling (BIM)) methods [2], [3], [4], information exchange standards [5], [6], [7], and formalizations of discipline-specific analysis methods [8], [9], [10], [11] now allow many different disciplines (e.g., structural and mechanical engineers) to have their concerns included in the early phases of a project [12], [13]. As a consequence, performance-based design supported by product models is becoming state-of-the-art practice [1] (Fig. 1). The number of performance criteria that can be analyzed from product models continues to increase and now include some architectural, many structural, mechanical (energy), acoustical, lighting, and other concerns. These VDC methods are enabling multi-disciplinary design teams to consider more performance criteria from more disciplines and life-cycle phases than possible with traditional, document-based practice. They contribute greatly towards better coordinated designs [14] and to creating Pareto-optimal designs [15] that are typically more sustainable than designs created by the traditional design process that involves design disciplines sequentially. In most cases several related product models form the basis of this performance-based design [16], [17]. These models also support the reuse of knowledge from project to project [18].

I.F.C. Smith (Ed.): EG-ICE 2006, LNAI 4200, pp. 186–205, 2006.
© Springer-Verlag Berlin Heidelberg 2006

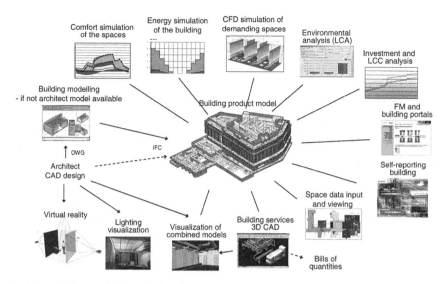

Fig. 1. Tools for analysis and visualization integrated through shared product models are emerging as cornerstones of integrated, performance-based, life-cycle focused facility design. The figure illustrates the current capabilities and offerings of mechanical design firm Granlund in Helsinki, Finland, Figure from [1].

A promise of virtual design and construction is that not only the traditional design disciplines, but also downstream disciplines (e.g., construction) can contribute to improve the design of a facility in a timely and effective manner. It supports an expansion of the concept of performance-based design from a traditional focus on the physical form of a facility and its predicted behaviors during facility operations (e.g., the performance of the structural system during an earthquake) to the concurrent design of a project's product (i.e., the facility itself) and the organizations and processes that define, make, and use it. The construction perspective is an important perspective to consider in this expanded performance concept of facility design. It considers the constructibility and therefore the economy in monetary, environmental, and social costs of a particular facility design and includes the performance-based design of the virtual and physical construction processes in the context of the facility's lifecycle.

However, construction knowledge has not yet been formalized to the extent necessary to consider construction input explicitly in the information models and systems used to represent and analyze the concerns of the various design disciplines in practice. Furthermore, a conceptual limitation of the modeling and analysis approaches used for the concerns of traditional design disciplines is that the underlying representation is typically a 3D product model. However, the explicit consideration of construction concerns in a performance-based design process requires not only the formalization of a wide range of construction knowledge to support computer-based analyses of productivity, safety, workflow, and other concerns, but also the addition of the time dimension to the 3D product model, since the time dimension is a critical factor in the consideration of construction concerns early in the design of a facility.

This paper addresses the large-scale integration problem of incorporating performance-based construction concerns in the design of facilities. It presents past, recent, and ongoing research efforts at the Center for Integrated Facility Engineering (CIFE) at Stanford University and elsewhere that focus on formalizing construction knowledge in support of performance-based facility design. It discusses the underlying representation and reasoning methods needed to incorporate construction concerns in intelligent building models. The paper also discusses the 'horseshoe' research method CIFE has developed to formalize experiential knowledge into model-based methods. The paper concludes with a vision for the use of model-based methods and organizational implications to incorporate construction concerns in the design of facilities and throughout a facility's lifecycle.

2 Construction as a Very Large-Scale Integration Problem

Construction is a critical part of the life-cycle of facilities and needs to be addressed as a very large scale integration problem as the scope and awareness of global concerns become focused on individual projects. Each project combines concerns and information from professional and other project stakeholders, lifecycle project phases, and economic, environmental, and social contexts in unique ways that need to be integrated for its successful realization. For example, the selection of a particular structural system and material for a building impacts construction costs and duration, use of materials and other resources, CO_2 emissions, performance of the building during natural and man-made hazards, the flexibility of the building to adapt to evolving uses, etc. Today's engineering methods and software enable project teams to optimize the performance of a facility for individual disciplines, but methods that address this large-scale integration problem in practice and facilitate the identification of Pareto-optimal solutions to construction problems are still in the research phase [15].

2.1 Overview of Approaches to Incorporate Construction Concerns into Facility Design

Approaches to incorporate construction concerns into facility design include human and process-based methods and automated methods that are based on a formal representation of construction knowledge, the facility design, and mechanisms to learn and update the construction knowledge.

Human and process-based methods. In today's practice constructibility input to design is provided with manual social processes, i.e., by bringing construction and design professionals together (Fig. 2), typically by involving construction professionals in the design phase [19]. Researchers and practitioners have explored how to improve constructibility for several decades [20], [21], [22], [23], [24], [25], including formal constructibility improvement tools [26]. Constructibility programs have gained acceptance for many types of projects [27]. A shortcoming of these human and process-based methods is that it is difficult for professionals to address the large-scale integration challenges of enhancing the constructibility of facility designs in the context of their lifecycle with the time, budget, and stakeholder attention available in the early project phases. To help overcome this difficulty, constructibility knowledge has been organized according to levels of detail of design decisions and the timing of constructiblity input [28], [29].

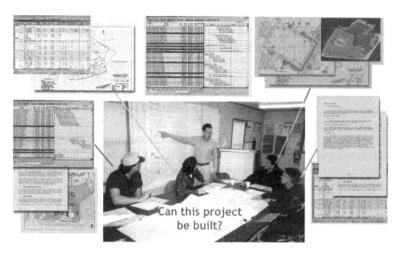

Fig. 2. Social integration process widely used in today's practice to improve the constructibility of projects. Professionals need to understand a wide range of project information by interpreting documents, make performance predictions in their minds, and share the predictions verbally and with sketches with the other professionals.

Social process supported by 4D visualizations. A first step towards larger scale integration is for the many professional and non-professional stakeholders to see what concerns others have during the lifecycle of a facility and understand the impacts of these concerns on the project design. Researchers and practitioners have developed 4D (3D plus time) and other visualization methods to visualize the planned construction process in the context of a facility's 3D model [30], [31], [32], [33], [34], [35]. In some cases, 4D visualizations have been related to discrete event simulations of the construction process [36]. 4D visualizations have proven cost-effective in practice [37] and find increasingly beneficial applications [38]. 4D visualizations support the constructibility reasoning of professionals, enhance the communication of construction information to project stakeholders, and support the collaborative development of more constructible design [39]. 4D visualizations enable project teams to understand and integrate more facility design and construction concerns more quickly and comprehensively than possible with today's engineering methods. They have proven effective in improving the economic (e.g., less rework and fewer design change orders), environmental (e.g., fewer wasted resources), and social performance (e.g., safer and more meaningful jobs on construction sites) of construction and reconstruction projects. Their usefulness depends, however, largely on the timing of the use of 4D visualizations and on the construction expertise of the participating professionals. The combination of lean construction methods and 4D visualizations promises to improve the timely use of 4D visualizations at the appropriate level of detail [40], [41].

Automated methods. To make the incorporation of construction input to design more timely (typically earlier), economical, and consistent, researchers have proposed several types of knowledge-based systems. Expert systems formalize constructibility knowledge [42], [43], [44], but these systems are cumbersome to apply if they are not explicitly related to a computer-based representation of the facility design that is shared with the design team. Therefore, constructibility review and construction planning systems have been developed that are based on a product model representing the facility design [45], [46], [47], [48]. To update and maintain this construction knowledge base automated learning methods have been proposed [49], [50] and methods to infer construction status and knowledge from documents are being developed [51].

The fundamental challenge for the development of these automated constructibility review and improvement methods is to make the various virtual construction elements self-aware in the context of other virtual construction elements, elements of design solutions, and other lifecycle concerns [52], [53]. The challenge is to find the right abstractions that support general, project-independent methods, but support construction professionals to find project-specific solutions quickly. In addition to formal product models [54], formal process models are needed to support the temporal aspects of construction knowledge [55], [56], [57], [58]. These abstractions need to address concerns arising at the operational or trade level of construction [59] and the strategic project management level [60]. Ontologies and hierarchical product and process models provide the underlying methodologies to formalize construction knowledge in a project-independent (i.e., general) way to support the powerful (rapid, consistent, widespread) application of this knowledge for a specific project [61], [62]. Construction knowledge for sub-domains has already been formalized, e.g., for steel construction [63], for reinforcement [64], and for concrete [28], [47] and for general constructibility concerns like tolerances [65].

2.2 Research: Self-aware Elements for Large-Scale Integration

Using these formal knowledge representation methods, a next, longer-term step towards large scale integration is for each element to "see" what affects its design and behavior. An "element" can be a *physical item* like a wall, a *process* like an activity, or an *organizational actor* like a company.

For example, a self-aware scaffold would recognize when the facility design has changed and check whether its design needs to change, or an activity in such a model would recognize when its sequence relationships to other activities have changed and compute the impact of the revised activity sequence on its production rate. Note that these self-aware elements are aware of what affects their own design and behavior, but do not need to be aware of the impacts changes in their design and parameters have on other elements. For example, the self-aware scaffold is not aware of the schedule impact of a change to its design. It knows only about when its design works in the context of the facility design and schedule. The self-aware schedule would compute the schedule impact of a change in the scaffold design. The self-aware activity knows how its production rate is affected by, among other things, the activity sequence it is part of, but does not know the overall cost impact of the change in production rate. A self-aware cost element would figure out the cost impact.

It is important that construction and facility elements are made self-aware in this manner, i.e., each element knows what affects its design and not what effect the design of a particular element has on other elements, to enable the flexible use of these elements for facility design and to make the maintenance of the knowledge encapsulated in these elements manageable. The knowledge encapsulated in self-aware elements that focus on the computation of the impact of their design on other project elements would be difficult to maintain since the knowledge base needed would typically come from many disciplines and the nature and magnitude of the impact cannot always be predicted a priori. For example, the cost impact of a sequence change may depend on other aspects of a construction schedule, e.g., access conditions to the site, which the activity cannot know about, but a cost element could include in its knowledge base. In my experience it becomes quickly an intractable problem to maintain, e.g., the knowledge about the possible impacts of a change in activity sequence because there are all kinds of conditions that affect the types and magnitudes of the impacts of such a change on other project elements and their performance in the context of the overall design of a facility and its organizations and processes.

To the extent to which self-aware "virtual elements" can be formalized and implemented as computational models and methods, the resulting computer model of the design of a project becomes intelligent and can actively support the concurrent efforts of the various construction disciplines (architects, structural engineers, builders, etc.) to integrate their concerns and information with everyone else's concerns and information. Such self-aware elements would also enable a pull-driven method for design, which should be more productive than the prevalent current push-driven design methods. For example, a construction activity that knows what building elements it is building and that knows what resources it consumes can react automatically to changes in the design of its building elements or to changed resource availability. It can automatically adjust its duration, its timing, its relationship to other activities, etc. and make this updated information available for other analyses, which can then be carried out when they are needed. In contrast, a push-driven design method would calculate the impacts of a design change just in case, regardless of whether a project stakeholder actually needs that information at the time.

Such a self-aware activity can support a construction team much more proactively and quickly with insights into the impact of changes and changed conditions than an activity that can only gain self-awareness through human interpretation. It is challenging, however, to formalize and validate the concepts needed for construction due to the large-scale integration needed and due to the unique nature and context of each project. The challenge is to find the appropriate level of formalization so that the conceptual model is general enough so that it can be applied in a number of situations, but is powerful enough to provide a useful level of intelligence or self-awareness in a specific situation on a construction project.

2.3 Examples of Self-aware Construction Elements

For example, the work in my research group has focused on formalizing the following construction-related concerns. This work is extending the conceptual basis of virtual construction elements to make them more intelligent and self-aware in the context of a project design:

- 'building components' to make them self-aware of their functional relationships to systems of one or several disciplines, e.g., to make a wall aware of its associations with and roles in a building's architectural, structural, and mechanical systems to enable, e.g., the automated checking whether a change in the structural design impacts the architectural function of a space [66], [67]
- subcontractors and their behavior to understand their allocation of resources to various projects so that the impact of schedule changes on the subcontractors' resource allocations can be understood and managed better [68]
- 'construction methods' to support automated (re)planning of projects given a facility design and available resources [69], [70]
- 'construction workspaces' to add them automatically to a given construction schedule to test the schedule for time-space conflicts that cause safety and productivity concerns [71]
- 'cost estimating items' to update the cost of constructing parts of a building automatically as the building design changes [72]
- 'construction activities' to make them self-aware in the context of other activities, the geometric configuration of a facility, and the state of completion of the facility at the timing of the activity [73]
- 'sequence relationships' between construction activities to make them aware of their role in a network of activities so that they can highlight opportunities for rescheduling when the schedule needs to be changed [74]
- 'design tasks' to embed them in a network of design tasks and make them aware of the information they depend on and the methods needed to execute them [75]
- 'design requirements' to relate them to each other, make them visible throughout the project lifecycle, and relate them to design solutions so that client requirements don't get lost or misinterpreted as a project progresses [76]
- 'decisions' to highlight the relationships between design options, decisions, and decision criteria [77].

Ongoing research in this area in my group focuses on formalizing the following elements: 'temporary structures' to understand and optimize their use during construction, 'detailed design specifications' to support the planning and handover of construction work in accordance with the design specifications, 'site workers and other resources' to model their role in the context of construction activities, available design information, and regulations affecting construction work, 'conceptual schedules' to provide continuity of overall schedule goals throughout the construction phase of a project, 'schedule uncertainty and flexibility' to highlight major schedule risks and assess the value of mitigating methods, 'building systems and controls' to check that a building is operated as designed and as built, 'building spaces and major components' to assess the energy performance of a building during early project stages, and 'material degradation' to understand the degradation of materials in the context of a facility's geometric configuration, use, and environment. Other researchers are extending the reach of computer-based models from the design and construction planning phase to the actual construction phase [78] and into facility operations [1], [79].

While many of these self-aware elements are still in the research stage, they will eventually enable project teams to consider many more lifecycle concerns for many more disciplines early in project design and throughout project development and help project teams develop integrated design solutions that perform better for more performance criteria. The result should be a seamless process of sustainable design, construction, and use of facilities. Significant research is still needed to formalize these self-aware construction elements in the context of design solutions of the many disciplines involved in a facility project and the economic, environmental, and social life-cycle context of facilities. Therefore, the second part of this paper describes the research method developed at CIFE in support of such research efforts. The goal of the method is to help researchers achieve research contributions that are scientifically sound and practically relevant and applicable in the experience-based, anecdotally-focused construction industry.

3 Formalizing Construction Knowledge Through Research in Practice: The CIFE Research Method

Formalizing construction knowledge requires three types of research efforts:

1) Research projects exploring new terrain in two different ways:
   (a) in practice, through careful participation or observation of preconstruction, construction planning, and construction work, researchers identify, document, and quantify a particular problem as best as possible, and
   (b) in the lab, through rapid prototyping and using test cases from past projects or text books, researchers determine the technical feasibility of a particular envisioned system or method.
   The two types of explorations often interact, i.e., the identification of a problem in practice might lead to the need for a new method, which can then be developed and tested first in the computer in the lab. Or, the availability of a new method might motivate observations in practice to see whether there really is a problem that would be addressed by the new method.
2) Pilot projects that pilot the use of a new method (often with new software tools or methods developed in a research effort of the first type) on a real project to learn about the value of the new method in practice and to learn about the needed improvements of the new method to address the challenges that engineers and constructors face every day.
3) Research efforts that take methods that have proven themselves in pilot projects to widespread use and develop guidelines for implementation.

   The distinction of research projects into these categories helps set the expectations of the researchers and practitioners and develop a research plan that all agree to.
   The research challenge with formalizing construction knowledge for use in model-based virtual environments as part of the performance-based design of facilities is that – unlike for other disciplines – lab experiments can rarely replicate the situations found in practice where the formal design methods need to apply. Therefore, construction sites and project offices become the research lab. There, it is difficult, however, to isolate a particular factor and study its effect on project outcomes, which makes it hard to formalize model-based design methods. To address this challenge, researchers need to triangulate results from field observations with theory in related

literature with predictions and insights from experts and with descriptions, explanations, or predictions from models developed from the observations, theory, and expert opinions. To support this research process, we have developed the 'horseshoe' research method at CIFE. The method supports researchers in building on the experiential knowledge and anecdotal evidence that can be gathered on construction sites in the context of existing theory and expert knowledge to carry out practically-relevant and scientifically-sound research.

3.1 Overview of CIFE Research Method

We call the method the 'CIFE horseshoe research method' (Fig. 3). Given the unique combination of large-scale integration challenges on construction projects the method cannot guarantee full repeatability of a research effort by different research teams addressing the same topic, but we have tried to make the method as explicit and replicable as possible given the nature of the domain studied. We have found that students who work with this method progress more quickly to defensible research results and can understand each others' work more easily, quickly, and fully.

While one can enter the steps in the horseshoe diagram showing the research process in Fig. 3 at any step, I will describe the method from the upper left corner around to the lower left corner. Throughout the remainder of Section 3 I will use an example from a recent research project in my group to illustrate the steps of this method – the development of a geometry-based construction process modeling method motivated by our experience in applying 4D models to plan the construction of part of Disney's California Adventure® theme park [73].

3.2 Observed Problem in Practice

In construction, the ultimate 'proof' of the value and soundness of a formalized concept or method is in its application in practice. This can only be explained if the problem that is addressed by a research project is clearly identified, described, and quantified. I have found that it is a significant effort to develop a crisp problem definition that sets up a research effort that may lead to a defensible contribution to knowledge. Typically, the problem is defined too broadly, making it difficult to quantify it, and to test a new method, concept, or system. For example, while often true, the problem "4D modeling is too time-consuming" is too vague and ill-defined. 4D modeling is too large an area for a single research project, and, while hinting at a quantifiable metric, time-consuming is too generic a metric to clearly articulate the problem. It is not clear for whom and for what task 4D modeling is too time-consuming. Are 4D models too time consuming to make for construction schedulers? Are they too time consuming for project engineers to update? Are they too time consuming for non-project stakeholders to understand? It is also not clear what type of 4D modeling is too time-consuming. Is 4D modeling at the master schedule level too time-consuming? Or is 4D modeling at the day-to-day construction level too time-consuming? Finally, the statement does not point out what domain is addressed, i.e., does the problem manifest itself for all types of projects anywhere in the world, or was the problem observed for the 4D modeling of a specific type of structure, for a specific type of stakeholder, in a specific project phase, in a particular area? A research project that starts with a problem definition that is too vague and too broad will probably not yield a productive research process and a strong research result because the criteria for success are not clear.

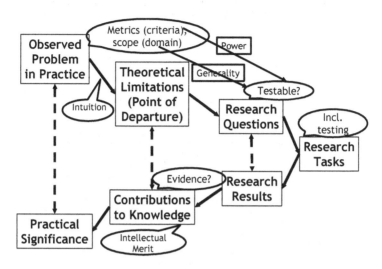

Fig. 3. CIFE 'Horseshoe' Research Process

In my experience, a specific observation of a problem in practice is a better starting point for a research problem statement. For example, the revision of the 4D model of the schedule for the construction of the lagoon in the Paradise Pier portion of Disney's California Adventure® theme park required about two work days per schedule alternative. The reason was that the lagoon was modeled as a single 3D CAD object, but the work to construct the lagoon consisted of four activities (excavate lagoon, place a clay layer for waterproofing, place reinforcement bars, place the concrete liner) for which the scheduler wanted to explore schedules with different 'chunks' of scopes of work, starting points, and workflow (direction of work). The scheduler wanted to understand the work on the lagoon and the relationship of that work and its sequence with the construction work around the lagoon. For example, in one schedule version, the scheduler wanted to break up the four lagoon activities into 6 work areas, in another version into 8 areas, and in another version into 22 areas, etc. For each version, this required about two days of work to remodel the lagoon in 3D with the right work areas, to generate and sequence the activities in the CPM schedule, to link the 3D objects with the activities, to review the 4D model, and to revise the schedule according to the insights gained in reviewing the 4D model. Hence, a problem statement for this problem could read as follows: "The construction scheduler cannot generate the 4D model to plan the construction of the lagoon in Disney's California Adventure® theme park fast enough (<1 to 2 hours) to support the owner's preconstruction project team with the insights about the workflow to build the lagoon and the lagoon work's interrelationships with work around the lagoon" (Fig. 4). While generating a solution to this problem would not be a sufficiently general contribution to knowledge this statement offers many more specific dimensions along which evidence for the existence of the problem can be sought and for which a general solution (approach, method) can be formalized. For example, does this problem manifest itself for all lagoon projects, or for all construction activities whose scope of work can be represented with one or a few 3D CAD objects, but requires many activities to complete?

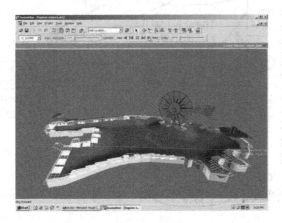

Fig. 4. Snapshot of a parametrically generated 4D model of the lagoon for the example project in this section [73]

My experience is that the precise articulation of a specific problem for a few specific projects is the fastest way to develop a precise problem statement that allows the researchers to look for relevant points of departure, formulate a sound research method and plan, and test their results.

3.3 Intuition

This is the least formalized step in the CIFE horseshoe research process. However, typically, an intuition is needed about how the problem could be addressed in a general, i.e., project-independent way. For example, for the research that eventually resulted from the lagoon case, the researcher's intuition was that discrete event simulation methods combined with geometric transformation mechanisms that could automatically generate the appropriate level of detail in the 3D model to match the desired level of detail in the schedule combined with a formalization of scheduling knowledge might yield a novel approach to 4D modeling that would solve the identified problem.

3.4 Points of Departure (Theoretical Limitations)

A well-defined problem statement and a clearly articulated intuition make it clear in what areas the researcher needs to look for the theoretical starting point for the research. The researcher needs to review the literature for three main reasons:

1) To complement the researcher's observations of practice and find further evidence for the existence, generality, and quantification of the problem. For the example research effort, this included reviewing the literature to find case studies by researchers and practitioners that identified the same problem.

2) To identify existing theory that is useful in addressing the problem. This is theory the researcher plans to use for the research, but that does not need extensions to make it useful to address the identified problem. For the example, this included, e.g., discrete event simulation, a method that the researcher planned to use, but not extend or contribute to in any fundamental way.

3) To identify existing theory that is partially useful in addressing the problem, but needs extensions to make it truly useful to help practitioners address the problem. The scientific contribution of a research effort needs to be shown in the context of existing theory that was insufficient, i.e., needed to be modified, to address the problem. For the example, this included geometric transformation algorithms from computer graphics that needed to be modified and tested for this problem.

3.5 Research Question(s)

Only one good question is needed to make a research effort worthwhile. In my experience, it also takes significant effort to develop a good research question or questions. Typically, initial research questions are too broad and the researchers have too many questions to make a successful research project possible in a reasonable timeframe. A key criterion for a good research question is whether it is testable or not. I.e., as soon as a research question is formulated the researcher should think about how evidence for the generality and the power of an answer (solution) to the question could be found or developed. Hence, it is critical that the research question(s) relate directly to the domain established in the research problem statement (to set up a future claim for the generality of the formalized solution) and to the metrics used to quantify the problem (to set up a future claim for the power of the formalized solution). Questions that can be answered with a 'yes' or 'no' answer are typically not good research questions. Many research questions start with 'what' or 'how'. A further challenge in formulating a research question is that it must be possible to know when one has answered the question, i.e., when the research is done. Any research project will create the foundation for new research, of course, but it is vitally important that a specific scope of work is identified in the question(s) for a particular research project. Finally, a research question needs to be formulated in way that makes any finding to the question an interesting answer. If not, the researcher might set himself up for failure or might anticipate (hope for) a specific result, which will likely cloud the researcher's objectivity and make the research biased towards a specific outcome, i.e., the researcher may look for, and therefore find, evidence for particular phenomena.

3.6 Research Tasks, Including Research Method and Plan

Depending on the research problem, the research tasks may include further literature study, interviews, surveys, case studies, observations of practice, participation in ongoing construction projects, ontology development, implementation of software prototypes, etc. to develop the research results and contributions to theory. An important consideration for research planning is the number of test cases (or interview or survey subjects) that will be needed to be able to argue for the generality of the research result(s). Testing of the research results is typically the most critical research task, and the researcher should consider and refine the test plan often during the research. A good test plan needs to be transparent, i.e., it needs to be plausible that another researcher would have come to the same conclusions if she had done the same test. Common test methods for the formalization of knowledge and methods for construction tasks include:

Variation studies on retrospective cases. Essentially all research projects that formalize new concepts and methods carry out validation studies based on retrospective or past cases. These studies are typically done in the computer in the lab and include varying various input parameters to test that the formalized concepts and implemented software prototype perform in a technically sound way.

Asking an expert panel. To validate the results of a research project, the researcher could show the results to experts and ask them whether the approach and results make sense and the important concepts in a domain are covered adequately. A better method is to give the experts several problems that can be addressed (solved) with the research results and ask them for their predictions of the solutions to these problems, then apply the method developed from the research results and compare the expert predictions with the predictions of the new method [80].

Charrette tests. The researcher develops a test that is representative of the engineering task(s) the research is trying to improve, but that practitioners or other researchers can carry out in an hour or two. Typically, a charrette test contrasts the performance of a traditional or typical method of performing the tasks with the method enabled by the concepts formalized in the research [81]. Ideally, a charrette test compares the output and process performance of the two methods and therefore sheds light on the impact newly developed concepts have on the execution of engineering tasks. The advantage of a charrette test is that it allows researchers to isolate certain factors that are difficult to isolate in practice. The disadvantage is that tasks rarely happen in practice in the way in which they are conceptualized in the charrette test. Nevertheless, charrette tests can provide excellent evidence for the power and generality of the research contributions because the researcher can design the tests so that the metrics for power identified in the problem statement are addressed and so that the generality of the research results can be tested (e.g., for how many project phases, stakeholders, building types, levels of detail, etc. a contribution applies). A difficult aspect of charrette tests is that a 'gold standard' (i.e., the correct output of a task or right way for doing a task) for test case tasks needs to be established. In some cases, this can only be done by consulting experts and taking their opinion as the gold standard. In other cases, a gold standard, i.e., the correct result of the test case tasks, can be calculated. It is usually quite easy to study process performance differences between the two methods with a charrette test. Typical process performance metrics include the durations to complete the tasks (or the number of tasks completed in a certain timeframe), the number of issues, concerns, or criteria that could be considered, the number of alternatives that could be developed, the number of stakeholders that could be included, the consistency of these process performances across different professionals. A challenge with the charrette test is that some training is required in the new method and that it is difficult to administer enough, but not too much training in the new method, so that the results of the test are not biased one way or the other. Another challenge is that to compare the performance of the two methods the same testers should be used, but once the testers have completed the tasks with one method they are likely to perform better the second time, even with a different method. Hence, the testing sequence and time between tests need to be designed carefully to minimize these effects.

Prospective or intervention case studies. The researcher (or someone else) applies the formalized concepts to a real, ongoing construction project and, if necessary, suggests changes to the project plan or design based on the insights from the application of the newly developed concepts. The project manager then implements or rejects the suggestions and the project or process outcome is observed. While anecdotal in nature, prospective cases demonstrate the value of formalized concepts vividly, and it is usually seen as convincing evidence of the power of the research contributions if the concepts work at the scale and under the time and organizational pressures of an ongoing project and if seasoned practitioners pay attention to the results generated from the application of the new concepts to a situation in practice [77].

For the case example research project, the researcher conducted – among other test cases – retrospective test cases using the construction of parts other than the lagoon of Disney's California Adventure®, such as the construction planning of the Seafood Restaurant. The resulting schedules and 4D models were then reviewed by a construction expert from Disney's project team that had built that part of the park. The expert assessed whether the resulting schedule was realistic and considered the important constraints. Additional test criteria included, e.g., comparisons of the time needed to generate a detailed construction schedule and 4D model from a 3D model and a Master schedule using the geometric construction process modeling method (implemented as a computer prototype) with the manual approach used on the actual construction project.

3.7 Research Results

The research results need to answer the research questions, of course. Most importantly, the results include evidence from the tests for the generality and power of the research contributions using the metrics from the problem definition. For the example research project, the results included the specification of the geometric construction process modeling method, the implementation of a computer prototype based on the method, and the results and insights from the tests.

3.8 Contributions to Knowledge

The research contributions, i.e., the tested and validated results of the research, extend prior work and contribute new concepts to theory. They become a new foundation for further research and improved practice. The main contribution to knowledge of the case example research is the formalization of a geometric construction process modeling method. The method has become a basis to study additional construction scheduling topics, such as studying the tradeoff between schedule uncertainty and flexibility and the planning of temporary structures.

3.9 Practical Significance

The research method asks researchers to address the practical significance of the planned or completed research. The practical significance or impact must be explained with the same metrics used to define the problem. Even though Fig. 3 shows this step as the last step, research projects often start here, i.e., with a vision for the desired impact of the research or a vision of how engineers should be able to carry a particular task. The remaining steps of the research method are then tailored to support the vision. The practical significance of the case example research project lies, e.g., in allowing construction engineers to base their work and their analyses on the same 3D models and project schedule information as other disciplines. They can, in this way, participate more effectively in the concurrent, performance-based design of facilities.

3.10 Discussion of Research Method

The research method works well for the formalization of new construction (and other) knowledge and methods to embed the newly formalized knowledge in facility definition, design, construction, commissioning, and operations phases. It forces the researcher to develop a scope of work and research plan that is manageable and executable and leads to scientifically defensible and practically relevant results. The researcher should, however, in a brief section embed the particular research effort in the larger picture of theory and practice surrounding the research topic. For the example project, this included a short discussion of concurrent engineering of the facility and its construction schedule at different levels of detail.

This research method works best when it is used in all phases of a research project, i.e., to define and select the focus of a research effort, to design a particular research effort, to manage it, and to report on it. In all phases, the researcher should advance the thinking on all steps as far as possible and proceed to the next research task based on the 'maximum anxiety principle' [82], i.e., tackle the task that has the greatest uncertainty or risk or lack of definition. I have found that a commonly used research method like the method presented in this paper is particularly important for construction research that has still a young research tradition.

For research efforts that aim mainly at deepening the understanding of current practice, in-depth case studies [83] are usually more appropriate than the presented method; see [84] for an example of the application of this type of research method.

4 Summary and Conclusions

In today's practice, the large-scale integration of design, construction, and other facility lifecycle concerns happens largely in the minds of engineering professionals and may therefore be slow, incomplete, inconsistent, and error-prone. Opportunities for improved lifecycle performance (e.g., lower use of natural and human resources for construction and operations) are often missed. Formalizing and integrating construction concerns into the facility design have proven particularly difficult. While it will be some time before construction project teams will be able to consider all the economic, environmental, and social concerns for all disciplines and stakeholders for all parts of a construction project throughout all the phases, it is important that better visualization, integration, and automation methods become available for engineering practice to improve the lifecycle performance of facilities. For the foreseeable future, these integration methods will blend formal computational models with visualizations and human cognition to leverage the expertise of humans and take advantage of existing and emerging computational modeling, simulation, and visualization tools. This paper has summarized underlying methods to represent this knowledge and make it available for concurrent engineering efforts of facilities and presented areas where such formalizations have been developed or are being researched.

As more formalizations become available, research will be needed to find mechanisms to integrate them across disciplinary concerns, lifecycle phases, and levels of detail of the facility, its development processes, and organizations. Practitioners who have focused on formalizing their knowledge and the knowledge of their firm for integrated and automated application are already able to capitalize on efficiency gains from the consistent and rapid application of this knowledge base at the right time in a construction project and are seeing opportunities to expand their involvement in earlier and later project phases. For example, the mechanical design firm Granlund has seen the opportunities to participate in the project definition and facility management phases increase significantly with the formalization of its knowledge base and with the use of product models to support the lifecycle of buildings [1]. The price for these opportunities is, however, the allocation of significant resources (about 5 to 10% of its staff in Granlund's case) to research and development activities, i.e., shifting repetitive design work to the development of formal computer-based methods.

Extensions of the presented work includes expanding the joint consideration of economic, environmental, and social performance goals and including the operations and maintenance phase in the continued development of methods to address this very large-scale integration problem. The goal is to make the construction phase much more efficient than it is today and enable students and professionals to understand the large scale integration problem they face when attempting to satisfy the concerns from all the stakeholders and contribute to methods that balance facility lifecycle costs and uses, maintenance costs, expected facility life, security costs, global, regional, and local concerns relative to impacts on energy, water, air, etc. Finally, tools to design appropriate integration mechanisms and methods to innovate on projects are needed to improve how we integrate the many stakeholder and lifecycle concerns from project to project to better their lives and the lives of their peers and children.

Acknowledgements

I am most indebted to my colleague at CIFE, John Kunz, for the many discussions and work sessions that have led to the formalization of the research method presented in this paper. I also thank the CIFE students for their inspiring work and the CIFE members for their support of the environment that has made the presented work possible.

References

1. Hänninen, R.: Building Lifecycle Performance Management and Integrated Design Processes: How to Benefit from Building Information Models and Interoperability in Performance Management. Invited Presentation Watson Seminar Series Stanford Univ. (2006)
2. Eastman, C.: General Purpose Building Description Systems. Computer Aided Design 8(1) (1976) 17–26
3. Bjork, B.C.: Basic structure of a proposed building product model. Computer Aided Design 21(2) (1989) 71–78
4. Eastman, C.M., Siabiris, A.: Generic building product model incorporating building type information. Automation in Construction 3(4) (1995) 283–304
5. Karola, A., Lahtela, H., Hänninen, R., Hitchcock, R., Chen, Q.Y., Dajka, S., Hagstrom, K.: BSPro COM-Server - Interoperability between software tools using industrial foundation classes. Energy and Buildings 34(9) (2002) 901–907
6. Lee, K., Chin, S., Kim, J.: A core system for design information management using industry foundation classes. Computer-Aided Civil and Infrastructure Eng. 18(4) (2003) 286–298
7. Eastman, C., Wang, F., You, S.J., Yang, D.: Deployment of an AEC industry sector product model. Computer Aided Design 37(12) (2005) 1214–1228
8. Rivard, H., Bedard, C., Ha, K.H., Fazio, P.: Shared conceptual model for the building envelope design process. Bldg. & Env. 34(2) (1999) 175–187
9. O'Sullivan, D.T.J., Keane, M.M., Kelliher, D., Hitchcock, R.J.: Improving building operation by tracking performance metrics throughout the building lifecycle (BLC). Energy and Buildings 36(11) (2004) 1075–1090
10. Shea, K., Aish, R., Gourtovaia, M.: Towards integrated performance-driven generative design tools. Automation in Construction 14(2) (2005) 253–264
11. Mora, R., Rivard, H., Bedard, C.: Computer representation to support conceptual structural design within a building architectural context. J. Comput. Civ. Eng. 20(2) (2006) 76–87
12. Howard, H.C., Levitt, R.E., Paulson, B.C., Pohl, J.G., Tatum, C.B.: Computer integration: Reducing fragmentation in AEC industry. J. Comput. Civ. Eng. 3(1) (1989) 18–32
13. Rivard, H., Fenves, S.J.: Representation for conceptual design of buildings. J. Comput. Civ. Eng. 14(3) (2000) 151–159
14. Hegazy, T., Zaneldin, E., Grierson, D.: Improving design coordination for building projects. I: Information model. J. Constr. Eng. & Mgt. 127(4) (2001) 322–329
15. Gero, J.S., Louis, S.J.: Improving Pareto optimal designs using genetic algorithms. Microcomputers in Civ. Eng. 10(4) (1995) 239–47
16. Turk, Z.: Phenomenologial foundations of conceptual product modelling in architecture, engineering and construction. AI in Eng. 15(2) (2001) 83–92
17. Kam C., Fischer M., Hänninen R., Karjalainen A., Laitinen J.: The product model and Fourth Dimension project. ITCon 8 (2003) 137–166
18. Demian, P., Fruchter, R.: Measuring relevance in support of design reuse from archives of building product models. J. Comput. Civ. Eng. 19(2) (2005) 119–136
19. Russell, J.S., Gugel, J.G., Radtke, M.W.: Comparative analysis of three constructibility approaches. J. Constr. Eng. & Mgt. 120(1) (1994) 180–195
20. O'Connor, J.T.: Impacts of Constructibility Improvement. J. Constr. Eng. & Mgt. 111(4) (1985) 404–410
21. Tatum, C.B.: Improving Constructibility During Conceptual Planning. J. Constr. Eng. & Mgt. 113(2) (1987) 191–207

22. Boeke, E.H. Jr.: Design for constructibility. A contractor's view. Concrete Constr. 35(2) (1990) 3p
23. Constructibility and constructibility programs. White paper: J. Constr. Eng. & Mgt. 117(1) (1991) 67–89
24. O'Connor, J.T., Miller, S.J.: Constructibility Programs: Method for Assessment and Benchmarking. J. Performance of Constructed Facilities 8(1) 1994 46–64
25. Glavinich, T.E.: Improving constructibility during design phase. J. Arch. Eng. 1(2) (1995) 73–76
26. Fisher, D.J., Anderson, S.D., Rahman, S.P.: Integrating constructibility tools into constructibility review process. J. Constr. Eng. & Mgt. 126(2) (2000) 89–96
27. Pocock, J.B., Kuennen, S.T., Gambatese, J., Rauschkolb, J.: Constructibility state of practice report. J. Constr. Eng. & Mgt. 132(4) (2006) 373–383
28. Fischer, M., Tatum, C.B.: Characteristics of Design-Relevant Constructibility Knowledge. J. Constr. Eng. & Mgt. 123(3) (1997) 253–260
29. Pulaski, M.H., Horman, M.J.: Organizing constructibility knowledge for design. J. Constr. Eng. & Mgt. 131(8) (2005) 911–919
30. Paulson, B.C.: Interactive Graphics for Simulating Construction Operations. J. Constr. Div. 104(1) (1978) 69–76
31. Cleveland, A.B. Jr.: Real-time animation of construction activities. Constr. Congr. I - Excellence in the Constructed Project (1989) 238–243
32. Retik, A., Warszawski, A., Banai, A.: Use of computer graphics as a scheduling tool. Bldg. & Env. 25(2) (1990) 133–142
33. Fischer, M., Liston, K., Schwegler, B.R.: Interactive 4D Project Management System. 2nd Civ. Eng. Conf. in the Asian Region (2001) 367–372
34. Fischer, M., Haymaker, J., Liston, K.: Benefits of 3D and 4D Models for Facility Managers and AEC Service Providers. 4D CAD and Visualization in Construction - Developments and Applications Issa, R.R.A., Flood, I., O'Brien, W. (eds.) Balkema (2003) 1–32
35. Chau, K.W., Anson, M., Zhang, J.P.: Four-dimensional visualization of construction scheduling and site utilization. J. Constr. Eng. & Mgt. 130(4) (2004) 598–60
36. Kamat, V.R., Martinez, J.C.: Comparison of simulation-driven construction operations visualization and 4D CAD. Winter Simulation Conf. 2 (2002) 1765–1770
37. Haymaker, J., Fischer, M.: 4D Modeling on the Walt Disney Concert Hall. tec21 38 (2001) 7–12
38. Fischer, M.: The Benefits of Virtual Building Tools. Civ. Eng. 73(8) (2003) 60–67
39. Liston, K., Fischer, M., Winograd, T.: Focused Sharing of Information for Multidisciplinary Decision Making by Project Teams. ITCon 6 (2001) 69–81
40. Khanzode, A., Fischer, M., and Reed, D.: Case Study of the Implementation of the Lean Project Delivery System (LPDS) using Virtual Building Technologies on a large Healthcare Project. 13th Annual Conf. of the Int. Group for Lean Constr. (2005) 153–160
41. Rischmoller, L., Alarcon, L.F., Koskela, L.: Improving value generation in the design process of industrial projects using CAVT. J. Mgt. in Eng. 22(2) (2006) 52–60
42. Hendrickson, C., Zozaya-Gorostiza, C., Rehak, D., Baracco-Miller, E., Lim, P.: Expert System for Construction Planning. Comput. Civ. Eng. l(4) (1987) 253–269
43. Fisher, D.J., Rajan, N.: Automated constructibility analysis of work-zone traffic-control planning. J. Constr. Eng. & Mgt. 122(1) (1996) 36–43
44. Poon, J.: Development of an expert system modelling the construction process. J. Constr. Research 5(1) (2004) 125–138
45. Darwiche, A., Levitt, R., Hayes-Roth, B.: OARPLAN: Generating Project Plans by Reasoning about Objects, Actions and Resources. AI EDAM, 2(3) (1988) 169–181

46. Cherneff, J., Logcher, R., Sriram, D.: Integrating CAD with construction-schedule genera-
    tion. J. Comput. Civ. Eng. 5(1) (1991) 64–84
47. Fischer, M.A.: Automating Constructibility Reasoning with a Geometrical and Topologi-
    cal Project Model. Comput. Syst. in Eng. 4(2-3) (1993) 179–192.
48. Chevallier, N., Russell, A.D.: Automated schedule generation. Canad. J. Civ. Eng. 25(6)
    (1998) 1059–1077
49. Nakasuka, S., Yoshida, T.: Dynamic scheduling system utilizing machine learning as a
    knowledge acquisition tool. Int. J. of Production Research 30(2) (1992) 411–431
50. Skibniewski, M., Arciszewski, T., Lueprasert, K.: Constructibility analysis: Machine
    learning approach. J. Comput. Civ. Eng. 2(1) (1997) 8–16
51. Brilakis, I., Soibelman, L., Shinagawa, Y.: Material-based construction site image re-
    trieval. J. Comput. Civ. Eng. 19(4) (2005) 341–355
52. Schmitt, G., Engeli, M., Kurmann, D., Faltings, B., Monier, S.: Multi-agent interaction in a
    complex virtual design environment. AI Communications 9(2) (1996) 74–78
53. Schnellenbach-Held, M., Geibig, O.: Intelligent agents in civil engineering. Int. Conf. on
    Comput. Civ. Eng. (2005) 989–998
54. Ito, K.: Utilization of 3-D graphical simulation with object-oriented product model for
    building construction process. Congr. on Comput. in Civ. Eng. (1988) 73–78
55. Froese, T.M., Paulson, B.C. Jr.: Integrating project management systems through shared
    object-oriented project models. Int. Conf. on Applications of AI in Eng. (1992) 69–85
56. Froese, T.: Models of construction process information. J. Comput. Civ. Eng. 10(3) (1996)
    183–193
57. Stuurstraat, N., Tolman, F.: Product modeling approach to building knowledge integration.
    Automation in Construction 8(3) (1999) 269–75
58. Halfawy, M., Froese, T.: Building integrated architecture/engineering/construction sys-
    tems using smart objects: Methodology and implementation. J. Comput. Civ. Eng. 19(2)
    (2005) 172–181
59. Shen, Z., Issa, R.A., O'Brien, W., Flood, I.: A trade construction knowledge module to en-
    able use of design component data in project management. Int. Conf. on Comput. Civ.
    Eng. (2005) 1595–1604
60. Lee, S.H., Pena-Mora, F., Park, M.: Dynamic planning and control methodology for stra-
    tegic and operational construction project management. Automation in Construction 15(1)
    (2006) 84–97
61. Ugwu, O.O., Anumba, C.J., Thorpe, A.: Ontological foundations for agent support in con-
    structibility assessment of steel structures - A case study. Automation in Construction
    14(1) (2005) 99–114
62. Udaipurwala, A.H., Russell, A.D.: Hierarchical clustering for interpretation of spatial con-
    figuration. Constr. Research Congr. - Broadening Perspectives (2005) 1137–1147
63. Anumba, C.J., Baldwin, A.N., Bouchlaghem, D., Prasad, B., Cutting-Decelle, A.F., Dufau,
    J., Mommessin, M.: Integrating concurrent engineering concepts in a steelwork construc-
    tion project. Conc. Eng. Research & Applications 8(3) (2000) 199–212
64. Navon, R., Shapira, A., Shechori, Y.: Automated rebar constructibility diagnosis. J.
    Constr. Eng. & Mgt. 126(5) (2000) 389–397
65. Milberg, C., Tommelein, I.: Role of Tolerances and Process Capability Data In Product
    and Process Design Integration. Contruction Research Congr. - Winds of Change: Integra-
    tion and Innovation in Construction (2003) 795–802
66. Clayton, M.J., Kunz, J.C., Fischer, M.A.: Rapid Conceptual Design Evaluation Using a
    Virtual Product Model. Eng. Applications of AI 9(4) (1996) 439–451

67. Clayton, M.J., Teicholz, P., Fischer, M., and Kunz, J.: Virtual components consisting of form, function, and behavior. Automation in Construction 8 (1999) 351–367
68. O'Brien, W.J.: Capacity Costing Approaches for Construction Supply-Chain Management. Ph.D. Thesis, Stanford Univ. (1998)
69. Fischer, M.A., Aalami, F.: Scheduling with Computer-Interpretable Construction Method Models. J. Constr. Eng. & Mgt. 122(4) (1996) 337–347
70. Luiten, G.T., Tolman, F., Fischer, M.A.: Project-modelling in AEC to integrate design and construction. Computers in Industry 35(1) (1998) 13–29
71. Akinci, B., Fischer, M., Kunz, J.: Automated Generation of Work Spaces Required by Construction Activities. J. Constr. Eng. & Mgt. 128(4) (2002) 306–315
72. Staub-French, S., Fischer, M., Kunz, J., and Paulson, B.: A generic feature-driven activity-based cost estimation process. Advanced Eng. Informatics 17(1) (2003) 23–39
73. Akbas, R.: Geometry Based Modeling and Simulation of Construction Processes. Ph.D. Thesis, Stanford Univ. (2003)
74. Koo, B.: Formalizing Construction Sequencing Constraints for the Rapid Generation of Scheduling Alternatives. Ph.D. Thesis, Stanford Univ. (2003)
75. Haymaker J., Kunz, J., Suter, B., Fischer, M.: Perspectors: composable, reusable reasoning modules to construct an engineering view from other engineering views. Advanced Eng. Informatics 18(1) (2004) 49–67
76. Kiviniemi A.: Product Model Based Requirements Management. Ph.D. Thesis, Stanford Univ. (2005)
77. Kam. C.: Dynamic Decision Breakdown Structure: Ontology, Methodology, and Framework For Information Management In Support Of Decision-Enabling Tasks in the Building Industry. Ph.D. Thesis, Stanford Univ. (2005)
78. Reinhardt, J., Akinci, B., Garrett, J.H.: Navigational models for computer supported project Management tasks on construction sites. J. Comput. Civ. Eng. 18(4) (2004) 281–290
79. Hammad, A., Garrett, J.H. Jr., Karimi, H.A.: Mobile Infrastructure Management Support System Considering Location and Task Awareness. Towards a Vision for Information Technology in Civ. Eng. Conf. (2003) 157–166
80. Christiansen, T.: Modeling Efficiency and Effectiveness of Coordination in Engineering Design Teams. Ph.D. Thesis, Stanford Univ. (1993)
81. Clayton, J., Fischer, M., Teicholz, P., Kunz, J.: The Charrette Testing Method for CAD Research, Applied Research in Architecture and Planning 2 Hershberger, R., Kihl, M. (eds.) (1996) 83–91
82. Kunz, J.: Concurrent Knowledge Systems Engineering. Working Paper 5, CIFE, Stanford Univ. (1989)
83. Yin R.: Applications of Case Study Research. 2nd Ed. Sage Publications (1994)
84. Hampson K.: Technology Strategy and Competitive Performance: A Study of Bridge Construction. Ph.D. Thesis, Stanford Univ. (1993)

Next Generation Artificial Neural Networks and Their Application to Civil Engineering

Ian Flood

Rinker School, College of Design Construction and Planning, University of Florida,
Gainesville, FL 32611, USA
flood@ufl.edu

Abstract. The aims of this paper are: to stimulate interest within the civil engineering research community for developing the next generation of applied artificial neural networks; to identify what the next generation of devices needs to achieve, and; to provide direction in terms of how their development may proceed. An analysis of the current situation indicates that progress in the development of this technology has largely stagnated. Suggestions are made for achieving the above goals based on the use of genetic algorithms and related techniques. It is noted that this approach will require the design of some very sophisticated genetic coding mechanisms in order to develop the required higher-order network structures, and may utilize development mechanisms observed in nature such as growth, self-organization, and multi-stage objective functions. The capabilities of such an approach and the way in which they can be achieved are explored in reference to the truck weigh-in-motion problem.

1 Introduction

Civil engineering, as with many disciplines, is fraught with problems that have defied solution using conventional computational techniques, but can often be solved by people with appropriate training and expertise. Examples include determining legal compliance of designs from drawings and specifications; identifying constructability problems from the design of a building; and measuring construction progress from site images. Automated methods of performing such tasks would help reduce design and construction costs, and improve or validate the efficacy of design and construction decision making. Classical artificial intelligence has targeted this class of problems by attempting to capture the essence of human cognition at the highest level, although progress has been frustratingly slow. This disappointment helped revive interest in computational devices that emulate the operation of the brain at the neuronal level (albeit in a highly abstract form) with the intent of achieving higher level cognitive skills as an emergent property.

Indeed, within the civil engineering discipline, artificial neural networks appear from publications statistics to be one of the great successes of computing. In the ASCE Journal of Computing, for example, over 12% of papers published from 1995 to 2005 (54 out of 445) have used the term "neural" as part of their title [1]. Furthermore, the distribution of these publications by year (see Figure 1) indicates that there

I.F.C. Smith (Ed.): EG-ICE 2006, LNAI 4200, pp. 206–221, 2006.

has been no decline in interest over this period. The citations indices similarly confirm the popularity of artificial neural networks: for example, according to the ISI Web of Knowledge [18] and summarized in Table 1, three of the top five most frequently cited articles from all issues of the ASCE Journal of Computing are on artificial neural networks, including the first and second placed articles in this ranking. This enthusiasm and the diversity of applications reported across all fields of civil engineering make this technology difficult to ignore.

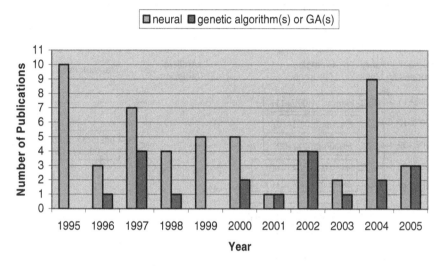

Fig. 1. Distribution of articles in the Journal of Computing using the terms: (1) "neural" and (2) "genetic algorithm(s)" or "GA(s)" in their title [1].

Table 1. The five most frequently cited articles in the Journal of Computing in Civil Engineering [18]

Article Title:	Number of Citations:
Neural Networks in Civil Engineering, Parts I and II [5]	131
Neural Networks for River Flow Prediction [11]	97
Genetic Algorithms in Pipeline Optimization [9]	74
Genetic Algorithms in Discrete Optimization of Steel Truss Roofs [13]	37
Damage Detection in Structures Based on Feature-Sensitive Neural Networks [17]	35

However, an analysis of the content of reported applications shows that progress in the development of applied artificial neural networks has not moved forward significantly since the earliest applications of the late 1980's and early 1990's. These networks were mostly simple vector mapping devices used for function modeling and pattern classification, the types of problem that are typically solved using methods such as multi-variate regression analysis. Although many new applications have been made in subsequent years (along with several improvements to the technique) these have still been mostly simple vector mapping problems. Clearly this is a long way from the goal of a computational device capable of emulating higher-level cognitive processes.

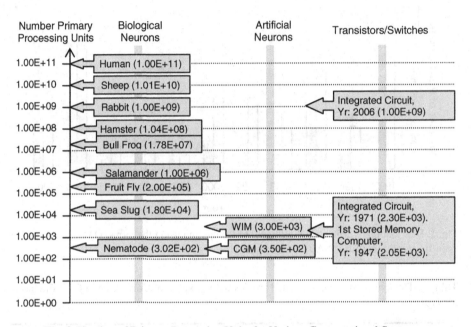

Fig. 2. Number of Primary Processing Units for Various Computational Systems

A comparison of progress with the most popular computational model, the general purpose electronic digital computer, reinforces this view that progress in the rate of development of artificial neural networks has been slow. The initial development of artificial neural networks dates back to the mid 1950's [15] whereas electronic digital computing is only about a decade or so older with the first stored memory device (the Small Scale Experimental Machine - SSEM) first operating around 1948. Given this, it might be expected that the two technologies would have reached a similar state of development. However, since its inception, the electronic digital computer has evolved steadily from a device comprising just a couple of thousand primary processing units (switches or transistors) into one comprising billions organized into a sophisticated structure of higher-order functional subsystems. Artificial neural networks, on the other hand, have failed to advance beyond simple applications that require rarely more than a few hundred primary processing units (neurons in this case) arranged with almost no higher-order structuring.

This point is illustrated in Figure 2 which compares complexity for various computational systems, including digital computers, artificial neural networks, and the brains of various animals. The chart measures complexity in the simple terms of the number of primary processing units that can be usefully employed by the system, and is scaled logarithmically with a range from around 300 units to 100 billion units. Using this simple measure, we can compare today's most complex integrated circuit to the brain of a rabbit (each comprising in the order of 10^9 primary processing units), while artificial neural networks have progressed little further than the brain of the nematode (comprising in the order of just a few hundred primary processing units). In fact, the vast majority of artificial neural network applications in civil engineering employ no more than a few tens of neurons. While there are examples of artificial neural networks in civil engineering that make useful employment of several hundreds of neurons (such as the CGM network for modeling transient-heat flow in buildings [7] or even thousands of neurons (such as the WIM network for truck Weigh-in-Motion [8]), the additional complexity in these devices is there simply to provide greater precision in results, not greater functionality.

Arguably, the number of primary processing units that can be employed usefully in a given system provides an overly simplistic measure of comparative complexity. After all, a neuron in an artificial neural network is usually a much more complicated processing device than a transistor. Likewise, a biological neuron is far more complicated than an artificial neuron. Also, it is likely that significant aspects of the computational mechanisms underlying biological neural networks are yet to be discovered and could be dependent on key processes that operate well below the level of individual neurons (see Bullock [2] for example). That said, the comparison of Figure 2 clearly demonstrates two important and related points: (i) that applied artificial neural networks, unlike digital computers, have failed to advance very far over their history; and (ii) that, according to the biological model, artificial neural networks have a great potential yet to be realized.

An obvious question at this stage is that if development has been so limited in the application of artificial neural networks to civil engineering then why has there been such a high and sustained level of interest in this technology? The answer is that artificial neural networks, notwithstanding their currently rudimentary form, are very good at solving vector mapping problems that are non-linear in form and comprise a fixed set of independent variables, a common class of problems in engineering. In this context, they frequently provide more accurate solutions than the alternative modeling techniques (such as multi-variate nonlinear regression analysis), and do not require the user to have a good understanding of the basic shape of the function being modeled. Still, solving direct vector mapping problems is no more than a primitive first step in the application of artificial neural networks if we dare aspire to the computational capabilities of the brain.

2 Higher Order Structuring

2.1 The Need for Greater Complexity

Not surprisingly, the biological model suggests that an increase in cognitive skills can be achieved by moving towards networks of greater complexity. Brain size alone is a poor indicator of intelligence of a species since larger organisms require more brain capacity for basic monitoring and control of the body; otherwise, we would have to conclude that the Blue Whale is the most intelligent species having a brain mass of 6 kg. The ratio of brain size to body size is also not a particularly accurate indicator of intelligence since the required brain capacity for basic monitoring and control of the body does not increase linearly with body size. This has led to the development of the so called encephalization quotient (EQ) as an indicator of intelligence in a species, being the ratio of the actual brain size of an organism to its expected brain size needed for basic monitoring and control of the body (see Jerison [10] for example). Even this measure can lead to some unexpected results in the ranking of species, and so it has been proposed that a measure of residual brain capacity (such as the difference between actual brain size and expected brain size for a species) is a better indicator of intelligence.

It could be argued that this search for a good indicator of intelligence is biased since we keep refining the method of measurement until we find one that places humans at the top of the scale. However, this should not be a problem here since our goal is to achieve greater human-like intelligence in our computational devices, and therefore we are not so much concerned with ranking as we are with the predictors of rank. Accordingly, these metrics indicate that an increase in cognitive capability requires an increase in the size of the neural network. It also seems that an increase in the number of neurons alone is not sufficient to provide greater cognitive skills, but that the neurons must also be formed into a structured system of higher-order functional units such as is found in the visual system [16].

The barrier in the development of applied artificial neural networks has not been an inability to construct systems comprising very large numbers of neurons. Networks comprising many millions of neurons are certainly feasible with today's technology whether they be implemented in hardware or as software simulations. Rather, the problem, at least in part, is knowing how to organize very large numbers of neurons into appropriate higher-order network structures. (Note, the term "higher-order structures" is used here to refer to the physical organization of a network, and should not be confused with the term "higher-order neural networks" which is commonly used in the literature to refer to devices composed of Sigma-Pi neurons.) The argument of this paper is that there is much that can be done to advance the scope of application and utility of artificial neural networks by focusing research on the development of higher-order structures.

2.2 Forms of Structuring

Higher-order neural network structures are composed of discrete network units each of which perform some sub-function and may contain any number of neurons. The way in which these units are organized defines the higher-order structure of the network and its collective functional behavior. The boundary of a unit may be identified by the mode of operation of the neurons and of their connections and/or by the connectivity of the neurons (the topology of the connections). In this sense, we are identifying patterns in the overall organization of the neural network.

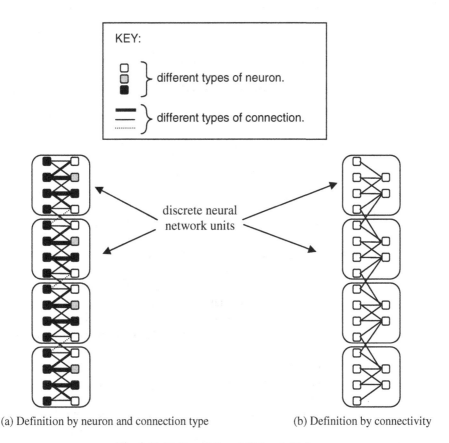

(a) Definition by neuron and connection type   (b) Definition by connectivity

Fig. 3. Definition of Neural Network Units

Figure 3(a) shows, as an example, a double array of neurons of different types and connections that form an underlying repetitive pattern. The repeating elements of the pattern mark the boundaries of the individual units, each of which will presumably perform a similar sub-function. Figure 3(b) shows a similar situation but with the connectivity identifying the boundaries of the elements. In either case, the definition of the boundaries of repeating units can be ambiguous. For example, referring to Figure 3, by moving the boundaries of the units down by two neurons in the arrays we still are able to identify collections of neurons and connections that are repetitive. Furthermore, the boundaries of higher-order units may range from very distinct to vague. This is illustrated in Figure 4 which shows three units defined by their connectivity. Units A and B are clearly demarked by a lack of cross-connections, whereas units B and C are less clearly differentiated since they have many cross-connections. It is also possible that some neurons are equally associated with two or more units, in which case the boundaries of those units would have to be considered to be overlapping.

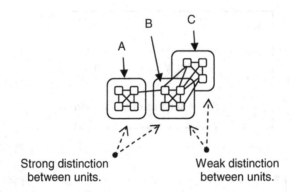

Strong distinction   Weak distinction
between units.   between units.

Fig. 4. Strength of Definition of Higher-Order Units by Connectivity

Three basic forms of higher-order network structuring can be identified, each of which is illustrated in Figure 5. Concurrently structured units are those that operate largely independently of each other, although often simultaneously. Serial units, on the other hand, are dependent on each other and process information along a sort of pipe-line. They may also operate simultaneously. Hierarchical organizations represent groupings of units that cooperate to perform some higher-order function. Each such grouping can be considered to be a higher-order network unit, and such groupings may extend up to many levels in the hierarchy. The boundary of a higher-order unit may be defined by a point of integration of the output of several units, although other determinants of cooperation may exist such as a recursive or circular feeding of information between a collection of units. Recursion as such can occur between units at any level in a hierarchy including the unit itself, as illustrated in Figure 6. Despite the vagueness in the concepts of higher-order structuring, it is an extremely important notion when it comes to developing complicated networks and in analyzing how they function.

2.3 Flexible Input Data Formats

For versatility in application, the structuring of a network will often have to allow for a variable format in the configuration of the input data. This is to allow the network to function in an environment where there is positional, temporal, and/or stochastic variance in the presentation of a problem at the network's inputs. Positional variance occurs when there is no fixed mapping of data sources to input neurons. Obviously a complete randomization of data input locations cannot work and so their must be some organization of the data spatially, however, this should only need be a relative positioning not an absolute positioning. Figures 7(b) to 7(d) illustrate the different types of positional variance that can occur for a two-dimensional spatially distributed set of inputs. Moreover, with the exception of rotation, all forms of positional variance can occur for a one-dimensionally distributed set of inputs. Next, temporal variance can occur when a series of values are input to one or more neurons over time. It can result from, for example, the use of arbitrary starting points in input data streams (translation), differences in the rate of data flow (scaling), and/or gradual shifts in the nature of a problem over time (distortion). Finally, stochastic variance, illustrated in Figure 7(e), is a corruption in the values of the data presented at the input neurons and can result from signal noise and/or missing data values.

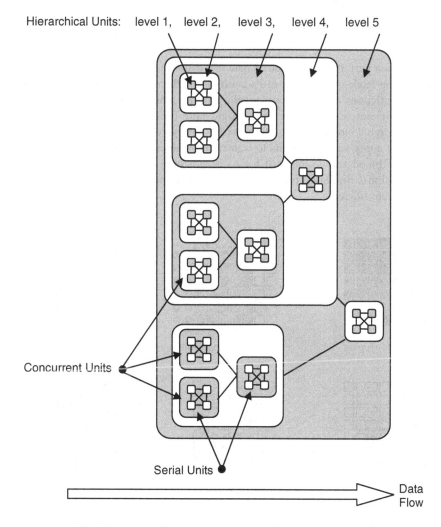

Fig. 5. Structural Forms in Higher-Order Network Configurations

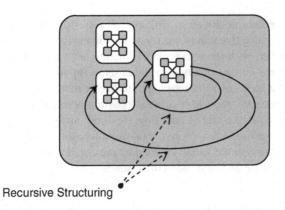

Recursive Structuring

Fig. 6. Recursion within a Structured Neural Network

The function performed by a neural network is not just dependent on its structure, but also on the mode of operation of its neurons and their connections. This proposal places no restrictions on these lower level operating modes. Studies might consider anything from neurons that act as simple logic gates (effectively making the network operate as a digital circuit) through to pulse frequency coded units. In addition, connections may apply simple weights to transmitted values or some more complicated function (see Flood and Kartam [6] for a classification of the alternative modes of neuron operation). Similarly, processes operating both below and above the neuronal level may be considered.

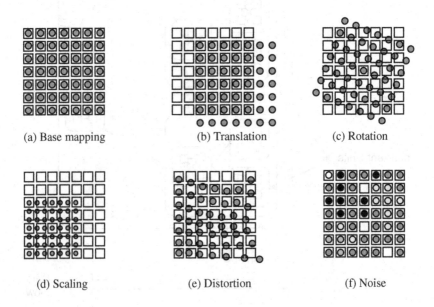

(a) Base mapping   (b) Translation   (c) Rotation

(d) Scaling   (e) Distortion   (f) Noise

Fig. 7. Types of Variance in the Configuration of Input Data

2.4 Development of Higher-Order Structures

Typically, artificial neural networks are developed using an appropriate learning mechanism in which the values of connection weights and other parameters, are developed in response to the input of examples of a problem. These examples may include target outputs, depending on the learning mechanism adopted. However, training schemes are only capable of developing single network units. While some training schemes can develop low-level structuring within a network (see for example Kohonen Networks [12] and the Cascade Correlation algorithm [4]) none are capable of developing the higher-order network structures proposed here.

To an extent, inspiration for the design of higher-order network structures can be gained by studying neural subsystems in the central nervous system. For example, the self-organizing subsystems that solve certain early processing tasks in the visual system (such as, line detection) are understood well enough to be replicated using artificial neural networks [16]. However, the organization of structures that perform higher-order operations than this, later in the visual system (such as face recognition) are not well understood. More importantly, most engineering problems do not have analogs in nature with readily available solutions provided by biological neural systems. Given this, and the fact that suitable higher-order structures cannot be derived by design for most problems, it seems that we must turn to techniques such as genetic algorithms for their development.

Interestingly, genetic algorithms are second only to neural networks in terms of their popularity as a civil engineering computing tool, and date back to at least 1987 (see, for example, Goldberg [9]). Referring to Table 1, of the top five most frequently cited articles in the ASCE Journal of Computing, while three were on neural networks the other two were concerned with genetic algorithms. Figure 1 also shows that, as with neural networks, interest in this technology has been sustained over the last decade.

While genetic algorithms and related optimization techniques have been used to develop artificial neural network applications for many years this has been restricted to developing single network units rather than the higher-order structures proposed here (see for example Cigizoglu et al., [3]). Development of higher-order structures poses a real challenge and requires the design of a sophisticated genetic coding system and corresponding set of objective functions. These must be designed to facilitate development of appropriate neural structures at: (i) the macro-level (the connectivity between the higher level neural units); (ii) the meso-level (the connectivity between the neurons within a given unit), and (iii) the micro level (the mode of operation of the individual neurons). One idea that may facilitate this is the use of some form of growth algorithm, whereby the genetic coding system defines a process for deriving a structure rather than defining the structure itself, more analogous to the way biological genetic systems operate. This may simplify, for example, the development of units that are repeated or have similar forms, and should reduce the amount of information required to encode each neural network. Inspiration for this approach may come from biological systems, fractal analysis, and chaos theory, all of which are concerned with the development of complicated structures and behaviors from relatively simple principles. The use of growth algorithms may be complemented by the concept of multi-stage objective functions which, rather than attempting to develop the neural network in a single evolutionary process, would have a series of intermediate objectives that prescribe for very simple versions of the problem through progressively more complicated and complete versions of the problem, each of which must be solved in turn. As each intermediate objective is solved, the resultant neural network codes would be used as the basis for solving the next stage in the problem. This would allow for development of primitive structures that are seminal in the development of a more complete solution.

Finally, the method of development might be designed to evolve automatic learning responses in a network when subjected to input data, including learning of connection weights and self-organization of the connection structure. In the longer term, other processing mechanisms operating at levels below and above that of the neuron may be considered, particularly as we come to understand these processes from biological studies.

3 Proposed Case Study

The main goal of this paper is to kindle a broader interest within the civil engineering computational research community for developing the next generation of artificial neural network applications, and to indicate where the author believes this research should be focused. At the same time, work is on-going at the University of Florida addressing exactly this issue. This work is based on the execution of a series of civil engineering oriented case studies in which an attempt is being made to develop higher-order neural network solutions to these problems and thus identify principles and effective approaches that may have broader application. The first case study chosen for this purpose is concerned with estimating truck attributes (such as velocity, axle spacings and axle loads) from the dynamic strain response of bridge girders during the passage of a truck (see, for example, Moses & Kriss [14]). This problem was selected as a preliminary case study since it cannot be solved satisfactorily with a single neural network unit, yet, it should not require a particularly complicated higher-order structure. In addition, a hand-crafted higher-order structure has already been developed that provides a partial solution to this problem [8] which can be used as a basis for comparison with the proposed evolved solution, in terms of the resultant structures, accuracy of results, and scope of application.

3.1 Truck Weigh-in-Motion

Figure 8 shows the existing hand-crafted structure for the network. The first unit was trained to determine the type of truck crossing the bridge according to the U.S. Federal Highway Administration (FHWA) system of classification (based on axle configuration and spacing). The input to this network was a sequence of noise filtered strain readings measured at a fixed location on a girder of a single span simply supported bridge. Following this network are a set of parallel units arranged into clusters of three. Each cluster was dedicated to determining the attributes of a specific truck type, and was selected based on the truck type output from the truck classifier unit. The units in each cluster were trained to estimate truck axle spacings, truck axle loads, and truck velocity from the filtered strain readings. The individual units used a form of radial Gaussian Basis Function neurons developed using a supervised training algorithm. A complete description of the development and performance of this neural network can be found in Gagarin et al. [8].

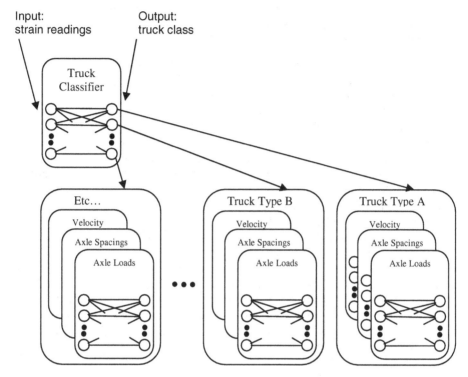

Fig. 8. Existing Hand-Crafted Higher-Order Neural Network Structure for the Truck Weigh-in-Motion Problem

This problem poses some real challenges to the development of a valid neural network based solution. In particular, the input data stream (the strain readings generated when a truck crosses the bridge) does not follow a fixed format for the following reasons:

1. The passage of other vehicles and the dynamic response of the bridge to traffic can add noise to the data stream, as illustrated in Figure 9(a). Noise, as such, makes the shape of the strain envelope less distinct and therefore more difficult to characterize.

2. Noise in the strain readings also makes it difficult to determine where the truck crossing event starts in the input data stream. Again, this is illustrated in Figure 9(a) where it can be seen that the first strain reading resulting from the truck crossing event is not easily determined. Adopting alternative starting points in the data stream as such represents a time-wise translation of the strain envelope.

3. The velocity of the truck crossing the bridge may range from just a few miles per hour to over 80 miles per hour, causing time-wise scaling of the data envelope as shown in Figure 9(b).
4. A truck may accelerate or decelerate while crossing the bridge, causing time-wise distortion of the data envelope as shown in Figure 9(c).
5. There are several other factors that affect the form of the input data stream but will be deferred to later studies – these include the effects of events such as simultaneous truck crossings and lane changes.

A second point of interest for this problem is that the data input to the neural network can be handled as a time series of values presented to a single neuron or concentrated cluster of neurons (analogous to the way hearing functions) rather than as a vector of parallel inputs of fixed size as considered in the original study. This is illustrated in Figure 10 in which the parallel input represents the approach adopted in the original study and the serial input represents the approach proposed here. The advantage of the serial approach is that it does not require predetermination of the number of elements (strain readings) in an input data set. However, it does create a challenge for the operation of the neural network since it now has to integrate inputted data over time. If significant ambiguity in the problem arises from, for example, deceleration of a truck as it crosses a bridge, then this may be alleviated by sampling strain at two or more different locations along the length of the bridge, and using separate parallel input neurons to the network for each of these data streams.

A third point of interest for this problem is that the number of values output from the network will vary depending on the type of truck crossing the bridge. That is, the number of axle spacings and axle loads that need to be predicted for any truck crossing event depends on the number of axles on the truck. Outputs might therefore be treated as a series of values generated at a single neuron, a selected value at a single neuron whereby another input is used to define which axle is of interest, or as a spatially oriented vector of values across an array of neurons where the length of the array is variable, for example.

This weigh-in-motion study is still in an early stage of development, but it serves to illustrate the sorts of issues that the next generation of artificial neural networks will need to tackle. In summary, these are: a need for sophisticated genetic coding system that can develop very large and highly structured network organizations (perhaps introducing the concepts of growth algorithms, fractal analysis, chaos theory, and multi-stage objective functions); an ability to handle unfixed formatting of input data (resulting from, for example, translation, rotation, and scaling either in a spatial or temporal framework); an ability to handle serial input data streams and the consequential need for the network to integrate these values over time; and an ability to handle variable output data structures.

With the lessons learned from this study, work will advance to more complicated problems, including dynamic decision-making for industrialized custom home design (the problem here being to make sure that all manufacturing resources are used effectively within a dynamic and uncertain market), and damage evaluation in structures determined from their dynamic response to signals such as acoustic pulses.

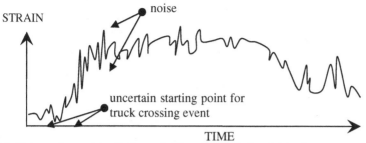

(a) Noise and uncertain starting point (translation) in input data stream

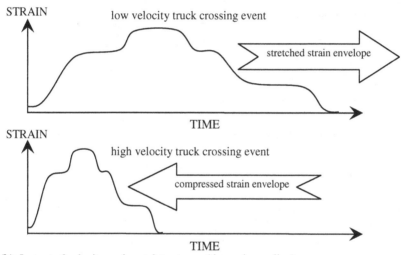

(b) Impact of velocity on input data stream (time-wise scaling).

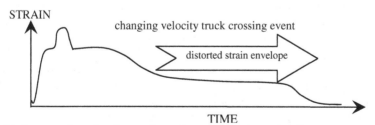

(c) Impact of acceleration on input data stream (time-wise distortion).

Fig. 9. Example Factors Affecting the Format of the Strain Data Envelope

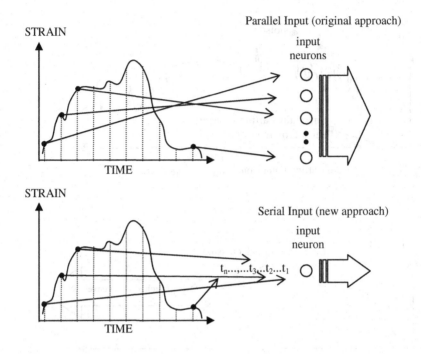

Fig. 10. Parallel Versus Serial Input Data Formats

4 Conclusions

For many years, artificial neural networks have enjoyed a significant and sustained level of interest in computer-based civil engineering research. They have provided a convenient and often highly accurate solution to problems within all branches of civil engineering. The concern raised here, however, is that the extent of this application has rarely ventured beyond rudimentary problems such as simple function modeling and pattern classification, using single unit neural network structures. Yet, biological neural systems suggest a far greater potential than this. In order to realize this potential, researchers must take on the challenge of developing networks that are vastly more complex than have been developed to date both in terms of size and structure, such as is found in the rich higher-order structuring and behavior of biological neural systems. Promising approaches to the development of these structures are genetic algorithms and related methods. While these will require the design of sophisticated genetic coding mechanisms, the potential payoff is considerable in terms of broadening the scope of application of neural computing to civil engineering. Research is, of course, a very long way from being able to replicate human cognitive skills using artificial neural networks, but the decision to take the next tentative step towards this goal is long overdue.

References

1. ASCE: Research Library. At: http://ascelibrary.aip.org/. (2006)
2. Bullock, T.H., Bennett, M.V.L., Johnston, D., Josephson, R., Marder, E., Fields R.D.: The Neuron Doctrine, Redux. Science, 310. (2005) 791-793
3. Cigizoglu, H.K., Tolun, S., and Öztürk, A.: Evolutionary Artificial Neural Networks in Hydrological Forecasting. In: proceedings of the World Water and Environmental Resources Congress 2003. Eds: Paul Bizier & Paul DeBarry. 118. ASCE. (2003) 149
4. Fahlman, S.E., and Lebiere, C.: The Cascaded-Correlation Learning Architecture. Rep. CMU-CS-90-100. Carnegie Mellon University, Pittsburgh, PA. (1990)
5. Flood, I., and Kartam, N.: Neural Networks in Civil Engineering, I: Principles and Understanding, II: Systems and Application. In: Journal of Computing in Civil Engineering, 8, (2). ASCE. (1994) 131-162
6. Flood, I., and Kartam, N: Systems. In: Artificial Neural Networks for Civil Engineers: Fundamentals and Applications. Eds: Kartam, N., Flood, I., & Garrett, J.H. Jr. ASCE. (1997) 19-43
7. Flood, I., Issa, R.R.A., and Abi Shdid, C.: Simulating the Thermal Behavior of Buildings Using ANN-Based Coarse-Grain Modeling. In: Journal of Computing in Civil Engineering, 18, (3). ASCE. (2004) 207-214
8. Gagarin, N., Flood, I. and Albrecht, P.: Computing Truck Attributes with Artificial Neural Networks. In: Journal of Computing in Civil Engineering, 8, (2). ASCE. (1994), pp 179-200
9. Goldberg, D.E., and Kuo, C.H.: Genetic Algorithms in Pipeline Optimization. In: Journal of Computing in Civil Engineering, 1, (2). ASCE. (1987) 128-141.
10. Jerison, H. J.: The Evolution of Intelligence. In: Handbook of Intelligence. Ed: Sternberg, R. J. Cambridge University Press. (2000)
11. Karunanithi, N., Grenney, W.J., Whitley, D., and Bovee, K.: Neural Networks for River Flow Prediction. In: Journal of Computing in Civil Engineering, 8 (2). (1994) 201-220
12. Kohonen, T.: Self-organization and associative memory. 3rd edn. Springer-Verlag, Berlin. (1989)
13. Koumousis, V.K., and Georgiou, P.G.: Genetic Algorithms in Discrete Optimization of Steel Truss Roofs. In: Journal of Computing in Civil Engineering, 8 (3). ASCE. (1994) 309-325
14. Moses, F., and Kriss, M.: Weight-in-Motion Instrumentation. Report FHWA/RD-78/81. Federal Highway Administration, McLean, VA. (1978)
15. Rosenblatt, F.: The Perceptron: A Probabilistic Model for Information Storage and Organization in the Brain. In: Psychological Review, 65, (6). Cornell Aeronautical Laboratory. (1958) 386-408
16. Sirosh, J. and Miikkulainen, R.: Topographic Receptive Fields and Patterned Lateral Interaction in a Self-Organizing Model of the Primary Visual Cortex. In: Neural Computation, 9. (1997) 577-594
17. Szewczyk, Z.P. and Hajela, P.: Damage Detection in Structures Based on Feature-Sensitive Neural Networks. In: Journal of Computing in Civil Engineering, 8 (2). ASCE. (1994) 163-178
18. Thomson Corporation: ISI Web of Knowledge, Web of Science. At: http://portal.isiknowledge.com/. (2006)

Evolutionary Generation of Implicative Fuzzy Rules for Design Knowledge Representation

Mark Freischlad, Martina Schnellenbach-Held, and Torben Pullmann

University of Duisburg-Essen, Institute of Structural Concrete,
Universitätsstraße 15, 45141 Essen, Germany
{mark.freischlad, m.schnellenbach-held,
torben.pullmann}@uni-due.de

Abstract. In knowledge representation by fuzzy rule based systems two reasoning mechanisms can be distinguished: conjunction-based and implication-based inference. Both approaches have complementary advantages and drawbacks depending on the structure of the knowledge that should be represented. Implicative rule bases are less sensitive to incompleteness of knowledge. However, implication-based inference has not been widely used. This disregard is probably due to the lack of suitable methods for the automated acquisition of implicative fuzzy rules. In this paper a genetic programming based approach for the data-driven extraction of implicative fuzzy rules is presented. The proposed algorithm has been applied to the acquisition of rule bases for the design of reinforced concrete structural members. Finally an outlook on the application of the presented approach within a machine learning environment for evolutionary design and optimization of complex structural systems is given.

1 Introduction

The preliminary structural design process is a knowledge intensive task. Knowledge based systems are suitable to support engineers within this process. These systems offer the possibility to represent knowledge in a transparent and comprehensible way. In order to handle the complexity, incompleteness and vagueness of experience knowledge, fuzzy reasoning mechanisms are of special interest [1].

In the following section the fundamentals of implicative rule bases as well as the differences between conjunctive-based and implicative-based inference are presented. In section 3 a new genetic programming based approach for the data-driven generation of implicative fuzzy rule bases is shown. The developed algorithm has been applied to the acquisition of rule bases for the design of reinforced concrete structural members. The results of these tests are demonstrated in section 4. An outlook on the application of the presented approach within a machine learning environment for evolutionary design and optimization of complex structural systems and concluding remarks is given in section 5. Conclusions and a short outlook are presented in the last section.

I.F.C. Smith (Ed.): EG-ICE 2006, LNAI 4200, pp. 222–229, 2006.
© Springer-Verlag Berlin Heidelberg 2006

2 Implicative Fuzzy Rule Bases

Two basic types of fuzzy rules can be distinguished [2]. A rule "If X is A_i, then Y is B_i" can be interpreted as a mapping of the domain U ($X \in U$) to the domain V ($Y \in V$). In this case each rule of a rule base encodes possible values for Y. This semantic is represented by using a conjunctive implication operator, e.g. the minimum or the product operator. As each conjunctive fuzzy rule defines a possible fuzzy point and a set of conjunctive fuzzy rules defines a fuzzy graph, rules have to be combined disjunctively. The aggregation of two (crisp) conjunctive rules is illustrated in figure 1a.

Furthermore, the rule can be interpreted as a logic implication. In this case the rule excludes impossible values of V. Consequential the implication operator has to satisfy the equivalence of the implication $A_i \rightarrow B_i$ and the proposition $\neg A_i \vee B_i$, e.g. the Kleene/Dienes or the Lukasiewicz operator. An implicative fuzzy rule base (FRB) represents a set of fuzzy constraints on the values of Y. Hence the rules have to be combined conjunctively. Figure 1b shows the aggregation of two implicative rules.

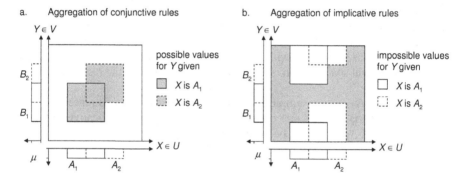

Fig. 1. Aggregation of (a) conjunctive rules and (b) implicative rules

The output fuzzy set of a conjunctive fuzzy rule base is the set of all possible values for Y given the situation X. By means of defuzzification a crisp value for Y can be derived. The output fuzzy set of an implicative fuzzy rule base represents an upper bound of possible values on V, according to the knowledge considered within the rule base. The determination of a crisp output value by means of defuzzification is not suitable in general. Often the incorporation of further knowledge is required. However, in combination with subsequent search mechanisms the elimination of impossible values can be very useful.

Within the scope of knowledge based decision support different areas for the application of the two types of rules bases can be stated. Depending on the structure of the knowledge to be represented the two reasoning mechanisms hold advantages and disadvantages. One advantage of conjunctive fuzzy rule bases is the possibility to directly obtain a crisp output value. To ensure the reliability a complete rule base and the consideration of all influence parameters is necessary. For the support of more complex decisions implicative rule bases are of advantage. In this case, not all influence parameters have to be considered in the rule base. On the one hand this leads to a reduced specificity of output sets. On the other hand the size of the rule base is reduced significantly. Hence the interpretability is increased.

3 Evolutionary Learning of Implicative Fuzzy Rules

In recent years many approaches for the data-driven generation of conjunctive fuzzy rule bases have been proposed, e.g. neuro-fuzzy systems and genetic fuzzy systems [3]. In former works the authors have presented a genetic programming based algorithm for the multi-objective optimization of fuzzy systems [4]. The optimization process is guided by the demands on the accuracy and the interpretability of fuzzy systems. Based on this approach the *R*ule Base *E*xtraction and *M*aintenance (REM) algorithm has been developed. The REM algorithm serves for the generation and optimization of both, conjunctive and implicative FRBs.

As pointed out in the previous section, the output fuzzy sets of both types of FRBs are interpreted in a different manner. Thus, the evaluation of the quality differs. In case of conjunctive FRBs a major measure of quality is the approximation error of the defuzzified outputs and the training data. Further measures have been proposed, e.g. the completeness of the rule base [5]. For the evaluation of implicative FRBs three measures are proposed: the specificity of the output fuzzy sets, the consistency of the output values of case examples with the corresponding output fuzzy sets and the congruency of the input space region covered by the case base with the region covered by the rule base.

Given a set of N_{CB} case examples (case base), where each case is represented by a set of M input variables and one output variable. Assume an implicative fuzzy rule base, consisting of N_R rules is to be evaluated.

3.1 Specificity

A major purpose of an implicative FRB is the elimination of impossible values of the output variable. A measure for this purpose is the mean specificity of the output fuzzy sets on the input vectors of the case base. The specificity is defined by

$$F_{SP} = 1 - \sum_{i=1}^{N_{CB}} \left(\int \mu_i(y) \, dy \, / \int dy \right) / N_{CB} \qquad (1)$$

where $\mu_i(y)$ is the membership function of the output fuzzy set on the *i*th case example. The higher the specificity, the more values of the output variable are (partly) excluded and consequently the more valuable is the rule base in terms of finding a crisp output value.

3.2 Consistency

By increasing the specificity the rule base might get inconsistent with the case knowledge. That means the output value of a case example is determined as an (highly) impossible value. The mean consistency of the rule base with the underlying case knowledge is derived by

$$F_{CS} = \sum_{i=1}^{N_{CB}} \mu_i(y_i) / N_{CB} \qquad (2)$$

where y_i is the output value of the ith case example and $\mu_i(y_i)$ is the corresponding degree of membership to the output fuzzy set on case i.

3.3 Congruency

An implicative FRB should only fire in those areas of the input domain that are covered by the knowledge contained in the case base. In order to determine the congruency two parameters, the fuzzy coverage of the input domain $\overline{U}$ ($\overline{x} \in \overline{U}$) by the case base $\mu_{CB}(\overline{x})$ and the coverage by the rule base $\mu_{RB}(\overline{x})$ are defined. The coverage of $\overline{U}$ by the case base is derived by

$$\mu_{CB}(\overline{x}) = \max(\mu_{C,i}(\overline{x})), \quad i = 1, 2, ..., N_{CB} \qquad (3)$$

with

$$\mu_{C,i}(\overline{x}) = \prod_{m=1}^{M} e^{-\left(\frac{x_m - x_{m,i}}{\sigma_m}\right)^2} \qquad (4)$$

and with $x_{m,i}$ is the value of the mth input variable of case i and σ_m is the predefined fuzziness of input variable m. By means of this fuzziness the degree of coverage in the neighbourhood of a case can be adapted.

The coverage of $\overline{U}$ by the rule base is derived by

$$\mu_{RB}(\overline{x}) = \max(\mu_{R,j}(\overline{x})), \quad j = 1, 2, ..., N_R \qquad (5)$$

where $\mu_{R,j}(\overline{x})$ is the degree of confidence of rule j.

The congruency measure is defined by

$$F_{CG} = 1 - \int \max(0, \mu_{RB}(\overline{x}) - \mu_{CB}(\overline{x})) d\overline{x} / \int \mu_{RB}(\overline{x}) d\overline{x} \qquad (6)$$

The overall accuracy fitness F_{AC} of the rule base is composed of the presented measures:

$$F_{AC} = (a_{SP} \cdot F_{SP} + a_{CS} \cdot F_{CS} + a_{CG} \cdot F_{CG}) / (a_{SP} + a_{CS} + a_{CG}) \qquad (7)$$

By means of the weight parameters a_{SP} (specificity), a_{CS} (consistency) and a_{CG} (congruency) the optimization process can be adapted in order to increase a preferred quality.

Besides this objective the REM algorithm takes into account the demands on the interpretability of the rule base represented by the number of rules [4]. The overall fitness of an individual is determined by the pareto-rank based approach presented by Fonseca and Fleming [6].

4 Evaluation

The developed approach was evaluated by applying it to a real world data set for the determination of the beam height of a beam slab. The input variables are the span length of the beam l_b, the span length of the slab l_s and the life load on the slab q. The output variable is the beam height h_b.

To demonstrate the advantages of implicative FRBs the goal was to find a rule base considering only the major influence parameters for this design decision. In order to simulate knowledge incompleteness only case examples covering parts of the input domain were chosen. The fuzzy coverage by the case base is shown in figure 2a.

The REM algorithm was run for 300 generations; the population size was set to 200 individuals. The maximum number of rules was set to 8 rules. The weight parameters of the accuracy fitness components were set to $a_{SP} = a_{CS} = a_{CG} = 1.0$.

The pareto-optimal rule base consisting of 6 rules is presented in further detail: Figure 2b shows the coverage of the input domain by this rule base. It is obvious that any rule fires significantly outside the fuzzy support of the case base, confirmed by the value of the congruency measure $F_{CG} = 0.978$.

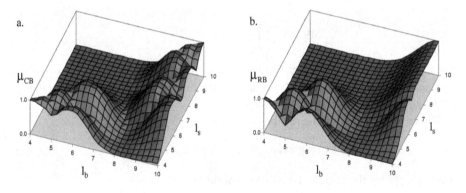

Fig. 2. Coverage of the input domain (a) by the case base and (b) by the rule base

Figure 3 shows the output fuzzy sets for three situations: $S_1(l_b = 5$ m, $l_s = 5$m), $S_2(l_b = 10$ m, $l_s = 7$ m) and $S_3(l_b = 8$ m, $l_s = 6$ m). For S_1 a highly specific fuzzy set is obtained (fig. 3a). This set is in accordance with the case examples $C_1(l_b = 5$ m, $l_s = 5$ m, $q = 3$ kN/m², $h_b = 35$ cm) and $C_2(l_b = 5$ m, $l_s = 5$ m, $q = 9$ kN/m², $h_b = 42$ cm). The fuzzy set obtained for S_2 is less specific (fig. 3b), representing well the larger range of the optimal beam height ($h_b = 65 \sim 82$ cm) for typical values of the life load ($q = 3 \sim 9$ kN/m²).

For S_3 almost any value of h_b is ruled out (fig. 3c). There was no case highly similar to this situation in the case base. Consequently almost any information can be provided and a nearly unrestricted search for a suitable output value has to be performed. Assumed a solution for this situation was found, the rule base can be extended based on this newly discovered knowledge.

The mean specificity for all case examples is $F_{SP} = 0.586$. The generated rule base is highly consistent with the underlying case base ($F_{CS} = 0.956$).

a.
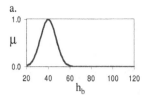

b.
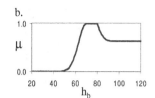

c.
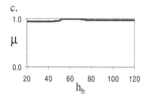

Fig. 3. Output fuzzy sets for (a) $l_b/l_s = 5m/5m$, (b) $l_b/l_s = 10m/7m$ and (c) $l_b/l_s = 8m/6m$

5   A Machine Learning Environment for Structural Design

The developed REM algorithm has been implemented into a machine learning environment for the conceptual design of concrete structures. Its core is a hybrid knowledge based evolutionary design system for the optimization of high-rise concrete structures. A prototype of this system has been presented in [7]. It was developed for the discrete optimization of the structural topology as well as the continuous optimization of single structural members. The structure is represented by a 3D product model based building information model. A finite element system for static and dynamic analyses of design candidates is included.

5.1   Knowledge Augmented Design Optimization

The conceptual design of high-rise structures includes major decisions on structural systems, e.g. the lateral load-bearing system, and subsidiary decisions on the dimension of structural members. In order to reduce the problem size two methods for the incorporation of design knowledge within the optimization process have been implemented.

Conjunctive fuzzy rule bases are used for direct determination of design variables. Thereby the number of optimization variables is decreased. Since the generation of reliable conjunctive fuzzy rule bases requires complete knowledge this method is restricted to secondary design decisions and ordinary conditions.

Highly complex design decisions are supported by implicative fuzzy rule bases. The search space of optimization variables is restricted according to the output fuzzy sets. Provided that knowledge is available for a given situation, the search space can be reduced significantly.

5.2   Machine Learning of Design Knowledge

In the scope of knowledge based systems the acquisition of knowledge by human experts is laborious and time-consuming. To overcome this problem the REM algorithm for the automated acquisition of design knowledge has been implemented into the presented design system. Figure 4 shows the architecture of the proposed machine learning environment.

The project data base contains the building information models of completed designs. When a new rule base is to be generated, the input and output variables have to be defined. Based on the project data base a set of case examples is retrieved. In the next step a fuzzy rule base is generated by the REM algorithm. The manual knowledge acquisition component serves human experts for checking and manipulation of

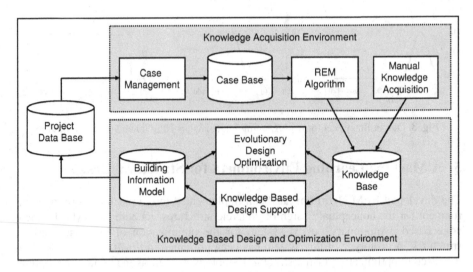

Fig. 4. Architecture of a machine learning environment for structural design

the knowledge base. Learned rule bases are incorporated within further optimization tasks as described in section 5.1.

Every time a project is completed, the project data base and, subsequently, the case base is updated. If necessary, the rule base is revised or extended.

6 Conclusions

In this paper a genetic programming based approach for the data-driven generation of implicative fuzzy rules was presented. Three measures for the evaluation of implicative fuzzy rule bases were proposed. The evaluation of the developed REM algorithm on a real world problem has shown that the generated fuzzy rule bases fulfill the demands on the specificity and the consistency with the knowledge of the underlying case base.

The application of REM within a machine learning environment for evolutionary design and optimization of complex structural systems was presented. Currently the authors investigate the impact of the proposed approach on the exploration of innovative design solutions.

References

1. Schnellenbach-Held, M., Freischlad, M.: Fuzzy Rule Based Models for Slab System Design. In: Schnellenbach-Held, M., Denk, H. (eds.): Advances in Intelligent Computing in Engineering. Proceedings of 9th International EG-ICE Workshop, VDI-Fortschritt-Berichte, Darmstadt (2002) 134-143
2. Dubois, D., Prade, H., Ughetto, L.: A new perspective on reasoning with fuzzy rules. International Journal of Intelligent Systems, V. 18 N. 5 (2003) 541-567

3. Cordón, O., Gomide, F., Herrera, F., Hoffmann, F., and Magdalena, L.: Ten years of genetic fuzzy systems: current framework and new trends. Fuzzy Sets and Systems, Vol. 141, Issue 1 (2004) 5-31
4. Freischlad, M., Schnellenbach-Held, M.: A machine learning approach for the support of preliminary structural design. Advanced Engineering Informatics, Vol. 19, No 4 (2005) 281-287
5. Jin, Y., von Seelen, W., Sendhoff, B.: On Generating FC^3 Fuzzy Rule Systems from Data Using Evolution Strategies. IEEE Transactions on Systems, Man and Cybernetics, Part B 29(4) (1999) 829-845
6. Fonseca, C. M., Fleming, P. J.: Genetic algorithms for multiobjective optimization: Formulation, discussion and generalization. In: Forrest, S. (ed.): *Genetic Algorithms: Proceedings of the Fifth International Conference.* San Mateo, CA: Morgan Kaufmann (1993) 416-423
7. Pullmann, T., Skolicki, Z., Freischlad, M., Arciszewski, T., De Jong, K., Schnellenbach-Held, M.: Structural Design of Reinforced Concrete Tall Buildings: Evolutionary Computation Approach Using Fuzzy Sets. In: Ciftcioglu, Ö., Dado, E. (eds.): Intelligent Computing in Engineering, Foundation of Design Research SOON (2003) 53-61

Emerging Information and Communication Technologies and the Discipline of Project Information Management

Thomas Froese

Dept. of Civil Engineering, 6250 Applied Science Lane, University of British Columbia, Vancouver, Canada
tfroese@civil.ubc.ca

Abstract. Project Management for architecture, engineering, and construction projects could include a well-defined sub-discipline of project information management. Not only might this improve project performance, particularly when new information and communication technologies are used, but it could also improve the technology transfer process—increasing the potential of new information and communication technologies being adopted into industry practice. An overall framework for project information management is presented that includes Information Management Plans consisting, in part, of a collection of Information Management Methods Statements. To enhance technology transfer potential, it is suggested that information and communication technology developers create templates for Information Management Methods Statements relating to their proposed systems.

1 Introduction

Recently, a design meeting was held for a local architecture, engineering, and construction (AEC) project. The project is typical of many such projects, except that it involves renovations and additions to one of the city's landmark buildings, situated on a stunning site, originally designed by one of her leading architects. The discussion focused around a cardboard model of the renovations. Although 3D computer models would offer significantly better visualization functionality—which is clearly vital for this project—and would also offer numerous other benefits, no such models have been developed. Why?

This simple anecdote is representative of one of the fundamental issues for the development of information and communication technologies (ICT) in AEC—how to transfer the technology to industrial practice. The answer to the question of why this project has not adopted technology that is readily available and undeniably beneficial is complex, but one part of the answer seems clear: it is *not* because of significant problems with the technology itself. In order to succeed in transferring new ICT to practice, then, the ICT research and development community must address issues that go beyond the technology, and address the corresponding changes to work practices. This paper suggests that an effective approach to address this challenge is through the development of the discipline of *project information management (PIM)*.

I.F.C. Smith (Ed.): EG-ICE 2006, LNAI 4200, pp. 230–240, 2006.

Information and communication have always been important to AEC projects, yet approaches for managing information have generally been informal and ad-hoc. We have argued that the AEC industry would benefit from the development of PIM as a well-understood and formalized sub-discipline of project management. Not only would this benefit projects in general, but it seems to be a necessary pre-condition to the successful implementation of more complex emerging ICT such as Building Information Models (BIM). In this paper, we consider PIM from a somewhat different perspective, and suggest that ICT researchers and developers could use a formalized PIM process as the primary tool for building the bridge from technology development to industrial practice.

This paper first presents a summary of our current thinking about a formalized sub-discipline of PIM (portions reported earlier in [5]-[8]) (as a research initiative, this represents early, conceptual work with validation and implementation effort yet to occur). The paper then discusses how elements of this approach to PIM could support the transfer of any new ICT technologies to improve its chances of successful adoption in industrial practice. The approach encourages a common approach to PIM, such that the more widely the approach is adopted, the higher the benefit for all.

2 Information Management as a Sub-discipline of Project Management

2.1 The Need to Formalize Project Information Management

Information and information management have always been recognized as important aspects of project management. But they have not been well-formalized—wide variations exist in the level and techniques used for managing project information. Some perspectives argue against an explicit PIM function: for example, suggestions that project management is inherently *all about* information and communication and cannot be sub-divided into a distinct information function sub-function; that information management is largely a technical support (staff) function rather than a project management (line) function; or that information management is a corporate, rather than a project-centric, function. However, we contend that the necessity, on one hand, for management tasks and technical expertise related specifically to information and ICT, while on the other hand, for tight integration with all aspects of project management, demands that PIM be treated as a critical, explicit function within the overall project management process.

This could be considered as very analogous to functions such as safety, risk, or quality, which have also been long recognized as important to project management in the construction industry; yet over time, these areas have evolved from loosely-defined project management objectives to distinct sub-disciplines with well-understood requirements, procedures, bodies of knowledge, and roles within the overall project management process. The same can be said for information management. For example, one chapter of the Project Management Institute's Project Management Body of Knowledge [13] defines a communications planning framework, yet this falls well short of a comprehensive approach to PIM. Information management seems far behind the areas of cost, schedule, scope, safety, risk, or quality as a well defined and understood sub-discipline of project management.

A number of efforts have been carried out within construction IT research related to information management practices. For example, Björk [3] defines a formal model for information handling in construction processes. Turk [15] explored the relationships between information flows and construction process workflows, and makes a distinction between base processes (the main value adding activities) and glue processes (that make sure that the materials and information can flow between the base processes) [16]. Mak [12] describes a paradigm shift in information management that focuses mainly in Internet-based information technologies. Betts [2] includes much work on the role of information technologies in the management for construction, with an emphasis on strategic management of the firm. These and other works have much to offer in the area of information management practices. This paper, however, takes a fairly specific perspective that has not been widely addressed: i.e., the development of specific information management practices as they relate to the management of individual construction projects in the context of emerging ICT.

We contend that improved PIM could improve performance on any construction project today. (Although we have no data to prove this statement, we suggest it is axiomatic that, first, information and communication is critical to the performance of AEC projects and, second, that any aspect of work processes can be improved through improved management practices. This leaves open the issue of how best to improve PIM and what level of PIM is appropriate). Yet improved PIM becomes much more significant as projects adopt more advanced, emerging ICT, such as building information models (BIMs). This is because of the increased complexity, required skills, and work tasks. Indeed, we contend that a careful consideration of how information management practices could adopt new ICT provides an essential bridge to move new ICT from development into industrial practice.

2.2 Elements of Project Information Management in Comparison with Quality Management

Fig. 1 illustrates the relationships between Project Management and PIM, including a comparison with quality management. In current practice, Quality Management, as a body of knowledge, can be thought of as a subset of both project management and production technologies. ISO 9001 [10] provides standardized specifications for quality management systems. For any individual AEC project, ISO 9001 is implemented in the Quality Management System, as documented by the Quality Manual. One component of the Quality Manual is a collection of individual Method Statements.

In a similar fashion, we suggest that PIM is a distinct sub-discipline of both Project Management and ICT (i.e., it represents the management of the project's information and ICT). Within the general body of knowledge of PIM, one possible way of implementing PIM is to develop a PIM protocol that provides a set of specifications for PIM systems. On an AEC project, the PIM protocol is implemented in the project's information management system (i.e., the socio-technical system that includes the people, work practices, technical systems, etc.—not just the software systems), which is documented in the Information Management Plan/Manual. This plan contains (among other things) a collection of specific information management methods statements (IMMS). Each IMMS generally addresses the treatment of specific information resources or ICT tools for specific project work packages or for specific project interfaces.

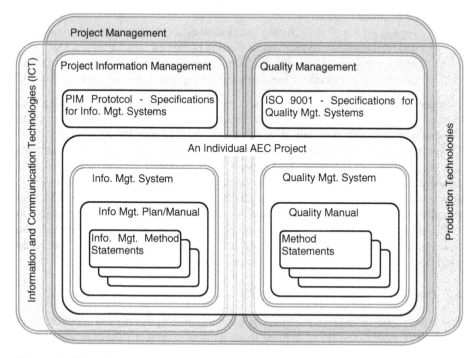

Fig. 1. An illustration (Venn Diagram) of elements of PIM relative to concepts of project management and quality management. The double grey lines represent fuzzy concepts (no exact delineations between concepts) while the single black lines represent specific, individual things.

3 A Framework for Project Information Management

A comprehensive list of all of the issues involved in the management of information systems for construction can grow very long indeed. To provide some structure to these issues, we propose that PIM be defined as ***the management of information systems to meet project objectives***. Though simple, this definition suggests a breakdown of PIM into four main topic dimensions: a *management process*, *project elements*, *information system elements*, and *objectives*. The following sections examine each of these topics.

3.1 A Management Process for Information Management

The management of information systems should follow general management processes:

- *Plan* all aspects of information system. This includes identifying the requirements and alternatives, designing a suitable solution taking into account all objectives and constraints and using appropriate analytical techniques and models, and adequately documenting the plan so that it can be communicated to all.
- *Implement* the plan, including issues such as securing the necessary authority and resources for the plan, implementing communication, training, etc.
- *Monitor/control* the results, including appropriate data collection relative to established performance measures and taking necessary corrective action.

Other generic management processes such as scope definition, initiating and closing the project, iterating through increasingly detailed cycles of the plan-implementation-monitoring sequence, etc. are all equally applicable.

Applying this generic management process, Fig. 1 suggested some of the elements that could make up a specific PIM process. For example, a widely-accepted PIM protocol could help raise the awareness of PIM and lead to a level of industry-wide commonality in the way that PIM is handled. The Information Management Manual would be the key document for representing the plan, supporting the implementation and control efforts. Information management methods statements would provide the basic building blocks for addressing specific applications of ICT tools, and could form the basis for transferring practices from one project to another.

3.2 Project Elements

The information management actions of planning, implementing and monitoring an information system should be applied to all parts of a project. This can involve the same project work breakdown structures used for other aspects of project management (e.g., breaking the project down by discipline, work package, etc.). However, there are perspectives on decomposing the work that are of particular relevance to information systems. Adopting a project processes model as a basis for structuring information management, the approach should address three primary elements: project tasks, information transactions, and overall integration issues. The process should define these elements, including identifying participants, project phase, etc. (this should correspond largely to an overall project plan and schedule, and thus it may not need to be done as a distinct activity). Then, for each of these elements, the information management process should analyze information requirements, design information management solutions, and produce specific information management deliverables, such as Information Management Methods Statements (this is generally at the level that various work packages must interact with each other, not into the details of how each participant performs their own work packages).

The model considers these elements across all project participants (spanning all participating companies, not just internal to one company), and the information management tasks should be carried out for each of these project elements.

3.3 Information System Elements

For each of the project elements to which we are applying our information management processes, there are a number of different elements of an information system that must be considered:

- *Information*: Foremost, we must consider the information involved in each of the project elements. First, the process should assess the significant information input requirements for each element, determining the type of information required for carrying out the tasks, the information communicated in the transactions, or the requirements for integration issues. With traditional information technologies, information requirements generally correspond to specific paper or electronic documents. With building information models and other newer information technologies, however, information requirements can involve access to specific data sources (such as specific application data files or shared databases) that do not correspond to traditional documents. Second, we must assess tool requirements by determining the key software applications used in carrying out tasks, communication technologies used for transactions, or standards used to support integration. Third, we must assess the significant information outputs produced by each task. This typically corresponds to information required as inputs to other tasks. After analysis, these results should be formalized in the information systems plan as the information required as inputs for each task, and the information that each task must commit to producing.

- *Resources:* the information management process should analyze the requirements, investigate alternatives, and design specific solutions for all related resources. These include hardware, software, networking and other infrastructure, human resources, authority, and third party (contracted) resources.

- *Work methods and roles:* the solution must focus not only on technical solutions, but equally on the corresponding work processes, roles and responsibilities to put the information system to proper use.

- *Performance metrics, specified objectives, and quality of service standards:* the information systems plan should include the specification of specific performance metrics that can be assessed during the project and used to specify and monitor information systems objectives and standards of service quality.

- *Knowledge and training:* the information systems solution will require certain levels of expertise and know how of people within the project organization. This may well require training of project personnel.

- *Communications:* implementing the information systems plan will require various communications relating to the information system itself, such as making people aware of the plan, training opportunities, procedures, etc.

- *Support:* information system solutions often have high support requirements, which should be incorporated as part of the information management plan.

- *Change:* the information management plan should include explicit consideration of change—how to minimize its impact, how to address unauthorized changes by individual parties, etc.

3.4 Information Systems Objectives

Solutions should be sought that meet the general project objectives of cost, time, scope, etc. However, there are a number of objectives that are more specific to the information system that should be taken into account:

- *System performance* is of primary concern, including issues such as efficiency, capacity, functionality, scalability, etc.
- *Reliability, security, and risks* form critical objectives for information systems.
- *Satisfaction of external constraints:* we have placed the emphasis on the project perspective, but the information management must also be responsive to a number of external influences. Of particular significance in alignment with organization strategies and information management solutions, including appropriate degrees of centralized vs. decentralized information management. Other external influence include client or regulatory requirements, industry standards
- *Life-cycle* issues should be considered. These include both the life cycle of the information (how to ensure adequate longevity to the project data), and of the information system (e.g., life-cycle cost analysis of hardware and software).
- *Interoperability* is key objective for many aspects of the information system.

4 The Technical Body of Knowledge: Project Systems and Areas of Expertise

The previous section outlines a very generic framework for information management. While this focus on the conceptual frameworks and management processes provides one leg to the practice of PIM, the other leg consists of the technical body of knowledge that underpins the information systems used throughout the construction industry. Ideally, there would be a well developed and widely understood body of knowledge for this discipline—but this does not seem to exist. At present, technical expertise is built up mainly through extensive industry experience with little in the way of unifying underlying theory or frameworks. Recent developments such as Master degree programs focusing on construction ICT (e.g., the European Masters program in Construction ICT, Rebolj and Menzel [14], or the ASCE Certificate in Computing [1]) are helping to contribute to a more formalize body of knowledge for both traditional and emerging construction ICT.

5 Organizational Roles: The Project Information Officer

Where possible, PIM requires a new, senior-level position within the project management team. We call such a position the *Project Information Officer (PIO).*

The overall responsibility of the PIO is to implement the information management as described previously. However, no single solution for implementing PIM will be ideal for all projects. Rather, ideal organizational solutions will depend on a number of factors, not the least of which are the size of the project and the relative complexity of the information systems to be used (for example, on small and simple projects, PIM may be a accomplished very informally as just on of the responsibilities of the overall project manager). We have gone into greater detail on possible organizational configurations for PIM and on the nature of the PIO position in other publications [5], [8].

6   Towards the Discipline of PIM

A long term, ideal implementation of project information systems may involve advanced ICT tools used throughout the project, extensive integration and interoperability, and wide-spread buy-in to a comprehensive PIM process. However, these changes to current practice are not preconditions to the beneficial application of PIM. We suggest that improved PIM, as outlined here, could benefit AEC projects under present conditions (with the most significant benefits to be seen on large and complex projects).

We have stated that this is preliminary, conceptual work that has yet to be implemented and validated. A likely validation process would begin with field studies of current best practices in PIM. Although we believe that the PIM framework as outlined here is novel, PIM itself is by no means new, and best practices for every aspect of the proposed PIM approach exist in current practice. Later work would attempt to fully develop the proposed approach and to implement it on actual projects along with benchmarking techniques to assess its impact.

7   PIM and ICT Development

Having summarized an approach for a formal PIM process, we return to the role of PIM in supporting the successful transfer of new ICT into industrial practice. Those involved in the research and development of ICT solutions for the AEC industry, whether for profit such as software vendors or not for profit such as universities, are of course interested in seeing those solutions used by industry. We characterize this as a technology transfer issue. Technology transfer has been modeled as involving the following five factors [17]:

- characteristics of the transferor, e.g., willingness to transfer technology, management systems, etc.,
- characteristics of the transferee, e.g., intent to learn technology, management systems, etc.,
- transfer environment, e.g., complexity of technology,
- learning environment, relationship and communication between transferor and transferee,
- value-added from the technology transfer, e.g., economic advancement, project performance, etc.

Clearly, the success of the transfer depends upon more than just the technical characteristics of the solution. In particular, there are significant non-technical challenges facing technology transfer of ICT in the AEC industry. For example, there is very little research capacity relative to the size of the industry, and perhaps even less ICT development capacity. The culture of the industry doesn't generally place a high value on ICT innovation, while the short-term, dynamic, virtual organizations that make up AEC projects eschew significant investment in developing large-scale ICT solutions and lead to high fragmentation. Some of the non-technical elements required for the successful transfer of any new ICT solutions include the following:

- Development of the entire socio-technical or "soft" system. E.g., for a new piece of software, the work practices required to use the software, the roles and responsibilities involved, training requirements, expected benefits and assessment techniques, etc.
- Interface with other systems. How the proposed system interacts with other social-technical systems (e.g., planning the interface between design systems and costing systems).
- Communication of the solutions from the system providers to potential end users.

Developers, preferably working along with the industry users, are well positioned to develop these elements of the overall solution. Indeed, technology transfer issues are frequently considered as part of research and development efforts. Current efforts, however, are relatively ad hoc, since there is little in the way of standard practice.

A PIM process as described above could improve this situation. It would provide a comprehensive structure and methodology for developing the complete socio-technical solution of new ICT systems, including interface issues. Furthermore, by using the same PIM framework that project teams use to develop their own PIM programs, the compatibility and communication of the overall solution would be improved, making it easier for industrial users to understand and adapt new ICT solutions. The common structure would also improve the reusability of new ICT systems between projects.

7.1 The Use of Information Management Methods Statement Templates to Enhance Technology Transfer

As described earlier, a *PIM Protocol* would provide a common, high-level structure or framework in the form of specifications for PIM systems. On individual projects, the PIM system would be documented in the *Information Management Plan/Manual*. This plan would be comprised, in part, of a series of *Information Management Methods Statements (IMMS)*, each of which would document the plan for addressing specific information issues (the use of a particular ICT tool or the production of a particular body of project information) for a particular work package or particular information exchange requirement. Although the IMMS's are unique and specific to an individual project, they would often be quite similar from one project to another, and could be derived from an *IMMS Template*.

It is these IMMS Templates that could provide the focus of ICT developers' efforts to bridge the gap from system development to industrial application. Developers could create IMMS Templates for their ICT solutions, thereby working through a comprehensive range of information management issues surrounding their systems, and creating an information management plan that could be readily adopted into any project's PIM process. This IMMS-based approach are consistent with the modular techniques taken by other research into the design and modeling of AEC processes [9], [11]. Where appropriate, they could utilize relevant techniques or standards such as the Unified Modeling Language (UML), the Industry Foundation Classes (IFC's), standard AEC enterprise models [4]. The IMMS would consist of sections such as the following:

- Name, description, and purpose of the method.
- Scope: description of the context and constraints for which the method is intended and is considered to be relevant.
- Roles and responsibilities: the parties involved and their respective roles with respect to the method.
- Procedures: who does what, when, and how.
- Resource requirements
- Expected results: costs, durations, benefits, as well as metrics and methods for assessing these metrics.
- Risks and responses

The development of a full example of an IMMS is beyond the scope of this paper, but one can imaging an IMMS relating to the use of 3D models for visualization during the preliminary design phases of a renovation project, as described in the opening scenario of this paper. It would contain specific methodology describing how the 3D visualization could be used to support this phase of the project, who would need to be involved and what their respective roles would be, what types of human and technical resources would be required, the costs and time required, the benefits to be expected, etc. All of this would be in a form that could easily be incorporated in the project's PIM process.

8 Conclusion

Project Information Management, as a formal sub-discipline of project management, could improve the performance of AEC projects. In addition, a formalized PIM process could provide the unifying structure to support ICT developers in addressing the wide range of issues involved in adopting new ICT systems into industrial practice, and in easing the process for industrial users to learn about and incorporate the new ICT. This paper has presented a conceptual framework for a formalized PIM processes and discussed its potential for supporting ICT technology transfer. In future work, we expect to further develop the approach and carry out implementation and validation work.

References

1. Arciszewski, T., Smith, I., & Melhem, H. *Progress Report*, ASCE Global Center of Excellence in Comp, http://www.asceglobalcenter.org/ProgressReports/ProgressReport.pdf Accessed April 28, 2006.
2. Betts, M. (Ed). Strategic Management of I.T. in Construction, Blackwell Science, 1999.

3. Björk, B-C. "A formalised model of the information and materials handling activities in the construction process," Construction Innovation, 2(3), pp. 133-149.
4. Ducq, Y., Chen, D. and Vallespir, B., "Interoperability in enterprise modelling: requirements and roadmap" Advanced Engineering Informatics, Vol. 18, No 4, 2004, pages 193-204.
5. Froese, T., "Help Wanted: Project Information Officer", 5th European Conf on Product and Process Modelling in the Building and Construction Industry, Istambul, Turkey, Sept. 8-10, 2004.
6. Froese, T., "Impact of Emerging Information Technology on Information Management", International Conference on Computing in Civil Engineering, ASCE, Cancun, Mexico, Paper #8890, July 12-15, 2005.
7. Froese, T. "Information Management for Construction," 4th International Workshop on Construction Information Technology in Education, Dresden, Germany, K. Menzel (Ed.), Institute for Construction Informatices, Technische Universitat Dresden, Germany, ISBN 3-86005-479-1, CIB Publication 303, pp. 7-16. Jul 18, 2005.
8. Froese, T. "Project Information Management for Construction: Organizational Configurations", Submitted to ASCE/CIB Leadership conference, Bahamas, May, 2006.
9. Haymaker, J., Kam, C. and Fischer, M. "A Methodology to Plan, Communicate and Control Multidisciplinary Design Processes", CIB W78 22nd Conference on Information Technology in Construction, Dresden, Germany, R.Scherer, P. Katranuschkov, S.-E. Schapke (Eds.), Institute for Construction Informatices, Technische Universitat Dresden, Germany, ISBN 3-86005-478-3, CIB Publication 304, pp. 75-82, Jul 19-21, 2005.
10. ISO. ISO 9001:2000. Quality management systems - Requirements. ISO, Switzerland, 2000.
11. Keller, M. and Scherer, R.J. "Use of Business Process Modules for Construction Project Management", CIB W78 22nd Conference on Information Technology in Construction, Dresden, Germany, R.Scherer, P. Katranuschkov, S.-E. Schapke (Eds.), Institute for Construction Informatices, Technische Universitat Dresden, Germany, ISBN 3-86005-478-3, CIB Publication 304, pp. 91-96, Jul 19-21, 2005.
12. Mak, S. "A model of information management for construction using information technology," Automation in Construction 10, pp. 257-263.
13. PMI. A Guide to the Project Management Body of Knowledge (PMBOK Guide), 2000 Edition, Project Management Institute: Newtown Square, PA, USA.
14. Rebolj D and Menzel K., "Another step towards a virtual university in construction IT," ITcon Vol. 9, pg. 257-266, http://www.itcon.org/2004/17
15. Turk, Z. "Communication Workflow Approach to CIC," Computing in Civil and Building Engineering, ASCE, pp. 1094-1101.
16. Turk, Z. "What Is Construction Information Technology," Proceedings AEC2000, Informacni technologie ve stavebnictvi 2000, Praha, CD-ROM.FIATECH (2004), Capital Projects Technology Roadmapping Initiative, FIATECH, Austin, USA.
17. Waroonkun, T., Stewart, R. and Mohamed, S. "Factors Affecting Technology Transfer Performance: Evidence from Thailand," 1st International Construction Specialty Conference, Calgary, Alberta, Canada May 23-26, 2006.

The Fishbowl™: Degrees of Engagement in Global Teamwork

Renate Fruchter

Director of Project Based Learning Laboratory (PBL Lab)
Department of Civil and Environmental Engineering
Stanford University, Stanford, CA 94305-4020, USA
fruchter@stanford.edu

Abstract. The challenge addressed in this study is to improve project based learning in cross-disciplinary, global teamwork that prepares the next generation of professionals for the global AEC market place. The paper presents the Fishbowl™ learning interaction experience as a pedagogical intervention method designed to support effective knowledge transfer from professionals to students. An innovative information and communication technology augmented (ICT) workspace mediates the communicative event among professionals and global learners and allows capture, transfer, and reuse of knowledge created during the Fishbowl™ pedagogical intervention. The assessment of the intervention and ICT workspace affordances, their impact on learning and team performance lead to the formalization of specific interaction zones and participants' degrees of engagement. The paper leverages this unique AEC global teamwork course as a testbed and tech transfer conduit. The paper concludes with examples of pilot projects in support of tech transfer of both the interaction experience and ICT workspace into industry practice.

1 Introduction

The challenge addressed in the study presented in this paper is to improve project-based learning (PBL) in cross-disciplinary, global teamwork based on role modeling methods by creating an innovative computer mediated learning experience. Architecture, engineering, and management students engaged in PBL courses typically apply the knowledge acquired in discipline centric classes to produce a product, but do not necessarily know how to acquire the cross-disciplinary communication competences that professional experts exercise in real life projects. These competences include exploration of alternatives to solve problems, inquiry, probing the boundaries between disciplines, and negotiation.

This goal is rooted both in market needs and a desire for innovation in AEC education. The globalization of economic activity is perhaps one of the defining constants of the rapidly changing market place. Increased competitive pressures shorten project lead times and use of concurrent engineering in cross-functional teams. The availability of communication technologies enables these cross-functional teams to be often geographically distributed. It is interesting to observe that the social context of

I.F.C. Smith (Ed.): EG-ICE 2006, LNAI 4200, pp. 241 – 257, 2006.

multi-stakeholder project teams and the advances in technology have increased the complexity of projects. Today, AEC projects are characterized not only by cross-disciplinary, multi-stakeholder teams, and the social context, but also by the global aspect of teams driven by diverse cultural perspectives, and the access to large volumes of digital information.

The mission of the Project Based Learning Laboratory (PBL Lab) at Stanford is to prepare the next generation of AEC professionals who know how to team up with professionals from other disciplines worldwide and leverage the advantages of innovative collaboration technologies to produce higher quality products, faster, more economical, and environmentally friendly. The goal is for these students to become leaders in global teamwork. The objective is a sustained effort in an integrated research and curriculum to develop, test, deploy, and assess radically new collaboration technologies, workspaces, processes, learning and teamwork models that support cross-disciplinary, geographically distributed teams.

This is accomplished through an authentic project-based learning (PBL) learning experience, an innovative information technology (ICT) infrastructure. The PBL Lab serves as a home for the PBL learning experience and a testbed to study the impact of ICT in global teamwork and learning [13]. The AEC Global Teamwork course established at Stanford in 1993 and run in collaboration with universities worldwide [4], offers an authentic PBL teamwork learning experience that enables students to identify discipline and cross-discipline objectives and thereby develop *know-why* knowledge in an interdisciplinary context. They exercise the theory and knowledge acquired in traditional discipline courses, i.e., *know-what* and *know-how*. The roles of each information technology as mediator for communication and cooperation within cross-disciplinary teams is justified and determined to support the diverse: (1) modes of learning and interaction over time and space, (2) needs to capture, share, and reuse information and knowledge, and (3) types of interactions among participants.

The paper presents the rationale and points of departure for this study, and discusses:

- The Fishbowl™ learning interaction experience as a pedagogical method designed to support effective knowledge transfer of communication skills from professionals to students.
- An information and communication technology augmented (ICT) workspace that (1) supports the communicative event among professionals and global learners, (2) promotes a better conceptual understanding of the professional practice, and (3) allows capture, transfer, and reuse of knowledge created during the Fishbowl™ sessions.
- The assessment of the Fishbowl™ learning experience, ICT augmented workspace affordances, their impact on learning, as well as the Fishbowl™ as a technology transfer conduit.

2   Rationale and Points of Departure

The study addresses three hypotheses.

Pedagogic Intervention. The first hypothesis is that knowledge is best transferred and acquired by students when (1) they are apprentices or are mentored [7] [13] and (2) the problem is not only situated [10][11][13][20] (3) but also identified and owned by the students. The Fishbowl^{TM} was designed to test this hypothesis. It was designed as a pedagogic intervention in the form of a *learning interaction experience* to support the knowledge transfer of communication skills from professionals to students. This *learning interaction experience* improves the students' cross-disciplinary, collaborative teamwork competences through learning by doing [3]. The *learning interaction experience* takes place in a geographically distributed multi-cultural global PBL setting, i.e., the AEC Global Teamwork course. The paper explores how students learn and interact in such a global Fishbowl^{TM} PBL setting.

ICT augmented workspace for PBL. *The second hypothesis asserts that a primary source of information behind design decisions is embedded within the verbal conversation among project members.* Capturing these conversations is difficult because the information exchange is unstructured and spontaneous. In addition, discourse is often multimodal. It is common to augment speech with sketches and gestures. Audio/video media can record these activities, but lack an efficient semantic indexing mechanism. The objective of the developed ICT augmented workspace is to *improve* and *support* the process of knowledge creation, capture, transfer, and reuse of the *learning interaction experience* interventions. The designed and deployed information and communication technology *(ICT) augmented workspace*: (1) supports the communicative event among professionals and learners during the creative process of concept generation and development, (2) promotes a better conceptual understanding of the professional practice based on cognitive apprenticeship and legitimate peripheral learning [13] principles, and (3) allows capture of the act, i.e., the creative discourse and problem solving sketching activity. Consequently the communicative act becomes a multimedia learning artifact. Such multimedia learning artifacts are used by learners for further reflection and knowledge reuse both in terms of product, i.e., solutions and ideas proposed by the professionals, as well as their communication process.

The third hypothesis asserts that any ICT that engages multiple participants in communicative events and tasks in an interactive workspace will determine (1) specific interaction zones, where participants engage in coordinated action in the physical workspace, and (2) degrees of engagement of the participants during the discourse. To explore this hypothesis we focus on:

- how to capture with high fidelity, and low overhead to the team members, the knowledge experience that constitutes conceptual design generated during informal events such as brainstorming or project review sessions?
- what interaction zones and degrees of engagement emerge in the proposed ICT augmented workspace, and
- how the Fishbowl^{TM} *learning interaction experience* is supported and impacted by the affordances of the ICT and the configuration of the interactive workspace in both collocated and geographically distributed learning environments? [6]

The theoretical points of departure for this study include: learning theory, interaction design and analysis, design theory and methodology, knowledge management, and human computer interaction.

The design of the AEC Global Teamwork course and ICT workspace are grounded in cognitive and situative learning theory. The cognitive perspective characterizes learning in terms of growth of conceptual understanding and general strategies of thinking and understanding [3]. The situative perspective shifts the focus of analysis from individual behavior and cognition to larger systems that include individual agents interacting with each other and with other subsystems in the environment [10][11]. Situative principles characterize learning in terms of more effective participation in practices of inquiry and discourse that include constructing meanings of concepts and uses of skills. Teamwork, specifically cross-disciplinary learning, is key to the design of the AEC Global Teamwork course. Students engage with team members to determine the role of discipline-specific knowledge in a cross-disciplinary project-centered environment, as well as to exercise newly acquired theoretical knowledge. It is through cross-disciplinary interaction that the team becomes a community of practitioners--the mastery of knowledge and skill requires individuals to move towards full participation in the socio-cultural practices of a larger AEC community. The negotiation of language and culture is equally important to the learning process--through participation in a community of AEC practitioners; the students are learning how to create discourse that requires constructing meanings of concepts and uses of skills.

This research builds on Donald Schon's concept of the reflective practitioner paradigm of design [18]. Schön defines the process of tackling unique design problems as *knowing-in-action*. To Schön, design is an *action-oriented* activity. However, when knowing-in-action breaks down, the designer consciously transitions to acts of reflection, termed *reflection-in-action*. Schön argues that, whereas action-oriented knowledge is often tacit and difficult to express or convey, what *can* be captured is *reflection-in-action*. This concept was expanded into a *reflection-in-interaction* framework to formalize the process that occurs during collaborative team meetings [9]. The act of reflection-in-action is viewed as a step in the knowledge creation and capture of the "knowledge life cycle" [8] – "creation, capture, indexing, storing, finding, understanding, and re-using knowledge." Knowledge that is created through dialogue among practitioners, or between mentors and learners represents an instance of what Nonaka's knowledge creation cycle calls "socialization, and externalization of tacit knowledge." [14]. The design of the Fishbowl ^{TM} and the ICT augmented workspace build on these constructs of the knowledge lifecycle and the "socialization, externalization, combination, and internalization" cycle of knowledge transfer.

The human computer interaction (HCI) scenario-based design approach [15] is used as a methodology to study the current state-of-practice, describe how people use technology and analyze how technology can support and improve their activities. The process begins with an analysis of current practice using *problem scenarios* that are transformed into *activity scenarios*, *information scenarios* and *interaction scenarios*. The final stage is *prototyping* and *evaluation* based on the interaction scenarios. The process as a whole from problem scenarios to prototype development is iterative.

3   The Fishbowl^{TM} Learning Interaction Experience

The AEC Global Teamwork course is used as the testbed for Fishbowl^{TM}. The course is based on the PBL methodology of teaching and learning that focuses on *problem* based, *project* organized activities that produce a *product* for a client. PBL is based on re-engineered *processes* that bring *people* from multiple disciplines together. It engages faculty, practitioners, and students from different disciplines, who are geographically distributed.

Situated Learning Context. The AEC Global Teamwork course is a two Quarter course that engages architecture, structural engineering, and construction management students from universities in the US, Europe and Japan, e.g., Stanford University, UC Berkeley, Cal Poly San Luis Obispo, Georgia Tech, Kansas University, University of Wisconsin Madison, in the US, Stanford Japan Center in Kyoto Japan, Aoyama Gakuin University in Tokyo Japan, University of Ljubljana in Slovenia, Bauhaus University in Weimar Germany, ETH Zurich and FHA in Switzerland, Strathclyde University in Glasgow, and University of Manchester, Manchester, in UK, KTH in Stockholm, Chalmers University and IT University in Goteborg, Sweden, and TU Delft in Netherlands, in Europe. The core atom in this learning model is the AEC student team, which consists of an architect, two structural engineers, and two construction managers from the M.Sc. level. This master builder atelier education model builds a number of bridges, between undergraduate and graduate students, between students and industry, and between academia and industry. Each year there are 4 to 12 AEC teams in the class. Each team is geographically distributed, and has an owner/client with a building program, a location, a budget, and a time line. The students have four challenges – cross-disciplinary teamwork, use of advanced collaboration technology, time management and team coordination, and multi-cultural collaboration. The building project represents the core activity in this learning environment. Each team has a building functional program, a budget, a site, a time for delivery, and a demanding owner. The project is based on a real-world building project that has been scoped down to address the academic time frame of two academic quarters. AEC teams model, refine and document the design product and process. The students learn to regroup as the different discipline issues become central problems and impact other disciplines. They use computer tools that support discipline tasks and collaborative work. The project progresses from conceptual design to a computer model of the building and a final report. As in the real world, the teams have tight deadlines, engage in design reviews, and negotiate modifications. A team's cross-disciplinary understanding evolves over the life of the project. The international structure of AEC teams adds the real-world collaboration complexity to the learning environment. A key focus is the effective use of IT resources to support instruction and learning outcomes. Typical project examples can be viewed in the project gallery of PBL under http://pbl.stanford.edu

The design of the **Fishbowl^{TM}** learning experience is inspired from the Medical School learning environment, where it is typical to see special operation rooms enclosed with glass walls where world expert surgeons operate on patients (e.g., open heart surgery) and medical students watch. This learning experience brought to mind the fishbowl metaphor. More importantly, the goal was to emulate this learning experience in the school of engineering, specifically in project-based design teamwork such as the AEC Global Teamwork offered by the Project Based Learning Laboratory (PBL Lab), at Stanford. In addition, the aim was to capture the activity during the project-based *Fishbowl^{TM}* session for future re-use, either by the student team in the *Fishbowl^{TM}* or by future generations of students. The result was what is now known as the *Fishbowl^{TM}* session.

Ownership of Problem by the Learners. The *Fishbowl^{TM}* session is a project review and problem solving session that is intended to act as a role modeling, mentoring and cognitive apprenticeship opportunity. The *Fishbowl^{TM}* session takes place after the teams have worked on their concepts for three months and face some key challenges. Each student team prepares for a *Fishbowl^{TM}* session in which they present in 10 min the status of their project and their key 3-5 challenges. They then hand over their project to a full team of AEC industry mentors who work on the student's project challenges in front of the students for an hour. Since the problems and challenges are defined by the students who work on that specific project, their level of attention and engagement is maximized. Consequently the students are highly motivated to acquire as much knowledge from the mentors as possible, as they observe them enact the cross-disciplinary dialogue that takes place in any typical building project.

Cognitive Apprenticeship Setting. The participants and their roles in the *Fishbowl^{TM}* session are as follows:

1. *the "fish learners"* i.e. the *A/E/C team* that brings *their project with current challenges* the *"patient with problems."* This is a geographically distributed team of students, e.g., an architect at Georgia Tech, a structural engineer at Stanford in the US and a second one at Bauhaus University in Germany, and a construction manager at Stanford and a second one at KTH in Sweden;
2. the AEC industry mentor team that represents the *"fish mentor."* This mentor team is typically composed of an architect, two engineers, and two construction manager from AEC companies (e.g., architect from HOK, structural engineers from Ove Arup and GPLA, and construction mangers from Swinerton Builders and Microestimation Inc.).
3. The rest of the students in the AEC Global Teamwork course. These are the *observers* watching the *"fish learners and mentors"* in the project review *Fishbowl^{TM}* session. They are from all the partner universities in the AEC class that are distributed worldwide. Consequently, there are collocated spectators who are in the PBL Lab, and geographically distributed spectators.

The students learn best practice skills for effective teamwork and cross-disciplinary collaboration as they observe how the industry mentors: redirect *fish learners'* questions from very specific "how-to" to broader project specific or process related questions, explore alternatives to solve problems, inquire about impacts across disciplines of specific discipline choices, probe the boundaries between disciplines, and negotiate. As a result of this learning experience, we observed that the students started to mimic the behavior of the industry mentors during the *Fishbowl^{TM}*. The interaction and the dialogue between team members during project meetings evolved from presentation mode to inquiry, exploration, problem solving, and negotiation.

4 The ICT Augmented Workspace

The interactive collaboration workspace created in the PBL Lab to support the *Fishbowl^{TM}* interaction was designed to address and validate the hypotheses of this study. The workspace consists of a hybrid software and hardware environment that includes:

- RECALL™ collaboration technology and knowledge capture, briefly described in section 4.1. [5],
- VSee™ technology (VSeelab.com) for parallel video streaming over the IE browser to enable the PBL participants to see all the remote sites, briefly described in section 4.2
- MS NetMeeting Videoconference for application sharing (e.g., RECALL™) with all the remote sites,
- a SmartBoard for direct manipulation and sketching through the RECALL™ application,
- a Webcam that enables the remote students to see the interactive workspace that covers the *Fishbowl™* area in the PBL Lab at Stanford. This includes the Smart-Board, the *mentor Fish* and *team Fish*;
- additional SmartBoard or projector and projection screen for the parallel video streams that enable the *mentor Fish* in PBL Lab to see all the remote students;
- a microphone for audio capture that feeds into the SmartBoard computer that runs RECALL™, and
- a teleconference bridge for high quality audio.

4.1 RECALL™

RECALL™ builds on Donald Schon's concept of the reflective practitioner. [17] [18] [19] It is a drawing application written in Java that captures and indexes the discourse and each individual action on the drawing surface. The drawing application synchronizes with audio/video capture and encoding through a client-server architecture. Once the session is complete, the drawing and video information is automatically indexed and published on a Web server that allows for distributed and synchronized playback of the drawing session and audio/video from anywhere at anytime. In addition, the user is able to navigate through the session by selecting individual drawing elements the user can jump to the part of interest. Fig. 1 illustrates the RECALL™ graphic user interface during real-time production that is during a communicative session. The participants can create free hand sketches or import CAD images and annotate them during their discourse. They have a color pallet, and a "tracing paper" metaphor that enables them to re-use the CAD image and create multiple sketches on top of it. The right side bar contains the existing digital sketch pad pages that enable quick flipping or navigation through these pages. At the end of the brainstorming session the participants exit and RECALL™ automatically indexes the sketch, verbal discourse and video. This session can be posted on the RECALL™ server for future interactive replay, sharing with geographically distributed team members, or knowledge re-use in other future projects. RECALL™ provides an interactive replay of sessions. The user can interactively select any portion of the sketch and RECALL™ will replay from that point on by streaming the sketch and audio/video in real time over the net.

The RECALL™ technology patented by Stanford University is aimed to improve the performance and cost of knowledge capture, sharing and re-use. It enables seamless, automatic, real-time video-audio-sketch indexing, Web publishing, sharing and interactive, on-demand streaming of rich multimedia Web content. It has been used in the AEC Global Teamwork course since 1999, as well as deployed in industry pilot settings to support:

- *solo brainstorming*, where a project team member is by him/her self and has a "conversation with the evolving artifact," as Donald Schon would say, using a TabletPC augmented with RECALL™ and then sharing his/her thoughts with the rest of the team by publishing the session on the RECALL™ server.
- *team brainstorming* and project review sessions, using a SmartBoard augmented with RECALL™
- *best practice knowledge capture*, where senior experts in a company, such as designer, engineers, builders, capture their expertise during project problem solving sessions for the benefit of the corporation.

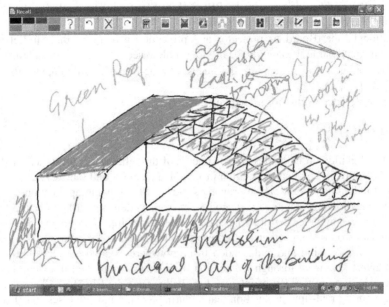

Fig. 1. RECALL™ User Interface and Sketch from an AEC Project

4.2   VSee™

The digital videoconferencing holds great promise to reduce travel cost and time for geographically distributed team members. Nevertheless, collocated face-to-face team meetings are still the most efficient and effective. One of the reasons is rooted in the multitude of cues that the participants leverage in a collocated face-to-face team meeting – such as seeing all participants at the same time, building a sense of community and shared understanding of the tasks and workspace, attention retention of participants, engagement, gesture language used to augment product descriptions or ideas. In order to support and strengthen the geographically distributed teamwork through videoconferencing each participant should be able to see all participants concurrently at low cost and low set-up overhead.

The VSee™ technology addresses this need. It allows dozens of students to take a class from different locations. The conceptual usage model of the VSee™ is that all participants can be seen and heard with minimal latency at all times. Unlike voice-activated switching in standard commercial solutions, the VSee™ allows the user to decide at whom to look. Unlike FORUM [12], the VSee™ does not require a student to explicitly request the audio channel before he/she can be heard. The experience in the AEC Global Teamwork course as well as the findings of [12] [16] suggests that keeping all channels open all the time is essential in creating spontaneous and lively dialogs. The VSee™ end point is implemented as a plug-in for Internet Explorer. Each student requires a web camera and a high-speed computer network connection. Students can see the instructor, mentors, and all the other students in a video grid on their computers. The mentors or instructor can see the students on a desktop computer or projected near life-size on a wall-size display. [1] [2].

For many years all the remote sites and PBL Lab participants were able to share data and use a teleconference bridge for high quality channel. (Fig. 2) Nevertheless, visibility was not available in any multipoint session, since NetMeeting Videoconference allows only for point-to-point video transmission and other commercial solutions

Fig. 2. PBL Lab Fishbowl™ ICT augmented workspace before the introduction of VSee™ SmartBoard running RECALL™ and NetMeeting for application sharing

Fig. 3. Stanford Fishbowl™ ICT augmented workspace with VSee™. Left SmartBoard runs RECALL™ and NetMeeting with application sharing to allow Stanford and remote participants to sketch and annotate. Right SmartBoard runs eight VSee™ video streams.

where costly and offered only voice-switched video and no data sharing. Fig. 2 shows the Fishbowl[TM] ICT augmented workspace setting before the introduction of VSee[TM]. VSee[TM] offered an important and valuable channel that allowed for all participants to be visible and create a sense of persistent presence. Fig. 3 illustrates the setting of the workspace after the introduction of VSee[TM].

4.3 Instrumenting the Fishbowl[TM] Augmented ICT Workspace for Data Collection and Analysis

The PBL Lab and the global learning workspaces create a network of hubs in which learners interact. This provides an innovative testbed to study the impact of collaboration technologies on teamwork, workspace, and engagement of learners. The data for this study was collected in the following way:

- indexed and synchronized sketch and discourse activities were captured through RECALL[TM]
- interactions, movement and use of collaboration technology within the PBL Lab workspace was captured with the video camera (Fig. 4)
- interaction and engagement of remote students was captured through a screen capture application that recorded all the concurrent VSee[TM] video streams for parallel analysis of interaction and engagement of all students at all sites.

A temporal analysis of the data was performed. It integrated the information about the speech-acts, discourse, movements, use of collaboration technologies and workspace in the PBL Lab. The result was a temporal spreadsheet with the following rubrics: (1) time stamp, (2) verbal discourse transcript, (3) mark-up of discourse transcript with specific the speech-acts (e.g., Question, Explanation, Negotiation, Exploration, etc.) (4) video snapshots of the PBL Lab configuration showing the participants movement in the physical space over time, (5) RECALL[TM] snapshots of key sketch actions, (6) screen snapshots of concurrent Virtual Auditorium streams of remote sites showing the participants movements in their physical space over time, (7) screen layout of applications on remote PCs, (8) field notes, and (9) data analysis observations.

5 Fishbowl[TM] Affordances

The analysis of the Fishbowl[TM] learning experience and ICT augmented workspace affordances was aimed to better understand the nature of engagement during the communicative events, the sharing of these workspaces, and the impact of the collaboration technologies. The analysis revealed three interaction zones and degrees of engagement, as well as and how the participants are using the collaboration technology and other artifacts to best explore and convey their ideas. The three interaction zones and corresponding degrees of engagement are (Fig. 4):

Zone 1 or Action Zone– is defined as the *action zone*, since it is in this zone that most speech-acts, interactions among participants and the digital content creation takes place, as the participants annotate, sketch, explore, explain, propose ideas.

Zone 2 or Reflection Zone – is defined as the *reflection zone*, since it is in this zone that the participants in the *Fishbowl™* reflect on proposed ideas that are presented in zone 1. In this zone questions are asked, clarifications are requested, and negotiation and what-if scenarios are brought up. As Donald Schon so well defines in the reflective practitioner, there is an interplay between zone 1 and zone 2 as the participants become reflective practitioners across disciplines building common ground and understanding of the artifact and its potential evolutions.

Zone 3 or Observation Zone – is defined as the *observation zone*, since the spectators either collocated in PBL Lab or remote participants observe the activities and discourse in the *Fishbowl™*. Rarely did participants from zone 3 move into zone 2 or zone 1 with a proposal or question.

The temporal analysis indicated the effect of the location of the collaboration technologies, more specifically the SmartBoard and RECALL™ in the use of the workspace and the disposition of the participants in the space. In the 30 *Fishbowl™* sessions studied, the presence of the SmartBoard and RECALL™ determines a circular disposition of the participants during most of the interaction with back and forth movement between the *mentor fish* and *student fish* participants between zone 1 and zone 2 in the PBL Lab. The student fish participants who were remote had a different

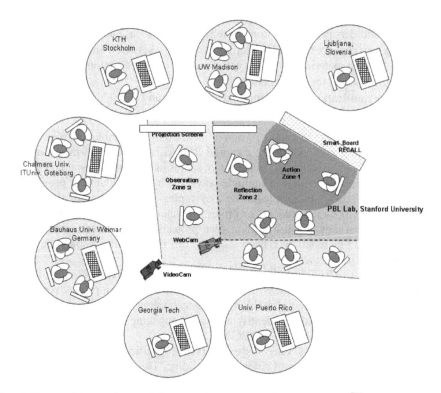

Fig. 4. Zones of interaction and degrees of engagement in the Fishbowl™ ICT augmented workspace of networked global hubs

disposition. They were sitting at tables in front of Tablet PCs in small groups (1-4 students per location) and had continuous control of their workspace and digital content that was shared. In such small group setting the collocated participants have a more confined space to control and negotiate vs. the PBL workspace that accommodates larger groups (e.g., 15-25 participants). The participants in zone 3 are students who work on similar projects as the ones in the $Fishbowl^{TM}$. They participate in the session as active observers. This provides valuable peer-to-peer learning and legitimate peripheral participation opportunities [Lave & Wenger, 1991]. Zone 3 students engage in the $Fishbowl^{TM}$ discussion by suggesting ideas related to the problem at hand, or posing questions about the proposed solutions in the $Fishbowl^{TM}$ by the mentors that they can relate to their own problems in their projects.

The observations and data collected from all the participants, i.e., learners and industry mentors, indicated that the value of new processes mediated by the $Fishbowl^{TM}$ experience and the ICT - RECALL^{TM} augmented workspace resides in the fact that it brings together and engages all participants (stakeholders) in:

- effective collaboration;
- rapid building of common ground. One of the effective actions was shared common ground building by having each participant share his/her priorities and challenges upfront made the session more effective;
- providing input from different discipline perspectives by having all the team members/ stakeholders present and contribute in real time ideas and solutions;
- enabling participants to manipulate, annotate and sketch on images, and interact with the project material that is discussed and track the discussion;
- joint exploration of ideas and concepts,
- effective identification of problems and key issues, and
- joint problem solving.

It is important to note that the $Fishbowl^{TM}$ is not only a valuable experience during the project meeting session, but the captured RECALL^{TM} rich multi-modal and multimedia content, i.e., sketch, discourse, video, serves as a learning resource over time. The session content is re-used in multiple ways: (1) by the students or project members in the $Fishbowl^{TM}$ project team, as they replay specific parts of the proposed solution details, (2) by the students or project members who were in zone 3 and find relevant ideas that they can use in their own projects, and (3) future generations of students or project members who have access to this rich learning resource leveraging knowledge from past projects.

The option to use RECALL^{TM} not only for project team meeting but also for asynchronous communication of ideas, issues, and solutions was adopted. Over time the participants started to feel comfortable with the technology and interact with the digital content, contributing ideas, identifying issues, engaging in collaborative problem solving, and proposing solutions in RECALL^{TM}.

With the introduction of VSee^{TM} we observed a significant improvement in communication and a strong sense of community building. This was caused by the fact that all sites were visible to each other. This enabled all participants to build a common ground understanding regarding the local conditions and configuration. In addition, all participants were able to observe the attention retention and level of engagement and respond to it. The temporal data analysis provided evidence that the concurrent video streams create a visibility environment that leads to new behaviors such as socializing acts, e.g., happiness to see each other, some of the participants decided to sing in front of the camera during breaks creating a sense of presence and friendship, point cameras to views from their window to share their impressions of the local weather (snow, sunny), or views of their pets at home (since some of the meetings were at very late hours for the European partners.

The comparison of the behaviors of the collocated PBL students vs. the remote students, shows the "spatial buffer effect" that is – the remote students felt more relaxed and free to move around, even leave their workspace for short periods of time, where as the collocated PBL student who were in the same room with the mentors behaved just as students would in a classroom. Consequently, even though the VSee video technology is based on the virtual auditorium metaphor of a lecture hall with large numbers of students visible to the instructor, the behavior of the students will be different and more relaxed than in a collocated context. When remote students need to discuss an issue, they feel free to have a private conversation. Such spontaneous private conversations are harder for the collocated participants in the PBL Lab where the mentors and instructor are. At that point the remote students' attention and level of engagement in the general discussion is very low. In addition, this private conversation is visible to all sites, and the mentors can intervene and ask if there is a need for clarifications.

6   The Fishbowl^TM as a Tech Transfer Conduit: Examples from Industry Pilot Projects

The Fishbowl^TM sessions offer an effective incubator and tech transfer environment for the industry mentors to experience hands-on new ICT and teamwork processes. As a result a number of successful pilot projects were launched by industry mentors engaged in the AEC global teamwork course to test the Fishbowl^TM and the augmented ICT workspace in a real industry environment. The following two examples illustrate the deployment of the Fishbowl^TM and augmented ICT workspace in collocated and geographically distributed industry setting.

Collocated Project Review Meeting with the Client. A retrofit project was launched by PG&E (Pacific Gas & Electricity Co. of California). A project review meeting with the client from PG&E and the key stakeholders who contributed to the project was organized in the ICT augmented workspace of the PBL Lab at Stanford for an industry Fishbowl^TM. The project review meeting engaged 12 participants from client PG&E, architecture firm, structural engineering firm, landscape architecture consultant, MEP subcontractor, facility management consultant. In preparation for this meeting two of the consultants put together a rich set of images and additional material to be presented and discussed during the meeting. The participants' indicated during the debrief after the session that being able to share, annotate, and manipulate the visual material was excellent. It was used to generate new ideas, discus and explore key concepts and points, that further generated good point, and questions what might work what might not work.

From a usability point of view the participants indicated that RECALL™ was easy to use in a professional environment. RECALL™ enables the participants to create rich content and share it with the rest of the team adding more value to the process, and come to the meeting prepared with questions and issues. Furthermore, having the ability of the team members to interact over the net since this is a server based technology facilitates rapid and timely content sharing that provides rich contextual rationale behind proposals or decisions. Fig. 5 shows a snapshot from the meeting in which two of the consultants identify a specific issue and explore alternative solutions annotating a preliminary CAD image. Note how seamless the RECALL™ on the SmartBoard is, as one of the participants is drawing with his finger directly on the interactive display.

The participants expressed a desire for every project review meeting to be held in the *Fishbowl*™ format using the ICT augmented workspace and RECALL™ – since it helps to see the results emerge, engage and work through the process. As a participant observed *"this experience allowed us to accomplish in three hours what would take in a typical project three months. It was amazing through how much material the team went over in such a short time. We spend time in substantial problem solving."* More than that, the process changed from sequential to concurrent interactions that provide timely input from other disciplines.

Fig. 5. Fishbowl™ ICT augmented workspace in use during an industry project review

Geographically Distributed Problem Solving and Expert Knowledge Capture. The ICT augmented workspace developed for the Fishbowl™ was deployed in an industry pilot project at Obayashi Corporation in Japan. The objective was to connect the different team members and experts located at Obayashi Headquarters in Tokyo and on a construction site in Mizunami (distance from Tokyo - couple of hours by bullet train). For that purpose, part of the Fishbowl™ ICT augmented workspace was emulated at both sites: a SmartBoard, with RECALL™ and speaker phone were installed in the Tokyo Headquarter. A TabletPC with RECALL™ and speaker phone were set up at the Mizunami construction site. This enabled the team members to have productive real-time distributed problem solving sessions with the company expert(s). This resulted in effective just-in-time problem solving. It allowed the team members to review the sessions and provided the company with a testbed environment to experiment with valuable knowledge capture in context, i.e., specific problems solved in the context of a project by corporate experts that can be reused in future projects. Fig.6 illustrates snapshots from such a distributed session from this project.

Obayashi Headquarter   **Obayashi Construction Site**
Tokyo   **Mizunami**

Fig. 6. Fishbowl^TM and ICT augmented workspaces in use during an industry pilot project at Obayashi Corporation in Japan showing a distributed problem solving session. The Obayashi Headquarters in Tokyo had a SmartBoard with RECALL^TM and the Mizunami construction site had a Tablet PC with RECALL^TM.

7 Conclusions

The paper described the *Fishbowl^TM*, as an engaging learning experience in support of global teamwork. It offers a pedagogical intervention method designed to facilitate knowledge transfer from industry mentors to students. It describes the interactive learning and ICT augmented workspace developed in support of creative, collaborative, synchronous, geographically distributed project meeting sessions. The *PBL Lab* offered a testbed for data collection and analysis of the impact of technology on behavior, learning, and team dynamics, as well as a tech transfer conduit of innovative ICT and teamwork processes from the research lab to industry.

This paper presents three additional take away messages.

1. *The act becomes the artifact.* Fishbowl^TM ICT workspace augmented by RECALL^TM mediates the creative concept generation process and supports the real-time capture of the communicative act in its original context including the dialogues, sketches, and annotated CAD/image artifacts. The RECALL^TM captured communicative act becomes a multimedia learning and knowledge transfer artifact. Such multimedia learning artifacts are used by learners for further reflection and knowledge reuse both in terms of product, i.e., solutions and ideas proposed by the mentors, as well as their collaboration process. This concept is not only valuable in an academic environment, but offers a model for corporations to capitalize on their core competence through knowledge capture and reuse.
2. *There are times when the process is more valuable than the product.* Often it is the successful new process that learners or corporations want to transfer or repeat,

rather than copy the product concept or details. Nevertheless, state-of-practice technologies focus on product or document archival. The Fishbowl™ ICT workspace augmented by RECALL™ offers an environment that captures both the product and process in context and allows future learners or practitioners to understand and reuse them.

3. *The tool equals the instrument.*   Ethnographic data collection in complex collaborative environments is typically tedious, time consuming, and intrusive. The approach taken in the PBL Lab is to integrate collaboration technologies that serve well two objectives - to support a specific design or collaboration task, and be a non-intrusive data collection instrument. In both cases, i.e., RECALL™ and VSee™, the technologies facilitate collaboration and communication and at the same time offer a channel for seamless and non-intrusive data collection for product, process, social behavior, and spatial analysis. RECALL™ system supports both the designer and the researcher since they can both use the same space (i.e., technology and knowledge archive). For the designer, the RECALL™ system is a tool that facilitates collaboration and improves productivity. For the researcher, RECALL™ is an instrument to collect data and study the team interaction and designer behavior, rationale, and work habits. VSee™ is a tool that serves two purposes at the same time – a visibility channel for the geographically distributed team members, and a data collection instrument for the research team.

The *Fishbowl*™ collaboration activities mediated by RECALL™ leads to effective (1) common ground building among team members, (2) knowledge transfer to participants that can not be present, and (3) knowledge reuse at a later time. Concurrent visual / video channels such as VSee™ enable to see all participants at the same time. This supports and strengthens the discourse and community building. Collaboration technologies that engage multiple participants in communicative events and tasks in an interactive work space define new interaction zones and lead to different degrees of engagement of the participants as shown in the *Fishbowl*™. These observations and formal models offer a blueprint for the design of the next generation interactive workspaces, collaboration technologies, and corresponding work practices and processes. One of the questions that the PBL research team is currently working on is "how to engage the students who are *observers?*" in the dialogue that takes place between *"fish mentors"* and students who are *"fish learners."*

Acknowledgements. This study was partially sponsored by the the PBL Lab, at Stanford, and the Wallenberg Global Learning Network (WGLN II). The author would like to thank PG&E, Obayashi Corporation, and all the PBL Lab academic and industry partners for their engagement and support. Last but not least, the author would like to thank the VSee Lab Inc. for the support of the Global Teamwork program at Stanford. For a complete list visit http://pbl.stanford.edu

References

1. Chen, M. *"Design of a Virtual Auditorium,"* Proceedings of ACM Multimedia. (2001)
2. Chen, M., *"A Low-Latency Lip-Synchronized Videoconferencing System,"* Proceedings of ACM Conference on Human Factors and Computing Systems, (2003).

3. Dewey, J. "Experience and Nature" Dover New York (1928, 1958)

4. Fruchter, R. *"Architecture, Engineering, Construction Teamwork: A Collaborative Design and Learning Space,"* Journal of Computing in Civil Engineering, October (1999), Vol 13 No.4, pp 261-270.

5. Fruchter, R. and Yen, S., *"RECALL in Action,"* Proc. of ASCE ICCCBE-VIII Conference, ed. R. Fruchter, K. Roddis, F. Pena-Mora, Stanford, August 14-16, CA. (2000)

6. Fruchter, R. *Bricks & Bits & Interaction,"* Special Issue on "Exploring New Frontiers on Artificial Intelligence," Eds. Takao Terano, Toyoaki Nishida, Akira Namatame, Yukio Ohsawa, Shusaku Tsumoto, and Takashi Washio, in Lecture Notes on Artificial Intelligence (LNAI) 2253, Springer Verlag, December 2001.

7. Fruchter, R. and Lewis, S., *"Mentoring Models,"* International Journal of Engineering Education (IJEE), Vol 19, Nr. 5. (2003), 663-671.

8. Fruchter R. and Demian, P. *"Corporate Memory,"* in "Knowledge Management in Construction," edited by Chimay J. Anumba; Charles Egbu; Patricia Carrillo, Blackwell Publishers. (2005)

9. Fruchter, R. and Swaminthan, S. *"Bridging the Analog and Digital Worlds in Support of Design Knowledge Life Cycle,"* Proc. ICCCBE-XI, (2006) Montreal, Canada.

10. Greeno, J.G., *"The Situativity of Learning, Knowing, and Research,"* American Psychologist, 53 (1998) pp5-26.

11. Goldman, S. & Greeno, J. G. *Thinking practices: images of thinking and learning in education.* In, Goldman, S. and Greeno, J. G. (eds) Thinking practices in mathematics and science learning. Lawerence Erlbaum Associates, Mahwah, NJ. pp1-13. (1998)

12. Isaacs, E. T. Morris, T. Rodriguez, and J. Tang *"A Comparison of Face-to-face and Distributed Presentations."* Proceedings of CHI, pages 354-361, (1995

13. Lave, J. & Wenger, E. Situated learning: legitimate peripheral participation. Cambridge, England. Cambridge University Press (1991)

14. Nonaka and Takeuchi, *The Knowledge-Creating Company,* Oxford University Press. (1995)

15. Rosson, M. B. & Carroll, J. M. Usability Engineering: Scenario-Based Development of Human Computer Interaction (2001).

16. Sellen, A. *"Remote Conversations: The Effects of Mediating Talk with Technology."* Human-Computer Interaction, pages 401-444. (1995)

17. Stiedel and Henderson, *The Graphic Languages of Engineering,* (1983)

18. Schön, D. A., , *The Reflective Practitioner* (1983)

19. Tversky, B., *"What Does Drawing Reveal About Thinking?"* Proceedings of Visual and Spatial Reasoning in Design. (1999)

20. Wenger, E. Communities of practice: learning as a social system. Systems Thinker June 1998. Available online at http://www.co-i-l.com/coil/knowledge-garden/cop/lss.shtml (1998)

Animations and Simulations
of Engineering Software:
Towards Intelligent Tutoring Systems

R. Robert Gajewski

Warsaw University of Technology, Faculty of Civil Engineering,
Division of Applied Computer Sciences in Civil Engineering,
Aleja Armii Ludowej 16, 00-637 Warszawa, Poland
R.Gajewski@il.pw.edu.pl

Abstract. The main objective of the paper is to present the state of art in the field of engineering software instruction and training. There are various approaches how to teach someone how to use an application. In the simplest illustrative approach training attempts to illustrate each screen and describe each task what is hardly possible in the case of complicated CAD/CAE software. An alternative is an exploration approach in which during training user can be asked to look at various functions. In many opinions the most effective way to learn software is scenario-based approach. Major types of authoring tools used for software simulations are described and discussed. Finally there are raised some open questions concerning costs of e-Learning and so called Civil Engineering Crisis.

1 Introduction

The types of content people are willing to have online are in majority of cases software instructions [18]. More and more online learning is related to Information Technology and also to engineering applications [8], [9], [10]. Nearly all organizations need to provide training to software users and the Web is one of the most effective tools to do this. What we are nowadays desperately looking for are multimedia-based materials [4], [5], [15], [17] - software animations and simulations. Most of the online content available is still "page turning courseware" (several years ago asynchronous kind of e-Learning has acquired a name "page-turner") and often does not yield the type of results possible from online learning. Web Based Training (WBT) is one of the most popular methods for instructing how to use software applications. One of the main aspects of WBT is software simulation.

The first objective of the paper is to acquaint with scenario and simulation based e-Learning. Simulation-based e-Learning (SIMBEL) is much more effective than classical asynchronous or scenario-based e-Learning (SbeL) [13]. The simplest definition of SIMBEL is "learning by doing". Software simulations, soft skills simulations (role-play, sales process simulations, business modelling) and hard skills simulations (troubleshooting, diagnostics, simulating physical systems) soon will be the crucial

I.F.C. Smith (Ed.): EG-ICE 2006, LNAI 4200, pp. 258–261, 2006.

part of future e-Learning environments [1]. Online animations and simulations are treated as next big wave in training. In the past simulations were extremely expensive. Nowadays tools used to prepare software animations are cheaper and this approach is gaining more interest.

The second objective is to show the potential of Intelligent Tutoring Systems (ITS) which can be used in software instruction and training. Computers have been used in education for more than 30 years. Computer Based Training (CBT) and Computer Assisted Instruction (CAI) systems were the first attempts to teach using computers. Instruction was not individualized to learners' needs. CBT and CAI do not provide personalized help and attention which could be received from a human tutor. Intelligent Tutoring Systems (ITS) and also Adaptive Training Systems (ATS) offer flexibility in presentation of material and ability to meet individualized students' needs. They are very effective in increasing students' performance and motivation.

2 Software Simulations and Training

Three major cases of use of software simulations are awareness training, full training course and performance support [7]. Typical full training course consist of three parts: animation, simulation and test. Web Based Training (WBT) is one of the most popular methods for instructing how to use effectively soft-ware applications. One of the main aspects of WBT are simulations. As presented in [12] there are five levels of software simulations (see Fig. 1). Full simulation includes all possible interactions [16]. This feature can be represented by wizards. Full simulation has usually limited number of choices in multiple paths.

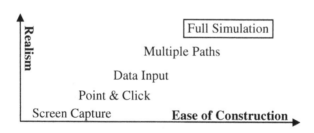

Fig. 1. Levels of Software Simulations

There are various approaches how to teach someone how to use an application [11], namely: illustrative, exploration and scenario-based approaches. The last one, in many opinions, is the most effective way to learn software. Scenario-based approach uses three fundamental interaction modes: show, teach and try [11] (see Fig. 2), but there are many arguments against it. One of them says that it is difficult to find meaningful and realistic activities for some types of software training and to develop scenarios which are directly related to specific jobs. Show modules are passive elements in which simulation level is a screen capture or point-and-click. Usually teach modules are point-and-click or data input simulations. Finally, try modules require data input, full simulations or multiple paths. More detailed information about SIMBEL and interaction modes can be found in [13] and [11].

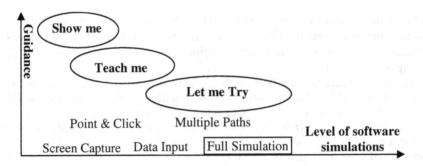

Fig. 2. Fundamental Interaction Modes

3 Authoring Tools - More Than Technology

There are three major types of authoring tools [14] used for software simulations: tools based on Hypercard tradition, tools based on Lotus ScreenCam and new generation intelligent tools understanding interaction between user and software. The extended review of such tools is presented in [3].

One of the first animated software tutorials was prepared by Simmons [19] in 2001 for ArchiCAD. Till now there are only several tutorials available on the market [10]. Software simulations are much more than only technology [6]. Three most important elements in design of simulations are wise choice of simulations, well targeted tasks and varying instructional techniques.

4 Concluding Remarks

One of the mayor conclusions concerns the costs. The average ratio for creating e-Learning content (development time versus finished hours) is 200. In the case of e-Learning simulations this ratio can be even higher - between 700 and 1300. The only chance to lower this factor is the use of appropriate authoring software enabling to create training content, to prepare simulations as tutorials, assessments and demos and to produce high quality documentation.

Intelligent Tutoring Systems could have positive impact on the solution of "Civil Engineering Crisis" [2] in the field of computing. In many cases engineers are reluctant to use computer programs because they are very complicated. Moreover, engineers being concentrated on clicking, dragging and dropping treat software as black box what can lead to incorrect use of software. Intelligent Tutoring Systems can help in crossing these barriers.

The question "how to make e-Learning interesting" [20] is still open. One of the biggest chances to do this is interactive software simulation, part of ITS. One particular animated tutorial cannot exactly fit to one particular user but it is much more effective than classical printed tutorials. There is a risk that very complicated tutorials can de as complicated as programs but it is easier to limit tutorial complexity on a certain level.

References

1. Aldrich, C.: Simulations and the future of learning: an innovative (and perhaps revolutionary) approach to e-learning, Pfeiffer (2003)
2. Arciszewski T., Civil Engineering Crisis, Leadership and Management in Engineering, 6(1), 26-30 (2006)
3. Chapman, B.: Online Simulations 2005: A Knowledge-Base of Custom Developers, Simulation Courses and Simulation Authoring Tools, Brandon Hall Research (2005)
4. Clark, R.C., Lyons, C.: Graphics for Learning: Proven Guidelines for Planning, Designing, and Evaluating Visuals in Training Materials, Pfeiffer (2004)
5. Clark, R.C.: e-Learning and the Science of Instruction: Proven Guidelines for Consumers and Designers of Multimedia Learning, Pfeiffer (2002)
6. Estabrook, S.: Making the Most of Software Simulations, Learning Circuits (2004)
7. Feldstein, M.: Desperately Seeking Software Simulations", eLearn Magazine (2004)
8. Gajewski, R .R., Morozowski, M.: Multimedialny podręcznik programu InRoads, In: III Sympozjum Ksztalcenie na Odległosc – Metody i Narzedzia (2005) 21-28 [in Polish]
9. Gajewski, R.R.: Czy i jak uczyc oprogramowania - narzędzia tworzenia animacji do symulacji oprogramowania i szkolen, In: Rozwój e-edukacji w ekonomicznym szkolnictwie wyzszym (2004) 191-203 [in Polish]
10. Gajewski, R.R.: Simulations and Animations for CAD/CAE Software, The 10th International Conference on Computer Supported Cooperative Work in Design, (CSCWD2006), May 3-5 (2006) Nanjing, P.R. China, to be published
11. Karrer, A., Laser, A., Martin, L.S.: Instruction and Feed-back Models for Software Training, Learning Circuits (2002)
12. Karrer, A., Laser, A., Martin, L.S.: Simulation Levels in Software Training, Learning Circuits (2001)
13. Kindley, R.: The power of simulation-based e-learning (SIMBEL), eLearning Developers Journal, 17 September (2002)
14. M. Paris, M.: Authoring Interactive Software Simulations for e-Learning, In: Proceedings of the 3rd IDEE International Conference on Advanced Learning Technologies ICALT'03 (2003)
15. Mayer, R.E.: Multimedia Learning, Cambridge University Press (2001)
16. Musslewhite, C.: Simulation classification system, Learning Circuits (2003)
17. Schecter, T., Fair, B., Managing the Matrix: Using Multimedia in Distance Learning Projects, Sidebars, British Columbia Institute of Technology, Learning Resources Unit
18. Shank, P.: Software Show N' S/Tell, Learning Circuits (2003)
19. Simmons T.M.: ArchiCAD v. 7.0. Step by step tutorial, Graphisoft (2001)
20. Wilson, E.: How to make e-Learning Interesting, The Sydney Morning Herald, 24 September (2002)

Sensor Data Driven Proactive Management
of Infrastructure Systems

James H. Garrett Jr., Burcu Akinci, Scott Matthews,
Chris Gordon, Hongjun Wang, and Vipul Singhvi

Department of Civil and Environmental Engineering
Carnegie Mellon University
Pittsburgh, PA 15213-3890

Abstract. In a paper presented at the ASCE International Conference on Computing in Civil Engineering in Cancun, Mexico, a vision was laid out for sensor data-driven, proactive management of infrastructure systems in which information and communication technology is used to more efficiently and effectively construct infrastructure systems, monitor their performance, and enable an intelligent operation of these systems. Since that time, a research center at Carnegie Mellon, the Center for Sensed Critical Infrastructure Research (CenSCIR), has been established with a mission to perform research towards this vision. The objectives for this paper are: 1) to discuss the motivation for such sensor-data driven proactive infrastructure management; 2) to identify and discuss the major research questions that need to be addressed by CenSCIR to achieve this vision; and 3) to present several CenSCIR projects that address some of these research questions.

1 Introduction

The U.S. infrastructure is a trillion dollar investment, defined broadly to include road systems and bridges, water distribution systems, water treatment plants, power distribution systems, telecommunication network systems, commercial and industrial facilities, etc. In spite of the enormous investments made in these systems and their importance to the US economy, we (as in government, industry and academia) are not being very good stewards of this infrastructure. The American Society of Civil Engineers (ASCE) recently announced the overall "grades" for our infrastructure systems as being a "D" [ASCE 2005]. By their nature, infrastructure systems are large-scale, networked systems with physical components that may be themselves networks of systems and whose health, due to use, environment, and abuse, can significantly deteriorate. Because of expense and growing local demand, these systems have expanded over decades in a more or less ad hoc fashion. Because of their size and highly interconnected nature, the operating conditions of the overall network are difficult to assess from local data. Often, local actions may give rise to unexpected global behavior.

There are numerous examples of where a lack of knowledge of the actual condition or state of an infrastructure system has led to failure of the system, or inefficient

I.F.C. Smith (Ed.): EG-ICE 2006, LNAI 4200, pp. 262–284, 2006.

operation from either a functional or environmental perspective. For example, the New York Times reported on March 15, 2006 the following story concerning a major oil leak in the trans-Alaska pipeline [Berringer 2006].

"The largest oil spill to occur on the tundra of Alaska's North Slope has deposited up to 267,000 gallons of thick crude oil over two acres in the sprawling Prudhoe Bay production facilities....The spill went undetected for as long as five days before an oilfield worker detected the acrid scent of hydrocarbons while driving through the area on March 2... officials from BP, the company pumping the oil, and from the Alaska Department of Environmental Conservation said they believed that the oil had escaped through a pinprick-size hole in a corroded 34-inch pipe leading to the Trans-Alaska Pipeline System. The pressure of the leaking oil, they said, gradually expanded the hole to a quarter- or half-inch wide. Most of the oil seeped beneath the snow without attracting the attention of workers monitoring alarm systems." [Berringer 2006]

On August 14, 2003, in less than 8 minutes, a blackout, twice as large as any in US history, affecting 250 power plants and 62 Gigawatts of generating capacity, left over 50 million people in the Northeast and Canada without electrical power [Mansueti 2004].

"Before it was over, three people were dead. One and a half million people in northern Ohio had no running water for two days. Twelve airports closed in eight states and one Canadian province. The estimated economic damage was $4.5-$10 billion." [Mansueti 2004]

In a more recent and local event for the authors, a section of an overpass on Interstate 70 near Washington, PA collapsed onto the Interstate and several vehicles collided with, or were struck by, the debris [Grata 2006]. The bridge had been recently inspected and given a rating of 4 out of 10, which indicates that it is structurally deficient, but was not in danger of imminent failure. Experts at the scene believe that the likely cause was a combination of "age, wear and tear in the structure, a history of being hit by trucks and very recently, another hit" [Grata 2006]. Extensive corrosion damage to the pre-stressing cables and reinforcing bars in the concrete beam is believed to have contributed to the failure; the extent of this damage was not detected during visual inspection. Fortunately, nobody was killed in this incident. However, Interstate 70 was closed for several days. "At this time of year, about 40,000 vehicles a day travel that section of I-70, about 25 percent of them trucks" [Grata 2006].

Recent technological developments make it feasible and practical to address the security, continuous monitoring, and rational and sustainable management of critical infrastructure systems, such as those described in the previous paragraphs using information and communication technology (ICT). A primary enabling technology is provided by sensor networks. These create tremendous opportunities to instrument distributed physical infrastructures with a large number of autonomous, heterogeneous, inexpensive sensors that locally process the measurements and communicate, under power constraints, the information to support intelligent decision making and control.

The objectives for this paper are: 1) to illustrate the need for such sensor-data driven proactive infrastructure management; 2) to briefly describe a new center that has been launched to explore these needs; 3) to identify and discuss the major research questions that need to be addressed by CenSCIR to achieve this vision; and 4) to present several projects within CenSCIR addressing some of these research questions.

2   Illustrations of the Need for Sensor-Data Driven Proactive Management of Infrastructure Systems

As stated in the previous section, ICT can be deployed to improve the performance and/or reduce the life-cycle cost and societal impacts of all life-cycle phases and over a broad range of physical infrastructure systems. The life-cycle of a constructed facility affords many opportunities for sensor-based decision support systems, from the first stages of construction through the demolition and reclamation of the construction materials. The next two subsections illustrate how we might take better advantage of sensors and other types of ICT to improve our construction, management and operation of infrastructure systems.

2.1   Opportunities in the Construction Phase

Construction is the first phase in the facility delivery process where ICT devices like sensors can actually be deployed and used. However, the data collected during this phase and downstream phases can also inform the upstream phases of planning and design. During construction, thousands of activities are conducted by hundreds of people to place a large amount of material and to configure hundreds of engineered-to-order components of a facility or component of an infrastructure system. There exists a number of opportunities for material to be misplaced or material quality to be affected. As such, there are a significant number of sensing opportunities on a construction site. Integrated microchips with multiple sensing functions, local storage, and integrated processing and communications can be used to track temperature, strain and acceleration on structural components [Sohn 1999; Tanner 2003; Sohn 2003]. Rewritable RFID tags can be used to track the minute-by-minute movements of equipment, material and engineered-to-order parts [Akinci 2004; Song 2004]. Embedded concrete maturity meters can be used to track the developing strength of cast-in-place concrete [Ansari 1999; Collins 2004; Gordon 2003]. Wide-area sensors, such as laser scanners, can be used to scan the physical locations of many components at the same time and help determine that they are properly placed and aligned [Stone 2004; Cheok 2000; Akinci 2004]. These sensor systems will have various forms of power supply; some will be wired to an emerging power grid, some will scavenge power from ambient vibrations [Sodano 2004], while others will be powered by battery or RF energy [Churchill 2003]. Sensors within this system will also need to detect their location and associate location with the measured values [Shaffer 2003; Tseng 2004; Brooks 2003]. These systems will also need to be self-monitoring and able to detect and correct errors in their sensed quantities before they are reported [Tanner 2003]. As these devices will be deployed in harsh environments, they must be robust and able to withstand rugged, dirty, and humid conditions. To be used on construction projects, these systems must also be cost-effective, preferably from a first-cost perspective, but definitely from a system's life-cycle perspective. Finally, these systems must be accessible and usable by humans at the home office, field office and workface [Reinhardt 2005].

2.2 The Operation and Maintenance Phase

The operation phase of a facility offers the greatest need for sensing and probably the most challenging context. From a management perspective, it is this phase that consumes the most energy and leads to air emissions. The loads on the facility (e.g., wind, live loads, snow), the performance of that facility, and the condition of the various components of that facility need to be monitored so as to permit its efficient and safe operation and to allow for efficient facility maintenance and management. The use of sensing to control the operation of the mechanical and electrical systems in a facility is not a new problem. Many companies manufacture and sell building control systems. However, the extensive use of sensing to determine the loads, performance, and condition of the structural and other functional elements is less common. For example, the integrity of roof and wall systems is paramount for maintaining a healthy facility and keeping water from entering the building envelope. Sensors could be deployed throughout a building façade and roof system to act as sentinels looking for penetrating moisture. Depending on the grid size of the sensor network, the system could simply locate the existence of a leak or more accurately pinpoint the specific location of the leak so that a very efficient and cost-effective maintenance process can be conducted. All too often in built facilities, small problems that could have been easily and cheaply fixed if detected early grow into costly and extensive maintenance efforts. If the actual loads experienced by a facility (from the environment and from usage) are tracked, the performance of the structure can be much better understood. Such knowledge could also significantly inform future design activities about the actual nature of the loads likely to be experienced. The challenges for deploying sensing during operation are: large scale; long time-frame; and harsh environment.

In addition to the 2003 North American Blackout and the I-70 bridge collapse described in Section 1, several additional examples further illustrate the opportunity to deploy ICT to improve the sustainable and secure operation of these systems.

Water Main Failures. Another example of a significant need for more condition information is the problem of water main breaks throughout the US and other countries. Water main breaks wreak extensive havoc on local residents and businesses. For example, a major break to a 36 inch (approx. 91.5 cm) diameter water main occurred in Pittsburgh in 2005, next to a dense collection of highrise buildings containing many businesses and condominiums [Lara-Cinisomo 2005]. The break caused floods in many of these buildings basements, leading to a loss of power and water to all residents, leading to businesses being closed and residents being displaced for several days [Lara-Cinisomo 2005]. According to Feiner and Rajani, "Direct inspection of all water mains is often prohibitively laborious and expensive"

[Feiner 2002]. According to the GAO, "In the United States, about 54,000 community water systems supply most of the nation's drinking water and about 16,000 wastewater treatment systems provide sewer service" [GAO 2004]. In this same report, a survey of water utilities indicates that one third of them stated that about 20% of their pipelines needed to be replaced. The GAO also states that "utilities reported that collecting accurate data about their assets provides a better understanding of their maintenance, rehabilitation, and replacement needs, which helps utility managers make better investment decisions" [GAO 2004]. This need for more information about water mains represents a significant opportunity for cost-effective applications of ICT.

Residential Sustainable Operation. There is a significant potential to improve the control and management of residential energy use using existing ICT. Networkable temperature sensors, energy meters and switches continue to decline in price and improve in sophistication, and could be integrated into energy monitoring and control systems that inform residents of how and where energy is being used and provide automation of many actions to affect their consumption [Williams 2006]. Unfortunately, only 27% of homes in the US had programmable thermostats in 2001, a basic step in energy management, and homeowners by and large still have very little information on energy use beyond their monthly meter reading [Williams 2006]. Delivering and deploying such ICT-based monitoring and control systems in the future can likely save money for homeowners as well as reduce energy use, as US consumers spend about US$1,400 per year on home energy, a significant incentive to improve home energy utilization [Williams 2006].

Sustainable Transportation Infrastructure. In the US, there are over 600,000 bridges in the National Bridge Inventory that must be inspected at least every two years. The current bridge inspection and management process utilizes a paper form-based condition rating method, whereby the rating of the entire bridge is based on the conditions of all of the elements, no matter the relative importance of the elements. ICT can be used to more efficiently and effectively construct infrastructure systems, monitor the performance of these systems (if one maintains accurate performance records, they should also inform future design decisions), assist in the management of the entire set of assets having to be managed within an allocated budget, and assist in the operation of these systems. Improved life cycle management tends to lead to less major renovation and reconstruction activities, leading to sustainable outcomes.

As an example of the kind of transformation that can occur, consider the 1106 km of the Portuguese system of toll roads operated by Brisa (the main part of the whole Portuguese network), which recently adopted an aggressive policy to deploy and use ICT in their operations over the entire toll road system [Bento 2004]. 57% of their tolls are collected electronically, currently using RFID tags. A 1-8Gb/s network over the entire road system and use it for data transmission, voice over IP (VOIP), office applications, distributed data storage, and roadway telematics (namely for the streaming of digital video generated in some 450 CCTV cameras along the motorway network). They use this ICT infrastructure to assist in the coordination of roadside operations, roadway incident monitoring, toll collection, and roadway incident/

condition monitoring. Brisa has seen significant and demonstrable economic benefit from their effective deployment of ICT. Thanks to a number of efficiency factors that have occurred after they intensified their ICT usage, their operating margin nears 75% (the world's highest amongst listed tolled operators). This is one reason why the company maintains their current level of ICT spending at around 1.5% of their operating revenues. As a matter of illustration, that compares with some 4% spent in road maintenance [Bento 2004].

3 CenSCIR: Center for Sensed Critical Infrastructure Research

In late 2005, the Center for Sensed Critical Infrastructure Research (CenSCIR, www. ices.cmu.edu/censcir) was created at Carnegie Mellon, building on components of the research activities in the departments of Electrical and Computer Engineering, Civil and Environmental Engineering, Engineering and Public Policy, in the College of Engineering, and the departments of Computer Science and Architecture. This section briefly describes the center vision and the research questions its members seek to address.

3.1 Vision and Mission for CenSCIR

To avoid costly failures and provide a 21st century infrastructure, the United States and other governments must build their critical infrastructures with a "nervous system" that collects and feeds data to places in the system that interpret it and allow better decision making. CenSCIR will perform research that will motivate the need for, provide design guidance for, and provide the clear justifications for implementers, operators and future designers of critical infrastructure systems to provide sensor data-driven awareness of the usage and condition of their systems (both for components and the entire network), and proactive, intelligent decision support and control of these critical infrastructure systems over their lifetime.

The mission of this center to: (1) perform research with industry and government partners to develop a thorough understanding of the data and decision support needs (human and autonomous control) in a variety of infrastructure contexts and the economic implications of delivering such support; (2) perform research on sensor devices, data models, data interpretation techniques, system behavior models, or decision support frameworks that addresses the needs of a specific critical infrastructure system; (3) perform research needed to develop sensor-data driven vertical decision support systems for specific critical infrastructure systems or components, from the sensors needed, to the data models used, to the algorithms and models applied to interpret that data, to the decision support needed to assist in the construction, operation and maintenance of an infrastructure system; (4) validate these systems using combinations of laboratory testbeds, actual infrastructure systems and simulations; (5) perform research to explore the common aspects of the developed vertical systems to determine if common approaches or tools may be developed from these more specific solutions; (6) explore new network theories concerning the value of the information provided by sensors distributed throughout a critical infrastructure network, how that information is best converted into decisions (centrally or locally), whether this decision making knowledge must be specified or can it be learned, and whether that information is able to improve the reliability, stability and quality of service of critical infrastructure systems; and (7) perform research to develop frameworks that assist developers of such sensor data-driven decision support systems to take advantage of the knowledge gained by this center when creating new vertical decision support systems for critical infrastructure contexts.

3.2 Research Questions to Deliver CenSCIR Vision

There are many research questions that will need to be addressed to achieve this CenSCIR vision. The set of research questions include:

What information needs to be measured about a system, and where and when it should be measured?

What types of sensor are needed, but do not exist in a form usable for infrastructure applications? How are they powered for long periods of time? How do they remain functional for the long lives of most infrastructure systems?

How should this extremely large amount of information be represented, stored, managed and exchanged and who is responsible for the stewardship of this data?

How does one predict global behavior, and more importantly the onset of an incident, from localized sensor information, and what inference algorithms are needed to infer the state of the system without having centralized knowledge of the network?

What types of action need to, and can, be taken to prevent abnormal behavior?

How does one translate the mass of information collected from the distributed sensor network into intelligent decision support that actually helps the operators of these systems make the best possible decision under the circumstances?

What are the economic conditions that make such an approach to delivering a "nervous system" for an infrastructure system economically viable? Does it make sense to use such a system only at certain times when problems are anticipated, such as at the end of life as in the I-70 bridge example? What can we do at construction time and what can wait until later?

These and many other questions need to be answered for the vision of CenSCIR to be achieved.

While specific technologies are becoming available for effective on-line gathering of real-time data in such systems and the case for their use is being made, it remains critically important to develop methods for differentiating between the information necessary to make decisions in order to prevent undesired performance, on one side, and the excessive data gathering, on the other. Such methods are not available, and this vacuum creates a major obstacle to enhanced performance of our critical infrastructures, and it creates a fantastic opportunity for research.

4 Several Illustrative Projects

In this section, three ongoing projects are presented to illustrate various approaches that are being taken to begin to answer the research questions presented in Section 4.

The first project, being conducted by Gordon, Akinci and Garrett described in Section 0, explores the need for, and mechanisms to deliver, sensor-based construction inspection planning. The second project, being conducted by Wang, Akinci and Garrett and described in Section 0, explores ways to more efficiently map between data models in an evolving modeling context by making the most use of previously acquired mapping knowledge. The third and final project, being conducted by Singhvi, Matthews and Garrett and described in Section 0, explores a utility maximization-based approach for sustainably controlling infrastructure systems based on sensor information and knowledge of users' and operators' utility functions.

4.1 Inspection Planning for Construction Site Inspection

4.1.1 Need
Construction inspectors need inspection planning assistance to ensure that inspection goals are properly specified, to identify and search among inspection alternatives, and to help intelligently deploy inspection resources. While reporting on best practice quality management strategies in the construction industry, the Commission on Engineering and Technical Systems (CETS) stated that inspection planning is needed to assure effective inspections, and that current practices for developing and verifying requirements for inspection are error-prone and can result in missed inspections [CETS 1991]. The need for inspection planning on construction sites, and the problems associated with poor specification and implementation of inspection requirements motivate the development of approaches to help plan inspections for construction projects.

Requirements for a formal representation of goals and plans for inspection have been established to address the need for inspection planning assistance. According to these requirements, construction inspection planning requires formal reasoning to develop and reduce inspection planning spaces. Development of inspection planning spaces requires development of inspection goals and feasible inspection plans that may be applied to address these inspection goals. Formally developed inspection goals specify the inspection action, building element, and property to be inspected; contain zero or many constraints on how they are to be addressed; are known to be applicable to projected weather conditions; and are refined to the lowest level of detail needed to support development of inspection plans. Formally developed inspection plans specify the method and corresponding sets of activities that are required to address inspection goals, and can be evaluated to determine if they meet constraints associated with inspection goals, and to determine how different inspection plans compare to each other for a given goal. Reduction of inspection planning spaces requires formal reasoning to simplify sets of inspection plans and search among sets of inspection plans that may be used to address the set of inspection goals selected for a construction site.

Inputs needed for these processes are detailed project models, including product and process information. This information, and knowledge of construction site conditions, prior inspections, and inspection knowledge, can be reasoned with to develop goals and constraints for inspection. Project models can also be reasoned with in addition to inspection knowledge and sensor knowledge to determine which measurements, actions, and technologies can be used to gather this information. Finally, knowledge of evaluation criteria can be reasoned with to reduce the set of feasible inspection plans for the set of inspection goals required for a construction site. Inspection plans developed for construction sites should be linked to the project model.

The formalism is component-based, in that inspection goals are associated with individual building elements, rather than a material or a construction activity. This reflects the emergence of research and current practice, such as in [Uzarski 1989; Demers 2002], into definition of goals for inspection and the collection and interpretation of inspection data for facilities at a component level. This section describes the formalism (e.g. a syntax and semantics) that we have developed for construction inspection planning to meet the requirements described above. The intent of the formalism described in this Section is to support systematic inspection planning through the development of a formal representation of inspection goals and inspection plans, and formal reasoning that can be used to develop inspection goals and inspection plans to support detailed comparison of various alternatives for performing inspections.

4.1.2 Motivating Example

Consider an office building that contains a cast-in-place concrete parking garage at its lowest levels. Among the components in the lower level are a slab-on-grade, *Slab S1*, and two lines of columns, *Columns 1-3*, and *Columns 4-6*. Columns *1-3* were placed and cured in the month of October, when air temperature averaged above 4° C. *Columns 4-6* were placed and cured during the month of November, when air temperatures averaged less than 4° C.

Among the inspection goals derived from specifications to inspect concrete quality in this example were the inspection of concrete strength, concrete workability, and concrete temperature, as shown in Table 1. In addition to these properties, the specifications listed single methods that were to be used to assess these properties. The specifications called for one test of concrete temperature per hour using ASTM C1064 when temperature was below 4° C and above 27° C, in addition to one test of concrete temperature from every set of samples taken for compressive strength

Table 1. Inspection goals drawn from specifications for the project in the motivating example are directly mapped to inspection methods by the specifications, and indicate if their existence is weather-dependent

Component	Inspection goal	Inspection method	Weather applicability
Slab S1, Columns 1-6	Inspect concrete workability	ASTM C143	All weather
Slab S1, Columns 1-6	Inspect concrete compressive strength	ASTM C39	All weather
Column 4 – 6	Inspect concrete temperature	ASTM C1064	Air temperature <4° C

testing. Concrete strength was specified to be inspected per ASTM C39 by taking samples of concrete as it was delivered to the site and then subjecting these samples to destructive compressive strength tests in a laboratory setting. Concrete was also specified to be tested daily for concrete workability by testing concrete slump per ASTM C143.

The specifications for this project partially address two main requirements for inspection planning: inspection goal development and inspection plan development. In terms of inspection goal development, (1) the specifications identify the properties to be inspected for components that are composed of concrete; (2) they specify the weather conditions in which the inspection goals are applicable; and (3) they contain informal mappings between measurable properties (e.g. slump) and the qualitative properties that these measurable properties can address (e.g. workability). Further inspection goal development concepts, not addressed by these specifications, are limits on the performance of inspection plans used to address these goals, such as the cost of addressing an inspection goal. In terms of inspection plan development, the specifications for this project contain direct mappings of inspection goals (i.e. concrete slump) to the inspection methods (i.e., ASTM C143) that can be used to address the goals. Because the specifications contained direct mappings among inspection goals and single inspection methods, the specifications did not permit a detailed comparison of different possible inspection methods for inspection goals.

Considering all possible methods for inspection goals as an alternative to ad hoc links between inspection goals and inspection methods opens up the possibility of improving the inspection planning process by enabling identification of alternatives that may be better suited for given contexts. For example, although the concrete maturity method was not mentioned as a possible method to test concrete strength, it has been recommended for schedule acceleration applications [Cable 1998]. Despite the potential for process improvement by considering multiple inspection methods, it is relatively difficult to search among the large amount of inspection methods. ASTM, for example, has standardized approximately 5,000 inspection methods. In the motivating example, the slab and six columns each had three inspection goals per component. Reviewing each of the 5,000 inspection methods to determine all of the methods that can address each of the inspection goals will result in 105,000 applicability tests. However, by classifying the inspection methods according to material, the search space is reduced to 3,000 applicability tests. By further classifying these inspection methods according to the inspected property, the search space is reduced to 147 applicability tests. This demonstrates the effectiveness of reasoning with more than one classification facet to limit the search for applicable inspection methods.

Once applicable inspection methods are identified, these may be reasoned with to develop and compare inspection plans. The number and cost of resources must be reasoned with to support cost-based comparisons of inspection plans. Some inspection methods, such as the concrete maturity method, require the use of sensors that are embedded on or within components. Such methods specify how to reason with component geometry or material quantities to determine the number of measurements or sensors that are needed to inspect a given component. Hence, it is necessary to reason about how to distribute sensors within a component and where to embed them to make necessary measurements.

The motivating case demonstrates concepts needed to develop inspection goals and inspection plans for construction sites. For example, some inspection goals are only applicable in particular weather conditions (e.g. cold weather conditions); some goals can be refined to more detailed inspection goals (e.g. a concrete workability inspection goal can be refined to a concrete slump inspection goal); and inspection goals can be addressed by one or more inspection methods. It also demonstrates the need for increased formalism to support consideration of multiple possible inspection methods and their associated resource distributions.

4.1.3 Overview of the Approach

Classical planning approaches develop plans that consist of activities that can transition a "world" from one or many initial states to one or many goal states. Hierarchical Task Network (HTN) algorithms start with one or more partially defined tasks, or "goal tasks", which are decomposed into more "primitive tasks" using methods or "complex tasks". Method-based planning incorporates methods, which are standardized procedures that are selected according to applicability conditions to automatically create plans [Sacerdoti 1977; Erol 1994].

The formalism discussed in this Section, and illustrated in Figure 1, builds on an HTN-style approach, whereby inspection goals are represented as partially defined activities. Inspectors begin with one or more project-specific inspection goals, but without any prior plan development. Project-specific inspection goals are matched to generic inspection goals, which may be associated with a decomposition hierarchy detailing more specific inspection goals. These more detailed goals are used to refine the project-specific inspection goals. Project-specific inspection goals are then matched with applicable generic constraints, which are then instantiated and associated back with the project-specific inspection goals. Project-specific inspection goals are then matched to inspection methods. These inspection methods can be instantiated to develop inspection plans, composed of the measurement actions, preparatory actions, and resources that are needed to address a given inspection goal.

Faceted classification is used to support reasoning about matching between project-specific inspection goals and generic inspection goals, constraints, and inspection methods. This approach entails developing one or more classification facets (e.g. classification by material, component type, etc.) and associating the nodes within of each facet (e.g. concrete, steel) with instances of each generic inspection goal, constraint, and inspection method. These pre-determined relationships are reasoned with to determine if an inspection method, generic inspection goal, or generic inspection goal constraint is applicable to a given construction site context.

Applicability reasoning begins when an applicability control object identifies an applicability facet to review first. It then traverses the tree of applicability nodes in this facet until a match is discovered, lists goals, constraints, or methods associated with a matching node, proceeds to the next facet; and processes this facet similarly. It then ensures that the listed goals, constraints, or methods are applicable to each matching node; and returns the set of applicable goals, constraints, or methods for a given inspection goal.

To demonstrate the formalism, we have implemented a Java-based prototype inspection planning system, called MakeSense. We use 3D design data generated in ArchiCAD 9.0 and output in .ifc format to develop project models. To initialize the prototype, a user imports a project model that has been saved in .ifc format and parsed into Java classes; loads files that describe inspection goal, method, and resource libraries; and loads files that describe specifications that have been represented in .xml format. These specifications are reasoned with to begin the inspection goal development process.

Research in individual inspections, such as the work described in [Bungey and Miller 1996], has identified criteria, such as resource cost, time, and accuracy, which may be used to compare inspection plans. Such evaluation criteria can guide search of available inspection plans. For example, Table 2 shows results of using such a cost function with search methods based on random search, genetic search, and simulated annealing, in comparison to exhaustive search.

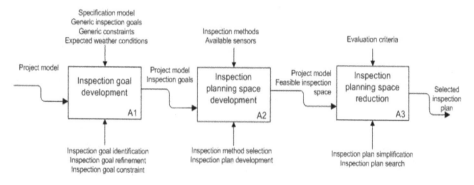

Fig. 1. Inspection planning process develops inspection goals and plans, and searches for plans to recommend from among the feasible plans that have been developed

Table 2. Comparison of search algorithms for to exhaustive search for plan selection

	Random Search	Genetic Search	Simulated Annealing
% Exhaustive search (motivating example)	83.4%	94.2%	94.9%
% Exhaustive search (large scale problem)	51.1%	76.17%	75.9%

In this example, the small problem corresponds to the set of combinations of inspection plans for a set of inspection goals in the motivating example, while the large problem corresponds to a similar ratio of applicable inspection plans, but for 100 goals. Genetic search and simulated annealing find near-optimal plans after evaluating 1,000 plans, while random search consistently finds good, but not near-optimal solutions. Stochastic search strategies can therefore be effectively applied in conjunction with existing evaluation criteria to identify comparatively good inspection plans from among the numerous possible combinations of inspection plans that can be created for a given set of goals. Furthermore, these results indicate that while applying the formalism results in large inspection planning spaces, these spaces can in fact be searched to identify relatively good inspection plans within a short amount of search time.

4.2 Flexible Data Exchange Under Changing Conditions

4.2.1 Need

The need for data exchange between computer applications has existed for decades. Data exchange among applications involves translating data, which is modeled specifically for one application, into data that can be understood by another application. It requires that the target data model represents the source data as accurately and completely as possible to minimize data loss during exchange [Fagin et al., 2003]. This requirement arises in many domains, where independent applications do not necessarily adopt the same data model (or schema). As we collect more and more data about the various components, subsystems, and systems related to critical infrastructure, this need for data exchange among models will only become more severe.

The Architecture, Engineering, and Construction (AEC) industry is recognized as a multi-disciplinary and multi-participant industry. Data created in the AEC domain include 2D and 3D drawings, contracts, specifications, standards, reports, etc. It is created by users like owners, designers, constructers and inspectors, through many different domain-specific applications. To enable interoperability between different software systems, data must be exchanged between multiple users or applications by some public data exchange standards such as Industry Foundation Classes (IFC) [IAI 2003a], ifcXML [IAI 2003b], CIMsteel [Eureka 2004], AEX [FIATECH 2004], etc..

Accordingly, there is a long standing requirement to match different data schemas (e.g., a task specific schema or a public data exchange standard). However, manual matching of data models is time-consuming, error-prone and tedious work. Given the rapidly growing number and scale of data models used in today's applications, manual matching is becoming a much harder task. In addition, the challenges associated with model matching become even more pronounced when a source or a target model is changing frequently, which often happens in the real world. For example, in the last three years, the IFC data exchange standard has undergone two major updates, Release 2x and 2x2, but most IFC-capable commercial software only supports Release 2.0 or worse only Release 1.5 [Steinmann 2004].

It has been demonstrated that a computer (IT)-enabled process can provide further help in this matching process. Some prior studies (e.g., [Doan et al., 2000, Madhavan et al., 2001, Mitra et al., 2000, Li & Clifton, 2000]) could perform a considerable part of schema matching automatically and output satisfactory results under certain circumstances. However, because of the complexity of applying human knowledge, there are two research areas that most of the prior studies do not address: 1) how to re-use previously existing matches; and 2) how to use domain specific knowledge [Rahm and Bernstein 2001].

4.2.2 Approach

In this research, we developed an approach that partially addresses the above issues to improve the data model matching process, by utilizing prior matching correspondence and constraints introduced by domain knowledge in a certain AEC domain (e.g., the Building Commissioning domain) where both the source and the target models are changing frequently.

Figure 2 illustrates the overall procedure of the proposed research, which aims at matching a Building Commissioning data model (i.e., BC data model [Akin et al. 2003]) to different releases of the IFC data exchange standard.

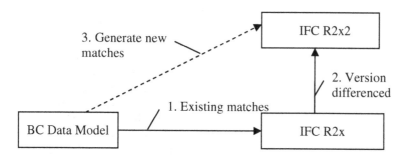

Fig. 2. Matching BC data model to various releases of the IFC

First, we have developed an approach that precisely detects differences between different versions of the same data schema. This approach includes two parts: 1) a taxonomy of version differences to identify which part of an element is changed; and 2) a semi-automated version matching approach that applies both linguistic-based and constraint-based schema matching algorithms to detect version differences based on the classifications. Because a new version of a schema usually builds upon its prior version and shares a lot of common contents, it is reasonable to assume that differences between releases will not be significant and a computer-aided approach can potentially perform in version matching much better [Amor and Ge 2002].

Second, we have integrated the version differences of the target model, with the existing matching results that map the previous version of the target model to a source data model, so as to deduce new matching results. The research has identified a collection of patterns, each of which combines a particular kind of existing matching result and a particular type of version difference to generate a new matching result. For example, the *Event* class declared in the BC data model represents an inspection activity in the commissioning process and in the IFC Release 2.0, it was matched to the class *IfcWorkTask* that is a unit of work. In the IFC Release 2x, the class *IfcWorkTask* was renamed to *IfcTask*. With those facts known, a *Rename* pattern is defined indicating that the new matching result is that *Event* is matched to *IfcTask* directly in the IFC Release 2x. Compared to scanning the entire IFC R2x schema to find the proper IFC class, updating prior matching results will save a considerable amount of effort and achieve better accuracy.

Finally, to utilize domain constraints, we have explored different types of constraints (i.e., those associated with form, function or behavior) introduced by domain knowledge that will affect the data matching process. A representation and reasoning approach has been developed to incorporate these constraints with fundamental data model matching algorithms proven by other studies, such as name matching, to improve the data model matching process.

4.2.3 Prototypes and Results

We developed a prototype of the version matching approach to validate this approach with real world test cases. When comparing the IFC R2x and the IFC R2x2, the approach can complete the matching process in a few seconds, compared to two weeks of manual work. In addition, the accuracy of our matching results is close to 97% at the class level and 99% at the attribute level. The approach can even find a few pairs of related classes that are hard to discover by the manual matching approach because the version change is deeply hidden in the schema.

Based on the precisely detected version matching results, another prototype was implemented to validate the upgrade patterns we created as part of this research, updating existing matching results associated with specific patterns of version differences. Compared to manual work, our approach significantly saves human effort while achieving comparable accuracy. Compared to other computer-aided approaches that compare two schemas from scratch, this approach achieves higher accuracy, because it utilizes previously proven correct human knowledge. For example, when matching the Building Commissioning (BC) data model to the IFC R2x, our approach achieves 95% accuracy, which is much higher than other computer-aided approaches that compare two models from scratch and much faster than the manual process. In addition, the approach also addresses some hard matching problems (e.g., 1:N matching) efficiently.

Currently, we are working on representing and reasoning with domain constraints. An ontology library has been developed and used to assist the schema matching problem. We are attempting to understand what types of constraints (i.e., Form, Function and Behavior) will bring the greatest impact to the matching process and in what ways each type will affect the results.

It is a significant challenge to perform data model matching efficiently. In this project, we have developed a semi-automated approach that addresses this challenge in a specific domain (HVAC), where the source or the target data model is being upgraded frequently. Human knowledge will be applied in terms of prior matching results and domain constraints, both of which are rarely addressed in prior studies.

Although the research is not completed yet, the current results have already demonstrated that re-using existing matching results will significantly reduce human workload on model matching and achieve comparable accuracy under some cases. On the other hand, our research also indicates that domain constraints show some significant potential to assist in the schema matching process.

4.3 Utility Based Decision Making in Building Infrastructure

4.3.1 Need

Meeting human preferences of comfort, safety and privacy are major factors in the success of many civil infrastructure projects. Currently, these factors are taken into account by incorporating available standards during decision making. In building operation, occupant's comfort is measured using standards like those from ASHRAE [ASHRAE 1980]. However, most of the standards represent approximation of these preferences, as in reality, they are unique for each individual and often are location and time dependent. The inability to integrate the seemingly important factor of human comfort in building operation is due to lack of a framework to quantitatively capture human comfort preferences. In this section, a project is presented that is aimed at developing a principled decision theoretic approach using utility theory to successfully integrate these preferences into automated building operations. The approach has been implemented as an intelligent lighting control system in three actual testbeds for testing under different scenarios.

4.3.2 Approach

Building operation is a complex activity, where building operators and occupants continuously interact with each other. Typically, the interaction is passive with little communication between the operator and the occupants. Historically, *maximizing occupant comfort* and *minimizing energy costs* have always been two primary objectives of intelligent buildings operation [Finley Jr. M. R. 1991; Flax B. 1991]. The trade-off between meeting occupant preferences for indoor environmental condition and reduction in energy usage leads to a difficult optimization problem and this optimization can be thought of as *decision-theoretic* — the objective is to *minimize* the expected cost of building operation and to *maximize* the occupant's expected comfort. The challenges to develop such a balanced control strategy are threefold. First, we need to identify the preferences of individual occupants in indoor environments continuously, as preferences change over time and as the occupants move in the building. The second challenge is to gather information about the immediate indoor and outdoor environment of the occupants. The third challenge is to optimize the trade-off between meeting occupants' preferences and reducing energy usage.

The proposed decision-theoretic approach described in [Singhvi 2005] optimally achieves occupants' light preferences and energy usage tradeoff by solving a multi-criterion optimization problem. We show that given the utility function for individual occupants and the operation cost, the proposed *coordinated illumination approach* can efficiently optimize the tradeoff in meeting occupant preferences and energy usage. Our control strategy integrates individual occupants' preferences and the real state of indoor and outdoor environment through a network of wireless sensors. The control strategies quantify the trade-off between meeting occupant preferences and the corresponding energy utilization. While decision-theoretic optimization provides a powerful, flexible, and principled approach for such systems, the quality of the resulting solution is completely dependent on the accuracy of the underlying utility function. To extend the coordinated illumination approach, we have developed a utility elicitation approach that addresses the needs and requirements of an intelligent building system. The main contributions of this work are:

- The development of a formal decision theoretic framework for integrating occupants' preferences in building operations;
- The development of an efficient *coordinated illumination* algorithm for optimizing the tradeoff between meeting occupants' preferences and reducing energy consumption;
- The development of a principled utility elicitation technique using minimal interaction and partial information provided by the occupants; and
- The extension of our approach to optimally exploit external light sources and optimally exploit spatial and temporal correlation for sensor scheduling.

4.3.2.1 Coordinated Illumination Algorithm. Utility functions are defined over a space, which is exponential in the number of variables on which the utility depends. Typically, such representation of utility function leads to an intractable optimization problem due to the exponential nature of the solution space. More tractable representations of the utility are possible if we make certain assumptions about the *additive independence* [Keeney 1976] among the variables. We assume that the occupant's utility function is derived from *sub-utility* components, which reflects preference of the occupant for various parameters in the indoor environment.

More formally we assume that the occupant's utility function **U** is linearly additive, i.e. there is a set of sub-utility functions $\boldsymbol{\varphi} = \left\{ \phi_1, \phi_2 \ldots \ldots \phi_k \right\} \in \left[0,1 \right]^k$, such that for any indoor environment state **s** in the building system the associated utility for the occupants is given by:

$$\Phi(\mathbf{s}) = \sum_{i=1}^{n} w_i * \phi_i(s_i) \tag{1}$$

Here, **s** is the indoor environment state defined as the vector of the various parameters s_i involved in defining the state. $\Phi(\mathbf{s})$ is the occupant's utility function representing the preference for the comfort in state **s**. ϕ_i is the sub-utility function and w_i are the associated weights in the utility function. The *operating cost utility function* is defined as Ψ, which decreases monotonically with the energy expended.

A building has multiple occupants with varying preferences and hence with varying utility functions $\Phi_1, \Phi_2 \ldots \ldots \Phi_n$. When cosidering occupant preferences and the operating cost, the goal is to tradeoff Ψ with the occupant utility $\sum_{i=1}^{m} \Phi_i$. A common technique of '*scalarization*' is then employed to solve this multi-criterion optimization problem by defining a *system utility* function, $U(\mathbf{s})$ which is defined as follows:

$$U(\mathbf{s}) = \sum_{i=1}^{m} \Phi_i(\mathbf{s}) + \gamma * \Psi(\mathbf{s}) \tag{2}$$

The optimal control strategy in this scenario is to find **s*** such that:

$$\mathbf{s}^* = \operatorname*{argmax}_{\mathbf{s}} U(\mathbf{s})$$

Note that the solution space is exponential with the number of acuatable state variables, and enumeration of all possible states is impossible. Singhvi et al. [Singhvi 2005] present an efficient algorithm to solve this exponential maximization problem exploiting the *zoning* principle in lighting design (i.e., not all lights affect all spaces).

The success of such a decision support system depends on how well the utility function Φ represents the preferences of the occupants. The main challenge is to estimate/elicit the shape of the *sub-utility* function ϕ_i (as defined in eq 1) and the associated weights w_i for each occupant. In Section 0 we propose an approach for a preference elicitation system that would estimate the utility functions using partial information from the users.

4.3.2.2 Preference Elicitation. To address the unique requirements and features required for a utility elicitation tool in building control, we have designed our tool in two stages. In the first stage called, **sub-utility elicitation**, we let the occupant visually explore the lighting space by letting them operate the lighting control, while constantly soliciting qualitative feedback. Since every occupant has a unique sub-utility function, it is important that each occupant provide the system with that information. This is the only stage where we actively solicit feedback from the occupants. The goal of this stage is to estimate the shape for all the sub-utility functions.

The second stage is called **weight elicitation**. In this stage, the system starts by using a set of weights to form the utility function for the occupants. The weights are chosen to meet certain constraints based on the domain knowledge and decision theoretic concepts. In this stage, the occupant does not have to interact with the utility elicitation system, however, if an occupant is unsatisfied with the current state of the system, he or she can use the lighting control to change the setting. The utility elicitation system observes the behavior and uses it to update the utility function. Over time, with minimal interaction from the occupant, the system would be able to learn a very close approximation of the utility function.

4.3.3 Results

We implemented the control strategy in a test bed at Carnegie Mellon University. We created the test bed (shown in Figure 3) to emulate the real situation at a smaller scale to test our lighting control strategy. Our test bed consists of twelve MICA2 motes and ten 60 Watt table lamps, arranged on a 146 in. by 30 in. table. The lamps are actuated by the X10 system, which wirelessly communicates with a single desktop PC, and uses power lines for controlling the lamps. Each lamp can be actuated to produce ten different light intensities. The lamps are arranged in a triangular pattern which corresponds to the zoning concept. Each of the seven triangular zones is affected by three lamps, and each lamp affects up to three regions. The motes are distributed over the different zones, and communicate with the base station over an ad hoc wireless network. We added 5 wall lamps to this test bed that acted as a source of external light.

We tested various scenarios in the test bed; a brief summary of the detailed results presented in [Singhvi 2005] is:

- For our setup, at $\gamma = 0.4$ (see Eq 2), the system saves about 30% of energy while allowing minimal loss of occupant utility;
- Coordinated lighting approaches do significantly better than a typical greedy algorithmic approach of actuating the lamps;

- Exploiting day-lighting (external lights) results in about 70% saving in electrical energy during daytime with minimal loss in occupant utility; and
- Scheduling sensors to take readings based on known sunlight distribution schedules leads to a utility maximization/energy minimization performance comparable to continuous sensing, but uses only 10 readings/sensor/day thus greatly extending the battery life of the sensors.

5 Closure

The U.S. infrastructure is a trillion dollar investment, and in spite of the enormous investments made in these systems and their importance to the US economy, we (as in government, industry and academia) are not being very good stewards of this infrastructure. To avoid costly failures and provide a 21st century infrastructure, the United States and other governments must build their critical infrastructures with a "nervous system" that collects and feeds data to places in the system that interpret it and allow better decision making.

In late 2005, the Center for Sensed Critical Infrastructure Research was created at Carnegie Mellon, building on components of the research activities in the departments of Electrical and Computer Engineering, Civil and Environmental Engineering, Engineering and Public Policy, in the College of Engineering, and the departments of Computer Science and Architecture. CenSCIR will perform research that will clarify the need for, provide design guidance for, and provide the clear justifications for the stewards of critical infrastructure systems to provide sensor data-driven awareness of the usage and condition of their systems and proactive, intelligent decision support and control of these critical infrastructure systems over their lifetime.

Fig. 3. Intelligent lighting control test bed

The three projects presented in this paper begin to address some of the research questions identified in Section 0. The first project related to inspection planning explores one approach to addressing the question: What information needs to be measured about a system, and where and when it should be measured? There are many additional issues that need to be addressed, but the need for formal representation and reasoning is apparent from this project and stochastic search processes appear to be feasible for supporting planning for sensor-based inspection. The second project related to semi-automated data model matching explores the research question: How does one manage, model, mine and exchange the large amount of data that will be generated by sensed infrastructure? This project illustrates how an evolving data modeling and exchange situation, which is the case in long-lived infrastructure applications, can be provided automated support that makes effective use of the mapping knowledge that has already been acquired. Over the infrastructure system life-cycle, such as system will be able to evolve as the data models evolve. The third project, related to utility maximization based on a sensed environment and knowledge of user and operator utility functions, addresses the research question: How does one translate the mass of information collected from the distributed sensor network into intelligent decision support that actually helps the operators of these systems make the best possible decision under the circumstances? The project explores the ways in which such sensor-based lighting control might be achieved with a dense set of environment sensors and an efficient utility optimization approach.

We must commit to deploy ICT during the construction phase, and utilize it over the whole life cycle of these infrastructure systems. This approach should allow us to have full life-cycle visibility of infrastructure, and enable us to move beyond the current "crisis mode" management paradigm that is pervasive within public agencies. While more robust networking and device technologies will need to be created to make ICT systems that deliver service over the lifetime of physical infrastructure assets (e.g., on the order of 50 years), this is a worthy goal and one that will have far-reaching cost, performance, and sustainability implications.

Acknowledgements

The project presented in Section 0 is funded by a grant from the National Science Foundation, CMS #0121549. NSF's support is gratefully acknowledged. Any opinions, findings, conclusions or recommendations presented in this paper are those of the authors and do not necessarily reflect the views of the National Science Foundation. The project described in Section 0 is sponsored by the National Institute for Standards and Technology. The support from, and interactions with, Kent Reed at NIST is also gratefully acknowledged. Finally, the project presented in Section 0 is sponsored by Pennsylvania Infrastructure Technology Alliance and PITA's support is also gratefully acknowledged.

References

Akin, O. Turkaslan-Bulbul, M.T. Gursel, I. Garrett, Jr. J.H. Akinci, B. Wang H. (2004) Embedded Commissioning for Building Design. *In: Proceedings of European Conference on Product and Process Modeling in the Building and Construction Industry*, 8-10 September, Istanbul, Turkey.

Akinci, B., Ergen, E., Haas, C., Caldas, C., Song, J., Wood, C.R., Wadephul, J. (2004) "Field Trials of RFID Technology for Tracking Fabricated Pipe," FIATECH Smart Chips Report, FIATECH, Austin, TX, February 25, 2004, http://www.fiatech.org/links.htm#smart.

Amor, A.W., Ge, C.W. (2002) Mapping IFC Versions. In: Proc of the EC-PPM Conference on eWork and eBusiness in AEC, Portoroz, Slovenia, 9-11 September, pp.373-377.

Ansari, F., Luke, A., Dong, Y., Maher A. (1999) Development of Maturity Protocol for Construction of NJDOT Concrete Structures, FHWA. Final Project Report, NJ 2001-017.

ASCE *Infrastructure Report Card*, http://www.asce.org/reportcard/2005. March 9, 2005.

ASHRAE (1980). American Society of Heating Refrigeration and Air Conditioning Engineers, Atlanta.

Bento, J. (2004) ICT@Brisa. Why does it work?, unpublished presentation made as part of the 2004 International Associated of Bridge and Structural Engineers Symposium, Shanghai.

Berringer, F. "Large Oil Spill in Alaska Went Undetected for Days" New York Times, March 16, 2006.

Brooks, R.R.; Ramanathan, P.; Sayeed, A.M. (2003). "Distributed target classification and tracking in sensor networks," *Proceedings of the IEEE*, 91(8):1163-71

Bungey, J.H. and Millard, S.G. (1996) "Testing of concrete in structures." Chapman and Hall.

Cable, J. (1998) Using NDT to Reduce Traffic Delays in Concrete Paving 1998 Transportation Conference Proceedings.

Cheok, G.S., Lipman, R.R., Witzgall, C., Bernal, J., and Stone, W.C. (2000) "Field Demonstration of Laser Scanning for Excavation Measurement," Proceedings of the 17th International Symposium on Automation in Robotics in Construction, Taipei, Taiwan.

Churchill, D.L., Hamel, M.J., and Townsend, C.P. (2003) "Strain energy harvesting for wireless sensor networks," Proceedings of the SPIE - The International Society for Optical Engineering (SPIE-Int. Soc. Opt. Eng) 5055, pp. 319-27.

Collins, V.A. (2004) Evaluation and Comparison of Commercially Available Maturity Measurement Systems, unpublished MS Thesis, Carnegie Mellon Department of Civil and Environmental Engineering, Pittsburgh, PA.

Commission on Engineering and Technical Systems (CETS). Inspection and Other Strategies for Assuring Quality in Government Construction. National Academy Press (1991).

Demers, Cornelia E.; Gregory, Rita A.; and Upton, Mark N. (2002) "Cost at Element Level" Journal of Infrastructure Systems Volume 8, Issue 4, pp. 115-121.

Doan AH., Domingos P., Halevy A. (2001) Reconciling schemas of disparate data sources: a machine-learning approach. In: Proc ACM SIGMOD Conf. pp.509-520

Erol, K., Nau, D., and Hendler, J. (1994) "UMCP: A Sound and Complete Planning Procedure for Hierarchical Task-Network Planning". In AIPS-94, Chicago, June.

The Eureka CIMsteel Project (2004). CIMsteel Integration Standards. Last accessed Nov 2004. http://www.cae.civil.leeds.ac.uk/past/cimsteel/cimsteel.htm

Fagin, R., Kolaitis P.G. and Popa, L. (2003) Data Exchange: Getting to the Core. Proceedings of the 22nd ACM SIGMOD-SIGACT-SIGART symposium on Principles of database systems, pp90-101, San Diego, California.

Feiner, Y. and B. Rajani. (2002) "Forecasting Variations and Trends in Water-Main Breaks." ASCE *Journal of Infrastructure Systems*, Vol. 8, No. 4, pp. 122-131

FIATECH (2004) Automating Equipment Information Exchange. Last accessed Nov 2004. http://www.fiatech.org/projects/idim/aex.htm.

Finley Jr. M. R., A. K., and R. Nbogni (1991). "Survey of intelligent building concepts." IEEE Communication Magazine.

Flax B., M. (1991). "Intelligent buildings." IEEE Communication Magazine.

Keeney, R., L., Raiffa, H. (1976). Decisions with Multiple Objectives: Preferences and Value Trade-offs. New York, Wiley

General Accounting Office (2004) "WATER INFRASTRUCTURE Comprehensive Asset Management Has Potential to Help Utilities Better Identify Needs and Plan Future Investments." Report No. GAO-04-461.

Gordon, C., Boukamp, F., Huber, D., Latimer, E., Park, K., and Akinci, B. (2003) "Combining Reality Capture Technologies for Construction Defect Detection: A Case Study," in Proceedings of EuropIA International Conference, pg. 99-108, Istanbul, Turkey.

Grata, J. (2005) "Road salt, hits from trucks likely led to bridge collapse." Pittsburgh Post-Gazette, December 29, 2005.

International Alliance for Interoperability (2003a) *Industry Foundation Classes*, Last accessed Nov 2004. http://www.iai-international.org

International Alliance for Interoperability (2003b) ifcXML Project. Last accessed Nov 2004. http://www.iai-na.org/.

Lara-Cinisomo, V. (2005) "Water main break stymies Downtown business." Pittsburgh Business Times, August 17, 2005.

Li, W., Clifton, C. (2000) SemInt: a tool for identifying attribute correspondences in heterogeneous databases using neural network. Data Knowledge Engineering 33(1):49-84.

Madhavan, J., Bernstein, P.A. and Rahm, E. (2001) Generic Schema Matching with Cupid. In: Proc the 27th VLDB Conference, Roma, Italy, 2001.

Malakooti, B. (2000). "Ranking and Screening Multiple Criteria Alternatives with Partial Information and use of Ordinal and Cardinal Strength of Preferences." IEEE Trans. on Systems, Man, and Cybernetics Part A 30: 355-369.

Mansueti, L. (2004) "Is Our Power Grid More Reliable One Year After the Blackout?" Conservation Update September-October 2004 Issue, U.S. Department of Energy - Energy Efficiency and Renewable Energy.

Mitra P., Wiederhold G. and Kersten M. (2000) A graph-oriented model for articulation of ontology interdependencies. In: Proc Extending Database Technologies, Lecture Notes in Computer Science, vol. 1777. Springer, Berlin Heidelberg New York, 2000, pp. 86-100

Rahm, E. and Bernstein P. A. (2001) A survey of approaches to automatic schema matching. The VLDB Journal, 10, 334-350.

Reinhardt, J, B. Akinci, and J. H. Garrett, Jr., (2005) A Navigational Model for Providing Customized Representations for Effective and Efficient Interaction with Mobile Computing Solutions on Construction Sites," *J. of Computing in Civil Engineering*, 19(2).

Sacerdoti, E.D. (1977) "A Structure for Plans and Behavior." American Elsevier, NY, NY.

Shaffer, J., and Siewiorek, D.P., Zhuang, W., Yeh, C.-H., Droegehorn, O., Toh, C.-K., Arabnia, H.R. (2003) "Locator@CMU a wireless location system for a large scale 802.11b network," International Conference on Wireless Networks - ICWN'03, CSREA Press, 666 pp. 61-65.

Singhvi, V., Krause A., Guestrin, C., Garrett, J., Matthews, S., (2005). Intelligent Light Control using Sensor Networks. SenSys, San Deigo.

Sodano, H.A., Park, G., Inman, D.J. (2004). "Estimation of electric charge output for piezoelectric energy harvesting," *Strain* (British Soc. Strain Meas), 40(2): 49-58.

Sohn, H., Dzwonczyk, M., Straser, E.G., Kiremidjian, A.S., Law, K.H. and Meng, T. (1999) "An Experimental Study of Temperature Effects on Modal Parameters of the Alamosa Canyon Bridge," *Earthquake Engineering and Structural Dynamics*, 28: 879-897.

Sohn, H. Park, G., Wait, J.R., Limback, N.P., Farrar, C.R. (2003) "Wavelet-Based Active Sensing for Delamination Detection in Composite Structures," *Smart Matls. and Structures*, 13(1):153-160

Song, J., Haas, C., Caldas, C., Ergen, E., Akinci, B., Wood, C.R., and Wadephul, J. (2004) "Field Trials of RFID Technology for Tracking Fabricated Pipe -Phase II." FIATECH Smart Chips Report, FIATECH, Austin, TX, June 30, 2004.

Steinmann, R. (2004) International Overview of IFC-Implementation Activities. Last accessed Nov 2004. http://www.iai.fhm.edu/ImplementationOverview.htm.

Stone, W.C. (2004) Performance Analysis of Next-Generation LADAR for Manufacturing, Construction, and Mobility, NISTIR 7117, Nat. Inst. of Stds. and Tech., Gaithersburg, MD.

Tanner, N.A., Wait, J.R., Farrar, C.R., and Sohn, H. (2003) "Structural Health Monitoring using Modular Wireless Sensors," *J. of Intelligent Matls., Sys. and Structures*, 14(1), 43-56.

Tseng, Y.C., Kuo, S.P., and Lee, H.W. (2004) "Location Tracking in a Wireless Sensor Network by Mobile Agents and Its Data Fusion Strategies" *Computer J.*, 47(4): 448-460.

Uzarski, D.R.; Tonyan, T.D.; and Maser, K.R. (1989) "Facility and Component Inspection Technology Concepts: Potential Use in U.S. Army Maintenance Management." USCERL Technical Report M-90/91.

Williams, E, Matthews, H.S., Breton, M., Brady, T., and M. Yao, (2006) Use of a Computer-Based System to Manage and Measure Energy Consumption in the Home, Proceedings of 2006 IEEE International Symposium on Electronics and the Environment.

Understanding Situated Design Computing and Constructive Memory: Newton, Mach, Einstein and Quantum Mechanics

John S. Gero

Key Centre of Design Computing and Cognition, University of Sydney, NSW 2006, Australia
john@arch.usyd.edu,au

Abstract. Situated design computing is an approach to the use of computers in design based on situated cognition. It is founded on two concepts: situatedness and constructive memory. These have the capacity to explain a range of design behaviors but have proven to be difficult to fully comprehend. This paper presents analogies with developments in physics that aim to assist in the comprehension of these foundational ideas. The ideas are drawn from the developments in the notions of space and observations in physics since, to a degree, they parallel the developments in constructive memory and situatedness.

1 Introduction

Design computing is the area of computing that deals with designing and designs. Designs and their representations have been the focus of considerable research that has resulted in a variety of representation models and tools. Designing involves the development and refinement of requirements and approaches, the synthesis of designs as well as the emergence of new concepts from what has already been partially designed. In this designing is not a subset of problem solving – problem solving is a subset of designing. This interaction between partial designs and the design process has been called "reflection" [1]. Reflection is the general term used to describe a designerly behavior that allows a designer to "see" what they have differently to what was intended at the time it was done. Current concepts in design computing make the modeling of this conception of designing very difficult. Other designerly behavior is equally difficult to model using current concepts of design computing. For example: how is that two designers when given the same set of requirements produce quite different designs? Why is that a designer when confronted with the same set of requirements at a later time does not simply reproduce the previous design for those requirements? How is that designers can commence designing before the requirements are fully specified? It is claimed that it is precisely these behaviors that distinguishes designing from problem-solving. Most models of design conflate it with problem solving [2], [3], [4] and are unable to represent the activities that produce these behaviors [5]. As a consequence our computational models of designing and the computational support tools for designing do not adequately match the behavior of designers and are insufficiently effective. This can be seen in the paucity of tools for the early stages of designers where all the critical decisions are taken.

I.F.C. Smith (Ed.): EG-ICE 2006, LNAI 4200, pp. 285–297, 2006.
© Springer-Verlag Berlin Heidelberg 2006

This paucity of tools for the most significant and influential parts of designing is not due to a lack of ingenuity in constructing such tools, it is due to a lack of knowledge about what such tools should embody. This paper claims that the concepts that provide the foundations for situated design computing and hence situated design computing itself provide the knowledge for the development of a new class of tools, tools that have the capacity to support these early stages of designing.

1.1 Situated Design Computing

A new approach is needed. Situated design computing is a new paradigm for design computing that draws its inspiration from situated cognition in cognitive science [6]. It is claimed that it can be used to model designing more successfully than previous approaches. It has the capacity to form the basis of a model that can represent and explain much of designerly behavior. In particular it is claimed it can model [7], [8]:

- how a designer can commence designing before all the requirements have been specified
- how two designers presented with the same specifications produce different designs
- how the same designer confronted with the same requirements produces a different design to the previous one, and
- reflection, ie, how a designer can change their design trajectory during the activity of designing.

As a consequence situated design computing warrants further investigation. Situated design computing is founded on three concepts that are new to design computing: acquisition of knowledge through interaction, constructive memory and the situation. The first concept draws its distinction from the source of the knowledge rather than the techniques used in its acquisition. The second introduces a novel notion of memory that conceives of memory primarily as a process rather than the current view of memory as a "thing in a location". The third introduces the notion of a gestalt view of the designer that influences what the designer "sees".

1.2 Interaction

Whilst it is clear that much of human knowledge is objective or third-person knowledge of the kind derived in engineering science, there is a category of everyday knowledge that depends on the person rather than deduction. This knowledge is developed based on first-person interaction with the world [9]. This class of knowledge is sometimes inappropriately encoded as third-person or deductive knowledge and when done so often causes a mismatch between the experience of the person who encoded the knowledge and a subsequent user of that knowledge.

A simple example of such encoding of personal knowledge can be seen even in the way objects are represented in a CAD system. Fig. 1(a) shows the screen image of a floor layout. Simply looking at the drawing of the floor layout gives no indication of how it has been encoded. The darkened line is the single polyline representation of the outline obtained by pointing to a spot on the boundary, but that representation could not be discerned from the image. Fig. 1(b) shows exactly the same outline but it is encoded differently, as indicated by the darkened polyline obtained by pointing to the same spot. The issue here is one of interpretation that has been missing in design computing. A common assumption is that the external world is there to be represented, ie that in some sense it has only one representation. This misses an important step: namely that of interpretation. Interpretation uses and produces first-person knowledge and is an interaction process.

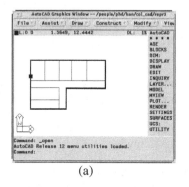 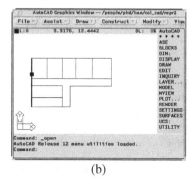

(a)   (b)

Fig. 1. The same image has different encodings (a) and (b) that depend on the individuals who created them rather than on any objective knowledge

Interaction is one of the fundamental characteristics of designerly behavior. The interaction can be between the designer and the representation of what is being designed, or it could be between the designer and other designers involved with designing what is being designed, or between the designer and other people, or even between the designer and their own concepts about the design – re-interpretation through reflection.

Interaction is the basis for the development of first-person knowledge, any one of the four kinds of interaction or any combination of them produces first-person knowledge. This first-person knowledge is what distinguishes one designer from another. Tow designers who share exactly the same third-person knowledge can behave markedly differently when presented with the same set of design requirements because of their different first-person knowledge.

1.3 Constructive Memory

Computationally memory has become to mean a thing in a location. The thing can take any form and the location need not be explicitly known. The thing can be accessed by knowing either its location or its content. There are a number of distinguishing characteristics of this form of memory:

- memory is a recall process
- there needs to be an explicit index (either location or content)
- the index is unchanged by its use
- the content is unchanged by its use
- the memory structure is unchanged by its use.

This is in contrast to cognitive models of memory, in particular constructive memory [10], which has the following distinguishing characteristics:

- memory is a reasoning process
- the index need not be explicit, it can be constructed from the query
- the index is changed by its use
- the content is changed by its use
- the memory structure is changed by its use
- memories can be constructed to fulfill the need to have a memory
- memories are a function of the interactions occurring at the time and place of the need to have a memory.

As can be seen constructive memory takes a fundamentally different view of memory than computational memory. Such memories are intimately connected to both the previous memories, called "experiences", the current need for a memory [11] and the current view of the world at the time and place of the need for that memory. Simple examples of some of these characteristics include the following.

Index need not be Explicit, It can be Constructed from the Query. Take the query: find an object with symmetry. There is no need for there to be an index "symmetry" in the memory. The system can use its experiences about symmetry to determine whether it can construct symmetry in objects in the memory. This concept allows for the querying of a memory system with queries it was not designed to answer at the time the original memories were laid down. This is significant in designing as there is evidence that designers change the trajectory of their designs during the process of designing and introduce new intentions based on what they "see" in their partial designs, intentions that were not listed at the outset of process. There are fundamental issues here that are not addressed by fixed index systems. For a novel query to be able to answered it first needs to be interpreted by the memory system using the experiences it has that might bear on the query.

Index Changed by its Use. If the same query is made a multiple times the response to the query should become faster, irrespective of whether it is a constructed index or not; this is a very simple example of how an index is changed by its use. A more profound and useful example of this phenomenon is when a memory is used to construct another, later, memory a new index is created that connects these two memories such that when either is used again the other is associated with it.

Content Changed by its Use. A trivial example is in the case above about symmetry, the index can now become part of the memory. A less trivial example would be in the case where an experience is used in constructing a new memory. The experience is changed by having a link to the newly constructed memory. The experience is no longer the same experience it was before it was used to construct the new memory. That experience can no longer be recalled without its role in the construction of a new memory being part of it.

Memory Structure Changed by its Use. In the example above, not only does the content change but also the structure of the memory. The link between the experience and the new memory changes the structure of the memory system itself such that the way these memories can be used is changed. Later experiences can change what was experienced before. The content of a constructive memory system is non-monotonic.

Memories can be Constructed to Fulfill the Need to Have a Memory. Take the case where a finite element analysis is carried out and both the cost of processing and the result are passed onto the design team leader. The team leader may query whether the analysis was cost effective. There is no memory of this but they can construct a memory in response. Later, if asked whether the analysis was cost effective they can respond directly through a recall-like process. If that query had not been asked earlier there would be no memory to respond to the later query.

Memories are a Function of the Interactions Occurring at the Time and Place of the Need to Have a Memory. Take the example of the cost effectiveness query above. If at the same time as that query being made another member of the design team states that his experience with the analysis group is that they always overestimate their costs, this changes the design team leader's construction of the memory to take account of some discounting of the cost provided.

One way to conceptualize a constructive memory system is as a global, continuously learning, associative system, where all later memories have the potential to include and affect all earlier memories while earlier memories affect later memories. The notion of past memories as fixed entities has to be modified such that all of the past can only be viewed "through the lens of the present", and the present is an encapsulation of the past. This brings us to the notion of a "situation".

1.4 Situatedness

The third concept that provides one of the foundations of situated design computing is situatedness. This is the notion that a designer works within a world of their own making. This world is based on their perception of the world outside and inside them and their behavior is a response to that world. This conceptualized world is called the "situation". Designers do not need to articulate the situation to behave in accordance with it, just as people do not either. Situations can be extrinsic or intrinsic. Extrinsic situations are available for observation, while intrinsic situations are emergent properties of behavior.

A well-known example of a situation is produced when people attempt to solve the nine-dot problem. Consider three rows of three dots equally spaced in both the horizontal and vertical directions. The problem is how to draw lines through all the dots under the following three constraints: using only four straight lines with the pen not leaving the paper. Most people cannot produce a solution. The reason is that they appear to view the world they are in as if the following situation prevailed: no line can pass outside the square (the convex hull) produced by the dots. No one told them this and most people don't even know that they have been working within this situation, but their behavior appears as if it is controlled by it.

Situations are the basis for expectations and interpretations and hence play a dominant role in designerly behavior. Situations can be explicit, ie, known to the designer and affect their behavior, or implicit, ie, not known to the designer and still affect the designer's behavior.

In order to understand situated design computing all three of interaction, constructive memory and situatedness need to be understood. Of the three constructive memory is the most contentious. The next sections draw analogies from the development of the concepts of space in physics that aim to assist in the development of the understanding of constructive memory. It might be argued that the development of the concepts of space in physics in itself is complex and difficult. It is claimed that the base concepts in physics are sufficiently widespread and well understood that they provide a foundation for enhancing the understanding that is trying to be achieved.

2   Understanding Constructive Memory Using Space Analogies

2.1   Newton and Absolute Space: Computational Memory

In 1687 Isaac Newton published *Philosophiae Naturalis Principia Mathematica* [12], in which he introduced the notion of absolute space as the "truest reference for describing motion". Newton's need for absolute space derived from the notion that objects had to have a reference against which some of their properties could be determined. He went on to say: "Absolute space, in its own nature, without reference to anything external, remains always similar and unmovable." This is similar to the view of cognitive realists who claim that there is an objective world that is independent of any observer and is the same for all observers. In Newton's world we could draw a grid and locate all objects on that grid. All object activity relates to that grid. The grid is unchanged by the objects and the grid does not change the objects: it is "always similar and unmovable".

This maps well onto our notion of computational memory, where the location of a thing is memory is absolute and does not depend either on the thing or the "observer" of the thing or whether you access the thing. If we access a thing in memory neither the object nor the memory system is changed, no matter how often we access it. There is no means of accessing it that can change the object. There is no means of accessing it that can change the memory system. It is absolute.

2.2   Mach and No Space: Linked Memory

In 1883 Ernst Mach published *Die Mechanik in ihrer Entwicklung* [13], in which he argued against Newton's absolute space in favour of a completely relative space, where objects are all relative to each other. Mach claimed that objects did not need an absolute space to sit in. They could simply be in relation to each other and that provided the reference that Newton argued was the basis of absolute space. This is equivalent to taking Newton's view as presented with the grid as the reference and removing the grid, leaving only the objects.

This maps well onto the notion of linked or network memory where all things in the memory system are accessed by associations with other things. In linked computational memory things have both absolute locations and links but the absolute locations need not be part of the visible accessing process. Semantic networks [14], [15] are examples of linked memory systems. Accessing a linked memory system does not change its contents, nor does it change anything about the next time you access it.

2.3 Newton and Mach and Gravity: Interacting Memory

In removing the grid Mach did not remove another of Newton's foundational concepts, namely that of interactions between objects: what he called gravitational attraction. In this view objects are linked together by the influence they exert on each other. The introduction of a new object potentially changes the location of some or all existing objects in the system. This influence is bi-directional. Here the newly introduced object's influence is dependent on the individual masses of the existing objects and their relative size in relation to it. A new object will have a small influence if its mass is small in relation to the nearby objects, which will then have a greater influence on it and vice versa. This applies to all the objects in the system.

Currently there is no memory system equivalent to this concept except aspects of constructive memory which we will call interacting memory. A thing in the memory is not only linked to other things in the memory but is influenced by them and influences them.

2.4 Einstein and Gravitational Space: Towards Constructive Memory

In 1916 Albert Einstein published the final version of his *Die Grundlagen der allgemeinen Relativitätstheorie* [16], in which he demonstrated that space was changed by objects (this laid the foundation for our current view of gravity) and that there was an absolute space-time continuum. In one sense this combined and extended the ideas of both Newton and Mach. In its simplest form the theory of relativity re-instated Newton's absolute space but did not keep it absolute. The space itself was now changed by the objects in it. Gravitation was no longer the "attraction of bodies to each other" but a "warping of space" by the masses of the objects.

Continuing with our grid metaphor, the grid was re-instated but was changed by the objects that were still interacting with each other but the interaction was caused by and caused a change in the grid. Imagine a two-dimensional grid with objects hovering over it. The Newtonian view was that the objects interacted with each other but did not change the grid. The Einsteinian view was that the objects changed the grid by warping it into the third dimension creating a three-dimensional grid and the third dimension played a role in gravity.

Consider this gravitational space view:

- a new object affects existing objects
- a new object warps the space around itself
- a new object affects an existing object with an effect that is a function of the size of the new object
- a new object affects an existing object with a relative effect that is a function of the distance to the existing object
- a new object affects an existing object with a relative effect that is a function of the distances and sizes of all the other existing objects
- no object sits by itself unaffected by other objects

Consider a memory system with the following characteristics:

- a new memory affects existing memories
- a new memory affects an existing memory with an effect that is a function of how important the new memory is
- a new memory affects an existing memory with a relative effect that is a function of how important the new memory is in relation to the importance of the existing memory
- a new memory affects an existing memory with a relative effect that is a function of the time distance of the activation of the existing memory to the new memory
- a new memory affects an existing memory with a relative effect that is a function of how important the new memory is in relation to the time distances of the activations and the importances of all the other existing memories
- no memory sits by itself unaffected by other memories.

This brings us close to the characteristics of a constructive memory system. Such a system is more complex than the gravitational space model described above, but one that shares a set of analogically similar characteristics. A constructive memory system has the potential to continuously change as it senses new input that becomes a memory. A constructive memory system has the capacity to meet Dewey's requirement [17] that: "Sequences of acts are composed such that subsequent experiences categorize and hence give meaning to what was experienced before." This is close to the understandings we have about designing. A significant addition to the gravitation space model that constructive memory brings is that new memories can be made entirely out of existing memories, no new sensing is required. This is one example of Schon's reflection.

3   Understanding Situatedness Using a Quantum Physics Analogy

In 1927 Werner Heisenberg published *Über den anschaulichen Inhalt der quantentheoretischen Kinematik und Mechanik* [18], one of the cornerstones of quantum mechanics, which displaced Newtonian mechanics, partly by replacing the Newtonian concept of atoms and fields and the implied concept of certainty with an emphasis on subatomic particles and uncertainty. Heisenberg's uncertainty principle states that it is not possible to simultaneously determine both the position and momentum of a particle. In quantum mechanics the location of particles is a function of their viewing. This is an intriguing concept that previously had been associated with social and behavioral science rather than physics.

The concept of situatedness is analogically related to this concept of not having a fixed location and momentum, rather the location or momentum is a function of the observation. A situation is like a particular worldview. A particular worldview affects the interpretation of the memory. The memory could support all manner of worldviews, which one it supports is only apparent when a situation is used to query or access it. Before the memory is queried with that situation it does not have any bias to that situation. The situation of the query biases the memory to produce interpretations that support that situation.

Let us commence with an example drawn from the behavior of a human designer. Take the drawings in Fig. 2. Fig. 2(a) shows the drawing produced by the designer at some point in the design.

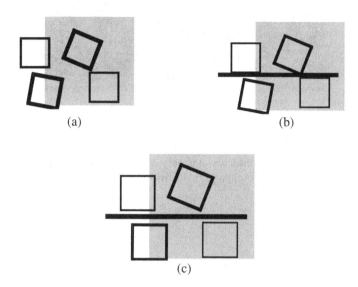

(a)

(b)

(c)

Fig. 2. (a) The original drawing, (b) the designer interprets the drawing as having a horizontal axis; this creates the "axis" situation; (c) the designer now moves the blocks based on the situation of the "axis"

In looking at this drawing the designer reinterpreted what he had drawn not as a series of blocks but as a horizontal axis connecting the blocks, Fig. 2(b). This is a new situation – a new worldview – with this he changes the meanings in his memory of the locations of the blocks and orients them with respect to the axis, Fig. 2(c). Schon and Wiggins [20] observed this behavior on numerous occasions in their studies as did other researchers [21]. This is an example of an extrinsic situation.

Take another example from human behavior. Suppose a designer is shown a picture of pile of rubble that is clearly that of a collapsed house, Fig. 3. The picture is labeled: "result of devastation by Cyclone Larry, 2006". The designer, through their interaction with the picture, is likely to have a response related to the damage caused by nature or similar.

However, if they were shown the same picture with the label: "example of safety issues in hand demolition of houses" the response would be quite different because the viewer had used a different situation on which to base their interpretation. If the picture was unlabelled it is unclear what the viewer would think without further knowledge about the viewer and the viewer's experiences.

Fig. 3. Picture shown to designer

This behavior is similar to the quantum physics behavior of the location or momentum being a function of the observer, not only of the observed the effect of this concept on memory systems is to change them from static memory to dynamic memory systems.

4 Discussion

Situated design computing is a paradigmatic change in the way we view computation in design. Typically design computing has taken the traditional computational stance of memory being a repository and the primary memory process being that of recall. This has served design computing well, but at the same time has restricted further conceptualization of its development. Novel concepts from cognitive science and in particular situated cognition have opened up new avenues for the development of design computing.

The three foundational concepts of situated design computing are:

- knowledge through interaction
- constructive memory
- situatedness.

Knowledge through interaction moves design computing from encoding third-person knowledge during a computational system's initial design to including first-person knowledge acquired during interactions continuously through the use and application of the system. The effect of this that such a computational system can adapt its behavior based on its use [22].

Constructive memory provides the foundation for all the activity in situated design computing. It turns memory into a dynamic process that reconfigures itself based on the "experiences" it has encountered. The trajectory of the development of our current understanding of space from physics provides a strong analogy with the development of our understanding of constructive memory [10].

Situatedness provides the foundation for a range of changes in a constructive memory system. It is one of the ways in which reinterpretation of existing memories can take place. In design computing it plays a role in emergence [23], [24], reflection [1] and reinterpretation [5]. Analogies from quantum physics map well onto the base concepts of the situation.

Situated design computing is a different paradigm for computing and the development of computational tools to support designing. It does not displace but rather augments the more traditional design computing paradigm. The distinction between the two is best summarized by treating design computing as embodying third-person knowledge and situated design computing embodies first-person knowledge. Design computing largely uses fixed structures both in the way it is conceptualized and in the way it is implemented. Situated design computing introduces dynamic structures throughout its conceptualization and implementation.

A variety of situated agents have been implemented based on situated design computing. These agents, to varying degrees, embody these concepts and have been applied to:

- designing in virtual environments [25], [26], [27]
- simulation [28]
- modeling expertise [29]
- design optimization [30]
- exploring creativity [31]
- learning [32]
- design ontologies [33], [34] and
- product model interoperability [35].

Projects that are based on situated design computing include:

- interpretation as a design process
- situated design prototypes
- situated design optimization
- design team behaviour.

Acknowledgements

This research is part of a larger project on developing situated design computing and is supported by a grant from the Australian Research Council, Grant No. DP0559885.

References

1. Schon, D. : The Reflective Practitioner: How Professionals Think in Action. Arena, Aldershot. (1995)
2. Pahl, G. and Beitz, G.: Engineering Design : A Systematic Approach (translated). Springer, Berlin (1999)
3. Coyne, R., Radford, A., Rosenman, M., Balachandran, M. and Gero, J. S.: Knowledge-Based Design Systems. Addison-Wesley, Reading (1990)
4. Suh, N.: Axiomatic Design. Oxford University Press, Oxford (2001)

5. Lawson, B.: How Designers Think, 4th edn. Architectural Press, London (2005)
6. Clancey, W.: Situated Cognition. Cambridge University Press (1997)
7. Lawson, B.: Acquiring design expertise. In Gero, J. S. and Maher, M. L. (eds), Computational and Cognitive Models of Creative Design VI. Key Centre of Design Computing and Cognition, University of Sydney, Sydney (2005) 213-229
8. Constructive Memory Group meetings. Key Centre of Design Computing and Cognition, University of Sydney, Australia (unpublished notes) (2006)
9. Wegner, P.: Interactive foundations of computing. Theoretical Computer Science (1998) 192: 315-351
10. Gero, J. S.: Constructive memory in design thinking. In G. Goldschmidt and W. Porter (eds), Design Thinking Research Symposium: Design Representation, MIT, Cambridge, (1999) I.29-35
11. Bartlett, F. C.: Remembering: A Study in Experimental and Social Psychology. Cambridge University Press, Cambridge (1932 reprinted in 1977)
12. Newton, I,: Sir Isaac Newton's Mathematical Principle of Natural Philosophy in His System of the World. trans. A. Motte and F Cajori, University of California Press, Berkeley (1934)
13. Mach, E.: The Science of Mechanics: A Critical and Exposition of its Principles. Open Court, Chicago (1893)
14. Shapiro, S. C.: A net structure for semantic information storage, deduction and retrieval. Proc. IJCAI-71, (1971) 512-523
15. Sowa, J. F.,:(ed) Principles of Semantic Networks: Explorations in the Representation of Knowledge. Morgan Kaufmann Publishers, San Mateo, CA (1991)
16. Einstein, A.: The foundation of the general theory of relativity. In H. A. Lorentz, A. Einstein, H Minkowski, H. Weyl, Principle of Relativity: A Collection of Original Memoirs on the Special and General Theory of Relativity, Dover, New York, (1923, republished in 1952) 109-164
17. Dewey, J.: The reflex arc concept in psychology. Psychological Review. 3 (1896 reprinted in 1981) 357—370
18. Heisenberg, W. : Über den anschaulichen Inhalt der quantentheoretischen Kinematik und Mechanik. Zeitschrift für Physik. 33 (1927) 879-893
19. Suwa, M., Gero, J. S. and Purcell, T.: The roles of sketches in early conceptual design processes. Proceedings of Twentieth Annual Meeting of the Cognitive Science Society, Lawrence Erlbaum, Hillsdale, New Jersey (1998) 1043-1048
20. Schon, D. A. and Wiggins, G.: Kinds of seeing and their functions in designing. Design Studies (1992) 13, 135-156.
21. Suwa, M., Gero, J. S. and Purcell, T.: Unexpected discoveries: How designers discover hidden features in sketches. In Gero, J. S. and Tversky, B. (eds), Visual and Spatial Reasoning in Design. Key Centre of Design Computing and Cognition, University of Sydney, Sydney, Australia (1999) 145-162
22. Gero, J. S.: Design tools as situated agents that adapt to their use. In W. Dokonal and U. Hirschberg (eds), eCAADe21, eCAADe. Graz University of Technology (2003) 177-180
23. Holland, J.: Emergence: From Chaos to Order. Perseus Books, Cambridge, MA (1999)
24. Stiny G.: Emergence and continuity in shape grammars. In U Flemming and S Van Wyk (eds). CAAD Futures'93. Elsevier Science (1993) 37-54
25. Maher, M. L. and Gero, J. S.: Agent models of 3D virtual worlds. ACADIA 2002: Thresholds, California State Polytechnic University, Pomona, (2002) 127-138
26. Smith, G., Maher, M. L. and Gero, J. S.: Towards designing in adaptive worlds. Computer-Aided Design and Applications (2004) 1(1-4): 701-708

27. Smith, G.J., Maher, M.L., and Gero, J.S.: Designing 3D virtual worlds as a society of agents. In M-L. Chiu, J-Y. Tsou, T. Kvan, M. Morozumi and T-S. Jeng (eds), Digital Design: Research and Practice, Kluwer (2003) 105-114.

28. Saunders, R. and Gero, J. S.: Situated design simulations using curious agents. AIEDAM (2004) 18 (2): 153-161

29. Gero, J. S. and Kannengiesser, U.: Modelling expertise of temporary design teams. Journal of Design Research (2004) 4(3): http://jdr.tudelft.nl/.

30. Peng, W. and Gero, J. S.: Concept formation in a design optimization tool. DDSS2006 (2006) (to appear)

31. Sosa, R. and Gero, J. S.: Design and change: A model of situated creativity. In Bento, C., Cardosa, A. and Gero J. S. (eds) Approaches to Creativity in Artificial Intelligence and Cognitive Science, IJCAI03, Acapulco (2003) 25-34

32. Gero, J. S. and Reffat, R.: Multiple representations as a platform for situated learning systems in design. Knowledge-Based Systems (20010 14(7): 337-351

33. Gero, J. S. and Kannengiesser, U.: The situated Function-Behaviour-Structure framework. Design Studies (2004) 25(4): 373-391

34. Gero, J. S. and Kannengiesser, U.: A Function-Behaviour-Structure ontology of processes. In J. S. Gero (ed), Design Computing and Cognition'06, Springer (2006) 407-422

35. Kannengiesser, U. and Gero, J. S.: Agent-based interoperability without product model standards. Computer-Aided Civil and Infrastructure Engineering (2006) (to appear)

Welfare Economics Applied to Design Engineering

Donald E. Grierson

Professor Emeritus, Civil Engineering Department, University of Waterloo,
Ontario N2l 3G1, Canada
Phone: +1 519/888-4567 (x2412); Fax: 519/888-4349
grierson@uwaterloo.ca

Abstract. The paper concerns design engineering problems involving multiple
criteria and, in particular, the development of a formal trade-off strategy that
can be employed by designers to mutually satisfy conflicting criteria as best as
possible. A Pareto-optimal exchange analysis technique is adapted from the
theory of social welfare economics as the basis for a search methodology to
identify good-quality compromise designs. The concepts are initially developed
for the two-criteria design problem so that the main ideas can be given a simple
geometric interpretation. Curve-fitting, equation-discovery and equation-
solving software are employed along with welfare economics analysis to find
competitive general equilibrium states corresponding to Pareto-optimal
compromise designs of a flexural plate governed by conflicting weight and
deflection criteria. The trade-off strategy is then extended to the design of a
multi-storey building governed by three conflicting criteria concerning capital
cost, operating cost and income revenue.

Keywords: multi-criteria design engineering, multi-goods welfare economics,
Pareto optimization.

1 Introduction

One of the difficulties in engineering design is that there are generally several
conflicting criteria, which forces the designer to look for good compromise designs by
performing trade-off studies between them. As the conflicting criteria are often non-
commensurable and their relative importance is generally not easy to establish, this
suggests the use of non-dominated optimization to identify a set of designs that are
equal-rank optimal in the sense that no design in the set is dominated by any other
feasible design for all criteria. This approach is referred to as 'Pareto' optimization
and has been extensively applied in the literature concerned with multi-criteria
engineering design (e.g., Osyczka 1984, Koski 1994, Grierson & Khajehpour 2002,
Cheng 2002). The number of Pareto-optimal designs so found can still be quite large,
however, and it is yet necessary to select the best compromise design(s) from among
them.

   Koski (1994) briefly reviews several methods for searching among Pareto optima
to identify one or more good compromise designs. The final selection generally
depends on the designer's personal preferences. This study employs a Pareto trade-off
analysis technique adapted from the theory of social welfare economics to identify

I.F.C. Smith (Ed.): EG-ICE 2006, LNAI 4200, pp. 298–314, 2006.

one or more competitive general equilibrium states of the conflicting criteria that represent good compromise designs; i.e., designs that represent a Pareto-optimal compromise between designer preferences for the various criteria. The trade-off strategy is initially developed for the two-criteria problem so that the concepts can be given a simple geometric interpretation. A flexural plate design involving conflicting weight and deflection criteria serves to illustrate the main ideas. Finally, a multi-storey building design involving conflicting capital cost, operating cost and income revenue criteria illustrates the extension of the trade-off strategy to the three-criteria design problem.

2 Welfare Economics

2.1 Utility Functions

In the theory of welfare economics involving multiple goods, utility functions are used to describe consumer preferences for different bundles of the goods. For an economy involving two goods, x_1 and x_2, a utility function assigns a number to every possible consumption bundle (x_1, x_2) such that more-preferred bundles get assigned larger numbers than less-preferred bundles. The utility assignment is 'ordinal' in that it serves only to *rank* the different consumption bundles, while the size of the utility difference between any two bundles isn't meaningful.

In the space of the two goods x_1 and x_2, a certain consumption bundle (x_1, x_2) lies on the boundary of the set of all bundles that the consumer perceives as being preferred to it. This implies that the consumer is indifferent to all bundles that lie on the set boundary itself, which is called an *indifference curve*. Utility functions are used to define (label) indifference curves such that those which are associated with greater preferences get assigned higher utility numbers. They can take on a variety of forms (a specific utility function form is considered later in the paper).

For any incremental changes dx_1 and dx_2 of the two goods along an indifference curve defined by utility function $u(x_1, x_2)$, there is no change in the value of the utility function. Mathematically, this may be expressed as (Varian 1992),

$$du = \frac{\partial u(x_1, x_2)}{\partial x_1} dx_1 + \frac{\partial u(x_1, x_2)}{\partial x_2} dx_2 = 0 \tag{1}$$

Equation (1) may be reorganized to find the slope of the indifference curve at point (x_1, x_2) as,

$$\frac{dx_2}{dx_1} = -\frac{\partial u(x_1, x_2) / \partial x_1}{\partial u(x_1, x_2) / \partial x_2} \tag{2}$$

which is known as the *marginal rate of substitution (MRS)* that measures the rate at which the consumer is willing to substitute good x_2 for good x_1. The negative sign indicates that if the amount of good x_1 *increases* then the amount of good x_2 *decreases* in order to keep the same level of utility, and vice versa.

2.2 Pareto Exchange and Competitive Equilibrium

Consider now a pure exchange economy in which two consumers A and B are seeking to achieve an optimal trade-off between goods x_1 and x_2 (Boadway & Bruce 1984). The total supply of good x_1 is x_1^*, while that for good x_2 is x_2^*. Suppose that consumer A's initial endowment consists of the entire supply of good x_1, as indicated by the distance between the origin 0_A and point x_1^* along the horizontal axis in Figure 1. Her initial utility level corresponds to the indifference curve labelled by the utility value u_A^0. If consumer A is offered a relative price of good x_1 in terms of good x_2 as given by the (absolute) value of the slope of the *terms-of-trade line (TL$_A$)* passing through her endowment point x_1^*, she will choose to trade x_1^*-x_{1A} units of good x_1 in exchange for x_{2A} units of good x_2 and, thereby, achieve increased utility level u_A^1 (note that utility increases with distance from the origin 0_A).

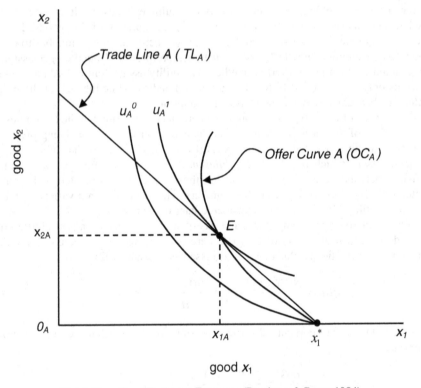

Fig. 1. Two-Good Exchange Economy (Boadway & Bruce 1984)

By offering consumer A different relative prices of good x_1 in terms of good x_2, her *offer curve (OC$_A$)* can be traced out by rotating the terms-of-trade line through her endowment point and drawing the locus of equilibrium points E chosen. The highest indifference curve (utility) available to consumer A at each relative price is tangent to the terms-of-trade line at its intersection with the offer curve. Offer curves indicate the willingness of consumers to exchange a certain amount of one good for a given amount of another good at any relative price. They can take on a variety of forms (a specific offer curve is considered later in the paper).

We can draw a similar diagram for consumer B by supposing that his initial endowment consists of the entire supply of good x_2. Upon doing that, the competitive equilibrium of the two-consumer and two-good exchange economy can be analytically examined by constructing the *Edgeworth box*[1] diagram in Figure 2, the horizontal and vertical dimensions for which are equal to the total supplies x_1^* and x_2^* of goods x_1 and x_2, respectively. The origins for consumers A and B are 0_A and 0_B, respectively. Their initial endowment points $A(x_1^*, 0)$ and $B(0, x_2^*)$ are both located at the lower right-hand corner of the box (note that consumer B's axes are inverted since they are drawn with respect to origin 0_B).

Of special interest is the *contract curve*[2], which is the locus of all allocations of the two goods such that the indifference curves of consumer A are tangent to those of consumer B. That is, the marginal rate of substitution between goods x_1 and x_2 for consumer A is equal to that for consumer B at each point on the contract curve, but not at any other point off the curve. This suggests the possibility for mutually beneficial trade.

The initial indifference curves u_A^0 and u_B^0 shown in Figure 2 form a lens-shaped area within which lie points that are Pareto superior to the initial endowment point and which can be reached by consumers A and B through trading goods x_1 and x_2 between themselves. Once they have traded to a point on the contract curve no further Pareto improvements are possible, since then one consumer can gain utility only at the expense of the other. That is, any point on the contract curve segment FG is a Pareto-optimal allocation of goods x_1 and x_2 between consumers A and B. But some points are better than others depending on the consumer; namely, all points from F up to almost E are unacceptable to consumer A because they do not lie on her offer curve and have smaller utility than desired, while the same situation applies for consumer B for all points from G up to almost E. It is only at the intersection point E of their offer curves OC_A and OC_B that consumers A and B are mutually satisfied with their highest attainable utilities u_A^1 and u_B^1, respectively. Point E is a *competitive general equilibrium* Pareto-optimal allocation of goods x_1 and x_2.

That point E lies on the contract curve follows from the fact the two offer curves at that point have the same marginal rate of substitution. In other words, as indicated in Figure 2, there exists a common terms-of-trade line $TL_A = TL_B$ that affords consumers A and B the opportunity to trade from their initial endowment to point E. This opportunity to directly proceed to a Pareto-optimal competitive general equilibrium state is exploited in the following concerning multi-criteria design engineering.

---

[1] Named in honor of English economist F. Y. Edgeworth (1845-1926), who was among the first to use this analytical tool.

[2] Each point on the contract curve is obtained by maximizing the utility of one consumer while holding that for the other consumer fixed: e.g., point G in Figure 2 is found by *maximizing* $u_A(x_{1A},x_{2A})$ subject to $u_B(x_{1B},x_{2B})=u_B^0$, $x_{1A}+x_{1B}=x_1^*$ and $x_{2A}+x_{2B}=x_2^*$.

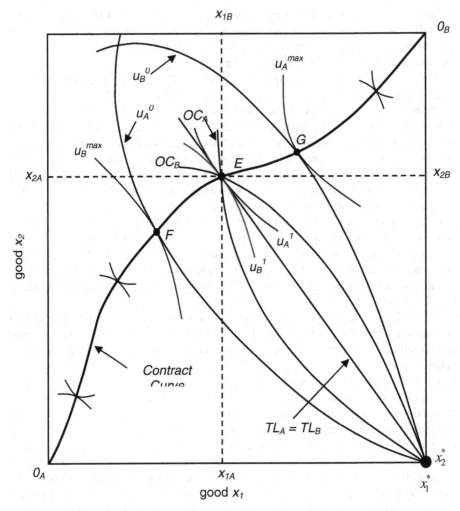

Fig. 2. Welfare Economics Edgeworth Box (Boadway and Bruce 1984)

3 Two-Criteria Design Engineering

3.1 Flexural Plate Design

Consider the simply-supported plate with uniformly distributed loading shown in Figure 3, for which length $L = 600$ mm, load $P = 0.4$ N/mm^2, material density $\rho = 7800$ kg/m^3, Young's modulus $E_{ym} = 206 \times 10^3$ N/mm^2, and Poisson's ratio $\nu = 0.3$ (Koski 1994). It is required to design the plate for the two conflicting criteria of *minimum weight (W-criterion)*, and *minimum deflection (Δ-criterion)* at midpoint M.

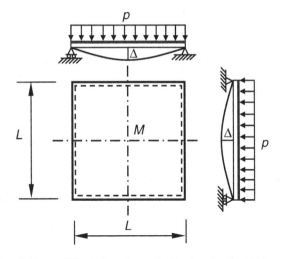

Fig. 3. Flexural Plate - Loading & Deflection (Koski 1994)

The analysis model for the plate is defined by the mesh of 36 finite elements shown in Figure 4(a). The plate thicknesses of the six zones indicated in the design model for the plate shown in Figure 4(b) are taken as the design variables. The (von Mises) stress σ_i (i =1, 2,..., 36) for each finite element is constrained to be less than or at most equal to 140 MPa, while the thickness t_j (j =1, 2,..., 6) for each plate zone is constrained to be in the range of 2-40 mm.

The two-criteria design optimization problem statement is:

$$
\left.
\begin{aligned}
&\text{Minimize: } F(\underline{t}) = [W(\underline{t}), \Delta(\underline{t})] \\
&\text{Subject to: } \sigma_i \le 140 \quad (i = 1,2,...,36) \\
&\qquad\qquad 2 \le t_j \le 40 \quad (j = 1,2,...,6)
\end{aligned}
\right\}
\tag{3}
$$

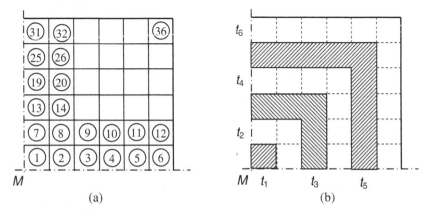

Fig. 4. Quarter-Plate (a) Analysis Model, (b) Design Model (Koski 1994)

A variety of optimization methods are available to find a Pareto-optimal design set for the problem posed by Eqs.(3). A genetic algorithm could be applied for solution (Khajehpour 2001, Grierson & Khajehpour 2002). Alternatively, Koski (1994) used sequential quadratic programming to find the ten Pareto-optimal designs having variously different plate thicknesses listed in Table 1. It can be seen from the last two table columns that no one design is dominated by any other design for both the W-criterion and the Δ-criterion.

Table 1. Pareto-Optimal Flexural Plate Designs (Koski 1994)

Design Point	T_1 (mm)	t_2 (mm)	t_3 (mm)	t_4 (mm)	t_5 (mm)	t_6 (mm)	W (kg)	Δ (mm)
1	20.6	19.7	18.4	16.4	13.8	8.6	39.4	2.73
2	26.1	20.8	18.4	16.4	13.8	8.6	40.0	2.50
3	30.2	26.1	20.6	16.4	13.8	8.6	42.4	2.00
4	31.0	28.9	24.7	19.4	14.1	8.6	46.8	1.50
5	37.3	34.3	26.8	22.1	16.3	9.8	53.3	1.00
6	40.0	37.1	30.2	24.0	18.3	10.8	58.8	0.75
7	40.0	40.0	36.4	27.8	21.0	12.8	67.6	0.50
8	40.0	40.0	40.0	32.6	24.6	14.4	75.6	0.375
9	40.0	40.0	40.0	40.0	33.5	20.5	90.8	0.25
10	40.0	40.0	40.0	40.0	40.0	40.0	112.3	0.1746

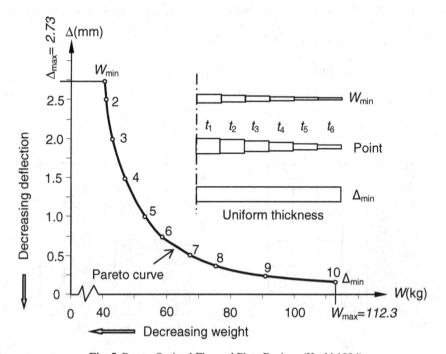

Fig. 5. Pareto-Optimal Flexural Plate Designs (Koski 1994)

The ten Pareto-optimal designs define the *Pareto curve* in Figure 5; in fact, any point along this curve corresponds to a Pareto design. Shown are sketches of the three Pareto designs corresponding to Wmin, *point 5* and Δmin. It remains to select a good-quality compromise design from among the set of Pareto designs in accordance with the preferences of the design team.

3.2 Pareto-optimal Design Compromise

A formal trade-off strategy based on the welfare economics analysis[3] presented earlier is applied in the following to identify a compromise plate design for which designer preferences concerning the conflicting W and Δ criteria are Pareto optimal (Grierson 2006).

To begin, normalize the data for the W and Δ criteria in the last two columns of Table 1 to be as given by the x_1 and x_2 values in the fourth and fifth columns of Table 2. Note that the largest value of each of the normalized criteria x_1 and x_2 is unity (i.e., $x_1^* = x_2^* = 1.0$). Then consider two designers A and B who are seeking between themselves to achieve an optimal trade-off of the two criteria x_1 and x_2 for the plate design. Suppose that designer A's initial endowment is the largest value $x_1^* = 1.0$ of criterion x_1, while that for designer B is the largest value $x_2^* = 1.0$ of criterion x_2. Similar to that in Figure 2, the competitive equilibrium of the two-designer and two-criteria trade-off exercise can be analytically examined by constructing the normalized Edgeworth box diagram in Figure 6, the horizontal and vertical dimensions for which are both equal to unity.

Table 2. Data for Flexural Plate Design Trade-Off Analysis

Design Point	Weight (W)	Deflection (Δ)	x_1 (W/Wmax)	x_2 (Δ/Δmax)	$(1-x_1)$	$(1-x_2)$
1	39.4	2.73	0.351	1.000	0.649	0.000
2	40.0	2.50	0.356	0.916	0.644	0.084
3	42.4	2.00	0.378	0.733	0.622	0.267
4	46.8	1.50	0.417	0.549	0.583	0.451
5	53.3	1.00	0.475	0.366	0.525	0.634
6	58.8	0.75	0.524	0.275	0.476	0.725
7	67.6	0.50	0.602	0.183	0.398	0.817
8	75.6	0.375	0.673	0.137	0.327	0.863
9	90.8	0.25	0.801	0.092	0.199	0.908
10	112.3	0.1746	1.000	0.064	0.000	0.936

In Figure 6, the origins for designers A and B are 0_A and 0_B, respectively, and their initial endowment points $A(1,0)$ and $B(0, 1)$ are both located at the lower right-hand corner of the box. Designer A's offer curve OC_A is a plot of the data points (x_1,x_2) in

---

[3] Here: consumers ≡ designers; goods ≡ criteria; x_1 ≡ W-criterion; x_2 ≡ Δ-criterion.

the fourth and fifth columns of Table 2 (i.e., a normalized plot of the Pareto curve in Figure 5), while designer B's offer curve OC_B is a plot of the data points $(1-x_1, 1-x_2)$ in the last two columns of Table 2.

4 Competitive General Equilibrium

It is observed in Figure 6 that there are *two* competitive general equilibrium points E_1 and E_2 defined by the intersections of the two offer curves OC_A and OC_B, the coordinates for which are found as follows. Upon applying curve-fitting/equation-discovery software to the data points (x_1, x_2) in the fourth and fifth columns of Table 2, designer A's offer curve OC_A is found to be accurately represented ($r^2 = 0.999$) by the function (TableCurve2D&3D 2005),[4]

$$17.15x_1^2 x_2 - 1.1x_2 - 1 = 0 \qquad (4)$$

Hence, designer B's offer curve OC_B is represented by the function,

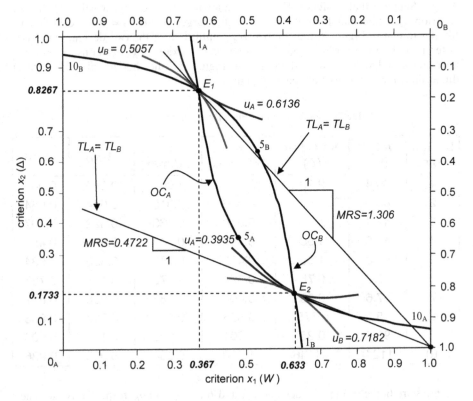

Fig. 6. Plate Design Edgeworth Box (Grierson 2006)

---

[4] Note that in Table 2 and Eqs.(4)-(8) the coordinates x_1 and x_2 are measured from the origin point O_A in Figure 6; i.e., $x_1 = x_{1A}$ and $x_2 = x_{2A}$, and therefore $(1-x_1) = x_{1B}$ and $(1-x_2) = x_{2B}$.

$$17.15(1\text{-}x_1)^2(1\text{-}x_2) - 1.1(1\text{-}x_2) - 1 = 0 \qquad (5)$$

Substitute for x_2 from Eq.(4) into Eq.(5) to obtain the function,

$$9.021x_1^4 - 18.042x_1^3 + 6.812x_1^2 + 2.209x_1 - 1 = 0 \qquad (6)$$

Equation (6) is solved to find the two roots $x_1 = 0.367$ and $x_1 = 0.633$ (MATLAB 2005), and then the corresponding roots $x_2 = 0.8267$ and $x_2 = 0.1733$ are found through Eq.(4). That is, as shown in Figure 6, the equilibrium points are $E_1(0.367, 0.8267)$ and $E_2(0.633, 0.1733)$.

4.1 Design Utility

Equilibrium point E_1 corresponds to a plate design intermediate to designs 2 and 3 in Table 1 that has weight $W = (0.367)(112.3) = 41.21$ kg and deflection $\Delta = (0.8267)(2.73) = 2.26$ mm, while point E_2 corresponds to a plate design intermediate to designs 7 and 8 in Table 1 that has weight $W = (0.633)(112.3) = 71.09$ kg and deflection $\Delta = (0.1733)(2.73) = 0.473$ mm. Each of these two designs is a Pareto-optimal compromise design of the plate. It remains to determine their utilities from the perspective of the designers' preferences for the two conflicting criteria.

This study adopts the commonly used *Cobb-Douglas* utility function (Varian 1992),

$$u(x_1, x_2) = x_1^c x_2^{1\text{-}c} \qquad (7)$$

where the exponent c is a function of the point at which the utility is being measured. Upon observing in Figure 6 that for both designers the marginal rate of substitution at any point (x_1, x_2) is $MRS = -x_2/(1\text{-}x_1)$, it is found from Eqs. (2) and (7) that the utility functions for designers A and B can be expressed as,

$$u_A(x_1, x_2) = x_1^{x_1} x_2^{1\text{-}x_1} \quad ; \qquad u_B(x_1, x_2) = (1\text{-}x_1)^{x_2}(1\text{-}x_2)^{1\text{-}x_2} \qquad (8a,b)$$

The utility levels u_A and u_B indicated in Figure 6 are found by evaluating Eqs.(8) for the two sets of (x_1, x_2) coordinates $(0.367, 0.8267)$ and $(0.633, 0.1733)$ corresponding to equilibrium points E_1 and E_2, respectively. As expected, utility u_A is greater at E_1 than at E_2 (i.e., $0.6136 > 0.3935$) because the plate weight W there is less (i.e., 41.21 $kg < 71.09$ kg). Conversely, utility u_B is greater at E_2 than at E_1 (i.e., $0.7182 > 0.5057$) because the plate deflection Δ there is less (i.e., 0.473 mm < 2.26 mm). Presuming that designer A is the advocate for the W-criterion, she will opt for the plate design at point E_1 because it provides her greatest utility $u_A = 0.6136$. However, as the advocate for the Δ-criterion, designer B will alternatively opt for the plate design at point E_2 because it provides his greatest utility $u_B = 0.7182$. This poses a dilemma, which may be overcome if the two designers agree to act together as a team that simply opts for the design having the maximum utility level u_{max} from all among all four utility levels associated with the two equilibrium points. That is, they would select the plate design at point E_2 having weight $W = 71.09\ kg$, deflection $\Delta = 0.473\ mm$, and utility $u_{max} = 0.7182$.

5 Three-Criteria Design Engineering

5.1 Multi-storey Building Design

Consider the design of a multi-storey building, such as that shown in Figure 7, having the following constraints and conditions: tax rate = 2% ; annual cost rate = 2% ; capital interest rate = 10% ; inflation rate = 3% ; geographical latitude = 40^0N ; angle with East = 0^0 ; land unit cost = $\$1000/m^2$; min-max annual lease rate = $\$100$-$\$360/m^2$; steel, concrete, reinforcement, forming, windows, walls, finishing, electrical, mechanical, and elevator unit costs = *local values* ; building colour = *dark* ; maximum height = *300m* ; minimum aspect ratio = *0.5* ; maximum slenderness ratio = 9 ; maximum footprint = *70m x 70m* ; core/footprint area ratio = 20% ; minimum lease floor area = $60,000m^2$; minimum core-to-perimeter distance = *7m* ; floor-to-ceiling height = *3m* ; energy cost = $\$140/MWhr$; desired inside temperature = 22^{0c} ; outside min-max temperature = -20^{0c} to 31^{0c} ; cold-hot daily temperature range = 10^{0c} ; desired inside relative humidity = *0.5%* ; cold-hot daily outside relative humidity = 50% -80%; dead load = $1.45\ kN/m^2$; gravity live load = $2.80\ kN/m^2$; wind pressure = *0.48kPa* ; clear sky percentage = 75% ; and rules of good design practice to ensure architectural and structural layouts are feasible and practical.

The design of the building is governed by a large variety of primary and secondary variables. The former include ten different structural types, two different bracing types, four different floor types for each of concrete and steel structures, four different window types, sixteen different window ratios, four different cladding types, from two to five times more column bays on the perimeter of framed tube structures than on the interior of the building, up to eight different core dimensions in each of the length and width directions for the building, and a large number of different regular-orthogonal floor plans having from three to ten column bays with span distances of *4.5 to 12m* in the length and width directions for the building.

It is required to design the building for the three conflicting criteria of *minimum initial capital cost (CC-criterion), minimum annual operating cost (OC-criterion)* and *maximum annual income revenue (IR-criterion)*. The initial capital cost consists of the cost of land, structure, cladding, windows, HVAC system, elevators, lighting, and finishing. The annual operating cost consists of maintenance and upkeep costs, the cost of energy consumed by the HVAC, elevator and lighting systems, and annual property taxes. The annual income revenue accounts for the impact that flexibility of floor space usage and occupant comfort level has on lease income.

Full details of the constraints, variables and objective criteria discussed in the foregoing may be viewed in Khajehpour (2001) and Grierson & Khajehpour (2002), as well as that the three-criteria building design optimization problem may be generally stated as:

$$Minimize: \{\ Capital\ Cost\ ;\ Operating\ Cost\ ;\ -Income\ Revenue\ \} \quad (9a)$$
$$Subject\ to: \{\ Specified\ Constraints\ \} \quad (9b)$$

where it is noted that minimizing the negative of income revenue is equivalent to maximizing its (absolute) value, as desired. Khajehpour (2001) and Grierson & Khajehpour (2002) applied a multi-criteria genetic algorithm to the optimization problem posed by Eqs.(9), to find the 139 Pareto-optimal (CC,OC,RI) design points listed in Table 3 (single-decimal accuracy is used solely for presentation purposes; in fact, e.g., maximum $CC = \$76,159,471$, $OC = \$5,211,802$ and $IR = \$18,713,876$). Each design point corresponds to a different building design. For example, the first (CC,OC,IR) design point in Table 3 corresponds to the building shown in Figure 7.

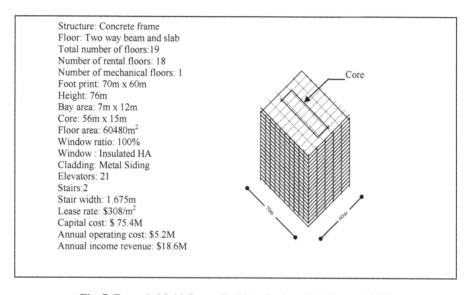

Structure: Concrete frame
Floor: Two way beam and slab
Total number of floors: 19
Number of rental floors: 18
Number of mechanical floors: 1
Foot print: 70m x 60m
Height: 76m
Bay area: 7m x 12m
Core: 56m x 15m
Floor area: $60480m^2$
Window ratio: 100%
Window : Insulated HA
Cladding: Metal Siding
Elevators: 21
Stairs:2
Stair width: 1.675m
Lease rate: $308/m^2$
Capital cost: $ 75.4M
Annual operating cost: $5.2M
Annual income revenue: $18.6M

Fig. 7. Example Multi-Storey Building Design (Khajehpour 2001)

The *Pareto surface* in Figure 8 is defined by the *139* design points in Table 3; e.g., see the point corresponding to the building in Figure 7. In fact, any point on this surface corresponds to a Pareto design. It remains to select a good-quality compromise design from among all possible Pareto designs in accordance with the preferences of the design team.

To begin, normalize the data in Table 3 using the maximum CC, OC and IR values noted in the foregoing (i.e., such that the largest value of each of the three normalized criteria is unity). Assign each of three designers A, B or C an initial endowment equal to the largest value of the normalized CC, OC or IR criterion, respectively. As an extension to that in Figure 6, the competitive general equilibrium of the three-designer and three-criteria trade-off exercise can be analytically examined by constructing the normalized *Edgeworth-Grierson cube* diagram in Figure 9.

In Figure 9, the origins for designers A, B and C are 0_A, 0_B and 0_C, respectively, and their initial endowment points $A(1,0,0)$, $B(0,1,0)$ and $C(0,0,1)$ are all located at the lower right-hand corner of the back face of the cube (see black dot). Designer A's offer surface OS_A is plotted from origin 0_A and is the same as the surface plot in Figure 8. Designer B's offer surface OS_B is a plot of the same data points but from origin 0_B, while designer C's offer surface OS_C is also a plot of the same data points but from origin 0_C.

Table 3.: Pareto-Optimal Multi-Storey Building Designs (Khajehpour 2001)

Design	CC	OC	IR	Design	CC	OC	IR	Design	CC	OC	IR
1	75.4	5.2	18.6	48	73.6	5.1	13.1	95	74.4	5.0	10.5
2	76.0	5.2	18.7	49	73.0	5.1	13.0	96	70.3	5.0	8.1
3	75.3	5.2	18.1	50	72.8	5.1	13.0	97	72.2	5.0	9.8
4	75.1	5.1	18.0	51	73.0	5.1	13.0	98	72.1	5.1	11.3
5	75.1	5.2	17.5	52	72.8	5.1	13.0	99	69.8	5.0	6.9
6	74.9	5.1	17.3	53	73.4	5.1	12.8	100	70.2	5.0	7.3
7	74.9	5.1	17.3	54	75.3	5.1	12.9	101	70.1	5.0	7.7
8	74.9	5.1	16.8	55	72.8	5.1	12.6	102	69.9	5.0	7.3
9	74.9	5.1	16.8	56	72.6	5.1	12.6	103	71.2	5.0	9.7
10	74.7	5.1	16.7	57	73.4	5.1	12.6	104	74.2	5.0	9.9
11	74.7	5.1	16.7	58	73.4	5.1	12.5	105	72.6	5.0	10.9
12	74.7	5.1	16.2	59	73.6	5.1	12.3	106	71.7	5.0	8.6
13	74.7	5.1	16.2	60	73.1	5.1	12.2	107	71.3	5.0	7.5
14	74.5	5.1	16.1	61	72.6	5.1	12.1	108	72.0	5.1	11.2
15	74.5	5.1	16.1	62	72.4	5.1	12.1	109	72.4	5.0	10.4
16	74.5	5.1	15.6	63	72.6	5.1	12.2	110	70.9	5.1	9.1
17	74.0	5.2	15.6	64	72.6	5.1	12.2	111	72.0	5.1	11.2
18	74.1	5.1	15.6	65	75.0	5.1	12.3	112	70.9	5.0	7.3
19	74.2	5.1	15.4	66	73.1	5.1	12.0	113	70.7	5.1	8.7
20	73.8	5.2	15.1	67	72.3	5.1	11.8	114	71.5	5.1	10.4
21	76.2	5.1	15.3	68	72.4	5.1	11.8	115	71.3	5.1	9.9
22	73.9	5.1	15.1	69	72.3	5.1	11.7	116	73.2	5.0	11.2
23	74.2	5.1	15.0	70	72.2	5.1	11.7	117	71.5	5.0	8.2
24	74.0	5.1	14.8	71	73.4	5.0	11.7	117	71.7	5.0	8.7
25	73.5	5.1	14.5	72	72.2	5.1	11.6	119	73.9	5.0	10.8
26	73.7	5.1	14.5	73	72.9	5.1	11.5	120	74.4	5.0	8.2
27	75.9	5.1	14.7	74	71.4	5.0	9.9	121	70.0	5.0	7.4
28	73.5	5.1	14.4	75	70.7	5.0	8.5	122	70.9	5.0	9.0
29	73.6	5.1	14.4	76	70.9	5.0	8.5	123	70.2	5.0	7.8
30	74.0	5.1	14.4	77	71.4	5.1	10.1	124	74.7	5.0	8.8
31	738	5.1	14.1	78	70.4	5.1	8.2	125	71.1	5.1	9.5
32	73.3	5.1	14.0	79	72.9	5.0	8.4	126	71.8	5.1	10.8
33	73.5	5.1	14.0	80	72.1	5.1	11.2	127	71.9	5.0	9.3
34	73.3	5.1	14.0	81	71.9	5.1	10.9	128	72.9	5.0	11.5
35	73.3	5.1	14.0	82	70.8	5.0	8.9	129	74.6	5.0	11.1
36	73.4	5.1	14.0	83	70.4	5.0	7.7	130	71.0	5.0	9.3
37	73.4	5.1	14.0	84	71.2	5.0	7.7	131	71.7	5.0	10.3
38	75.7	5.1	14.1	85	71.3	5.0	9.4	132	75.0	5.0	9.3
39	73.8	5.1	13.7	86	70.5	5.0	8.1	133	71.7	5.1	10.5
40	73.1	5.1	13.5	87	71.1	5.0	9.0	134	70.5	5.0	8.5
41	73.3	5.1	13.5	88	70.4	5.0	7.6	135	70.7	5.0	8.1
42	73.1	5.1	13.5	89	70.6	5.0	8.1	136	71.4	5.0	8.1
43	73.1	5.1	13.5	90	71.5	5.0	9.9	137	73.6	5.0	10.2
44	73.2	5.1	13.5	91	71.1	5.0	9.4	138	71.8	5.1	10.8
45	73.2	5.1	13.5	92	71.9	5.1	10.8	139	748	5.0	11.7
46	73.6	5.1	13.5	93	74.2	5.0	11.3				
47	75.5	5.1	13.5	94	71.6	5.1	10.3				

CC = Capital Cost ; **OC** = Operating Cost ; **IR** = Income Revenue ($ Million)

Interestingly, as indicated in Figure 9, it is found that there are only *two* competitive general equilibrium points E_1 and E_2 defined by the intersections of the three offer surfaces OS_A, OS_B and OC_C, the coordinates for which are found as described in the following. Upon applying curve-fitting/equation-discovery software to the normalized data points derived from Table 3, the surface they form is found to be accurately represented ($r^2 = 0.981$) by the function (TableCurve2D&3D 2005),

$$x_3 = a + b/x_1 + c/x_2 + d/x_1^2 + e/x_2^2 + f/(x_1 x_2) + g/x_1^3 + h/x_2^3 + i/(x_1 x_2^2) + j/x_1^2 x_2) \quad (10)$$

where the constant values are $a = -4654$, $b = 1505$, $c = 12168$, $d = -2654$, $e = -13086$, $f = 2349$, $g = 1634$, $h = 3925$, $i = 1073$ and $j = -2261$, and the parameters x_1, x_2 and x_3 correspond to the CC, OC and IR axes, respectively. The offer surface for each of the three designers is defined by plotting Eq.(10) from each of the three origins. MATLAB software (2005) is applied to plot Eq. (10) from the three origins 0_A, 0_B and 0_C, in turn, to form a superimposed image of the three offer surfaces OS_A, OS_B and OS_C that is exactly as shown in Figure 9. The two competitive general equilibrium points E_1 and E_2 where the three curves simultaneously intersect are readily identified, and the data cursor command finds their coordinates to be as given in Table 4 for all three origins.

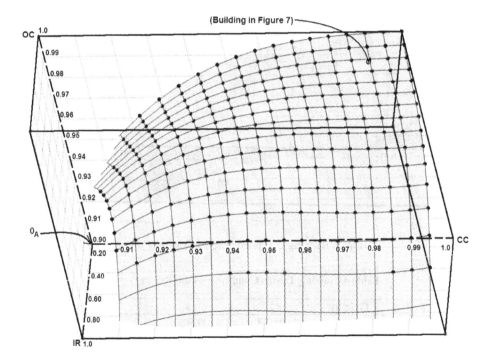

Fig. 8. Pareto-Optimal Building Designs (Khajehpour 2001)

Equilibrium point E_1 corresponds to a building that is very nearly the same as design *102* in Table 3 having capital cost CC = *$69,853,181*, operating cost OC = *$5,003,072* and income revenue IR = *$7,255,537*. Alternatively, equilibrium point E_2 corresponds to a building similar to design *137* in Table 3 having CC = *$73,644,156*,

OC = *$5,032,577* and IR = *$10,176,632*. Each of these two designs is a Pareto-optimal compromise design of the building. It remains to determine their utilities from the perspective of the designers' preferences.

Here, the Cobb-Douglas utility function accounting for three conflicting criteria is,

$$u(x_1, x_2, x_3) = x_1^e x_2^f x_3^{1-e-f} \qquad (11)$$

where the exponents e and f are each a function of the point at which the utility is being measured.

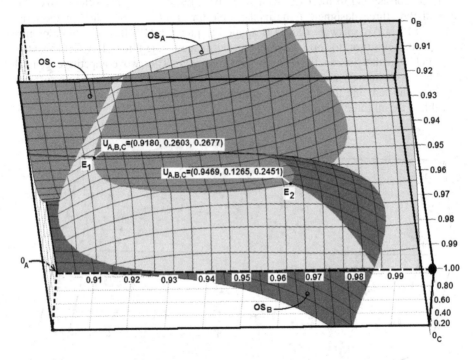

Fig. 9. Building Design Edgeworth-Grierson Cube

Table 4. Building Design Competitive Equilibrium Points

Point	Origin	x_1	x_2	x_3
	O_A	0.9173	0.9668	0.8794
E_1	O_B	0.0827	0.0332	0.8794
	O_C	0.0827	0.9668	0.1206
	O_A	0.9671	0.9450	0.5112
E_2	O_B	0.0329	0.0550	0.5112
	O_C	0.0329	0.9450	0.4888

There is no change in the value of the utility function for any incremental changes of the three criteria along an indifference surface defined by Eq.(11), i.e.,

$$du = (\partial u / \partial x_1)dx_1 + (\partial u / \partial x_2)dx_2 + (\partial u / \partial x_3)dx_3 = 0 \qquad (12)$$

from which the expressions for marginal rates of substitution are readily derived as,

$$dx_2 / dx_1 = -(\partial u / \partial x_1)/(\partial u / \partial x_2) - ((\partial u / \partial x_3)dx_3)/((\partial u / \partial x_2)dx_1) \quad (13a)$$

$$dx_3 / dx_1 = -(\partial u / \partial x_1)/(\partial u / \partial x_3) - ((\partial u / \partial x_2)dx_2)/((\partial u / \partial x_3)dx_1) \quad (13b)$$

$$dx_2 / dx_3 = -(\partial u / \partial x_3)/(\partial u / \partial x_2) - ((\partial u / \partial x_1)dx_1)/((\partial u / \partial x_2)dx_2) \quad (13c)$$

Setting one-at-a-time the incremental changes dx_1, dx_2 and dx_3 equal to zero in the second term of Eqs.(13), the values of the exponents in the utility function Eq.(11) associated with each of the three origins for designers A, B and C are found to be as given in Table 5.

The utility levels u_A, u_B and u_C indicated in Figure 9 at equilibrium points E_1 and E_2 are found by evaluating Eqs.(11) for the six sets of coordinates (x_1,x_2,x_3) in Table 4 and the exponents in Table 5. If acting alone, designer A will opt for the building design at point E_2 because it provides her greatest utility $u_A = 0.9469 > 0.9180$. However, designer B will alternatively opt for the building design at point E_1 because it provides his greatest utility $u_B = 0.2603 > 0.1265$. Similarly, designer C will also opt for the building design at point E_1 because it provides her greatest utility $u_C = 0.2603 > 0.1265$. The foregoing dilemma is overcome when the three designers agree to act together as a team that, for example, opts for the design having the maximum utility level from among all six utility levels associated with the two equilibrium points. That is, select the building design at point E_2. Perhaps more reasonably, however, the design team will opt for the design for which two of the three utility types is greatest. In which case, select the building design at point E_1.

Table 5. Exponents for Utility Function Eq.(11)

Origin	e	f	$1-e-f$
0_A	$x_1/(2-x_1)$	$(1-x_1)/(2-x_1)$	$(1-x_1)/(2-x_1)$
0_B	$(x_2-1)/(x_2-2)$	$-x_2/(x_2-2)$	$(x_2-1)/(x_2-2)$
0_C	$(1-x_3)/(2-x_3)$	$(1-x_3)/(2-x_3)$	$x_3/(2-x_3)$

6 Concluding Remarks

For a given multi-criteria design problem, this study demonstrates that only a very few of the theoretically infinite number of designs forming the Pareto front represent a mutually agreeable trade-off between the competing criteria. This finding, which depends primarily on the shape of the Pareto front, is under ongoing investigation. At present, the research is in its early stages and prompts fewer conclusions than it does questions, some of which are as follows. Does the form of the utility function have an influence on the results? When there are multiple competitive general equilibrium points, what is the veracity of selecting the design to be that at the particular equilibrium point having the maximum utility level from all among all utility levels associated with all equilibrium points? What is the veracity of selecting the design at the equilibrium point for which the number of the different utility types is maximum? Can the methodology be applied to design problems involving four or more conflicting criteria? These and other lines of enquiry will be pursued by the ongoing research program.

Acknowledgments

This study is supported by the Natural Science and Engineering Research Council of Canada. For insights into social welfare economics thanks are due to Kathleen Rodenburg, Department of Economics, University of Guelph, Canada. For help in preparing the paper text and the engineering example thanks are due to Yuxin Liu, Joel Martinez and Kevin Xu, Faculty of Engineering, University of Waterloo, Canada.

References

Boadway, R., and Bruce, N. (1984). *Welfare Economics*. Basil Blackwell, Chapter 3, 61-67.
Cheng, F.Y. (2002). "Multi-Objective Optimum Design of Seismic-Resistant Structures". *Recent Advances in Optimal Structural Design*, Edited by S.A. Burns, ASCE-SEI, NY, Chapter 9, 241-255.
Grierson, D.E. (2006). "Pareto Analysis of Pareto Design." *Proceedings of Joint International Conference on Computing and Decision Making in Civil and Building Engineering*, Edited by H. Rivard, E. Miresco and M. Cheung, June 14-16.
Grierson, D.E., and Khajehpour, S. (2002). "Method for Conceptual Design Applied to Office Buildings." *J. of Computing in Civil Engineering*, ASCE, NY, 16 (2), 83-103.
Khajehpour, S. (2001) *Optimal Conceptual Design of High-Rise Office Buildings*. PhD Thesis, Civil Engineering, University of Waterloo, Canada, pp 191.
Koski, J. (1994). "Multicriterion Structural Optimization." *Advances in Design Optimization*, Edited by H. Adeli, Chapman and Hall, NY, Chapter 6, 194-224.
Osyczka, A. (1984). *Multicriterion Optimization in Engineering*. Ellis Horwood, Chichester
Varian, H.R. (1992). *Microeconomic Analysis*. Third Edition, W.W. Norton & Co., NY.
___MATLAB, Version 7.0 (2005). *Automated Equation Solver*. The MathWorks, Inc.
___TableCurve2D&3D, Version 5.01 (2005). *Automated Curve-fitting and Equation Discovery*. Systat Software, Inc., CA.

A Model for Data Fusion in Civil Engineering

Carl Haas

Professor,
Canada Research Chair in Sustainable Infrastructure, and
Director of Centre for Pavement and Transportation Technology
Department of Civil Engineering
University of Waterloo
200 University Avenue West
Waterloo, Ontario, Canada N2L 3G1
Phone: 519-888-4567 x5492
chaas@civmail.uwaterloo.ca

Abstract. Thoroughly, reliably, accurately and quickly estimating the state of civil engineering systems such as traffic networks, structural systems, and construction projects is becoming increasingly feasible via ubiquitous sensor networks and communication systems. By better and more quickly estimating the state of a system we can make better decisions faster. This has tremendous value and broad impact. A key function in system state estimation is data fusion. A model for data fusion is adapted here for civil engineering systems from existing models. Applications and future research needs are identified.

1 Introduction

Data fusion may be defined as the process of combining data or information to estimate or predict entity states [16]. In civil engineering the entity state may be the structural health of a bridge, the productivity of a construction project, or the traffic flow in a section of a traffic network. State may be multidimensional. An element of a system or the whole system may be the entity. Need and application typically drive further definition.

To be useful a data fusion model should facilitate understanding, comparison, integration, modularity, re-usability, scalability, and efficient problem solving or design. It may take the form of a process model, a formal mathematical model, a functional model, etc. For data fusion, a functional model is desirable, because it may be both flexible and general.

This short abstract presents a simple functional data fusion model adapted from Steinberg [16]. Examples of the application of data fusion methods in civil engineering are interspersed in the explanation of the model. Then, common tools for data fusion are identified. Finally, a conclusion and a recommendation are presented.

However, before expanding on the model, it is worth exploring how data fusion relates to system identification and case based reasoning. System identification methodologies have been used to identify characteristics of structural systems using measurement data [12,13,14]. It is a type of inverse problem solving. Case based reasoning is similar in the sense that attribute values are used to identify the closest

I.F.C. Smith (Ed.): EG-ICE 2006, LNAI 4200, pp. 315–319, 2006.
© Springer-Verlag Berlin Heidelberg 2006

past case to that observed [2]. Typically, multiple inverse solutions may exist. A classic case is the inverse kinematic solution for a multi-degree-of-freedom manipulator where none, one, a finite number greater than one, or infinite solutions may exist depending on the link configuration. A redundant manipulator may result in an infinite solution space, for example. In data fusion, it is typical and generally necessary to begin with one or a finite number of a-priori models of the solution space for the level at which the fusion occurs. Where system identification is used to define a population of candidate models, data fusion can be used to reduce the population via refined input estimates or complementary measurement data. Where case based reasoning is being used, data fusion may improve the case fit by better estimating attributes and by adding attributes. In all cases, Smith's admonishments in, "Sensors, Models, and Videotape," should be observed [15].

2 Model for Data Fusion

Figure 1 presents a functional data fusion model adapted from Steinberg [16] for civil engineering applications. Data input sources, a data fusion functional domain, and a human/computer interface are defined.

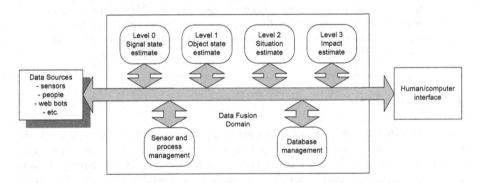

Fig. 1. Model for data fusion in civil engineering adapted from Steinberg, et al [16]

In addition to sensors, data sources may include people, web bots, or data already fused at a lower level. Typical sensors sources include strain gauges, inductive loop detectors (ILD's), radio frequency identification (RFID) tags, range cameras, etc. People provide data via probe vehicles, on site assessments, etc. Web bots may troll the web to provide data from sources with varying degrees of reliability. And data fused at some lower level such as at the automated incident detection (AID) algorithm functional level may be a source to be fused with probe vehicle reports, for example.

Within the data fusion domain six functions are defined. They are not necessarily sequential. In fact, sensor management may be interpreted as a meta-level activity. Database management serves the other functions and the objective to archive some data. Four functional levels for data fusion are defined that correspond to typical hierarchical levels associated with aggregation, semantic content, and/or decision level in an operational environment.

Level zero data fusion is signal state estimation. A signal is hypothesized, and data sources are fused to construct the signal. Examples include: (1) stacking in signal processing to improve the quality of the signal itself, (2) merging of range point clouds from range scans to improve the signal's representation of the hypothesized entity [5], (3) better boundary detection in piles of construction aggregates or mine tailings by fusing Canny edge detection and watershed algorithm processed signals [6], and (4) accurate, autonomous, machine vision based crack detection by fusing digital and range image data [4].

Level one data fusion is object state estimation. Inferences from two or more observations of the object state are combined to improve the accuracy of and confidence in the object state estimate. Examples include: (1) proximity based RFID tag locating using constraint set techniques, fuzzy logic, and Dempster-Shafer theory [1], (2) pavement crack mapping by logically combining manual clues with machine vision based processes, (3) integrating range image data and CAD data to better describe infrastructure entities, (4) integrating point based data (from ILD's) with link based data (from portal to portal reads of AVI tags on vehicles) to improve estimates of traffic flow, and (5) integrating output data from diverse AID algorithms to detect traffic incidents at a higher rate with less false alarms.

Level two data fusion is situation state estimating (assessment) based on inferred relations among entities. Examples in transportation, structural health monitoring, and construction are briefly described. At the TMC (traffic management centre) level, level one traffic flow state estimates from several related links may lead quickly to an assessment of a major traffic congestion situation. Typically, TMC floor level human controllers integrate data via shouts, gestures and brief conversations, in a manner similar to the communication that occurs on the floor of a stock exchange or in a situation room in a battle environment. Integration may be facilitated via web based collaboration schemes and graphical visualization of related data. Whatever the extent of automation, the situation assessment may trigger a **sensor management function** such as channeling remote video camera output of the situation to TMC monitors or redeployment of probe vehicles. A rather simplistic example of level two data fusion for structural health monitoring is pavement condition assessment. Data on cracking, rutting, roughness etc. of AC pavements, when fused via a multi-layer quadtree and an expert system, yielded reasonably accurate assessments of the pavement structural health and the likely cause [4]. For construction project state estimation, it is becoming increasingly clear that fusion of data sources such as GPS readers, RFID tags [1], digital images, range images [5], heat sensors, strain gauges, etc. creates the opportunity to automatically estimate project state (situation) variables such as productivity, object location, process maturity (such as concrete curing), machine activity (as in automated earth moving), and test completion. In fact the opportunities are overwhelming and companies such as Trimble^{TM} and Leica^{TM} (major positioning and spatial data equipment makers) are moving quickly to fill the gap. However, rigorous implementation of data fusion tools at this level becomes increasingly challenging.

Even more challenging is **data fusion at level three**, the impact assessment level. And yet, there is active research in this area. Bridge scour impact assessment tools have been developed based on risk analysis. They assess for example, the risk associated with structural failure based on situation state estimates at for instance the bases of the piers and the traffic situation. Widely used construction project impact assessment tools have been developed via weighted indices such as the PDRI (project definition rating index) from the CII (Construction Industry Institute) located in Austin Texas. In fact a CII research team is currently working on the concept of project situation "lead indicators". This is a very applied and very relevant form of data fusion for estimating impact.

At all levels of data fusion, **database management** is a key service function. It maintains entity attribute information, it manages data relationships (such as spatio-temporal, part/whole, organizational, causal, semantic, legal, emotional, etc.) [16]. It facilitates associative and relational operations. Data structures are used such as hierarchical (e.g. quadtrees), object oriented, and relational. Integrating and managing multi-modal data is a particular challenge.

3 Data Fusion Tools

Functional tools for data fusion include alignment, correlation, averaging and voting. For example, cross correlation matrices for data sources indicate the extent to which they may be considered independent which is a critical input to evidence based reasoning [10]. Evidence based reasoning methods for data fusion include logic, fuzzy logic, Bayesian reasoning, and Dempster-Shafer theory methods based on belief functions [1]. Incorporating human beliefs and perceptions is valid but non-trivial.

4 Conclusions and Recommendations

As stated earlier, to be useful a data fusion model should facilitate understanding, comparison, integration, modularity, re-usability, scalability, and efficient problem solving or design. Validating its usefulness in facilitating understanding will take time, however an opportunity exists to use the model in the author's and others' teaching and to find out whether it helps. Applicability of the model to problem solving will also take time and many cases to analyze. Usefulness of the model for facilitating comparison, integration, modularity, re-usability, and scalability may be validated via future research and development in construction project state estimation for decision support and knowledge management. The model may also be used to facilitate development of state estimation methods for infrastructure and transportation system management.

References

1. Caron, Razavi, Song, Haas, Vanheeghe, Duflos, and Caldas, "Models For Locating RFID Nodes," for ICCCBEXI, Montreal, Canada, June 14-16, 2006.
2. Cheng, Y. and Melhem, H. G., "Monitoring bridge health using fuzzy case-based reasoning", Advanced Engineering Informatics, Vol. 19, No 4, 2005, pages 299-315.
3. Haas, C., "Evolution of an Automated Crack Sealer: A Study in Construction Technology Development," Automation in Construction 4, pp. 293-305, 1996.
4. Haas, C., and Hendrickson, C., "Integration of Diverse Technologies for Pavement Sensing," the NRC's Transportation Research Record, No. 1311, pp. 92-102, 1991.

5. Kim, C., Haas, C., Liapi, K., and Caldas, C., "Human-Assisted Obstacle Avoidance System Using 3D Workspace Modeling for Construction Equipment Operation," J. Comp. in Civ. Engrg., Volume 20, Issue 3, (May/June 2006), pp. 177-186.

6. Kim, H., Haas C.T., Rauch, A., and Browne, C., "3D Image Segmentation of Aggregates from Laser Profiling," Computer-Aided Civil and Infrastructure Engineering, pp. 254-263, Vol. 18, No. 4, July 2003.

7. Kim, Y.S., and Haas, C., "A Model for Automation of Infrastructure Maintenance using Representational Forms," Vol. 10/1, Automation in Construction, pp. 57-68, Sept. 2000.

8. Kwon, Bosche, Kim, Haas, and Liapi, "Fitting Range Data to Primitives for Rapid Local 3D Modeling Using Sparse Range Point Clouds," Automation in Construction 13, January 2004, pp. 67-81.

9. Mahmassani, H.S., Haas, C., Logman, H., Shin, H., and Rioux, T., "Integration of Point-Based and Link-Based Data for Incident Detection and Traffic Estimation," Center for Transportation Research, Bureau of Engineering Research, University of Texas at Austin, research report no. 0-4156-1, March, 2004

10. Mahmassani, H., Haas, C., Peterman, J., and Zhou S., "Evaluation of Incident Detection Methodologies," Center for Transportation Research, Univ. of Texas, Report No. 1795-2, Austin TX, Oct. 1999.

11. McLaughlin, J, Sreenivasan, S.V., Haas, C., and Liapi, K., "Rapid Human-Assisted Creation of Bounding Models for Obstacle Avoidance in Construction," Journal of Computer-Aided Civil and Infrastructure Engineering, vol. 19, pp. 3-15, 2004.

12. Robert-Nicoud, Y., Raphael, B. and Smith, I.F.C. "System Identification through Model Composition and Stochastic Search" J of Computing in Civil Engineering, Vol 19, No 3, 2005, pp. 239--247

13. Robert-Nicoud, Y., Raphael, B. and Smith, I.F.C. "Configuration of measurement systems using Shannon's entropy function" Computers & Structures, Vol 83, No 8-9, 2005, pp 599-612.

14. Saitta, S., Raphael, B. and Smith, I.F.C. "Data mining techniques for improving the reliability of system identification" Advanced Engineering Informatics, Vol 19, No 4, 2005, pp 289-298.

15. Smith, I., "Sensors, Models and Videotape," proc.s, 2005 ASCE International Conference on Computing in Civil Engineering, Soibelman, and Peña-Mora - Editors, July 12–15, 2005, Cancun, Mexico.

16. Steinberg, A.N., and Bowman, C.L., "Revisions to the JDL Data Fusion Model," pp. 2-1--2-18. in Hall, D.L., and Llinas, J., "Handbook of Multisensor Data Fusion," CRC Press, 2001.

Coordinating Goals, Preferences, Options, and Analyses for the Stanford Living Laboratory Feasibility Study

John Haymaker[1] and John Chachere[2]

[1] Center for Integrated Facility Engineering, Construction Engineering and Management,
Civil and Environmental Engineering, Stanford University
haymaker@stanford.edu
[2] Management Science and Engineering, Civil and Environmental Engineering,
Stanford University
chachere@stanford.edu

Abstract. This paper describes an initial application of Multi-Attribute Collective Decision Analysis for a Design Initiative (MACDADI) on the feasibility study of a mixed-use facility. First, observations of the difficulties the design team experienced communicating their goals, preferences, options, and analyses are presented. Next, the paper describes a formal intervention by the authors, integrating survey, interview, and analytic methods. The project team collected, synthesized, and hierarchically organized their goals; stakeholders' established their relative preferences with respect to these goals; the design team formally rated the design options with respect to the goals; the project team then visualized and assessed the goals, options, preferences, and analyses to assist in a transparent and formal decision making process. A discussion of some of the strengths and weaknesses of the MACDADI process is presented and opportunities for future improvement are identified.

1 Introduction

To achieve multidisciplinary designs, Architecture, Engineering, and Construction (AEC) professionals need to manage and communicate a great deal of information and processes. They need to define goals, propose options, analyze these options with respect to the goals, and make decisions [1]. This is a social process [2]; they need to coordinate these processes and information amongst a wide range of team members and stakeholders. AEC professionals have difficulty doing this today.

This paper describes observations of these processes on the feasibility study for the Living Laboratory project: a new student dormitory and research facility being planned for the Stanford University campus. It describes the ways in which goals were defined, options were proposed and analyzed, and decisions were made. It also describes some of the difficulty the team had communicating and coordinating these processes and information.

The paper then describes a process called MACDADI: a Multi-Attribute Collective Decision Analysis for the Design Initiative. The authors designed and implemented MACDADI with the help of the design team towards the end of the feasibility stage. The authors and project team collected, synthesized and hierarchically organized the

I.F.C. Smith (Ed.): EG-ICE 2006, LNAI 4200, pp. 320–327, 2006.

project goals; the stakeholders established their relative preference with respect to these goals; the authors collected and aggregated these preferences; the design team analyzed the design options with respect to the goals; and the design team, stakeholders, and authors visualized and assessed these goals, options, preferences, and analyses to assist in a transparent and formal decision making process. The methods of Decision Analysis3, including adaptations for group decision-making and multi-attribute decisions [4], inspired our analysis but we did not adhere strictly to those methods. Finally, we discuss some of the strengths and weaknesses of the implemented MACDADI process, and we identify opportunities for future improvement.

2   Observations on the Living Laboratory Feasibility Study

During the fall of 2005, Stanford University hired a design team consisting of architects, structural, mechanical, electrical, and civil engineers, and construction consultants to perform a feasibility study for a mixed use dormitory and research facility on campus. The Stanford stakeholder constituency consisted of project managers, housing representatives, a cost engineer, the University architects, an energy manager, student representatives, and several professors and researchers with interests ranging from innovative water treatment, to renewable energy strategies, to innovative structural solutions, to design process modeling. This section describes the processes this project team used to define their goals, propose options, analyze these options with respect to their goals, and make decisions. It also describes some of the difficulties they had communicating and coordinating these processes.

Goals: From the moment the dorm was first proposed, stakeholders began defining and refining project goals. A class focused on the project and issued a statement of goals as a final report, a presentation to the Provost to get funding for the project contained additional goals, and the request for proposals to design teams contained another collection of project goals. As the feasibility stage progressed goals were further proposed and refined through meetings and e-mail exchanges. However, as the feasibility stage neared its conclusion, these goals remained distributed throughout the project team and in several documents. A definitive and comprehensive set of goals had not been collected and made public for the design team and stakeholder to collectively assess and share, and the relative preferences amongst goals had not been established. For example, there remained doubt amongst the stakeholders as to whether the project was intended to demonstrate the financial viability of building sustainably, thereby emphasizing first and lifecycle cost, or whether the project should emphasize other goals, and let these project costs fall where they may.

Options: Throughout the process, the design team proposed many design options. For example, the design team proposed options for energy reduction (i.e., passive ventilation, high performance fixtures, daylighting solutions, and light dimming) and energy production (i.e., photovoltaic panels, heat recovery from water, a fuel cell, and even methane gas from a membrane bioreactor). As the design progressed, the design team aggregated several of these options into two alternatives: a baseline green alternative, and a living laboratory alternative. Although the alternatives were intended to only give a sense of the feasibility of the project, several other options and alternatives that were discussed were omitted and not formally documented. For example a well, which would both close the water loop and serve as a source of heat and cooling for the dorm, was not mentioned in the draft of the report, even though several project stakeholders felt this option still had merit.

Room Type Decision Matrix	NSF/Bed (Factor)	Efficiency (Sustainability)	Social Interaction	Future Flexibility	Popularity	Fit w/ Campus Housing Plan	
Singles	110 (1.0)	2	3	2	4	?	1 = Worst
Doubles	100 (1.1)	4	5	5	3	?	5 = Best
Divided Doubles	110 (1.0)	2	3	2	4	?	
Triples	95 (1.16)	4	1	5	1	?	
Quads	90 (1.22)	4	3	5	1	?	
Suites	135 (0.82)	1	1	1	5	?	
Mix	110?	?	?	?	?	?	

Fig. 1. An analysis of room types

Analyses: In some cases the design team performed formal and explicit analyses of options and alternatives, such as for the carbon emissions of the different energy sources they were investigating. In other cases the analyses were less formal, albeit explicit. For example, Figure 1 shows a matrix the design team constructed to help them perform and communicate the analysis of room types with respect to a collection of goals they deemed relevant. In other cases the analyses were neither formal nor explicit, such as for the analysis of the roof deck for its impact on the social life of the dorm. In this case, the roof deck was generally determined to be desirable, without explicit analyses as to why this was the case.

Decisions: As the design progressed the project team made many decisions, and recorded these in the meeting minutes. However, many other decisions, such as the choice of which room types to pursue, or whether to include the well, were not explicitly recorded or accessible to the stakeholders as the design progressed, but rather most decisions and their rationale were stored in the heads of the participants. The design team did not have a consistent, formal means to record decisions in a way that described how the decision helped address project goals.

3 The MACDADI Process

As the feasibility study drew to a close, the authors saw that a more collective and explicit coordination of goals, preferences, options, and analyses may help the team arrive at consensus and better communicate their rationale. This section describes the MACDADI process that the authors designed and implemented for the project. Figure 2 diagrams the process, and the subsequent text describes each step.

1. Design Team and Stakeholders define the Project Goals: The architect and authors reviewed and synthesized many of the project documents and compiled a comprehensive set of hierarchically organized goals. To develop consensus, these goals were then sent out to the project stakeholders and feedback was collected on how well these goals matched each stakeholder's understanding. Another round of synthesis incorporated this feedback into a cohesive set of project goals. At the highest level, the goals state that the project should be the Most Desirable Housing on Campus, serve as a Living Laboratory for Research, achieve a high level of Measurable Environmental Performance, and should be Economically Sustainable. The project goals can be found across the top of the Matrix shown in Figure 3.

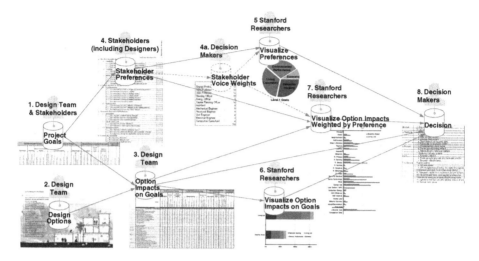

Fig. 2. The MACDADI process involved seven steps, described below. The process is diagrammed using the Narrative formalism [5] in which processes are described as dependencies (arrows) between representations (barrels). The reasoning agent (in this case human) required to construct each representation is shown above each barrel.

2. Design Team identifies Design Options: With input from the stakeholders, the design team proposed several design options, ranging from architectural solutions such as a green roof and clerestory windows, to mechanical solutions such as solar hot water and photovoltaic arrays, to structural alternatives, such as optimized wood framing and an earthquake resistant steel framing system. The design team coupled these many options into two primary alternatives: Baseline Green and Living Laboratory. The options are shown in the left-hand columns of the Matrix. The MACDADI process did not ostensibly impact the process of choosing options.

3. Design Team assesses Options' Impacts on Goals: Using the matrix the design team next assessed the impact of each of these options on the project goals. Other projects [6] have similarly used a matrix to evaluate project options with respect to goals, although the matrix shown in Figure 3 is perhaps more comprehensive in terms of the number of goals assessed. The assessment rated each option's impact on each goal with a numeric score. In this case the architects first completed the entire matrix, then consulted with the specialty engineers to validate their scores. Some assessments, for example the impact of the photovoltaic array on the Low/No Carbon Per kwh goal, are reinforced by rigorous analysis in the appendices of the feasibility report. Other assessments, such as the impact of the large roof deck on Dynamic Social Life, are more qualitative and rely on the assessing designer's experience and intuition.

4. Stakeholders report Preferences: Step 1 established the stakeholders' goals. However, that effort provided no indication of the relative importance among these sometimes-competing goals. To determine their relative perceived importance, each stakeholder was asked to represent their preferences by allocating 100 points amongst the lowest level (detailed) goals. Lower level goal preferences were summed to

Stanford Green Dorm / MACDADI Matrix	The Most Desirable Housing			A Living Laboratory for Research			Measurable Environmental Performance			Economically Sustainable
	Community	Learning	Indoor environmental quality	Experimentation	Demonstration		Zero carbon	Closed water cycle	Material resources	

Baseline Green
- Shared "Information Center" (foyer) and entry
- Solar orientation for passive solar design
- Radiant slab heating
- Optimized 24" O.C. wood framing
- Natural ventilation for passive cooling
- High-performance light and water fixtures
- Fly ash or slag, low-cement concrete
- First floor location for building systems lab
- Large roof deck at second level
- Electric car garage
- 85% daylit interior

Living Laboratory
- 100% daylit interior
- Steel structure w/concrete-filled metal deck
- FSC-certified wood
- 5 kw fuel cell
- Solar hot water system
- Greywater heat recovery
- 80 Kw Photovoltaic array
- Dimmed lighting in dorm rooms
- Evening lighting setback
- Highest-performance lighting and water fixtures
- Building systems monitors
- Rainwater collection
- Greywater and blackwater collection
- Stormwater Features and Native Landscaping
- Sustainable finish materials (interior and exterior)
- Extensive green roof, 2 to 4 inches of soil, 1400 sf
- Triple-paned, double low-e windows
- Three foot clerestory pop-up at upper, north-facing rooms
- Ventilation atrium on first floor

Fig. 3. The Matrix: The top rows of the Matrix show the current consensus on project goals. Each high-level goal is broken down into lower-level goals that, if achieved, would positively impact the higher-level goal. The left side of the Matrix shows the options, aggregated into two alternatives. The body of the matrix contains an evaluation of each option with respect to each goal (3 = high positive impact, 0 = no impact, -3 = high negative impact).

approximate the preferences for higher-level goals. 23 responses were received: 3 from students, 3 from engineering faculty, 8 from designers, 1 from the school of engineering, 3 from housing, and 3 from facilities. Responses within each stakeholder group were averaged to approximate an aggregate perspective for each constituency. To calculate a single aggregate perspective, all stakeholder constituencies were weighted equally (i.e., the collection of student voices count equally to the collection of faculty voices). Figure 2 (see 4A) shows that a decision maker can optionally add weights to the voices of different constituencies, if certain groups' opinions are considered more relevant to a collective decision analysis, although on this project that was not deemed desirable.

5. Analyze Preferences: Having collected the goals, options, assessments, and preferences, this information was visualized in illuminating ways. For Example, Figure 4A shows that taken together, the stakeholders feel the dorm should equally balance the goals related to Living Laboratory, Desirable Housing, and Environmental Performance. According to the stakeholders, Economic Sustainability, which includes first cost, lifecycle cost, and project schedule, is important, but is not as critical to project success. Follow-up discussions revealed some confidence that a profoundly innovative project might offset first costs through fund raising, and as long as the project is ready for a fall move in schedule is not important (which year the move in occurs is not critical). Figure 4B shows the relative preferences of all of the lowest level goals, broken out by different Stakeholder constituencies. For example, reducing energy use and creating that "wow" factor are considered to be extremely important, while project schedule, and research in vehicle energy are not considered to be as important for this project. Closer inspection reveals the relative importance of each of these goals for individual stakeholder constituencies.

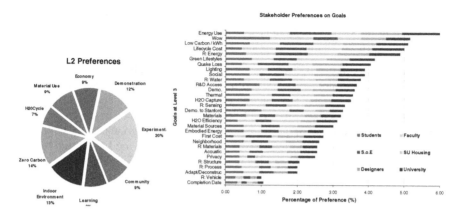

Fig. 4. Visualization of stakeholder preferences amongst goals. **A.** Level 2 preferences show relatively equal preference among level 1 goals of Environmental Performance (green) Desirable Housing (red-brown) ad research (orange-yellow) with lesser emphasis on Economy (blue). **B.** Preferences for Level 3 goals, broken down by stakeholder.

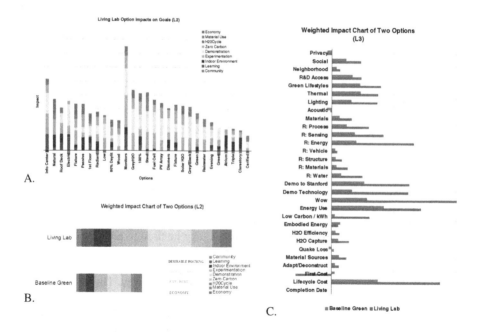

Fig. 5. Impacts of Options on goals. **A.** Impacts of the Living Lab Stakeholder Value of the Baseline Green and Living Laboratory Options. **B.** Weighted Impact of the Living Laboratory and Baseline Green Alternatives broken down to level two goals. **C.** Weighted impact of the Baseline Green and Living Laboratory alternatives on each level three goal.

6. Analyze Options' Impacts on Goals: Visualizations of the designers' assessments of the impact of each option on each goal without taking preference into account can also be constructed. For example Figure 5A shows the assessment of the impact of each option in the Living Laboratory alternative on each goal. The overall impact is broken down to show the impact on each level two goal.

7. Analyze Options' Impacts, Weighted by Preference: Combining designer's assessments of the design options' impact from Step 3 with stakeholder preference data collected in Step 4 generates a prediction of the *perceived* costs and benefits of each design option – or a measure of overall Stakeholder Value. For example, Figure 5B compares the average value of the Baseline Green and Living Laboratory Options for all stakeholders, showing that these stakeholders find the Living Lab option to be far more valuable. Figure 5C illustrates the relative value of the Baseline Green and Living Laboratory Options with respect to the Level 3 goals.

4 Conclusions and Next Steps

While still in the early development stages, MACADADI has shown benefits, and more value is expected as the research advances. Step 1 helped to formally define project goals, leading to a statement of objectives that clarified team members' discussions. These goals became the organizing framework for the feasibility study. Step 3 guided the design team to methodically and explicitly analyze the design options (resulting in over 900 explicit analyses) and helped them communicate the impacts of each option more clearly to the stakeholders. Step 4 helped the project team gain explicit understand of the importance of each of these goals, and through step 6, helped them to assess consensus of the various stakeholders. Step 7 lent confidence that the proposed design addresses stakeholders' preferences equitably, and can guide the team when additional options are explored in future phases. The charts stimulated explicit discussion about the importance of the economic goals and of the reasons for the electric vehicle, and lend specificity to a range of claims in the final report -- twelve charts are included to explicitly communicate goals, preferences, assessments, and project value [7].

This MACDADI process deserves further research. In Step 1, we collected goals by culling documents, synthesizing these goals, and asking stakeholders for comments. More collaborative methods of developing project goals should lead to greater goal congruence. Case-based reasoning methods may be used to retrieve goals developed on previous projects. Standard templates for capturing goals [8], and incorporating more specific definitions of goals and ranges could also be incorporated. In Step 2, the design team aggregated options into only two alternatives. The ability to systematically propose and manage more options and collect these options into more alternatives [9] is also an area of future research. In Step 3, the design team systematically assessed the impact of each option on each goal. However, there may be benefit in using more precise and uniform methods to assess the impact of options on goals. Additionally these analyses should capture the emergent effects of complementary or conflicting design options. It may also prove valuable to investigate a link that would consider the stakeholders preferences (step 4) when assessing impacts on goals. In step 4, there is potential for finer grained preferences that can account for nonlinear preferences (for example, cost is not important until a certain threshold is reached) or interactions among preferences. In step 5 – 7, there are opportunities for more advanced tools to visualize the rich information MACDADI produces. Finally, there are opportunities to develop frameworks that represent and manage these processes at lower levels of detail10. Moving forward, we intend to first develop a web based application that will enable the method to be deployed on many projects in order to popularize the method. We also intend a more detailed literature review to establish the relationships to prior work, and more clearly define the roadmap for future development outlined above.

References

1. Gero, J. S. (1990). Design Prototypes: A Knowledge Representation Schema for Design, AI Magazine, Special issue on AI based design systems, M. L. Maher and J. S. Gero (guest eds), 11(4), 26-36.
2. Kunz, W., & Rittel H. (1970). Issues as elements of information systems. Working Paper No. 131, Institute of Urban and Regional Development, University of California at Berkeley, Berkeley, California, 1970.
3. ASTM International (1988). Standard Practice for Applying the Analytic Hierarchy Process to Multiattribute Decision Analysis of Investments Related to Buildings and Building Systems, ASTM Designation E 1765-98, West Conshohocken, PA, 1998.
4. Keeney R., and Raiffa, H., (1976). "Decisions with Multiple Objectives: Preferences and Value Tradeoffs," John Wiley and Sons, Inc.
5. Haymaker J., Fischer M., Kunz J., and Suter B. (2004). "Engineering test cases to motivate the formalization of an AEC project model as a directed acyclic graph of views and dependencies," ITcon Vol. 9, pg. 419-41, http://www.itcon.org/2004/30
6. BNIM (2002). "Building for Sustainability Report: Six scenarios for the David and Lucile Packard Foundation Los Altos Project", http://www.bnim.com/newsite/pdfs/2002-Report.pdf
7. EHDD (2006). Stanford University Green Dorm Feasibility Report, In production
8. Kiviniemi, A. (2005). "PREMISS - Requirements Management Interface To Building Product Models" Ph.D thesis, Stanford University.
9. Kam, C. (2005). "Dynamic Decision Breakdown Structure: Ontology, Methodology, And Framework For Information Management In Support Of Decision-Enabling Tasks In The Building Industry." Ph.D. Dissertation, Department of Civil and Environmental Engineering, Stanford University, CA.
10. Gentil, S. and Montmain, J. (2004). "Hierarchical representation of complex systems for supporting human decision making", Advanced Engineering Informatics, 18,3,143-160.

Collaborative Engineering Software Development: Ontology-Based Approach

Shang-Hsien Hsieh and Ming-Der Lu

Department of Civil Engineering, National Taiwan University, Taipei 10617, Taiwan
shhsieh@ntu.edu.tw, mdlu@caece.net

Abstract. This paper proposes an ontology-based approach to facilitate collaboration between domain experts and software engineers for development of engineering software applications. In this approach, ontologies are employed to serve as the knowledge interfaces for encapsulating the analysis units identified from the solution workflow of the targeted problem domain and for reducing the collaboration complexity and knowledge coupling between domain problem solving and software engineering in software development. In this paper, the proposed ontology-based software development approach is discussed. Some considerations are given to the software development environment needed for realization of the proposed approach. In addition, the application of the proposed approach is demonstrated using an engineering software development example and a prototype ontology-based software development environment.

1 Introduction

Engineering software development requires knowledge from both the engineering problem domain and software engineering domain. In the past, because it was often more difficult for software engineers to learn engineering domain knowledge, successful engineering application development often relied on domain experts with good knowledge of software development. However, with the rapid advancement of information technology, it has become increasingly difficult for domain experts to keep up with the most advanced software development technologies. Furthermore, the growing scale and complexity of today's engineering problems have resulted in increasingly large and complicated engineering applications that are much more difficult to extend and maintain. Therefore, successful development of engineering applications with good software flexibility, extensibility, and maintainability demands more and more on good collaboration between domain experts and software engineers.

In recent years, the object-oriented technology [1] has emerged as a state-of-the-art software development technology to address the reusability, extensibility, and maintainability issues in development of large-scale and complex software applications. It also provides a solution for domain experts and software engineers to collaborate on software development. In the object-oriented approach (see Fig. 1), the functional requirements of the targeted software application are specified by the domain expert and often expressed in use cases. The software engineer then works collaboratively with the domain expert to design the object-oriented software architecture for the targeted application so that all of the use cases can be supported. Finally, the

I.F.C. Smith (Ed.): EG-ICE 2006, LNAI 4200, pp. 328–342, 2006.

implementation of the object classes in the software architecture is carried out and the targeted application is built by the software engineer.

The key challenge in this software development process is on the design of the object-oriented software architecture not only because the design requires good collaboration between the domain expert and the software engineer but also because the reusability, extensibility, and maintainability of the targeted application depend largely on the design. Often, to achieve good object-oriented design, the domain expert needs to have sound background on the object-oriented design method so that he or she can help laying out a good overall architecture for the software at the first place. Furthermore, through the object-oriented design process, the domain knowledge for problem solving explicitly specified in the use cases is transformed and represented in a more implicit manner by the designed set of object classes and their interactions. This also means that the domain knowledge and the software technology are more strongly coupled in the object-oriented design. As a result, the software architecture as well as its components cannot be easily reused, extended, and maintained without good knowledge on both the problem domain and the underlying object-oriented design.

This paper proposes an ontology-based approach for collaborative engineering software development. In previous researches on ontology-based technologies, ontologies are often used for knowledge representation, knowledge interchange, or knowledge integration in the collaborative design processes [2, 3, 4]. In the present approach, ontologies are employed to represent explicitly the important concepts and their relationships of the targeted engineering problem domain. They also serve as knowledge interfaces between software components of the targeted engineering software application.

In the collaborative software development process using the proposed ontology-based approach (see Fig. 2), the domain expert first identifies and defines the domain knowledge ontologies as well as analysis units involved in the solution workflow of the problem domain, with the consideration of the nature of domain knowledge, previously defined domain knowledge ontologies, and available analysis units. Then, the software engineer works collaboratively with the domain expert to design the analysis units before he or she carries out the implementation of the analysis units. Finally, following the solution workflow, the domain expert integrates the analysis units along with associated ontologies to construct the application.

Although the design of the analysis units requires good collaboration between the domain expert and the software engineer, the collaboration complexity and knowledge coupling for the design task has been greatly reduced because each analysis unit is well encapsulated by the ontologies and is responsible for solution of only a subset of the entire targeted problem. In addition, because the major knowledge for problem solving is expressed explicitly in the proposed approach using a set of ontologies and analysis units, the targeted application and its components (i.e., the analysis units) can be more easily reused, extended, and maintained.

The remaining of this paper is organized as follows. Section 2 presents the proposed ontology-based approach for collaborative software development. In Section 3, requirements on the software development environment needed for realization of the proposed approach are discussed. Also, an example environment is demonstrated using an ontology-based engineering application integration framework prototyped in this work. Section 4 uses an engineering software development example to demonstrate the application of the proposed approach. Finally, some conclusions are drawn in Section 5.

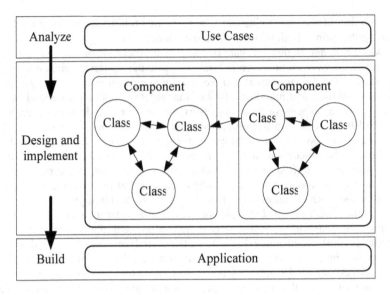

Fig. 1. Object-oriented software development approach

2   The Proposed Ontology-Based Approach

An ontology-based approach is proposed here for collaborative engineering software development. This approach employs ontologies to represent explicitly the important concepts and their relationships needed in the solution of the targeted engineering problem. Also, to reduce the collaboration complexity and knowledge coupling for the software design tasks performed cooperatively by domain experts and software engineers, this approach decomposes the targeted software system into a logical set of components, called analysis units, based on the solution workflow and using the ontologies as the knowledge interfaces to encapsulate the analysis units.

Section 2.1 discusses the basic concepts of the ontologies, and analysis units used by the proposed ontology-based approach. The collaborative software development process supported by the proposed approach is then described in Section 2.2.

2.1   Ontologies and Analysis Units

The terminology "ontology" originates from the domain of Philosophy and refers to the study of existence. In recent years, the concept of ontology is used in software engineering for representation of concepts and their relationships of a problem domain (or knowledge domain). It is also used for providing a vocabulary of terms and relations in knowledge engineering to facilitate construction of a domain model [5].

An ontology is defined by Gruber [6] as "an explicit and formal specification of a conceptualization." The meaning of conceptualization may be apprehended as abstraction in object-oriented design. The purpose of either ontology conceptualization or object-oriented abstraction is to specify the relations between the common concepts

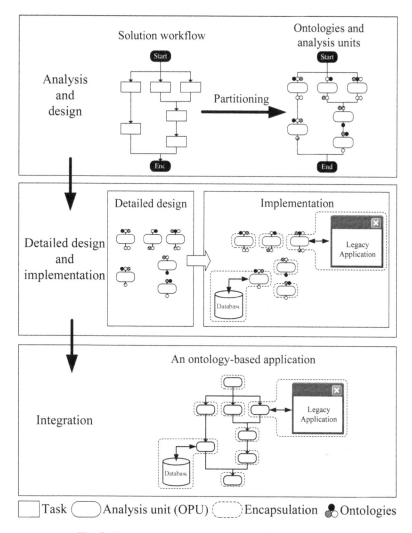

Fig. 2. Ontology-based software development approach

among different information objects (e.g., classes of objects) in a formal and explicit way of expression using vocabularies with logical axiom.

A popular recent application of ontologies is in providing better organization and navigation of web sites and improving the accuracy of searches in the World Wide Web. In this application, ontologies are employed to provide "a shared understanding of a domain" [7] so that semantic interoperability can be supported in the Web environment. A markup language called Web Ontology Language (OWL) [8] has been developed to facilitate application of ontologies on the Web. In engineering field, most of physical concept ontologies of a particular problem domain are usually employed to integrate engineering models that are formed based on domain theories [9]. In addition, for construction of ontologies for a domain, the method proposed by Noy and McGuinness [10] may be used.

Here, we use the reinforced concrete (RC) section shown in Table 1 as an example to demonstrate what the ontology of the RC section may look like, using the OWL vocabularies and notations. As shown in Fig. 3, based on the general knowledge of an engineer, one can explicitly express the knowledge concepts about a RC section into a set of classes (e.g., Section, Concrete, Steel, Location, etc.) and establish the relations between the classes using a set of properties (e.g., SubClassOf, Domain, MateralIs, hasSteel, etc.). In Fig. 3, the elliptical and rectangular boxes are used for representation of classes (or concepts) and user-defined properties (or relations), respectively. The texts denoted on the arrows indicate the properties already defined in OWL. For example, the Section class represents the concept of a structural section and the RectangleSection class, representing the concept of a rectangular section, is a subclass of the Section class. The hasSteel property is established to relate the objects of the Section class to those of the Steel class, stating that a Section object may have some Steel objects (i.e., steel bars). Similarly, the LocateAt property relates the objects of the Location class to those of the Steel class and describes the locations of steel bars inside the RC section. It should be noted that the ontology shown here (in Fig. 3) is not meant to be a comprehensive one and, depending on the purpose of the ontology, different set of classes and properties as well as different levels of details may be developed.

Table 1. Some general knowledge about a RC section

Illustrative figure	Some knowledge concepts
	(1) Cross section geometry (e.g., rectangular or circular shape) (2) The sizes and material properties of steel bars and stirrups (3) Locations of steel bars (4) Spacing of stirrups (5) Material properties of concrete

In addition, the ontology model is more general and reusable than the Entity-Relation Model (E-R Model), commonly used in design of relational database, in defining the relations between the information objects needed for solving the domain problem. In the E-R Model, the Entity, Attribute, and Relation concepts are used to represent the data model. An entity usually can have many attributes and the relations between two different entities are established through the common attributes they share. An attribute must belong to a certain entity and cannot exclusively represent the concept of a relation. However, in the ontology model, both the classes and

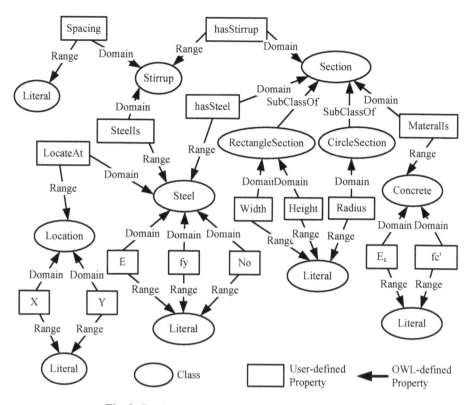

Fig. 3. Graph representation of a RC section ontologies

properties can represent independent concepts and relations in one domain and can be also be reused to represent the same concepts and relations in another domain.

For solving the targeted domain problem, engineering software usually follows a solution workflow that can be decomposed into a logical set of analysis or information processing tasks. There is no fixed rule for the decomposition and the sizes of the tasks can vary in a wide range. For processing of these tasks, a set of corresponding software components is usually designed in the engineering software. In this paper, the term "analysis unit" is used to refer to the software component and an analysis unit may contain an integrated set of analysis units.

Furthermore, the decomposition creates interfaces between analysis units. These interfaces are called knowledge interfaces in this paper because they represent not only the knowledge for associating the analysis units but also the knowledge that can be processed by the analysis units. In the proposed approach, ontologies are employed to define the knowledge interfaces.

Every analysis unit deals with two groups of ontologies: input ontologies and output ontologies (See Fig. 4). The output ontologies are resulted from the processing of the input ontologies by the analysis unit. Therefore, an analysis unit is also called an Ontologies Processing Unit (OPU) in this paper.

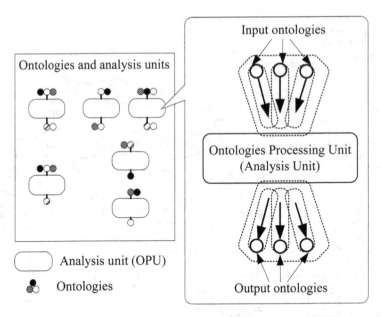

Fig. 4. Each analysis unit processes input ontologies and delivers output ontologies

2.2 Collaborative Software Development Process

The collaborative software development process supported by the proposed ontology-based approach can be divided into the following three main stages (see Fig. 2):

1. **Analysis and Design Stage.** Because the work involved in this stage mainly requires the domain knowledge, the domain expert should take the major responsibility. First, the domain expert lays out the solution workflow for the software application. Then, he or she decomposes the workflow into a set of analysis tasks and, for each of the tasks, designs the knowledge interfaces (in terms of input and output ontologies) of the corresponding analysis unit. The identification of analysis units and the design of their ontology interfaces are two key tasks in this software development approach for software flexibility, extensibility, and maintainability. Also, they are interrelated tasks and should be performed jointly in an iterative process with the following considerations:

   - The ontologies designed should reflect the important concepts in the domain knowledge and should serve as a knowledge unit to be processed by an analysis unit.
   - Reuse of the ontologies already defined in other applications of the same problem domain should be considered.
   - There may be software components, applications, or tools available in the problem domain for full or partial solution of some analysis task in the workflow or even some available analysis units developed in other applications of the same problem domain. How to reuse or take advantage of these components, applications, tools, or analysis units should be considered.

- For designing ontology interfaces of the identified analysis units, the domain expert may seek help from an ontology engineer for better accuracy and usability of the designed ontology interfaces.

Once the ontology interfaces of the analysis units are designed and expressed by an ontology language (e.g., OWL), the processing logics of the analysis units also needs to be described and defined by the domain expert.

For domain experts with no or very little basic software engineering background, partition of the solution workflow into an appropriate set of analysis units may not be an easy task for them. In this case, they should ask for assistance from their software engineer partner to have a better understanding about the implication of software complexity associated with the knowledge interfaces and analysis units. They can also ask their software engineer partner to review the partition and provide suggestions in the iterative partitioning process.

2. **Detailed Design and Implementation Stage.** Following the design of the ontologies and analysis units from the previous stage, the software engineer performs detailed design and implementation of the analysis units at this stage. Although the software engineer needs to work collaboratively with the domain expert to have enough domain knowledge for completing the detailed design tasks, he or she can work on each analysis unit without knowledge of other analysis units as well as the solution workflow of the software application because the complexity of each analysis unit is well encapsulated by its interface ontologies. This also means that the complexity of collaboration between the software engineer and the domain expert depends only on the limited domain knowledge needed for the detailed design of an analysis unit. In addition, as long as the designed input and output ontologies of an analysis unit are implemented accordingly, the implementation of an analysis unit may reuse existing software libraries, components, packages, applications, or analysis units.

3. **Integration Stage.** Following the solution workflow, the domain expert integrates the analysis units and their associated ontologies to build the targeted engineering software application. If new analysis units are needed for extending the application, the domain expert can go through the previous stages again to develop them and then integrates them with the existing ones to build the extended application.

It should be noted that the collaborative software development process proposed above is not a waterfall (or sequential) one but an incremental and iterative one. During the development process, any of the analysis units may be extended and the solution workflow may be adjusted to reflect evolutionary changes of ontologies, analysis task partitions, and analysis logic. Also, the domain expert and the software engineer collaborate in iterative processes for re-analysis, re-design, re-implementation, and re-integration of the developing software.

3 Environment for Ontology-Based Software Development

To ease the tasks of both domain experts and software engineers in the collaborative software development process presented in Section 2.2 as well as to facilitate the management, sharing, and reuse of ontologies and analysis units, a user-friendly software development environment for ontology-based applications is needed. This section discussions some basic considerations on the requirements of such a software development environment. In addition, discussions are given to an example software development environment for ontology-based applications prototyped in this work.

3.1 Requirements Analysis

The software development environment should include a runtime environment that acts as a bridge between ontology-based applications and the underlying operating system to support dynamic integration of analysis units and execution of ontology-based applications. Also, it should provide services and tools to aid domain experts and software engineers in collaborative software development tasks.

Some basic considerations on the requirements of the software development environment are discussed as follows:

- The runtime environment should support management of ontologies so that any ontology, once published by a domain expert, can be easily queried, shared, and reuse among domain experts and software engineers.
- The runtime environment should support management of analysis units so that any analysis unit, once implemented and deployed by a software engineer, can be easily queried, shared, and reuse among software engineers and domain experts.
- The runtime environment should support management of ontology-based applications so that domain experts can easily deploy ontology-based applications and end users can easily execute ontology-based applications.
- Software tools for defining and editing ontologies are needed. They should also provide the function for publishing ontologies to the runtime environment.
- Software tools to ease the tasks of detailed design and implementation of analysis units are needed. They should also provide the function for deploying analysis units to the runtime environment.
- Software tools are needed for facilitating the integration of analysis units with their associated ontologies into ontology-based applications. Some forms of visual editors should be very helpful for the tasks of manipulating analysis units in the process of building or rebuilding ontology-based applications. The tools should also provide the function for deploying ontology-based applications to the runtime environment so they can be shared with end users.
- Software tools are needed to encapsulate the software development details and complexity from end users and help the end users in managing and executing ontology-based applications.

3.2 Prototype Environment

A software development environment, called the OneApp Environment (Ontology-based Engineering Application Integration Environment), is prototyped using the .NET framework on Windows in this work to serve as an example for proof of the concept. Because the development of the OneApp Environment is not the focus of this paper, only a brief discussion on the functionality and application of the Environment is provided here.

Fig. 5 illustrates the major components of the OneApp Environment, their applications in the proposed software development process, and their interactions with domain experts, software engineers, and end users. Among all the components in the OneApp Environment, the OneApp runtime environment is the most essential one because it serves as the bridge between all other OneApp components and the Windows operating system and supports all of the basic functions in the OneApp Environment, such as management of ontologies, analysis units, and ontology-based applications.

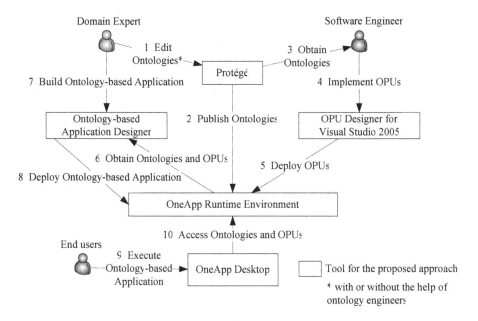

Fig. 5. The OneApp Environment to support the proposed approach

Besides the OneApp runtime environment, four other major tools are provided by the OneApp Environment to facilitate the collaborative software development of ontology-based applications:

- **Protégé Ontology Editor.** Protégé is a free, open-source ontology editor developed at Stanford [11]. It provides many well-developed functions for modeling ontologies. It can also export ontologies into a variety of formats including OWL.
- **OPU Designer for Visual Studio 2005.** This tool is designed for helping the software engineer in obtaining the ontologies defined by the domain expert, implementing the analysis units, and deploying the analysis units to the OneApp runtime environment. The tool is developed as a Visual Studio add-in to take advantage of the powerful built-in development functions in Visual Studio and to shorten the time for learning the tool.

- **Ontology-based Application Designer.** This tool provides a visual editor to help the domain expert in building ontology-based applications. It helps the domain expert to integrate the analysis units with their associated ontologies in a way similar to visual programming (see Fig. 6).
- **OneApp Desktop.** This tool is designed to serves as a friendly interface for end users to manage and execute ontology-based applications in the OneApp Environment (see Fig. 7).

4   Demonstration Example

A simple engineering analysis example is used here to demonstrate how the proposed ontology-based approach can be applied to development of engineering software applications and how the prototype ontology-based software development environment presented in Section 3.2 can be used to assist the development tasks. The example selected is the analysis of plastic hinge properties for a Reinforced Concrete (RC) column using the analysis method proposed in [12]. When the column member is under a constant axial force, the analysis method can account for both flexural and shear failure models of the member to compute the plastic hinge properties in terms of the moment-rotation curve. In this example, the first author of [12], Dr. Y. C. Sung,

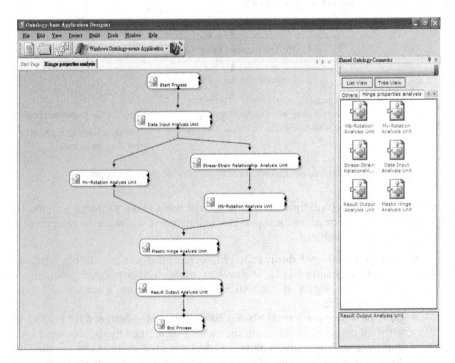

Fig. 6. Ontology-based Application Designer for the domain expert to construct ontology-based applications

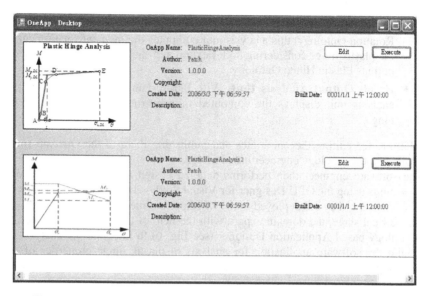

Fig. 7. OneApp Desktop for end users to manage ontology-based applications

was invited to act as the domain expert and the second author of this paper played both the role of the ontology engineer to help define the required ontology interfaces and the role of the software engineer.

Following the software development process described in Section 2.2, the domain expert first defines the solution workflow for the analysis problem and partitions the workflow into six analysis units with eight ontologies (see Fig. 8). The ontologies are expressed in the format of OWL. A brief description about each analysis unit and its associated ontologies is provided below:

- **Data Input Analysis Unit.** This analysis unit obtains the necessary data for the analysis of plastic hinge properties from the user through a graphical user interface. Its output ontologies include Section Ontology, Column Ontology, and Axial Force Ontology.

- **Stress-Strain Relationship Analysis Unit.** This analysis unit inputs Section Ontology, computes the stress-strain relationships for both the concrete and steel in the column, and outputs Concrete Stress-Strain Ontology and Steel Stress-Strain Ontology.

- **M_b-Rotation Analysis Unit.** After inputting Section Ontology, Column Ontology, Axial Force Ontology, Concrete Stress-Strain Ontology, and Steel Stress-Strain Ontology, this analysis unit computes the moment-rotation capacity of the column considering only flexural failure of the column. The output is M_b-Rotation Ontology.

- **M_v-Rotation Analysis Unit.** This analysis unit inputs Column Ontology, Section Ontology, and Axial Force Ontology, computes the moment-rotation capacity of the column considering only shear failure of the column, and outputs M_v-Rotation Ontology.

- **Plastic Hinge Analysis Unit.** After inputting M_b-Rotation Ontology and M_v-Rotation Ontology, this analysis unit computes the moment-rotation curve of the plastic hinge considering both flexural and shear failures of the column. It outputs Plastic Hinge Ontology.
- **Result Output Analysis Unit.** With Plastic Hinge Ontology as input, this analysis unit displays the computed moment-rotation curve of the plastic hinge.

In addition, for editing the ontologies and publishing them to the OneApp runtime environment, the ontology engineer uses Protégé.

The software engineer then performs detailed design and implementation of the analysis units using the OPU Designer for Visual Studio 2005. Once the analysis units are implemented, they are deployed to the OneApp runtime environment.

At the final stage, the domain expert, with the help of the ontology engineer, uses the Ontology-based Application Designer (see Fig. 6) to integrate the analysis units and build the software application for analysis of plastic hinge properties of a RC column. Fig. 9 shows the final product of the application.

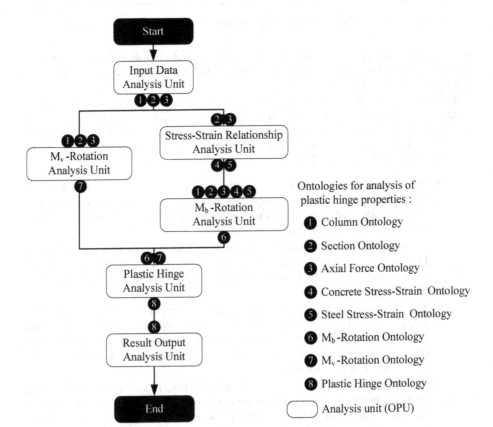

Fig. 8. Solution workflow for analysis of plastic hinge properties

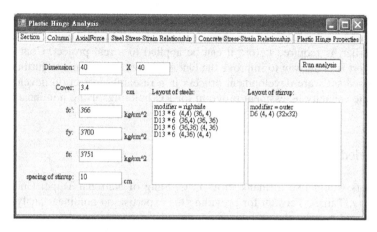

Fig. 9. An example ontology-based application for analysis of plastic hinge properties of a RC column

5   Conclusions

An ontology-based approach for collaborative software development has been proposed in this paper to facilitate collaboration between domain experts and software engineers in developing engineering software applications. In this approach, ontologies are employed to represent important concepts and their relationships for problem solving. They also serve as knowledge interfaces for encapsulation of software components so that knowledge coupling between the software components can be reduced. Furthermore, based on the decomposition of the solution workflow, this approach decomposes the software development task into a set of smaller development tasks to reduce collaboration complexity in the software design tasks performed cooperatively by domain experts and software engineers. With the support of the OneApp software development environment prototyped in this work, the application of the proposed approach has been demonstrated using an engineering software development example.

To better support the three-stage collaborative software development process of the proposed approach, the prototyped OneApp environment needs to be further implemented to provide more complete services and tools for editing of ontologies, design and implementation of analysis units, integration of analysis units into applications, management and sharing of ontologies, analysis units, and ontology-based applications, etc. Moreover, research is still needed for development of a more effective and efficient software development environment for ontology-based applications. For example, an ontology-based reasoning system discussed in [9] can be used to facilitate integration of the analysis units and their associated ontologies by the domain expert for building the targeted engineering software application.

Further research is needed to validate the proposed software development process in a real engineering software development project. The simple example demonstrated in this paper is not sophisticated enough to serve as a good validation example for the proposed software development process. Also, the OneApp environment is currently prototyped with only basic and limited functionalities for carrying out the presented demonstration example. Further implementation of the OneApp environment, as discussed earlier, is required before it can be applied to a real project. Continuous research effort is underway to improve the OneApp environment and to further validate the proposed software development process in a research project on development of an aseismic capacity assessment program for RC buildings using nonlinear pushover analysis.

Acknowledgements

The authors would like to thank Prof. Y. C. Sung of National Taipei University of Technology, Taipei, Taiwan, for providing his expertise on nonlinear analysis of RC structures and helping the authors to validate the proposed ontology-based software development process.

References

1. Booch, G.: Object-Oriented Analysis and Design with Applications. 2nd edn. Addison Wesley, Redwood City California (1994)
2. Gu, N., Xu, J., Wu, X., Yang, J., and Ye, W.: Ontology based semantic conflicts resolution in collaborative editing of design documents. Advanced Engineering Informatics, Vol. 19, No. 2 (2005) 103-111
3. Garcia, A. C. B., Kunz, J., Ekstrom, M., and Kiviniemi, A.: Building a project ontology with extreme collaboration and virtual design and construction. Advanced Engineering Informatics, Vol. 18, No. 2 (2004) 71-83
4. Kim, T., Cera, C. D., Regli, W. C., Choo, H, and Han, J.: Multi-Level modeling and access control for data sharing in collaborative design. Advanced Engineering Informatics, Vol. 20, No. 1 (2006) 47- 57
5. Studer, R., Benjamins, V. R., and Fensel, D.: Knowledge Engineering: Principles and Methods. Data and Knowledge Engineering, Vol. 25, No. 102 (1998) 161-197
6. Gruber T. R.: A translation approach to portable ontology specifications. Knowledge Acquisition, Vol. 5, No. 2 (1993) 199-220
7. Antoniou, G. and van Harmelen, F.: A Semantic Web Primer. The MIT Press, Massachusetts London (2004)
8. McGuinness, D. L. and van Harmelen, F..: OWL Web Ontology Language Overview. W3C Recommendation, World Wide Web Consortium: http://www.w3.org/TR/owl-features/ (2004)
9. Yoshioka, M., Umeda, Y., Takeda, H., Shimomura, Y., Nomaguchi, Y. and Tomiyama, T.: Physical concept ontology for the knowledge intensive engineering framework. Advanced Engineering Informatics, Vol. 18, No. 2 (2004) 95-113
10. Noy, N. F. and McGuinness, D. L.: Ontology Development 101: A Guide to Creating Your First Ontology. Technical Report KSL-01-05, Knowledge Systems Laboratory (2001)
11. Noy, N., Fergerson, R., and Musen, M.: The knowledge model of Protege-2000: Combining interoperability and flexibility. The 2nd International Conference on Knowledge Engineering and Knowledge Management (EKAW'2000), Juan-les-Pins France (2000)
12. Sung, Y. C., Liu, K. Y., Su, C. K., Tsai, I. C., and Chang, K. C.: A Study on Pushover Analyses of Reinforced Concrete Columns. Structural Engineering and Mechanics, Vol. 21, No. 1 (2005) 35-52

Optimizing Construction Processes by Reorganizing Abilities of Craftsmen

Wolfgang Huhnt

Berlin University of Technology, Institute for Civil Engineering,
Gustav-Meyer-Allee 25, D-13355 Berlin, Germany
huhnt@tu-berlin.de

Abstract. This paper is focused on the consideration that construction processes can be executed in a better way if specific interfaces are avoided. The interfaces regarded in this paper result form the decomposition of construction activities into subsets. This decomposition is influenced by traditional outlines of professions. As a consequence of this decomposition, interfaces in the process occur. The interfaces can be weighted by costs. Different decompositions can be determined where the sum of the project costs is varying. The goal is to determine a decomposition of the overall process where the total of the costs results in a minimum. This decomposition represents the optimum concerning interfaces between project participants. As a consequence of a different decomposition, abilities of craftsmen need to be reorganized so that this paper describes a theoretical background that can be used in civil engineering for reorganizing abilities of craftsmen with the aim to minimize project costs.

1 Introduction

It is well known that craftsmen with different professions are necessary for the execution of construction projects. Even if a construction project is executed by a general contractor, the general contractor subdivides the set of construction activities of the project into subsets. Subcontractors tender for the execution of selected subsets.

The process of forming packages of construction activities is influenced by traditional outlines of professions. However, there are interdependencies between construction activities that are assigned to different subsets and – as a consequence - executed by different parties. These interdependencies are called interfaces in the context of this paper. Interdependencies between construction activities that are assigned to the same subset are called transitions in the context of this paper.

Costs can occur at interdependencies between construction activities. And, in general, these costs differ depending on whether the interdependency is an interface or a transition. In general, responsibilities change at interfaces so that the progress of work needs to be documented clearly. Furthermore construction activities before and after an interface are generally executed by different parties so that traveling costs can occur. In addition, experienced project manager know where expensive interfaces occur due to rework that is required because the necessary quality is not achieved. Thus, calculating project costs need to consider costs at interfaces and transitions.

I.F.C. Smith (Ed.): EG-ICE 2006, LNAI 4200, pp. 343–347, 2006.

In this paper, the question of whether a specific interdependency between activities is an interface or a transition is not preset as it is usually done in construction projects. Thus building subsets of construction activities becomes a result of an algorithm. The algorithm determines decompositions with minimal project costs. The process can be executed with minimal project costs if the craftsmen would have exactly the abilities required to execute the construction activities contained in a specific subset.

In this paper, a small example is used to explain the considerations. This example is introduced in section 2. Section 3 describes costs that can be assigned to activities, interfaces and transitions. Based on this input, an algorithm can search for a decomposition of the set of construction activities into subsets with minimal project costs. This is explained in section 4.

The considerations presented in this paper are a first step into an area which is at present time not supported by information technology. Outlines of professions of craftsmen are modified based on experiences only. The author of this paper knows that there is a lot of work to do until the considerations presented in this paper will result in a software tool that can be used in practice to compute outlines of professions in such a way that construction processes can be executed with minor project costs. In section 5, some aspects of further research in this area are explained.

2 Forming Subsets of Construction Activities

A given schedule of a construction project consists of two different types of information, descriptions of activities including start and end of each activity and relations between activities. Figure 1 shows an example of such a schedule. Relations between activities are shown. These relations describe technological interdependencies between the construction activities.

The example in figure 1 shows a situation where usually drywall construction worker fill the expansion joints of the drywall construction with silicone. They also fill and sand each other joint of the drywall construction. Priming is usually executed by someone who needs that priming, house painters for painting and tilers for tiling. Therefore, a traditional decomposition of construction activities would result in three subsets where drywall construction work, filling expansions with silicone and filing and sanding would be a subset, priming for painting and painting would be another subset, and priming for tiling and tiling would also be a subset.

3 Costs

It is state of the art to assign costs to construction activities. This is usually done when a project is calculated. In addition, costs can be assigned to interdependencies between construction activities. Figure 2 shows such costs. The mentioned costs result from traveling costs. The construction activities partly stop and start at noon. In such cases, traveling costs are involved due to the departure of project participants assigned to the ending activity and the arrival of project participants assigned to the succeeding activity.

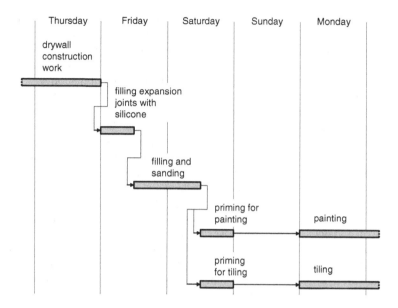

Fig. 1. Extract of a Schedule

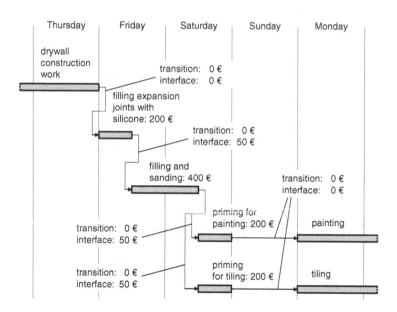

Fig. 2. Costs of Activities, Transitions and Interfaces

4   Calculation Subsets of Construction Activities

Calculating subsets of construction activities belongs to the class of np-complete problems. This calculation can be regarded as an assignment of resources to construction activities where all those activities that are assigned to the same resource form a subset. The assignment of resources to activities with minimal costs has to be determined. The solution space consists of all possible assignments between resources, in this case construction workers, and activities. To guarantee the optimum, each combination has to be checked and the sum of costs has to be determined. A combination with minimal costs is the optimum. The example shown in figure 1 has 4*4*4 possible solutions if house painters, tillers and drywall construction worker are considerer as ressources. Np-complete problems are treated in Computer Science. In general, they can have more than one solution i.e. more solutions have minimal costs.

In Computer Science, branch and bound techniques have been developed to reduce the effort of computation. This technique is based on the consideration of subdividing the solution space into disjoint subsets (branches) and checking whether all solutions in such a subset cannot be better than an already determined solution (bound). [1] If a subset can be checked, the solutions in that subset do not need to be considered so that the effort of computation can be reduced. The disadvantage of branch and bound techniques is that there is no theoretical proof that in any case the effort of computation can be reduced. An additional concept to reduce the effort of computation is the use of heuristics. [2] Heuristics determine a solution in an efficient way, but in general it cannot be proven that the solution determined is the optimum.

For the problem discussed in this paper, an analysis of the complete solution space, a branch and bound technique and a heuristic have been implemented. The pilot implementation is able to find an optimum for the example shown in figure 1 and the weights shown in figure 2. Several optima exist with minimal costs of 1.000 €. An optimum is the assignment of drywall construction workers to all activities so that they have to execute activities that are usually executed by house painters and tilers.

The computed results can be regarded as a proposal for continuing education. The results show whether it is possible to optimize a process by the assignment of specific resources to activities; and continuing education has to be initiated if specific coworker shall execute activities that are usually not part of their work. However, such considerations require the analysis of several projects because the investment into a specific continuing education cannot be returned in a single project. But the concept described in this paper shows that the decision of investing into continuing education can be quantified so that a comprehensible basis is available for such decisions.

5 Conclusions and Outlook

The approach presented in this paper is specifically focused on building subsets of construction activities where these subsets are not preset. The subsets are calculated with the aim to achieve minimal project costs.

A lot of investigations have been done in the past to optimize construction processes. However, these investigations can be subdivided into two groups, those who modify a schedule so that for instance start and finish dates of specific activities are modified and those who do not change a given schedule. An example of the first group of optimization procedures is the determination of the minimal project duration under multiple resource constraints [3]. This paper is focused on a specific optimization technique which is from the second type where a given schedule is accepted as it is. Within such a given schedule, several optimization possibilities exist. Start and end date of activities can be regarded as prescribed time frames for the execution of these activities, and a given set of resources can be assigned to the activities with the aim to achieve a balanced use of these resources [4]. However, the optimization procedure presented in this paper is focused on finding an optimal mapping of craftsmen to construction activities where the abilities of the craftsmen are not preset.

As a consequence of applying the approach presented in this paper to real projects, abilities of craftsmen need to be rearranged. Activities always require abilities of construction workers to execute these activities. Entrepreneurs have to think about the abilities of their co-workers, and continuous education already takes place in construction companies. The approach presented in this paper tries to quantify the benefit of specific continuous educations so that an entrepreneur can come to the decisions whether a specific continues education will help or not on a comprehensible basis.

A lot of investigations are still necessary to analyze the complete potentials of the approach presented. In general, decisions that specific co-workers should learn specific additional abilities are not based on a single project. Representative projects need to be analyzed where core abilities should not be questioned. At present time, the approach presented is discussed with associations in Germany that are responsible for the content of teaching apprentices. These contents of teaching need to be reviewed permanently so that at the end construction workers are available at the market that are able to execute construction projects in an efficient way.

References

1. Branch and Bound. http://en.wikipedia.org/wiki/Branch_and_bound, 20.2.2005
2. Heuristic (computer science).
   http://en.wikipedia.org/wiki/Heuristic_%28computer_science%29, 20.2.2005
3. Jiang, G., Shi, J.: Exact Algorithm for Project Scheduling Problems under Multiple Resource Constraints. Journal of Construction Engineering and Management, ASCE, September 2005, 986-992
4. Huhnt, W.: Disposition von Mitarbeitern im Kontext der Ausführung von Bauleistungen. Technische Universität Berlin, Institut für Bauingenieurwesen, 2004

Ontology Based Framework Using a Semantic Web for Addressing Semantic Reconciliation in Construction

Raja R.A. Issa and Ivan Mutis

M.E. Sr., Rinker School of Building Construction, University of Florida,
Gainesville, FL 32611
{raymond-issa, imutis}@ufl.edu

Abstract. Exchanging, sharing, or integrating information in the construction industry involve the reconciliation of multiple data formats, structures and schemas with minimal human interaction. Approaching interoperability employing these models via integration, mapping or merging information forces the use of purely syntactical layers with heavy human intervention. Enhancing our ability to explicitly model information and to find methods that consistently harmonize the construction participant's use of common language will allow interoperability to be moved to new levels of flexibility and automation by structuring the information upon semantic levels. These levels must have conceptualization aspects to define knowledge, a common vocabulary that covers the syntax, symbols, grammars, and axiomatization that captures inference. Our approach inherits the ability of enhancing interoperability at the syntactic and semantic levels of the Semantic web and proposes onto-semantic schemas, which are ontology constructs of concepts of the construction domain. The approach produces an analysis from the primitives to more refined concepts using the Semantic Web's power of representation with the purpose of enhancing interoperability at semantic and syntactic levels.

1 Introduction

There is a need in the construction industry to develop strong interaction between customers, contractors, and owners using emerging information technologies to facilitate communication and collaborative use of project information. A successful implementation of these technologies will enable integration of project information based on real time and transparent information transfers to facilitate decision-making or to work on concurrent engineering.

The problems encountered in the exchanging, sharing, transferring, and integrating of information when actors interact with one another include the lack of coordination, inconsistencies, errors, delays, and misinformation. These problems make the exchanging, sharing, transferring, or integrating of information among construction participants burdensome with their high costs and need for human intervention. The consequence is a reduction in the productivity and efficiency of current interoperability activity. This problems were analyzed in a recent labeled domain, construction informatics [1]. Hence, construction participants in order to interoperate are forced to either partially solve coordination errors or to totally rework the representation of information, to manage the resulting project delays, and to use additional resources.

I.F.C. Smith (Ed.): EG-ICE 2006, LNAI 4200, pp. 348–367, 2006.
© Springer-Verlag Berlin Heidelberg 2006

A solution is proposed to diminish the need for human intervention in exchanging, sharing, transferring, and integrating information. To make this possible construction participants have to comprehend *what* to exchange, share, transfer and integrate. In addition, they must know *how* and *when* the action has to occur. '*What*' refers to the content of information that one construction participant queries from another, '*how*' refers to the process that is going to be used, and '*when*' refers to the instances when a transaction is suitable to be executed.

One approach that tackles the need for human intervention within the exchanging, sharing, transferring, and integrating of information is that of executing agreements, in partnering or alliance models, between construction participants. The agreements are pursued by sharing referential models, taxonomies from each partner, and standards, among others. These agreements are created to answer the queries concerning the content of information, the process and the instances when the interoperability is going to be unexecuted. However, current approaches have not been successful in finding agreements within the construction community. For example, in the case involving the content of information that one construction participant queries from another, the efforts that only address issues at the platform and syntactic levels by using mapping strategies between taxonomies are hampered by the fact that a new mapping has to be created for each new element of the taxonomy. Thus, the mappings do not have inferenceability capabilities. In addition, they lack the flexibility to adjust the mapping based on context.

New studies have been aimed at the development of ontologies [2] and the establishing of agreements among agents in order to interoperate and to allow them to share a common syntax that allows them to can query each others data. These efforts aim to work at the semantic levels of the construction participant's information. In these approaches, essentially the actors have to establish mechanisms to understand that the elements of the information in their conceptual model have the same meaning for them. Hence, the construction participants must know what operations they are going to execute in exchanging, sharing, transferring, and integrating information. The main drawbacks of these approaches are that the implementation is extremely expensive and that it is valid only for a particular group of construction participants that reach agreements with each other. In addition, most firms lack the level of technology sophistication needed to implement solutions that are based on their interoperation agreements [3].

Key to the newer, semantic approach has been the emergence of the Semantic Web with its vision of providing enhanced information management based on the exploitation of machine-processable metadata. Documents retrieved from the Semantic Web are annotated with meta-information, which defines their content in a way that specifies the machine-processable components. Ontologies are a key enabling technology for the Semantic Web. They offer a way to cope with heterogeneous representations of Web resources. An ontology can be taken as a unifying structure for representing information that is able to be a common representation among the construction participants, and that is able to define the semantics of the participant's concepts. Ontologies attempt to model the content of information. Our approach inherits the ability of enhancing interoperability at the syntactic and semantic levels of the Semantic web.

This research assumes that the members of the community agree on a common knowledge and an explicit ontology, but in contrast to the aforementioned efforts, our strategy takes into account inferenceability and extensibility aspects. The first aspect considers computation by inference of the concepts of the common ontology. The second aspect consists of the proposed ontology which will allow for the addition of concepts as well as the reuse of already developed concepts from other sources e.g. concepts from IFC Industry Foundation Classes (IFC) schemas and taxonomy standards. Our strategy proposes onto-semantic schemas, which are ontology constructs of construction domain concepts. The result of this approach is to produce a framework that contains an analysis from the primitives to more refined construction concepts in the domain thorough the use of Semantic Web's representations with the purpose of enhancing interoperability at both the semantic and the syntactic levels. The goal of this strategy is to revaluate current efforts on mapping and on information integration in order to achieve effective interoperability by reducing human intervention. This research advances the current state of the art on semantic interoperability by proposing the enrichment of an ontology to approach reconciliation. Other efforts in semantic interoperability support particular processes within engineering practice such as functional design knowledge [4] or infrastructure management based on finding semantic agreements [5].

In summary, there is a need to create a framework in which dissimilar construction participants are able to exchange, share, transfer, and integrate information in order to fulfill collaboration goals. This approach aims to reach common knowledge by using explicit ontologies through the Semantic web at semantic and syntactic levels.

Next, a brief analysis of the problems encountered in exchanging, sharing, transferring, and integrating information will be presented. The analysis is narrowed to the reconciliation of two sources of information. The subsequent section illustrates some approaches to model the information in the construction industry. The last section discusses the *onto-semantic* framework approach and its relation to the semantic web.

2 Reconciliation Problem

Our research recognizes multiple sources for the problem of creating the *effective* exchanging, sharing, transferring, and integrating of information. Roughly, the sources of the problem are the different methods used to represent information [6], different levels of specification of the concepts in the domain, and the various levels of systematization or sophistication of the construction participants' systems.

The *effective* exchanging, sharing, transferring, and integrating of information is embraced within the *semantic interoperability* paradigm. *Semantic interoperability* can be defined as the *understanding* of the information concept representations by domain's agents. This paradigm can be approached from two different standpoints: *information systems* and *problem domain*. The former is addressed within the sphere of *computer science* domain, and the latter is addressed by the domain problem field. The approaches have different input to find semantics. A solution from the problem domain perspective embraces "informal" methods such as surveys to find semantic definitions, such as mechanisms that prescribe models that structure manageable pieces of work and that are aimed at agents that have the same view of the models (e.g. framework models that propose break down of construction products). Solutions from the information systems attempt to find axiomatizations and models of information [7]. These solutions use complex algorithms and other tools from fields such as artificial intelligence. These solutions from each perspective have been driven independently by specialists from the domain and from the scientific community. However, we envision that research on semantic interoperability should encompass inputs from both perspectives. For example, while a project participant is interested in how to make other participants *understand* the information furnished by each agent in any interoperability activity, and computer scientists and knowledge engineers are interested in how to model a domain, inputs from both perspectives are necessary to develop an approach within the semantic interoperability arena. An attempt to compromise these perspectives is done through the elaboration of product models, which are conceptual models that define schemata in which concepts of the construction domain could fit, and that attempts to embrace the whole information during the life cycle of a building. However, these type of models do not define the associated nature of the construction domain, but employ a modeling technique that domain experts use to reshape the world [8]. The authors claim that a conceptual model will lack effectiveness, reliability, and reusability if the solutions are not addressed from the two standpoints. Approaches that only work on finding conceptual models without a close induction or collaborative input of the concepts by domain experts will lack reusability and effectiveness among domain agents because of their poor semantics.

New efforts on *semantic interoperability* by the authors are aimed at finding frameworks based on inducting a conceptual model from the *intention* of the construction participants in order to reduce human intervention [9, 10]. Our claims approach semantics by associating it with perception of the agents, which are states and direct agent's assertion to the form that characterize a concept. The idea of intentionality has been rooted in the integrationists' paradigm [11, 12].

With the purpose of finding strategies within this paradigm, part of our research is aimed at finding insights into the problem of reconciling concept representations within the construction domain. The reconciliation problem deals with finding common semantics about sources of information from different agents. Reconciliation is the step of exchanging, sharing, transferring, and integrating information. To illustrate the problem complexity, consider the following example, which employs simple knowledge representations but uses highly demonstrative schemes. The example is narrowed to ontology-based information integration.

The results of this research will not be a panacea that solves the reconciliation problem described by the database community as one of the obstacles on the road toward solving the heterogeneity of information problem [13]. Preliminary results indicate that further effort should be spent on another step of semantic interoperability, which is the reconciliation of two or more sources of information. For this purpose, this research appeals to the ontological tradition by distinguishing the natural ontological categories that are implicit on knowledge representations and it proposes a strategy to map two sources of information.

The following example selects *ad hoc* ontologies of certain construction businesses. The ontologies represent the construction business model. A graphical view of one of these ontologies is presented in Figure 1. The reader is reminded that ontologies are often associated to taxonomic hierarchies of classes and the subsumption relations, but they are not limited to this structure, which could be easily confused with the rigid inheritance structures from object oriented representations. Ontologies acquire knowledge about the world and frame them into categories and add terminology and constrain them with axioms from the traditional logic. The nodes in Figure 1 represent the Concepts.

Figure 1 shows some specializations of the root, represented by Concept 1, such as {*Functional Areas, Administration and Buildings*}, which in turn are represented by Concepts 2, 3, and 4 respectively. The concept *Project Manager*, Concept 7, has instances like {*Bill O'Connell, General Manager, New Hall UF*} with fixed attributes like {*Name, Hierarchy, Project In Charge*}. Thus, the instance Project Manager O'Connell has the values *Name=Bill O'Connell, Hierarchy = General Manager, Project in Charge=New Hall UF*. Additionally, there is a set of relations among concepts like {*is a, part of, has*}. The relations are generally denoted as *PartOf(Project Manager, Administration)*. These relations describe associated defined relations within classes, inheritance relations, and instances of properties.

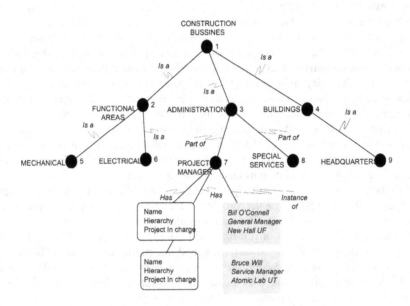

Fig. 1. Structure of an ad-hoc, construction business ontology

The assumption is that an actor exchanges, shares, transfers, and integrates information between two construction businesses, i.e. construction participant queries information from the two-construction businesses.

If we reduce the analysis to finding mappings that express semantic equivalence between the two ontologies, then the problem can be expressed in terms of how an actor can semantically map one concept from one ontology to the "most semantically <u>similar</u>" concept of the other ontology. It is clear that mappings for this example are semantic relations between two or more concepts.

Graphically, it is possible to bring up a clear picture of the complexity in finding a reconciliation of two *ad hoc* ontologies. The two ontologies are shown in Figure 2. The nodes in the ontology represent concepts that have levels of specializations from their parent's concept. In addition, the reader can identify relationships among concepts, for example, *Isa*(CPVC, Plastic Pipe Fittings) among the Concepts 11 and 9, as shown in the Figure 2b. The '*is a*' relation corresponds to a semantic link between two concepts that have a subsumption relation and the '*has*' relation corresponds to a semantic link that constrains the subsumption relation to a directional relation of containment [14, 15].

Suppose the construction participant queries information about the availability and costs of specific items from different material suppliers, say pipes for internal water distribution in buildings. Specifically, he/she will need to perform a semantic mapping between the most similar concepts of the two ontologies that contain the information sought. The problem is in how a construction participant can semantically map a concept or concepts of one data representation to another concept of the other data representation. Moreover, as the reader can intuitively notice in Figures 2a and 2b, the question of how a construction participant will be able to semantically map more complex matchings when one concept is semantically similar to a concatenation of two or more concepts needs to be addressed.

In addition, the mappings shown in Figure 2 resemble one to one and complex matching problems that are studied within the database community [16]. For instance, consider a one to one semantic match at one of the levels. With the aid of auxiliary information such an "expert knowledge", Concept 7 from Figure 2a (*Steel Pipe, Black Weld, Screwed*) is matched with Concept 8 from Figure 2b (*Metal, Pipes & Fittings*). Note that although they <u>semantically</u> match, they fully <u>syntactically</u> mismatch.

Consider the case where the construction participant queries more detailed information from two businesses. In other words, the participant needs to map specific instances from one source to another source. For example, the user semantically maps Concept 4 (*Plumbing Piping*) in Figure 2a to Concept 6 (*Pipes and Tubes*) in Figure 2b. But this mapping does not resolve the aforementioned query. The user should follow the subsumption relations of the concepts. This approach is similar to following down the hierarchy of a taxonomy. A taxonomy is a central component of an ontology [17]. Assume the expert's information asserts that a Polymer Pipe between 1" and 1"1/2 diameter is suitable for internal water distribution, e.g. PVC (Polyvinyl Chloride) pipe. Hence, the user matches Concepts 9 (*PVC 1¼"*) from Figure 2a to Concepts 10 (*PVC*) and Concept 13 (*1¼" Diameter*) from Figure 2b. Observe that Concept 10 is concatenated with Concept 13 to perform the match. Concept 10 is a more general concept than Concept 13 and vice versa, Concept 13 is more specific than Concept 10. This is a complex type of match that includes a joint of two concepts and a specialization of one concept.

Figures 2a and 2b, illustrate how relations between concepts of two representations can match at a *more general or specific* level; or they can *overlap;* or they can *mismatch*. Based on these types of intuitively semantic relationships, these relations will have what is called a level of *similarity* [16, 18]. Thus we could say that the concept of Concept 4 (*Plumbing Piping*) in Figure 2a is similar to Concept 4 (*Building Service Piping*) in Figure 2b.

In addition, it is important to remark that the relation between concepts such as *Isa*(HVAC, Mechanical) is also a possible map to other concepts of the ontology. These mappings between different types of elements make the process more complex.

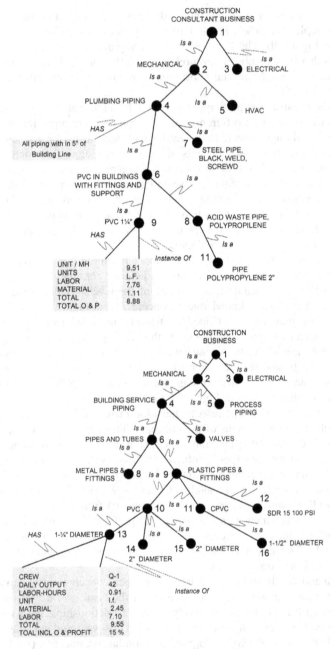

Fig. 2. (a) and (b), *Ad hoc* construction business ontologies

Note that the concepts from Figure 2a and Figure 2b partially match; they semantically match although they are *not equal*. In fact, the concepts are *similar* and *over-
lapped* into *more specific* elements of the concepts. It easy to observe that expert

knowledge would be needed if the reader intends to perform reconciliation among other nodes that represents the same concept (e.g. node 7 in Figure 2a and 8 and Figure 2b).

In summary, the above example illustrates the reconciliation problem of two ontology representations that are from the same domain, but have different terminologies. This semantic heterogeneity is a conflict that is categorized as ambiguous reference [19], because the same term has a different meaning in different ontologies. In addition, the example shows that one concept of one ontology has similar but not exactly the same semantics as the other ontology, and that two concepts with a similar meaning can be structured differently in different ontologies.

3 Reconciling Conceptual Models for Interoperability

As mentioned earlier, *effective semantic interoperability* requires expert domain intervention in implementing the conceptual model and, furthermore, in aiding reconciliation between the output of two agents' sources of information. In other words, human intervention is necessary for interoperability operations in order to aid the reconciliation of sources of information. In the former example, if the mapping operation must be manually performed, the operation becomes ineffective and expensive. There are some efforts in the database community that attempt to diminish human intervention in mapping operations [20] and they can roughly be classified into two basic approaches:

1. *Developing a standard.* All representations must conform to a common vocabulary or set of rules. The standards are common conceptual models for specific domains that could be structured in multiple forms of knowledge representation structures i.e. basic taxonomies, schemas, or intended models. In the construction domain, the need to provide interoperability has persuasively motivated the use of standards as a driver to perform integration. For example, the IAI through its IFC representation [21] has developed the specification of attributes to fully represent the area of work in the construction domain. The information, which is present in any computer application with structured data, can be mapped into IFC data files. In this way, IFC data files provide a neutral file format that enables AEC computer applications to efficiently share and exchange information. The ifcXML and aecXML initiatives, which are based on the IFCs, seek to exchange information in XML formats. For example, aecXML provides XML schemas to describe information specific for data exchange among the participants involved.

   Other standards play the role of classification models or *taxonomies*. They are simple conceptual models for representing entities. Examples of this type of model are classifications such as MASTERFORMAT and the newly proposed Overall Construction Classification System (OCCS) . The latter is a classification system that organizes data from the conception to the whole life cycle of the construction project in a common language [22].

   These standards are models used to improve communication and interoperability among the construction participants. They are employed as higher layer models from which more *consistent*, specific models are derived and used in custom construction applications i.e. *ad hoc* construction estimate databases based on CSI. The consistency maintained in the higher layer models diminishes the disagreement over the meaning, interpretation, or intended use of the same concept among construction participants. However, higher layer models fail in their lack of comprehensive information [23, 24]. They attempt to include the whole set of concepts of the construction domain into a model, but this is an incommensurable task that results in lack of comprehensive of construction concepts when agents attempt to use them within their applications. Organizations in charge of creating standards will need to continuously expand them requiring a time-consuming consensus building process. Another resulting problem is that the lack of comprehensiveness of concepts in the model forces community members to find methods to include the missing concepts or set of concepts in custom, fashion conceptual models i.e merging concepts from higher layer models into custom fashion concepts. The resulting products are models that are not possible to reconcile with other construction participant's applications. Therefore, standards cannot be a general solution to reconcile conceptual models built in schemas, taxonomies or other type of knowledge representation structures.

A domain community generates multiple competing standards, which defeats the purpose of having a common model in the first place. Then, the members of the community are forced to reconcile *manually* their own conceptual model, which is based on a higher layer model, to other conceptual models. As mentioned earlier, this is an expensive process and the results do not satisfied the needs of the agent's domain. For an illustration, consider the attempt to find a reconciliation of two competing standards in the construction industry. The mapping of the two standards, CSI (Construction Specification Institute) and UNIFORMAT II (ASTM standard), is illustrated in Figure 3. Figure 3 illustrates the difficulty involved in reconciling at the syntactic level two upper level taxonomies in the construction industry, which are based on different competing standards.

UNIFORMAT		MASTERFORMAT		Cardinality	Granularity Level
01	Foundations	02	Sitework	M:N	Structure
		03	Concrete		
		04	Masonry		
		05	Metals		
02	Substructure	02	Sitework	M:N	Structure
		03	Concrete		
		04	Masonry		
		07	Thermal and Moisture Protect		
03	Superstructure	03	Concrete	M:N	Structure
		05	Metals		
		06	Woods and Plastics		
04	Exterior Closure	02	Sitework	M:N	Structure
		03	Concrete		
		04	Masonry		
		05	Metals		
		06	Woods and Plastics		
		07	Thermal and Moisture Protect		
		08	Doors and Windows		
		09	Finishes		
		10	Specialties		
05	Roofing	05	Metals	M:N	Structure
		06	Woods and Plastics		

Fig. 3. An example of the reconciliation of two competing standards at their upper levels

2. ***Semi-Automatic approaches.*** These approaches are prototype implementations
    that run on an agent's systems and that attempt to map or merge two or more
    conceptual models built in relational schemas or basic ontologies. The proto-
    types are semi-automatic since they need an input from the prototype's user to
    complete the mappings. These approaches are not domain specific [16]. The
    reason is that they employ single matching criteria. As a consequence, these pro-
    totypes have less accuracy and they have limited applicability.

4 Onto-semantic Framework

This research is based on a framework that uses an architecture that sets stages to
reconcile construction domain concepts. The framework employs an ontology as its
knowledge representation structure in order to implement an information integration
strategy (see Figure 4). The designed architecture resembles ontology translation, but
uses a centralized global ontology as a repository. The approach is focused on finding
semantic relations between two ontologies to build Onto-semantics, which are generated ontology constructs elaborated via Web Services. Onto-semantics are ontological
concepts of the construction domain. This strategy allows the delineation of more
specific problems for ontology specifications using Semantic Web Services.

The framework is aimed at finding a strategy to solve the problem of reconciliation
based on ontology. As was mentioned earlier in section two, this research identifies
the semantic heterogeneity between information sources as a reconciliation problem.
The objective is to use the ontology constructs as a knowledge base that recognize the
semantic mismatch explicitly. This is performed by structuring ontology constructs
that define construction concepts for specific semantics.

The purpose of developing these constructs is to define semantics, which have a
form of ontology constructs. The constructs will articulate the input sources by using
inference algorithms. The outputs of the articulation are the relationships among the

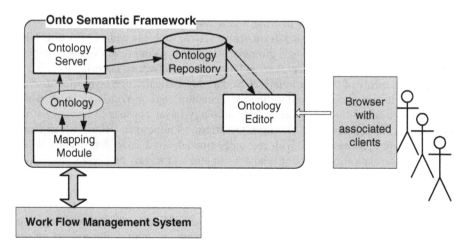

Fig. 4. Onto-semantic Framework

concepts of the sources or input. This result is expected to aid in the analysis of the structure of poorly specified ontologies as well as in defining a framework for other future implementations using these outputs.

There are three steps involved in implementing this approach. The first involves the layers of the semantic web that this strategy employs, the second illustrates the main components of the framework, and the third articulates the reasoning explaining the concepts on which the approach is based.

Semantic Web Service

Currently efforts aimed at enabling the exchange of information from distributed sources are focused on *web service solutions* such as Apache AXIS –Apache eXtensible Interaction System. These efforts endorse solutions based on layers such as *networking* (e.g. using the http network protocol), *messaging* (e.g. using Simple Object Access Protocol –SOAP, XML messaging based protocol), *description* (e.g. Web Service Description Language (WSDL), which describes the characteristics of the existing Web services: details in WSDL files and Web service interfaces), *discovering* (e.g. using UDDI, which locate service description), *publication* (e.g. UDDI, which make service descriptions available for publication), and *flow* (e.g. Business Process Execution Language, which combines web services to form composite web service process.)

Some of the advantages of using web services include the capability of working on loosely coupled applications; of working on agnostic operating systems; of working independent of Application Program Interfaces (API) and the possibility of combining and linking existing web services. However, these services produce multiple proliferations of results in the search process, which give unsatisfactory results in meeting the user's goal of finding specific semantics. This process fails at the *discovery* layer level.

This limitation inhibits the development of the applications needed in exchanging, sharing, transferring, and integrating information tools in a specific domain, thus making it difficult to implement applications with a shared common layer among actors. *Semantic Web Services* provide a new non-domain–specific approach, which makes knowledge representation formalisms visible and extensible by using URI-based vocabulary and XML-based grammar. Its goal is to improve understandings of concept representations between users and a system or between systems.

The components of the *Semantic Web Service* are basically one layer of a *Mark up language*, which represents well-defined information, and an *Ontology* layer, which represents the semantic composition. The *Ontology* Layer consists of arbitrary structured data in XML and it has a Resource Description Framework (RDF), which facilitates automated processing of Web resources (metadata); a RDF data model, which describes interrelationships among resources in terms of named properties and values; web ontology, which describes the common meaning among metadata definitions; RDF schema, which describes the mechanism for declaring the properties and relationships [25].

Most of the ontologies of the semantic web use RDF and the web ontology language (OWL), which are semantic networks that build taxonomies for conceptualisms with inheritance and membership relations, named *definitional networks* [26]. In addition, other ingredients of the semantic web services play a significant role such as the *Agents*, which create multiple inputs of data from diverse sources and process the information, and *Proofs*, which ensures data transfer when data comes with semantics. New developments derived from RDF are RDFS, which provide better support for semantic definitions, and DAML+OIL which extends RDF with richer definitional primitives, that facilitate modeling semantics by using ontological constructs.

Our approach develops the ontological constructs in RDFS which facilitates the creation of relation: *rdf:property*, *rdf:subpropertyof*, and *rdf:domain*. The use of RDFS facilitates the representation of the construction domain ontology.

Producing Semantics Through the Use of Ontology

The main contribution of this framework is to generate the necessary understanding to enable reconciliation processes via semantic web services. The development of the prototype will allow the delineation of more specific problems for ontology specifications. In addition, this approach will leverage implementation of consistent applications for information exchange with the use of data representation ontologies in the construction domain.

Ontologies have a taxonomy component and a set of inference rules that can be manipulated by an application for more effective human understanding. The use of ontology, as it used in *web services*, provides a large extent of flexibility and expressiveness; has the ability to express semi-structured data and constraints; and supports types and inheritance. Although the development of the logic layer is currently in research, one of the advantages of this type of data representation is that it allows for the manipulation of concepts or knowledge structures in the taxonomy, in order to determine the semantics via inference rules. Another advantage is that it allows the use of ontology to semantically annotate WSDL constructs in order to sufficiently explain the semantics of the data types used in the *Web Service* as well as their functionality. This step falls at the discovery and publication layers. The idea is to use the ontology to categorize registries based on the domains maintaining properties of each registry and the relationship between registries. This additional information on the ontology leverages the employment of process models, giving dynamic features to the service.

The proposed architecture uses a central ontology for providing specifications of construction concepts. The twofold use of ontology, as an element to support the construction concepts and/or as data representation, promises a viable solution to this problem in the construction domain, specifically with the description of explicit or declared semantics. This solution takes advantage of the rich functionality of the ontology, i.e the specification hierarchy.

The application of *semantic web services* provides a solution-approach due to the nature of its components. For instance, any definition of a useful concept will exist through the collaboration of agents and it will have a specific Universal Resource Identifier (URI) and other additional semantic links. These concepts must adequately associate the names of entities in the universe of discourse with human-readable text describing the meaning of the labels of concepts, and formal axioms that constrain the interpretation and the use of the terms.

The Onto-semantic Framework

The architecture of the framework is shown in Figure 4. This framework contains existing *Browsers* that are the interfaces to generate the information in the *Ontology Editor*. The *Editor* generates new ontology and allows the agents to brows through it. The *Ontology Repository* contains the high level layers ontologies, which are specified by the Ontology Web Language OWL[27] , and also provides functions to retrieve, update and validate them. The *Ontology Server* has an important role in the framework because it controls the other components of the framework.

The *Ontology Server* uses the specifications, provided by the *Ontology Repository*, to allow the search for specific mappings, requested by the *Mapping Module*, in order to map the two ontology data representations. As was mentioned previously, these requested searches are based on specific classification systems or semantic interpretations included in the ontology. The *Workflow Management System* supplies these two data representations. However, due to the potential incompleteness of any of the ontology data representations and/or the general layer ontology, the mappings might not be performed. Therefore, as shown in Figure 5, an additional *Knowledge Base Repository* is necessary to assist the process.

As shown in Figure 5, the *Mapping Module* contains a *Semi-Automatic Mapping Engine* that coordinates the requests and responses with the ontology server. It also has a *Wrapper* that serves as a translator for the specific format used by the *Work Flow Management System*. The *Mapping Module* receives the input that consists of two ontology data representations. The *Mapping Engine* should also allow for eventual human intervention when the *Ontology Server* does not provide any mapping results. The *Engine* generates ontology representation instances or mapped data, expressed in XML format, as a final result of the process. This result is intended for the specific use of the internal application of the *Workflow Management System* and it will enable readable human interpretation of the XML format that can be expressed in other formats such as HTML.

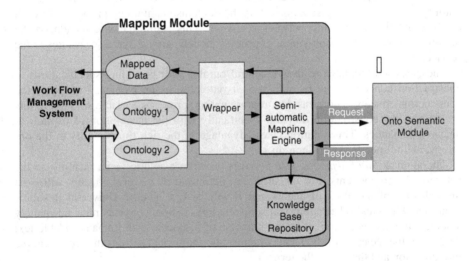

Fig. 5. Semantics Mapping Module

Inference Algorithms

The capabilities of the framework to make inferences are aimed at solving the reconciliations problem of concepts from the external source over the main, global ontology. The algorithms take into account the following factors: syntactical and semantic patterns, similarity indexes, and frequency of queries. This strategy undertakes the definitions of typical mapping patterns to derive a comprehensive understanding of what needs to be mapped and how it should be performed (see Figure 5). This process is supported with the use of search engines and, the aid of a knowledge base library.

Global Ontology and Agents

This research proposes the development of a global ontology that promotes interoperability and consistency. The consistency is promoted by virtue of the existence of a central repository of construction concepts. In other words, the approach will attain coherence in the ontology components by virtue of a high-level layer of concepts and operations. This layer will be leveraged, as much as possible, from standards such as IFC [21]. The assembly of the ontology into the *Semantic Web Service* approach is intended to mitigate the complexity of the meaning of concepts. This reduction of dissimilar concept interpretations is attained by the interaction between the Ontology representation and the agents, i.e. between the Onto-Semantic Framework and the browsers of multiple clients. In addition, a strategy is implemented for registering the *reactions* of *multiple agents* to the interpretation of the concepts that are presented to them through the *ontology;* these interpretations depend on the specific domain or application data. This reaction is called the intention of the agents or construction participants. In this case, the agents are construction participants that are cognitive agents, which interact with the browsers as clients. The record of the agent's interaction will then be utilized to anticipate data redundancy and other data conflicts in future concept reconciliations in the mapping module. The incremental clarification of the agent's intention over the construction concepts will reduce semantic differences on the interpretation of the concepts. In this framework, the domain experts are represented by the cognitive agents.

Fundamentals of the Strategy

It is suggested that the reader navigates through the explanations of the fundamentals in order to capture the full understanding of the onto-semantic framework. The following sections explain the construction concepts analysis used in the framework, which is implemented in the inference algorithms, and the server management system.

For this purpose, a *conceptual framework* to analyze concepts to aid the interpretation of cognitive agents is employed. The objective of this strategy is to take into account the interaction of the expert's domain and his/her world. This *conceptual framework* differs from the traditional, passive models that process information, since the framework incorporates external clients or construction participants as cognitive agents and their social role in the construction organization. As was previously mentioned, construction participants are multi-agents that interact with the system via the browser clients using web services. The approach employs ontological levels to represent construction concepts and wraps them in a *framework* with the purpose to approximate those concepts to the cognitive agent's intentions. In this section, the fundamentals of the approach are explained and justified through examples.

Construction Concepts

Concepts are abstract, universal notions, of an entity of a domain that serves to designate a category of entities, events, or relations. A concept that is used in the construction community comprises geometric features, components or parts, additional or assembled items, and functional characteristics. *Details* and *conditions* are expressions used to define characteristics of any concept employed in a construction project by the construction industry community. A concept is represented in two forms either as a physical construct or as an abstract expression.

Concept <u>details</u> are modes of describing a concept with features (e.g. geometrical) and ontological aspects (e.g. dependency relations). For example, the concept details that describe the component 'hung' of an entity 'window' are a part of the entity 'window' and have functional characteristics which cannot exist independently; 'hung' needs a 'window' to perform the locking and handling functions necessary that allow an agent to open or close the 'window'.

The approach is based upon the assumptions that <u>any concept in a region of space-time has no intrinsic meaning</u>. Accordingly, this research aims at finding semantics to determine how an entity, which is an abstract, universal notion, is related to other entities. The semantics takes into account additional relationships such as situational conditions. The conditions identify a separate piece of the 'world' in which the construction concept is involved. For any concept, specific situations, which are bounded in a space-time region, are considered and are labeled as situational conditions. Situational conditions include state of affairs, which embrace the entity's location, position, site, place, and settings; status condition, which is the stage of the concept (e.g. completed, installed, delayed) during its life in the time-space region; and the relationships with other products or context relations (e.g. set by, part of). In this research context relations are strictly locative to the object the concept describes. This means that the space or region of analysis is limited to the closest location of the objects.

Situational conditions help the analysis handle states of affairs and context relations. As an illustration, Figure 6 depicts a construction concept 'wood frame window'. It shows the conditions of the visual symbol representation and the possible situational condition (e.g. relative position of the wood window in the wall, and the window settings). Figure 6 sketches the construction concept context relations and indicates the state of affairs of this particular entity.

For example, in Figure 6, 'place in' is a context relation of the 'wood frame window' to another physical concept; the wood window is vertically placed in the wall. The wood window and wall represent construction concepts, and 'placed in' represents the relationship between these two concepts.

Representations attempt to describe an extension of a concept in the real world. The representations themselves are simple *metaphors* that give meaning to some concept. Concept representations are not merely elaborations of signs in the mind, but are extended to something physical, such as the context space, in order to be realized or instantiated [28]. This means that representations of concepts cannot fully describe the meaning of the concepts if their relationships to the other concepts are not taken into account. These relationships are termed *contextual relations*.

Contextual relations attempt to identify a possible agent's relations to other locative objects or construction concepts, which might influence the current concept interpretation, and to link such relation to other concepts. This line of characterization of the interpretation has roots in the semiotic tradition [29]. The *contextual relations* rest on the cognitive agent's purpose in interpreting a concept. This strategy takes *contextual relations* in the consideration of a valid construction participant's interpretation.

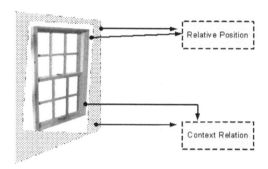

Fig. 6. Contextual relations

Conceptualization on the Domain

In order to introduce the *conceptualization* notion, one has to keep in mind that the *intension* of a concept in the construction domain can be stated by its *details and situational condition relations*. Construction concept *details* comprise their geometric features, their components or parts, additional or assembled items, and their functional characteristics. The concept *conditions* are the *situational conditions* or *state of affairs*, which embrace the concept location, position, site, place, and settings; the *status condition,* which is the stage of the concept (e.g. completed, installed, delayed), and its *relationships* with other products or *context descriptions* (e.g. set by, part of).

A *conceptualization* accounts for all intended meanings of a representation's use in order to denote relevant relations [30]. This means that a *conceptualization* is a set of informal rules that constraint a piece of a physical construct concept or an abstraction concept. An actor or observer uses a set of rules to isolate and organize relevant relationships. These are the rules that tell us if a piece of such a concept remains the same independently of the states of affairs. Guarino further clarifies the *conceptualization* notion, which refers to a set of conceptual relations defined on domain space that describe a set of *state of affairs*, by making a clear distinction between a set of state of affairs or possible worlds and intended models [31].

A *conceptualization* of any physical construct or abstract notion in the construction domain must include details that will independently describe the construction concept from its states of affairs. *Situational conditions* will be needed to describe some extensions of the concepts in order to reflect common situations or relevant relations to the *states of affairs*.

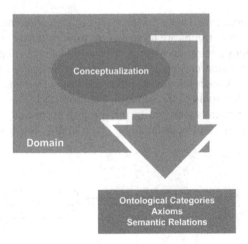

Fig. 7. *Conceptualization* on a domain

*Conceptualization*s are described by a set of informal rules used to express the intended meaning through a set of domain relations. These meanings are supposed to remain the same even if some of the *situational conditions* change [30]. One particular set of rules, which describes an extension to the world, is called the intended model. These descriptions are implemented in a language that has specific syntax. The syntax can be a natural language (e.g. English words), a programming language (e.g. LISP syntax), or any visual representation or topology that aids in communication (e.g. electrical symbols for drawings). An intended model uses a particular interpretation of the language to elaborate representations and create the constraints. The syntax of the languages composes what is called a vocabulary. This vocabulary is used to define the intended models. The models fix a particular interpretation of such a language [31] . The intended models, or the models that partially commit to a certain situation, weakly describe a state of affairs by an underlying *conceptualization*, but at least describe the most obvious or primitive situations. These means that we recognize the existence of other states of affairs which are not register in the model by any conceptualization.

For a better illustration of the *conceptualization* concept, consider Figure 7, which schematically depicts a *conceptualization* into a specific domain, and indicates components that help define a *conceptualization*. The components are minimal ontological definitions of the entity, logical axioms that use the syntax and vocabulary of a language, and additional semantic relations, which help describe several states of affairs.

For an example in the construction domain, see Figure 8. The *conceptualization* of this 'wood frame window' involves an explicit description of the ontological definition. Additional description of the concept intension, which comprehends context relations and other constraints that do not change with the states of affairs of the concept (e.g. the relationship 'set by' and 'on' of a detail do not change with the position of the product), will help to define 'wood frame window' for further interpretation.

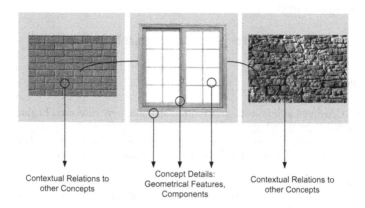

Fig. 8. Context relations and details for conceptualization

From Figure 8 it easy to notice that the details and condition are specifications of the *intension* of the concept. Specifications or ontological refinement processes are explicit formalizations of the concept conditions and the concept details. Reifying a concept denotes an understanding of the *conceptualization* of the representation. Reifying a concept, in computer science and artificial intelligence, means to make a data model for a previously abstract concept, i.e. to consider an abstract concept to be real. Reification allows a computer to process an abstraction as if it were any other data.

Explicit formalization of concepts is by definition an ontology specification [7, 30, 32]. As it is illustrated in Figure 8, *conceptualization*s become extractions of the domain knowledge and are specified by ontological categories, relationships, and constraints or axioms. Categories are forms of classifications of the ways cognitive agents see the world. *Conceptualization*s, through the use of relationships and constraints or axioms, attempt to formally define cognitive agent's views or their perception of the world according to the nature of the concepts themselves and the categories that cognitive agents use.

5   Summary

This paper explains a step of the semantic interoperability paradigm, particularly the reconciliation problem. It explains the complexity involved in exchanging, sharing, transferring, and integrating knowledge and the need for human intervention to aid these processes. An Onto-semantic framework is proposed as an approach for semantic interoperability that allows two sources of information represented each one in an ontology to be reconciled. Our approach inherits the ability of enhancing interoperability at the syntactic and the semantic levels of the Semantic web, and a central, global ontology to support reconciliation of two ontologies. Further development of this research is aimed at finding better strategies to induct the global ontology with the goal of finding conceptualizations based on the intention of the cognitive agent.

Acknowledgements

This work is partially supported by NSF research grant ITR-0404113.

References

1. Turk, Z. Construction informatics: Definition and ontology. Advanced Engineering Informatics. (2006). 20(2): 187-199
2. Jeusfeld, M.A. and A.D. Moor. Concept Integration Precedes Enterprise Integration. In: 34th Annual Hawaii International Conference on System Sciences (HICSS-34). Island of Maui, Hawaii: IEEE. (2001) 10
3. Veeramani, R., J.S. Russell, C. Chan, N. Cusick, M.M. Mahle, and B.V. Roo, State of Practice of E-Commerce Application in the Construction Industry, in E-Commerce Applications for Construction. 2002, Construction Industry Institute, The University of Texas at Austin: Austin, TX. 138
4. Kitamura, Y., M. Kashiwase, M. Fuse, and R. Mizoguchi, Deployment of an ontological framework of functional design knowledge. Advanced Engineering Informatics. (2004). 18(2): 115-127
5. Peachavanish, R., H.A. Karimi, B. Akinci, and F. Boukamp, An ontological engineering approach for integrating CAD and GIS in support of infrastructure management. Advanced Engineering Informatics. (2006). 20(1): 71-88
6. Partridge, C. The role of ontology in integrating semantically heterogeneous databases, in LADSEB-CNR, T.R. 05/02, Editor. 2002, National Research Council, Institute of Systems Theory and Biomedical Engineering. (LADSEB-CNR): Padova - Italy. 24
7. Zúñiga, G.L. Ontology: its transformation from philosophy to information systems. In: Proceedings of the international conference on Formal Ontology in Information Systems. Ogunquit, Maine, USA: ACM Press. (2001) 187 - 197
8. Turk, Z. Phenomenological foundations of conceptual product modelling in architecture, engineering, and construction. Artificial Intelligence for Engineering. (2001). 15(2): 83 - 92
9. Mutis, I.A., R.R.A. Issa, and I. Flood. Semantic Structures of Construction Product Conceptualization. In: International Conference on Computing Decision Making in Civil and Building Engineering. Montreal, Canada: Submitted. (2006)
10. Mutis, I.A., R.R.A. Issa, and I. Flood. Conceptualization of Construction Industry Organizations Via Ontological Analysis. In: International Conference on Computing Decision Making in Civil and Building Engineering. Montreal, Canada: Submitted. (2006)
11. Gibson, J.J. *The Theory of Affordances*. In: R. Shaw and R.J. Brachman, (eds.): Perceiving, acting, and knowing. Toward and ecological psychology. Vol. 1. Lawrence Erlbaum Associates: Hillsdale, New Jersey. 492 (1977)
12. George Lakoff and M. Johnson, Philosophy in the Flesh: the Embodied Mind and Its Challenge to Western Thought. First ed. New York: Basic Books, Perseus Books Group. 624 (1999)
13. Garcia-Molina, H., J.D. Ullman, and J. Widom, Database systems: the complete book. New Jersey: Prentice-Hall (2002)
14. Brachman, R.J. On the epistemological status of semantic networks. In: N.V. Findler, (eds.): Associative Networks: Representation and Use of Knowledge by Computers. Academic Press: New York, NY. 3 - 50 (1979)
15. Woods, W.A. What's in a link: Foundations for semantic networks, in Representation and Understanding: Studies in Cognitive Science, D.G.B.a.A.M. Collins, (Eds.). Academic Press: New York, N.Y. (1975) 35-32

16. Doan, A. Learning to Map between Structured Representations of Data, in Computer Science & Engineering. 2002, University of Washington: Seattle, Washington. 133
17. Noy, N.F. and D.L. McGuinness, Ontology Development 101: A Guide to Creating Your First Ontology', R. KSL-01-05 and SMI-2001-0880, Editors. 2001, Stanford Knowledge Systems Laboratory Technical.: Stanford, Ca. 25
18. Giunchiglia, F. and P. Shvaiko, Semantic Matching. 2003, University of Trento, Department of information and Communication Technology. 16
19. Ding, L., P. Kolari, Z. Ding, S. Avancha, T. Finin, and A. Joshi, Using Ontology in the Semantic Web. 2004: Department of Computer Science and Electrical Engineer. University of Maryland. Baltimore, MD. 34
20. Rahm, E. and P.A. Bernstein, A survey of approaches to automatic schema matching. The VLDB Journal. (2001). 10: 334-350
21. International Alliance for Interoperability (IAI). Industry Foundation Classes (IFC). Accessed in April, (2006). Last Update (2005)
22. OmniClass Construction Classification System. web. Accessed in July 4. Last Update
23. Amor, R. Integrating Construction Information: An Old Challenge Made New. In: Construction Information Technology 2000. Reykjavik, Iceland: International Council for Building Research Studies and Documentation. (2000) 11-20
24. Zamanian, M.K. and J.H. Pittman, A software industry perspective on AEC information models for distributed collaboration. Automation in Construction. (1999). 8(3): 237 - 248
25. Sivashanmugam, K., J.A. Miller, A.P. Sheth, and K. Verma, Framework for Semantic Web Process Composition, in Large Scale Distributed Information Systems. 2003, Computer Science Department. The University of Georgia: Athens GA. 42
26. Sowa, J.F. Knowledge Representation: Logical, Philosophical, and Computational Foundations. 1st. ed, ed. K.R. Theory. Pacific Grove, CA: Brooks Cole Publishing Co. 594 (1999)
27. World Wide Web Consortium (W3C). Web Ontology Language (OWL). Accessed in April, (2006). Last Update (2004)
28. Emmeche, C. Causal processes, semiosis, and consciousness, in Process Theories: Cross disciplinary Studies in Dynamic Categories., J. Seibt, (eds.). Dordrecht, Kluwer. (2004) 313 - 336
29. Luger, G.F. Artificial intelligence : structures and strategies for complex problem solving. Fourth ed. Harlow, England: Pearson Education Limited. 856 (2002)
30. Guarino, N. Understanding, building and using ontologies. International Journal Human-Computer Studies. (1997). 46: 293 - 310
31. Guarino, N. Formal Ontology and Information Systems. In: FOIS'98. Trento, Italy: IOS Press. (1998) 12
32. Gruber, T.R. A Translation Approach to Portable Ontology Specification. Knowledge Acquisition. (1993). 5(2): 199 - 220

GPS and 3DOF Tracking for Georeferenced Registration of Construction Graphics in Outdoor Augmented Reality

Vineet R. Kamat and Amir H. Behzadan

Department of Civil and Environmental Engineering
The University of Michigan
Ann Arbor, MI 48109, USA
{vkamat, abehzada}@umich.edu

Abstract. This paper describes research that investigated the application of the Global Positioning System (GPS) and 3DOF angular tracking to address the registration problem in visualization of construction graphics in outdoor Augmented Reality (AR) environments. AR is the overlaying of virtual images and computer-generated information over scenes of the real world so that the user's resulting view is enhanced or augmented beyond the normal experience. One of the basic issues in AR is the registration problem. Objects in the real world and superimposed virtual objects must be properly aligned with respect to each other, or the illusion that the two coexist in augmented space is compromised. In the presented research, the global position and the 3D orientation of the user's viewpoint (i.e. longitude, latitude, altitude, heading, pitch, and roll) are tracked, and this information is reconciled with the known global position and orientation of superimposed CAD objects. The result is an augmented outdoor environment where superimposed CAD objects stay fixed to their real world locations as the user moves freely on a construction site. The algorithms are implemented in a prototype platform called UM-AR-GPS-ROVER that is capable of interactively placing 3D CAD models at any desired location in an outdoor augmented space.

1 Introduction

Augmented Reality (AR) is the overlaying of virtual images and computer-generated information over the scenes of the real world so that the resulting view is enhanced or augmented beyond the normal experience. In other words, AR allows users to see and navigate in the real world, with virtual objects superimposed on or blended with their view. Virtual Reality (VR), on the other hand deals with only computer-generated models in a totally synthetic environment in which the surrounding real world is not contributing to the simulation process.

Figure 1 presents a snapshot of an AR-based simulation for an airport terminal extension project. While the existing terminals and ongoing airport operations are used as real background, CAD models of the crane and the new steel frame, the only the simulation objects under study, are superimposed over this view.

I.F.C. Smith (Ed.): EG-ICE 2006, LNAI 4200, pp. 368–375, 2006.
© Springer-Verlag Berlin Heidelberg 2006

Fig. 1. AR View of a Virtual Crane Erecting a Virtual Steel Frame on a Real Jobsite

AR can be classified into two categories: indoor and outdoor. In indoor AR, the user takes advantage of a prepared environment in which movements are usually restricted to a finite space. For a domain such as construction, however, indoor AR has limited applications because most construction activities are performed in outdoor, unprepared environments (e.g. heavy construction, roads, bridges, dams, buildings, etc.).

The main requirement of outdoor AR is the need to accurately track the user's viewpoint, and to respond correctly to variations in user's movement and environmental characteristics (e.g. irregular terrain, lighting conditions, etc.). The AR system must be capable of generating an accurate representation of augmented space in real time so that the user experiences a world in which virtual objects stay fixed to their intended location and seamlessly blend with real entities (Azuma et al. 1997).

2 Technical Approach for Registration

Figure 2 presents a schematic view of the hardware components selected for this study and the configuration in which they were assembled. A video camera models the behavior of the eye where all the real objects in the surrounding space appear in perspective view. It continually captures a view of the real world and transmits the images to the AR platform's laptop computer. A head-mounted display (HMD) device worn by the user is also connected to the computer's video port.

In addition, there are two important pieces of equipment connected to the user to keep track of viewpoint movements and basically provide the input for the registration computations. These are the GPS receiver and a 3-DOF orientation tracker. The GPS receiver provides the AR platform with real-time position data in the global space in the form of longitude, latitude, and altitude (Rogers et al. 1999, Roberts et al. 2002, Dodson et al. 2002). The orientation tracker on the other hand, provides the platform with three important pieces of information called yaw, pitch, and roll angles which represent the orientation of the user's viewpoint in a 3D space.

Using these six pieces of information, the AR platform is capable of computing the user's viewpoint in real-time, and based on the computed viewpoint position and three-dimensional orientation, can correctly register virtual CAD objects in the user's viewing frustum. The CAD objects are drawn inside a standard OpenGL perspective viewing frustum. The OpenGL viewing frustum is reconciled with the truncated viewing pyramid of the video camera that captures the real world view. Accurate, real time alignment (registration) of the two viewing frustums (real and virtual) leads to a realistic augmented view where both real and virtual objects coexist. The AR platform transmits images of this augmented environment to the user's display.

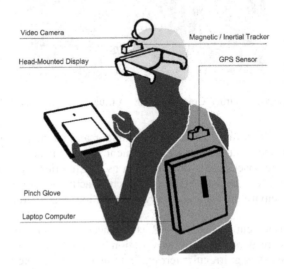

Fig. 2. Hardware Setup for Mobile Outdoor Augmented Reality

The most important requirement for realizing an AR scene in which virtual models appear to coexist with objects in the real environment is real-time knowledge of the relationship between the models, real world objects, and the video input device (Barfield and Caudell 2001). Registration in AR means accurate overlapping of the real and virtual object coordinate frames. Once accurate registration is achieved and maintained, CAD models placed in the augmented space are correctly located and oriented in the real world regardless of where in the augmented space they are viewed from. Registration of virtual objects in the real environment requires accurate tracking of the user's viewpoint position and orientation.

3   Georeferenced Registration of CAD Models

Figure 3 depicts the technical approach used in the presented work to register virtual CAD models in a user's view and superimpose them in the final augmented scene. The GPS receiver obtains the user's global position in real-time. The global position of the virtual object (i.e. the excavator in this figure) can be read from an existing database file. Knowing these two point locations, the relative distance (R) and the corresponding heading angle (α) between the user and the virtual object is calculated. The distance calculation is based on Vincenty's method (Vincenty 1975).

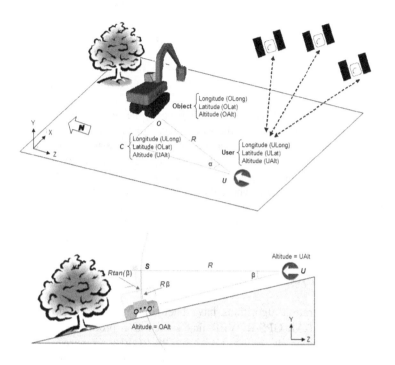

Fig. 3. Registration Method Used in the Presented AR System

Now, considering the fact that the CAD objects are placed in a perspective view, transformation matrices can be effectively used to manipulate them. In order to do that, each object is first subjected to a translation matrix by which it is translated into the depth of view by an amount equal to the distance between the two points (R). Then a rotation matrix is applied to the object by which it is rotated about the Y axis (vertical axis) by an amount α. These two transformations update the virtual object's location based on the user's last incremental move (Behzadan and Kamat 2005).

In Figure 3, the basic assumption in calculating the horizontal displacement (R) and angle (α) is that both the user and the CAD object have the same elevation (altitude). Thus, in case the object is located at a different elevation, further adjustments are needed requiring additional computation steps. In Figure 3, for example, the object has a lower altitude than the user. In this case, the relative pitch angle between the user and the object (β) must be calculated using properties of triangle USO.

The virtual object is then rotated about the X axis by an amount equal to the calculated angle β. Referring to Figure 3, note that by doing this the object is being rotated along the SO' curve. For simplicity, a good assumption can be translating the object along the cord SO in which case the final position of the virtual model ends up to be the point O instead of O'. A final adjustment may be needed to be made on the object's initial side roll angle to represent the possible ground slope (equal to β in Figure 3). In the present work, a pure rotation equal to the slope is being applied to the object around its local X axis so that it lies completely on the ground. In other words, the ground plane is assumed to be uniform and parallel to line segment UO'.

Using the translation along SO instead of rotation along SO' may cause a minute positional error that can be safely neglected for the purposes of this study without experiencing any adverse visual artifacts. However, for wider range applications (e.g. objects that are to be placed far away from the user), the length of OO' can be large in which case, instead of translating the object from S to O, a rotation equal to β must be applied to transform the object from S to O'.

Thus, when a mobile AR user turns his/her head, the relative change in the viewpoint orientation is obtained using the three pieces of data coming from the 3-DOF orientation tracker. Applying the reverse transformations in the amount of the computed angles to the virtual objects in the form of a rotation matrix leads to a final augmented view wherein the objects' locations are unaffected by the user's head movement. In addition, in cases where the user both moves and rotates the head at the same time, all the computation steps in the described procedure are continually repeated (i.e. distance and relative yaw angle calculation, and pitch adjustment) so that the final composite AR view remains unaffected as a user moves freely.

4 UM-AR-GPS-ROVER Augmented Reality Platform

The designed registration algorithms have been implemented in an AR prototype platform named UM-AR-GPS-ROVER that serves as a proof-of-concept, and also helps validate the research results. UM-AR-GPS-ROVER comprises of two main components working in parallel. In addition to the software components of the tool, supporting hardware devices are appropriately integrated in an AR backpack to provide the necessary sensing, input, and output capabilities. The connected hardware devices primarily consist of a GPS receiver, a 3DOF orientation tracker, an HMD, and a laptop. Figure 4 shows the AR backpack setup used in UM-AR-GPS-ROVER.

A Trimble agriculture grade GPS receiver and a digital compass module are used for user position and orientation sensing respectively. An i-glasses SVGA Pro 3D HMD is used as the wearable display. A Unibrain Fire-i digital firewire camera is also used as the video input device which is connected in front of the HMD. The GPS receiver is placed in the backpack with the antenna coming out of it. The digital compass is connected inside the hard hat in an isolated chip box. In the presented work, a HP Pavilion laptop (placed inside the AR backpack) with 3 GHz CPU speed and 512 MB RAM is used as a mobile computing power.

The software components of UM-AR-GPS-ROVER are independent interconnected modules that can be easily replaced or updated as necessary. The platform is basically implemented as a set of four loosely-coupled interacting modules. The first module captures a live video stream from the real world using the video input device. The second module is mainly a data collector for the GPS receiver and orientation tracker and communicates with the sensors via USB ports. This module provides the input for the third module which is the transformation module in which the viewpoint's movements are registered and appropriate transformations are applied to the virtual objects. The fourth module is a graphical module that reads graphical data from indicated CAD files and places each model in the user's view in real time.

Fig. 4. AR Backpack Setup of UM-AR-GPS-ROVER

5 Validation

To validate the research results and registration algorithms, UM-AR-GPS-ROVER platform was used to place several static and dynamic 3D CAD models at several known locations in outdoor augmented space. In particular, the prototype was successfully tested in many outdoor locations at the University of Michigan north campus using several 3D construction models (e.g. buildings, structural frames, pieces of equipment, etc.). Figure 5 presents two such snapshots in outdoor AR.

Fig. 5. Structural Steel Frames Registered in Outdoor AR Models Courtesy of Mr. Robert R. Lipman (NIST)

6 Future Work and Challenges

One of the important issues in almost any outdoor AR application is occlusion. Incorrect occlusion happens when a real object is placed between the user's view and the virtual object(s) in augmented space. In such a situation, as the distance between the real object and the user is less than that between the virtual object(s) and the user, the real object should theoretically block (at least partially) the user's view of the superimposed models that lie behind it. This idea is presented in Figure 6.

The stick of the virtual excavator in this figure is partially occluded at two locations by the light pole and the stick of a real excavator, both of which are closer to the user in augmented space than the virtual excavator. Thus, under ideal circumstances, the hidden portions of virtual models should not be visible in the composite AR output as shown in the figure. That is, however, not currently the case because UM-AR-GPS-ROVER draws the pixels of all virtual models after painting the captured video image as a background.

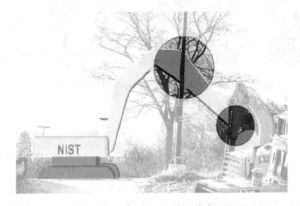

Fig. 6. Dynamic Occlusion Problem in Outdoor Augmented Reality CAD Models Courtesy of Mr. Robert R. Lipman (NIST)

One of the solutions being explored for this issue is using a combination of rapid geometric modeling of the surrounding environment or other depth sensing techniques (e.g. stereo cameras) and utilizing the graphics processor's z-buffer to draw the appropriate set of pixels in each composite AR frame. In other words, if this depth of real objects is greater than the depth of virtual object(s) for a given view, the real object does not occlude any virtual objects. In the opposite set of circumstances, appropriate corrections should be made to user's view to take into account the existence of an occluding real object.

7 Conclusions

The primary advantage of graphical simulation in AR compared to that in VR is the significant reduction in the amount of effort required for CAD model engineering. In order for AR graphical simulations to be realistic and convincing, real objects and augmented virtual models must be properly aligned relative to each other. Without accurate registration, the illusion that the two coexist in AR space is compromised.

Traditional tracking systems used for AR registration are intended for use in controlled indoor spaces and are unsuitable for unprepared outdoor environments such as those found on typical construction sites. In order to address this issue in the presented research, the global outdoor position and 3D orientation of the user's viewpoint are tracked using a GPS sensor and a 3DOF orientation sensor. The tracked information is reconciled with the known global position and orientation of CAD objects to be overlaid on the user's view.

Based on this computation, the relative translation and axial rotations between the user's eyes and the CAD objects are calculated at each frame during visualization.

The relative geometric transformations are then applied to the CAD objects to generate an augmented outdoor environment where superimposed CAD objects stay fixed to their real world locations as the user moves about freely on a construction site. Designed algorithms have been validated using several 3D construction models in a number of outdoor locations.

Acknowledgements

The presented work has been supported by the National Science Foundation (NSF) through grant CMS-0448762. The authors gratefully acknowledge NSF's support. Any opinions, findings, conclusions, and recommendations expressed in this paper are those of the authors and do not necessarily reflect the views of the NSF.

References

1. Azuma, R. (1997). "A Survey of Augmented Reality." Teleoperators and Virtual Environments, 6(4), 355–385.
2. Barfield, W., and Caudell, T. [editors] (2001). Fundamentals of Wearable Computers and Augmented Reality, Lawrence Erlbaum Associates, Mahwah, NJ.
3. Behzadan, A. H., and Kamat, V. R. (2005). "Visualization of Construction Graphics in Outdoor Augmented Reality", Proceedings of the 2005 Winter Simulation Conference, Institute of Electrical and Electronics Engineers (IEEE), Piscataway, NJ.
4. Dodson, A. H., Roberts, G. W., and Ogundipe, O. (2002). "Construction plant control using RTK GPS." FIG XXII International Congress, Washington, D.C.
5. Rogers, S., Langley, P., and Wilson, C. (1999). "Mining GPS data to augment road models." In Proceedings of the 5th International Conference on Knowledge Discovery and Data Mining, 104-113, San Diego, CA.
6. Roberts, G. W., Evans, A., Dodson, A., Denby, B., Cooper, S., and Hollands, R. (2002). "The use of Augmented Reality, GPS, and INS for subsurface data visualization." FIG XXII International Congress, Washington, D.C.
7. Vincenty, T. (1975). "Direct and inverse solutions of geodesics on the ellipsoid with application of nested equations." Survey Review, 176, 88-93.

Operative Models for the Introduction of Additional Semantics into the Cooperative Planning Process

Christian Koch and Berthold Firmenich

Bauhaus University Weimar, CAD in der Bauinformatik, Coudraystr. 7,
99423 Weimar, Germany
{christian.koch, berthold.firmenich}@bauing.uni-weimar.de

Abstract. According to the State-of-the-Art building information is described by standardized evaluated models like the Industry Foundation Classes (IFC). Instances of evaluated models are stored either in documents that are frequently cloned and exchanged or in model servers that allow for sharing data between the actors involved. In the cooperative planning process this approach has considerable shortcomings concerning model complexity, implementation effort, model transformation and model semantics. In the solution approach presented the currently used evaluated models are complemented by additional semantics. This approach is denoted as operative modeling. Here, the native application models remain unchanged and the operations applied are stored using a simple, standardized language. Because operative models have additional semantics they could be a solution in the fields of data exchange, version management and long term compulsory archiving.

1 Introduction

In the computer-supported planning process building information is described by instances of evaluated models. Due to the iterative and distributed nature of the planning process several versions m_i of the building instance are created and exchanged between the actors involved. According to law, these versions have to be archived over long periods of time.

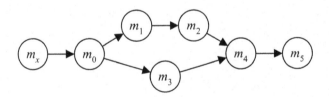

Fig. 1. Version graph of a building instance with m_x as the empty version

In order to be able to manage versions it is indispensable to compare and merge versions. It is distinguished between two types of versions – revisions (Fig. 1: m_1, m_2) and variants (Fig. 1: m_2, m_3).

Information exchange, long term compulsory archiving and version management have considerable shortcomings when applied to traditional evaluated models. The

I.F.C. Smith (Ed.): EG-ICE 2006, LNAI 4200, pp. 376–382, 2006.

objective of this contribution is to introduce an alternative modeling approach denoted as operative modeling. Here, the native application models remain unchanged and the operations applied are described using a simple, standardized language. The application of operative models in the planning process is shown below.

2 State-of-the-Art

According to the State-of-the-Art software applications used in the building planning process create and modify structured object sets called application specific model instances.

2.1 Standardized Object Models

The standardization of evaluated object models is one attempt to support the distributed planning process. An example for this approach is the introduction of the Industry Foundation Classes (IFC) and the physical file exchange format STEP. In practice, however, the cooperation on the basis of evaluated object models is an error-prone process. Firstly, a common model that covers all the disciplines tends to be too complex. If it existed, an application would have either to implement the standardized model and consequently would have to be newly developed or the standardized models and the application specific models would have to be transformed into one another. The latter is characterized by a cumulative information loss because standardized models and application models cannot be mapped completely onto one another [1].

Information exchange, long term compulsory archiving and version management have considerable shortcomings when applied to traditional evaluated models. The reason for this assumption is that the private object attributes have to be processed by tools which are not aware of the semantics of this information. Moreover, evaluated object models can be very complex in the case of deeply nested object relationships. The interpretation of such models is a difficult task.

2.2 Distributed Cooperation on the Basis of Standardized Object Models

The traditional approach of cooperation consists of the exchange of documents. This approach has not proven in practice for reasons already mentioned. As a solution, Document Management Systems (DMSs) are used to manage document versions in the planning process. In DMSs the history of a document is managed. Only revisions but not variants of a building instance can be stored (Fig. 2) [2].

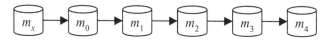

Fig. 2. Versions of a building instance on the basis of document history in DMSs

Furthermore, DMSs cannot support the merging of building instance versions since the semantics of the document are generally not known [3].

Like DMSs, model servers are used for sharing building information in the distributed planning process. A shared model server running on the Internet manages object model instances in database systems. An example for IFC models is the IFC Model Server Development Project [4]. Using the SOAP communication service IFC instances are completely or partially exchanged between client applications and server. However, available servers do not support versions of IFC instances yet and the already mentioned problems remain unsolved. Recently, an approach to manage IFC instance versions was described in [7].

For the addressed reasons standardized object models in conjunction with DMSs and model servers have considerable shortcomings in the iterative and distributed planning process of buildings.

3 Solution Approach

3.1 Operative Modeling

In the traditional approach a model instance is represented by objects and attributes. In contrast, an operative model instance is described by the operations applied in the design process that finally lead to the respective building state. While an object model is evaluated, an operative model is unevaluated [1].

In order to exemplify the concept of operative building modeling a comparison with solid modeling can be drawn. While a BRep model instance describes the topology and geometry of a solid boundary in an evaluated form, a CSG model instance consists of an unevaluated description of a solid. The building shown in figure 3 shall be abstracted as a solid model instance.

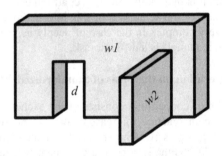

Fig. 3. An abstracted building model instance

Using the *Scheme* language of the solid modeler ACIS [5] the applied operations for defining an unevaluated CSG model instance can be formulated in just four lines of code:

```
1:  (define w1(solid:block 0 0 0 10 0.5 5))
2:  (define w2(solid:block 8 3 0 9 0 3.5))
3:  (define d(solid:block 2 0 0 4 0.5 3))
4:  (solid:subtract(solid:unite w1 w2)d)
```

In contrast, using the BRep modeling approach the evaluated description of the building's solid results in a *SAT* file (Standard ACIS Text) containing almost 300 lines of code:

```
1: 700 0 1 0
2: 22 ACIS/Scheme AIDE - 7.0 11 ACIS 7.0 NT 24 Thu May
           27 09:50:57 2004
3: 1 9.9999999999999995e -007 1e -010
4: body $-1 -1 $-1 $1 $-1 $2 #
5: lump $-1 -1 $-1 $-1 $3 $0 #
6: transform $-1 -1 1 0 0 0 1 0 0 0 1 5 0.25 2.5 1
           no_rotate no_reflect no_shear #
7: shell $-1 -1 $-1 $-1 $-1 $4 $-1 $1 #
8: face $5 -1 $-1 $6 $7 $3 $-1 $8 reversed single #
...
285: straight-curve $-1 -1 $-1 4 -1.75 -2.5 0 -1 0 II #
286: point $-1 -1 $-1 4 -0.25 -2.5 #
287: point $-1 -1 $-1 4 -3.25 -2.5 #
288: End-of-ACIS-data
```

This example demonstrates that an operative description (1) is compact and (2) can be interpreted by users – not only by applications.

3.2 Mathematical Description

Formally, the operative building instance is described by a sequence of applied operations that are contained in one of the sets A, S, C or R with

$$
\begin{aligned}
A &:= \{a_i \mid a_i \text{ is an operation that adds an object}\} \\
S &:= \{s_i \mid s_i \text{ is an operation that selects or unselects objects}\} \\
C &:= \{c_i \mid c_i \text{ is an operation that modifies selected objects}\} \\
R &:= \{r_i \mid r_i \text{ is an operation that removes selected objects}\}.
\end{aligned}
\tag{1}
$$

The set O of all operations is defined as

$$
O := A \cup S \cup C \cup R. \tag{2}
$$

A change δ_{ij} is a sequence of operations $o \in O$

$$
\delta_{ij} := \langle o_0, o_1, o_2, \cdots, o_{n-1} \rangle \tag{3}
$$

and can be represented as an edge (m_i, m_j) in the version graph. The version graph is a rooted tree with m_x as its root node (Fig. 4). It should be noted that in this approach the tree structure is preserved, even in the case of merges. In contrast to a version graph (Fig. 1) this structure is denoted as delta tree.

An instance version m_j is a node in the delta tree and can be formulated as a root path starting at the root node m_x. A path is denoted either by a sequence of n edges or a sequence of $n+1$ nodes.

$$
rootpath(m_j) := \langle \delta_{x0}, \delta_{01}, \cdots, \delta_{ij} \rangle = \langle m_x, m_0, m_1, \cdots, m_i, m_j \rangle. \tag{4}
$$

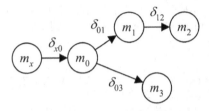

Fig. 4. Delta tree of a building instance

For example, the version m_2 of the model instance in figure 4 is described by the root path $rootpath(m_2) = \langle \delta_{x0}, \delta_{01}, \delta_{12} \rangle = \langle m_x, m_0, m_1, m_2 \rangle$.

3.3 Simple Language for Operative Modeling

Analogous to object models operative models have to be described by a language. The operative modeling language (OML) has a substantial similarity to a command language because an operative model instance consists of operations. An operation is described by an identifier (*id*) and a list of arguments (*arglist*). The type of arguments is either *number, string, boolean*, persistent identifier (*poid*) or *list*. An extract of the language's grammar in Backus-Naur-Form (BNF) is shown below:

```
operation ::= id arglist
arglist   ::= arg(,arg)+
arg       ::= number | string | boolean | poid | list
```

Because of its additional semantics OML is considered to have advantages in the areas of data exchange and version management. OML is also useful in the area of long term compulsory archiving because it does not change as frequently as evaluated models do [1].

Extending OML by variables, procedures and control structures results in a simple operative programming language called OPL. Via OPL both users and programmers can communicate with the planning system in an ad-hoc way.

3.4 Distributed Cooperation on the Basis of Operative Models

A prerequisite for the use of operative models in the planning process is the definition of standard operations. Once defined, the standard operations can be used to exchange the building information.

Additionally, existing applications need to be slightly adapted. While the evaluated application building model remains unchanged the application itself has to be extended by journaling functionality. A journaling mechanism is responsible for recognizing changes applied to the model instance and describing these changes by a standardized language. The operations that advance the instance version m_i to m_j are serialized in a journal file as a δ_{ij} change.

A sequence of journal files represents a version of a building instance. Consequently, building information is exchanged by journal files. Contrary to the traditional data exchange of evaluated models the proposed approach has the advantage of a non-accumulating information loss: Instead of sequential information transformations the exchanged δ_{ij} changes are only interpreted and thus remain unchanged [1]. The exchanged information not only describes the result but to some extent also the intent of design. Operative modeling is applicable in both a versioned and unversioned environment. Besides the journaling mechanism an application has to be extended by an interpreter capable of applying standard operations of the journal file.

Planner A

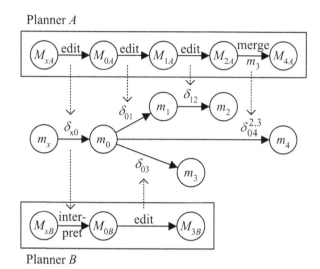

Planner B

Fig. 5. Distributed workflow on the basis of the operative modeling approach

Figure 5 illustrates the distributed workflow between planner A and planner B on the basis of operative building models. It is assumed, that the planners use different planning applications. Planner A starts designing and creates his native model instance M_{0A}. The operations issued are automatically journaled as changes δ_{x0}. Subsequently, planner B generates his native model instance M_{0B} by interpreting and applying the changes δ_{x0}. Then, both planner A and planner B synchronously edit the building instance m_0. These changes are automatically recorded as $\langle \delta_{01}, \delta_{12} \rangle$ and $\langle \delta_{03} \rangle$ respectively.

As described in [6], unevaluated operative models have advantages in the context of comparing and merging versions. The reason is that an operative instance explicitly stores the semantics of differences between versions. This results in advantageous diff and merge algorithms that operate on the delta tree. For example, merging versions of the building instance means to join the respective sub-paths into a new one on the basis of the existing changes [6]. Specifically, merging the variants m_2 and m_3 by planner A results in the version m_4 that is described as the sequence of changes $\langle \delta_{x0}, \delta_{04}^{2,3} \rangle$ in figure 5.

4 Conclusions

Evaluated standardized models like the IFC have considerable shortcomings in the context of the cooperative planning process. This paper presents a new solution approach that complements evaluated models by the semantics of unevaluated operative models. The operative modeling approach has advantages in the area of data exchange, version management and long term compulsory archiving. However, a lot of research topics remain open: for example the definition of suitable standardized operations for each discipline of the cooperative planning process.

Acknowledgement

The authors gratefully acknowledge the support of this project by the German Research Foundation (DFG).

References

1. Firmenich, B.: A Novel Modelling Approach for the Exchange of CAD Information in Civil Engineering. In: Proceedings of the 5th ECPPM. A.A. Balkema, Leiden, London, New York, (2004) 77 pp.
2. Beer, D. G., Firmenich, B., Beucke, K.: A System Architecture for Net-distributed Applications in Civil Engineering. In: Proceedings of the Joint International Conference on Computing and Decision Making in Civil and Building Engineering, Montreal, Canada, (2006), - accepted paper
3. Firmenich, B., Koch, C., Richter, T. and Beer, D. G.: Versioning structured object sets using text based Version Control Systems. In: Proceedings of the 22nd CIB-W78. Institute of Construction Informatics, Dresden, (2005) 105 pp.
4. http://cic.vtt.fi/projects/ifcsvr/ (2006-02-28)
5. Corney, J., Lim, T.: 3D modeling with ACIS. Saxe-Coburg. Stirling, (2001)
6. Koch, C., Firmenich, B.: A Novel Diff and Merge Approach on the Basis of Operative Models. In: Proceedings of the Joint International Conference on Computing and Decision Making in Civil and Building Engineering, Montreal, Canada, (2006), - accepted paper
7. Nour, M., Firmenich, B., Richter, T., Koch, C.: A versioned IFC Database for Multi-disciplinary Synchronous Cooperation. In: Proceedings of the Joint International Conference on Computing and Decision Making in Civil and Building Engineering, Montreal, Canada, (2006), - accepted paper

A Decentralized Trust Model to Reduce Information Unreliability in Complex Disaster Relief Operations

Dionysios Kostoulas[1], Roberto Aldunate[2],
Feniosky Peña-Mora[3], and Sanyogita Lakhera[4]

[1] Graduate Student, Department of Computer Science,
Newmark Civil Engineering Laboratory, Room 3119, MC-250, 205 North Mathews Avenue,
University of Illinois, Urbana-Champaign, IL 61801
kostoula@uiuc.edu
[2] PhD Student, Construction Management and Information Technology Group, Department of
Civil and Environmental Engineering, Newmark Civil Engineering Laboratory, Room 3119,
MC-250, 205 North Mathews Avenue, University of Illinois , Urbana-Champaign, IL 61801
aldunate@uiuc.edu
[3] Associate Professor of Construction Management and Information Technology,
Department of Civil and Environmental Engineering, Newmark Civil Engineering Laboratory,
Room 3106, MC-250, 205 North Mathews Avenue , University of Illinois, Urbana-
Champaign, IL 61801
feniosky@uiuc.edu
[4] Graduate Student, Information Technology Group,
Department of Civil and Environmental Engineering, Newmark Civil Engineering Laboratory,
Room 3119, MC-250, 205 North Mathews Avenue, University of Illinois ,
Urbana-Champaign, IL 61801
lakhera2@uiuc.edu

Abstract. The vulnerability of urban areas to extreme events is a vital challenge confronting society today. Response to such events involves a large number of organizations that had no past interactions with each other but are required to collaborate during disaster relief efforts. Participants from diverse teams need to form an integrated first response group to effectively react to extreme situations. Civil engineers are expected to play a key role in collaborative first response groups, because of their structural expertise, as complex disasters in urban areas are usually followed by structural damage of critical physical infrastructure. The establishment of trust is a major challenge in extreme situations that involve diverse response teams. Although the means of communication are available (e.g., ad-hoc networks), first responders are hesitant to interact with others outside of their organization because of no prior experience of interactions with them. Moreover, the spread of inaccurate information in cases of complex disaster relief operations increases uncertainty and risk. Participants must be given the ability to assess the trustworthiness of others and information propagated by them in order to enforce collaboration. In this paper, we propose a decentralized trust model to reduce uncertainty and support reliable information dissemination in complex disaster relief scenarios. Our model includes a distributed recommendation scheme, incorporated into an existing membership maintenance service for ad-hoc networks, and a nature-inspired activation spreading mechanism that allows trust-based information propagation. To evaluate the effectiveness of our method in reducing information unreliability in

I.F.C. Smith (Ed.): EG-ICE 2006, LNAI 4200, pp. 383–407, 2006.

complex disaster areas, we tested it through software simulations and by conducting a search and rescue exercise involving civil engineers and firefighters. Results indicate fast and robust establishment of trust and high resilience to the spread of unreliable information.

1 Introduction

Extreme Events (XEs), (i.e., rare and significant occurrences in terms of their impacts, effects or outcomes), have become the main focus of many research studies, especially after the "9/11" terrorist attack [1] and the recent Asian tsunami disaster [2]. XEs include natural disasters, such as earthquakes, hurricanes and floods, as well as intentional disasters, such as fires or terrorist attacks. Specifically, the way urban areas respond to XEs has been reported to be a vital challenge confronting society today [3, 4, 5, 6]. The current response system to such emergent occurrences has proved to be inadequate and needs to be improved [3, 4, 7, 8].

During XEs, a large number of organizations that do not work together on a regular basis are forced into different kinds of interactions. These interactions lead to the formation of the so-called hastily formed networks [30].For example, a Canadian research team in a study of a massive fire near Nanticoke, Canada identified 346 organizations that were on site, i.e., at the scene of the fire [9]. If the efforts among the different organizations involved in complex disaster scenarios are not well coordinated, the actions made by one organization can generate problems for others [10]. Therefore, the members of different organizations are expected to collaborate and coordinate their response efforts into an integrated team, where they would share common goals but have distinct roles determined by their agency, rank and location in relation to the disaster.

Civil engineers should be considered as key participants in this integrated response group. Extreme events in urban areas are inevitably followed by structural and functional damage of critical physical infrastructure. According to FEMA [11] complete knowledge and accurate structural and hazard information about the incident site may not be readily available in a disaster relief effort. Yet, because of civil engineers' knowledge and experience on structural analysis, their role needs to be extended beyond infrastructure life-cycle management and sustainability to also involve first response to XEs; particularly, the engineers and contractors involved in the original design and construction of the critical physical infrastructure.

The diversity of groups forming an integrated multi-agent first response team that includes civil engineers as well as the lack of experience from pre-disaster interactions inhibit collaboration and limit the effectiveness of the response team. Although the means for communication exist, first responders are hesitant to communicate and interact with others outside their own organization [12]. The different groups of responders involved in relief operations, police officers, firefighters, medical personnel, civil engineers, city hall personnel, public and private agencies, among others [13], bear distinct roles in the disaster environment and may have different backgrounds of skills, interests, knowledge and experience, as well as policies and protocols on how to organize and prioritize activities. Moreover, inaccuracy of communicated information increases hesitance to collaborate. According to Heide [12], initial actions in disaster relief efforts are undertaken based on vague and inaccurate information. Therefore, every interaction between first responders has an inherent risk, because of uncertainty. Yet, first responders, including civil engineers, must be given the ability to assess the trustworthiness of others in order to reduce uncertainty.

However, trust establishment in first response teams is not trivial. Traditional means of trust management by centralized approaches would not work in large-scale disaster scenarios. Quarantelli [9] argues that disasters have implications for many different segments of social life and the community, each with their own preexisting patterns of authority and each with the necessity for simultaneous action and autonomous decision-making, and, therefore, it is impossible to create a centralized authority system. Disasters being considered in this paper may require a large number of human resources being deployed into the disaster zone from a municipal, state, federal or even international level, and responding in an on-demand manner. A strictly centralized and hierarchical model, as the command and control model currently applied in response efforts, would be unwieldy in such an environment, as human and artificial resources outgrow the resource capabilities of a possibly existent central agency. Prieto [7], for example, points out that the current command and control model has shown limited effectiveness in complex disaster contexts. In addition, Shuster [14] argues about the inadequacy of centralized approaches to cope with systems comprised of large number of individuals or elements and points out the need to establish a complex system approach due to the increasing number of resources. For the above reasons, decentralized mechanisms for establishing trust and providing reliable communication are needed to supplement any centralized approaches, currently supported by the response system, in order to increase the effectiveness of the response process, even in cases of large-scale disaster contexts.

Furthermore, history-based approaches to engineer trust in a multi-agent first response group would also fail. As shown earlier, organizations involved in complex relief operations do not have a shared history and they may interact for the first time during the disaster. It is clear that first responders cannot rely on their past experience to assess the reliability of others. Under these conditions, reputation through word-of-mouth could be used.

In this paper, we propose a decentralized reputation-based trust model to establish trust and provide reliability in communication between civil engineers and other first responders involved in complex disaster relief efforts. Our model uses a decentralized recommendation scheme to allow participants of a first response network to evaluate the trustworthiness of other participants. This scheme is piggybacking on a membership maintenance protocol that is supposed to run on the system. To efficiently support reliable information dissemination we build a reputation-based nature-inspired activation model on top of the recommendation scheme. To validate our method, we tested it through software simulations and by conducting a search and rescue exercise involving civil engineers and firefighters.

The remainder of this paper is organized as follows: The following section describes our System Model. In the Nature-Inspired Systems Section, some background knowledge on systems inspired by biological paradigms is presented. The section Related Work, presents previous work in the trust management area. Section Decentralized Trust Model provides details and analysis of the proposed model and the next section presents how information dissemination is provided based on the trust model.

Sections Simulation Results and System Evaluation provide results from the validation of our model based on simulations and on a search and rescue exercise respectively. To conclude, the last section of this paper summarizes contributions and describes future work.

2 System Model

In the previous section, we argued that although the communication capabilities are available, first responders from diverse groups do not interact because of the lack of trust. In this section, we present the communication model on which we have based our trust model.

Several infrastructure-based initiatives have been undertaken in order to effectively support communication between the participants involved in disaster relief operations, [15, 16]. However, disasters studied in this paper have low probability. Furthermore, the time and location of some of these events are difficult if not impossible to predict. Therefore, maintaining an infrastructure with constant availability and capable of supporting relief efforts solely in anticipation of a disaster would be prohibitively expensive. Instead, a communication platform for relief operations needs to support interaction in a dynamic, on-demand manner, between participants who may randomly enter or abandon the disaster site at any time. An ad-hoc network, (i.e. a network whose functionality does not rely on any existent infrastructure), is the ideal candidate to meet this challenge since it allows network devices to dynamically connect and be part of the network for the duration of a communications session only. Infrastructure networks may provide backbone connectivity between actors and remote analysts, scientists, or task coordinators, if they are available.

In this paper, we assume robust communication through mobile ad-hoc networks (MANETs), (i.e. a peer-to-peer infrastructure-less communication networks formed by short-range wireless enabled mobile devices). Each first responder equipped with an IT-based mobile device as well as any IT component that may perform independently of a physical actor in the first response system plays the role of a node in the mobile ad-hoc network. Nodes participate in a dynamic network which lacks any underlying infrastructure. Communication takes place hop-by-hop. In other words, each node acts as wireless router. Nodes may route packets through neighbors (i.e. nodes with which they have direct communication) to reach an intended destination. This allows the network to accommodate high mobility and frequent topology changes. Any appropriate communication protocol can be used to provide such wireless communication capabilities among first responders and civil engineers. We assume IEEE 802.11b/802.11g and AODV (Ad Hoc On-Demand Distance Vector) since they are widely used standard protocols. AODV is a routing protocol for ad-hoc mobile networks with large numbers of mobile nodes. The protocol's algorithm creates routes between nodes only when the routes are requested by the source nodes, giving the network the flexibility to allow nodes to enter and leave the network at will. Routes remain active only as long as data packets are traveling along the paths from the source to the destination. The communication protocol provides connectivity, data transmission and routing among the mobile devices.

3 Nature-Inspired Systems

Several computer system design approaches have taken inspiration from the collective behavior of social animals, and particularly insects. Problems solved by these insects' affinities appear to have counterparts in both engineering and computer science. The computational and behavioral metaphor for solving distributed problems that takes its inspiration from the biological paradigms provided by social insects (e.g., ants), is usually referred to as Swarm Intelligence (SI).

Insects' societies are organized in a completely decentralized manner [17], similar to the distributed systems used in the area of computer science. For example, in the case of bees, the queen is not in control of the whole colony, but only of a small part of the nest [18]. Moreover, insects act on local information and make single decisions based on simple rules defined locally. This is similar to decentralized computer networks, where no global knowledge of the network is assumed and operation is based on local information provided by a single node, and its neighbors. In addition, a network of positive and negative feedbacks is built between insects. This organization scheme makes such social systems very robust [17]; in the same way decentralized communication protocols make distributed systems robust.

Complexity in social insects emerges at the level of a group. Simple behaviors of individuals interact in a manner that produces a range of interesting complex behaviors. This is usually referred to as self-organization of insects' societies. At a global level, structure or order appears because of interaction between lower-level entities. However, the behavioral rules or the rules for interaction among entities are implemented on a local basis. Self-organization helps social insects to easily adapt to changes in their environment [17]. Adaptability and self-organization is needed in some cases of distributed systems as well. For example, if the load of the system shifts rapidly from one region to another, as in the case of an adaptive grid computing environment when the computing needs of a node suddenly increase, then the system needs to easily adapt to changes in the computing environment, as social insects do by self-organization. Social insects' behavior may be used to model many problems of distributed computing due to the similarities found in the way insects' colonies and distributed systems are organized.

In this paper, the behavior of ants, and specifically ants' division of labor process, is examined. Under threatening situations ants secrete a specific pheromone to inform other ants. Not all the ants react in the same way to the levels of pheromone perceived (i.e., division of labor). The heterogeneous response to alarm pheromone avoids cascading effects. In general terms, it is a binary decision problem which can be described by the rule: an ant will react to an alarm pheromone (adopting a "alarm" behavioral pattern and also secreting pheromone) based on the amount of pheromone present in the area where the ant is moving and its threshold level.

Models inspired by entomology and particularly by ants, have been widely used to address optimization and routing problems in computer networks. Moreover, ant-based techniques have been used for designing distributed applications that work without the use of a central authority, similar to the decentralized trust management scheme presented in this paper. Agassounon [19] presents a swarm-based system for distributed information dissemination and retrieval. The proposed model uses mobile, autonomous agents that interact with each other in order to provide a distributed complex process.

When agents are used as sensors for counting small objects, it is found that cooperation decreases the standard deviation of the sensed data with the decrease proportional to the number of collaborating agents. In other words, the greater the number of agents that cooperate to sense data, the more narrowly data is distributed around the average. It is found that the sensing information retrieved from a single collaborating agent has as much precision as that retrieved from all agents individually, when no cooperative action takes place and information is processed afterwards [19].

Furthermore, Yingying et al. [20] have proposed a multi-robot cooperation algorithm that can organize different numbers of robots to cooperate on a task according to the task difficulty. The algorithm is based on the labor division process of the ant society. The main idea is that more difficult tasks possess a higher pheromone amount than easier ones. Robots are attracted by pheromone and, as a result, more robots are engaged in difficult tasks. Similar to this scheme, higher rated first responders in our trust model possess higher stimulus, attracting more actors to adopt information provided by them.

Ants' division of labor belongs to a class of processes that can be analyzed using spreading activation models. In spreading activation models, members of the network are represented as nodes in a mathematical graph and relationships between members are represented as graph edges, connecting the related nodes. Nodes that are connected with an edge in the mathematical graph are called neighbors. Information spreading across the network is modeled as a Boolean state. A node has either adopted the information or it has not. A node becomes activated, i.e. adopts the information, using some activation function, which is typically based on some threshold [21]. In a sense, a node decides whether or not to adopt the information depending on the trend followed by its neighbors. If a node's neighbors appear to get activated, the node itself gets also activated. Ants use a threshold-based activation function to decide whether or not to adopt an alarm based on the trend followed by other ants, which is represented by the amount of pheromone associated with the alarm. As more ants adopt the alarm, the amount of pheromone increases.

4 Related Work

Trust is subjective. It can be viewed as the trustor's perception of the trustee's reliability, or else as the subjective probability by which the trustor relies on the trustee.

Trust may be based on various factors, including personal experience and reputation. In the absence of personal experience, as in the case of diverse first response groups, reputation systems can be used to form trust. In such systems, recommendations in the form of ratings are used to provide subjective feedback about the reliability of other nodes. Reputation is considered here as a collective measure of trustworthiness based on ratings. If a complete set of ratings is used to measure reputation, i.e., if ratings from all nodes in the network are used, then reputation is objective. However, this is not usually the case in distributed systems where memory constraints do not allow the maintenance of a complete set of ratings at each node. Therefore, in the sense of distributed systems reputation is rather subjective.

Hung [22] argues also about the need to find peripheral, i.e., word-of-mouth based, mechanisms, like recommendation systems, to assess trustworthiness in the absence of past experience. He mentions that when people first meet, the lack of personal knowledge about the interacting parties hinders their ability to engage in deliberate assessment, even when they have high motivation to do so. This forces people to use simple heuristics based on the peripheral cues embedded in the interaction environment.

Recommendation (or reputation) systems have been widely used for trust establishment in distributed systems. Marti et al. [23] proposed a reputation system for ad-hoc networks. In their system, a node monitors the transmission of a neighbor, to make sure that the neighbor forwards others' traffic. For example, a civil engineer that processes information on the state of a building through a police officer checks to see if this information is being further propagated. If the neighbor does not forward others' traffic, it is considered as uncooperative, and this uncooperative reputation is propagated throughout the network. A similar approach is being used in our decentralized trust model. However, we do not consider misbehavior in the sense of uncooperative behavior, but rather as unreliable behavior in the sense of providing inaccurate information.

Liu et al. [24] introduce a distributed reputation-based trust model to detect threats and enhance the security of message routing in ad-hoc networks. The rating scheme [24] resembles our approach in the way trust reputation is calculated by averaging on reported ratings. However, their scheme uses discrete trust values, whereas our rating system is based on continuous trust scores in the real interval [0,1].

CONFIDANT [25] detects malicious nodes in ad-hoc networks by means of observation and reports about non-trustworthy nodes. The ability to make direct observations is assumed and although second-hand, i.e. word-of-mouth based, observations are used, only first-hand, i.e. direct, observations based ratings are reported. In our system, ratings that are solely based on rumor spreading may also be taken into account as long as their reliability has been assessed and weighted

Many reputation systems use complex probabilistic approaches (Bayesian systems) to compute reputation scores. Bayesian systems take binary ratings as input (i.e. positive or negative) and are based on computing reputation scores by statistical updating of beta probability density functions [26]. As opposed to these systems, our rating mechanism uses a simple weighted average approach to compute ratings.

Our rating mechanism bears resemblance to the approach proposed by Abdul-Rahman and Hailes [27]. Both systems use conditional transitivity of trust, i.e. trust is considered transitive only under certain conditions, for assessing trustworthiness by applying two distinct trust ratings: a direct trust rating and a recommender one. However, the reputation system in of Abdul-Rahman and Hailes [27] uses discrete trust values. In comparison, continuous rating values are applied in our system. Moreover, Abdul-Rahman and Hailes [27] provide no insight on how ratings are stored and communicated.

5 Decentralized Recommendation Scheme

In this section we present a distributed recommendation scheme to allow the assessment of trustworthiness in diverse groups of first responders, including civil engineers, where no past experience exist between participants from different groups. The main properties of a reputation-based trust system are the trust relationships between the entities of the system, the representation of trust, how trust reputation is built and updated, and how the recommendations of others are integrated. We study each of these properties in the following sections

Trust Relationships

A trust relationship exists between two nodes in the system if one of the nodes holds a belief about the other node's trustworthiness. However, the same belief in the reverse direction need not exist at the same time. In other words, trust relationship is unidirectional.

Two different types of relationships exist in our trust model. If node i has a belief about the reliability of node j, then there is a direct trust relationship. Direct trust relationship may rely on first hand information based on direct observations, on second hand information provided through recommendations or on a combination of both. If node i has a belief about the reliability of node j to give recommendations about other nodes' trustworthiness, then there is a recommender trust relationship. Recommender trust relationship is always based on first hand observations.

Trust relationships can be modeled as a trust graph. Nodes are represented by vertices, direct trust relationships by straight line edges and recommender trust relationships by dotted line edges on the trust graph. Since trust relationships are non-symmetrical, edges are directed. If a trust relationship of any type (direct or recommender) exists between two nodes in the network then the appropriate edge (straight or dotted line respectively) appears between the two corresponding vertices on the trust graph. An example from our disaster scenario is shown in Figure 1. The structural engineer and the police officer are each represented by a vertex on a trust graph. The structural engineer has a perception of trust for the police leader based on previous interactions between these two nodes and therefore a straight line edge, directed from the structural engineer to the police officer, connects their corresponding vertices on the trust graph. Moreover, we assume that from the past interactions the structural engineer also has knowledge on the trustworthiness of the police leader as a recommender. For this reason, a directed dotted line edge connects the two vertices on the trust graph that stand for the structural engineer and the police leader respectively.

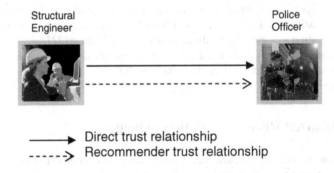

Fig. 1. An example of modeling trust relationships on a trust graph

Trust Transitivity

In our decentralized model, we consider conditional transitivity of trust similar to Abdul-Rahman and Hailes [27]. This means that transitivity is not absolute. It may hold only under certain conditions. In Figure 2, we consider one more trust relationship from our disaster scenario in addition to those demonstrated in Figure 1. Then, with reference to the examples in Figures 1 and 2, the conditions that allow transitivity in our model are:

a) The police leader explicitly communicates his perceived trust for the firefighter to the structural engineer, as a recommendation.
b) The structural engineer trusts the police officer as a recommender.
c) Trust is not absolute. The structural engineer may trust the firefighter more or less than the police leader does.

Structural   Police   Fire Fighter
Engineer   Officer

Fig. 2. An example of trust relationships

Trust Representation

Trust values in our system are in the [0,1] space for both direct and recommender trust relationships. Two different types of values are used in accordance to the different types of trust relationships: direct trust value is relevant to the direct trust relationship and recommender trust value is relevant to the recommender trust relationship. Direct trust value shows the probability θ with which node i thinks node j will be reliable. This outcome is drawn independently each time an observation is being made or a recommendation from some third node on the trustworthiness of j is received. Each independent outcome is then used to update the direct trust rating of i for j. In a similar way, the recommender trust rating shows the probability θ with which node i thinks node j will provide reliable recommendations. This outcome is also drawn independently based on direct observations after a recommendation is received. Recommender trust value is used to update the recommender trust rating.

Trust Ratings

In our model, node i maintains two ratings about another node j. The direct trust rating represents the opinion of i about j's reliability as an actor on the first response network. We represent the direct trust rating that node i has about node j as a variable $D_{i,j}$. The recommender trust rating represents the opinion of i about the trustworthiness of j as a recommender. We represent the recommender trust rating that node i has about node j as a variable $R_{i,j}$.

The direct trust rating of i for j is based either on i's observations of j's behavior or on recommendations received by i concerning the trustworthiness of j. A recommendation is the communicated direct trust rating of some other node for j. It has the form of a report containing a direct trust rating value for j. A combination of both direct observations and recommendations may also be used. Initially, the direct trust rating $D_{i,j}$ is set to NULL if no information exists for j, or else an initial direct trust value is assigned in order for the trust model to be initiated.

A high direct rating for j indicates that j behaves in a reliable manner in the first response system, while a low rating indicates misbehavior of j. Whenever i makes an observation of j's behavior or a recommendation is received, the direct trust rating $D_{i,j}$ gets updated.

The recommender trust rating of i for j is based on the directly observed reliability of j as a recommender in the trust model. Initially, there is no knowledge about j as a recommender. Therefore, the recommender trust rating $R_{i,j}$ is set to NULL. With repeated interaction, the subjective quality of a recommender's recommendations can be judged with increasingly greater accuracy since $R_{i,j}$ is updated each time j provides i with some recommendation on the trustworthiness of some third node.

If only one's own experience is considered in order to form attitudes concerning some other node in the network, then the reputation system is more likely to be reliable, since it will not be vulnerable to possible false ratings provided by others. However, in that case, the potential of learning from experience made by others goes unused [26]. Moreover, as shown earlier in this paper, at first the lack of personal knowledge about the interacting parties hinders the ability to engage in deliberate assessment, even if there is high motivation to do so. Therefore, a peripheral route attitude formation needs to be used. In other words, ratings from others need also to be taken into account. If the ratings given by others are considered, then the reputation system may be more vulnerable to false ratings, i.e. false praises or false accusations. However, since more information is available, the detection of behavior is faster. The goal is to make the trust model both robust and efficient. For this reason, direct trust ratings may be based on recommendations provided by other nodes in the system but the recommendation trust rating is being used to weight each of the recommendations based on personal experience only.

Direct trust rating calculation

A simple calculation of direct trust rating for j on node i could just take the average trust value of node j, based on recommendations concerning j from all nodes reporting to i as well as any possible direct observation of i for j's behavior. In that case, a direct observation is equivalent to a recommendation for j provided by i itself. However, such an approach would not make our model resistant to false recommendations, as described earlier, since all direct trust rating reports (i.e. recommendations) would be taken equally into account regardless of the reliability of the recommender.

Alternatively, we can infer the reliability of direct trust rating reports by using the reporting node's recommender trust rating as a means of determining the quality of the report. In other words, a recommendation for j received from a reporting node k with a high recommender trust rating $R_{i,k}$ by i would be weighted more than a reported trust rating from some other reporting node of i that has a lower recommender trust rating when calculating direct trust rating of i for j. Direct observations in that case are given the maximum weight of 1. In other words, we assume that for any node i, $R_{i,i} = 1$.

The direct trust rating of node i for node j will be the weighted average of all remembered reported direct trust ratings and/or first hand trust ratings for j that i receives or forms based on direct observations, respectively. We use the term remembered because memory constraints do not allow the maintenance of a complete set of ratings at each node. Therefore, at each point of time only the most recently received reported ratings are remembered. If N nodes report a direct trust rating for j and are remembered, then the direct trust rating of i for j will be

$$D_{i,j} = \frac{\sum_{t=1}^{N} R_{i,t} * D_{t,j}}{\sum_{t=1}^{N} R_{i,t}}.$$

We use 10 as the maximum value of N for validating our scheme later in this paper.

Figure 3 shows an example of how direct trust rating is calculated in the case of our disaster scenario. At some point of time t_1, the structural engineer who has knowledge only of the reliability of the police leader as an actor in the system, wants to know about the reliability of the firefighters' leader in order to make a trust-based decision on whether or not to adopt information regarding the state of a building that is being forwarded by the firefighter. The police leader and the rescuer, who are both neighbors of the structural engineer in the first response network and at time t_1 have a direct trust rating for the firefighter, provide recommendations to the structural engineer concerning the trustworthiness of the firefighter's leader. We assume that the structural engineer has already received other recommendations in the past (i.e., at some time before time t_1) by both the police leader and the rescuer and, therefore, he has, at time t_1, some recommender trust rating value assigned to them.

We see that the low direct trust rating of the police officer for the firefighter has a greater affect on the calculated direct trust rating of the structural engineer since the police officer, in this example, is more reliable as a recommender compared to the rescuer.

Recommender trust rating calculation

Each time node i receives a recommendation from some node k on the trustworthiness of node j, node i evaluates the quality of this trust report and updates the recommender trust rating for k. The evaluation of the reliability of the recommendation is based on the deviation of the reported direct trust rating from the weighted average of all other reported direct trust ratings that are remembered. In other words, if N nodes have already reported a direct trust rating for j at the time k reports one and this is still remembered by the system, the recommender trust rating of i for k will be:

$$R_{i,k} = 1 - \left| D_{k,j} - \frac{\sum_{t=1}^{N}(R_{i,t} * D_{t,j})}{\sum_{t=1}^{N} R_{i,t}} \right|.$$

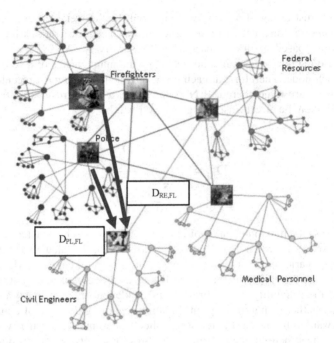

At time t1:
$D_{SE,FL}$ = NULL
$D_{PL,FL}$ = 0.5
$D_{RE,FL}$ = 0.9
$R_{SE,PL}$ = 0.9
$R_{SE,RE}$ = 0.4
Direct trust rating calculation $D_{SE,FL}$:
$D_{SE,FL}$ = (($R_{SE,PL}$ * $D_{PL,FL}$) + ($R_{SE,RE}$ * $D_{RE,FL}$)) / ($R_{SE,RE}$ + $R_{SE,PL}$) = ((0.9*0.5) + (0.4*0.9)) / (0.4+0.9) = (0.45 + 0.36) / 1.3 = 0.62

Fig. 3. An example of calculating direct trust ratings based on recommender trust ratings and recommendations

If the deviation is small then the evaluated recommender is considered trusted. Also, if node k provides node i with recommendations for nodes other than j as well, then the recommender trust rating of i for k will be calculated as the average deviation found for all the different nodes for which node k acts as a recommender.

Distributed Recommendation Reporting

Direct trust report distribution takes place in order for nodes to receive information about other nodes in the system in the form of recommendations since direct knowledge is not always possible. We use a simple approach to distribute trust reports.

At each point of time, every node p that wants to send some information is able to communicate with only a subset of nodes in the network, which could be either nodes that are within p's communication range at this point of time, and therefore can hear any message broadcasted by p, or nodes that p is aware of and can unicast information to, using the capabilities of the underlying ad-hoc routing protocol. We refer to all nodes that node p can communicate with at some point of time as the logical neighbors of p. The topology of the ad-hoc network is continuously changing over time as nodes are moving in and out of other nodes' communication range. Moreover, nodes fail making themselves inaccessible in the communication network. For the above reasons, the group of nodes that node p can access at each point of time is continuously changing. A group membership protocol allows us to model the availability of logical neighbors, called also group members of node p. To do so, we use a heartbeat-style group membership protocol.

In a heartbeat-style membership protocol, like the one proposed by Friedman and Tcharny for ad-hoc networks [28], each node p periodically multicasts a heartbeat message (incremented sequence numbers) to all other nodes in its group list, i.e. to all of its logical neighbors. These heartbeats are used to proactively learn about new prospective logical neighbors, as well as for failure detection. The latter is achieved by timing out on the time since the last heartbeat was received by node p from a neighbor q; this results in p deleting q from its membership list.

The basic heartbeat-style membership protocol has each node p periodically (a) increment its own heartbeat counter; (b) select some of its logical neighbors (defined by the membership list of p), and send to each of these a membership message containing its entire membership list, along with heartbeats. Each node receiving the message merges heartbeat values in the received message with its own membership list.

To support trust recommendations distribution, we piggyback on the membership protocol allowing each node p to periodically send trust recommendations along with the membership information. This is done by including the direct trust ratings for the group members, if available, together with the heartbeats when sending the membership list.

A node q that receives such information from p will use it not only to update its own membership list by merging the heartbeat values, but also to update its recommender trust rating for p as well as its direct trust ratings for all other nodes that a recommendation is included for in the reported membership list.

To update its recommender trust rating for p, q compares all its previous direct trust ratings with those reported by p and for any common entries in the two lists it calculates the deviation between the two trust values available for the same node. Doing so for all nodes for which both a direct trust rating existed before and a new one has been included by p in the reported message, it then calculates the average of deviations for all nodes as specified earlier in this section.

After updating the recommender trust rating for p, the receiving node q will then update its list of direct trust ratings. The new recommender trust rating for p will act as the weight for its reported recommendations to q. Each node keeps in memory only the last 10 recommended ratings received for some other node and only those ratings are taken into account when calculating the direct trust rating for this node. That means that if node p includes a rating for node r in its list sent to q and the heartbeat for r in this list is higher than any of the heartbeats in the last 10 reports received for r, then the recommendation of p for r, will be taken into account. The new direct trust rating for r will be calculated as the weighted average of the last 10 received ratings for r.

6 Trust-Based Information Dissemination

In the previous section, we described a decentralized recommendation system that allows nodes in the first response network to remove uncertainty and obtain knowledge on the reliability of other nodes through the distribution of word of mouth reputation ratings over the network.

Table 1. Pseudo code for our activation model

```
Trust-based Activation Model (on a non-initiator node)
Input: threshold (t), neighborhood list (L)

Initially:
activated = FALSE;   // am I activated?
activated_fraction = 0; // trust fraction of neighbors being activated
trust_rating_sum = 0; // sum of all direct trust ratings
    for each nodeID such that (find (nodeID,L) == TRUE) do
    // for each node in the neighborhood list do
    trust_rating_sum += direct_trust_rating(nodeID);
    // add the direct trust rating for it in the sum of ratings
    enddo
initiator = NULL;   // am I the initiator of contagion process?
Once every protocol period:
if (activated == FALSE)
    for each nodeID such that (find (nodeID,L) == TRUE) do
    // for each node in the neighborhood list do
      if activated(nodeID) == TRUE // if node is activated
      activated_fraction += direct_trust_rating(nodeID);
      // add the direct trust rating for it to the activated fraction
      endif
    enddo
if ((activated_fraction / trust_rating_sum) > t) activated = TRUE;
// if activated trust portion of nodes exceeds the threshold get activated
```

However, apart from the assessment of the reliability of the participants in a first response network, we also want our trust model to control information dissemination based on the reliability of information. Information reliability can be associated to that of the nodes that adopt it. The decision on whether or not to adopt the information will then be trust-based. In other words, a node would make a decision whether or not to adopt information and further disseminate it based on the trustworthiness of information, which is implied by the reputation of the nodes that have already adopted it. In that way, unreliable information would be filtered, allowing only trusted information to get propagated. This would cause information overload to be substantially reduced improving the efficiency of communication. We propose an activation model, similar to the one used by ants for division of labor, for efficient and reliable information spreading on top of the decentralized reputation scheme.

As seen earlier in this paper, under threatening situations ants secrete a specific pheromone to inform other ants. However, not all the ants react in the same way to the levels of pheromone perceived. Information or alarm spreading is a binary decision problem. An ant will react to an alarm pheromone (i.e. an ant will adopt an "alarm" behavioral pattern) based on the amount of pheromone present in the area where the ant is moving and its threshold level.

Assuming that the greater the amount of pheromone perceived, the more the ants that have adopted the information, we can translate the behavior of ants using a spreading activation model. In such a model, a node becomes activated, i.e. adopts the information, using some activation function, which is typically based on some threshold. This activation function in the case of nodes could be a simple fraction activation function that takes as parameters the number of neighbors of a node and their activation states. If, for some threshold fraction t, the number of active neighbors is t or more of a node's total number of neighbors, the node itself gets activated next. The threshold t is different for different nodes.

In the case of information dissemination in a first response network, the alarm is some critical information propagated over the network. Moreover, since we want the decision on whether or not to adopt the information to be trust-based, the amount of pheromone and the threshold level will be related to direct trust ratings. We slightly modify the fraction activation function in order to have direct trust ratings affect the activation decision. We call our modified activation function trust fraction activation function. This modification causes highly trusted nodes to be more influential in causing the activation of their neighbors. If, for some threshold fraction t, the number of activated neighbors multiplied by their respective direct trust ratings on the node that wants to make an activation decision is equal or greater than t of the sum of all direct trust ratings on the same node, then this node will get activated next. The threshold t is different for different nodes and is a function of the average of all direct trust ratings for a node's neighbors.

Tables 1 and 2 show pseudo code, and the analogy between ants' alarm activation and our trust-based information dissemination model, respectively.

Table 2. Analogy between ants' activation model and our trust activation model

Ants activation model	Trust-based activation model
Ants are represented by nodes in the network graph.	Each first responder who is equipped with an IT-based mobile device as well as each IT component that performs independently of a physical actor in the first response system is represented by a node in the network graph.
Pheromone paths are represented by graph edges, connecting the related ants.	Trust relationships between the first response network entities are represented by graph edges, connecting the related nodes.
Information spreading across the network is modeled as a Boolean state.	Information spreading across the network is modeled as a Boolean state.
A node/ant becomes activated, i.e. adopts the information/alarm, using fraction activation function, which is based on some threshold, different for each node/ant.	A node becomes activated, i.e. adopts the information, using trust fraction activation function, which is based on a direct trust ratings related threshold, different for each node.

7 Experimental Results

To evaluate the performance of our trust system, we develop a simulation environment in Visual C. Each actor in the simulated disaster network is represented by a node in the network used for our experiments. The connectivity of the network is modeled using the modified group membership protocol for mobile ad-hoc networks described earlier in this paper. Our simulation model performs in rounds of execution.

We consider networks of 100 or 1000 first responders, including civil engineers, which move randomly within the disaster area. The mobility of actors is handled by the membership protocol that updates the membership lists used from our trust model for communication. Our trust system runs independently of the movement of first responders in the disaster area.

We test the performance of our trust scheme as follows:

Detection of unreliability. We consider the case when a portion of the nodes in the system are misbehaving, i.e. behave in an unreliable manner, and we assume that another portion of nodes has initially classified them as not trustworthy. We simulate our trust scheme and measure how many rounds of the recommendation protocol are needed in order for all the nodes in the system to be able to classify the misbehaving nodes as not trustworthy. Figure 4 shows results for a network of 110 nodes when 10 nodes are misbehaving. The number of nodes that has originally detected them is 10. We notice that only a small number of rounds are needed in order for all nodes in the system to be able to classify the misbehaving nodes as unreliable.

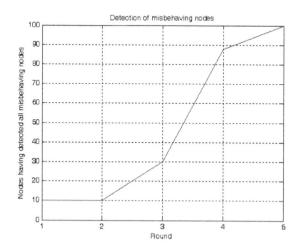

Fig. 4. Rounds of the recommendation protocol needed to have 10 misbehaving nodes detected in the whole network, starting from 10 spontaneous detections

Robustness. We consider the case when a portion of the nodes in the system are misbehaving and, in addition to this, another portion is lying about their trustworthiness, i.e. it provides false praises. Assuming that there are a number of nodes that have been able to detect misbehavior, we measure again how many rounds of the rating protocol are needed in order for all the nodes in the system to be able to classify the misbehaving nodes as not trustworthy. The performance of the trust system in that case would be an indication of its robustness to false rating and liar strategies. Figure 5 shows results for a scenario similar to that in figure 4 and for 10 nodes reporting false praises for the misbehaving nodes in the recommendation protocol. We see that our protocol is robust to such liar strategies because of the use of the recommender trust rating that gradually reduces the impact of unreliable recommenders in the system.

Trust-based information spreading to reduce information overload. To study the effect of trust in the spreading of information we consider a network of 1000 nodes, where 250 nodes are initially activated, i.e. spontaneously insert some information in the network. In figure 6, we measure the number of rounds needed for all the nodes to get activated when the trustworthiness characteristics of the initially activated group

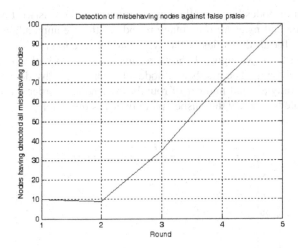

Fig. 5. Rounds of the recommendation protocol needed to have 10 misbehaving nodes detected in the whole network, starting from 10 spontaneous detections and having 10 nodes providing false ratings

of nodes change from a generally not trusted group with an average reputation of 0.25 to a generally trusted one with an average reputation of 0.75. It is clear that an activation that is initiated from more trusted nodes spreads faster in the whole network. Similarly, figure 7 shows the number of activated nodes after 1000 rounds for different average trust levels of an initially activated portion of nodes. This figure clearly indicates how our trust model reduces information overload in the system, since in cases of unreliable information it only spreads to a limited number of nodes.

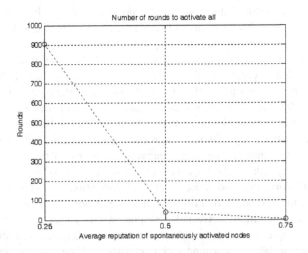

Fig. 6. Number of rounds required for all nodes to get activated

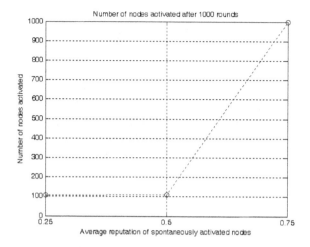

Fig. 7. Number of nodes activated after 1000 rounds

8 System Evaluation

To test the usefulness of our trust model we incorporate our proposed schemes in a prototype disaster support application and simulate a disaster scenario.

MASC: A prototype implementation

Our model is implemented as part of the Mobile Ad-Hoc Space for Collaboration (MASC) application [29]. MASC is an application, built using Microsoft Embedded Visual C++ for Windows and Windows CE, which runs on handheld devices enabled with short range wireless communication. It aims to provide robust and efficient collaboration among first responders, using a short range wireless communication platform. MASC provides all users that run the application with a shared view of the disaster area (Figure 8). Users may enter information into the system regarding the stability of the buildings in the disaster area. The stability information is represented by colored flags where each color stands for a stability state (stable, caution or unstable). According to the stability state of a building, as perceived by some mobile user running the MASC application, the building is marked with the corresponding colored flag and this information becomes available to all users in the communication range of the information provider, through the shared view. Apart from information regarding the stability of buildings, users may share pictures taken by using a digital camera attached to the short-range wireless enabled handheld device. To access a particular picture, a user should first click on the representation of the user that shares this picture on the ongoing shared view of the disaster area and then request the shared image by clicking on the appropriate button that pops-up. Clicking on the representation of a particular user on the shared view, allows also other users to access some profile information regarding this user.

Fig. 8. An example of a shared view in MASC application

We further extend the capabilities of the MASC application in order to provide data on the reliability of the information entered by users regarding the stability of buildings. The reliability of the information is evaluated based on the reliability of the nodes adapting the information. We call each action of marking the representation of a building on the shared view with a colored flag, in order to report on its stability state, a task. Therefore, the trustworthiness of the users adopting a task indicates the reliability of the information provided through this task. The trustworthiness of each user is provided by our trust model that runs under the application environment. Users periodically exchange recommendations on the trustworthiness of other users as described earlier in this paper. All recommendations are transferred as files (Figure 9) that contain the name of the user being evaluated and the corresponding direct trust rating being reported for this user. When users receive recommendations, they calculate their direct trust ratings for the users these recommendations are about; based on the direct trust ratings being reported and the recommender trust ratings they have for the users that provide the recommendations. When a user clicks on the representation of a task on the shared view, the trust-based activation model runs comparing the direct trust rating fraction with the user's trust threshold. The result of the algorithm's run is provided through a pop-up window that indicates whether or not the corresponding information should be trusted or not. The threshold used is user-specific.

```
<?xml version="1.0" encoding="UTF-8" ?>
- <Recommendation>
    <Recommender>Sergio</Recommender>
    <Posx>45</Posx>
    <Posy>88</Posy>
    <Recommended>Roberto</Recommended>
    <Dirtrust>5</Dirtrust>
  </Recommendation>
```

Fig. 9. An example of a trust recommendation in MASC application

The Evaluation Setting

In the previous section, we presented evaluation results from computer simulations conducted to test the effectiveness, efficiency and robustness of our proposed approach. In this section, we show results from a disaster scenario simulated to test the usability of our methods. For this reason, we use our prototype implementation, MASC, to support a real rescue exercise. The evaluation of our system was carried out in parallel to the development of a rope rescue exercise conducted by the Illinois Fire Service Institute (IFSI) of the University of Illinois at Urbana-Champaign at its training facilities.

The setting of the rope rescue exercise included one metal building, where rescue took place, a group of apprentice firefighters that were acting as rescuers, a group leader that participated in the system evaluation, and three students testing the system on site. They assumed the roles of rescuers, and structural expert.

Our prototype implementation, MASC, was installed on the different computing devices used in the exercise: 1 HP Jornada Pocket PC 568 using a Socket WLAN, 2 Compaq IPaq 3950 with a Compaq WL110 wireless card and 1 Compaq IPaq 5550 with an integrated wireless card. The network protocol used for communication was TCP/IP.

The Rope Rescue Exercise

The exercise began after digital information about the disaster area had been dispersed to all participants that were part of the system testing team, i.e. the rescue group leader, the student playing the role of the structural engineer and the other two students playing the roles of rescuers. During the two hour exercise the members of the MASC evaluation team that were not part of the actual rescue group that was being trained, dynamically moved and located according to the movement of the trained first responders without interfering with their activities, as shown in Figures 10 and 11.

The role of the structural engineer was to indicate the simulated stability of the disaster area, i.e. of the metal building where the rope rescue exercise was taking place. Information regarding the stability of the building was entered into the system by the structural engineer in the form of colored flags indicating the level of stability. This information was available in the shared view provided by the MASC application.

The role of one of the rescuers, rescuer A, was to act as a mobile field user and to take pictures of the building. Those pictures had to be shared with the structural engineer and other members of the testing team. Then, the structural engineer would use them in order to better evaluate the stability of the building. Specifically, the mobile field user took pictures of the physical infrastructure, using a digital camera attached to the short-range wireless enabled handheld device. As the pictures of the physical infrastructure were stored in MASC, they were transparently available for any actor using a handheld device running the MASC application. For instance, at any time during the exercise when the structural engineer is required to see the pictures of the building, taken by the field agent, the engineer would click on the representation of the field user, in the ongoing shared view of the scenario, provided by the application. By selecting one of the items in the popup list, the structural engineer would see the corresponding picture on his/her short-range wireless enabled PDA.

The second rescuer, rescuer B, was acting also as a mobile agent, but his/her role in the simulated scenario was to provide inaccurate, unreliable information in order to test the resistance of our trust model to such information.

Finally, the rescue group leader was using the MASC application in order to have access to structure information provided by the structural engineer so that he/she could direct his/her rescue group accordingly.

Fig. 10. The rescue group takes part in the exercise and a student located according to the movement of the group uses MASC application to validate our trust model

Fig. 11. The rescue group leader uses the MASC application to access structural information

An Illustrative Example

The rope rescue exercise was carried out in normal conditions supported by the MASC application and our trust system, incorporated in the application, was tested for the following:

Detection of unreliability. To test how our system reacts when actors using the MASC application provide unreliable information, we let rescuer B provide incorrect information about the stability of the building. For this test we have all four actors being part of the application testing team to be within each other's communication range. Therefore, the information entered by rescuer B would appear in the shared view of all four actors, including the structural engineer. We also let the structural engineer already have a good reputation for providing structural information. This reputation has been gained through interactions with the rescue group leader, to whom he provides information on stability of buildings, and the provision of recommendations from the group leader to the other actors forming the testing group. On the other hand, we assume that no information on the trustworthiness of rescuer B is available before he/she provides the information. Therefore, when the structural engineer accesses the information provided by rescuer B, which is incorrect, he/she first rates rescuer B low and sends recommendations through the trust reputation model, and second he/she provides the correct information about the stability of the building. Since, after those actions taken by the trust model, the reputation of the structural engineer will be high while that of rescuer B will be low, the information provided by the engineer will be the one adopted by the other actors, as required.

Robustness. To test the robustness of our system, we use a test setting similar to the one used for detection of unreliability, but instead of having rescuer B report false information on stability, we let him/her report a false accusation, i.e. a false low rating for the structural engineer. In other words, in this test we let rescuer B be unreliable not as an actor but as a recommender. The other actors that receive the false rating for the structural engineer from rescuer B will use it in order to evaluate rescuer B as a recommender. To do so, they are going to compare the reported rating with the direct trust rating they already have for the structural engineer. However, as we mentioned above, the reputation of the structural engineer is high and therefore the deviation between the old and the newly reported rating will also be high, causing rescuer B to be rated low as a recommender. Therefore, our trust system proves to be robust to false ratings.

Reducing information overload. After the rescuer B has been detected as not trustworthy any information on stability of buildings provided by him/her will be ignored by other actors as long as this information does not get adopted by other, preferably trustworthy, actors as well. It is clear that because of this feature of our trust system information overload is reduced, since unreliable information is filtered out. This factor, among other, proves the usability of our system in disaster scenarios, where information overload is one of the major concerns.

9   Conclusions

The vulnerability of urban areas to extreme events is one of the most vital problems confronting society today. Significant human and economical costs associated with

XEs emphasize the urgent need to improve the efficiency and effectiveness of first responses. Any attempt to effectively support communication between civil engineers and other first responders in disaster scenarios should among others provide trust management between actors. This paper proposes a distributed trust model that aims to establish trust among first responders. A reputation-based scheme, incorporated in a group membership protocol, is being used for this purpose. A nature-inspired activation mechanism is also proposed for trust-based information dissemination on top of the trust model. Experimental results indicate fast and robust establishment of trust and high resilience to the spread of unreliable information.

Simulations have significantly contributed to the construction decision making process by providing 'what-if' scenarios. Discrete Event Simulation (DES) has been one of the primary means of simulation, focusing on construction operational details. Considering the similarity between construction operations and queuing theory, DES, which is good at representing queuing theory, would be an appropriate method to represent construction operations.

References

1. New York Times, 9/11/2001, "The 9/11 Report", Web Page: http://www.nytimes.com/indexes/2001/09/11/
2. Newsweek, 1/4/05, "Tsunami Report", Web Page: http://www.msnbc.msn.com/id/6777595/site/newsweek/?ng=1
3. Mileti D., "Disasters by Design: A Reassessment of Natural Hazards in United States" Joseph Henry Press, Washington D.C, 1999.
4. Tierney K., Perry R. and Lindell M., "Facing the Unexpected: Disaster Preparedness and Response in the United States" The National Academies Press.
5. Columbia/Wharton Roundtable, "Risk Management Strategies in an Uncertain World" IBM Palisades Executive Conference Center, April 2002.
6. Godschalk D., "Urban Hazard Mitigation: Creating Resilient Cities." Natural Hazards Review, ASCE, August 2003, pp. 136-146.
7. Prieto R., "The 3Rs: Lessons Learned from September 11th" Royal Academy of Engineering, Chairman Emeritus of Parsons Brinckerhoff, Co-chair, New York City Partnership Infrastructure Task Force, October 2002.
8. National Science and Technology Council: Committee on the Environment and Natural Resources, "Reducing Disaster Vulnerability through Science and Technology" July 2003.
9. Quarantelli E.L, "Major Criteria for Judging Disaster Planning and Managing and Their Applicability in Developing Societies" Newark, Delaware: Disaster Research Center, University of Delaware.
10. Comfort, L., "Coordination in Complex Systems: Increasing Efficiency in Disaster Mitigation and Response". Annual Meeting of the American Political Science Association, San Francisco, USA, 2001.
11. FEMA, "Federal Response Plan Basic Plan" October 22, 2004, URL: http://www.fema.gov/rrr/frp/
12. Erik Auf der Heide, "Disaster Response – Principles of Preparation and Coordination", St. Louis, Mosby, 1989.
13. FEMA, "Federal Response Plan" Federal Emergency Management Agency, 9130.1-PL. April, 1999.
14. Shuster P., "The Disaster of Central Control" Complexity, Wiley, Vol. 9, No. 4, March-April 2004

15. MSCMC, NCSA, Multi-Sector Crisis Management Consortium, Web page: http://www.mscmc.org/, 2003.
16. Lee R. and Murphy J., "PSWN Program Continues to Provide Direct Assistance to States Working to Improve Public Safety Communications", Homeland Defense Journal, Vol. 1, Issue 22, December 2002.
17. L. Leonardi, M. Mamei, F. Zambonelli, "Co-Fields: A Unifying Approach to Swarm Intelligence", 3rd International Workshop on Engineering Societies in the Agents' World, Madrid, September 2002.
18. S. D' Silva, "Collective Decision Making in Honey Bees: Selection of Nectar Sources and Distribution of Nectar Foragers Through Self-organization", Spring 1998 Colloquium, Michigan State University, Department of Physics & Astronomy, March 1998.
19. W. Agassounon, "Distributed information retrieval and dissemination in swarm-based networks of mobile autonomous agents", IEEE Swarm Intelligence Symposium, Indianapolis, USA, April 2003.
20. D. Yingying; H. Yan, J. Jingping, "Multi-robot cooperation method based on the ant algorithm", IEEE Swarm Intelligence Symposium, Indianapolis, USA, April 2003.
21. D. J. Watts (2000) "A simple model of fads and cascading Failures" Working Papers 00-12-062, Santa Fe Institute, December 2000.
22. Y.T. Hung, A. Dennis, L. Robert, "Trust in Virtual Teams: Towards an Integrative Model of Trust Formation", Proceedings of the 37th Annual Hawaii International Conference on System Sciences (HICSS'04), Big Island, Hawaii, January 5 - 8, 2004.
23. S. Marti, T. Giuli, K. Lai and M. Baker, "Mitigating routing misbehavior in mobile ad hoc networks," Proceedings of The Sixth International Conference on Mobile Computing and Networking 2000, Boston, MA, August 2000.
24. Z. Liu, A. W. Joy, and R. A. Thompson, "A dynamic trust model for mobile ad-hoc networks". In 10th IEEE International Workshop on Future Trends of Distributed Computing Systems (FTDCS'04), pages 80–85, 2004.
25. S. Buchegger and J-Y. Boudec, "Performance analysis of the confidant protocol: Cooperation of nodes". In IEEE/ACM Symposium on Mobile Ad Hoc Networking and Computing (MobiHoc), Lausanne, June 2002.
26. S. Buchegger and J-Y Le Boudec, "A Robust Reputation System for P2P and Mobile Ad-hoc Networks", 2nd Workshop on the Economics of Peer-to-Peer Systems, Harvard, June 4-5, 2004.
27. A. Abdul-Rahman and S. Hailes, "A distributed trust model". In ACM New Security, 1997.
28. R. Friedman and G. Tcharny, "Evaluating Failure Detection in Mobile Ad-Hoc Networks", Technical Report, Technion, Israel Institute of Technology, CS-2003-06.
29. R. Aldunate, S. F. Ochoa, F. Pena Mora and M. Nussbaum, "Robot Mobile Ad-hoc Space for Collaboration to Support Disaster Relief Efforts Involving Critical Physical Infrastructure", J. Comp. in Civ. Engrg., Volume 20, Issue 1, pp. 13-27 (January/February 2006).
30. Peter J. Denning, "Hastily Formed Networks", Communications of the ACM, Volume 49, Issue 4, April 2006.

MGA – A Mathematical Approach to Generate Design Alternatives

Prakash Kripakaran[1] and Abhinav Gupta[2]

[1] Ecole Polytechnique Fédérale de Lausanne, Switzerland
prakash.kripakaran@epfl.ch
[2] North Carolina State University, USA
agupta1@ncsu.edu

Abstract. Optimization methods are typically proposed to find a single solution that is optimal with respect to the modeled objectives and costs. In practice, however, this solution is not the best suited for design as mathematical models seldom include all the costs and objectives. This paper presents a technique - Modeling to Generate Alternatives (MGA), that instead uses optimization to generate good design alternatives, which the designer may explore with respect to the unmodeled factors. The generated alternatives are close to the optimal solution in objective space but are distant from it in decision space. An application of this technique to design of moment-resisting steel frames is illustrated.

1 Introduction

Computational approaches based on optimization are increasingly used to assist engineers in the design process. Optimization methods [1,2] that attempt to find the best solution for a specific mathematical model of the actual design problem assume that all the costs and objectives in the real problem are included in the mathematical model. However, designers are seldom able to model all the factors in a practical design problem due to the following reasons:

- Presence of certain objectives that cannot be numerically defined. e.g., aesthetics.
- Difficulty in quantifying the relative importance of different objectives in a multi-objective optimization problem. For example, an engineer may prefer to have supports at certain locations on a pipe in a support optimization problem. However, quantifying this preference relative to minimizing the total cost of the supports may be difficult.

Moreover, optimization models typically involve simplifications in the cost model. Complexities in the calculation of the costs of certain aspects of the problem are often left out from the optimization formulation. For instance, the cost of a rigid connection in a steel frame design problem may be specified as a constant value. In practice, the cost depends on various factors such as the type of the connection, the member type for the beam and the column, and the weld length. Due to these reasons, the solution obtained from optimization, while possibly good, is seldom the best solution for design.

I.F.C. Smith (Ed.): EG-ICE 2006, LNAI 4200, pp. 408–415, 2006.

This drawback is not a limitation of optimization but is a problem that arises from the manner in which optimization is employed for design. It can be overcome if optimization is instead used to generate good design alternatives. The designer can explore these alternatives with respect to the unmodeled factors and then choose one for final design. The designer may also wish to tinker with some of these alternatives and generate a better alternative.

Over the last decade, researchers have attempted to address these issues using multi-objective optimization (MOO) approaches [3]. Particularly in the domain of evolutionary computing, numerous methods [4,5] have been developed to generate the pareto fronts for MOO problems . The solutions on the pareto front can form a set of alternatives that the designer may explore with respect to the unmodeled factors. MOO approaches have been successfully applied to various problems such as conceptual design of buildings [6]. However, MOO approaches have some drawbacks. They tend to be computationally expensive. Also, their performance often depends on appropriately setting the values of certain parameters such as fitness sharing and niching parameters in multi-objective genetic algorithms. The engineer, who may not be familiar with the intricacies of the optimization algorithm, may experience difficulty in finding the correct values for these parameters.

This paper presents Modeling to Generate Alternatives (MGA), an approach to generate good design alternatives with single-objective optimization techniques. MGA aims to generate solutions that are close to the "optimal" solution in the objective space but are distant from it in the decision space. The technique is compatible with any single-objective optimization method. The parameters that need to be adjusted by the designer in MGA involve the primary objective in the design problem. In this paper, this technique is illustrated for the design of moment-resisting steel frames. MGA is used to identify alternatives with different locations of beam-column rigid connections in the frame. The alternatives are compared with the optimal solution with respect to different parameters like wind load capacity and the location of rigid connections. The alternatives are observed to form a good set of solutions that show better performance than the optimal solution with respect to the unmodeled factors. Professional knowledge and judgment of the designer are required to choose the design for implementation.

2   Structural Optimization

Mathematical models of real-world problems often involve some degree of approximation in the costs and the objectives. Consequently, the solution generated using traditional single-objective optimization techniques may not perform satisfactorily with respect to the unmodeled parameters. For illustration, consider the following formulation of a design optimization problem:

$$\text{Maximize } z = x + 2y, \text{ subject to } x + y \leq 30; \ y \leq 20; \ x, y \geq 0 \qquad (1)$$

The decision space for the problem is shown as the shaded region in Fig. 1a. Mathematical optimization correctly produces z = 50 at point A as the solution.

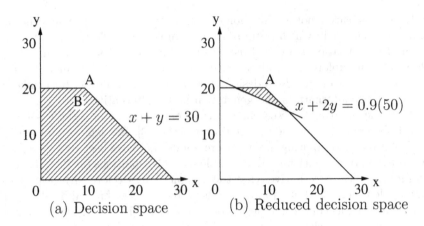

Fig. 1. Illustration of MGA

Thus, if all parameters of the real problem are present in this formulation, the optimal design would be $(x, y) = (10, 20)$. However, the premise is that there may be features that are not completely captured by the model. When those issues are considered, point A may be less desirable overall than a point, call it B, originally deemed inferior (to the optimal solution) by the model. The issues that are involved in finding B and establishing the computer assistance needed in this process are discussed below.

- Generating all feasible solutions suggests that one has no confidence in the model - one would expect that point B optimizes the objective function nearly as well as A, so only these good solutions need to be examined. This idea is illustrated in Fig. 1b, where only those solutions that are within 10% of the optimal are retained.
- Available solutions should represent a cross-section of good solutions, so that, if B is not actually among them, perhaps one of them is close enough from which to begin "tinkering."
- Since a decision maker can reasonably consider only a small number of designs, a subset of these good solutions should be presented for inspection.

In the following section, a formal description of the technique - Modeling to Generate Alternative (MGA), and its application to generate alternatives for a generic optimization problem are given.

3 Modeling to Generate Alternatives (MGA)

MGA [7,8] is an extension of single and multiple-objective mathematical programming techniques with an emphasis on generating a set of alternatives that are "good" but "as different as possible." By generating these good yet different solutions, a decision maker can explore alternatives that satisfy unmodeled objectives to varying degrees. These alternatives may be assessed by the decision maker either subjectively or quantitatively. MGA requires the designer to specify a metric, δ_{ij} to measure the distance between solutions i and j. For problems with real valued decision variables, δ can be the euclidean distance between the solutions.

For the purpose of illustration, assume that the objective function for a minimization problem is represented by Z. Let solution a be evaluated as the optimal solution using an appropriate optimization method. $Z(a)$ represents the value of the objective function at solution a. To generate alternatives, a distance metric δ is defined. A new optimization formulation is created to evaluate the solution b that maximizes δ_{ab}. The formulation also imposes an upper-bound constraint on $Z(b)$. This formulation is shown below.

$$\text{Find } b \text{ that Maximizes } Y = \delta_{ab} \text{ s.t. } Z(b) \leq kZ(a) \tag{2}$$

k is a constant with a value that is greater than 1. If the original optimization formulation Z minimizes cost, then k specifies an upper-bound on the cost for the alternative. For obtaining an alternative with a cost that is utmost 20% greater than that of the optimal solution, $k = 1.2$. Note that for a maximization problem, the constraint is specified as $Z(b) \geq kZ(a)$ and the value for k is between 0 and 1. For the problem given in Figure 1a, the difference metric δ can be defined as the euclidean distance between the optimal solution $a = (10, 20)$ and the alternative $b = (x, y)$ in the decision space, i.e., $\delta_{ab} = \sqrt{(x - 10)^2 + (y - 20)^2}$. Additionally, the following lower-bound constraint - $x + 2y \geq 0.9(50)$, is imposed to ensure that the alternative is close to the optimal solution in objective space.

To generate additional alternatives, the objective function is modified to maximize the distance of the new solution from all previously-evaluated alternatives. Such a formulation can be formally specified as follows.

$$\text{Find } c \text{ that Maximizes } Y = \sum_{j=1}^{n} \delta_{cj} \text{ s.t. } Z(c) \leq kZ(a) \tag{3}$$

n is the total number of previously-evaluated alternatives including the optimal solution. The same optimization method that is used for solving the original formulation can also be used to generate the alternatives.

4 Design of Moment-Resisting Steel Frames

In this section, MGA is illustrated for the design of moment-resisting steel frames. Rigid beam-column connections have significantly higher labor costs associated with their fabrication and erection than hinged connections. Computational approaches [9] that combine optimization with MGA are used to identify the optimal locations of rigid connections in a frame. This paper focuses on the MGA techniques used in the study.

4.1 Problem Description

Let R represent the total number of rigid beam-column connection locations in the frame. Let c_i represent the decision variable corresponding to the presence or absence of a rigid connection at location i, $i = 1..R$. $c_i = 0$ represents the presence of a hinged connection at location i and $c_i = 1$ represents the presence of a rigid connection at location i. Then, the binary string $< c_1, c_2, c_3, c_4...c_R >$ represents the decision variables that correspond to the various beam-column connections in the frame. The total cost C of the frame is expressed as,

$$\text{Total cost, } C = C_s \sum_{j=1}^{m} w_j(p_j)l_j + C_r \sum_{i=1}^{R} c_i \tag{4}$$

p_j represents the product type of member j in the frame. For a given set of c_i, p_j are evaluated using a heuristic algorithm. m is the total number of members in the frame. w_j is the weight in tonnes per unit length of the product p_j. l_j is the length of member j. C_s is the cost per tonne of steel. C_r is the cost of a single rigid connection. In order to determine C_r, detailed discussions were held with practicing engineers, steel fabricators and erectors. The actual cost of a connection depends upon various factors such as the type of connections, amount of welding, web stiffening and doubler plates. However, a fixed cost of $900 per connection is adopted. This value is representative of the average value for fabricating a rigid connection in the state of North Carolina. Furthermore, $C_s = \$600$ per tonne of steel is assumed.

The constraints for the design problem are prescribed by the strength and serviceability requirements specified in the Manual of Steel Construction, Load and Resistance Factor Design [10]. These constraints are not described here for brevity. Readers are referred to [9] for a complete description of the problem and the optimization approach.

4.2   Optimization Formulation

The objective function for the problem is given as

$$\text{Minimize } Z = C + aP_s + bP_d \tag{5}$$

P_s and P_d correspond to the violations of the strength and serviceability constraints respectively. a and b are penalty factors. A Genetic Algorithm-based optimization approach [9] is used to perform a trade-off study between the number of rigid connections in the frame r_{req} and its total cost C. The trade-off curve is generated by obtaining the optimal cost solution for different specified number of rigid connections. The trade-off curve illustrates that the cost of the structure is minimum for $r_{req} = 10$ and it gradually increases for values lesser or greater than this number. The engineer may want to explore alternatives to the least cost solution ($r_{req} = 10$) as well as to the solutions for $r_{req} = 8$ and 12. In the following sections, the application of MGA to generate the alternatives is illustrated and the alternatives generated for $r_{req} = 12$ are examined.

4.3   MGA for Moment Frame Design

The "Hamming distance," which is a metric used in binary computation, is used in Equation 2 as the difference metric for the MGA. If two solution strings are represented as $< c_{1a}, c_{2a}, .., c_{Ra} >$ and $< c_{1b}, c_{2b}, \ldots, c_{Rb} >$, then the distance between the two solutions δ_{ab} is given as follows.

$$\delta_{ab} = \sum_{i=1}^{R} |c_{ia} - c_{ib}| \tag{6}$$

The Genetic Algorithm, which is used to optimize the objective function given by Equation 5, is also used to identify alternatives by optimizing the objective function Y given in Equation 3. The value of k is taken to be 1.1, i.e., the search is for alternatives whose cost does not exceed the cost of the optimal solution by more than 10%. Using this formulation, alternatives are generated for $r_{req} = 12$, i.e., frames with exactly 12 rigid connections. These alternatives are evaluated on the basis of the following criteria - (1) preferences for certain locations to place rigid connections, and (2) margins against excessive lateral loads.

4.4   Results and Discussion

Although many alternatives can be generated, in this paper we compare only two alternatives with the corresponding optimal solution on the trade-off curve for discussion purposes. The optimal cost solution as well as the alternatives - $MGA1$ and $MGA2$, for $r_{req} = 12$ are given in Figure 2. The figure also gives the costs of these solutions. The locations of the rigid beam-column connections are indicated by the small circles. The two alternatives along with the cost-optimal solution constitute a small solution set which the designer can explore with respect to the unmodeled objectives. All three solutions have significantly different locations for rigid connections. While the model assumes that all the rigid connections have the same cost, this is not entirely true in practice. The connection cost varies depending on the products used for the corresponding beam and column which in turn determine the required weld specifications, fabrication and stiffness requirements. In the optimal solution, the largest beam with a rigid connection uses W18 × 35. In $MGA1$, the largest beam with a rigid connection is member 19 with W24 × 68. In $MGA2$, the largest beam with a rigid connection uses a W24 × 62. Also, $MGA1$ has most of its rigid connections on the innermost bay while $MGA2$ and the optimal solution have the rigid connections primarily in the outermost bays. Since rigid connections require on-site welding, the engineer may consider construction aspects to determine if one alternative is superior to another. Specifically, engineers may prefer to have the rigid connections on the inner bays in a space-limited urban setting.

Another parameter that an engineer may consider to evaluate the relative quality of the alternatives is associated with the ability of the frames to withstand increased lateral loads. While the generated solutions satisfy the strength constraint, it is possible that certain solutions have greater margins than others. An elegant way to compare the alternatives on this basis is by gradually increasing the wind load and identifying the wind load at which the different design alternatives fail. It is observed that the optimal solution fails when the wind load is increased by 95%. On the other hand, the alternatives $MGA1$ and $MGA2$ fail

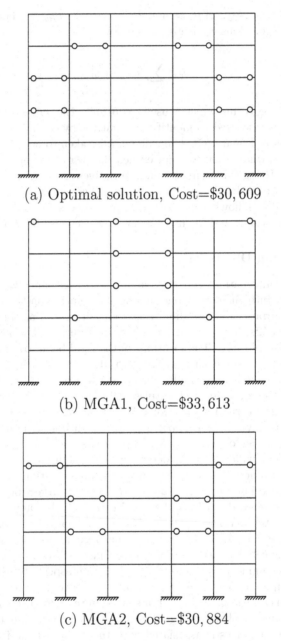

(a) Optimal solution, Cost=\$30, 609

(b) MGA1, Cost=\$33, 613

(c) MGA2, Cost=\$30, 884

Fig. 2. Optimal solution and alternatives for $r_{req} = 12$

when the wind load is increased by 123% and 117%, respectively. The optimal solution, $MGA1$, and $MGA2$ have costs of \$30, 609, \$33, 613 and \$30, 884 respectively. Expertise and judgment can be used to evaluate the alternative that is best suited for final design.

5 Summary

This paper presents a mathematical technique - Modeling to Generate Alternatives (MGA), to generate design alternatives. MGA requires an appropriate distance metric to be identified before it can be applied to a particular design problem. The technique works by modifying the objective function in the original optimization formulation to maximize the difference metric. The technique is generic and can be used in conjunction with other optimization methods. The alternatives generated using MGA form a good subset of solutions that the designer can explore to identify the final design for implementation. This paper illustrated the method for locating rigid beam-column connections in the design of moment-resisting steel frames.

References

1. Deb, K., Gulati, S.: Design of truss-structures for minimum weight using genetic algorithms. Finite Elements in Analysis and Design **37** (2001) 447–465
2. Hasancebi, O., Erbatur, F.: Layout optimization of trusses using simulated annealing. Advances in Engineering Software **33** (2002) 681–696
3. Shea, K.: An approach to multiobjective optimisation for parametric synthesis. In: Proceedings of International Conference on Engineering Design, Glasgow. (2001)
4. Kicinger, R., Arciszewski, T., De Jong, K.D.: Evolutionary computation and structural design: A survey of the state-of-the-art. Computers and Structures **83** (2005) 1943–1978
5. Marler, R.T., Arora, J.S.: Survey of multi-objective optimization methods for engineering. Structural and Multidisciplinary Optimization **26** (2004) 369–395
6. Grierson, D.E., Khajehpour, S.: Method for conceptual design applied to office buildings. Journal of Computing in Civil Engineering **16** (2002) 83–103
7. Baugh Jr., J.W., Caldwell, S.C., Brill Jr., E.D.: A mathematical programming approach to generate alternatives in discrete structural optimization. Engineering Optimization **28** (1997) 1–31
8. Gupta, A., Kripakaran, P., Mahinthakumar, G., Baugh Jr., J.W.: Genetic Algorithm-based decision support for optimizing seismic response of piping systems. Journal of Structural Engineering **131** (2005) 389–398
9. Kripakaran, P.: Computational approaches for decision support in structural design and performance evaluation. PhD thesis, North Carolina State University (2005)
10. AISC: Manual of Steel Construction - Load Resistance Factor Design. 3rd edn. AISC (2001)

Assessing the Quality of Mappings Between Semantic Resources in Construction

Celson Lima[1], Catarina Ferreira da Silva[1,2], and João Paulo Pimentão[3]

[1] CSTB, Centre Scientifique et Technique du Bâtiment, Route des Lucioles BP209,
Sophia-Antipolis cedex, 06904 France
{c.lima, catarina.ferreira-da-silva}@cstb.fr
[2] LIRIS, Laboratoire d'InfoRmatique en Image et Systèmes d'information,
Université Claude Bernard Lyon 1, 43, boulevard du 11 novembre 1918,
F-69622 Villeurbanne, France
[3] UNINOVA, Institute for the Development of New Technologies, Quinta da Torre,
Caparica, 2825-114, Portugal
pim@uninova.pt

Abstract. This paper discusses how to map between Semantic Resources (SRs) specifically created to represent knowledge in the Construction Sector and how to measure and assess the quality of such mappings. In particular results from the FUNSIEC project are presented, which investigated the feasibility of establishing semantic mappings among Construction-oriented SRs. The paper points to the next lines of inquiry to extend such work. In FUNSIEC, a 'Semantic Infrastructure' was built using SRs that were semantically mapped among them. After quite positive results from FUNSIEC, the obvious questions arose: how good are the mappings? Can we trust them? Can we use them? This paper presents FUNSIEC research (approach, methodology, and results) and the main directions of investigation to support its continuation, which is based on the application of fuzzy logics to qualify the mappings produced.

1 Rationale: Why FUNSIEC?

The second generation of the WWW is emerging based on the addition of "meaning" to data and information, provided by the development of new semantic-oriented tools and resources. Web services are now gaining a semantic layer that allows the development of 'web service crawlers' capable of understanding what exactly a given web service does. Semantics is undoubtedly the cornerstone of the whole evolving web.

The first generation of the web was essentially focused on the creation and publication of content with humans as the main consumers. Subsequently, the immense sea of digital content available became attractive enough to be exploited by automatic tools. This requires (and is essentially based on) the formal definition of meaning and its respective association with the information published on the web. The work carried out by the Semantic Web group has prepared the ground on this subject. We are getting closer to the futuristic vision of the Web's creator, Tim Berners-Lee [4].

I.F.C. Smith (Ed.): EG-ICE 2006, LNAI 4200, pp. 416–427, 2006.

Experts say that the use of Semantic Resources[1] (SRs) can contribute to transforming that vision into reality. Formalisms are required to support the definition of the real meaning of web *resources* (both information and services) allowing them to be published and to be precisely understood by other agents (more specifically, software agents). This is where semantic interoperability then comes into play. Semantic interoperability enables systems to process information produced by other applications in a meaningful way (i.e. in isolation or combined with their own information) As such, it represents an important requirement for improving communication and productivity.

The European Construction sector has been offered several results produced by international initiatives at standardisation level related to interoperability and semantic matters. In order to move next steps in these directions, the FUNSIEC project aimed at evaluating the feasibility of creating an Open Semantic Infrastructure for the European Construction Sector (OSIECS). Such an infrastructure was to be built by selecting publicly available, dedicated construction semantic resources available from results produced by international initiatives and European funded projects.

Essentially, FUNSIEC looked for an answer to the following question: is it possible to establish (semantic) mappings between SRs tailored to construction needs? The driving quest of FUNSIEC was to know if it would be possible to use in an integrated way (some of) the SRs already available to Construction. Additionally, FUNSIEC aimed to enhance the semantic interoperability of those SRs.

FUNSIEC, supported by its own methodology, designed and partially implemented the OSIECS Kernel, which is essentially a human-centred tool to produce the OSIECS meta-model and the OSIECS model. Through them, it is possible to evaluate the establishment of mappings among SRs, either in a purely research-oriented perspective or in an (embryonic) business-oriented way.

This paper is structured as follows. Section 2 presents FUNSIEC goals and methodology. Section 3 describes the FUNSIEC results. Section 4 discusses the application of fuzzy logics to qualify the mappings. Section 5 briefly summarises the related work. Section 6 draws some conclusions and points out future work and expectations.

2 FUNSIEC Scenario and Methodology

The core subject in FUNSIEC work was semantic. Semantic Resources are available in many forms and flavours even though they are still used (and exploited) in a very embryonic level. The European Construction sector is not an exception, and it has been offered several results produced by international initiatives at standardisation level (e.g. CEN eConstruction workshop, IFC model, International Framework Dictionary, LexiCon, Barbi, bcXML language, e-COGNOS ontology, etc.).

Taking these results into account, the FUNSIEC project analysed the feasibility of building an Open Semantic Infrastructure for the European Construction Sector (OSIECS). Such an infrastructure was to be built by selecting semantic resources

---

[1] Term coined in the SPICE project to refer to controlled vocabularies, taxonomies, ontologies, etc.

devoted to construction, so exploiting public results produced by prominent international initiatives and European funded projects. A methodological approach was devised to support this work. The innovation of OSIECS is on the semantic mappings established among the existing semantic resources.

It is worth emphasizing that in the context of FUNSIEC, semantic mapping tackles the relations between ontological entities (e.g. concepts, relations, axioms) from two ontologies representing the same domain. Several methods can be used to find mappings, such as terminological, structural, extensional (i.e. based on instances) or semantic methods. Those methods come from different disciplines such as data analysis, machine-learning, language engineering, statistics or knowledge representation. Their applicability depends on the features (e.g. labels, structures, instances, semantics) to be compared. They also depend on the expected type of results. Some of them were used in FUNSIEC.

2.1 The FUNSIEC Scenario

The scenario used to support the development of OSIECS (figure 1) is formed by three domains of construction activity (scenarios), namely IFC, eConstruct, and e-COGNOS. The IFC scenario is basically focused on the exchange of design-related information. It is an extension of CAD drawing exchange (that primarily focuses on exchange of geometry) into semantic representation of construction objects and relationships.

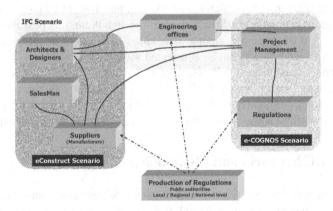

Fig. 1. The FUNSIEC Scenario

The eConstruct scenario is focused on e-procurement of construction products. It shows the use of electronic catalogues to support both design and sales process. So the designer wants to try different products in his project and the salesman uses the catalogue to show different alternatives to his clients.

The e-COGNOS scenario focuses on Knowledge Management (KM) practice related to ontologies for regulations (Lima et al 2003b). It relies on the e-COGNOS Knowledge Management Infrastructure (e-CKMI) as well as on the e-COGNOS

Ontology Server (e-COSer). For instance, the project manager feeds the system with knowledge about regulations, in this case the *url* of regulatory bodies. During a project, he is informed about the publication of new regulations and then he uses the e-CKMI to verify if his projects have to be changed according to the new regulations regarding accessibility matters for disabled people.

2.2 The FUNSIEC Methodology

Literature presents a long list of methodologies developed in the field of ontology engineering, including methodologies for building ontology, ontology reengineering, ontology learning, ontology evaluation, ontology evolution, and techniques for ontology mapping, merging, and alignment [5] [6]. It is generally accepted that a consensual methodology is difficult to establish due to the lack of maturity of the field and the difficulty to develop a methodology adaptable to different applications, sectors and settings [1].

The choice of the "proper" methodology to be used is very much dependent on the nature and characteristics of the targeted domain and its various applications. FUNSIEC developed its own methodology[2] which used the strengths of several established methodologies. In brief, it comprises the following phases: (*i*) *Domain Scoping*: characterisation of the domains covered by OSIECS; (*ii*) *SRS Identification*: SRs used to form OSIECS are identified and selected based on the analysis of features relevant to the FUNSIEC context; (iii) *Conversion and Similarities*: handles syntax-related problems as well as semantic heterogeneity and detection of correspondences among SRs (the ultimate result of this phase is the OSIECS Kernel); (*iv*) *OSIECS Meta-model and Model*: mapping tables produced by OSIECS Kernel representing the *meta-level* and the *level* itself; (*v*) *Testing & Validation*: assessment of the OSIECS *Triad* (Kernel, Meta-model and Model); and (*vi*) *Maintenance*: it is about correcting and updating OSIECS during its working life, which includes the work reported here.

3   The FUNSIEC Results

The FUNSIEC methodology is the first output of the project. The OSIECS *Triad (i.e.,* the OSIECS Kernel, the OSIECS meta-model, and the OSIECS model) are the major results of the application of the methodology.

3.1 The FUNSIEC Kernel

The OSIECS Kernel produces the OSIECS meta-model and OSIECS model. The Kernel is composed by the Syntax Converter, the Semantic Analyser, the Converter, the Detector of Similarities, and the Validator. Experts are required to provide the right inputs to the OSIECS Kernel and, as such, make the best use of it.

The formalism adopted to represent both the OSIECS meta-model and model is the OWL language, for two important reasons: *i*) its rich expressiveness; and *ii*) the

---

[2] For more information on FUNSIEC methodology, see [1].

explicitness of OWL semantic representation that enables implementation of automatic reasoning. Additionally, OWL is the ontology representation format promoted and recommended by the Semantic Web group.

The operation of the OSIECS Kernel is shown in figure 2. This is the transformation process leading to the creation of both OSIECS meta-model/model. Experts are invited to verify the results produced by the Syntactic Converter and the Semantic Analyser, as well as to help validating the lists produced by the Detector of Similarities (the FUNONDIL system).

The whole process is performed in a semi-automatic way. Both the Syntactic converter & Semantic Analyser work co-operatively to create the 'production rules' (for each language used to represent SRs) to guide the Converter. The Converter is software based on the JavaCC compiler which takes the production rules and generates the Transformers. Transformers are software tools used to convert SRs (frm their native language) into OWL. The Detector of Similarities (based on Description Logics) is another software tool which compares the SRs and identifies the existing mappings among them. Finally, the Validator is used by the experts to assess and validate the quality of the mappings created.

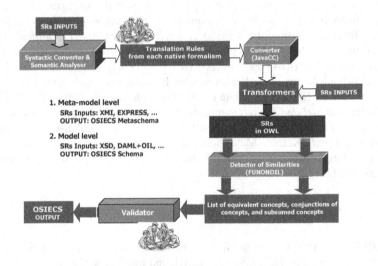

Fig. 2. Generating OSIECS meta-model and model

The OSIECS meta-model and OSIECS model are inter-connected indirectly via the SRs they represent. Their usage depends on the level of representation required when dealing with SRs. The OSIECS meta-model is used by the creators of semantic resources where different sources have to be combined and mappings made between them. For instance, the work currently conducted by TNO, concerning the development of the New Generation IFC, is a potential candidate to take advantage of the OSIECS meta-model since it already maps the IFC (Kernel only) and the ISO 12006-3 meta-schemas [2]. Shortly, both OSIECS meta-model and model are sets of tables respectively mapping meta-schemas and schemas of the SRs forming OSIECS.

3.2 OSIECS Meta-model and Model

Mapping semantic entities, even if they are formally defined, is a considerable challenge involving several aspects. Quite often it requires identifying pair-wise similaritybetween entities and computing the best match for them [9]. There are many different possible ways to compute such similarity with various methods designed in the context of data analysis, machine learning, language engineering, statistics, or knowledge representation. FUNSIEC relied on semantic methods [8] to deal with the problem. The basic assumption behind semantic methods is that they aim at discovering relations between (pairs of) entities belonging to different schemata based on the meaning of those two entities. The following SRs were used as input to OSIECS: *bcBuildingDefinitions* taxonomy, e-COGNOS ontology, ISO 12006, and the IFC model. The first step in the development of OSIECS is mapping their meta-schemas. More information on this is available in [7].

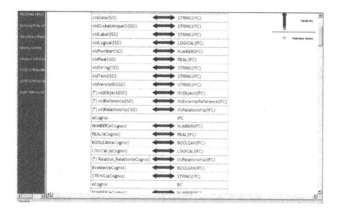

Fig. 3. Partial representation of OSIECS meta-model showing equivalence mappings between ISO 12006-3 and IFC kernel; and between e-COGNOS and IFC

The OSIECS Kernel uses the 'reasoning services' of FUNONDIL to determine and identify semantic correspondences, i.e., the relations between pair of entities belonging to different SRs. The FUNONDIL inference engine uses two ontologies as input (O and O') and a set of axioms (A), producing a set of inter-ontology axioms (A') that represents the mappings.

Three types of mappings are considered, namely *equivalence, subsumption* and *conjunction. Equivalence* means that the concept A is 100% equivalent to the concept B, considering the semantic expressed in each SR. *Subsumption* has a rank relation that defines the relation subconcept → superconcept between concept A and concept B. The *conjunction mappings* are the result of the mappings obtained in the previous stage.

The mapping search is performed between each pair of SRs producing semantic correspondences considered equivalents and non-equivalents. The former refers to absolute equivalences among the entities mapped. The latter refers to mappings in which only a part of the concepts of the SRs is common. This is the case of subsumption and conjunction.

For illustrative purposes only, bcXML to bcXML were mapped in order to help assess the operation of the OSIECS Kernel. As expected, to map a SR to itself produces equivalences (and only equivalences) between the same concepts. In addition, results for subsumption and conjunction are also presented, but this means only redundant information, because if $A \sqsubseteq B$ and $B \sqsubseteq A$ then A is equivalent to B. This exercise helped us to be sure that the mapping process was working properly.

4   Quantifying FUNSIEC Mappings Using Fuzzy Logics

FUNSIEC work was very human-centred in the sense that the participation of experts was essential to guarantee the quality of the results. The reason is the very essence of the work: semantics. The SRs currently available were not developed in the context of the need to establish mappings with other SRs. Only the experts know the meaning of things. This scenario has been slowly changed with the advent of the semantic web and the related elements (standards, tools, etc.). Ontologies written in one standard format (OWL) are likely to be more easily mapped among themselves.

As previously explained, FUNSIEC relied on *semantic methods* to tackle the semantic heterogeneity problem. It is worth noting that these methods being semantically exact do only provide an absolute degree of similarity for entities considered equivalent. Therefore, the continuation of FUNSIEC work depends on the quality of the mappings produced, which needs to be measured.

Part of the problem relies on the way of defining the quality of the mapping. How can we say that the "quality" of something is between 0 (bad) and 1 (perfect)? Fuzzy Logic theory [3] provides a qualitative approach to this inherently vague idea. In fact, instead of relying exclusively on quantitative approaches, Fuzzy Logic represents these concepts using *linguistic variables* whose values are *terms* that represent the concept (e.g. *bad, acceptable, good, and excellent*). These terms are then mapped onto Fuzzy Sets that are extensions to the classic sets theory where the membership function can allow values between 0 and 1, thus denoting a degree of membership instead of the biblical dichotomy of '*true* or *false*'.

4.1   Modelling Information for Quantifying Mappings Using Fuzzy Logic

A mapping is a binary relation between a C_1 concept of a SR and a C_2 concept of another SR. In a non-equivalent mapping, the C_1 concept is not 100% equivalent to the C_2 concept. We define a linguistic variable *non-equivalent mappings* (shortly *nem*) associated to the set of terms $D_{(nem)} = \{$*non-acceptable, acceptable, good, strong*$\}$.

In order to model membership functions we define three input variables (E_1, E_2, E_3) that are intended to represent the commonalities between two mapped concepts. E_1 is related to the property (object property or data type property considering the OWL notation) and the respective range type. In other words, we define it as the number of shared properties. E_2 is the number of 'lexical entries' shared by the two concepts.Lexical entries [10] are terms equivalents to a given concept which are used to enrich ontological concepts. They can be used, for instance, to provide a long list of terms that can be used to refer to a single concept (e.g. the concept *Actor* could be referred to by employee, person, driver, engineer, etc.).

E_3 captures the similarity between concept annotations[3]. An annotation contains natural language terms and expressions, which means that (part of) the annotation content can be labelled or tagged as an expression representing a rich semantic content. Let n be the number of terms in the expression e, n is called the order. By extension a term is an expression of order $n = 1$. It is clear that n cannot be a meaningless term. The meaningless terms are *the, to, of, for*, etc., also called 'stop-list'.

The input variables are *fuzzified* with four linguistic terms: *non-acceptable, acceptable, good*, and *strong*. If the definition of two mapped concepts, C_1 and C_2, do not share a property then the similarity related to the two concepts properties is *non-acceptable*. If C_1 and C_2 share one or two properties then the similarity related to the two concepts properties is *acceptable*. If C_1 and C_2 share two to five properties then the similarity related to the two concepts properties is *good*. Finally, if C_1 and C_2 share more than four properties then the similarity related to the two concepts properties is *strong*. A similar argument is applied to the E_1 and E_2 input variables. For instance, if C_1 and C_2 share two to six 'lexical entries', then the similarity of the two concepts regarding their 'lexical entries' is *good*. If C_1 and C_2 associated annotations share more than six expressions then the similarity between both concept annotations is *strong*.

For illustrative purposes only, figure 4 depicts the membership functions for E_3. For instance, if $E_3 = 3$, i.e., C_1 and C_2 annotations share a term and an expression of order 2, then the similarity between C_1 and C_2 annotations is *acceptable* to a degree of membership of 0.66 and is *good* to a degree of membership of 0.33. Based on that conclusion, the similarity between a pair of concepts is qualified, which allows to infer how good the mappings are.

Table 1. Summary of the assignment of the fuzzy linguistic terms, where z is *integer*

		Linguistic terms			
		Non-acceptable	*Acceptable*	*Good*	*Strong*
Input variables	E_1	0	*{1,2}*	*{2,...,5}*	*{4,...,z}*
	E_2	0	*{1,2,3}*	*{2,...,6}*	*{4,...,z}*
	E_3	0	*{1,...,4}*	*{3,...,7}*	*{6,...,z}*

Table 1 provides an example of definition of the fuzzy linguistic terms. It shows intervals where the number of properties (for E_1), of lexical entries (for E_2), and of similar terms in the annotations (for E_3) define the *class* they belong to. The numbers filling in the table are expected to be provided by the experts validating the mapping process. Based on this table, validation rules are created and can be automatically applied to assess the quality of the mappings produced.

---

[3] Concept annotation here is defined following OWL context, meaning comments, free text associated to a given concept. This annotation can also hold the *definition of a concept in natural language*.

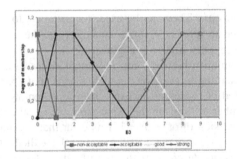

Fig. 4. Fuzzy membership function of E3

4.2 The FUNSIEC Vision – The Whole Picture

The FUNSIEC vision is shown in figure 5. It includes the OSIECS triad together with the SRs and the respective tools used to manage them, namely the eConstruct tools (bcXB, RS/SCS, and TS), the IFC tools (IFCViewer and IFCEngine), the e-COGNOS tools (e-CKMI and e-COSer), and the LexiCon Explorer.

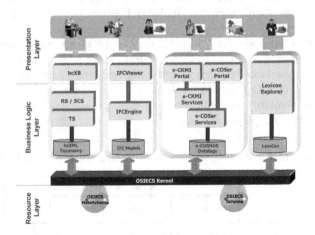

Fig. 5. The FUNSIEC Vision

The vision is that the OSIECS Kernel, supported by both OSIECS Model and Meta-model, acts as a bridge between the different tools providing richer possibilities of using the SRs in a transparent way. For instance, an expert looking for knowledge (using the e-COGNOS tools) about problems related to the fire resistance of a given product can, at the same time, find the information about alternative products and their suppliers, prices, etc., using the eConstruct tools in a totally open way. The OSIECS Kernel is responsible for translating the need of the expert in the respective bcXML query, sending it to the bcXML server and getting back the right answers. Another example is for a designer developing a CAD drawing (IFC compliant) and, at the same time, needing to know about the regulations that must be followed in his/her project. In this case, OSIECS Kernel provides the link between the IFC tools and the e-COGNOS tools.

5 Related Works

The techniques of mapping[4] are used to facilitate the interoperability among heterogeneous Semantic Resources. Currently, mapping of ontologies is used in several fields ranging from machine learning, concept lattices and formal theories to heuristics, database schema and linguistics.

In the literature we can find either similar or very different approaches to the one adopted in FUNSIEC. For a good source on related work, please see [11]. Briefly four are referred to here:

- Su [13] approached ontology mapping using mapping methods based on extension analysis. The mapping discovery approach is based on ontological instances, supported by text categorization and Information Retrieval techniques. It creates a "feature vector" for instances of concepts and assigns a 'similarity value' for each pair of concepts. The process is completed by experts that accept/reject the mappings.
- MAFRA is a framework for distributed ontologies in the Semantic Web [11], where ontologies to be mapped are normalised to a uniform representation – in their case RDF(S) – thus avoiding syntax differences and making semantic differences between the source and the target ontology more apparent.
- The Anchor-PROMPT [12] is an ontology merging and alignment tool with a sophisticated prompt mechanism for possible matching terms. Its alignment algorithm uses two ontologies and a set of anchored-pairs of related terms[5]. The alignment produced is then refined based on the ontology structures and users feedback.
- The work presented by Garcia et al. [14] is focused on the early phase of the design process using extreme collaboration in an environment similar to the used by NASA. It includes the creation of a Product-Organisation-Process ontology (using an Excel spreadsheet) in a collaborative process involving the team of designers. They argue that large ontologies (in their view IFC is an example) are not that useful and, as such, small and really common ontologies can help to reduce significantly the time required to complete the design process.

As might be expected, FUNSIEC has similarities and differences when compared to other works. For instance, it shares the use of IR techniques and need for approval from end users considered in [13] and has used the same approach as in MAFRA regarding the normalisation of ontologies. FUNSIEC is different when compared with IF-MAP because we believe that the previous agreement advocated is currently far from reality, since organisations use what they have to hand. Regarding ONION,

---

[4] There is also a multitude of terms expressing similar works in this area, such as *mapping, alignment, merging, articulation, fusion, integration,* and *morphism.*

[5] These are identified using string-based techniques or defined by the user.

FUNSIEC does not agree with the assertion that ontology merging is inefficient, costly and not scalable. Indeed, the continuation of FUNSIEC tackles efficiency from a quality perspective. Good mappings are likely to be useful whilst bad ones are to be useless.

6 Conclusions and Future Work

This paper presents the results achieved by the FUNSIEC project and describes the application of fuzzy logic to assess the quality of the mappings offered by both O-SIECS meta-model/model. These are created by the OSIECS Kernel, which is a semi-automatic software tool to transform SRs from their native language into OWL and map them. The FUNSIEC quest was about answering the question of *how feasible it is to establish semantic mappings among semantic Construction-oriented resources?*

After getting a positive answer, the next step was the qualification of the mappings produced by OSIECS kernel, which has been initially tackled with the application of fuzzy logics to evaluate the quality of OSIECS mappings. This is done taking into account the "commonalities" found between the mapped concepts as well as the semantics associated to them. FUNSIEC is exploring a way of modelling fuzzy membership functions and of assigning membership based on the available information about the mapped concepts. The preliminary results allow us to say how good the mappings are, from a conceptual point of view. Only real cases will ratify these results or prove that we cannot rely on them.

FUNSIEC intends to define appropriate rules to support the reasoning process of a fuzzy inference engine. The next task is to define an appropriate *defuzzification* method in order to obtain quantified mappings, targeting the implementation, evaluation and assessment of this approach in a real scenario from Construction sector. The conceptual part of this work has been carried out in the context of a Ph.D. thesis that will be finished by the end of 2006.

References

1. Barresi, S., Rezgui, Y., Lima, C., and Meziane, F. Architecture to Support Semantic Resources Interoperability. In Proceedings of the ACM workshop on Interoperability of Heterogeneous Information Systems (IHIS05), Germany, ACM Press, November 2005, page 79-82.
2. Lima, C.; Ferreira da Silva, C.; Sousa P.; Pimentão, J. P.; Le-Duc, C.; Interoperability among Semantic Resources in Construction: Is it Feasible?, CIB-W78 Conference, Dresden, Germany, July 2005.
3. Zadeh, L.A., Fuzzy Sets. Information and Control, 1965. 8: p. 338-353.
4. Berners-Lee, T., Hendler, J., Lassila, O.: The Semantic Web: A new form of Web content that is meaningful to computers will unleash a revolution of new possibilities. Scientific American, May (2001).
5. Corcho, O., Fernando-Lopez M., and Gomez-Perez A. Methodologies, tools and languages for building ontologies. Where is their meeting point?ï *Data and Knowledge Engineering 46(2003)*ï ï 41-64.ï

6. Fernandez-Lopez, M. Overview of methodologies for building ontologies. In *Proceedings of the IJCAI-99 workshop on ontologies and problem-solving methods (KRR5)*, Stockholm, August 2, 1999.

7. Lima, C., Storer, G., Zarli, A., Ferreira da Silva, C. (2005). "Towards a framework for managing standards-base semantic e-Resources in the European Construction Industry." Construction Research Congress 2005, ASCE, Chicago, EUA (vol and page numbers).

8. Benerecetti, M., Bouquet, P. & Zanobini, S. 2004. Soundness of Semantic Methods for Schema Matching. In P. Bouquet & L. Serafini (eds), *Workshop on Meaning Coordination and Negotiation (MCN-04) at the 3rd International Semantic Web Conference; Working notes, Hiroshima, Japan, 8 November 2004.*

9. Euzenat J., Le Bach T., Barrasa J., Bouquet P., De Bo, J., Dieng R. et al.: D2.2.3: State of the art on ontology alignment – Knowledge Web project, realizing the semantic web, IST-2004-507482 Programme of the Commission of the European Communities (2004)

10. Lima, C. P., Fiès, B., Lefrancois, G., Diraby, T. E. (2003). The challenge of using a domain Ontology in KM solutions: the e-COGNOS experience. In: 10TH ISPE 2003, Funchal, Por-tugal. International Conference on Concurrent Engineering: Research and Applications, p. 771-778.

11. Kalfoglou, Y. and Schorlemmer, M. (2003). Ontology mapping: the state of the art. The Knowledge Engineering Review, 18(1):1–31, 2003.

12. Noy, N. and Musen, M. (2001) Anchor-PROMPT: Using non-local context for semantic matching. In Proc. IJCAI 2001 workshop on ontology and information sharing, Seattle (WA US), pages 63–70, 2001. http://sunsite.informatik.rwthaachen. de/Publications/ CEUR-WS/Vol-47/.

13. Xiaomeng, S., (2003). Improving Semantic Interoperability through Analysis of Model Extension. In Proc. of CAiSE'03 Doctoral Consortium Velden, Austria, 2003. http:// www.vf-utwente. nl/~ xsu/paper/docConstCRD.pdg.

14. Garcia, A. C. B., Kunz, J., Ekstrom, M. and Kiviniemi, A., *Building a project ontology with extreme collaboration and virtual design and construction*, Advanced Engineering Informatics, Vol. 18, No 2, 2004, pages 71-85.

Knowledge Discovery in Bridge Monitoring Data: A Soft Computing Approach

Peer Lubasch, Martina Schnellenbach-Held,
Mark Freischlad, and Wilhelm Buschmeyer

University of Duisburg-Essen, Institute of Structural Concrete,
Universitätsstraße 15, 45141 Essen, Germany
{peer.lubasch, m.schnellenbach-held, mark.freischlad,
wilhelm.buschmeyer}@uni-due.de

Abstract. Road and motorway traffic has increased dramatically in Europe within the last decades. Apart from a disproportionate enlargement of the total number of heavy goods vehicles, overloaded vehicles are observed frequently. The knowledge about actual traffic loads including gross vehicle weights and axle loads as well as their probability of occurrence is of particular concern for authorities to ensure durability and security of the road network's structures.

The paper presents in detail an evolutionary algorithm based data mining approach to determine gross vehicle weights and vehicle velocities from bridge measurement data. The analysis of huge amounts of data is performed in time steps by considering data of a corresponding time interval. For every time interval a population of vehicle combinations is optimized. Within this optimization process knowledge gained in the preceding time interval is incorporated. In this way, continuously measured data can be analyzed and an adequate accuracy of approximation is achieved. Single vehicles are identified in measured data, which may result from one or multiple vehicles on the bridge at a given point of time.

1 Introduction

Traffic loads are usually obtained from so-called weigh-in-motion (WIM) systems [1]. For these systems two types can be distinguished: Pavement and bridge systems.

In the case of pavement systems, weighing sensors are embedded in or mounted on the pavement. To achieve high accuracy and partly overcome dynamic effects induced by crossing vehicles to single sensors, multiple-sensor WIM (MS-WIM) arrays were developed. The accuracy class of A(5) according to the COST323 specifications [2] could be achieved, using a high number of up to 16 sensors [3]. Sensor noise and inaccuracy [3] as well as sensor stability in calibration and operation [4] are remaining critical problems within the application of this type of systems. For maintenance and installation traffic lanes have to be blocked. This may lead to inconvenience on the part of the road users.

Bridge WIM (B-WIM) systems use an instrumented bridge as measuring device. By means of appropriate algorithms deformations recorded at well chosen locations

I.F.C. Smith (Ed.): EG-ICE 2006, LNAI 4200, pp. 428–436, 2006.

are backward analyzed to identify single vehicles and the associated gross vehicle weight, vehicle velocity, axle spacings and axle loadings. These systems have proven to reach high accuracy: Accuracy class of A(5) could be obtained for bridges with very smooth pavement [5] in combination with sophisticated algorithms to analyze measured data. Since the systems are invisible on the road surface, drivers of heavy goods vehicles can hardly avoid crossing the weighing scale. In consequence the total traffic flow is recorded. Moreover, compared to pavement based systems the system's durability is increased whereas costs for installation and maintenance are reduced.

B-WIM systems were initially studied in the USA by the Federal Highway Administration (FHWA) in the 1970's. A first approach based on the evaluation of influence lines was introduced in 1979 [6]. An approach making use of techniques from the field of soft computing can be found in [7]: Given measured data as input, a first artificial neural network (ANN) is used to obtain the type of a vehicle passing over the bridge. Subsequently a second, vehicle type specific ANN is used to determine the vehicle's velocity, axle loads and spacings. For this procedure, ANNs need to be formulated for all kinds of possible vehicles. A major drawback of the system is its limitation to the evaluation of single vehicle events.

In the following an evolutionary algorithm based approach for the acquisition of knowledge about traffic (i.e. vehicles and associated attributes of interest) from bridge monitoring data will be introduced in detail. Single vehicles are identified from data recorded during presence of one or multiple vehicles on the bridge at a given point of time.

The overall knowledge discovery process consists of three consecutive steps [8]: (i) data pre-processing, (ii) data mining and (iii) data post-processing. The step of data pre-processing addresses methods of data preparation. Within the framework of the presented approach, measured data is pre-processed by a digital filtering to reduce noise as well as dynamic effects and data is normalized to eliminate thermal influences. The data post-processing step covers all aspects of how the gained knowledge is eventually treated. Gained knowledge (i.e. identified gross vehicle weights) is used for studies on changes of traffic loads and composition. The approach presented in this paper covers the data mining step of the knowledge discovery process.

2 Bridge Monitoring

The Institute of Structural Concrete Essen performed long-term measurements for more than twelve months at the superstructure of a post-tensioned concrete bridge according to figure 1. The box girder bridge was built in the seventies of the last century and consists of two independent superstructures having two lanes each. Figure 1 also shows the calculated influence line for a 5 axle articulated vehicle of a gross weight of 40 t for an underlying linear temperature difference. Changes of strains ($\Delta\varepsilon_p$), concrete and ambient air temperature (T_c, T_a) were measured at several points of the cross section. Furthermore, displacement sensors (Δw_c) were attached to the top plate as also presented in [9] and [10].

Having more than one vehicle on the bridge the vehicles' influence lines superpose and the combination of single vehicles is recorded. Therefore, measured data may contain single vehicles or combinations of vehicles.

Possible recordings for two vehicles following each other in short distances are demonstrated in figure 2. For the j th vehicle the variable $t_{0,j}$ denotes the time of occurrence, v_j the velocity and G_j the gross weight. By means of the illustrated superposition of two influence lines shall be demonstrated that the identification of single vehicles from the recording is a non-trivial task and analytical methods are not sufficient.

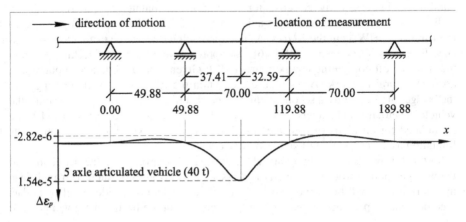

Fig. 1. Location of measuring and static influence line

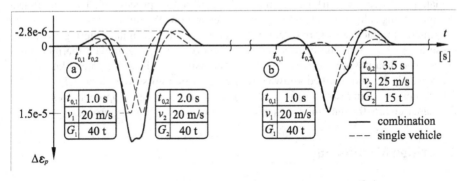

Fig. 2. Superposition of vehicle influence lines

The appearance of a single vehicle varies with its gross weight and its velocity, the type of vehicle and the level of basic stress. The gross weight scales the recording over the $\Delta\varepsilon_p$-axis whereas the vehicle's velocity is reflected in the duration of recording since measurement is performed over time. Axle configurations and spacings as well as the distribution of axle loads determine the static influence line. The level of basic stress describes physical non-linearities of the cross section and is caused by permanent and temperature actions and in the case of prestressing the level of basic stress is also dependent on the height of initial strain (also see [10]).

3 Data Analysis

The analysis of measured data is performed by the evaluation of time intervals of length $\bar{t}$ and in steps of $\Delta t \ll \bar{t}$. The i th interval begins at $t_{\text{Beg}(i)}$ and ends at $t_{\text{End}(i)}$ (see figure 3). For effectiveness, periods without any traffic are detected and excluded from the process of optimization.

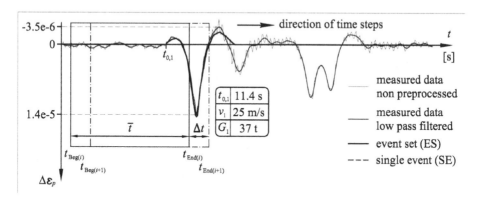

Fig. 3. Time intervals and time steps

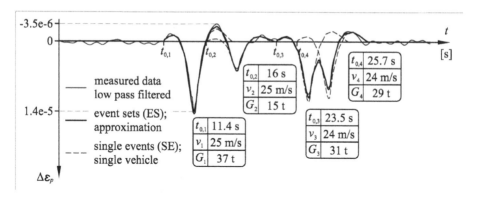

Fig. 4. Event sets (ES) and single events (SE)

For each time interval the data is analyzed by means of an evolutionary algorithm. Within the overall procedure, the traffic situation to be investigated is represented in an object-oriented manner: Event sets (ES) represent combinations of vehicles and are set up of single events (SE), which describe real vehicles. A SE is completely declared by its time of occurrence t_0, its velocity v, its gross weight G and the type of vehicle S (e.g. 3 axle rigid vehicle, 5 axle articulated vehicle among others). ES are made up of one or several SE and are used to approximate measured data within the optimization. Figure 4 shows four SE setting up ES.

As a very important element ES and associated SE from the former time interval are taken into account for the initialization of the current populations (SE and ES) as can be seen from figure 5. Since steps of $\Delta t \ll \bar{t}$ are carried out, the environment changes from the previous time interval i-1 to the current time interval i only slightly. The time intervals always overlap. Thus, the situation to be investigated does not alter significantly by performing a time step. The use of ES and their SE from the former time interval to build up the current populations leads to a dynamical adaptation of the algorithm to the optimization task. This way, knowledge gained from the former time step is not rejected but transferred to the time interval of consideration towards an improved optimization.

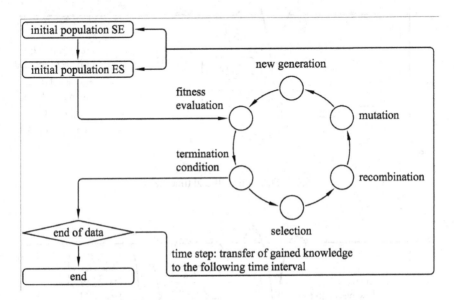

Fig. 5. Approach's scheme

The fittest ES of the last generation of the i th time interval is referred to as $ES_{win(i)}$. A SE being part of $ES_{win(i)}$ for a predefined number of time steps is assumed to describe a real vehicle and gets assigned to the database containing the discovered single vehicles. Such a SE was optimized within several evolutionary algorithm runs and thus holds a high accuracy of approximation.

3.1 Initialization

For initialization of the SE- as well as the ES-population the specialties of knowledge transfer to the current time interval i have to be taken into account. If a former time interval i-1 and a winner $ES_{win(i-1)}$ exist, the two populations are partly initialized on the basis of the $ES_{win(i-1)}$ and its SE_j. The index j refers to the j th SE of an ES.

Since ES are made up of SE, the ES-population is generated after initialization of the SE-population (also see figure 5). Consequently a SE can belong to several ES.

For the initialization of the SE-population four basic types of SE can be distinguished:

SE_1: Elitism: Every SE_j of $ES_{win(i-1)}$ is kept unchanged.

SE_2: On the basis of the SE_j of $ES_{win(i-1)}$ new SE are generated for the analysis of the current time interval i. The elements t_0, v, G and S of a new SE are designed by choosing one SE_j of $ES_{win(i-1)}$ and applying the mutation-operator to its elements. By this procedure, the elements of the SE_j of $ES_{win(i-1)}$ serve as a starting point and are changed slightly for the generation of a new SE to stay close to the former –good– solution.

SE_3: Initialization of SE, which are compatible to the SE_j of $ES_{win(i-1)}$ but occur to a later point of time. These SE are necessary to cover vehicles that follow and occur in the current time interval i.

SE_4: Random initialization of SE featuring preferably a t_0 in the last third of the current time interval i. For this initialization the gaussian function $z(t)$ as defined in figure 6 is utilized. By means of a value $\mu = t_{End(i)} - 1/6 \cdot \bar{t}$, which is adapted every time interval, and a predefined value σ, influence is taken on the initialization of the element t_0.

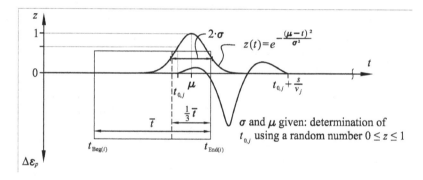

Fig. 6. Random initialization of SE

For initialization of ES one or several SE are combined. The selection of SE to form an ES requires a check of compatibility to assure that the SE fit to each other. Two SE, which are very similar in terms of their time of occurrence t_0 and velocity v are unsuitable for combination. Furthermore, the fitness of SE is considered for the initialization of ES. SE comprising a good fitness are chosen more frequently to form ES.

For the initialization of the ES-population four variants can be noted:

ES_1: Elitism: A certain number of $ES_{win(i-1)}$ including their SE is kept unchanged.

ES_2: By principal use of the elements from the group of SE_2 $ES_{(i)}$ are generated having the same number of SE as $ES_{win(i-1)}$. By this means, these $ES_{(i)}$ differ only slightly from $ES_{win(i-1)}$.

ES_3: Initialization of $ES_{(i)}$ being similar to $ES_{win(i-1)}$ but including one additional SE. These $ES_{(i)}$ are mainly generated by use of SE_3 to cover vehicles that follow and occur in the current time interval i.

ES_4: Pure random initialization of $ES_{(i)}$.

3.2 Evolutionary Operators

ES are set up of SE; whereas SE, which represent real vehicles, are made up of the code elements t_0, v, G and S. Following this architecture the evolutionary operators are either applied on the level of the ES or on the level of the SE. The application of an operator on the ES [SE] causes changes of the ES [SE] itself or its SE [code elements t_0, v, G and/or S]. As usual, before applying evolutionary operations a selection is performed: Parents ES_{par1} and ES_{par2} are selected according to their particular fitness value f for generating an offspring ES_{off}. Subsequently, the two genetic operators recombination and mutation are applied distinguishing between ES and SE.

Recombination: This operator is applied to an ES with an adaptive probability of $p_{1,rcb}$. If the operator is not applied, the prime parent ES_{par1} is directly transferred for mutation as ES_{off}. In case of recombination every $SE_{par1,j}$ of ES_{par1} is checked for application of this operator ($p_{2,rcb}$). The recombination of a selected $SE_{par1,j}$ is either performed in the level of SE or on the SE's code elements ($p_{3,rcb}$). The direct recombination of a SE signifies either the replacement of the chosen $SE_{par1,j}$ by a compatible $SE_{par2,j}$ or adding a $SE_{par2,j}$ to ES_{par1} to form ES_{off}. For recombination of the chosen $SE_{par1,j}$'s elements a compatible $SE_{par2,j}$ is selected. Subsequently an interchange of single elements t_0, v, G and S between the chosen $SE_{par1,j}$ and $SE_{par2,j}$ is carried out.

Mutation: An adaptive probability $p_{1,mut}$ decides about mutation of every single ES_{off}'s SE. Similar to the recombination, this operator is either directly applied to the chosen offspring's SE or the SE's code elements ($p_{2,mut}$). In case of directly mutating the SE it is either replaced by a compatible SE from the SE-population, removed from ES_{off} or as a third variant a compatible SE from the SE-population is added to ES_{off}. Mutating the SE's code elements signifies changes in either t_0, v, G or S.

Adaptation: Within the first generations the recombination according to $p_{1,rcb}$ is performed frequently on quite huge populations. Simultaneously the probability $p_{1,mut}$ is held little in order not to vary the individuals too much. After a certain number of generations the population sizes as well as the probability of recombination are reduced drastically. At this point mutation is raised to be carried out intensely with little variations on already quite good solutions.

3.3 Fitness Evaluation

Since discrete time intervals are considered for data analysis the fitness evaluation within the evolutionary algorithm is always carried out on the current time interval i (see figure 5). The fitness is calculated for every ES and subsequently assigned to the SE assembling the ES. Due to the definition of the fitness function, the individual's fitness is to be minimized for a better solution. This signifies that a SE being part of several ES – as basically possible from the initialization – is to be assigned the maximal fitness of its ES (worst fitness possible).

Primarily the fitness is determined by rating the ES's approximation performance of measured data. For this purpose the normalized mean squared error *NMSE* according to equation (1) is calculated for every ES.

Besides this most important fitness measure additional criteria are covered within the fitness evaluation: A SE, which could contribute to a better approximation of ES but features an improper time of occurrence t_0, shall not be rejected by assigning a bad fitness value. Thus, for the ES and its SE a certain neighborhood surrounding every discrete value of real data is examined for a better approximation. In the case of making use of such a neighborhood-value a worse fitness is assigned depending on the distance of the used value to the real data value. Within the following evolutionary operations the SE still comprises a more or less good fitness value, can be modified towards a better approximation and may be selected for initialization of ES.

During the evolution process the t_0 of SE are changed. This may lead to ES consisting of SE (representing vehicles) with unrealistic distances among them. To overcome this inconsistency, the fitness value of such ES is modified by means of a soft constraint based penalty function.

$$NMSE = \frac{\frac{1}{M} \cdot \sum_{m=1}^{M} \left(\Delta\varepsilon_{p,a,m} - \Delta\varepsilon_{p,r,m} \right)^2}{\left(\max\left[\Delta\varepsilon_{p,a,m} ; \Delta\varepsilon_{p,r,m} \right]_{m=1}^{M} - \min\left[\Delta\varepsilon_{p,a,m} ; \Delta\varepsilon_{p,r,m} \right]_{m=1}^{M} \right)^2} \qquad (1)$$

M total number of discrete data points of consideration
m discrete data point
$\Delta\varepsilon_{p,a,m}$ strain approximation, event sets
$\Delta\varepsilon_{p,r,m}$ real values of strain, measured data

4 Conclusions and Future Work

The extraction of compact and useful knowledge from vast amounts of data often exceeds the possibilities of traditional data analysis methods. Techniques from the field of knowledge discovery as well as soft computing methods are increasingly allowing for a better data analysis. In this paper an evolutionary algorithm based approach, which can be applied for the data mining step of an overall knowledge discovery process in measurement data, was presented. By means of the developed algorithm gross vehicle weights and vehicle velocities can be concluded from measured changes of strains without any installations on the top of the bridge or on the pavement.

The approach was implemented in C++ and a repetitive evaluation of data recorded during 10 test drives with a vehicle of known loads and geometry has shown that the approach is suitable for accurate results: Over 51 runs of the algorithm for the aforementioned test vehicle's gross weight of $G = 27.7$ t a minimum error of -5.7 % and a maximum error of 3.9 % could be obtained for the total number of 510 samples. The mean standard deviation for the gross vehicle weight and the mentioned 51 samples was calculated to be 1.2 t.

As a further development artificial neural networks (ANN) are currently tested to identify axle locations from displacement measurements (Δw_c) being carried out on the top plate. This procedure seems very promising for the analysis of smeared signals from light axles (especially from light axle groups) to locate the single axles. Eventually ANN will complement the overall approach towards more accurate results as well as the additional information about axle configurations and axle loads.

Additional repetitive test drives with two different test vehicles were carried for future accuracy evaluation. A classification according to the COST323 specifications [2] is intended.

References

1. WAVE: Weigh-in-Motion of Axles and Vehicles for Europe. General Report of the 4th FP Transport, RTD project, RO-96-SC, 403, ed. B. Jacob, LCPC, Paris. (2001)
2. COST 323: European Specification on Weigh-in-Motion of Road Vehicles. EUCO-COST/323/8/99, LCPC Paris (1999)
3. Jacob B., O'Brien E.J.: Weigh-in-Motion: Recent Developments in Europe. Proceedings of the 4th International Conference on WIM, ICWIM4, Taipei (2005) 2-12
4. Opitz R., Kühne R.: IM (Integrated Matrix) WIM Sensor and Future Trials. Proceedings of the 4th International Conference on WIM, ICWIM4, Taipei (2005) 61-71
5. Brozovič, R., Žnidarič, A., Vodopivec, V.: Slovenian Experience of using WIM Data for Road Planning and Maintenance. Proceedings of the 4th International Conference on WIM, ICWIM4, Taipei (2005) 334-341
6. Moses, F.: Weigh-in-Motion System Using Instrumented Bridges. ASCE, Transportation Engineering Journal, Vol. 105, No. 3, (1979) 233-249
7. Gagarin, N., Flood, I., Albrecht, P.: Computing Truck Attributes with Artificial Neural Networks. ASCE, Journal of Computing in Civil Engineering, Vol. 8 (2), (1994) 179-200
8. Fayyad, U., Piatetsky-Shapiro, G., Smyth, P.: From Data Mining to Knowledge Discovery in Databases. AI Magazin (1996) 37-54
9. Lutzenberger S., Baumgärtner W.: Evaluation of measured Bridge Responses due to an instrumented Truck and free Traffic. Bridge Management 4, ed. Ryall, Parke, Hardening, Thomas Telford, London (2000).
10. Schnellenbach-Held, M., Lubasch, P., Buschmeyer, W.: Evolutionary Algorithm based Assessment of Traffic Density Changes. IABSE Symposium, Budapest (2006)

Practice 2006: Toolkit 2020

Chris Luebkeman and Alvise Simondetti

Arup Foresight, Innovation and Incubation, 13 Fitzroy Street,
London W1T 4BQ, United Kingdom
{Chris.Luebkeman, Alvise.Simondetti LNCS}@arup.com

Abstract. This paper discusses advances in computing applications as they affect the design of the built environment. It presents emergent applications selected from Arup's current projects, examines the changing nature of the profession, and concludes with a preview of research leading to Arup's vision of the designer's toolkit for 2020. Applications address in the first instance the vertical integration of the supply chain and secondly the horizontal integration across the different design disciplines. This paper further explores the link between three dimensional representation and analysis, automated generation of representation, computational design optimization and Realtime synthetic environments. Of integral importance to the continuing innovation in computational applications are of course the people who use the tools. We describe four emergent specialists: toolmakers, custodians, math modellers and PhD candidates embedded in the practice. The vision of Toolkit 2020 ranges from persisting hardware limitations for near-realtime design regeneration, to the emergence of open source freeware.

1 Introduction

Arup is a firm of some seven thousand designers spread across the world. The firm has been traditionally known in the built environment for its structural consultancy work from the Sydney Opera House onward. However, Sir Ove Arup in 1970 was quick to point out that innovation in design occurs in a multidisciplinary practice [1]. In 2006, forty years from the date when Sir Ove Arup founded it, the firm employs specialists consultant in many design disciplines including acoustics, lighting, fire, flow of air and water, flow of people through spaces and during evacuation, flow of goods through airports, manufacturing plants and hospitals, traffic and vehicle movement. Professor William Mitchell[1] points out that research in computing applications is most successful when experiments occur in non-trivial scenarios. Advances in computing applications as they affect the design have consistently occurred at the confluence between information technology and creative practices [2], [3]. Both authors made a conscious decision to move to consultancy, accepting the challenges of putting invention and innovation into practice, on the assumption that the real world challenges are seldom trivial [4]. The focus of this paper is on work occurring in

---

[1] William J. Mitchell, Professor of Architecture and Media Arts and Sciences at MIT. Mitchell is currently chair of The National Academies Committee on Information Technology and Creativity.

I.F.C. Smith (Ed.): EG-ICE 2006, LNAI 4200, pp. 437–454, 2006.

between the practice of architecture and engineering, in between academia and practice and finally between implicit design intuition and the explicit rule-based generation of design [5], [6].

2 Computing Applications in Practice at Arup

The first series of applications address the vertical integration of information in the supply chain with the contractor and fabricators at one end, and the clients and building owner at the other. Contemporary examples of the application are driven by the client aspiration to improve their asset management, to counteract the time restrictions of market forces and minimise the risk of human error.

A second area of application addresses the horizontal integration across the different design disciplines (or physics of a building) to create multi-performance simulation environments. Current Arup projects include an urban environment, a highway widening and a transport interchange where natural light reaches the platform level thirty five metres below ground. The openness necessary to allow the penetration of light through the building creates a design challenge for both the intelligibility of public announcement as well as fire and smoke propagation.

Further reference is made to the link between three dimensional design representation and analysis beyond its application in structural steel design. Presented are applications in construction programming (4D) of complex construction sites.

Most generation and iteration of three-dimensional design representation occurs manually, whereas here we focus on emergent automated (not automatic) applications. Parametric relational modelling has already been used in the design of sport venues, characterised by geometrical complexity and a fixed opening date.

Regenerative modelling with Visual Basic scripting was applied for example to the design and documentation of an Olympic swimming centre so that the 3D representation and documentation becomes a by-product of three thousand lines of script. Computational optimisation has also been applied to projects that include panelization and rationalization of curved surfaces, optimization of building envelope, and sizing of structural members. Automated design methods such as computational optimisation for the sizing of structural members, present the need for Realtime synthetic environments to enable us to understand results which, through static images, alone would remain unreadable.

Immersive Environments was used to demonstrate design to non-specialist project stakeholders. Realtime[2] synthetic environments is applied to operationally driven design facilities such as healthcare facilities and infrastructure design with a large number of stakeholders; SoundLab[3] is applied in the design of performance venues.

2.1 Vertical Integration of the Supply Chain

The Building Information Model (BIM) are becoming the integrator of the supply chain all the way along the line from the client to the fabricator. Such geometrical

---

[2] Realtime is an interactive environment, is an innovative design tool developed by Tristan Simmonds at Arup, based on cutting edge computer graphics technology.

[3] SoundLab, with 12 channel ambisonic 3D sound system, is an innovative design tool developed by Arup to give an auralization of sound in performance spaces and other building types remote from the spaces themselves.

models have become the interface for the complete database of project information. Although they are already commonplace in the aerospace and automotive industry, they are just emerging in the construction industry where they often challange the standard contractual arrangement of the supply chain.

As an example, we will introduce three projects, each of which integrates a different portion of the supply chain. The Sydney Opera House Opera Theatre Refurbishment integrates design representation with facilities management, Westland Road Tower integrates architectural, structural, mechanical and quantity surveyor disciplines in a single model during the schematic design phase of the project; Serpentine Pavilion 2005 integrates design and fabrication.

Sydney Opera House – Opera Theatre Refurbishment. For this project, Arup developed a 3D model of the existing structure of the Sydney Opera House, Opera Theatre Refurbishment [7] using Bentley software. Each entity in the model contains the complete set of the original information, with some three thousand architectural drawings, one thousand structural, and several hundred services and associated subcontractor drawings.

Each entity in the model, in addition to geometrical modeling information, holds a description of where the drawing information for that entity originated. Tags containing the entity's unique number from the original drawings, list all the structural and architectural drawings used to create that entity (the list can contain as many as 10 different drawings). All existing drawings are in TIFF format and can be retrieved by double clicking on the entity in the 3D model.

In addition, each entity in the 3Dmodel is directly linked to the Opera House Facilities Management database. By double clicking on an entity in the 3D model, using its unique entity number, the Facilities Management master spreadsheet opens. This spreadsheet in turn is then hyperlinked to all other spreadsheets that are used for the daily running of the building. Similarly, by double clicking the entity number in the spreadsheet, it will open Bentley Structural, create a report, find the tag within the 3D model and show the location of the entity in one view.

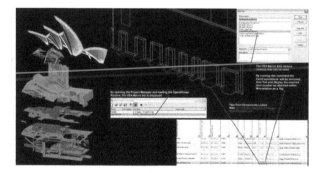

Fig. 1. Sydney Opera House, Opera Theatre Refurbishment linked to the facility management spreadsheets

This BIM, directly linked with the Building Management System (BMS), is currently being used by the building owner. The long-term purpose of this 3D model is to assist the team responsible for the proposed internal Opera Theatre refurbishment in attempting to realize the architect Jorn Utzon's original concepts.

Westland Road Tower. Arup participated in the creation of a BIM for Westland Road Tower in Hong Kong. Here the BIM was used for co-ordination and clash detection, within a project that was constrained by a very short timeframe. The tower is three hundred meters tall, seventy nine floors, and the client allocated a short six month design development period to produce Structural Tender Drawings. The tower client, Swire Properties [8], decided to use Digital Project[4] software.

Approximately twenty-five team members including the client's project manager, Quantity Surveyor, Architect, Structural, Mechanical, Electrical and Public Health (MEP) engineer and a 3D consultant were trained by the software supplier, Gehry Technologies [9]. A Model Coordinator was resident in the project office.

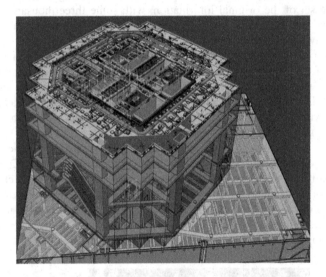

Fig. 2. BIM Westland Road Tower in Hong Kong (courtesy of Gehry Technologies)

This represents a pilot project for the client and for the industry as a whole. We are waiting for this tower to be completed and a few other applications before we can measure the full scale of the success.

Serpentine Pavilion 2005. Arup used Visual Basic (VB) scripting language and AutoCAD software for the twelve week Design Development phase of the Serpentine Pavilion 2005 [10] in London's Hyde Park. The driver for this method of working was to reduce the risk of mistakes in this project with a short time frame. The 3D geometrical model was a graphical instance of the script, the design rules evolved during the design development. This process insured that at least 80% of the model had no risk of mistakes. Doors and special edge conditions were added manually. The

---

[4] Digital Project software is the customization of Dassault's CATIA for the construction industry by Gehry Technologies.

success of the project was to build the pavilion with zero mistakes that were unacceptable to the client. To communicate with the fabricator, Arup used the geometrical model along with a text file containing the joint number and coordinates. Based on Arup's rules and joint information, the German fabricator rebuilt their model and checked it against the Arup three-dimensional model. When they started assembly on site, Arup produced a 2D drawing upon request.

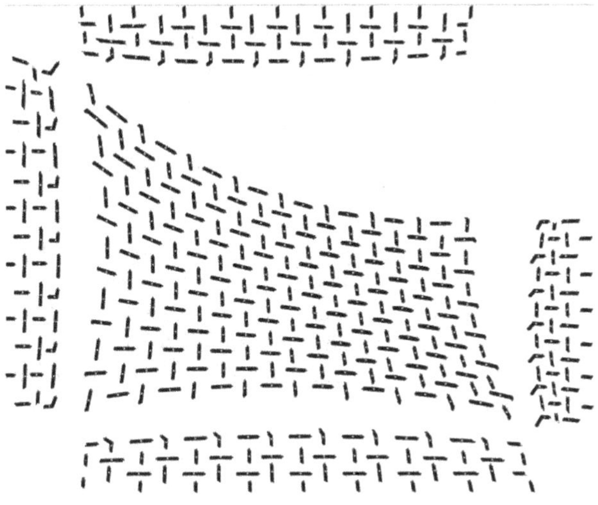

Fig. 3. Serpentine Pavilion 2005

Serpentine Pavilion 2005 proves the success of this methodology for a temporary pavilion. Now, the industry has to apply these techniques to the realm of permanent buildings.

2.2 Horizontal Integration of Disciplines

This section refers to 3D models which are used to integrate the results of the analysis of different design disciplines contributing to the design development of a project. These models are used to achieve a higher level of integration of the individual design disciplines and to communicate performance based designs to the public, the client and the regulatory authorities.

Three projects are highlighted here, each of which has been used to communicate with different bodies: *Florence High Speed Train Station* simulations were used to communicate with regulatory authorities, *Ancoats Village* urban digital prototype has been used for seven years by the client, a regeneration agency, to communicate with stakeholders, and *M1 Widening* was used in the project's public exhibition and public consultation.

Florence High Speed Train Station. Together with architect Norman Foster and Partners five disciplines at Arup [11] completed distinct design analysis of the Florence High Speed Train Station. The station box is some thirty five metres underground and some four hundred metres in length, dictated by the length of the new Pendolino high speed train. The design of the project was driven by the aim of bringing natural light thirty-five metres underground to the platform level via holes in the slabs.

Designers focussed mainly on two operational conditions of the building: one under standard conditions and another in the case of an extreme event. The worst case scenario was exemplified by the coach of a train arriving at the station on fire, aligned with the hole in the slab, with the doors open, some smoke coming out, a loud speaker announcement that the whole station must be evacuated, and finally with the windows of the coach exploding and more smoke invading the station.

The architect produced a geometrical model of the architectural surfaces that incorporated the bare-faced concrete structural model and the complex geometry of the steel roof. This three-dimensional model, evolving at each design iteration, was then used as a basis to create a RADIANCE[5] three-dimensional computational lighting simulation to demonstrate and refine the natural lighting levels at platform level. The same models were also used as a basis for a Computational Fluid Dynamics analysis, using STAR CD[6] to map the smoke propagation during the extreme event described above, and as the basis for a STEPS[7] people evacuation model of the entire station to support communications with the Italian Regulatory Authorities. Finally, the same model was used as the basis for a three-dimensional computational acoustical simulation, or auralization, of the station which demonstrated to the client and the authorities the intelligibility of emergency announcements with alternative acoustical insulation and public address systems.

Fig. 4. Florence High Speed Train Station multidisciplinary simulation

---

[5] The Radiance open source software is a distributed raytracing package developed by Greg Ward Larson, then at the Lawrence Berkeley National Laboratory (LBNL) in Berkeley, California.

[6] The computer program STAR CD by cd-adapco.

[7] The computer program STEPS (Simulation of Transient Evacuation and Pedestrian movementS) has been developed by Mott MacDonald to simulate the movement of people during emergency evacuation scenarios.

Upon completion of this discreet simulation, Arup produced a multidisciplinary simulation that integrated the results of all discrete disciplines described above in one specific extreme event condition. To produce this multidisciplinary model, all analysis results were extracted directly from the different analysis models and were then manipulated in spreadsheets and imported into a common modelling environment, using 3D Studio MAX[8], via custom-made scripts. The model explicitly demonstrated the co-ordination of disciplines to the client; for example, no passengers evacuating the space were still on the mezzanine bridge by the time the smoke engulfed that level.

Ancoats Village Urban Regeneration. Arup has created and maintained an integrated area wide ICT information system for Ancoats Village Urban Regeneration [12], which, for the past seven years, has remained continuously accessible on the internet to all stakeholders (planners, developers, politicians, designers, engineers, transport and utility service providers and individual citizens). This has enabled more inclusive decision-making and has supported a more sustainable practice. The urban prototype, based on a live accurate geometrical model, has been used as a basis for visual assessment studies, transport modeling, pedestrian simulation, noise studies, wind loading, lighting simulation, and remote sensing technology studies.

Fig. 5. Ancoats, Manchester, City Modelling

M1 widening. Motorway design has changed since the M1, Britain's north to south motorway, was built in the mid 1960s. Designers now have to make sure that local communities are protected from over-development and that measures to minimise any impact on the surrounding environment are carried out. This is a process known as 'mitigation'. Arup [13] has created a synthetic environment to introduce the public and the project stakeholders to the key issues related to the M1 widening project:

---

[8] 3D Studio MAX by Discreet, Autodesk.

visual and acoustical performance of the noise barriers; new road surface to mitigate noise; retaining structures to mitigate impact on existing mature vegetation and light spillage mitigation.

Fig. 6. M1 Widening for Public Consultation

The two km^2 synthetic environment, represented a stretch of motorway that was eighty-three kilometres in length which had been earmarked for widening. The synthetic environment was created from: three-dimensional design data of the proposed widening design, Ordinance Survey maps, LiDAR[9] data pre-processed for the ground surface building outlines and tops of trees, three types of aerial photography to create the models texture, accurate acoustical and lighting surveys, accurate lighting simulation, accurate auralization and indicative traffic flow. Three-dimensional tool In-Roads[10] design data, in turn integrated several design disciplines including bridges, civil and geotechnical engineering.

With a large number of disciplines coming together for the first time in one environment, despite a virtual one, the challenge was to integrate different stages of design development and different entities which would have otherwise looked awkward in a single visual representation. For example the DTM captured by an aircraft flying overhead necessarily wouldn't include data under the existing motorway overbridges; however in the M1 synthetic environment it would have looked awkward to leave an empty gap in the model.

2.3 Construction Programming

Construction programming is undergoing a rapid transformation where a project program is linked two-ways with the three-dimensional geometrical model to create an

---

[9] Light Detection and Ranging (LIDAR) is an airborne mapping technique which uses a laser to measure the distance between the aircraft and the ground. This technique results in the production of a cost-effective digital terrain model (DTM).

[10] InRoads software by Bentley.

interactive graphical simulation of the construction process. Also referred to as four-dimensional modeling, this form of representation improves the analysis and communication of a proposed construction sequence or construction program and enables all stakeholders to contribute with greater understanding and confidence.

Linking three-dimensional representation with the construction program is possible in object based modeling and was not in the traditional layer based modeling. This idea is emerging most evidently in structural steel, but also in electrical design and it will probably spread to the other design disciplines.

Channel Tunnel Rail Link. For the Channel Tunnel Rail Link project, Arup piloted the use of four-dimensional models for the new St Pancras International Terminal using Timeliner[11] software. The pilot project included the modeling of over six-thousand activities in worksite areas and construction elements. This provided an accurate overview of the trade contractors planned works on the project, months in advance, and was used to identify contractor interfaces and generate time related clash reports.

Heathrow Terminal 5. At Heathrow Terminal 5 Arup piloted four-dimensional modeling of the Control Tower. The model was developed off the critical path of the project. Arup is currently using four-dimensional modeling on the T5 Interchange which is in the pre-construction phase. The interchange model is reputed to have been very successful in bringing groups of planners from different contractors closer together, proving to be a major aid in facilitating improvement in the program.

Fig. 7. Heathrow T5 Construction Programming

2.4 Regenerative Modeling

Arup uses two techniques for regenerative modelling: parametric relational modelling, and automated modelling.

Parametric relational modelling, also referred to as "live intelligent modelling" is a three-dimensional model that contains information about entities as well as their

---

[11] Timeliner by NavisWorks.

relationships or dependencies. This differs from the large majority of three-dimensional modelling where entities are independent. Early results from Arup pilot projects have demonstrated dramatic improvements in the design cycle latency, which drops sometimes from weeks to seconds. To create them, the math modeller, an emergent professional specialist, needs to develop the skills necessary to be able to extract hierarchies and entities for the design and define them in terms of parameters, dependencies and constrains. It is common with beginners to observe an iterative process of identification of parameters and dependencies, where the over-constrained model grinds to a halt and has to be rebuilt from scratch several times.

Arup conducted a post-mortem study on Selfridges Birmingham Pedestrian Bridge with RMIT[12] university [14]. The bridge's three-dimensional parametric relational models with architectural and structural dependencies and constraints were built using CATIA[13], thereby reducing the interdisciplinary design cycle latency. Construction constraints, including the maximum available tube diameter that changed during fabrication, were introduced in the model thus demonstrating that such a radical change in the diameter value dramatically reduce the design cycle time.

Arup also used parametric relational model methods on live projects, using Digital Project[14] software, in the structural design studies for critical portions of the Beijing National Stadium for the 2008 Olympics. On more recent stadium roof designs, once the parameters and constraints were defined, Arup used computational design optimization techniques to improve the selected parameters.

For the concept design of complex geometry, high-rise building façades, Arup has implemented the "user feature" tool within the software, which allows the definition of the design principles of a detail to be automatically adapted at each instance within the design.

Finally, automated modeling defines a technique that is becoming popular with designers, architects and engineers alike whereby the three-dimensional model becomes a by-product of the design process. The technique involves the identification of a set of rules that define the geometry. The rules are then scripted in a simple program that takes the values of the parameter as input and automatically generates a three-dimensional model in near real-time. The ability to adjust the defining parameters and rules for the dependencies of the entities and automatically update the entire three-dimensional data design model allows for radical changes at a late stage in the design process.

This technique was used, for example, to develop the design and eventually automatically produce the structural tender drawings for the superstructure of the Beijing National Swimming Centre[15]. Close to three thousand lines of VB script automatically regenerated approximately eighty percent of the roof's cellular moment frame geometries in 3D, which included some twenty-five thousand beams. In this design, cells are twelve-sided and fourteen-sided polyhedrons according to Weaire-Phelan foam theory. The three-dimensional model was generated using Microstation[16] software.

[12] Royal Melbourne Institute off Technology (RMIT) SIAL.
[13] CATIA, by Dassault is a standard software of Aerospace industry.
[14] Digital Project is a parametric relational modeling software , by Gehry Technologies.
[15] Beijing National Swimming Centre by PTW Architects of Melbourne and Arup with China State Construction International Design. Under Construction. May 2006.
[16] Microstaion by Bentley.

One can imagine that this design technique will be more suitable to some designers then others, however it will become necessary in all those instances were a manual design process will take too long or it will carry too much risk of human error.

2.5 Interactive Synthetic Environment

Arup has developed an interactive synthetic environment called Realtime, which gives users the possibility of exploring their designs in three dimensions by roaming through them at will without the limitation of a pre-set path. All the project stakeholders have access to a preview of the project, avoiding expensive delays, misunderstanding or design mistakes. Realtime brings this capability to the projects quickly, reliably and without the need for expensive equipment. Successful applications have included a diverse range of time-critical projects.

Realtime synthetic environments are a development of cutting edge 3D graphics technology that fit the needs of clients, designers, planners and other professionals involved in the built environment. Three-dimensional models are brought to life with surface texturing, lighting and other visual detailing. Realtime environments can run on current specification computers with operation by a simple intuitive game console.

Realtime enhances the communication between the client technical team and the user and support staff by providing a realistic experience on the innovative proposed spatial design that affects current practice.

Arup's Realtime simulation provides three general modes of navigation: fly-around mode allows the user to go anywhere in the three-dimensional model, including through walls and slabs; first-person mode uses gravity and collision detection to give a 'human' dimension to the space. The user experiences the space with a view at head height moving along the floor or up stairs, with a pre-programmed natural walking bounce effect, constrained to accurate walking speed. Finally, a third-person mode adds an avatar to the first person experience. The avatar, male or female, allows the user to understand the scale of the design.

Realtime has, for example, been used by the firm to demonstrate the development of St Helens and Knowsley hospitals near Liverpool with architect Capita Symmonds to a group of stakeholders including patients and nurses, doctors and local council. The project includes a one million square foot newly built acute care hospital. It will be one of the largest building in town and the first of such buildings to be built within the community in this generation. An accurate understanding of the spatial relationship between the building and the surrounding community was crucial in order to reach a consensus among those present. All stakeholders in a project benefit from using Realtime, as it accurately represents and co-ordinates proposed three-dimensional geometry.

Similarly, Realtime was used to demonstrate a proposed design for the M1 Widening of a section of motorway near Nottingham described above. Here, Realtime took centre stage at the public consultation where people were encouraged to explore the proposed motorway widening design. Not surprisingly, the accurate and explicit demonstration of such a highly controversial project proved popular when presented to the public.

Fig. 8. St Helens and Knowsley hospitals screen-shot from Realtime

Finally, Realtime has become a necessity to a design team when the design is auto-mated following a set of rules. This proved to be the case with the Beijing National Swimming Centre described above. The design team used Realtime to double check the geometry generated by the script. The interactive synthetic environment became a powerful tool for debugging the script and developing the design itself.

Fig. 9. Beijing National Swimming Centre screen shot from Realtime

Arup has also developed an interactive synthetic environment called SoundLab, used in design development and in the demonstration of three-dimensional audio mate-rial or auralization. The purpose of the SoundLab environment is primarily for the direct communication of acoustic concepts to other members of a design team, and for the acoustic design of performing arts venues and auditoria, but it touches on all as-pects of the project. By recreating the spatial acoustic field of existing venues, and by comparing this directly to new design options developed out of three-dimensional geometric models, we are now able to immerse clients and project colleagues in the project during the design phase. SoundLab for example has been used for the design of public announcement systems for the Florence Station project mentioned above.

3 Preliminary Findings – The Designer's Desktop 2020

Within the context of current Arup practice described above, we carried out an international survey as part of an ongoing study on the changing nature of the practice. We interviewed two dozen colleagues from around the world, including clients, automotive designers, architects, engineers, software manufacturers and academics. The study will also include interviews with those who are more cynical about future developments, in order to present a balanced view. It must be reinforced here that purpose of the study is not to predict what the future of practice might be in 2020, which nobody knows, but to develop a map of possible futures. Such foresight studies are common for example in government.

This paper moves on to examine the changing nature of the profession; the toolmakers and custodians as emergent specialists in the digital design landscape; and the roles people are playing across academic disciplines and commercial practice.

To conclude we present a discussion of the social, technological, economic, environmental and political implications of designer's desktop 2020. Mention is made of hardware limitations for near-Realtime design regeneration, as well as the emergence of open source freeware and of the various initiatives used to create a (minimum) common denominator for exchanging data.

We began our interviews by presenting our colleagues with the following diagram, and by describing four arbitrary contexts in which designers might operate in 2020. We asked the participants to position their practice today with a dot on the diagram and to draw an arrow representing their perception of the context in which their practice might operate in 2020.

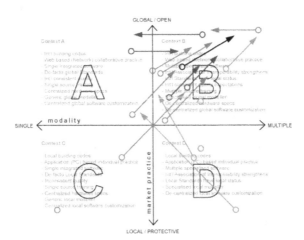

Fig. 10. Four contexts of practice range from the local and protective market with a single mode of practice *(Context C)* to the global and open market with a multiple mode of practice *(Context A)*

Developers appear to be pushing towards a design environment based on a single tool as a measure to reduce interoperability costs. Aerospace managers, who opted for a single tool design environment a while back in order to improve the quality of their life-dependent products, appear now to be grappling with a designer-led mini-revolution, driven by efficiency. Designers, on the other hand, are breaking away from the well-established single tool design environment towards an environment based on whatever tool does the job.

To canvas general perceptions of the designer's desktop 2020 we used a straw poll method. We asked our colleagues to rate, on a four point scale (from strongly disagree to strongly agree), the following ten arbitrary scenarios "I believe that by 2020 …". The table below shows the scenarios and the mean response. The preliminary sample was seventeen.

Table 1. Scenario Straw Poll

I believe that by 2020 … [sample of 17]	Strongly disagree 0%	Strongly agree 100%
we will work digitally directly from bldg site		82%
we will be designing design systems		73%
2D Documentation will disappear	41%	
Multidisciplinary BIM will be commonplace		78%
we will all work in open source	36%	
we will use programs to stimulate versus simulate		76%
algorithms will learn from their user		72%
we will regularly simulate a dozen physics at once		81%
Modeling animated architecture will be commonplace		71%
Immersive environments will be used for design reviews		69%

These preliminary results, often conflicting and inconclusive, nevertheless highlight the highest level of agreement for a 2020 practice commonly using multidisciplinary integrated modelling which is accessible on the out of the office. This is in contrast with the current two-dimensional line drawing stored on individual computers. In addition, very much like the paperless office myth, it is suggested that 2D documentation will not disappear, immersive environments will have to wait even longer to get out of the university labs and unexpectedly for us, open source software will not be taking over.

Further recurring views were that tools will be assembled by project; we will be playing parameters in real-time; there will be an easier entry level to software, nobody will be teaching it; designers will be doing more programming and the motivation will be in seeing what can be done.

To assess the world around us we used the STEEP framework to ask colleagues "what do you envision lies ahead?". The driving forces and implications that resulted from these questions allowed us to focus on the factors which might influence our future, some of which are extrapolations and some of which are speculation. The following table shows the key excerpts gained from the interviews.

Table 2. Desktop 2020 drivers and implications STEEP framework

	Drivers/ Implications
Social	It's not just a matter of learning AutoCAD.., but more a way of transforming the idea of the design office, so that it is intimately connected with crafting these tools themselves. The quality of products in Aircraft engineering is so much higher than the quality of buildings. It may well be that we need to turn to alternative sources other than the traditional civil engineering department or architecture department. Transitory employment between employers will become a real challenge for employers
Technological	Search and access of knowledge will become a bit easier. The viewer will be on the web somewhere. You are going to have access to pretty much anything, anywhere you are. Biological modeling is going to drive the next ten years.
Economical	All sectors at all levels of the industry always must necessarily remain incentivised. Reducing waste in the construction industry can easily pay for the enhanced work at the front end. If the value proposition gets redefined, then the fee structures will change. Project insurance will be like decennial insurance, so none of the designers indemnify themselves, the client will actually indemnify the project
Environment	Buildings need to be designed in such a way as to diminish energy consumption. Green architecture is probably as big a force for design integration as all the other stuff.
Political	A more open, co-operative agreement, rather than an adversarial type of contract. It's embedded in the American psyche that every state gets to do whatever it wants. It is all depending on trust between the designer, the contractors and the clients. The openness of European countries will drive change. In New Orleans to replace three hundred thousand housing units the traditional design bid is not going to work. Three-dimensional objects are required by planners

4 Conclusions – The Changing Profession

Having described what we are designing across Arup, and having interviewed some two dozen professionals we conclude as follows.

The industry needs new specialists, and if academia doesn't provide them, the industry will have to resort to setting up private academies. This has already occurred in the past, for example in the mid 1960s, the Istituto Europeo di Design was set up in Milan when the Italian Academic community couldn't meet the demands for emerging specialists including the Industrial Designers. Similarly, a few years back, the Interactive Institute in Ivrea was set up by Telecom Italia outside Turin, when once again Italian Academia failed to address the need for multidisciplinary design education based on advances in computation.

Don't be mistaken, the blame is not only to be placed on academia, our professional practices will also have to develop attractive careers for these new specialists if we are to reduce the current outflow of highly valuable professionals towards setting up their own shop. A trend that would not be negative for the industry as a whole if it wasn't for the inevitable consequence that the specialist, once on their own, are forced to manage the process as opposed to practising it, generally with the result that they interrupt their research.

We identified four new emergent specialists in our profession: the tool maker, the math modeller, the custodian and the embedded PhD. What follows is an attempt to profile each of them.

Toolmakers might be individuals that create programs or scripts to generate geometry and could have a fundamental understanding of first principles of design as well as a solid background in computer science or graphics programming. They would be a central resource to the office or the region and would spend short and sharp periods of time (from 2 weeks to a month) with each project team. Tool making is a part-time activity that combines very well, but not necessarily, with design itself. When not helping the project team, toolmakers would be given time to reprogram relevant code written for specific projects in a more generic way for re-use throughout the firm. These individuals would also be given time to connect with the programming community outside the firm, and would regularly present their novel work at technical conferences.

The BIM Custodian, also known as the BIM Master, or BIM Co-ordinator, would be an individual with solid experience in 3D modelling, preferably in several different software packages. He or she would have an understanding of how to set up modelling protocols with a broad grounding in construction techniques and a good understanding of multiple design disciplines.

The BIM co-ordinator is a fulltime position working on one or a few projects depending on size. Similar to the current Project Manager, the BIM co-ordinator is one of the foremost specialists dealing with the client and public relations. He/she would be on the move and work with a powerful graphics laptop. Because of the nature of the work the BIM co-ordinator would also have exceptional interpersonal and team management skills.

Regenerative modelling requires an individual with formal education in design and in computation or equivalent experience. The Math Modeller work involves the abstraction of mathematical rules that determine geometry; the ability and design experience to foresee architectural, engineering and/or construction constraints that might occur during the design of the project. These individuals would have an interest in the emerging field of explicit capturing of geometrically engineered detailing knowledge for reuse on other projects. This is an emerging area that promises to transform the engineering practice and these individuals would need to be given the time and resources to interact with the software developer, university and external communities.

The Embedded PhD would involve a PhD candidate sitting within the professional practice. The PhD thesis topic would be determined through discussion with the host practitioner. The financial structure of the mechanism would be designed to maximise the relevance of the candidate's research work in conjunction with the new professional practice. His or her PhD advisor would provide research methods, protocols and computational software and hardware tools to support the work. This is in striking contrast with other more common mechanisms, like many of the engineering doctorates, where a PhD candidate sits in the professional practice and uses the workplace as a test-bed for his or her PhD research – research whose topic is determined in discussion with an academic advisor. One practical implication of this one mechanism versus the other is the employability of the individual in the professional practice at the conclusion of their PhD; where in one case the work is relevant to the professional practice, in the other case it might not be.

Object based modelling is very rapidly taking over line drafting. Currently, the main remaining obstacles seem to be hardware computational power and a large stock of outdated training courses and software boxes filling the firm's IT asset depreciation lists.

Based on the evidence presented, interoperability, currently heralded as the number one enemy of the profession, shall largely sort itself out with more of the general functions of the software moving to the operating system. Interoperability issues related to the specialist functions of the software might be dealt with by a network of people, including the toolmakers, as opposed to standards.

Following in the footsteps of the one-dimensional world of text, the preferred viewing platform for the databases of three-dimensional information might also be the Web, accessible anywhere and searchable in a "Google Earth" style.

Based on our synthetic environment experience, one of the implications will be the ability of these individuals to work towards targets and overcome the drive to resolve too many design details, too early in the project, a tendency that appears more common in three dimensional modelling when compared to two-dimensional drawing and drafting. By 2020 most of us will have retired and necessarily, in the developed world at least, we will be replaced by those that today spend hours playing computer games. As highlighted by Harvard Business Review's study, these individuals will make better managers as they develop faster at decision making skills. Explicit representation of the design process, together with the enhanced ability to make many small and fast decisions will change the way we design.

References

Personal Interviews for Designer's Desktop 2020 study with: Professor Mark Burry, RMIT, Melbourne; Reed Kram, Kram Design, Stockholm; Charles Walker, Arup; Jeffrey Yim, Swire Properties, Hong Kong; Martin Riese, Gehry Technologies, Hong Kong; Axel Killian, MIT, Boston; Jose Pinto Duarte, Lisbon University, Lisbon; Joe Burns, Thorn Tommasetti, Boston; Mark Sich, Ford Motor Company, Michigan; Phil Bernstein, Autodesk, Boston; Lars Hesselgren, KPF, London; Mikkel Kragh, Arup; Bernard Franken, Franken Architekten, Frankfurt; Martin Fisher, Stanford University, San Francisco; Tristram Carfree, Arup; Mike Glover, Arup; Duncan Wilkinson, Arup.

Discussions for 3D Documentation study with: Richard Houghes, Stuart Bull, Steve Downing, Simon Mabey, Valerio Giancaspro, Dominic Carter, Neill Woodger, Tristan Simmonds, Martin Self, Martin Simpson, Dan Brodkin, J Parrish, Nick Terry, BDP, London.

1. Sir Ove Arup: Key Speech, delivered in Whinchester (1970), internal publication (2001).
2. Mitchell, William J.: et al eds., Beyond Productivity: Information Technology, Innovation, and Creativity. Washington, D.C.: The National Academies Press. 2003.
3. Shmitt, G.: Micro Computer Aided Design, New York, John Wiely & Sons, (1988)
4. Glymph, J.: et al. A parametric strategy for free-form glass structures using quadrilateral planar facets, Automation in Construction 13 (2004) 187– 202, Elsevier.
5. Burry, M.: Between Intuition and Process: Parametric Design and Rapid Prototyping, in Architecture in the Digital Age (ed. Branko Kolarevic) Spon Press, London, (2003)
6. Frazer, J.H.: An Evolutionary Architecture, Architectural Association, London, (1995)
7. Bull Stuart, Arup, Sydney Opera House, Theatre Refurbishment, Unpublished paper, Jan 2005
8. Yim, J.: Swire Properties. Unpublished interview notes. March 2006
9. Ceccato, C.: Gehry Technologies, Unpublished interview notes. March 2006
10. Self, M.: et al. in Dan Brodkin and Alvise Simondetti, 3D Documentation Report, Arup internal publication, Nov 2005
11. Clark, E., Woolf, D., Graham, D., Patel R., Shaw J., Simmonds, T., Simondetti, A.: unpublished project notes, February 2002
12. Mabey S.: Arup, unpublished project notes, February 2003
13. Simmonds, T., Simondetti, A.: et al. Arup, Project notes. March 2006
14. Maher, A. and Burry, M. The Parametric Bridge: Connecting Digital Design Techniques in Architecture and Engineering, ACADIA 2003 Proceedings Indianapolis (Indiana) pp. 39-47

Intrinsically Motivated Intelligent Sensed Environments

Mary Lou Maher[1], Kathryn Merrick[2], and Owen Macindoe[1]

[1] Key Centre of Design Computing and Cognition, University of Sydney,
Sydney, NSW 2006, Australia
[2] School of Information Technologies, University of Sydney and
National ICT Australia
mary@arch.usyd.edu.au, kathryn.merrick@nicta.com.au,
macindoe@mail.usyd.edu.au

Abstract. Intelligent rooms comprise hardware devices that support human activities in a room and software that has some level of control over the devices. "Intelligent" implies that the room is considered to behave in an intelligent manner or includes some aspect of artificial intelligence in its implementation. The focus of this paper is intelligent sensed environments, including rooms or interactive spaces that display adaptive behaviour through learning and motivation. We present motivated agent models that incorporate machine learning in which the motivation component eliminates the need for a benevolent teacher to prepare problem specific reward functions or training examples. Our model of motivation is based on concepts of "curiosity", "novelty" and "interest". We explore the potential for this model through the implementation of a curious place.

1 Introduction

The development of intelligent rooms has been dominated by the development of sensor configurations, effectors, and software architectures that specify protocols for interpreting and responding to sensor data. A current practical application is the home automation package, in which computational processes monitor activities within the home and respond by turning lights on and off, locking and opening doors, and other actions usually performed by the inhabitants of the home. Home automation systems are possible with sensors and effectors that are programmed to respond as expected.

Another approach to sentient rooms, or intelligent rooms, is to use an agent approach to the computational processes and allow the agent to reason about the use of the room so that it can facilitate human activity. This research started with the intelligent room project [4, 5] and has progressed in several directions, from sensor technology and information architectures, to possible agent models for intelligent reasoning [14]. While these systems go beyond the home automation systems to proactively support human activities, they still respond as programmed. Similarly, in the developments in sentient buildings, with a goal of building self-aware systems, computational processes monitor various environmental and human activity parameters to maintain a model of the state of the building and to maintain a comfortable environment for the inhabitants.

I.F.C. Smith (Ed.): EG-ICE 2006, LNAI 4200, pp. 455–475, 2006.

An alternative to facilitating practical human activities is to create sentient places that respond to human movement and interaction. These kinds of places use sensors to monitor the location and movement of the inhabitants, and then respond by changing images and sound through a transformation of the sensor information [8]. While these changes and transformation respond to human interaction, the computational processes are fixed and the responses are pre-programmed.

Throughout the literature, we see that a variety of agent models have been developed with different ways of mapping sensor input to effector output, from simple rule-based reactive agents through to complex cognitive agents that maintain, and reason about, an internal model of the world. The development of the agent models and the implementation of specific agent applications is domain specific and time consuming. The question then is, what kind of agent model would be able adapt its response to and learn from changing patterns of usage?

Psychologists have theorised that, in humans and animals, adaptable learning processes for developing competency at multiple tasks over extended periods of time may be the result of intrinsically motivated behaviour [7, 39]. More recently, the notion of self-motivation has also been adopted as a means by which artificial systems may achieve greater autonomy through selecting their own focus of attention. A number of algorithms have been developed to model various forms of motivation for use in both planning [24] and learning systems [17, 28, 30]. While these algorithms have demonstrated potential to display lifelong, adaptable, multitask planning or learning, there are no unified architectures for their usage or clear understanding of the impact of self-motivation on artificial systems.

This paper presents a schema for defining computational models of motivation and uses this schema to define three agent models for intrinsically motivated learning as a basis for intelligent environments that adapt learned behaviours from patterns of usage derived from their sensor data. Rather than specifying a set of competencies or goals with an external reward, we look at computational models of novelty and curiosity that allow the agent to respond to unexpected changes in the patterns of sensor data, that is, in the patterns of usage of the room. With a model of curiosity as a filter for the agent to process the sensed environment, we have developed a model for an intelligent sensed environment that is intrinsically motivated to learn new behaviours as the sensors and sensor data patterns change. We develop the idea of a curious place as a platform for exploring these agent models in order to gain an insight into the role of self-motivation in artificial systems and, specifically, in intelligent sensed environments.

2   A Computational Schema for Intrinsic Motivation

In natural systems, motivation plays three key roles: direction, organisation and activation. The directing function steers an individual's behaviour towards or away from specific goals, the organising function influences the combination of behavioural components into coherent, goal-oriented behavioural sequences and the activating function energises action in pursuit of goals [12]. Psychologists have developed a range of motivation theories which describe different aspects of these roles. There are also many different computational models of motivation, ranging from action selection architectures inspired by biological systems [5, 11] through to motivated goal creation schemes based on cognitive theories of the mind [1, 20, 24, 29, 31]. These models have been used to focuses an agent's reasoning and action around a particular subset of its perception, to generate goals or to trigger other processes that satisfy or stimulate its motivational mechanism. A survey of the interaction the motivation component has with the model of the environment and the agent is shown in Figure 1 and elaborated in Table 1 and Table 2.

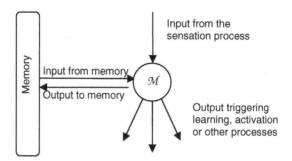

Fig. 1. The interaction of motivation with other agent processes

Table 1. Information available to the motivation process

Input from Sensation Process		Input from Memory	
Observed state	$O_{(t)}$	Sequence of observed states	$O_{(t-1)}$
Current events	$E_{O(t)}$	Sequence of events	$E_{(t-1)}$
Example	$X_{(t)}$	Sequence of examples	$X_{(t-1)}$
		Sequence of rewards	$R_{(t-1)}$
		Set of current goals	$G_{(t-1)}$
		Set of behaviours	$B_{(t)}$
		Set of actions	$A_{(t)}$
		Set of plans	$P_{(t)}$

Table 2. Information produced by the motivation process

Output to Memory		Output to other processes	
New sequence of observed states	$O_{(t)}$	Observed state	$O_{(t)}$
New sequence of events	$E_{(t)}$		
New sequence of examples	$X_{(t)}$	Example	$X_{(t)}$
New sequence of rewards	$R_{(t)}$	Reward	$R_{(t)}$
New set of goals	$G_{(t)}$	Goals to pursue	$G_{(t)}$
		Behaviour to execute	$B_{(t)}$
		Action to execute	$A_{(t)}$
		Plan to execute	$P_{(t)}$

The motivation process takes information from the sensed environment and its own memory to trigger learning, planning, action or other agent processes. Sensors provide the agent with information describing the state of its environment. A sensation process transforms this data into structures more appropriate for reasoning. These structures include observed states which focus attention on subsets of the sensed environment and events which provide information about the dynamics of the environment by representing the difference between sequential observed states. The agent may also sense examples of behaviours performed by other entities in their environment. In addition to these structures, the agent's memory may include rewards, goals, behaviours, actions and plans, depending on the higher reasoning processes employed by the agent.

3   Agent Models of Intrinsically Motivated Learning

An agent model assumes that the reasoning process is in response to the agent's sensors and the agent's response includes changes to the world and itself through the activation of its effectors. In developing these models for intrinsically motivated intelligent sensed environments we first focus on the role that motivation can play in learning systems. Machine learning algorithms fall into three main categories [25], supervised learning, reinforcement learning and unsupervised learning. In the following sections we discuss the implications of each of these types of learning for self-motivated learning agents in intelligent environments.

3.1   Motivated Reinforcement Learning Agents

Reinforcement learning [33] uses rewards to guide agents to learn a function which represents the value of taking a given action in a given state with respect to some task. A reinforcement learning agent is connected to its environment via perception and action as shown in Figure 2(a). On each step of interaction with the environment, the agent receives an input that contains some indication of the current state of the environment. The agent then chooses an action as output. The action changes the state of the environment and the value of this state transition is communicated to the agent through a scalar reinforcement signal. The agent's behaviour should choose actions that tend to increase the long-run sum of values of the reinforcement signal. This behaviour is learnt over time by systematic trial and error.

In motivated reinforcement learning agents, shown in Figure 2(b), the sensation process S differs from that of a standard reinforcement learning agent by producing both an observed state $O_{(t)}$ and a set of events $E_{O(t)}$. The observed state is stored in memory M and overwritten at each time step, while the set of events is forwarded to the motivation process. The motivation process M stores the set of current events $E_{O(t)}$ using a cumulative process to incorporate it into the set of all events $E_{(t)}$ in memory. The motivation process then reasons about the set of all events to produce a reward signal $R_{(t)}$ which is forwarded to the learning process. The learning process L performs a reinforcement learning update such as the Q-learning update [35] to incorporate the previous action and the current observed state into the behaviour $B_{(t-1)}$ in memory. Finally, the activation process A uses some exploration function of the Q-learning action selection rule to select an action $A_{(t)}$ to perform from the updated behaviour $B_{(t)}$. The chosen action $A_{(t)}$ is stored in memory and triggers a corresponding effector $F_{(t)}$ which makes a change to the agent's environment.

Motivated reinforcement learning agents aim to use systematic trial and error to maximise the long term sum of rewards produced by their motivation process. In motivated reinforcement learning agents, the reward received in each state may change over time, resulting in the emergence, stabilisation and disappearance of multiple mappings from states to actions within the behaviour B. An extension to the standard motivated reinforcement learning model might save multiple behaviours in memory so that each mapping from states to actions need only be learned once and is not forgotten [30]. Either way, over time, motivated reinforcement learning agents are able to learn to solve multiple tasks.

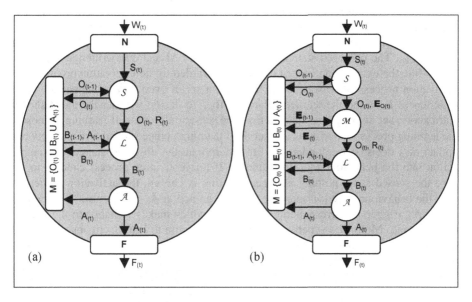

Fig. 2. Role of motivation in learning agents: (a) an agent model for reinforcement learning, (b) an agent model for motivated reinforcement learning

Reinforcement learning without motivation has been used in intelligent environments to control robotic assistants [35]. These assistants accept variable levels of user reward input including joystick control or way-point specification and learn tasks such as security patrols or item retrieval. Where standard reinforcement learning requires the user to specify a reward signal for training, motivated reinforcement learning is less obtrusive as it does not. However, as all reinforcement learning requires trial and error to learn, motivated reinforcement learning may still be too disruptive to consider as a means of controlling equipment such as projectors or lights. Instead, motivated reinforcement learning agents might be used to control virtual "critters" [19], characters which can move around a room and interact with a virtual image of human inhabits. Rather than using pre-programmed behaviours like Krueger's critters, "motivated critters" might learn such tasks as moving, climbing, jumping and falling, depending on which of these events they find interesting. These "curious critters" would make the environment more interesting and entertaining for its human occupants.

3.2 Motivated Supervised Learning Agents

Supervised learning uses examples of correct behaviour to guide agents to learn a function which represents a mapping between observations and correct actions with respect to some task. A supervised learning agent is connected to its environment via perception and action as shown in Figure 3(a). On each step of interaction with the environment, the agent receives an input that contains some indication of the current state of the environment and, optionally, an example of the correct action to take when in that state. When an example is not provided, the agent chooses an action as output.

A motivated agent model that incorporates a supervised learning control strategy is depicted in Figure 3(b). In this model, the sensation process S is responsible for disassembling $S_{(t)}$ into the observed state $O_{(t)}$, an example $X_{(t)}$, if it is provided, and a set of events $\mathbf{E}_{O(t)}$. The observed state is stored in memory $\mathbf{M}$ and overwritten at each time step, while the example and set of events is forwarded to the motivation process. The motivation process M reasons about the current set of events $\mathbf{E}_{O(t)}$, the agent's prior experiences $\mathbf{E}_{(t-1)}$ and the learned behaviour $B_{(t-1)}$ to decide whether the agent should learn about, act on or ignore the data from the sensation process. If learning is chosen, the learning process L performs a supervised learning update such as a neural network update or a decision tree update [25] to incorporate the observed state and example action into the behaviour $B_{(t-1)}$ in memory. If the motivation process chooses to ignore the sensed state, nothing is done. If acting is chosen, the activation process A uses the behaviour stored in memory to select an action $A_{(t)}$ to perform. The chosen action $A_{(t)}$ triggers a corresponding effector $F_{(t)}$ which makes a change to the agent's environment. Motivated supervised learning agents aim to learn from observations to minimise the percentage error between the actions they perform and the example actions that would be presented given the same observed state.

While a reinforcement learning agent can use trial and error to recognise that it has effectors which produce the same effect as devices it cannot control, for a supervised learning agent to be feasible in real time it requires an "example-sensor" to automatically sense examples by mapping sensor changes to actions. For example, when a human turns on a light, the agent's light sensor would sense the new light level in the room and the example-sensor would map the change in light level to the light-on action and provide it to the agent as an example. An example-sensor is feasible because, for an agent with a finite number of sensors and effectors, there is a finite number of mappings required to make an example-sensor.

Supervised learning without motivation has previously been used to learn services such as turning on lights and turning on music in intelligent environments [3]. Brdiczka et al experimented with Find-S, Candidate Elimination and decision tree learning to adapt situation networks according to feedback from a supervisor. Unlike standard supervised learning agents, motivated supervised learning agents can choose to act on or ignore observed states which are not accompanied by example actions. This means that, in comparison to standard supervised learning, motivated supervised learning agents can potentially have more sensors to collect contextual information which the agent cannot affect. The motivation process can then elect not to act in situations where it is not confident that it knows what to do. In motivated supervised learning agents, the motivation process acts as a filter for observations and examples, so attention can be autonomously focused on learning specific tasks. With the development of appropriate motivation functions, motivated supervised learning agents have the potential to derive contextual information from observations and examples, rather than requiring a separate perception process dedicated to this task. In a curious place, this may be useful for distinguishing task boundaries when there is more than one person in a room, performing multiple tasks. As is the case with reinforcement learning, it may be possible to develop hierarchical behaviours to represent these tasks.

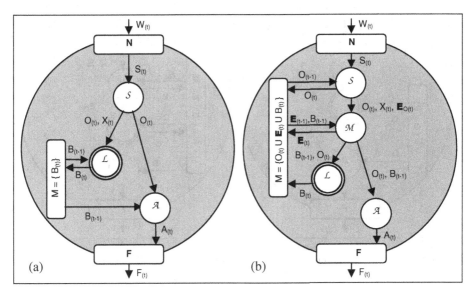

Fig. 3. Role of motivation in learning agents: (a) an agent model for supervised learning, (b) an agent model for motivated supervised learning

3.3 Motivated Unsupervised Learning Agents

Unlike supervised and reinforcement learning algorithms which learn functions mapping observed states to actions, unsupervised learning algorithms aim to identify patterns or important features in observed data. Agent models that incorporate unsupervised learning strategies with and without motivation are shown in Figure 4. In motivated unsupervised learning, the sensation process S is responsible for disassembling $S_{(t)}$ into the observed state $O_{(t)}$ and a set of events $\mathbf{E}_{O(t)}$. The motivation process $\mathcal{M}$ acts as a filter to decide which observations will be passed on to the learning process for learning. The learning process $\mathcal{L}$ then performs an unsupervised learning update such as a neural network update, k-means update or data mining update to incorporate the observed state and example action into the world model $\mathbf{O}_{(t-1)}$ in memory. Unlike motivated reinforcement and supervised learning, motivated unsupervised learning does not build behaviours which represent mappings from input to predefined output values such as actions. Rather, the output values produced by unsupervised learning represent important patterns or features in the input data such as clusters, principal components or repeated patterns in temporal data. The activation process $\mathcal{A}$ uses a set **R** of predefined behavioural rules to act on the features identified by the learning process. The chosen action $A_{(t)}$ triggers a corresponding effector $F_{(t)}$ which makes a change to the agent's environment.

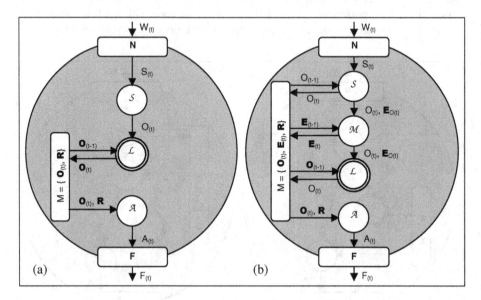

Fig. 4. Role of motivation in learning agents: (a) an agent model for unsupervised learning, (b) an agent model for motivated unsupervised learning

In a curious place, predefined behavioural rules might be represented as a mapping from sensors to actions, similar to the example-sensor described for supervised learning agents. Maher et al. [21] proposed a similar technique with sequential pattern mining to identify repeated patterns in sensor data, then act by associating actions with these observed states. While behavioural rules are fixed, the actual actions performed by the agent change over time as new observed states cause the learning process to identify different features or patterns as being important. This use of the unsupervised learning architecture is essentially similar to the supervised learning architecture, but enables the use of unsupervised techniques such as data-mining.

Because behavioural rules in the unsupervised learning architecture are fixed and only fire when triggered, not necessarily at every time-step, this architecture can also be used to create agents in which the sensed world and the effectible world are mutually exclusive. One such existing application is Microsoft's Lumiere project [15] which served as a basis for the Office Assistant in the Microsoft Office '97 suite of programs. This application augments the Microsoft Office environment with an intelligent helper by monitoring the stream of data produced by user actions and using Bayesian learning to model user needs.

In a curious place, agents for which the sensed and effectible worlds are mutually exclusive may be used to visualise information about human inhabitants. In a "window on the mind" approach, an unsupervised learning technique could be used to model user actions, then display its model to humans using a set of predefined behavioural rules defining actions to visualise the user model. However, like motivated supervised learning agents, the motivation process can be used as a filter to allow the agent to focus its attention on particular parts of its environment when there are multiple activities being performed around it. For example, in the visualisation scenario, the agent might choose to visualise a user model of just one user when, in fact, there are multiple users in the room. A standard unsupervised learning agent would simply generalise over all users.

4   A Curious Place: Curious Information Displays

The idea of a curious place promises a kind of sensed environment that is interested in the people that inhabit it and that may in turn be interesting to the inhabitants. A curious place builds on the technology that allows us to consider places that cross the boundaries between the physical world and the digital world. This technology includes rooms and buildings with embedded sensors and effectors that are monitored and activated by computational processes. A curious place departs from the current capabilities of sentient buildings and rooms by incorporating models of curiosity and learning within the computational processes.

We envisage that curious places might become the models for such spaces as intelligent rooms, entertainment arcades or data centres. A curious place as an intelligent room would be capable of observing the actions of its inhabitants, identifying novel or interesting actions to learn about using unobtrusive techniques such as supervised learning, then modifying the physical environment to meet the changing needs of its users. In contrast, a curious place as an entertainment arcade might include characters or augmented reality displays which can directly interact with occupants via active learning methods such as reinforcement learning. Characters and displays would be capable of actively seeking novel stimuli, to provoke interaction with and entertain users. Finally, a curious place as a data centre would be capable of observing the actions of its inhabitants, or even a wider space such as an entire building or the internet, identifying novel or interesting phenomena to learn about using techniques such as data mining, then modifying a digital environment to reveal these finding to users.

In this section we introduce a Curious Information Display as an application that transforms a physical space into a curious place. Traditional information displays such as posters and billboards present a fixed image to observers. More recently, with the advances in large screen display technologies, digital displays are becoming more prevalent as an alternative means of presenting information. Digital displays have allowed the amount of information being presented to be increased by attaching the display to a computer which executes software to change the contents of the display automatically. However, as yet, the full power of a digital display has not been realised, with the most common scenario using a database of images and displaying images at random or in a predefined order.

In scenarios where the display of digital information is more familiar, such as web-browsers for the display of information from the world-wide-web, novel interaction algorithms have been developed to automatically personalise the digital space [9]. Similarly, intelligent tutoring systems use artificial intelligence algorithms to tailor learning material to the individual needs of students [12]. Large digital information displays in public spaces have the same capacity for the use of novel techniques to improve the usefulness of the displays, however the public, multi-user nature of these displays calls for new algorithms to improve the ability of such displays to impart information. We propose Curious Information Displays as a means of creating digital displays that can attract the interest of observers and impart information by being curious and learning about the ways in which observers respond to information being displayed.

4.1   A Curious Information Display

Here we describe a curious place as an information gallery that attracts people by changing its information display. The purpose of this implementation is to explore the models described for motivated learning and to experiment with the parameters that control the motivation processes. Our Curious Information Display Environment includes a display on a large rear projection screen. The display comprises a matrix of displayed information items (IIs) and aesthetic items (AIs). Information items may be definitions, an image from the web, or video from a webcam. Aesthetic items are solid colour images. Each item can be displayed in a 1x1, 2x2 or 4x4 cell as shown in Figure 5. The layout of the Curious Information Display is implemented using an hierarchical tree data structure in which non-leaf nodes represent 1x1, 2x2 and 4x4 resolution displays and leaf-nodes represent IIs or AIs. Each non-leaf or leaf node has a number of properties describing the item or items they hold:

```
<node>        →      <id><itemType><source><keyword><colour>
<id>          →      [1, 999]
<itemType>    →      II | AI
<source>      →      web | database | webcam
<keyword>     →      curious | design | agent | computing
<colour>      →      red | orange | yellow | black
```

The information or aesthetic item being displayed is determined using the following rules about the properties of the leaf nodes:

```
if <itemType> = AI then
        display <colour>
else
        if <source> = web
                display <keyword> from web
        else if <source> = database
                display <keyword> from database
        else
                display webcam
```

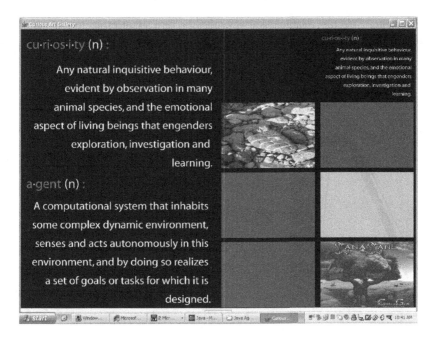

Fig. 5. The layout of the curious information display

Both non-leaf and leaf node properties can be sensed by a software agent that reasons about ways in which to modify the display by modifying these properties. These properties are sensed according to the following grammar:

```
DisplayState          →    <non-leaf nodes><leaf node><leaf nodes>
<non-leaf nodes>      →    <node><non-leaf nodes> | ε
<leaf nodes>          →    <node><leaf nodes> | ε
<leaf node>           →    <node>
<node>                →    <id> II web <keyword> | <id> II database
                           <keyword> | <id> II webcam | AI <colour>
<id>                  →    [1, 999]
<keyword>             →    curious | design | agent | computing
<colour>              →    red | orange | yellow | black
```

This grammar stipulates that every display has at least one leaf node and zero or more non-leaf nodes. In addition, only the properties relevant to a particular item type are sensed for any node at any time. For example, if the current source is webcam, the colour property will not be sensed. Likewise, if the current item type is AI then the source property will not be sensed as only colour is relevant. Leaf node properties describe the individual item held in the leaf node. Leaf node properties change whenever the item in the leaf changes. Non-leaf node properties summarise the properties of their child leaf nodes. Non-leaf node properties only change when all the child items achieve the same value for some property. This allows an agent to recognise the formation of clusters of like information or aesthetic items in the display.

The Curious Information Display is located in a meeting room equipped with sensor and effector hardware that allows a software agent to monitor and modify the environment. Effectors enable a software agent to turn a rear projector on and off and change the application it displays. Floor sensors provide information about the location of people in the room near the display and near the doorway. The layout of the room in which the display is mounted is shown in Figure 6.

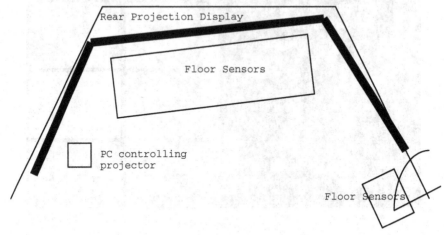

Fig. 6. Layout of the Curious Information Display Environment

4.2 A Sensor Model for Curious Places

Traditionally, representing the state of a world using attributes uses a fixed length vector and a sensed state $S_{(t)}$ is constructed by combining sensations into a vector $S_{(t)}$ = $<s_{1(t)}, s_{2(t)}, \ldots , s_{K(t)}>$. Two sensed states are compared by comparing the values of sensations that have the same index in each sensed state. While this representation is effective in many environments, it becomes inadequate in complex, dynamic environments where it is not known what objects may need to be represented. While relational representations can be used for variable length state spaces in reinforcement learning [10], such a representation is not viable for use in learning algorithms such as neural networks which require an attribute based representation. One solution to this problem uses placeholder variables in an attribute based representation to take the value of new objects. However, such representations place a hard limit on the number of new objects that can be introduced and hold large amounts of redundant data when old objects are removed. As an alternative to fixed length vectors, we propose that a state W_w of an environment may be represented as a string from a context-free grammar (CFG) [23].

In worlds represented by CFGs , a sensed state may be produced as follows. Agents have a set $\mathbf{N} = \{N_1, N_2, \ldots N_{|N|}\}$ of several different sensors. We call the data provided by a single sensor N_n a sensation. Each sensor may return sensations of varying length. Each sensor N_n assigns a label L to each sensation $s_{n(t)}$ such that that two sensed states can be compared using the values of elements with the same label. The world state at time t, $S_{(t)}$, is the tuple of all the agent's sensor inputs, $S_{(t)} = <s_{1(t)}, s_{2(t)}, \ldots s_{L(t)}, \ldots>$.

In general, an agent has several different types of sensors so information from a single sensor does not capture the complete state $S_{(t)}$. An iconic memory structure [21] associated with each sensor N_n holds sensations until they are replaced with new sensations or until a short time has passed. When each sensor has associated iconic memory, each time new data becomes available to one sensor, the information stored in the iconic memory of the other sensors can be used to construct the sensed state $S_{(t)}$ of the current world state $W_{(t)}$. This sensor model is illustrated in Figure 7.

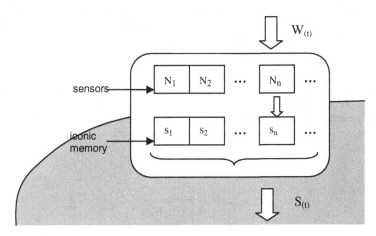

Fig. 7. A sensor model for learning agents in curious places. The sensor N_n has just registered information about the world state $W_{(t)}$ at time t. The new sensation s_n is combined with sensations from iconic memory for the other sensors to produce the sensed state $S_{(t)}$.

While CFGs can represent any environment which can be represented by a fixed length vector, they have a number of advantages over a fixed length representation. Using a CFG, only objects that are present in the environment need be present in the state string. Objects that are not present in the environment are omitted from the state string rather than represented with a zero value as would be the case in a traditional, fixed length vector representation. This means that the state string can include any number of new objects as the agent encounters them, without a permanent variable being required for that object. Fixed length vectors are undesirable in dynamic environments where it is not known what objects may occur as it is unclear at design time what variables may be needed. Similarly they are undesirable in complex environments where there are many objects which only appear infrequently as the state vector would be very large but only a few of its variables would hold values in any state.

Learning algorithms such as neural networks, both supervised and unsupervised, can be modified to accept CFG representations of states by initialising each neuron as an empty vector and allowing neurons to grow as required. Likewise, a table-based reinforcement learner can be modified to use a CFG representation by storing strings from the CFG in the state-action table in place of vectors.

4.3   Sensors and the Sensation Process for the Curious Information Display

The sensor and effector architecture of The Sentient, shown in Figure 8, consists of five separate sensor subsystems: The C-Bus subsystem for sensing and controlling lighting, the Teleo subsystem for sensing movement via pressure pads embedded in the floor, the Bluetooth subsystem for sensing and controlling Bluetooth devices, the virtual subsystem for controlling software running on machines in the Sentient and any virtual worlds or other software-based data sources that the Sentient is coupled with, and the camera subsystem which is an independent subsystem that provides a video stream via network cameras. All of these subsystems have device monitors, which are software that records the sensor inputs arriving from the subsystems into tables in MySQL-based context database. These tables are polled by the agent to enable it to sense the state of the room. The agent is also able to command the effector systems of the room by writing requests for actions into an action queue in the context database. These actions are carried out by device monitors, software daemons that poll the action queue and pass on the requested action to the appropriate effector subsystem.

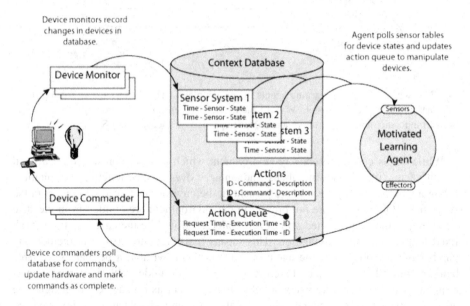

Fig. 8. The Sentient's sensor and effector architecture

In order to sense the movement of people in The Sentient, the Curious Information Display uses a portion of The Sentient's Teleo sensor subsystem. The Sentient's Teleo sensor subsystem, shown in Figure 9, is responsible for handling data from The Sentient's pressure pads. The pressure pad relays are connected to a series of Teleo modules, programmable sensor modules created by MakingThings (http://www.makingthings.com). A series of Analog Input Modules are wired to the relays along with a Teleo Intro Module which interfaces via USB with a PC on which the Teleo subsystem's device monitor is running. The device monitor makes use of the C Teleo Module API to receive pressure pad activation messages sent by the Teleo Intro Module whenever the weight pressing down on a pad changes and stores the activation data in a table in a database that records context information detected by the Sentient's sensor systems. The agent polls the table at regular intervals to retrieve the sensor inputs.

We are using just the sensors directly in front of the rear projection screen as shown in Figure 6. The sensors are polled at intervals and the sensor state at that time is returned. The Curious Information Display can also sense information about the state of its display. This includes the type of media being displayed on each grid cell. Thus, a sensed state displaying a single 9x9 image of type 1 will look like:

```
S((pic1itemType1)(pic1source:1)(pic1keyword:2))
```

A world state containing a combination of 1x1 and 2x2 images might look like:

```
S((pic141itemType:1)(pic141keyword:2)(usb/3.7.0value:1)(usb/3.6.0
value:1)(pic144itemType:1)(pic144source:3)(pic11itemType:1)
(pic11source:3)(usb/3.5.0value:1)(pic13itemType:1)
(pic13source:3)(pic141itemType:1)(pic142source:2)(usb/3.4.0value:
1)(pic120itemType:2)(pic120source:4)(pic120colour:1)(pic141source
:2)(pic142keyword:2)(pic143source:3))
```

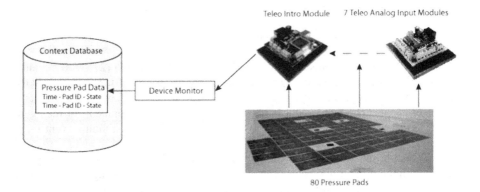

Fig. 9. The Sentient's Teleo sensor system

4.4 The Motivation Process

The motivated agent model for our Curious Information Display incorporates two forms of motivation as reward signals: extrinsic motivation and intrinsic motivation. The purpose of the extrinsic motivation is to reward the agent when people come close to the display, based on the triggering of the floor sensors near the display. We interpret people coming close to the display as a sign that the display is interesting to them and thus reward the agent.

While the extrinsic reward signal is present when people are attracted to the display, when people are not attracted to the display an agent using only an extrinsic reward signal would have no learning stimulus to direct its behaviour. Rather it would fall back on its random exploration strategy. The purpose of the intrinsic reward signal is to motivate the agent to develop interesting displays in periods where people are not attracted to its current display. Thus the intrinsic motivation signal motivates the agent to attempt to capture people's attention while the extrinsic motivation signal motivates the agent to maintain people's attention.

Extrinsic Motivation: The extrinsic motivation function for the Curious Information Display assigns a positive reward each time the agent encounters a world state in which one or more of the floor sensors directly in front of the display are triggered:

$$R_{ex(t)} = \begin{cases} 1 \text{ if } \exists \text{ floorSensor} \mid \text{floorSensor} \notin \text{doorFloorSensors \& floorSensorValue} = 1 \\ 0 \text{ otherwise} \end{cases}$$

Intrinsic Motivation: We use a motivation function based on the computational model of curiosity created by Saunders and Gero [26]. Saunders and Gero implemented a computational model of interest for social force agents by first detecting the novelty of environmental stimuli then using this novelty value to calculate interest. The novelty of an environmental stimulus is a measure of the difference between expectations and observations of the environment where expectations are formed as a result of an agent's experiences in its environment. Saunders and Gero model these expectations or experiences using an Habituated Self-Organising Map (HSOM) [22]. Interest in a situation is aroused when its novelty is at a moderate level, meaning that the most interesting experiences are those that are similar-yet-different to previously encountered experiences. The relationship between the intensity of a stimulus and its pleasantness or interest is modelled using the Wundt curve [2].

An HSOM consists of a standard Self-Organising Map (SOM) [27] with an additional habituating neuron connected to every clustering neuron of the SOM. A SOM consists of a topologically structured set U of neurons, each of which represents a cluster of events. The SOM reduces the complexity of the environment for the agent by clustering similar events together for reasoning. Each time a stimulus event $E_{(t)} = (e_{1(t)}, e_{2(t)}, \dots e_{L(t)} \dots)$ is presented to the SOM a winning neuron $U_{(t)} = (u_{1(t)}, u_{2(t)}, \dots u_{L(t)} \dots)$ is chosen which best matches the stimulus. This is done by selecting the neuron with the minimum distance to the stimulus event. The winning neuron and its eight topological neighbours are moved closer to the input stimulus by adjusting their weights. The neighbourhood size and learning rate are kept constant so the SOM is always learning. The activities of the winning neuron and its neighbours are propagated up the synapse to the habituating layer. The synaptic efficacy, or novelty, $\mathcal{N}_{(t)}$, is then calculated using Stanley's model of habituation [32]. Habituation has the effect of causing synaptic efficacy or novelty to decrease with subsequent presentations of a particular stimulus or increase with subsequent non-presentations of the stimulus. This represents forgetting by the HSOM and allows stimuli to become novel, and thus interesting, more than once during an agent's lifetime.

Once the novelty of a given stimulus has been generated, the interest I of the stimulus is calculated using the Wundt curve in Equation 1. The Wundt curve provides positive feedback $\mathcal{F}^+$ for the discovery of novel stimuli and negative feedback $\mathcal{F}^-$ for highly novel stimuli. It peaks at a maximum value for a moderate degree of stimulation as shown in Figure 10, meaning that the most interesting events are those that are similar-yet-different to previously encountered experiences.

$$I(2\mathcal{N}_{(t)}) = \mathcal{F}^{+}(2\mathcal{N}_{(t)}) - \mathcal{F}^{-}(2\mathcal{N}_{(t)}) = \frac{F_{max}^{+}}{1+e^{-\rho^{+}(2N_{(t)} - F_{min}^{+})}} - \frac{F_{max}^{-}}{1+e^{-\rho^{-}(2N_{(t)} - F_{min}^{-})}} \qquad (1)$$

F_{max}^{+} is the maximum positive feedback, F_{max}^{-} is the maximum negative feedback, ρ^{+} and ρ^{-} are the slopes of the positive and negative feedback sigmoid functions, F_{min}^{+} is the minimum novelty to receive positive feedback and F_{min}^{-} is the minimum novelty to receive negative feedback. The interest value I is used as the reward $R_{in(t)}$ which is passed from the motivation process $\mathcal{M}$ to the learning process $\mathcal{L}$ as follows:

$$R_{in(t)} = \begin{cases} I(N_{(t)}) \text{ if } E_{(t)} \text{ not empty} \\ 0 \text{ otherwise} \end{cases}$$

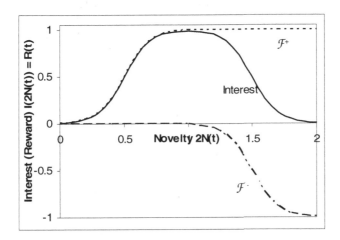

Fig. 10. The Wundt curve is the difference between positive and negative feedback functions

4.5 The Learning Process

The learning process uses the reward generated by the motivation process to drive a reinforcement learning algorithm: the Q-learning algorithm with an ε-greedy exploration strategy. We use ε=0.1 and a learning rate of 0.9. The reinforcement learning algorithm chooses an action to perform based either on expected reward or randomly to explore other potential rewards. Over time, the learning algorithm generates a behaviour policy mapping sensed states to actions that increases its potential reward.

4.6 Actions and the Activation Process

The agent modifies the display area using the following actions to manipulate the leaf nodes in the underlying data structure. Certain changes will cause an affect that can be sensed while others will not. For example, changing the colour of a leaf node currently displaying a webcam image will have no sensed affect. However, changing the source of a leaf node to 'database' while a webcam image is displayed will cause the agent to sense an image from the database rather than the webcam image.

```
A₀ = Change itemType of <leaf node> to <itemType>
A₁ = Change source of <leaf node> to <source>
A₂ = Change keyword of <leaf node> to <keyword>
A₃ = Change colour of image <leaf node> to <colour>
A₄ = Change resolution of <non-leaf node> to <res>
```

The agent cannot directly sense changes to the resolution of a node but it can sense this indirectly as a change in the configuration of nodes and leaf nodes it can sense.

4.7 Reflexes

In a real-world application, a number of practical issues arise. Firstly, the bulb on the projector has a limited lifetime and must be conserved where possible by switching the projector off. We propose to use two reflex actions to achieve this. Reflex actions use rules of the form if <condition> then <action> to decide how to respond to changes in the environment.

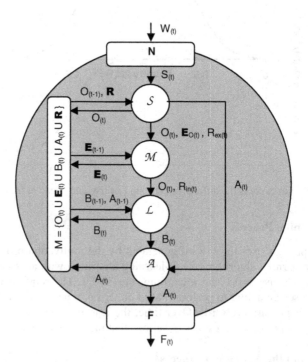

Fig. 11. A motivated reinforcement learning agent incorporating both intrinsic and extrinsic rewards and reflexes

```
if S_R1(FloorSensors:On) then
      A_R1(Projector,On) and A_R2(Agent,On)
if S_R2(TimeFloorSensorsOff:>5mins) then
      A_R3(Projector,Off)
      A_R4(Agent,Off)
```

A second issue arises because the room in which the Curious Information Display is to be displayed is also used as a space for teaching, seminars and meetings within a university. To make it possible for these activities to use the rear projection screen, we propose the following reflexes to allow people to turn the agent off:

```
if S_R3(SoftwareCommand:On) then
      A_R2(Agent,On)
if S_R4(SoftwareCommand:Off) then
      A_R4(Agent,Off)
```

These reflexes will respond to a Pause button on the application. An updated agent model showing how reflexes are implemented is shown in Figure 11. It differs from the standard motivated reinforcement learning framework in that the sensation process S also retrieves a set of rules **R** from memory which can trigger an immediate action $A_{(t)}$ by the activation process $\mathcal{A}$.

5 Conclusion

This paper has presented three agent models for intrinsically motivated learning as a basis for intelligent environments that adapt learned behaviours from patterns of usage derived from their sensor data. We develop the process of intrinsic motivation as computational models of novelty and interest that allow the agent to be rewarded for responding to unexpected changes in the patterns of sensor data, and apply these models to an application called a curious place. A model of curiosity acts as a filter for the agent to process the sensor data to determine a reward for achieving the current state of the environment. Our observations of the curious information display reveal that the model of intrinsic motivation causes the information display to generate different repeating patterns of changes in order to attract people to the sensor in front of the display, and then the display repeats a specific pattern in order to continue to receive the extrinsic reward until people move away from the sensors in front of the display. Our plans for the curious place application include the development of a parallel virtual place that can be used as a platform for motivated reinforcement learning and motivated supervised learning. This allows us to move in the direction of adaptable intelligent rooms that learn how to use the room by observing how people use the room.

Our current model of intrinsic motivation is based on novelty and interest. This has provided a good starting point for a curious place that responds to the movement of people in the place by generating novel and interesting patterns of behaviour. We plan to augment the "curiosity" approach to motivation with a "competency" approach. This combination may be more suitable for applications in which the intelligent room responds to changes in sensor technology by being curious about new events that meet the criteria of "similar yet different" because it is rewarded for repeating behaviours that generate reward for novelty, and also being motivated to develop new learned behaviour patterns that receive reward for developing competencies. Experiments with the different models of motivated learning will help determine the triggers for different kinds of motivation, the scenarios for different types of learning (reinforcement, supervised, and unsupervised), and the parameters for learning and forgetting, essential features for adaptive behaviour.

References

1. Aylett, R.A. et al., Agent-based continuous planning, in: Proceedings of the 19th Workshop of the UK Planning and Scheduling Special Interest Group (PLANSIG 2000), 2000.
2. Berlyne, D. E. Aesthetics and psychobiology. Englewood Cliffs, NJ: Prentice-Hall, (1971).
3. Brdiczka, O., Reignier, P., Crowley, J., Supervised learning of an abstract context model for an intelligent environment, in: Proceedings of the Joint sOc-EUSAI Conference, Grenoble, (2005).
4. Brooks, R.A., Coen, M., Dang, D., DeBonet, J., Kramer, J., Lozano-Perez, T., Mellor, J., Pook, P., Stauffer, C., Stein, L., Torrance, M., Wessler, M.: The Intelligent Room Project. In: Proceedings of the Second International Cognitive Technology Conference (CT'97). Aizu, Japan (1997) 271-279.
5. Canamero, L., Modeling motivations and emotions as a basis for intelligent behaviour, in: Proceedings of the First International Symposium on Autonomous Agents, ACM Press, New York, (1997).
6. Coen, M.H.: Design Principles for Intelligent Environments. In: Proceedings of the Fifteenth National / Tenth Conference on Artificial Intelligence / Innovative Applications of Artificial Intelligence. Madison, Wisconsin, United States (1998) 547–554.
7. Deci, E., & Ryan, R. Intrinsic motivation and self-determination in human behaviour. New York: Plenum Press, (1985).
8. Gemeinboeck, P., Negotiating the In-Between: Space, Body and the Condition of the Virtual, in: Crossings - Electronic Journal of Art and Technology, 4(1), (2004).
9. Dieterich, H., Malinowski, U., Khme, T. and Schneider-Hufschmidt, M. State of the art in adaptive user interfaces. In: M. Schneider-Hufschmidt, T. Khme and U. Malinowski (Editors), Adaptive User Interfaces: Principle and Practice, North Holland (1993).
10. Dzeroski, S., De Raedt, L., & Blockeel, H. Relational reinforcement learning. Paper presented at the International Conference on Inductive Logic Programming, (1998).
11. Gershenson, C., Artificial Societies of Intelligent Agents, Bachelor of Engineering Thesis, Fundacion Arturo Rosenblueth, (2001).
12. Graesser, A., VanLehn, K., Rose, C., Jordan, P. and Harter, D. Intelligent tutoring systems with conversational dialogue. AI Magazine, 22(4): 39-52, (2001).
13. Green, R.G., Beatty, W.W. and Arkin, R.M., Human motivation: physiological, behavioural and social approaches. Allyn and Bacon, Inc, Massachussets, (1984.).
14. Hammond, T., Gajos, K., Davis, R. and Shrobe, H., An Agent-Based System for Capturing and Indexing Software Design Meetings, in: Gero, J.S. and Brazier F.M.T., eds., Agents in Design 2002, Key Centre of Design Computing and Cognition, University of Sydney (2002) 203-218.
15. Horvitz, E., J. Breese, Heckerman, D., Hovel, D. and Rommelse, K., The Lumiere project: Bayesian user modelling for inferring the goals and needs of software users, in: Proceedings of the Fourteenth Conference on Uncertainty in Artificial Intelligence, (1998), 256-265.

16. Kandel, E.R., Schwarz, J.H. and Jessell, T.M., Essentials of neural science and behaviour. Appleton and Lang, Norwalk, (1995).

17. Kaplan, F., & Oudeyer, P.-Y. Motivational principles for visual know-how development. Paper presented at the Proceedings of the 3rd international workshop on Epigenetic Robotics : Modeling cognitive development in robotic systems, Lund University Cognitive Studies, (2003).

18. Kohonen, T. Self-organisation and associative memory. Berlin: Springer, (1993).

19. Krueger, M., Gionfriddo, T., et al., Videoplace: an artificial reality, in: Human Factors in Computing Systems, CHI'85, ACM Press, New York, (1985).

20. Luck, M., & d'Inverno, M. Motivated behaviour for goal adoption. Paper presented at the Multi-Agent Systems: Theories, Languages and Applications - Proceedings of the fourth Australian Workshop on Distributed Artificial Intelligence, (1998).

21. Maher, M.L., Merrick, K., and Macindoe, O. Can designs themselves be creative?, in: Computational and Cognitive Models of Creative Design, Heron Island, (2005), 111-135.

22. Marsland, S., Nehmzow, U., & Shapiro, J. A real-time novelty detector for a mobile robot. Paper presented at the EUREL European Advanced Robotics Systems Masterclass and Conference, (2000).

23. Merceron, A. Languages and Logic: Pearson Education Australia, (2001).

24. Norman, T. J. and Long, D., Goal creation in motivated agents, in: Intelligent agents: theories, architectures and languages, Springer-Verlag, (1995).

25. Russel, J. and Norvig P., Artificial intelligence: a modern approach, Prentice Hall Inc, (1995).

26. Saunders, R., Gero, J.S.: Designing for Interest and Novelty: Motivating Design Agents. In: de Vries, B., van Leeuwen, J., Achten, H. (eds.): CAADFutures 2001. Kluwer, Dordrecht (2001) 725-738.

27. Saunders, R., & Gero, J. S. Curious agents and situated design evaluations. In J. S. Gero & F. M. T. Brazier (Eds.), Agents In Design (pp. 133-149): Key Centre of Design Computing and Cognition, University of Sydney, (2002).

28. Schmidhuber, J. A possibility for implementing curiosity and boredom in model-building neural controllers. Paper presented at the The International Conference on Simulation of Adaptive behaviour: From Animals to Animats, (1991).

29. Schmill, M. and Cohen, P., A motivational system that drives the development of activity, AAAMAS, ACM, Bologna, (2002).

30. Singh, S., Barto, A.G., and Chentanez, N.: Intrinsically Motivated Reinforcement Learning. http://www.eecs.umich.edu/~baveja/Papers/FinalNIPSIMRL.pdf Accessed 7/4/2004 (2004)

31. Sloman, A. and M. Croucher, Why robots will have emotions, in: Proceedings of the 7th International Joint Conference on Artificial Intelligence, Vancouver, (1981).

32. Stanley, J. C. Computer simulation of a model of habituation. Nature, (1976) 261, 146-148.

33. Sutton, R.S. and Barto, A.G., Reinforcement learning: an introduction, MIT Press, (2000).

34. Wang, Y., Huber, M. et al., User-guided reinforcement learning of robot assistive tasks for an intelligent environment, in: Proceedings of the IEEE/RJS International Conference on Intelligent Robots and Systems, IEEE, Las Vegas, (2003).

35. Watkins, C., Learning from delayed rewards, PhD Thesis, Cambridge University, (1989).

36. White, R. W. Motivation reconsidered: The concept of competece. Phychological Review, (1959), 66, 297-333.

How to Teach Computing in AEC

Karsten Menzel[1], Danijel Rebolj[2], and Žiga Turk[3]

[1] Professor, University College Cork, College Road, Cork, Ireland
K.Menzel@ucc.ie
[2] Professor, Faculty of Civil Engineering, University of Maribor,
Smetanova 17, 2000 Maribor, Slovenia
danijel.rebolj@uni-mb.si
[3] Professor, Faculty of Civil and Geodetic Engineering, University of Ljubljana,
Jamova 2, 1000 Ljubljana, Slovenia
ziga.turk@itc.fgg.uni-lj.si

Abstract. Information Technology in Architecture, Engineering and Construction (IT in AEC) is a niche topic, lacking critical mass in most faculties. Researchers are creating critical mass by intense international collaboration. The same is true for education about IT in AEC where critical mass can be achieved in a similar way. In 2004 the first students entered a new postgraduate program called IT in AEC. The program was developed by seven European universities. It is unique not only because of its content, which covers various related IT-topics, but also in the ways in which it is organized and executed and the didactic methods used. It is based on a commonly agreed upon curriculum and is delivered using distant learning technologies. In the first part, the paper describes the reasons for developing the new program. The second part of the paper describes the development process of the program and its content. The final part of the paper describes the learning environment and the underlying teaching-learning scenario(s).

1 Motivation

Information Technology in Architecture, Engineering and Construction (IT in AEC - sometimes called Construction Informatics) is a mature research field with a dynamic and growing body of knowledge (Turk 2006). However, the impact of this research on construction practise has been limited (Froese 2004). This impact is generally achieved through three mechanisms: (1) developments for products (such as software that embeds that knowledge), (2) standardisation and best practise that prescribe the knowledge to be used in the industry and most importantly (3) education (Turk 2004).

Generally, curricula development has not followed the results of research- the range of information and communication technology in Civil Engineering curricula, for example, is incomplete and often restricted only to skills related to the use of technology (Table 1 "Heitmann"). Therefore, traditional teaching and learning scenarios need to be re-shaped, interconnected and extended to meet the needs for specialized civil engineers with deeper IT knowledge in the AEC-industry (Table 1 "ASCE").

Surveys and experiences show (Rebolj and Tibaut 2005) that the share and content of courses related to computer science and IT in undergraduate civil engineering

I.F.C. Smith (Ed.): EG-ICE 2006, LNAI 4200, pp. 476–483, 2006.
© Springer-Verlag Berlin Heidelberg 2006

curricula varies considerably from university to university. Typically, there are general introductory courses, programming courses and specialized courses in IT applications like the design of building models, technical drawings, finite element and heat loss programs for the determination of physical behaviour, information systems to support construction management and systems for enterprise resource planning.

However, the courses are mostly unconnected and only provide limited knowledge about the specific aspects of computer science and IT. There is little understanding of the holistic potentials of today's Information and Communication Technology (ICT). Such a way of learning about discrete, unconnected software tools does not properly educate in an area where the potential of IT in construction is the largest- in integrating the fragmented profession. *"Often there are gaps in modern engineering curricula regarding computer science. ..... However, many engineers believe inaccurately that computing is only a skill to be acquired on the job, not also a science to be learnt in an academic setting. Nevertheless, most will agree that there is a growing lack of correlation between what is taught ..... and how engineers use computers in practice."* (see Smith 2003).

According to the research conducted by Heitmann (Heitmann et al. 2003), academic requirements for all engineering programs at the Bachelors' level should contain the following IT-related abilities to:

- use common computer tools to produce documents and make presentations
- develop system models to carry out calculations and simulations
- design and maintain simple Internet work-space to support interactive teamwork
- computer-based tasks using object oriented programming and expert systems
- use professional computer codes to prepare data, obtain reasonable results from calculations and to verify results from an "engineering" point of view

The Master level should add the ability to:

- understand algorithms, their limitations and requirements, prepare the data for processing with professional code and analyze obtained results.

Due to the concern of appropriate computing component in the curriculum, the 'American Society of Civil Engineers' Task Committee on Computing Education of the Technical Council on Computing and Information Technology conducted a series of surveys in 1986, 1989, 1995 and 2002 to assess the current computing component in the curriculum of civil engineering. Key findings of the latest study (Abudayyeh 2004) include:

- the relative importance of the top four skills (spreadsheets, word processors, computer aided design, electronic communication) has remained unchanged;
- the importance and use of geographic information systems and specialized engineering software have increased over the past decade;
- the importance and use of equation solvers and databases have declined over the past decade;
- programming competence is ranked very low by practitioners; and
- the importance and use of expert systems have significantly decreased over the past decade.

Table 1. Competence Areas and Importance of Educational Programs in AEC

	Heitmann[1]	Smith	ASCE[2]
Common tools (spreadsheet, word processing, presentation)	Bachelor	n.a	High
Programming and underlying concepts	n.a	B ~)	Low
Computer Graphics, Geometric Modelling (GIS, CAD)	B*)	B ~)	High
Data preparation, evaluation	B*)	n.a.	n.a.
Modelling, calculation, simulation (databases, equation solvers)	B*)	B ~)	Medium
Expert systems (search techniques, case-based reasoning, model based reasoning)	B*)	B ~)	Medium
eWork, CSCW (web-technology, workflow management)	B*)	+)	High
Understand, evaluate, and select algorithms	Master	+)	n.a.
Design, operation, maintenance of software (Requirements engineering, complexity analysis, reverse engineering)	n.a.	B ~)	n.a.
Ubiquitous Computing	n.a.	+)	n.a.

Legend: B=Bachelor Curriculum *)=application level ~)=Smith 2000 +)=Smith 2003
[1]=see Heitmann et al. (2003)   [2]=see Abudayyeh et al. (2004)

2 A Postgraduate Course Pool on It in AEC

Undergraduate courses and graduate programs in the area of IT in AEC are offered in Germany, Sweden, Denmark, The Netherlands, Switzerland, Austria, the USA, Canada, Australia and New Zealand. Turk and Delic (2003) developed a proposed core undergraduate curriculum for the Bologna undergraduate model. At the graduate level, team oriented, project-based scenarios are used in several distant education courses in AEC. The best known course is Stanford University's P^{5}BL Course, which has been running for a decade now (Fruchter 1996). Most of these courses force students to use information and communication technology in order to learn the underlying principles of IT in AEC (Menzel 1997, Bento et al. 2004). The maturity of these topics has been proven by some excellent books with engineering examples and exercises (Pahl and Damrath, 2000; Raphael and Smith 2003).

IT in AEC is still a new topic which is rapidly developing. In research, critical mass was achieved by creating international, collaborative research projects. To spread the research results through teaching, the same kind of critical mass had to be achieved. Forces have been joined to develop an international, multi-institutional postgraduate program (ITC Euromaster 2005). In 2001 seven European universities started to develop a new postgraduate program in IT in AEC. The program development was funded by the European Commission through the Socrates / Erasmus framework between 2001 and 2003, and again in 2004 and 2005 when two more partners joined in the program dissemination process.

The main purpose of the ITC-Euromaster project was to develop a commonly agreed upon curriculum in IT in AEC. This curriculum should complement the existing portfolio of teaching programs and should meet the growing demand for IT-skills in the AEC-sector. The development of the individual modules in the curriculum was coordinated by a responsible partner institution. Teaching materials have been prepared in digital form, conforming to available e-learning standards, such as SCORM.

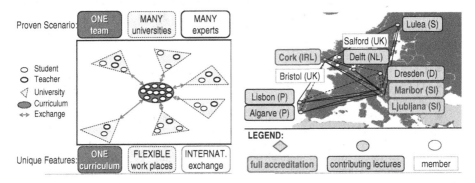

Fig. 1. Scenarios, Features and Partners of the ITC-Euromaster Network

Based on the results of a skill audit, the review of existing courses at partner institutions, as well as market research and analysis, a course structure has been developed consisting of 12 subjects (as listed in Table 2).

Table 2. ITC-Euromaster course pool: the commonly agreed upon curriculum in IT in AEC

INFORMATION MANAGEMENT	KNOWLEDGE
1. The role of construction informatics	6. Knowledge management
2. Data structuring and databases	7. Engineering artificial intelligence
3. Information modelling and retrieval	COMMUNICATION
4. Geometric modelling & visualization	8. Mobile computing in AEC
5. Software engineering	9. Computer mediated communication
	BUSSINESS / MANAGEMENT
	10. Virtual enterprises
	11. Virtual construction
	12. eBusiness and Data Warehouses

The curriculum is focused on students who have finished their undergraduate studies with a university degree in civil, building or structural engineering as well as architecture. The program graduates will earn a new "European Masters in Construction Information Technology" academic degree, which shall enable them to continue with the relevant PhD study or immediately start to work in the industry as civil engineers with a specific focus on Information Technology. Organization of the program called for inventive methods as well. Students enroll at each university, but they enter a single virtual class with teachers coming from the different partner universities (see Figure 1).

3   Applied Principles of Teaching and Learning

Since the creation of the Open University (London, 1969), "open", "distance", "flexible", "resource based" and "distributed learning" have expanded dramatically all over the world. The era of "e-learning" begun in the late 1990s. E-learning is now at the top of the agenda of public and private universities worldwide because it has the potential to change education and training radically, to open new ways of learning and to increase the ability of people to acquire new skills. From the functional point of view, there is a widespread acceptance that a holistic e-learning solution comprises three key elements (Henry 2001):

- CONTENT: adds to the knowledge, skills and capabilities of the human capital.
- TECHNOLOGY: comprises the infrastructure, management systems and learning technologies.
- SERVICES: include consulting, support, as well as design and build services.

Design, construction and implementation of eLearning scenarios aim at the optimal support of different methodological scenarios. In Figure 2, teaching methods are classified into three main groups: Presentation, Interactive Teaching and Independent Work. These teaching methods can be applied to realize different teaching strategies, such as 'didactica magna' (Comenius 1657), 'visiual instruction' (Pestalozzi 1801), 'advance-organizer strategy' (Ausubel 1960), 'basic-concept strategy', 'establishment of networks' or 'teaching of schemata'. (For complete references see Steindorf 2000).

Kolb (1984) provides one of the most useful, descriptive models of the learning process. This model suggests that there are four consecutive stages: Concrete Experience (CE), Reflective Observation (RO), Abstract Conceptualization (AC) and Active Experimentation (AE). Based on these four stages, a four-type definition of learning styles was developed for which Kolb introduces the terms: Diverging (CE/RO), Assimilating (AC/RO), Converging (AC/AE) and Accommodating (CE/AE).

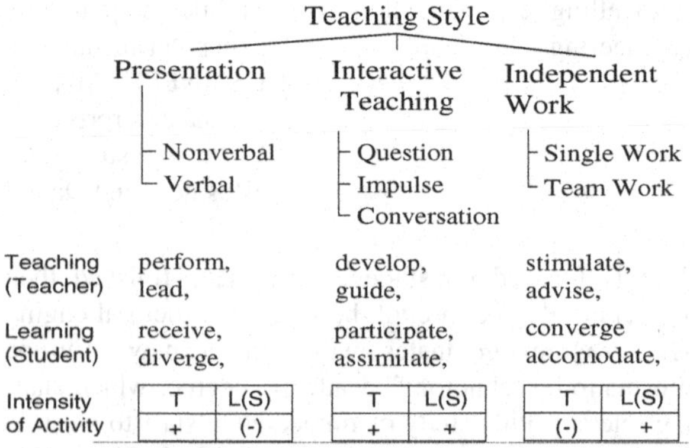

Fig. 2. Teaching Styles and related Teaching-Learning Activities

The different teaching methods and teaching-learning principles depicted in Figure 2 require different levels of ICT-support. For each of the categories, different levels of ICT-needs can be defined and are applied to the ITC-Euromaster framework as summarized in Table 3.

4 Collaborative Networks Within the ITC Euromaster Framework

Teaching on the Web induces different social and collaborative processes to the traditionally time-bound, place-bound and role-bound education models. Collaborative learning tries to create a virtual social space that must be managed for the teaching and learning needs of the particular group of people inhabiting that space. In an ideal collaborative learning set up, a student will know a great deal about his fellow students and faculty before he begins working through the material. He will be prompted with questions that have been very carefully designed to encourage him to link the material he is learning to his own knowledge and experience, as well as stimulate him to interact with other students and faculty. This model will use a database underlying the course management system to link people and information in new ways that will help them to understand the community of learners they have joined, as well as affect their relationship to the material itself.

Therefore, the current ITC Euromaster e-learning environment consists of two components: the Course Management System (CM), which is the entry point to the program (ITC Euromaster 2005), and a Virtual Classroom (VC).

Table 3. ICT-support for different teaching styles within the ITC-Euromaster network

Teaching Method: *ICT-support*	Presentation	Interactive Teaching	Independent Work	ITC Euromaster
Degree of Organization	scheduled, prepared	scheduled, improvisation	not scheduled, spontaneous	
PRESENTATION (nonverbal, verbal)				
Presentation devices	(++)	(++)	(-/+)	VC
Shared information	Read	Read / (Write)	Read / Write	CM
Audio transmission	UniDirectional	BiDirectional	BiDirectional	VC
Video transmission	BiDirectional	(+)	BiDirectional	VC
QUESTION:				
eForum	(--)	(-/+)	(++)	CM
eVoting	(--)	(+)	(--)	n.a.
Chat	(--)	(-/+)	(++)	VC
IMPULSE				
Messenger	(--)	(-/+)	(++)	*)
CONVERSATION / TEAM WORK				
Red-lining capability	(--)	(++)	(++)	*)
Shared application	(--)	(--)	(++)	VC

Legend: (--) not required; (-/+)recommended; (+) highly recommended; (++) required
VC: Virtual Classroom; CM: Course Management System; *) Windows components

The main function of the CM is to enable access to teaching and learning material, as well as other relevant functions (e.g. forums) and information (teacher and student list, timetables etc.) from the Internet. The course management system is based on the Modular Object-Oriented Dynamic Learning Environment (Moodle 2005).

The VC is supported by the ClickToMeet videoconferencing system, enabling teachers to directly communicate with their classes. A participant list, chat, audio and video control, document sharing, application sharing and a whiteboard are the basic parts of the VC. Both systems are interlinked and can be used as an integrated system.

There is no specific electronic tool or methodology to assess the students' learning progress or teachers' performance. Lecturers organize short tests during lectures and seminars. Students deliver the test results by eMail as electronic documents. Essays and theses are delivered by the students and commented on by the supervisors in an electronic way. Final exams are prepared by the responsible lecturer. The exam itself is organized locally by each participating university and monitored by local lecturers or teaching assistants. At the end of each teaching period, questionnaires are handed out to the students in order to get their feedback with regards to teaching style, performance of the ICT-infrastructure, etc.

5 Conclusion

At the University of Maribor and the University of Ljubljana, the program was accredited in 2004. Since January 2006 University College Cork has become the tenth member of the ITC-Euromaster network and accredited a 12-month postgraduate master program on "Information Technology in Architecture, Engineering, and Construction," using most of the ITC-Euromaster modules (http://zuse.ucc.ie/master). The other partner institutions are either in the accreditation process; or, the integration of the new program into existing programs is in progress.

The ITC-Euromaster network and the ITC@EDU workshop series are the two basic elements to support sustainable, further development of the ITC-Euromaster framework. The ITC-Euromaster network has managed the programme organization since 2004. It has organized the transition process from an EU-project into a self-sustaining course pool after the EU-funding period ended in the middle of 2005. All project members have signed a common "course pool agreement".

Since 2002 the ITC@EDU workshop series has proven to be a stable platform to maintain the discussion process amongst the members of the ITC-Euromaster network, external advisors and international partners. Through the workshops, the internal discussion process is stimulated and feedback is given by external experts to the network members, contributing to the continuous refinement and improvement of the program content, the "delivery" mode and the ICT-infrastructure of the ITC-Euromaster program.

With ten partners and three program accreditations developed out of the EU-funded ITC-Euromaster project, our network has developed the necessary critical mass to promote "IT in AEC" as an interdisciplinary scientific discipline. It has also substantially contributed to the sufficient transfer of knowledge and technology from academic institutions into the different areas of the AEC- and FM industry.

References

1. Abudayyeh, Cai, Fenves, Law, O'Neill, Rasdorf. (2004). "Assessment of the Computing Component of Civil Engineering Education 2004." *J. Comp. in Civ. Engrg.*, 18(3), 187-195
2. Adelsberger, Collis, Pawlowski, (Eds.) (2002). "Handbook of information technologies for education and training." Springer, Berlin - Heidelberg.
3. Bento, Duarte, Heitor, Mitchell, (Eds) (2004). "Collaborative Design and Learning – Competence Building for Innovation". Praeger Publisher, Westport CT, USA.
4. Froese (2004). "Help wanted: project information officer." *eWork and eBusiness in Architecture, Engineering and Construction* A.A. Balkema, Rotterdam, The Netherlands.
5. R. Fruchter (1996). "Multi-Site Cross-Disciplinary A/E/C Project Based Learning" in: *Proceedings of the Third Conference on Computing in Civil Engineering* S. 126 ff, American Society of Civil Engineers, New York, 1996 (ISBN 0-7844-0182-9).
6. Heitmann, Avdelas, Arne. (2003). *Innovative Curricula in Engineering Education.* Firenze University Press.
7. Henry P. (2001), "E-learning technology, content and services", *Education + Training*, MCB University Press, USA, 43(4), (251)
8. ITC EUROMASTER (2005), *The programme portal.* <http://euromaster.itcedu.net>.
9. Kolb (1984). "Experiential Learning: experience as the source of learning and development." New Jersey: Prentice-Hall.
10. Menzel, Garrett, Hartkopf, Lee. "Technology Transfer in Architecture and Civil Engineering by Using the Internet - Illustrated Through Multi-National Teaching Effort" *Forth Congress on Computing in Civil Engineering.* ASCE. New York. 1997. (224-231).
11. Moodle (2005), *The Moodle homepage.* <http://moodle.org>.
12. Pahl and Damrath. "Mathematische Grundlagen der Ingenieurinformatik"; Springer, Berlin, Heidelberg. 2000.
13. Raphael and Smith. "Fundamentals of Computer Aided Engineering". John Willey. 2003.
14. Rebolj and Menzel (2004). "Another step towards a virtual university in construction IT", *Electroic Journal of Information Technology in Construction*, 17(9), 257-266 <http://www.itcon.org/cgi-bin/papers/Show?2004_17> (Oct. 20, 2005).
15. Rebolj and Tibaut (2005). "Computer Science and IT in Civil Engineering Curricula" *Proceedings of the IVth International Workshop on Construction Information Technology in Education.* TU Dresden. (35-42) (ISBN 3-86005-479-1).
16. Smith and Raphael (2000). "CA course on fundamentals of computer-aided engineering" *Computing in Civil and Building Engineering (VIIIth ICCCBE).* American Society of Civil Engineers, Reston, VA, USA, pp 681 ff.
17. Smith, I.F.C. (2003). "Challenges, Opportunities, and Risks of IT in Civil Engineering: Towards a Vision for Information Technology in Civil Engineering." American Society of Civil Engineers, Reston, VA, USA (on CD) (1-10).
18. Steindorf, G. "Grundbegriffe des Lehrens und Lernens". Klinkhardt. Bad Heilbunn. 2000.
19. Turk (2006). "Construction Informatics: Definition and Ontology", accepted paper, Advanced Engineering Informatics.
20. Turk Z (2004). "Construction Informatics Themes in the Framework 5 Programmed." *5th European conference on product and process modelling in the building and construction industry - ECPPM 2004*, A.A. Balkema: Taylor & Francis Group. (399-405).
21. Turk and Delic (2003): "Undergraduate Construction Informatics Curriculum." *Concurrent engineering - The vision for the future generation in research and application*, A.A. Balkema: Taylor & Francis Group. (1185-1191).

Evaluating the Use of Immersive Display Media for Construction Planning

John I. Messner

Assistant Professor of Architectural Engineering,
Director of the Computer Integrated Construction Research Program,
The Pennsylvania State University, University Park, PA 16802, USA
jmessner@engr.psu.edu

Abstract. The aim of this research is to develop construction planning and plan review processes within virtual environments that result in the consistent development of more innovative and higher quality construction plans. During the early stages of this research, we have shown that affordable immersive display systems can be constructed and effectively used to allow construction project teams to better visualize product and process information in a stereoscopic 3D environment. The Immersive Construction (ICon) Lab has been built at Penn State University as a test bed facility for 3D and 4D CAD model visualization. This paper presents an overview of three case study projects performed in the immersive display system with a focus on construction planning activities. Early results illustrate that a project team can identify innovative solutions to construction challenges when performing a plan review of 3D and 4D virtual prototypes in an immersive environment.

1 Overview

The ability to visualize and experience facility design and construction plan information is critical to providing valuable information and feedback into design and construction process decisions. Many valuable research efforts have focused on the development of solutions for modeling the product and process information for a facility. An overview of many of these modeling efforts can be found in Eastman[1] and Lee et al.[2]. Additional research has also focused specifically on the development of information models and modeling methods related to construction planning with 4D CAD models along with the addition of space planning information to the models. Several valuable references include Koo and Fischer[3], Akinci et al.[4], and Dawood et al.[5] along with others.

While significant research has focused on the development of models that can benefit the Architecture/Engineering/Construction (AEC) Industry, much less research has focused on the most appropriate methods for displaying and interacting with the models. This research focuses on the impact of the display media used to visualize 3D and 4D models during the construction planning and plan review process. The use of immersive display systems allows the project team to gain an increased sense of immersion within the 3D or 4D model, sometimes referred to a sense

I.F.C. Smith (Ed.): EG-ICE 2006, LNAI 4200, pp. 484–491, 2006.

of presence[6]. This additional immersion can provide an experience where people feel embedded in the design, and they can gain a better sense of scale since they can visually navigate the model at full scale.

2 Virtual Facility Prototyping

The goal of this paper is to summarize the research results and observations gained from performing three construction plan reviews using virtual facility prototypes.

Many industries are implementing virtual prototyping to increase design iterations and allow for a more effective review of product designs prior to fabrication[7]. Many definitions exist for virtual prototyping and there are significant variations in these definitions[8]. Chua et al.[9] define virtual prototyping as 'the analysis and simulation carried out on a fully developed computer model, therefore performing the same tests as those on the physical prototype.' A common thread in most definitions includes the development of a computer model and the visual display of the model in some form of virtual environment which allows a user to interact with the model in a more natural manner. For this paper, virtual facility prototypes will be defined as having two components: 1) the computer model (e.g., 3D or 4D model), and 2) the immersive display media used for viewing and interacting with the model(s).

2.1 Computer Model

For the three case study projects, both 3D and 4D models were developed for each project. These models were developed in various 3D CAD applications. The models were then converted to formats that could be easily loaded into the immersive display systems. The conversion varied based on the display media used. It is important to note that the conversion of 4D models into the immersive display system for stereoscopic visualization was not always accomplished since it is a more tedious process to address the temporal data. Some construction plan reviews included a combination of immersive 3D models and non-immersive (or non-stereoscopic) 4D models.

2.2 Media of Display

This research focused on the use of immersive display systems that allowed users to be placed within a 3D stereoscopic virtual model. Two different immersive display systems were used in this research. The first system is a CAVE (Cave Automatic Virtual Environment) in the SEA Lab at Penn State[10]. This display system has five projected surfaces (four walls and a floor) and includes active stereoscopic visualization, position tracking, and specialized audio (see Fig. 1).

The second display system is a 3 screen immersive projection display system in the Immersive Construction (ICon) Lab[11]. The 3 screen display uses 6 projectors for passive stereo visualization which allow users to be immersed within a 3D or 4D virtual model of a facility at full scale (see Fig. 2). The display system can be operated from a single computer with a Windows XP operating system or from a four computer Linux cluster. This allows for the use of standard applications developed for a Windows operating system along with immersive virtual reality applications. This lab was developed to provide an affordable and relatively easy to use immersive display system to members of the Architecture, Engineering and Construction (AEC) industry. The cost of the display system was significantly lower than the cost of a typical CAVE display.

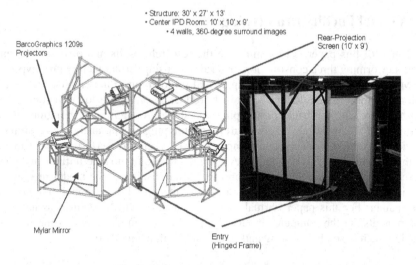

Fig. 1. SEA Lab CAVE at Penn State

Fig. 2. Rendering of three screen display system in the Immersive Construction (ICon) Lab at Penn State

3 Construction Plan Review Case Studies with Virtual Facility Prototyping

The three case studies summarized below have been performed over the past three years. A summary of the case studies are shown in Table 1 and a brief description of each is provided in the following sections. Note that references are provided for readers who wish to learn more about any individual case study.

Table 1. Virtual Facility Prototype Case Study Projects

Case Study Project	Model	Media
Westinghouse AP600 Nuclear Power Plant – Room 12306	3D CAD & 4D CAD Model	SEA Lab CAVE
Stuckeman Family Building (University Park, PA)	3D Building Information Model & 4D CAD Model	ICon Lab
The Village at Shirlington, Shirlington, VA	3D CAD & 3D CAD Model	ICon Lab

3.1 AP 600 Nuclear Power Plant Prototype

The first case study was one part of a larger project aiming to investigate the value of virtual mock-ups for nuclear power plant design, construction, and operation. A virtual prototype of the Westinghouse AP 600 was developed from the 3D CAD design of the facility (see Fig. 3). Two construction planning experiments were then performed to assess the value of the virtual facility prototype.

Fig. 3. Construction plan review meeting for Room 12306 displayed in the SEA Lab CAVE

Prototype Components. The prototype for a room (Room 12306) within the AP 600 plant was developed which included a 3D virtual model in the SEA Lab CAVE. In addition, a 4D interface was developed to allow for the sequential display of the construction components within the modular design.

Summary of Results. Two experiments were performed in the virtual facility prototype. The first was an investigation of the ability for two groups of two graduate students to develop logical construction schedules for Room 12306 in the CAVE. The results show that relatively inexperienced participants could use the mock-up to develop reasonable schedules for the room.

The second experiment focused on evaluating the value of reviewing construction plans in 4D within the prototype. For this experiment, four experienced construction superintendents (in two teams of two people) were provided with the paper drawings in isometric format for Room 12306. They were then asked to develop a schedule for the room and identify constructability issues. Following the development of their schedule, they traveled to the SEA Lab and reviewed their schedule in the CAVE. It was interesting to note that prior to the CAVE review, the two teams only identified a total of 2 constructability issues. Following the review, they identified a total of 10 module boundary suggestions and a total of 9 weld location change suggestions. The superintendents also rated the CAVE review as high on ease of use and the planners gained confidence level in their schedules based on pre and post surveys. By interactively developing a schedule within the CAVE, the planners were able to reduce their previous schedule duration of 35 days to a revised schedule of 25 days, primarily through the identification of opportunities to perform multiple activities at the same time without physical space interferences [12].

3.2 Stuckeman Family Building Prototype

The Stuckeman Family Building is a 4 story building on the Penn State campus at University Park, PA opened in August 2005. During construction, a virtual facility prototype of the building was developed to gain feedback from future occupants, and to aid in the construction planning and plan communication for the project.

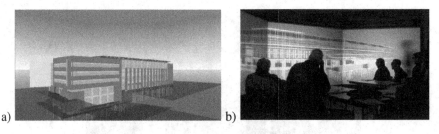

a)   b)

Fig. 4. Stuckeman Family Building: a) 3D Model; and b) construction planning meeting in ICon Lab

Prototype Components. The virtual facility prototype for this building project included a 3D Building Information Model that was developed in Graphisoft ArchiCAD along with a 4D CAD model developed in Common Point Project 4D. The 3D model was converted into VRML format for stereoscopic display in the ICon Lab using the BS Contact Stereo application. The modeling team worked with the construction project manager to develop a 4D model of the project.

Summary of Results. Participants from the project team reviewed the virtual prototype during a construction progress meeting. Following the meeting, the participants completed a survey on their perspectives. Results from the survey show that 85% of the participants felt that they had a better understanding of the building design following the 1 hour meeting in the ICon Lab. 85% of the participants also felt that the virtual prototype could be a valuable communication tool on the project and that it could help avoid delays in construction[13, 14].

3.3 The Village at Shirlington Prototype

The Village at Shirlington is an eleven story condominium building project under construction in Shirlington, VA. The project fell behind schedule early in the excavation phase due to contaminated soil and unexpected groundwater. Therefore, the project team decided to participate in a detailed schedule review meeting to seek opportunities for accelerating the remaining schedule.

a)   b)

Fig. 4. Shirlington Village Project: a) 4D CAD model; and b) construction plan review meeting in ICon Lab

Prototype Components. The prototype included a 3D CAD model that was converted into VRML format for display within the ICon Lab. A 4D CAD Model was also developed in Common Point Project 4D from the contractor's baseline schedule.

Summary of Results. The prototype was reviewed by seven members of the project team during a project team meeting that occurred within the ICon Lab. During the meeting, a time study was performed to evaluate the types of discussion that occurred during the meeting using discussion categories outlined by Liston et al.[15] Following the meeting, these values were compared to an average value taken from 17 more traditional project meetings. The results show that the type of conversation that occurred in the meeting in the ICon Lab included a greater percentage of evaluative and predictive discussions with less time spent on descriptive and explanative discussion[16]. In post meeting interviews, the participants stated that they felt the team members were able to 'play off of other's comments and could look at different alternatives.' The team was able to identify opportunities to recover 14 working days of schedule time, primarily through changes in the scaffolding method for the masonry work, and the team gained an improved level of confidence in the schedule.

4 Summary of Case Study Findings

The three case studies show that virtual facility prototypes can provide value in the construction planning and plan review process. The quantification of the value is difficult to determine due to the challenges of documenting specific value added on each project and isolating the specific benefits of the prototype. The following observations regarding the value of virtual facility prototyping were documented in the case studies performed:

1. Participants were engaged by the virtual environments which caused them to spend more time performing their review than they would typically spend in another meeting;
2. Participants found it relatively easy to identify potential conflicts and constructability issues within the space;
3. Participants engaged in more in-depth conversions as illustrated by the shift in communication types from descriptive and explanative, to an increased amount of time spent on evaluative and predictive communication;
4. Participants gained an increased level of confidence in the project schedule (frequently achieved after identifying several logic revisions in the schedule);
5. Participants were able to identify innovative planning solutions that they had not previously identified using traditional media; and
6. The prototypes made the review and development of schedules easier for inexperienced construction planners.

It is important to note that it is difficult to specifically study the value of the virtual facility prototype for construction planning on actual case study projects. For the three case studies presented in this paper, the participants in the study were required to perform the review after completing a paper based review of the project schedule which potentially impacted the results obtained in the virtual facility prototype.

5 Future Research

Research is continuing on the evaluation of the use of immersive display systems for construction planning. If the process can be systematized to take full advantage of the virtual prototype, the author believes even more benefit can be gained. For example, if the participants can be provided with a review process which specifically focuses on the information available to them in the virtual prototype, then they could gain increased benefits. Another goal is to continue to isolate the various attributes of the virtual facility prototype, along with other factors (e.g., people, physical place, and planning process), in more controlled experiments. Additional research is also needed to improve the use of product and process information in immersive display environments. Throughout these case studies, information was frequently lost when converting from one format to another and these issues associated with application interoperability and data transfer can significantly hinder the value of the developed prototype since it does not contain detailed information.

While it remains difficult to specifically quantify the additional value of displaying 3D and 4D models within immersive display systems, it is clear that the display media can impact the communication between team members and aid in the identification of innovative planning solutions.

Acknowledgements

The author would like to thank all the participants and researchers who aided in the development and execution of the case study projects with specific reference to members of the CIC Research Program and the SEA Lab. The author would also like to thank the U.S. Department of Energy and the National Science Foundation (Grants 0343861 and 0348457) for supporting this research. Any opinions or recommendations expressed in this paper are those of the author and do not necessarily reflect the views of the NSF or the DOE.

References

1. Eastman, C. M. (1999). *Building product models: Computer environments supporting design and construction*, CRC Press LLC, Boca Raton, FL.
2. Lee, A., Marshall-Ponting, A. J., Aouad, G., Wu, S., Koh, I., Fu, C., Cooper, R., Betts, M., Kagioglou, M., and Fischer, M. (2003). "Developing a vision of nD-enabled construction." The Center of Excellence for Construct IT, Salford, U.K.
3. Koo, B., and Fischer, M. (2000). "Feasibility study of 4D CAD in commercial construction." *Journal of Construction Engineering and Management, ASCE*, 126(4), 251-260.
4. Akinci, B., Fischer, M., and Kunz, J. (2002). "Automated generation of work spaces required by construction activities." *Journal of Construction Engineering and Management, ASCE*, 128(4), 306-315.
5. Dawood, N., Sriprasert, E., Mallasi, Z., and Hobbs, B. (2002). "4D visualisation development: real life case studies." *Distributing Knowledge In Building, CIB W78 International Conference*, The Aarhus School of Architecture, Denmark.
6. Slater, M., and Wilbur, S. (1997). "A Framework for Immersive Virtual Environments (FIVE): speculations on the role of presence in virtual environments." *Presence: Teleoperators and Virtual Environments*, 6(6), 603-616.
7. Schrage, M. (2000). *Serious play: how the world's best companies simulate to innovate*, Harvard Business School Press, Boston, MA.
8. Wang, G. G. (2002). "Definition and review of virtual prototyping." *Journal of Computing and Information Science in Engineering (Transactions of the ASME)*, 2(3), 232-236.
9. Chua, C. K., Leong, K. F., and Lim, C. S. (2003). *Rapid prototyping: principles and applications*, World Scientific Publishing Co. Pte. Ltd., Singapore.
10. Shaw, T. (2002). "Applied Research Lab at Penn State University, Synthetic Environment Applications Lab (SEA Lab)." May 7, (www.arl.psu.edu/facilities/facilities/sea_lab/sealab.html), Accessed: Dec. 20, 2002.
11. Computer Integrated Construction Research Program. (2005). "Immersive Construction (ICon) Lab." (http://www.engr.psu.edu/ae/cic/facilities/ICon.aspx), Accessed: February 28, 2006.
12. Yerrapathruni, S., Messner, J. I., Baratta, A., and Horman, M. (2003). "Using 4D CAD and immersive virtual environments to improve construction planning." *CONVR 2003, Conference on Construction Applications of Virtual Reality*, Blacksburg, VA, 179-192.
13. Gopinath, R. (2004). "Immersive virtual facility prototyping for design and construction process visualization," M.S. Thesis, Architectural Engineering. The Pennsylvania State University, University Park.
14. Gopinath, R., and Messner, J. I. (2004). "Applying immersive virtual facility prototyping in the AEC industry." *CONVR 2004: 4th Conference of Construction Applications of Virtual Reality*, Lisbon, Portugal, 79-86.
15. Liston, K., Fischer, M., and Wingrad, T. (2001). "Focused sharing of information for multidisciplinary decision making by project teams." *ITcon*, 6, 69-82 (http://www.itcon.org/2001/6).
16. Maldovan, K., and Messner, J. I. (2006). "Determining the effects of immersive environments on decision making in the AEC Industry." *Joint International Conference on Computing and Decision Making in Civil and Building Engineering*, Montreal, Canada, (Submitted for Review).

A Forward Look at Computational Support for Conceptual Design

John Miles[1], Lisa Hall[2], Jan Noyes[3], Ian Parmee[4], and Chris Simons[4]

[1] Cardiff School of Engineering, Cardiff University, UK
[2] Institute of Biotechnology, University of Cambridge, UK
[3] Department of Experimental Psychology, University of Bristol, UK
[4] Faculty of Computing, Engineering & Mathematics, University West of England, UK

Abstract. Future research needs for computational support for conceptual design are examined. The material is the result of the work of a so called *design cluster*. The cluster has, through a series of workshops, defined what it believes are the salient areas in which further research is needed. The work has a strong people centred approach as it is believed that, for the near future, it is only through a combination of man and machine that acceptable designs will be achieved. The cluster has identified 5 key areas and 39 sub-classes. The discussion focuses on the key areas and how these link to future research requirements in people centred computation for conceptual design.

1 Introduction

In 2004, two of the UK's research councils, the Arts and Humanities Research Council and the Engineering and Physical Sciences Research Council launched an initiative to set the future design research agenda for the UK. The initiative invited bids for funding for *"design clusters"*. These were to be groups of people who would spend a year looking at a design area of their choosing. The bids were to contain information regarding the people within the proposed cluster, the programme of work and the deliverables. Bidders were encouraged to be as innovative as possible in their thinking and also multi-disciplinary. It was hoped that by bringing together people from diverse backgrounds that generic ideas and concepts would emerge.

The call for proposals was aimed as widely as possible, including, for example, choreographers, as well as more traditional design backgrounds. A substantial number of bids were received from which 20 clusters were selected. One of the successful clusters is entitled *Discovery in Design: People Centred Computational Issues* and the work and findings of this cluster form the subject matter of this paper.

2 The Discovery in Design Cluster

The cluster consists of both academics and industrialists. As well as people from traditional engineering backgrounds such as aerospace, military, chemical and civil engineering, there are people who work on drug discovery, art based product design, user interfaces, design of new materials, biochemical sensors and software. The

I.F.C. Smith (Ed.): EG-ICE 2006, LNAI 4200, pp. 492–499, 2006.

cluster has a central core of people who have experience in the usage of software techniques to support conceptual design and particularly in the use of evolutionary computation. Also, the cluster has an emphasis on human factors. Cluster members believe that the way forward is a cooperative blending of knowledge and skills between humans and computers. To ensure that human factors are fully considered, the cluster also contains psychologists and social scientists. The objective of the cluster was defined as "*... to identify primary research aspects concerning the development of people-centred computational design environments that engender concept and knowledge discovery across diverse design domains*".

The cluster's main way of working has been through a series of four workshops. The participants have been the members of the cluster (typically some 15 of these attended each workshop) plus invited guests. Most of the guests have strong track records in design. However there were one or two people from other disciplines such as, for example, a specialist in detecting and predicting trends relating to lifestyles and fashions. In all cases the guests gave presentations on their particular speciality. Further information about the cluster can be found at www.ip-cc.org.uk.

3 The Workshops

The first three workshops were used to explore ideas and concepts and to highlight problems and weaknesses in terms of conceptual design and its computational support. These were then explored as potential areas for future research. This was an essentially divergent process. The fourth workshop was convergent, with the work of the previous workshops being analysed and synthesized.

4 Previous Work

There have been a number of other efforts to develop *roadmaps* for future research directions. For the construction industry alone notable examples which have concentrated on IT are the FIATECH / NSF initiative [3] and Amor at al [1]. Predicting what technologies will succeed in the future is always difficult and often the real breakthroughs come from something which cannot be foreseen. This is always a problem with roadmapping exercises. The cluster has largely avoided the pitfall of prediction and has limited itself to identifying areas of need. Also the focus of the cluster is different in that it is multi-disciplinary.

5 The Key Areas

In the following, the main findings of the cluster are presented in terms of future research needs. The cluster identified five key areas for future work, based on an examination of the deficiencies in current technologies, these being :-

- Understanding humans
- Representation
- Enabling environment for collaboration and user interaction

- Two way knowledge capture
- Search and exploration

In addition to the 5 key areas, a large number of sub-classes were identified. These are also discussed below.

5.1 Understanding Humans

There was a discussion about whether or not this area should be included. This is a well established area of research in which many research teams are working. Hence, some people thought that it was outside the cluster's remit and that by including knowledge capture and enabling environments, human factors were sufficiently considered. However the majority argued that without understanding human needs, abilities and reactions to different developments, the proposed research would never fulfil its aims. It should be appreciated though that the cluster's suggestion is limited to the specific context of conceptual design assisted by computational support rather than a more global understanding of human behaviour.

The basic argument is that, computational tools have to fit in with human capabilities and needs. For example humans are very good at pattern matching and assimilating visual information, although from any image they typically only take in 30% of the information. Research is needed to better define human abilities, especially in relation to design. The design studies undertaken to date (e.g. [6]) have shown that designers have problems with cognitive overload and bias and tend to stick with their initial thoughts and decisions. One obvious area for research is to ensure that the sort of software environments that are envisaged, will help designers to avoid these problems. The other research need is that of communication between the user and the computer, not in the terms described in the knowledge capture section but more in the fundamental area of what sort of tools and interface strategies are best suited to transferring information. Finally the cluster unanimously agreed that any software should ideally be exciting and interesting to use. This is something which design software has so far largely failed to achieve.

5.2 Representation

The term representation caused the cluster problems because it means different things to different people. Some within the cluster argued against representation being a research area because they considered that it is a part of search and exploration. However once the cluster had fixed on a common definition, it was agreed that representation should be included. The cluster's definition of representation is that it includes all areas within the software where the properties and characteristics of the problem domain are described. If the specific example of genetic algorithms is considered, the genome and the coding strategy are a part, as is the fitness function. Also as Zhang & Miles [7] show, for certain classes of problems, crossover and mutation can affect the form of the final solution and so, some cases can be considered to be a part of the representation.

To date, much of the representation used, especially in genetic algorithms, has relied on the ability of the problem to be expressed as a series of parameters. As Zhang et al [8] show, for some classes of problems such as topological search, this can be limiting. Also, one of the forthcoming challenges for design software is for it to be able to tackle complex, multi-participant, multi-objective, highly constrained problems. These will require far more complex representation strategies than are currently used. The work of some cluster members on software design has shown that there are many areas for which the development of the relevant software techniques is still in its infancy. Even for the more mature domains, there are significant challenges in terms of representation techniques. For example, for topological search, there is yet to be an established a generic form of representation which can handle a multiplicity of highly complex shapes. Without this, true topological search is not possible.

Although representation has not been a significant limiting factor to date, as work progresses in other areas, the limitations of current strategies will start to hinder progress and the need for further work in this area will become more apparent.

5.3 Enabling Environment for Collaboration and User Interaction

The cluster looked at current technology and what is required if it is to be advanced to cope with the complexity of many common design problems. A lot of work has been done on creating techniques to find areas of high performance within design spaces and to present the results to the designer [4] but these methods have so far largely been applied to single designer or single discipline problems and the communication is largely in one direction, from the computer to the designer. Many design problems are multi-disciplinary and involve large design teams. These people are typically tackling complex design problems with massive, multi-dimensional search spaces which contain areas of non-linearity and discontinuities. Enabling designers to understand the form of search spaces is a necessity and a major challenge.

The complexity of these search spaces is such that any attempt to communicate the findings of a design search to the users is likely to lead to significant problems with cognitive overload. The obvious solution to this is for the design software to select what information the user sees but this runs the risk of the designer not being given sufficient information to understand the results. Additionally, this is straying into the area of the decision being made by the software rather than the designer. As previous experience with other design software, such as knowledge-based systems, has shown, this is bad practice. It deskills the design process and designers don't like this. Also computers lack world knowledge and so don't have the common sense to detect even the most basic of errors in their reasoning and so are not competent decision makers.

There is also the question of communication in the other direction, from the designer to the computer. Designers possess a wealth of knowledge, some of which is procedural and hence difficult to access but if it could somehow be made available to the computational search this would be very useful.

The need to protect Intellectual Property Rights (IPR) could mitigate against greater collaboration. In many industries such as construction, an organisation's main asset is its IPR. Therefore protecting this, while also enabling other members of the design team to access the resources that they need, is vital. The IPR can be expressed in a number of ways. For example it may be encapsulated within software or it may be the staff who possess it. Either way, if true multi-organisation collaborative design with software support is to be possible, ways have to be found of enabling design consortia to have access to the IPR in a manner which allows access to it to be controlled.

5.4 Two Way Knowledge Capture

If computational intelligence is to be able to support design undertaken by large multi-disciplinary teams, methods are required for capturing the design team's objectives and related constraints. Currently each designer or discipline tends largely to work in isolation in terms of setting the constraints for their part of the work. Objectives are defined by the client and the design team. The interaction between the constraints and the objectives is complex and when they are combined, nobody is able to fully predict what their impact will be. Anecdotal evidence from the UK defence industry on submarine design (Biddle per comm.) suggests that the impact can be substantial. Unfortunately because this has been work undertaken by the defence industry, it has never been published.

What is needed are methods for eliciting and modelling each discipline's constraints. These will come in a variety of forms and will need to be handled within the design software. This concept can be extended beyond constraints to all forms of knowledge relating to a design. Intuitively it would seem that the greater the degree of shared knowledge and understanding between the software and the designer(s), the better will be the outcome of the design process. Also this should help the system to "understand" the designers' objectives and provide better support. If the system truly does have an understanding of what the designer is trying to achieve, it could suggest new ideas. This is examined more in the following section.

5.5 Search and Exploration

Search and exploration are basic features of any viable conceptual design tool. The potential complexity of multi-discipline, multi-objective search spaces has already been discussed but what has yet to be covered is the difficulty of searching such spaces in a sensible manner. As computational support for design tackles ever more complex and obtuse domains then the search will become more difficult. With multi-objective search, there are techniques such as Pareto analysis for selecting areas of high performance but only for a limited number of objectives. Parmee & Abraham [4] present a method which avoids these limitations. However, there is still a concern that, with substantial numbers of objectives and constraints, trade offs in the search process will render the results meaningless. The implications of such searches need to be thoroughly investigated to either prove or assuage these fears.

The cluster spent some time looking at innovation and creativity. Undoubtedly the successful economies of the future will be those that are the most innovative and creative in terms of commercial products. Creativity typically arises from moments of inspiration, which are often the coalescing of random and previously unconnected thoughts. The cluster discussed whether it would be possible to assist with this process using computational support using, for example, a "nonsense" generator. This could be attractive but also could be extremely wearing if one had to spend hours considering random nonsense. The idea of "jump out" agents was considered, these being agents which somehow leave the current search space and look elsewhere for solutions [2]. Another idea was contradiction; going against the accepted wisdom.

Linked to the ideas discussed in Two Way Knowledge Capture is the concept that, if the system could understand what the designer is trying to achieve, then it could search for relevant ideas and information, very much in the way that the semantic web anticipate needs. This for example could be in the form used by Amazon: people who designed one of these also looked at..... or it could be more like a Google search.

6 Sub-classes

In addition to the above 5 key areas of work, a considerable amount of time was spent looking at how the areas could be broken down into sub-classes. At the end of workshop three, some 350 potential subjects for sub-classes were identified. These ranged from statements made by some of the guest speakers such as one designer saying he has a "butterfly mind" to categories such as "team integration". Cluster members were asked to place each of the 350 potential sub-classes into one of the 5 key areas. Inevitably there was some divergence of opinion but the exercise made it possible to identify groupings within the potential sub-classes. The analysis was used at the start of workshop four to reduce the 350 down to 39 sub-classes.

An exercise was then undertaken to analyse these 39 sub-classes and determine how they relate to the five key areas in terms of importance. This was achieved by plotting two dimensional graphs with the graph axes being two of the key areas, giving in total ten graphs. The purpose of the exercise was to make the cluster members think about the relevance of the 39 categories in relation to each of the key areas and also to provide a visual aid to stimulate discussion. An example of a graph is shown in fig.1.

The exercise was useful because it stimulated discussion and it gives an indication of the potential difficulty of the research within each of the key areas. The amount of information obtained from this exercise is so large and complex that its analysis is incomplete but Parmee et al [5] have extracted the sub-classes that lie in the upper quartiles of the four graphs of each key area and identified the sub-classes that occur most often. These are shown in table 1. Note that the sub-classes are not exclusive to a given key area. This is an important finding and one which is still being analysed.

7 Discussion

The multi-disciplinary nature of the cluster was very beneficial and the interaction brought out some interesting concept and ideas. The body of information produced by the cluster is large and contains useful pointers as to the way forward. The cluster has identified that there is a huge amount of research yet to be undertaken before we can provide comprehensive software environments to support most areas of conceptual design. Some of the work to be done is fairly straight forward but much of it will require a substantial amount of fundamental research.. The cluster has focussed on areas where current approaches are lacking and identified the shortcomings. The workshops have produced a huge amount of information and this is still being analysed, especially with regard to the sub-classes. The work has been so rewarding and information rich that the members of the cluster have come together to form the Institute for People Centred Computation (www.ip-cc.org.uk). This will inherit the intellectual property of the cluster and continue its work both in looking at future research requirements but also in delivering the research.

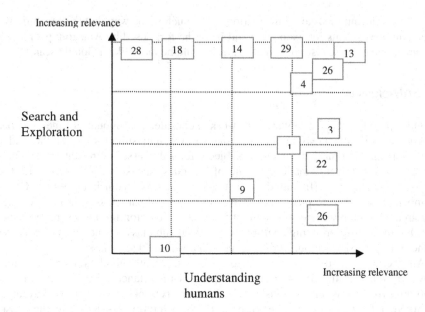

Fig. 1. An example of relating the 39 sub-classes to the key areas. (Some of the 39 categories have been omitted in the interests of clarity). [5]

Table 1. The more significant sub-classes in relation to the key areas

Two Way Knowledge Capture	Search & Exploration	Enabling Environments	Representation	Understanding Humans
Co-operation and collaboration; Capturing/ Extracting knowledge; Enabling Computational Technologies; Emergence; History and Traceability; Representation.	User support; Creativity and Innovation; Modelling; Emergence; History and Traceability; Capturing/ Extracting Knowledge; Data Issues.	User-centric Issues; Co-operation and Collaboration; Useability; Interface; Creativity and Innovation; Multi-users and Multi-user Interaction; Capturing / Extracting Knowledge.	Visualisation / Senses Stimulation; Form; Modelling; Capturing / Extracting Knowledge.	Usability; Visualisation / Senses Stimulation; User Interaction; Validation and Risk; Creativity and Innovation; Interface; User-centric Issues; User Support; End User Issues; Learning; Form; Co-operation and Collaboration.

A Forward Look at Computational Support for Conceptual Design   499

8 Conclusions

A cluster consisting of people from diverse backgrounds has come together to look at the requirements for software support for conceptual design. The starting point of the cluster was that the work needed to be people centred and nothing that has arisen in the workshops has caused this assumption to be questioned. The cluster has identified five key areas in which further research is needed. Beneath these five areas are thirty nine sub-classes which relate to one or more of the key areas. The cluster has identified a significant body of research that needs to be undertaken to enhance the current technology of computational support for conceptual design.

Acknowledgements. The work of the cluster was supported by the UK's Engineering and Physical Sciences Research Council and the Arts and Humanities Research Council. Also thanks to the guest speakers for their contribution, notable Tom Karen, Simeon Barber, Chris Jofeh and Pat Jordan.

References

1. Amor, R, Betts, M & Coetzee, G, 2002. IT for research: Recent work and future directions, ITcon, 7, 245-258.
2. Cvetkovic D & Parmee I, 2002. Agent-based support within an interactive evolutionary design system, AIEDAM, 16(5), 311-342.
3. Vanegas J, 2004. http://www.ce.gatech.edu/research/NSF-FIATECH_Charrette/index13.htm
4. Parmee I & Abraham J, 2004, Supporting implicit learning via the visualization of COGA multi-objective data, Proc EVOTEC, IEEE, 395-402.
5. Parmee I, Hall A, Miles J, Noyes J & Simons C, 2006. Discovery in Design: Developing a People Centred Computational Approach, Design2006, Dubrovnik Croatia.
6. Ullman D, Stauffer L & Diettrich T, 1987, Preliminary results of an experimental study of the mechanical design process, Waldron M (ed), NSF workshop on the Design Process, Ohio State Univ, 143-188.
7. Zhang, Y & Miles J, 2004. Representing the problem domain in stochastic search algorithms, in Schnellenbach-Held, M & Hartmann, M. (eds) Next Generation Intelligent Systems in Engineering, EG-ICE, Essen, 156-168.
8. Zhang Y, Wang K, Shaw D, Miles J, Parmee I & Kwan A, 2006, Representation and its Impact on Topological Search in Evolutionary Computation, Hughes R (ed), ICCCBE XI, Montreal Canada,

From SEEKing Knowledge to Making Connections: Challenges, Approaches and Architectures for Distributed Process Integration

William J. O'Brien[1], Joachim Hammer[2], and Mohsin Siddiqui[1]

[1] Department of Civil, Architectural and Environmental Engineering, University of Texas at Austin, 1 University Station C-1752, Austin, TX 78712-0273, USA
{wjob, mohsink}@mail.utexas.edu
[2] Department of Computer & Information Science & Engineering, University of Florida, Box 116125, Gainesville, Florida, 32611-6125, USA
jhammer@cise.ufl.edu

Abstract. Integration and coordination of distributed processes remains a central challenge of construction information technology research. Extant technologies, while capable, are not yet scalable enough to enable rapid customization and instantiation for specific projects. Specifically, the heterogeneity of existing legacy sources together with firms' range of approaches to process management makes deployment of integrated information technologies impractical. This paper reports on complementary approaches that promise to overcome these difficulties. First, the SEEK: Scalable Extraction of Enterprise Knowledge toolkit is reviewed as a mechanism to discover semantically heterogeneous source data Second, a schedule mapping approach is presented that integrates firms' diverse individual schedules in a unified representation. The mapping network is supported by a Process Connectors architecture that also incorporates SEEK components. While this paper focuses primary on schedule process integration, the Process Connectors architecture is viewed as providing a broad solution to discovery and integration of firms' process data.

1 Introduction

This paper reports on ongoing research by the authors to create new mechanisms for distributed process integration in the construction domain. The construction industry poses significant challenges for integration given a large number of firms of varying sophistication and a corresponding variety of data formats including a range of proprietary legacy sources. Technical integration challenges are made more difficult given the business climate that includes short-term associations and differing business practices (especially practices that involve firms operating on different levels of detail, making integration and constraint propagation exceptionally challenging.) These collective integration challenges exceed the ability of extant systems to create scalable, rapidly deployable information systems that address coordination needs. Our paper reports on a broad architecture for distributed process integration in the construction domain, with specific emphasis on (1) rapid discovery of legacy data, (2) mapping and management of discovered data to support process coordination.

I.F.C. Smith (Ed.): EG-ICE 2006, LNAI 4200, pp. 500–518, 2006.

Discovery of firms' data remains a significant challenge to integration. Despite the advent of modern data standards, firms have a broad choice of how to implement these standards, particularly for process information. Further, many firms use a variety of applications for process management, and many of these are homegrown or dated applications. To semi-automatically discover legacy data whatever the source, we review the SEEK (Scalable Extraction of Enterprise Knowledge) toolkit. SEEK provides a generalized, structured approach using data reverse engineering and wrapper development technologies to overcome the challenges of integrating information resident in heterogeneous legacy sources. SEEK exploits meta-data (schema, application code, triggers, and persistent stored modules) to develop a semantically-enhanced database schema for the data source. The data source owner is presented with a graphical user interface (GUI) displaying the correspondence matches and allowing the owner to approve or change the matches. This allows rapid generation of a wrapper that translates the legacy source data to a common schema.

Building from the wrapper generated by SEEK (or related toolkits), discovered data must be made available for process coordination. We present a mapping approach and architecture (Process Connectors) for process coordination building on integration of schedules. Schedules provide a skeleton for including additional resource information and related constraints, enabling rich process description and coordination support. The Process Connector architecture includes a stub at each firm (incorporating a SEEK generated or related wrapper to legacy sources) as well as a bridge component between stubs to enable process coordination. Stubs and bridge components can be invoked within a web services framework, enabling rapid and scalable connections to suit projects of any size. The core of the bridge component is a tree-based mapping approach to integrate schedules of varying levels of detail that have m:n matches among sets of activities (i.e., schedules that do not fall neatly into a simple hierarchical arrangement). A tree-based approach provides flexibility to add a variety of constraint information at each leaf node based on data availability at the firm/stub. The bridge can make recommendations for process coordination among firms using simple algorithms, or provide collective information to more complex decision support applications. Recommendations are received by the participating firms through their respective stubs and can be approved or rejected (thereby invoking a new coordination process at the bridge). Process Connectors thus support distributed process coordination across firms.

Collectively, the SEEK toolkit and the Process Connectors architecture provide a modular, scalable, and rapidly deployable infrastructure to aid process coordination across a wide variety of firms that have heterogeneous data sources and business processes that operate at different levels of detail. As such, this research lays the groundwork for extending existing standards based approaches to data and process integration, bringing the benefits of distributed process coordination to a widespread audience.

This paper first reviews the challenges of distributed process integration from both a semantic and process integration perspective, reviewing the current state of the art and defining challenges in section two. Section three of the paper reviews the SEEK architecture and approach to extracting semantically heterogeneous data from diverse legacy sources. While SEEK provides an approach to discovery heterogeneous data, section four of the paper presents a schedule mapping approach to integrating heterogeneous schedules. Section five brings together the SEEK and schedule mapping approach within the Process Connectors architecture, presenting a broad and scalable approach to process integration. Some concluding remarks are made regarding implementation and future research.

2 Distributed Process Integration: Review and Challenges

The challenges of integration are many and exist at several levels, from simple data exchange to process coordination and ultimately to inter-organizational cooperation. Much research has focused on semantic integration, which is a necessary building block for any electronic integration. Development of information standards such as the Industry Foundation Classes (IFC, [1]) and CIMSTEEL (CIS/2, [2]) have taken years of effort on the part of industry and academia to develop. These efforts have seen some fruition in deployment, particularly with the respect to the interchange of CAD/design data [3]. A natural application of standards for semantic integration is supporting information exchange between applications across project phases – from early design through construction execution and ultimately supporting operations. Such integration among applications can support rapid testing and feedback from all parties during design (e.g., circle integration [4]). The current view of such life-cycle approaches to information is Building Information Modeling (BIM) [5], which envisions all parties sharing data seamlessly across the life of the project.

Maturation of standards such as the IFC has also seen increased understanding of their limitations. Several researchers and practitioners [6-8] argue that a single integrated project database and supporting data standard will be not be able to contain or reconcile the multiple views and needs of all project participants. They predict that rather than a single standard, multiple protocols will evolve over time. These protocols will require significant translation to enable data exchange. Similarly, authors O'Brien and Hammer have argued that the number of firms in the construction supply chain – involving hundreds of subcontractors and suppliers – makes it unlikely that all firms will be able to subscribe to a single data exchange standard [9]. O'Brien and Hammer have further argued that the varying level of detail firms operate at also makes integration more difficult [10]. Whatever the advances of industry standards, semantic heterogeneity is likely to remain a significant barrier to integration.

There are further challenges to integration of process information. Even limiting process integration to a specific phase of the project – construction execution – there are additional challenges beyond semantic information as process descriptions and constraints must be expressed and processed. Put simply, integrating processes involves more than integrating data as several data items need to be processed simultaneously and in an interconnected fashion to adequately describe a process. Further, process descriptions can range from simple descriptions (such as a business rule like quantity multiplied by unit cost equals total cost) to complex descriptions (like schedules with predecessor/successor information and resource loading) to lengthy workflow descriptions (like handling an order from initial acceptance through fulfillment). The authors are hesitant to call workflow more complex than schedule integration; the differences are primarily in emphasis. Workflow can be more transactional in nature and translate to information technology implementations; there are certainly some mature research models in the area [11]. Schedule integration is less defined, in part because schedule definition in practice is an evolving process and we lack good representation and solution models for such a dynamic view of scheduling [12, 13].

Process integration requires both semantic integration and a shared description of the process(es) to be integrated (see [14] for a related discussion). With respect to process representation, Froese has produced a seminal summary of high level process descriptions in the construction domain [15]. However, Froese notes that these high level descriptions must be significantly enriched for implementation. It is safe to say that definition and testing of process related definitions in data standards such as the IFC is less mature than that for engineering products; some initial evaluations are being conducted several researchers [16, 17].

Probably the most extensive test of process integration standards to-date was conducted by Law and his students [18], who used the Process Specification Language (PSL) developed by NIST [19] as a neutral data format for exchange of information between engineering and construction applications. Law's group found PSL to be reasonably rich, but that it required extension to support specific project management components. Further, the group spent considerable effort manually developing wrappers to each legacy source as well as specifying mediators [20] to translate information for specific service applications. The intent of Law's research was to provide a framework for integration of applications as services rather than provide a complete test of process integration, so it is unclear the extent to which that group addressed challenges in integrating multiple views of the same processes or constraint representation beyond those needed for the specific service definitions.

We can illustrate these challenges in a small scheduling example. Fig. 1 depicts a mapping between the schedule of a construction manager (Centex) and a subcontractor (Miller) (further details of the case are found in [21]). Note that these mappings exist on several levels (e.g. 1st floor to 1st floor and activity number to activity number). Further, although it the mappings in Fig. 1 are easy for a human to interpret, both the ID numbers activities names are semantically different at the instance level. It is hard to automatically generate a match or mapping even if the underlying data schemas of the scheduling applications are machine interpretable. Most important, note that the mappings between individual activities do not have a 1:1 or even 1:n mapping in many cases (i.e.., mappings 3, 6, 9 and 11 in Fig. 1). A 1:1 one mapping (such as mapping 2) makes it easy to coordinate these two schedules as the set of information for each activity is largely the same (i.e., start finish dates, duration, etc.). Similarly, a 1:n mapping could be considered a hierarchical mapping where the construction manager's or general contractor's schedule summarizes a more detailed breakdown by the subcontractor. Such a hierarchical level makes it easier to integrate differing views of the schedule. For example, the LEWIS system specifies four different levels of detail within a schedule [22]; as long as subcontractors plan as a pre-specified level integration is reasonably straightforward. It is more difficult to integrate activities that have an m:n mapping as such a mapping implies potentially very different perspectives on the part of the firms. In the case of Miller in Fig. 1, their scheduling system is build from cost codes where every activity corresponds to a standard cost code, directly supporting continuous refinement of the firm's estimating database [21]. This is a very different view than the spatial and set related view of the schedule evinced by the construction manager Centex. We lack a formal mechanism to represent such a complex mapping, although elements of such mapping to support process integration are presented below. Similarly, propagating constraints between these complex mappings is difficult as each activity may have different resources, making it difficult to summarize resource constraints and share them with others. The mapping problem can become more complex as we add multiple firms.

Mapping No	Centex Schedule		Miller Schedule	
	ID	Activity Name	ID	Activity Name
1	1st Floor Activities		1st Floor	
2	1210	Install Light Fixtures	32	0200-Fixtures
3	1040	In-wall Elec. Rough in	33	0300-Wiring Devices
	1255	Install Electrical Finishes	34	0500-Cable Tray
	1020	Overhead Elec. Rough in	35	0602-Above Grade
			36	0900-Single Conductor Wire
			37	1300-Equipment
4	2nd Floor Activities		2nd Floor	
5	2090	Install Light Fixtures	39	0200-Fixtures
6	2045	In-wall Elec. Rough in	40	0300-Wiring Devices
	2110	Install Electrical Finishes	41	0500-Cable Tray
	2020	Overhead Elec. Rough in	42	0602-Above Grade
			43	0900-Single Conductor Wire
			44	1300-Equipment
7	3rd Floor Activities		3rd Floor	
8	3090	Install Light Fixtures	46	0200-Fixtures
9	3045	In-wall Elec. Rough in	47	0300-Wiring Devices
	3110	Install Electrical Finishes	48	0500-Cable Tray
	3020	Install Electrical Finishes	49	0602-Above Grade
			50	0900-Single Conductor Wire
			51	1300-Equipment
10	Structure/Exterior/Roof		Roof	
11	0211	Elec. Rough in/Equipment/Tie-in	53	0300-Wiring Devices
			54	0602-Above Grade
			55	0900-Single Conductor Wire
			56	1300-Equipment

Fig. 1. An example of mapping activities between two firms' schedules for the same construction tasks (electrical work). The sets of activities associated with each other are in many cases m:n sets, suggesting a disjoint view of the work plan.

The example of Centex and Miller can be generalized. The different parties involved in a construction project view their responsibilities at a level of detail that is most useful for their business objectives. A general contractor (GC) or construction manager (CM) in most cases owns the master schedule and uses critical path networks [23] often representing activities for a subcontractor at an aggregated level of detail. The GC is usually unaware of the specific resource and capacity constraints of the subcontractors, which can limit the ability of the GC to coordinate schedules [24]. A subcontractor models its responsibility in the project at a finer level of detail for proper control and management relying on critical path networks for the overall project view. Site staff usually relies on bar charts and activity lists for detailed planning of specific site tasks [25]. These views are typically disjoint and the current research does not provide the necessary theory and tools to integrate these independent views. Lacking such a representation, constraint propagation is similarly difficult. It is thus no surprise that, as Smith notes, most scheduling optimization algorithms operate within a static and single party view of temporal and resource constraints [12].

To summarize, we can generally state that current technology and understandings show that, with effort, it is possible to manually integrate processes across software for specific applications. Further, research demonstrates that data standards such as the IFC and specific process related data standards such as PSL and OZONE (a scheduling ontology) [26] can be used to support integration, although extensions may need to made for specific applications. More research is needed, however, to allow more general and reusable integration efforts. With respect to semantic information, there needs to be mechanisms to rapidly understand the data inside a source application so multiple uses can be made of the data and wrappers/mediators can be generated (semi-)automatically. Existing process models and standards can be similarly used to develop specific instantiations. However, we lack methods to link disjoint views of processes, particularly with respect to level of detail. Similarly, we need generalized methods to represent and propagate process constraint information.

3 SEEK: Extraction from Heterogeneous Sources

Before process coordination is possible, firm data sources must be made available. In theory, data standards, APIs, and emerging web service functionality should make this process easier. Certainly, data standards and application functionality to support data sharing makes the integration process more feasible now than previously. That said, significant issues remain in accessing legacy sources that have proprietary specifications and perhaps poor documentation. Similarly, modern applications that subscribe to data standards such as the IFC may have users that interpret the standards differently or operate at different levels of detail. There needs to be a mechanism to rapidly discover and make available legacy source data available both rapidly in terms of building links to data and interpreting what data is contained in legacy source applications. Towards that end, the SEEK: Scalable Extraction of Enterprise Knowledge toolkit was developed by authors O'Brien and Hammer and described in detail in several publications [9, 10, 27]. SEEK is reviewed briefly below.

A high-level view of the core SEEK architecture is shown in Fig. 2. SEEK follows established mediation/wrapper methodologies (e.g., TSIMMIS [28], InfoSleuth [29]) and provides a software middleware layer that bridges the gap between legacy information sources and decision makers/support tools. As such, SEEK links to legacy applications, which can range from proprietary applications to modern applications supporting APIs and data standards such as the IFC. SEEK works as follows. During run-time, the *analysis module* processes queries from the end-users (e.g. decision support tools and analysts) and performs knowledge composition including basic mediation tasks and post-processing of the extracted data. Data communication between the analysis module and the legacy sources is provided by the *wrapper component*. The wrapper translates SEEK queries into access commands understood by the source and converts native source results into SEEK's unifying internal language (e.g., XML/RDF).

Prior to answering queries, SEEK must be configured. This is accomplished semiautomatically by the *knowledge extraction module* that directs wrapper and analysis module configuration during build-time. The wrapper must be configured with information regarding communication protocols between SEEK and legacy sources, access mechanisms, and underlying source schemas. The analysis module must be configured with information about source capabilities, available knowledge and its representation. We are using a wrapper generation toolkit [30] for fast, scalable, and efficient implementation of customized wrappers. To produce a SEEK-specific representation of the operational knowledge in the sources, we are using domain specific templates to describe the semantics of commonly used structures and schemas. Wrapper configuration is assisted by a source expert to extend the capabilities of the initial, automatic configuration directed by the templates in the knowledge extraction module. Use of domain experts in template configuration is particularly necessary for poorly formed database specifications often found in older legacy systems. Furthermore, the knowledge extraction module also enables step-wise refinement of templates and wrapper configuration to improve extraction capabilities.

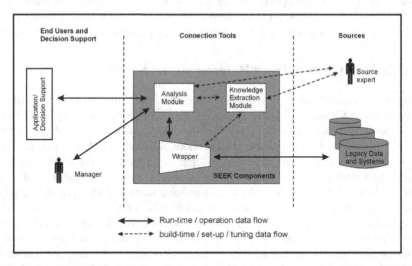

Fig. 2. High-level SEEK architecture and relation to legacy sources and querying applications. The SEEK wrapper and analysis components are configured during build-time by the knowledge extraction module. During run-time, the wrapper and analysis component allow rapid, value-added querying of firm's legacy information for decision support applications.

The authors believe that the modular structure of the architecture provides a generalized approach to knowledge extraction that is applicable in many circumstances. That said, it is useful to provide more details of the current implementation architecture with respect to discovery. Such a more detailed schematic is shown in Fig. 3, which can be seen as a more detailed view of the knowledge extraction module in Fig. 2. SEEK applies Data Reverse Engineering (DRE) and Schema Matching (SM) processes to legacy database(s), to produce a source wrapper for a legacy source (shown in Fig. 2). The source wrapper will be used by another component wishing to communicate and exchange information with the legacy system. We assume that the legacy source uses a database management system for storing and managing its enterprise data.

First, SEEK generates a detailed description of the legacy source, including entities, relationships, application-specific meanings of the entities and relationships, business rules, data formatting and reporting constraints, etc. We collectively refer to this information as enterprise knowledge. The extracted enterprise knowledge forms a knowledgebase that serves as input for subsequent steps outlined below. In order to extract this enterprise knowledge, the DRE module shown on the left of Fig. 3 connects to the underlying DBMS to extract schema information (most data sources support at least some form of Call-Level Interface such as JDBC). The schema information from the database is semantically enhanced using clues extracted by the semantic analyzer from available application code, business reports, and, in the future, perhaps other electronically available information that may encode business data such as e-mail correspondence, corporate memos, etc.

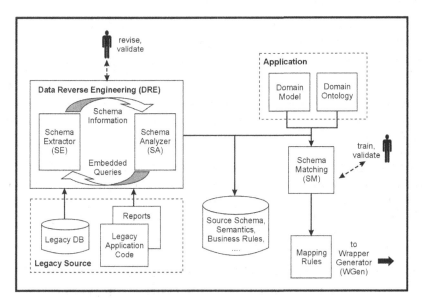

Fig. 3. Schematic diagram of the conceptual architecture of SEEK's knowledge extraction algorithm. The structural and semantic information contained in a firm's legacy applications are semi-automatically discovered and refined in a bootstrapping process with input from firm domain experts. The source description is supplied to a wrapper generator tool to automatically construct the SEEK wrapper component.

Second, the semantically enhanced legacy source schema must be mapped into the domain model (DM) used by the application(s) that want(s) to access the legacy source. This is done using a schema matching process that produces the mapping rules between the legacy source schema and the application domain model. In addition to the domain model, the schema matching module also needs access to the domain ontology (DO) describing the model. Third, the extracted legacy schema and the mapping rules provide the input to the wrapper generator (not shown in Fig. 3), which produces the source wrapper.

With these three steps, SEEK is thus able to discover semantically rich legacy data and supply this description to a wrapper generator [30] and value-added mediator. The wrapper generator allows automatic generation of source wrapper while the mediator can be configured by the source description. The generated wrapper can be refined in a bootstrap manner by human domain experts and the entire build-time process can be re-run to rapidly generate new wrappers should the sources change (for example, by adding an additional or upgrading an existing application program).

The current SEEK prototype is capable of extracting schema information including relationships and constraints from relational databases and inferring semantics about the extracted schema elements and any business rules from the accompanying (Java) application/report generation code using the database. With some modifications, non-relational databases or other programming languages (C++, PHP, TKL/TK) can be supported. The SEEK prototype has been tested using sample databases and report interfaces developed as part of a class project in an introductory database course at

UF. We are currently evaluating SEEK against the THALIA integration benchmark consisting of 44+ University course catalogs and 12 challenge queries [31]. In addition, we validated the SEEK approach using legacy data sources from the construction and first responder domains. For example, in the case of the Gainesville Fire Department, the SEEK software was used to extract schema and source descriptions for a variety of legacy sources (e.g., regional utilities, property appraiser, telephone companies) and helped produce translations between the legacy sources and the information contained in a emergency dispatching system, improving the information available to first responders. Using SEEK drastically reduced the time it took to make the data in the legacy sources available for sharing with the fire department [32]. Construction data was used mainly to develop the SEEK algorithms initially and to validate their correctness. We also used sample data from a construction project for most of the SEEK demonstration scenarios. SEEK is continuing development with application data from construction and other domains.

4   Making Use of Discovered Information: Mapping an Overlay of Networks for Distributed Schedule Process Integration

SEEK allows discovery of firms' information in a rapid manner. However, further steps are required to make use of that data to support process integration. As noted above, we must move from semantic integration to process representations that support differing levels of detail and constraint propagation. For this purpose we use schedule integration as a specific example of process integration as schedules are a central component of the construction process; coordinating firms' schedule under changing conditions also remains one of the most difficult tasks to accomplish manually due to the large number of possible solutions and conflicting constraints [13, 24]. Schedules, although varying in representation across firms, are also well enough understood that a flexible shared representation can be generated and customized for specific instantiations. This section describes such a representation, building from the example in section 2 of this paper.

The basic issue with schedule process integration is that firms' differing representations must be linked so that useful coordination tasks can be accomplished. For example, it is useful for managers to know that differing representations are reconciled with respect to time activities are to take place. Similarly, it is useful to be able to explore and recommend alternatives when changes in schedule are needed. As a starting point for schedule integration, we model different schedules for the same set of construction tasks using the traditional network based approaches. We represent two such schedules in Fig. 4 where each node in the graph represents and activity and the arrows represent precedence constraints between the activities. Although they represent schedules for the same construction tasks, these networks have a different topology. This heterogeneity of the networks stems from the different level of detail used in modeling the activities, different constraints of the owner of the schedule and the different software systems used by the participants.

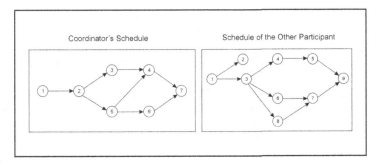

Fig. 4. Different schedule networks for the same physical tasks as viewed by the participants. The coordinator's schedule is a subset of a larger project-wide schedule.

Mapping is the process of identifying corresponding subsets of activities within these independent schedule networks, provides a mechanism to reconcile these disjoint views. (See extended discussion of mappings in [33]). Mapping allows formal capture of information that can help coordinate any changes in either schedule. This process can result in 1:1, 1:n and m:n matches between activities where each such match is called a schedule mapping. We can recursively discover smaller subsets within these mappings to establish the smallest possible mappings. Fig. 5 shows the final result of the mapping process down to the smallest possible decomposition; Fig. 7 shows the steps in the decomposition.

Fig. 5 also depicts precedence relationship of the mapping network, using the precedence information of the coordinating firm as a model. The coordinating firm is

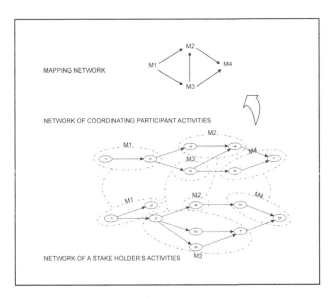

Fig. 5. Mapping network viewed as a result of an overlay of two networks that are the internal process representation of the coordinating firm and project stake holder

typically the CM or GC on the project that is responsible for the master schedule. As such, the coordinating firm directs precedence constraints for any given schedule mapping. In this sense, the mapping process can be considered conceptually as an *overlay of networks* where the coordinating firm determines the overall precedence constraints of the final mapping network. A set of mapping networks is developed for a project (e.g., a project with one CM/GC or coordinator and ten subcontractors would contain ten mapping networks). The designation of a single coordinating firm helps to ensure consistency when building and relating multiple mapping networks.

Each mapping in the mapping network is a collection of matching activity identifiers extracted from the respective schedules and the time period associated with these activities. Fig. 6 shows two sets of start and finish dates as a representation of the time periods associated with some activities of the coordinator and the corresponding activities for the other participant. The coordinator sets the time period the other firm must fall within. Note that each box in Fig. 6 may contain multiple; the set in each box represents the mappings (e.g., mapping M3 in Fig. 5).

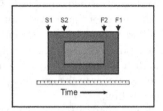

Fig. 6. Single mapping showing two pairs of start and finish dates. The outer start and finish dates represent the limits set by the coordinating firm.

As noted above, the mapping process recursively discovers smaller mappings. This process is illustrated in Fig. 7, progressing from a simple mapping of overall schedule to overall schedule down to the smallest possible mappings. However, a network representation at any point in the decomposition does not provide information about the previous decomposition. As such, potentially useful information might be lost. Referring back to the Centex and Miller example in Fig. 1, an intermediate decomposition can be activities grouped by floor. Such an association could be useful for reasoning about trade space constraints [34] and we would not want to lose that information in subsequent decompositions. As decomposition is hierarchical, we can utilize a tree based representation [35] to retain information about each step in the decomposition. The tree representation of each mapping is shown in Fig. 7 aside the network representation. During construction of the tree, it is possible to record additional information about the decomposition in a descriptive fashion that supports further reasoning and analysis.

Once the tree based representation of the mapping network is created, all the intermediate networks from the final tree based representation can be generated automatically by combining stored information for the children of any node at any level. Summarizing all the nodes up to the root level will give us the initial mapping. The only way to record this information using networks alone would be to store all

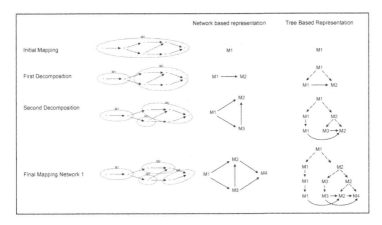

Fig. 7. Tree and network based representations of the mapping network. Note that the tree representation retains information about the previous decompositions, allowing reasoning at multiple levels of detail in the hierarchy.

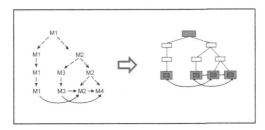

Fig. 8. A tree based representation of the schedule mappings, incorporating precedence information at the leaf node or smallest possible decomposition level

intermediate networks separately. Using a tree based structure to store the mapping information allows better consistency and reduces redundancy of information.

Note that information about precedence constraints are also stored as a part of the schedule mappings, shown as links between the leaf nodes of the tree. Fig. 8 depicts the network and tree based representations in more detail, incorporating pictorially the individual mapping shown in Fig. 6 in the leaf nodes of the tree. A further advantage of the tree-based representation is that it is easy to represent in a computer and add information at each node. Thus once a basic framework has been established, other constraints (e.g. resource constraints) associated with the activities of the participants can also be attached to these mappings. Ideally, once the mapping is established, associating additional organizational process information from existing legacy systems should can be accomplished semi-automatically using SEEK or related technologies.

There are multiple screens in the user interface for our mapping application. The screen shown in Fig. 9 allows a user to go through the step by step process of

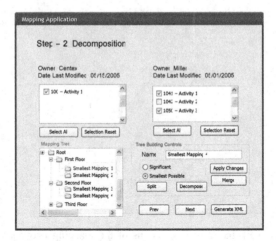

Fig. 9. Prototype interface for mapping function

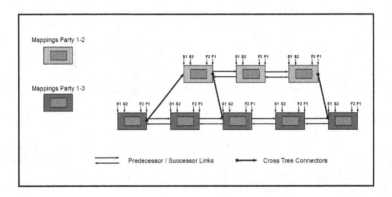

Fig. 10. Cross tree connectors showing precedence connections between two mapping trees

decomposing a mapping and building a hierarchy of decomposition (Fig. 7 and Fig. 8). The Mapping Tree area updates as the user performs various operations on the tree.

Building from schedule mappings between pairs of firms, it is possible to generate connections between mapping networks using precedence information from the coordinating firm. This is shown in Fig. 10, showing links (cross tree connectors) between the leaf node level of two pairs of mapping networks. These cross tree connectors supplement the predecessor/successor information contained within each individual mapping network, allowing coordination of all firms' schedules on the project. Reasoning about coordination can be done at several levels. At a simple level, once mappings are created and populated with constraint information, it is possible to validate that there are no conflicts. A more complex use of the mappings is to explore alternatives should there be a conflict or a change in schedule. Schedule optimization is also possible. Once the mappings are created and linked, the resulting representation can be used for a variety of schedule coordination activities.

5 Process Connectors: Enabling Distributed Process Integration

A tree based mapping provides a rich and flexible framework with which to facilitate integration and coordination of distributed schedule information. To implement these mappings in a scalable manner, we propose the Process Connectors architecture as illustrated in Fig. 11. (More detailed discussion of the architecture can be found in [36]). The architecture consists of bridge and stub components. The stubs are attached to the existing information systems of the firm and translate data into the internal data format of the architecture. The bridge component directs the information extraction by the stub components, performs basic analysis and supports the connection of processes across firms. While only two firms and a single bridge are shown, conceptually the architecture scales to many firms. Each firm has a single stub that services each of its multiple projects, and the bridge component exists as a (web) service that is activated when needed to support mapping or analysis. Functionality of the bridge and stubs are briefly described below.

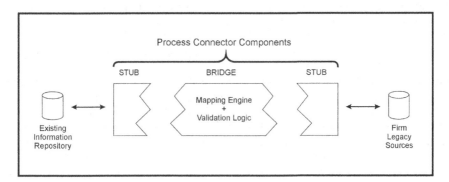

Fig. 11. Process Connectors architecture, depicting bridge and stub components between firms' existing legacy sources. Stubs represent firms' schedule process data, and the bridge performs mappings and supports process coordination.

Stubs are a main component of the process Connectors Architecture as they are the connection point between a firm's internal data and applications and connection to other components and analysis. Stubs are key to scalability as they translate firms' process information to a data format for sharing and further processing. We envision a stub will include elements of the SEEK architecture [9] and will support extraction of information from existing legacy systems. Stubs must also send information back to the firm (for example, a revised schedule for confirmation and acceptance by subcontractor management). Thus beyond being a wrapper or translator, they must also contain some application logic to process information as well as an interface for setup and confirmation. With respect to the SEEK architecture (Fig. 2), a stub uses the wrapper component but extends the functionality of the analysis module beyond mediation to include some analysis and conditional processing. In a larger context, we view the role of the stub as a gateway to the firm's internal data and process information.

For each stub, there is an initial, one-time setup or build-time process. In this process, the firm management must map the firm data and processes to the internal data format of the stub/process connectors architecture. When data is not available in an understood data format, automated discovery processes such as those in the SEEK toolkit (reviewed above) can be used to speed setup. In any case, managers must verify the translation; ideally this can be done by operations managers with limited input from technical staff. Of course, incremental improvements or changes can be made by reinitiating the build-time process. The authors believe the setup-once/use many times aspect of the stub provides a key aspect of scalability for the Process Connectors architecture.

In contrast, the bridge component, which manages and utilizes schedule mappings, requires a build-time activity for every project. As every project has a unique schedule, the mapping network must be established for each pair of firms (i.e., the coordinating firm, likely the CM/GC, and affiliated subcontractors). This will necessarily be accomplished by a user interface to the bridge component; further implementations may build this into existing software such as MS Project to make the application seem transparent to the user. The mapping process is not viewed by the authors as excessively burdensome as much related work in generating the schedule must be performed by humans already. The mapping process progresses one step further and formalizes the schedule coordination process. If firms work together repeatedly, they may also be able to build re-usable templates. As needed, the mapping process can be repeated if there are changes in definition of activities by either party.

Once a mapping is established, the bridge component can conduct basic analysis to support schedule coordination. The first such analysis function is a validation process that ensures dates and related constraints for sets of mapped activities are not in conflict (e.g., scheduled dates may not coincide and hence the mapped schedules are invalid). Validation is an automatic process. If the matches are invalid, the bridge can send a report to the affected firms about such invalid mappings and/or conduct analysis and propose recommendations for a new schedule. These reports are processed by the stub component. Overall, we expect the analysis performed in the bridge to be limited to relatively simple changes such as a right shift and validation of subsequent mappings. Thus a further capability of the bridge and stubs is to support data collection for external analysis (e.g., the LEWIS system that combines 4D analysis and subcontractor resource constraints for detailed trade coordination [22]). Analysis systems such as LEWIS would need only one connection to the Process Connectors architecture; the bridge and stub components could relay proposed schedule changes to affected firms for confirmation or rejection.

Fig. 12 depicts the Process Connectors architecture in more detail, showing sections of the internal architecture. A mapping repository that contains all the mapping networks is logically stored at the coordinating firm stub as this centralizes information about mapping networks. All stubs will contain elements of the SEEK architecture (depicted only in the right-side stub); as noted above, the analysis module in the Process Connectors architecture is more capable than that originally envisioned for SEEK. However, in each case the analysis functions are modular and will be customized for each instantiation depending on the sophistication of host firm. In an extreme case, a stub may provide most of the functionality of a scheduling application if a firm uses few or no electronic tools. In the majority of envisioned cases, the stub will work with existing applications that contain process data, and be configured with rules and knowledge about how the firm makes process decisions and data about constraints during the build-time phase. The bridge component is connected to each stub. While shown as a single block in Fig. 12, physical implementations of the bridge will likely contain many modules that can be called upon as needed. At a minimum, these modules will support the mapping process and perform simple analysis to aid schedule coordination. More sophisticated analysis can be supported by external applications. In this sense, the Process Connectors architecture acts as a data collection source for the entire project, making it easier to gather information heretofore difficult to collect.

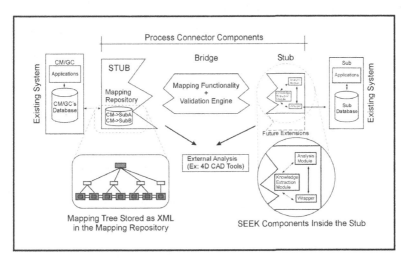

Fig. 12. Internal details of the Process Connectors architecture, depicting from left to right: Existing legacy systems at the coordinating firm, a stub that contains analysis functionality as well as a mapping repository to store mapping networks, a bridge component to construct and manipulate mappings, a stub attached to another firm that depicts the incorporation of SEEK components, and legacy systems at the stub. The Process Connectors architecture also supports links to external analysis tools.

The discussion of mapping networks and their implementation within the Process Connectors architecture has focused primarily on a single coordinating firm whose schedule is mapped to a set of firms participating in the same activities. This corresponds to the activities of a CM/GC and a number of subcontractors. Of course, this does not represent all the firms on a project as each subcontractor will likely have one or more tiers of suppliers. In this case, the authors believe it is possible to augment the Process Connectors architecture to enable coordination across all firms. We can view each subcontractor as a coordinating firm for its suppliers, mirroring the mapping network setup between the CM/GC and the subcontractors. This enables local instantiation of stubs and bridges for these firms. This recursive structure is shown in Fig. 13. Constraints on the part of the suppliers can be reflected in constraints of the subcontractor, and in turn incorporated into the subcontractor's schedule coordination with the CM/GC. Constraints of suppliers to suppliers can also be represented in this way, with the Process Connectors architecture being recursively implemented down each tier of the project supply chain. Such a recursive structure is easier to implement than direct links between a central hub at the CM/GC and all firms in the supply chain both in terms of number of links and in terms of following the contractual structural typical of projects.

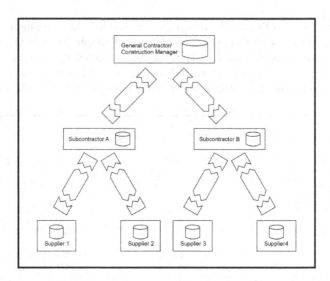

Fig. 13. The Process Connectors architecture can be extended recursively down the supply chain, following contractual arrangements. Constraint information from lower tiers is summarized at each local coordinating firm to make it available to the higher tiers.

6 Conclusions

In the broadest sense, the Process Connectors architecture is an implementation level view of a generalized services architecture for process coordination in the construction supply chain. Stubs provide both the translation of semantically heterogeneous data to a sharable data format and representation of a firm's process information. As such, they can be viewed as a general-use, value-added gateway between firm's internal data/processes and external data/process coordination needs. The example of schedule coordination, although very important to construction applications, need not be the sole distributed process that is supported by the Process Connectors architecture. Other applications for the bridge, such as order processing, can be envisaged. An insight of the schedule mapping representation is that firms' internal process representations, though dissimilar, may have enough common elements such that a generalized representation can maintain links between them. In the same sense we describe the schedule mapping process as an overlay of networks, it is possible to consider process coordination between firms as an overlay of processes. This is an important shift of thinking as distributed process coordination is often viewed as imposition of a single view of the process on many firms. Rather, our view opens new lines of research by considering process coordination as the intersection of multiple processes.

Acknowledgements

The authors would like to thank the faculty (Sherman Bai, Joseph Geunes, Raymond Issa, and Mark Schmalz) and students (Sangeetha Shekar, Nikhil Haldavnekar,

Huanqing Lu, Oguzhan Topsakal, Bibo Yang, Jaehyun Choi, and Rodrigo Castro-Raventós) involved on the SEEK project for their contributions to the research and understanding behind this chapter. The authors also thank the National Science Foundation who supported SEEK research under grant numbers CMS-0075407 and CMS-0122193. Additionally, the authors wish to thank the faculty (Ron Wakefield and Nashwan Dawood) and students (Ting-Kwei Wang, Jungmin Shin, and Tim Apsia) involved on the Process Connectors project. The authors wish to thank the National Science Foundation for its support of the Process Connectors research under grant numbers CMS-0542206 and CMS-0531797.

References

1. IAI: End user guide to Industry Foundation Classes, enabling interoperability in the AEC/FM industry. International Alliance for Interoperability (IAI). (1996)
2. Crowley, A.: The development of data exchange standards: the legacy of CIMsteel. Vol. 2001. The CIMsteel Collaborators (1999) 6 pages
3. Kam, C., Fischer, M.: Product model and 4D CAD final report. Center for Integrated Facility Engineering, Stanford University (2002) 51 pages
4. Fischer, M.A., Kunz, J.: The Circle: Architecture for Integrating Software. ASCE Journal of Computing in Civil Engineering 9 (1995) 122-133
5. Goldberg, H.E.: AEC From the Ground Up: The Building Information Model. Cadalyst (2004)
6. Zamanian, M.K., Pittman, J.H.: A software industry perspective on AEC information models for distributed collaboration. Automation in Construction 8 (1999) 237-248
7. Turk, Z.: Phenomenological foundations of conceptual product modeling in architecture, engineering, and construction. Artificial Intelligence in Engineering 15 (2001) 83-92
8. Amor, R., Faraj, I.: Misconceptions about integrated project databases. ITcon 6 (2001) 57-66
9. O'Brien, W.J., Issa, R.R., Hammer, J., Schmalz, M., Geunes, J., Bai, S.: SEEK: Accomplishing enterprise information integration across heterogeneous sources. ITcon - Electronic Journal of Information Technology in Construction - Special Edition on Knowledge Management 7 (2002) 101-124
10. Hammer, J., O'Brien, W.: Enabling Supply-Chain Coordination: Leveraging Legacy Sources for Rich Decision Support. In: Geunes, J., Akçali, E., Pardalos, P.M., Romeijn, H.E., Shen, Z.J. (eds.): Applications of Supply Chain Management and E-Commerce Research in Industry. Kluwer Academic Publishers, Boston/Dordrecht/London (2005) 253-298
11. Borghoff, U.M., Schlichter, J.H.: Computer Supported Cooperative Work: Introduction to Distributed Applications. Springer-Verlag, Heidelberg, Germany (2000)
12. Smith, S.: Is Scheduling a Solved Problem? : Proceedings of the First MultiDisciplinary Conference on Scheduling: Theory and Applications (MISTA), Nottingham, UK (2003) 16 pages
13. O'Brien, W.J., Fischer, M.A., Jucker, J.V.: An economic view of project coordination. Construction Management and Economics 13 (1995) 393-400
14. Ducq, Y., Chen, D., Vallespir, B.: Interoperability in enterprise modelling: requirements and roadmap. Advanced Engineering Informatics 18 (2004) 193-203
15. Froese, T.: Models of construction process information. ASCE Journal of Computing in Civil Engineering 10 (1996) 183-193
16. Danso-Amoako, M., O'Brien, W.J., Issa, R.: A case study of IFC and CIS/2 support for steel supply chain processes. Proceedings of the of the 10th International Conference on Computing in Civil and Building Engineering (ICCCBE-10), Weimar, Germany (2004) 12 pages
17. Pouria, A., Froese, T.: Transaction and implementation standards in AEC/FM industry. Proceedings of the 2001 Conference of the Canadian Society for Civil Engineers, Victoria, British Columbia (2001) 7 pages

518   W.J. O'Brien, J. Hammer, and M. Siddiqui

18. Liu, D., Cheng, J., Law, K., Wiederhold, G., Sriram, R.: Engineering information service infrastructure for ubiquitous computing. ASCE Journal of Computing in Civil Engineering **17** (2003) 219-229
19. Schlenoff, C., Gruninger, M., Tissot, F., Valois, J., Lubell, J., Lee, J.: The Process Specification Language (PSL): Overview and Version 1.0 Specification. NIST (2000) 83 pages
20. Wiederhold, G.: Weaving data into information. Database Programming and Design **11** (1998)
21. Castro-Raventós, R.: Comparative Case Studies of Subcontractor Information Control Systems. M.E. Rinker, Sr. School of Building Construction. University of Florida (2002)
22. Sriprasert, E., Dawood, N.: Multi-constraint information management and visualization for collaborative planning and control in construction. ITcon, Special Issue on eWork and eBusiness **8** (2003) 341-366
23. Antill, J.M., Woodhead, R.W.: Critical Path Methods in Construction Practice. Wiley, New York (1990)
24. O'Brien, W.J., Fischer, M.A.: Importance of capacity constraints to construction cost and schedule. ASCE Journal of Construction Engineering and Management **126** (2000) 366-373
25. Mawdesley, M., O'Reilly, M.P., Askew, W.: Planning and controlling construction projects: the best laid plans. Longman, Essex, England (1997)
26. Smith, S.F., Becker, M.A.: An Ontology for Constructing Scheduling Systems. Working Notes of the 1997 AAAI Symposium on Ontological Engineering. AAAI Press (1997) 10 pages
27. Hammer, J., Schmalz, M., O'Brien, W., Shekar, S., Haldavnekar, N.: Enterprise knowledge extraction in the SEEK project part I: data reverse engineering. Department of Computer and Information Science and Engineering, University of Florida (2002) 30 pages
28. Chawathe, S., Garcia-Molina, H., Hammer, J., Ireland, K., Papakonstantinou, Y., Ullman, J., Widom, J.: The TSIMMIS Project: Integration of Heterogeneous Information Sources. Tenth Anniversary Meeting of the Information Processing Society of Japan. Information Processing Society, Tokyo, Japan (1994) 7-18
29. Bayardo, R., Bohrer, W., Brice, R., Cichocki, A., Fowler, G., Helal, A., Kashyap, V., Ksiezyk, T., Martin, G., Nodine, M., Rashid, M., Rusinkiewicz, M., Shea, R., Unnikrishnan, C., Unruh, A., Woelk, D.: Semantic Integration of Information in Open and Dynamic Environments. MCC (1996)
30. Hammer, J., Breunig, M., Garcia-Molina, H., Nesterov, S., Vassalos, V., Yerneni, R.: Template-based wrappers in the TSIMMIS system. Twenty-Third ACM SIGMOD International Conference on Management of Data, Tuscon, Arizona (1997) 532-543
31. Hammer, J., Stonebraker, M., Topsakal, O.: THALIA: Test Harness for the Assessment of Legacy Information Integration Approaches. Proceedings of the 21st Int'l Conf. on Data Engineering (ICDE2005), Tokyo, Japan (2005) 2 pages
32. O'Brien, W.J., Hammer, J.: A case study of information integration problems in first responder coalitions. Proceedings of the 3rd International Conference on Knowledge Systems for Coalition Operations (KSCO-2004), Pensacola, Florida (2004) 145-149
33. Siddiqui, M., O'Brien, W.J., Wang, T.: A mapping based approach to schedule integration in heterogeneous environments. Proceedings of the Joint International Conference on Computing in Building and Civil Engineering, Montreal, Canada (2006) 10 pages
34. Thabet, W.Y., Beliveau, Y.J.: SCaRC: Space-constrained resource-constrained scheduling system. ASCE Journal of Computing in Civil Engineering **11** (1997) 48-59
35. Cormen, T.H., Leiserson, C.E., Rivest, R.L., Stein, C.: Introduction to algorithms. MIT Press, Cambridge, Mass. (2001)
36. Wang, T.-K., O'Brien, W.J., Siddiqui, M., Hammer, J., Wakefield, R.: Process Connectors: Mapping Distributed Processes in the Construction Supply Chain. Proceedings of the Joint International Conference on Computing in Building and Civil Engineering, Montreal, Canada (2006) 10 pages

Knowledge Based Engineering and Intelligent Personal Assistant Context in Distributed Design

Jerzy Pokojski

Institute of Machine Design Fundamentals, Warsaw University of Technology,
Narbutta 84,
02 524 Warsaw, Poland
jerzy.pokojski@simr.pw.edu.pl

Abstract. The work focuses on the problem concerning the application of the KBE approach and its tools in distributed environments. The first period of applying industrial KBE systems did not only show their significant advantages but also revealed their shortcomings. This problem is especially disturbing with distributed design where the potential of communication is very limited. Every design process is closely connected with the designer's knowledge. In general, this is a very individual knowledge which is stored in the designer's personal memory. The work represents an attempt of integration of KBE and IPA (Intelligent Personal Assistant), [11], which is the designer's personal knowledge repository.

1 Introduction

The work focuses on the problem concerning the application of the Knowledge Based Engineering (KBE) approach [7] and its tools in distributed environments [16].

Today's CAD systems enable building geometric models which allow parametric modeling at large scale. Often, when applying the conception of parametric modeling a vision of the product's further development is implied – a vision of the various versions being planned [1, 4, 7, 12]. Along with these CAD systems it became also possible to record the calculation procedure and design rules (Knowledge Based Engineering) [1, 7, 12, 15, 16] (fig. 1). It is difficult to find publications of complete description of industrial KBE applications. One reason is that firms are not willing to publish their know-how (work [4] is an exception). Second, KBE applications in general are large and complicated. Additionally the CAD systems can be integrated with external calculation processes as well as with external information and knowledge sources [12]. All these modeled and integrated components (forming the computerized form of procedural or declarative knowledge) retrieve the parameterized geometric objects (modeled in the CAD system) and thus support in the same process the generation, evaluation and modification of the objects [1, 7]. Consequently, models of a geometric construction based on the recorded knowledge can be created very quickly (it is also possible to create non-geometric models, for instance simulation models or FEM models [12]).

I.F.C. Smith (Ed.): EG-ICE 2006, LNAI 4200, pp. 519–528, 2006.

The first period of applying industrial KBE systems did not only show their significant advantages but also revealed their shortcomings [7, 12]. The two most significant ones have to be mentioned: 1) the lack of an universal work methodology with the KBE systems and 2) the lack of an universal methodology for the storing and managing of the engineer's knowledge implemented in the KBE system. In both cases the shortcomings are decisively influenced by the individual development of a branch, a firm or even a single person.

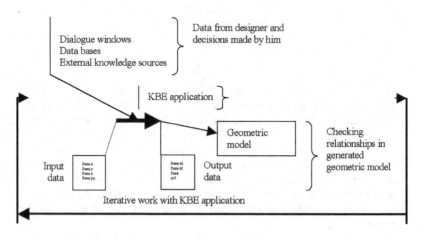

Fig. 1. Knowledge Based Engineering application and its structure

It became obvious that the process of building KBE applications also reflects the engineer's knowledge which he assigned to the tools and the representation in the KBE system. This process is realized by many a'priori conditions which are never completely articulated in an accessible form. Because of that the resulting applications may differ from each other (concerning the same domain) by their various structures and their final version. That means that they can only be fully understood by their creators. Any attempts of non-insiders to comprehend the process involves the risk of misunderstandings. This problem is especially disturbing with distributed design where the potential of communication is limited.

Every design process is closely connected with the designer's knowledge [2 –3, 7-11]. In general, this is a very individual knowledge which is stored in the designer's personal memory. As it comprises formally articulated knowledge, implemented procedural knowledge and computer tools it makes its exploitation a relative individual matter. Only in rare cases the designer articulates and stores his knowledge explicitly for external users. Moreover, the knowledge is not static, it develops individually, it is constantly enriched and continuously integrated with the current available knowledge. Thus it creates the work basis for the activities of a professional designer. Considering the inadequacy of the human mind the recording of the designer's knowledge on external storage units would make a future access to one's own knowledge easier (both resolved and accidentally obtained) and also provide a more efficient management of the knowledge. At the same time new knowledge elements could be better articulated and become available for external partners which would be especially helpful in the case of distributed design. The practical consequence of the above observations is the building of a personal notepad (Intelligent Personal Assistant) [11] which is computerized and functions to store and manage the designer's personal knowledge. After having fulfilled the mentioned requirements the notepad can become the designer's knowledge repository [11].

The ideas presented in this paper are based on concepts which the author explicitly laid out in his book (Pokojski, J., 2004 [11]). Among them can be found: the nature of individual and team based design, a survey of software concepts of the IPA (Intelligent Personal Assistant), a general model of the IPA concept, issues of integrating, a survey of engineering knowledge representations, design process modeling, knowledge modeling, relationships between the IPA and optimization as well as its implementation. It would be beyond the scope of this article to summarize the whole book. Consequently, merely the most basic points are taken up in the introduction of the paper, i.e. the role of knowledge in the design and knowledge based engineering in the IPA-context.

In many practical tasks we see that a professional's knowledge finds its representation in various forms and that all the knowledge elements are connected to each other even when they are recorded in different representations [11]. This fact becomes especially obvious when we cooperate with designers who can look back on a relatively long professional career. When such an experienced engineer reveals the work proceedings of a design task he is mostly able to explain the sources of the knowledge on which he based certain design steps. In general those sources are recorded multimedially as texts or drawings.

Often the designer is also able to make clear how the knowledge developed in the past. If there was any kind of knowledge representation – procedural or declarative recordings – then the designer is enabled to show its evolution as well. If we want to record a single designer's knowledge in a personal knowledge repository, the knowledge and its structuring have to be decomposed. For that purpose it is necessary to predict the data structures which allow to record the dynamics of the knowledge development. Equally important is the proper classification of each knowledge element as it makes its possible reuse in the future easier. Concerning the notebook we require that the picture of the recorded knowledge depicts as truly as possible its real form. Consequently, the proposed tool has to become one of the most essential means used by a designer. At the same time it should guarantee fast access to the required knowledge element if necessary. As another important function the notebook offers the possibility of managing the available knowledge in design processes of distributed design. There the notebook, i.e. the personal assistant, apart from providing single, detailed knowledge elements should also present a wider context of the selected knowledge or information element and make clear how it is connected with other knowledge elements from its user's point of view. The work represents an attempt of KBE and IPA (intelligent personal assistant) integration in distributed design.

2 The Use of the KBE Application in the Design Process

The applied concept assumes that a design process exploiting KBE tools is carried out in a linear way as a sequence of certain design steps (fig. 2) [12, 15]. Merely in some cases alternatives to the steps are offered. The concept also ignores an iterative realization of design steps.

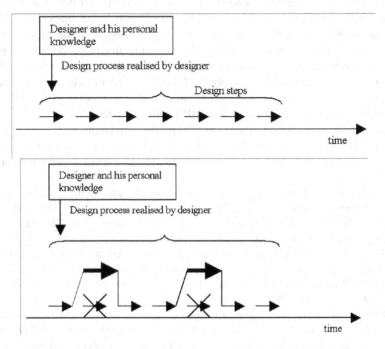

Fig. 2. Design process without (upper) and with (lower) KBE application

It has to be stated that in general real modeled processes are not completely articulated. Only certain stages and typical paths are known. As each step is manually indicated we often have to cope with surprising outcomes: correctly or incorrectly calculated steps as well as meritorically correct or incorrect ones [12]. Of course, we can try to solve most of the tasks correctly by testing certain mathematical models but this attempt requires a lot of work. Moreover, with diversity tasks which are characterized by a big complexity such an undergoing wouldn't be economically worth while doing. Apart from that it is quite difficult to build an universal formalism for that task. Because of those difficulties we mostly decide to apply the engineer's traditional way of proceeding in the design in the KBE system. This means, first the calculations are done then the constraints and the relations between the parameters of certain objects are checked. Any time a condition in the step is not fulfilled it is indicated to the user in an appropriate way. It has to be pointed out that every single step has its geometric consequences in the form of partial models. They can be visualized and their mistakes can become obvious. The designer does not only obtain information whether the implemented constraints are formally fulfilled, he can also virtually observe the model which he generated and study the constraints he took into account. After some experience with partial models the designer is mostly able to build modules which indicate that a certain relation is not fulfilled. They reflect the evaluation stages which naturally appear with manual calculation.

As presented here the most common way of implementing KBE applications is to make automatic the design steps and to perform trial modeling on the evaluation basis of the achieved relations. The realization of a design step is initiated manually and directly controlled by the engineer. The selection of the following steps is also done by the designer himself. In principle the whole design process is equipped with tools for an interactive support of single steps which are accompanied by the visualization of their geometric consequences. Thus the complete application is given an elastic structure which adequately reflects the knowledge the designer has applied in a relatively exact but realistic way. If aiming at a wider complexity the presented approach can be further automatized. However, the user should be aware of the fact that any automation is very labour intensive and not necessarily economic.

3 The Personal Assistant as a Knowledge Repository in the KBE System

In general, the knowledge owned and used by a designer represents a quite rich and coherent complexity which has been accomplished over many years [10, 11, 12]. As it would take a lot of effort and be very difficult to garner that knowledge entirely, selections and simplifications have to be made. It is advantageous when the selections and simplifications are carried out by the designer himself because he best knows the status of certain problems. One of the most basic classifications is to assign the knowledge elements to different activities carried out by the designer. Mostly the engineers opt for those activities which were used by them and whose appearance, evaluation and vanishing they know.

Each activity is accompanied and defined by its knowledge support whereby the knowledge may have various forms and sources. It is possible to establish a recording of the computer knowledge elements on which the designer founds his activities. When the knowledge elements are connected to a certain activity the evolution process of the knowledge for a given activity becomes obvious. If the activity is carried out with the help of KBE tools, we are able to trace back the process of the knowledge evolution which stands behind each version of the given tool – from the designer's personal perspective (fig. 3). We can then look back and try to comprehend the wider context and endeavour to pursue the individual path of development. The insight acquired by the retracing is especially helpful in the case of distance-cooperation as we may regard it a kind of a condensed professional biography of the designer which makes us better understand his work proceedings. For projects in distributed design which are realized by many cooperators this is advantageous because of the possibility to integrate the personal knowledge stores of the team members (fig. 4). Each participant can then suitably manage the process and offer his knowledge elements for the other team members. This brings about a better mutual understanding of the participants at a distance especially with various cultural backgrounds.

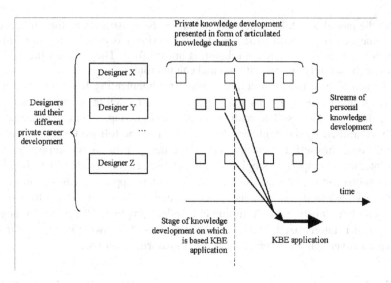

Fig. 3. KBE application and its knowledge resources

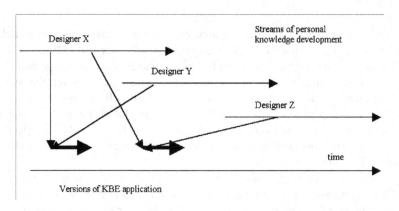

Fig. 4. Streams of personal knowledge developments and KBE applications

4 The Integration of Personal Assistant Storages in Distributed Design

KBE applications are an effective means to reduce the time of realizing design projects significantly. However, it is a very time consuming task to build KBE applications. When we want that applications of this class be used by engineers who are not in one way or the other involved in their creating then understanding the knowledge which forms the basis of the used application becomes an essential problem. Such a situation is easily foreseeable with teams cooperating over distance. In most cases the team members never meet and as a consequence never work face-to-face. As a result the application may have to be used by designers who are not its author and are also far away from him. This means that non-authors only exploit the results of the respective application. In both cases the users might want to or even have to understand the knowledge behind the design process whose model is to a certain degree the KBE application. In any case this model cannot easily be depicted or restored. Similar situations may also occur when different firms start to cooperate. It can happen that potential partners use different approaches in their projects and at least at the beginning of the cooperation are not able to completely understand each other's procedure.

The presented applications of a designer's personal assistant which contain the recordings of the evolution of the knowledge that was the bases for the establishing of further modules and their KBE application version may turn out to be helpful with finding information which explains its functioning . The advantage of this concept is that the personal assistant's content can be recorded parallel to the process by which the KBE application is created. At the same time it is possible to integrate the various knowledge elements with each other. Additionally, management providing modeled and controlled access to the information can be implemented.

4.1 Example

The example below shows how a computer application supports the design process of a car gear box [11, 12, 13, 14]. In each of the variants the gear box has the same structure (fig. 5), that means the same system of wheels, shafts, bearings, clutches etc. The wheels and clutches are KBE models. Because of that the complete geometric model of the gear box can be changed according to selected and calculated parameters. The presented KBE application is integrated with a calculation module which is equipped with a data base and a Case-Based- Reasoning module. Additionally, the designer can add to each gear box variant – manually or automatically – design rationale information concerning the respective variant. The module for the calculation of the tooth wheels is integrated with a multi-criteria optimization system.

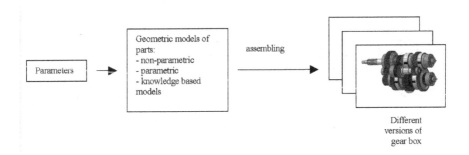

Fig. 5. Exemplary KBE application

The complete computer environment presented above was developed in 2001/2002 by M. Okapiec and G. Witkowski and submitted for a diploma, supervised by J. Pokojski [14]. The dissertation contains software and know-how from the supervisor

....

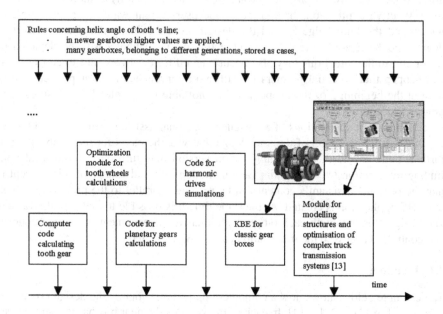

Fig. 6. Structure of IPA layer of author

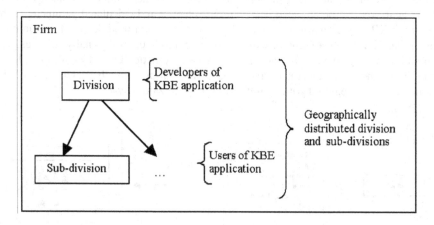

Fig. 7. Structure of industrial cooperation

and the consultants A. Wąsiewski and S. Skotnicki. Figure 6 depicts the structure and the extremely abstracted layer of the IPA content (belonging to the supervisor). The two kinds of information – 1) description of the KBE application and 2) personal knowledge development reflected in an IPA content – illustrate how a KBE application actually works.

5 Conclusions

The paper lays out an approach of how the process of creating a KBE application can be stored in a wider range. A well elaborated application is compiled of many modules. They may have been carried out in several version and are meant for use in distributed design.

Many KBE applications can be understood when we analyze their objects, attributes, activities, etc. But we should remember that even the knowledge of single automotive units contains huge amounts of data and information. A car transmission system for example can have thousands of attributes. Nowadays there are attempts of building models by using formal knowledge [5, 6, 7]. But they turn out to be too restrictive. Moreover, the level of these attempts is not necessary for potential users of KBE applications. In most cases they need a general informal description of the problem, explaining the foundations of a KBE application.

The example described in the work only wants to illustrate the concept. It is not an example which was originated in an industrial reality. The approach to the concept itself, however, results from actual experience gained while implementing it industrially in other domains [15], (fig. 7).

References

1. CATIA – manual, (2005)
2. Clarkson, J., Eckert, C. (ed.): Design Process Improvement. A review of current practice. Springer – Verlag, London (2005)
3. Fujita, K., Kikuchi, S.: Distributed Design Support System for Concurrent Process of Preliminary Aircraft Design. Concurrent Engineering: Research and Applications 11 2 (2003) 93- 105
4. Glymph, J., Shelden, D., Ceccato, C., Mussel, J., Schober, C.: A parametric strategy for free-form glass structures using quadrilateral planar facets. Automation in Construction 13 (2004) 187-202
5. Gu, N., Xu, J., Wu, X., Yang, J., Ye, W.: Ontology based semantic conflicts resolution in collaborative editing of design documents. Advanced Engineering Informatics 19 (2005) 103-111
6. Kim, T., Cera, C.D., Regli, W.C., Choo, H., Han, J.: Multi-level modeling and access control for data sharing in collaborative design. Advanced Engineering Informatics 20 (2006) 47-57
7. Managing Engineering Knowledge, MOKA - project. Professional Engineering Publishing Limited, London (2001)
8. Moran, T.P., Carroll, J.M.: Design Rationale, Concepts, Techniques, and Use. Lawrence Erlbaum Associates, Publishers, Mahwah, New Jersey (1996)
9. Nahm, Y., Ischikawa, H.: Integrated Product and Process Modeling for Collaborative Design Environment. Concurrent Engineering: Research and Applications, 12, 1(2004) 5- 23
10. Pokojski, J., (Ed.): Application of Case-Based Reasoning in Engineering Design. WNT, Warsaw (2003) (in Polish)
11. Pokojski, J.: IPA (Intelligent Personal Assistant) – Concepts and Applications in Engineering. Springer-Verlag, London (2004)
12. Pokojski, J.: Expert Systems in Engineering Design. WNT, Warsaw (2005) (in Polish)

13. Pokojski, J., Niedziółka, K.: Transmission system design – intelligent personal assistant and multi-criteria support. In: Next Generation Concurrent Engineering – CE 2005, ed. by M. Sobolewski, P. Ghodous, Int. Society of Productivity Enhancement, NY (2005)   455-460

14. Pokojski, J., Okapiec, M., Witkowski, G.: Knowledge based engineering, design history storage, and case based reasoning on the basis of car gear box design. In: AI-Meth 2002, Gliwice (2002) 337-340

15. Pokojski, J., Skotnicki, S.: Knowledge-based Engineering in support of industrial stairs geometric model generation. XV Conference „Methods of CAD" , Proceedings, Institute of Machine Design Fundamentals, Warsaw University of Technology (2005) 299-306

16. Sriram, R.D.: Distributed and Integrated Collaborative Engineering Design. Sarven Publishers (2002)

Model Free Interpretation of Monitoring Data

Daniele Posenato[1], Francesca Lanata[2], Daniele Inaudi[1], and Ian F.C. Smith[3]

[1] Smartec SA, via Pobiette 11,CH-6928 Manno, Switzerland
posenato@smartec.ch, inaudi@smartec.ch
[2] Department of Structural and Geotechnical Engineering
University of Genoa Via Montallegro, 116145 Genoa, Italy
[3] Ecole Polytechnique Fédérale de Lausanne (EPFL)
GC G1 507, Station 18, CH-1015 Lausanne, Switzerland
Ian.Smith@epfl.ch

Abstract. No current methodology for detection of anomalous behavior from continuous measurement data can be reliably applied to complex structures in practical situations. This paper summarizes two methodologies for model-free data interpretation to identify and localize anomalous behavior in civil engineering structures. Two statistical methods i) moving principal component analysis and ii) moving correlation analysis have been demonstrated to be useful for damage detection during continuous static monitoring of civil structures. The algorithms memorize characteristics of time series generated by sensor data during a period called the initialisation phase where the structure is assumed to behave normally. This phase subsequently helps identify anomalous behavior. No explicit (and costly) knowledge of structural characteristics such as geometry and models of behaviour is necessary. The methodologies have been tested on numerically simulated elements with sensors at a range of damage severities. A comparative study with wavelets and other statistical analyses demonstrates superior performance for identifying the presence of damage.

1 Introduction

Structural health monitoring engineers may employ sensors to perform non-destructive in-situ structural evaluation. These sensors produce data (either continuously or periodically) that are analyzed to assess the safety and performance of structures [1]. For static monitoring, damage can be identified through comparing static structural response with predictions of behavior models [2]. However, models can be expensive to create and may not accurately reflect undamaged behavior. Difficulties and uncertainties increase in the presence of complex civil structures so that well defined and unique behavior models often cannot be clearly identified [3]. Furthermore, multiple-model system identification may not succeed in identifying the right damage [4]. Despite important research efforts into interpretation of continuous static monitoring data [5], no reliable strategy for identifying damage has been proposed and verified for broad classes of civil structures [3][6]. Another approach is to evaluate changes statistically [7]. This methodology is completely data driven; the evolution of the data is estimated without information of physical processes [8-10].

The objective of this paper is to propose methodologies that discover anomalous behavior in data generated by sensors without using behavior models. The paper is

I.F.C. Smith (Ed.): EG-ICE 2006, LNAI 4200, pp. 529–533, 2006.

organized as follows: Section 2 provides a description of the numerical simulation used to compare the damage detection performances of several algorithms and a description of the moving PCA and correlation analysis. Section 3 explains how PCA and correlation analysis are used in structural health monitoring. Section 4 presents a comparison between these algorithms and continuous wavelet analysis [11-12]. More details of this work are contained in [13].

2 Numerical Simulation and Model-Free Data Interpretation

2.1 Numerical Simulation

Due to difficulties in retrieving databases from real structures with a range of damage severities, a finite element model [3] of a two-span continuous beam in healthy and in damaged states has been used to evaluate the efficiency of algorithms to detect damage [5]. A thermal load simulates structural behavior under varying environmental conditions. The response is measured by means of a 'virtual' monitoring system composed of twelve elongation sensors, see Figure 1.

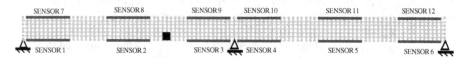

Fig. 1. The FE model used to test the methodology showing where sensors are placed (bold lines) and positions of simulated damage (black squares). The beam is 10x0.5x0.3 m, and each cell of the mesh is 10x8 cm.

2.2 Moving Principal Components Analysis (MPCA)

PCA is based on an orthogonal decomposition of the process variables along the direction that explains the maximum variation of the data (the components that contain most of the information) [15]. PCA is applied to the data in this study in an effort to reduce the dimensionality of the data and to enhance the discrimination between features of undamaged and damaged structures. To reduce the dimensionality of the space, values of sensor measurements are projected onto the eigenvectors that correspond to the largest eigenvalues of the covariance matrix. Most of the variance is contained in the first few principal components while the remaining components are defined by measurement noise.

Each eigenvalue expresses the variance in time associated with the corresponding eigenvector. Orthogonal eigenvectors are time-invariant: the eigenvector associated with the maximum eigenvalue represents the spatial behavior corresponding to the time function with the maximum variance. Normally the PCA is computed on the covariance matrix computed for all the measurements in the time series. To reduce the computation time and thus the time required to detect new situations in the time series, a Moving Principal Components Analysis is used [13], in which a window (a subset of time series) containing only a fixed number of last measurements is used. The original aspect of this method is that the covariance matrix of all the sensors is computed only for a moving window of constant size. This means that after each measurement session, the covariance matrix and principal components are computed only for the points inside the active window.

2.3 Moving Correlation Analysis

This method is used to calculate the correlations between all sensor pairs for a reference period with the aim to quantify the tendency of sensors to change in similar ways. During the reference period, variations of correlations are calculated for each sensor pair. After the initialisation phase, all correlations are calculated at each step to determine the presence of anomalies in the evolution of values. Anomalous behaviour is observed through correlations lying outside the thresholds defined during the initialisation phase. In normal conditions the correlation should be constant or stationary. However, when damage occurs, correlations between the sensors change. To follow the evolution of the time series more effectively, a moving window of fixed size is employed [13]. After each session of measurement the correlation is calculated only for the points inside the moving window. The size of the window has to guarantee stability of average values, to ensure rapid damage identification and to reduce the effects of noise.

3   Application to Structural Health Monitoring

Application of MPCA and Moving Correlation to structural health monitoring involves an initial phase (called initialization) where the structure is assumed to behave in an undamaged condition. The aim of this initialization period is to estimate the variability of the time series and to define thresholds for detecting anomalous behavior [14]. This period is normally one or two years because in this manner most variations in the behavior of structures due to periodic environmental and load changes is recorded. In order to make all the further recorded behavior comparable with those recorded during the initialization, the schedule of measurements is unchanged during monitoring. Once thresholds have been fixed, the parameters of the process are monitored (main eigenvectors for MPCA and Correlations on all sensor pairs for the Moving Correlation) to ensure that they are inside predefined ranges. For identifying damage location, the rule has been used that candidate damage zones are close to sensors that have the status parameters exceeding a threshold.

4   Results

A comparative study between the proposed algorithms and wavelet transform (CWT) [11-12], short-term Fourier transform (STFT) [12], the instance-based method (IBM) [15-16] has been carried out for several damage scenarios [13]. In this paper, only comparisons with using the algorithms and CWT for damage between sensors 2 and 3 (4 cells with reduction to 20% of original stiffness) are presented, see Figure 1. Results show that the algorithms proposed in this paper identify anomalous behaviour more effectively than CWT. Figure 2 shows plots of the eigenvectors related to the two main MPCA eigenvalues. The moment when damage occurs and its location are visible in both plots. Specifically, one of the eigenvectors (eigenvector 11) indicates a new state when it becomes stable, while the other (eigenvector 12) indicates when the damage occurred. The location of the damage is detected by the fact that within the main eigenvectors, there are one or more rapidly changing components that are associated with sensors close to the damage. In Figure 2 the location of the damage can be detected by sensor 3 (Sn3), which is the closest to the damage location since its variation bigger than the other sensors.

Figure 3 shows Moving Correlation results for sensor pairs that are closest to the damage. The moment when damage occurred and when the behavior of the structure can be considered to be stable are visible. Figure 4 shows CWT results. The moment when damage occurred is not visible and there is no information regarding whether the anomaly is due to a new temporary situation or due to permanent damage.

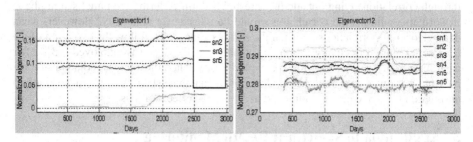

Fig. 2. MPCA plots of eigenvectors related to the two main eigenvalues. They show the moment when damage occurs and its location. One eigenvector (eigenvector 11, only values of sensors 2, 3 and 5 are presented) gives an indication of the new state of the structure when it becomes stable while the second eigenvector (eigenvector 12) gives an indication of the damage. The symbols, sn1, sn2, ... are the components of the eigenvector referred respectively to sensor 1, sensor 2, etc.

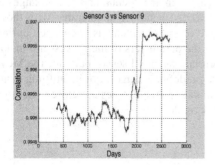

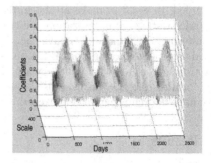

Fig. 3. Diagnostic plots of Moving Correlation calculated from measurements of two sensors close to the damage. Calculations were performed using a moving window of one year.

Fig. 4. CWT calculated from the difference between results of the two sensors normalized, closest to the damage. The Gauss wavelet with a scale of 1024 has been used.

5   Conclusions

Moving Principal Component Analysis and Moving Correlation are useful tools for identifying and locating anomalous behavior in civil engineering structures. These approaches can be applied over long periods to a range of structural systems to discover anomalous states even when there are large quantities of data. A comparative study has shown that for quasi-static monitoring of civil structures, these new methodologies perform better than wavelet methods. While these methodologies have good capacities to detect and locate damage, they also require less computational resources. Another important characteristic is adaptability. Once new behavior is identified, adaptation allows detection of further anomalies. The next step of the research is to apply the proposed methodology to a database of measurements taken from full-scale structures.

References

1. A. Bisby, An Introduction to Structural Health Monitoring. ISIS Educational Module 5 (2005)
2. Y. Robert-Nicoud, B. Raphael, O. Burdet & I. F. C. Smith, Model Identification of Bridges Using Measurement Data, Computer-Aided Civil and Infrastructure Engineering, Volume 20 Page 118 - March 2005
3. F. Lanata Damage detection algorithms for continuous static monitoring of structures PhD Thesis Italy University of Genoa DISEG, (2005)
4. S. Saitta, B. Raphael, I.F.C. Smith, Data mining techniques for improving the reliability of system identification, Advanced Engineering Informatics 19 (2005) 289–298
5. A. Del Grosso, D. Inaudi and F. Lanata Strain and displacement monitoring of a quay wall in the Port of Genoa by means of fibre optic sensors 2nd Europ. Conf. on Structural Control Paris, (2000)
6. A. Del Grosso and L. Lanata, Data analysis and interpretation for long-term monitoring of structures Int. J. for Restoration of Buildings and Monuments, (2001) 7 285-300
7. J. BROWNJOHN, S. C. TJIN, G. H.TAN, B. L. TAN, S. CHAKRABOORTY, "A Structural Health Monitoring Paradigm for Civil Infrastructure", 1st FIG International Symposium on Engineering Surveys for Construction Works and Structural Engineering, Nottingham, United Kingdom, 28 June – 1 July 2004
8. H. M. Jaenisch, J. W. Handley, J. C. Pooley, S. R. Murray, "DATA MODELING FOR FAULT DETECTION" , *2003 MFPT Meeting.*
9. F. Lanata and A. Del Grosso, Damage detection algorithms for continuous static monitoring: review and comparison 3rd Europ. Conf. On Structural Control (Wien, Austria), 2004
10. Sohn, H., J. A.Czarneski and C. R. Farrar. 2000. "Structural Health Monitoring Using Statistical Process Control", Journal of Structural Engineering, 126(11): 1356-1363
11. C. K. CHUI, Introduction to Wavelets, San Diego, CA: Academic Press, p.264, 1992
12. I. Daubechies, Ten Lectures on Wavelets, Philadelphia: Soc. for Indust. and Applied Mathematics, p. 357, 1992
13. Daniele Posenato, Francesca Lanata, Daniele Inaudi and Ian F.C. Smith, "Model Free Data Interpretation for Continuous Monitoring of Complex Structures", submitted to Advanced Engineering Informatics, 2006
14. M. Hubert and S. Verboveny, "A robust PCR method for high-dimensional regressors", Journal of Chemometrics, 17, 438-452.
15. Kaufman and Rousseeuw, 1990,L. Kaufman and P.J. Rousseeuw. Finding Groups in Data: An Introduction to Cluster Analysis. Wiley, New York, 1990.
16. Mahamud and Hebert, S. Mahamud and M. Hebert. Minimum risk distance measure for object recognition. Proceedings 9th IEEE International, 2003 Conference on Computer Vision (ICCV), pages 242–248, 2003.

Prediction of the Behaviour of Masonry Wall Panels Using Evolutionary Computation and Cellular Automata

Yaqub Rafiq, Chengfei Sui, Dave Easterbrook, and Guido Bugmann

University of Plymouth, Drake Circus, Plymouth, PL4 8AA, UK
{mrafiq, csui, deasterbrook, gbugmann}@plymouth.ac.uk

Abstract. This paper introduces methodologies that not only predict the failure load and failure pattern of masonry panels subjected to lateral loadings more accurately, but also closely matches deflection at various locations over the surface of the panel with their experimental results. In this research, Evolutionary Computation is used to model variations in material and geometric properties and also the effects of the boundary types on the behaviour of the panel within linear and non-linear ranges. A cellular automata model is used that utilises a zone similarity concept to map the failure behaviour of a single full scale panel *'the base panel'*, tested in the laboratory, to estimate variations in material and geometric properties and also boundary effects for any unseen panels.

1 Introduction

Due to the highly composite and anisotropic material properties of masonry, it has been difficult to accurately predict the behaviour of masonry panels. The research presented in this paper proposes a numerical model updating technique that studies the behaviour of masonry panels subjected to lateral loading within the full linear and non-linear ranges. The method uses evolutionary computation (EC) techniques to model variations in geometric and material properties over the entire surface of the panel. The EC search produces factors *'the corrector factors'*, which reflect the collective effects of the above mentioned variations. These factors are then used to vary the value of flexure rigidity at various locations over the entire surface of the panel. The modified flexure rigidity are then used in a specialised non-linear finite element analysis (FEA) program to predict the failure load, failure pattern and load deflection relationships over the full linear and non-linear ranges. The EC exploration also includes the effect that boundary types may have on the response of panels to lateral loading. Results obtained from the non-linear FEA are compared with the experimental results from a full scale panel tested in the laboratory. Finally a cellular automata (CA) is used to map information obtained from the single full scale panel 'base panel'[1] to an 'unseen panel'[2] for which an estimate in variations of material and geometric properties and boundary effects is produced. A non-linear FEA is then used to predict the failure behaviour of these unseen panels.

---

[1] The base panel is a panel for which displacement values are known at various load levels and locations over the surface of the panel and for which failure load and failure pattern are also known.

[2] The unseen panel is a panel for which the above parameters are normally not known.

I.F.C. Smith (Ed.): EG-ICE 2006, LNAI 4200, pp. 534–544, 2006.

The generality of the methodologies proposed in this paper was tested on several *'unseen panels'* with and without openings and the results were found to have a reasonable match with their experimental results. A sample of this study is presented later in this paper.

2 Modelling and Measurement Error

Robert-Nicoud et al. [1] define modelling error $|e_{mod}|$ as the difference between the predicted response of a given model and that of an ideal model representing the real behaviour accurately. Raphael and Smith [2] categorised the modelling error into three components, e_1, e_2 and e_3. The component e_1 is the error due to the discrepancy between the behaviour of the mathematical model and that of the real structure. The component e_2 is introduced during the numerical computation of the solution to the partial differential equations representing the mathematical model. The component e_3 is the error due to the assumptions that are made during the simulation of the numerical model.

For masonry wall panels, assumptions regarding the choice of boundary conditions are very difficult to justify. This is because the true nature of the panel boundaries either for a wall tested in the laboratory or real structures does not comply with the known boundary types (fixed, simply supported etc.). Hence the error resulting from the use of incorrect boundary types would be relatively large.

Another factor that greatly affects the behaviour of masonry wall panels is the existence of a large error due to the component e_2 [2] introduced during the numerical computation process. This error is mainly due to the uncertainty in modelling the material and geometric properties of highly composite anisotropic material such as masonry. Yet, there is not an agreed material model for masonry to represent the true anisotropic nature of this material. There are very few commercial packages for modelling masonry structures. Adding to this error is the modelling complexity due to the propagation of cracks over the surface and along the depth of the masonry panel, when performing a non-linear FEA.

The value of e_1 for steel and reinforced concrete structures may be reduced through the use of more precise mathematical models [2]. However, due to the extremely complex nature of masonry material, the applicability of this approach in practice would be almost impossible. Another reason for ambiguity in this error is the lack of sufficient laboratory test data and the high cost of these tests for masonry wall panels. The majority of tests performed on masonry wall panels only report the failure load of the panel, as this is the major design requirement, and the failure pattern, which is the crack pattern observed during the laboratory tests. Recording load and deflection information is generally limited to a single location, the location of maximum deflection, over the entire surface of the panel. This makes it extremely difficult to understand the true behaviour and the boundary effects on the response of masonry wall panels. One of the mostly cited published data available is the data from 18 full scale masonry wall panels tested in the University of Plymouth (UoP) by Chong [3] that reports load deflection data at 36 locations over the surface of the panel. These data are used in this research.

2.1 Error Due to Incorrect Support Types (University of Plymouth Test Panels)

The panel's vertical sides were supported on a steel angle connected to the test frame abutment to simulate a simply supported support type, and the base of the wall was enclosed in a steel channel packed with bed joint mortar at both sides (refer to Fig.1 for support details). It was assumed that a combined effect of the support details and the self weight of the wall might provide sufficient restraint to the base of the panel to simulate a fixed support type.

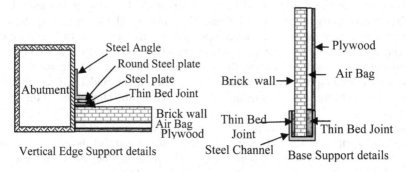

Fig. 1. Test panel support details

From Fig. 1, one can argue that the vertical edges of the panel are not truly simply supported and there is some degree for restrain to rotation. Similarly the base of the panel is by no means fully fixed and allows some degree of rotation. Due to the flexible nature of the edge support, some degree of movement perpendicular to the plane of the wall was observed at the right hand support.

2.2 Error Due to Applied Loads and Deflection Measurements

The load was applied to the wall by means of an air bag and it was assumed to be uniformly distributed over the entire surface of the panel. This assumption may be true when the air bag is fully inflated, but not at the lower load levels.

Fig. 2 shows the location of the measurement points (36 points in total) on the face of the base panel. This panel was a solid single leaf clay brick masonry wall panel. Linear Variable Differential Transformers (LVDTs) were placed at each gridline intersection to measure the wall movement perpendicular to the plane of the wall. Due to the unevenness of the surface, inherent to masonry panels, irregularities were observed in the deformed shape of the panel surface, as shown in Figs. 2(b) and 3. The reason for these irregularities could be the slippage of the LVDTs from their intended location and/or inaccuracy in the LVDT readings.

Controlling the load levels at each load increment and maintaining a uniform load over the entire surface of the panel by means of the airbag is another source of error that needs to be considered.

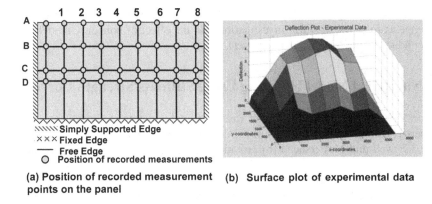

(a) Position of recorded measurement   (b) Surface plot of experimental data
points on the panel

Fig. 2. Position of recorded measurement points and 3D deformed shape of panel SBO1

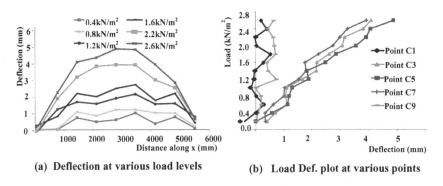

(a) Deflection at various load levels   (b) Load Def. plot at various points

Fig. 3. Measured load deflection along grid line C at various load levels

2.3 Numerical Computation Finite Element Modelling

As mentioned earlier, it was very difficult to find a FEA package that accurately models the masonry material properties, crack propagation and failure characteristics. Therefore, in this study a specialised non-linear FEA program, developed by Ma and May [4], was used. This FEA program was purely developed for research on masonry wall panels, but it lacks essential flexibility of the FEA packages.

In this analysis the following essential aspects were considered:

♦ The non-linearity failure criteria include both tension cracking and compression crushing of the masonry.

♦ The wall thickness was divided into 10 equal slices to monitor the crack propagation through the depth of the panel.

♦ Due to the inflexibility of this FEA program, degrees of freedom are only allowed to be either free or restrained. Spring stiffness was not provided to model support flexibility.

♦ The vertical edges of the panel were modelled as simply supported and the base edge as fully fixed. A full investigation into the effect of boundary modelling was conducted (not reported in this paper).

3 Numerical Model Updating Using Stiffness/Strength Correctors

A comprehensive literature review of various model updating methods is presented by Robert-Nicoud et al [1]. Siatta et al [5] discusses the reliability of system identification. Friswell and Mottershead [6] provide a survey of model updating procedures in structural damage detection research, using vibration measurements. Recent papers published in this area include [7, 8, 9, 10, 11, 12], and Cheng and Melhem [13] used fuzzy case-based reasoning for monitoring the bridge health. The majority of the research on model updating process involves computing sets of stiffness coefficients that help predict observed vibration modes of structures. The location and extent of damage are inferred through a comparison between the stiffness coefficients of damaged and undamaged structures.

Zhou [14] and Rafiq et al. [15] developed a numerical model updating technique that more accurately predicts the failure load and failure pattern of masonry wall panels subjected to lateral loading. They introduced the concept of stiffness/strength correctors which assigns different values of flexural rigidity or tensile strength to various zones within a wall panel. Stiffness/strength corrector values were derived from the comparison of laboratory measured and the finite element analysis computed values of displacement.

Zhou [14] used a number of experimental panels with different geometric properties and aspect ratios, and panels with and without opening, for which the stiffness correctors were determined. It was discovered from a comparison of the contour plots of corrector factors on these panels, that there appeared to be regions, termed 'zones', with similar patterns of corrector factors which are closely related to their relative positions from similar boundary types. In other words, zones within two panels appear to have almost identical corrector factors if they are located the same distance from similar boundary types.

Based on this finding, Zhou et al. [16] developed methodologies for zone similarity techniques. In order to achieve a more reasonable and automatic technique for establishing this zone similarity between the base panel and any new panel, a cellular automata (CA) model was developed to propagate the effect of panel boundaries to zones within the panel. The CA assigns a unique value, the so called 'state value' for each zone within the base panel and an unseen panel, based on their relative locations from various boundary types. The CA then identifies similar zones between the two panels by comparing similar state values of the two panels. Zones on two panels are considered to be similar if they are surrounded by similar boundary types and having similar distances from similar boundary types. Obviously the exact match is not possible. Therefore to find a good match for a zone on an unseen panel with a zone on the base panel, each zone on the unseen panel is compared with every zone on the base panel and the errors between the state value of a zone on the new panel and all zones on the base panel are calculated. The zone on the base panel with minimum error value is selected as the closest similar zone.

Although the proposed methodologies improved the predicted failure load and failure pattern of a number of unseen panels, two important issues were not given enough attention:

(i) the effects of panel aspect ratio on the response of the panel;
(ii) the load deflection relationships

The research presented in this paper, has concentrated more these issues.

3.1 Methodologies for Reducing Error in Laboratory Data

Figs. 2(b) and 3 demonstrated the existence of irregularities in the load deflection data recorded in the laboratory and a need for minimising (correcting) error in the experimental data. A finite element analysis method approximates the numerical solution for differential equations, which are generally based on the displacement function that closely matches the theoretical deformed shape for structural elements. For plate and shell type structures, these displacement functions generally produce deformed shapes which are similar to the Timoshenko [17], analytical solutions of the differential equations for isotropic plates and shells within the elastic load level.

In order to compare results obtained from a non-linear finite element analysis and those obtained from the laboratory tests, an in depth investigation was carried out to reduce the error in the laboratory data to reflect the real response of the panel under the action of a uniformly distributed lateral load.

3.2 Three Dimensional Surface Fitting

The first step was to carry out a regression analysis on the 3D laboratory load deflection data to fit a surface that is a closer representation of an ideal response of laterally loaded panel under ideal conditions, which matches with the FEA model as closely as possible. On the 2D linear load deflection data, the objective was to minimise the local irregularities in the deformed shape of the panel as depicted in Fig 3.

In this investigation the following three different regression formulae were used:

1. A polynomial function:

$$F_1 = \sum_{i=1}^{m} A_i * (ABS \ (X \ / L_x \ ))^i \quad and \quad F_2 = \sum_{i=0}^{n} B_i * (Y \ / \ L_y \ )^i \tag{1}$$

$$W = A_0 * F_1 * F_2 + C$$

2. A trigonometric function:

$$F_1 = COS^2 (ABS \ (\pi /2 *(X / (0.5 * L_x) * A_1))^{A_2}) \quad and \quad F_2 = \sum_{i=1}^{m} B_i * (Y \ / \ L_y \ )^i \tag{2}$$

$$W = A_0 * F_1 * F_2 + C$$

3. A Timoshenko like function [17].

Where: W is the deflection normal to the panel surface; Lx, Ly are panel length and height respectively; A_1, A_2 A_i, B_i and C are constant. Constant C is needed to model any movement in panel boundaries during the experiment; X and Y represent co-ordinates of point i.

Fig. 4. shows that all three models give a good fit for the experimental data while maintaining symmetry about the centreline of the panel. A more detailed investigation proved that the Timoshenko type surface gives a better fit with the experimental data at all measured points.

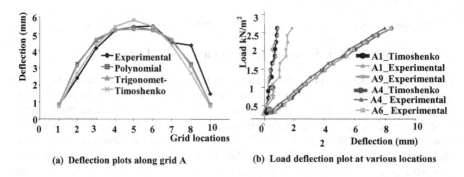

(a) Deflection plots along grid A   (b) Load deflection plot at various locations

Fig. 4. Load def. plots using various regression models (for nodal positions refer to Fig. 2a)

3.3 Evolutionary Computation Refined by Regression to Derive Correctors

In this research, the Genetic Algorithm (GA) was used to directly derive the corrector factors at various locations over the surface of the panel. At first the panel was divided into 36 locations to cover all measurement points (see Fig. 2). For simplicity a symmetrical half model was used. Corrector factors at each location were assigned to a GA variable (20 different variables for the symmetrical half model). Corrector factors, identified by the GA, were used to modify the flexural rigidity at each location on the panel. The objective function of the GA was designed to minimise the error between the modified experimental deflection (def_3D) and the deflection obtained by the FEA (def_FEA), over the entire surface of the panel.

Although the GA was able to find models that improved the predicted deflected shape of the panel, due to compensatory effects of many variables it was difficult to identify a suitable model. At this stage a regression analysis was used to refine corrector factors, selected from a number of the GA runs, to obtain a set of corrector factors that represent a best fit for the experimental deflected shape of the panel. Table 1 gives details of corrector factors derived by the GA and refined by regression. It should be noted that these corrector factors are used in the FEA to modify the flexural stiffness by multiplying these factors to the global elastic modulus (E) at each zone.

Table 1. Corrector factors derived by the GA and refined by Regression

	1	2	3	4	5	6	7	8	9
A	0.697	0.981	1.125	1.152	1.153	1.152	1.125	0.981	0.697
B	0.704	1.016	1.174	1.204	1.205	1.204	1.174	1.016	0.704
C	0.716	1.076	1.258	1.292	1.294	1.292	1.258	1.076	0.716
D	0.749	1.237	1.484	1.531	1.533	1.531	1.484	1.237	0.749

3.4 Boundary Modelling

A careful study of the corrector factors in Table 1 revealed that the flexural rigidities were mainly modified around the panel boundaries with relatively small changes inside the panel. It was therefore necessary to investigate the effect that boundary types may have on the behaviour of the panels.

At this stage it was decided to conduct a parametric study by changing the boundary types at the panel supports, and the GA was allowed to obtain corrector factors that produced a best fit with the modified experimental deformed shape. In this paper only the effect of the boundary at the base of the panel is discussed.

At first, the same boundary conditions as shown in Fig. 1 were assumed. The results from the FEA showed a kink around load level of 1.0 kN/m^2 (Fig. 5).

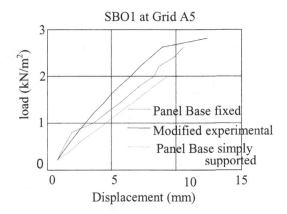

Fig. 5. Comparison of modified experimental with the predicted deflection using Table 1 Corrector Factors – Base simply supported and Fixed

Careful study revealed that this kink was due to the development of tensile cracks parallel to the bed joints, produced by the hogging moments along the panel base. As the tensile strength of the masonry is low parallel to the bed joints compared with that perpendicular to the bed joints, the first crack appears at a very low load level near the fixed support, which causes a kink in the load deflection curve. As this kink was not visible in the experimental load deflection data, it caused some concern.

The obvious choice for the next step was to change the boundary condition at the base of the panel to a simply supported type that allows full rotation of the base support. This eliminated the kink, but the stiffness of the panel was naturally reduced. This was reflected in the load deflection plots, (see Fig 5 for details).

A close investigation of Fig. 5 also revealed that the gradient of the predicted curves at various load levels were different from those of the experimental curves. In order to obtain a suitable set of corrector factors, it was decided to modify the objective function of the GA to include the gradient effect. The following errors were used:

1. Deflection error: minimise deviation of the FEA deflection values from the target values over the entire surface of the panel.
2. Gradient error: minimise deviation of the gradients of the FEA load deflection curve between two adjacent load levels.
3. Load error: minimise deviation of the FEA failure load from the target failure load.

A study of the corrector factors, derived from the simply supported base, revealed an increase in the corrector factors around the base of the panel. By changing the base of the panel to a fixed support (Table 2 corrector values) the opposite effect was observed. A close look at the results clearly strengthened the initial findings that the base of the panel is neither simply supported nor fixed, but there is only some degree of fixity at this edge.

Table 2. Corrector factors derived by the GA with panel base fixed

	1	2	3	4	5	6	7	8	9
A	1.283	1.278	1.278	1.278	1.278	1.278	1.278	1.278	1.283
B	1.187	1.182	1.181	1.181	1.181	1.181	1.181	1.182	1.187
C	0.927	0.921	0.920	0.920	0.920	0.920	0.920	0.921	0.927
D	0.223	0.218	0.218	0.218	0.218	0.218	0.218	0.218	0.223

From Fig. 5 it was observed that the FEA predicted failure load for simply supported base was much below the measured failure load. Although the failure load for the fixed base model was relatively increased, it was still below the measured values.

A closer look at this revealed that the decrease in failure load was due to the lower values of tensile strengths perpendicular to bed joints.

The results of the full boundary investigations revealed that:

♦   Boundary conditions shown in Fig 2 give closer results than the other models.
♦   Corrector factors derived by the GA for this model (refer to Table 2) give a better load deflection match at various locations over the surface of the panel.
♦   Changing the tensile strength perpendicular to the bed joints by 50% improved the predicted failure load of the panel.

4 Case Study

The corrector factors not only modeled the boundary effects, but also modelled variation in the material and geometric properties. One of the objectives of this research was to use these corrector factors to predict the behaviour of unseen panels with and without openings and panels for which the boundary conditions are different from the base panel. In this investigation it is important to note that the corrector factors derived in Table 2 are used to estimate the correctors at various locations on any unseen panel (Panel SBO2 in this study). The Cellular Automata 'zone similarity' technique [14,16] was used to estimate corrector values for unseen panels. These corrector factors are then used in a non-linear FEA to predict the load deflection, failure load and failure pattern for the unseen panel.

To assess the validity of the numerical model updating techniques presented in this paper a panel which was the same size as the base panel SBO1, but with an opening at the middle of panel (Panel SBO2), was investigated. Results of this investigation are presented in Fig 6.

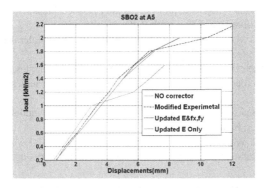

Fig. 6. Load deflection curve at maximum deflection location (A5)

5 Further Work

The challenging task is to extend this model updating technique to panels tested elsewhere, under different laboratory conditions and using different material constituents for construction of the panels. We were able to locate test data for a limited number of panels, tested elsewhere and would greatly appreciate the offer of further data, particularly on load deflection and tensile strength information for any type and size of masonry panels from researchers around the world.

The plan would be to investigate the suitability and generality of the corrector factors derived for the base panel to predict the failure criteria for as many unseen panels as possible.

6 Conclusions

This is perhaps the first time that an attempt has been made to use a numerical model updating technique to study the behaviour of a highly composite anisotropic material such as masonry within the full linear and non-linear range.

The research presented in this paper introduces a numerical model updating technique that has the potential to be extended to masonry material, which is highly composite and anisotropic.

In this research, corrector factors from a single panel tested in the laboratory were used for a number of unseen panels with different boundary types, size and configurations. The results produced more accurate prediction of the behaviour of the laterally loaded masonry wall panels.

References

1. Robert-Nicoud, Y., Raphael, B., and smith I. F. C. (2005). " System Identification through Model Composition and Stochastic Search", ASCE Journal of Computing in Civil Engineering, Vol. 19, No. 3, pp 239-247.
2. Raphael, B., and Smith, I. F. C., (2003) Fundamentals of computer aided engineering, Wiley, New York.
3. Chong, V. L. (1993). The Behaviour of Laterally Loaded Masonry Panels with Openings. Thesis (Ph.D). University of Plymouth, UK.
4. Ma, S. Y. A. and May, I. M. (1984). Masonry Panels under Lateral Loads. Report No. 3. Dept of Engineering, University of Warwich.
5. Saitta, S., Raphael, B. and Smith, I. F. C., "Data mining techniques for improving the reliability of system identification", Advanced Engineering Informatics, Vol. 19, No 4, 2005, pages 289-298.
6. Friswell, M. I., and Mottershead, J. E. (1995). Finite element model updating in structural dynamics, Kluwer, New York
7. Brownjohn, J. M. W., Moyo, P., Omenzetter, P., and Lu, Y. (2003). "Assessment of highway bridge upgrading by testing and finite-element model updating." J. Bridge Eng. 8 (3), 162–172.
8. Castello, D. A., Stutz, L. T., and Rochinha, F. A., (2002). "A structural defect identification approach based on a continuum damage model." Comput. Struct. 80, 417–436.
9. Teughels, A., Maeck, J., and Roeck, G. (2002). "Damage assessment by FE model updating using damage functions." Comput. Struct. 80, 1869–1879.
10. Modak, S. V., Kundra, T. K., and Nakra, B. C. (2002). "Comparative study of model updating studies using simulated experimental data."Comput. Struct. 80, 437–447.
11. Hemez, F. M., and Doebling, S. W. (2001). "Review and assessment of model updating for non-linear, transient dynamics." Mech. Syst. Signal Process. 15 (1), 45–74.
12. Hu, N., Wang, X., Fukunaga, H., Yao, Z. H., Zhang, H. X., and Wu, Z. S. (2001). "Damage assessment of structures using modal test data." Int. J. Solids Struct., 38, 3111–3126.
13. Cheng, Y. and Melhem, H. G., "Monitoring bridge health using fuzzy case-based reasoning", Advanced Engineering Informatics, Vol. 19, No 4, 2005, pages 299-315.
14. Zhou, G. C. (2002). Application of Stiffness/Strength Corrector and Cellular Automata in Predicting Response of Laterally Loaded Masonry Panels. School of Civil and Structural Engineering. Plymouth, University of Plymouth. PhD Thesis.
15. Rafiq, M. Y., Zhou G. C., Easterbrook, D. J., (2003). "Analysis of brick wall panels subjected to lateral loading using correctors." Masonry International 16(2): 75-82.
16. Zhou,G. C., Rafiq M. Y. Easterbrook, D. J., and Bugmann, G. (2003). "Application of cellular automata in modelling laterally loaded masonry panel boundary effects", Masonry International. Vol. 16 No 3, pp 104 -114.
17. Timoshenko, S. P., Woinowsky-Krieger, S. (1981) Theory of Plates and Shells,2nd Edition, McGraw-Hill.

Derivational Analogy: Challenges and Opportunities

B. Raphael

Assistant Professor, Department of Building, National University of Singapore,
4 Architecture Drive,
Singapore 117566
bdgbr@nus.edu.sg

Abstract. Transformational analogy is currently more widely employed than derivational analogy in CBR applications, even though the latter has significant advantages over the former. The main reason for the reluctance to use derivational analogy is the complexity of representation. Other factors include issues related to retrieval and difficulties in system validation. Means of addressing these issues are described in this paper. Unique opportunities offered by the approach are illustrated with examples.

1 Introduction

There are two approaches to case based reasoning (CBR) namely, transformational analogy and derivational analogy. In transformational analogy, similar past solutions are retrieved and adapted in order to propose new solutions [1]. In contrast, derivational analogy involves the application of the reasoning steps that were used to perform tasks in the past [2]. Most CBR applications that have been developed during the last two decades follow the transformational analogy approach. This is evident from recent publications in this area. In the proceedings of the latest international conference on CBR (ICCBR 2005), thirty six papers discuss issues related to transformational analogy [3]. Only one paper mentions about derivational analogy. Similarly, in the proceedings of ECCBR 2004, there are forty one papers that discuss applications of transformational analogy, but there are none related to derivational analogy [4].

The apparent lack of interest in derivational analogy might be partly due to the lack of knowledge among researchers about this approach. However, the primary reason is that it is more difficult to implement. This paper discusses difficulties and challenges in the implementation of derivational analogy and proposes solutions to overcome them.

2 Complexity of Representation

The issue of representing solutions in transformational analogy has been extensively studied [5]. Popular options include semi-structured forms such as text and images, and structured forms such as attribute-value pairs, tables and objects. Representing cases in these forms does not require considerable expertise and the task of storing cases are routinely performed by non-programmers using graphical user interfaces (GUI). On the other hand, representing methods is inherently more complex. In

I.F.C. Smith (Ed.): EG-ICE 2006, LNAI 4200, pp. 545–553, 2006.

Figure 1, differences between the two representations are illustrated using an example in steel plate girder design. In the first representation, only the solution is stored in the form of attribute-value pairs. The reasons for choosing the values are not known. In the second representation, the method used to arrive at the solution is described. Even in this simple example, the complexity of representation of methods compared to solutions is evident.

Early works on derivational analogy were in the context of autonomous problem solving [2,6]. Cases are automatically captured by recording actions during the course of problem solving. In such situations, representation is not a serious issue. A case consists of essentially a sequence of operators that are selected from a pre-defined set, along with additional information related to the choice of these operators. Carbonnell recommends recording the following pieces of information in each step of the solution process: Task decomposition (sub-goal structure), Alternatives considered and rejected, Rationale, Dependencies and Final solution. It is clear that all the above information can be easily captured by autonomous problem solvers. However, in complex domains where cases are not automatically generated, these details are rarely available. Furthermore, complete domain knowledge is not available in the form of a predefined set of operators. Usually, solutions are computed using complex methods that contain a number of activities that are unique in each case. Representing methods containing these activities requires special purpose languages. This issue is largely ignored even in more recent research in derivational analogy [11, 12].

Variable	Value	Units
thicknessOfWebPlate	10	mm
depthOfWeb	2200	mm
...		

a. Representation of solution (transformational analogy)

Choose lowest available value for thicknessOfWebPlate (10 mm)
Compute depthOfWeb using the empirical formula
1.1*sqrt(maxBendingMoment*1e6/
        (allowBendingStress*thicknessOfWebPlate))
Round off depthOfWeb to the nearest 100 mm

b. Representation of method (derivational analogy)

Fig. 1. Comparison between representations

2.1 Representing Methods

There are many options available for representing methods that are used to formulate solutions in cases. A simple option is production rules. A case contains a sequence of rules that represent actions taken in a particular context. These rules should not be interpreted as generic knowledge that can be applied in any context such as in a rule based system. Instead, they should be viewed as case-specific actions which may or may not be repeated in a new situation. Knowledge related to when and where they can be reused need to come from an external source.

Not all activities can easily be represented as rules. More complex schemes have been developed in order to alleviate the drawbacks of rules. An object oriented representation was employed in CADREM [7,8] in which case-specific methods are organised in the form of an abstraction hierarchy. A generic class called an MREM (memory reconstruction method) encapsulates common features of all types of methods and is at the root of the hierarchy. Specialized classes are derived from generic classes in order to represent different types of methods. More than 40 MREM classes have been identified in [7]. Out of these, about five are required in most applications. Commonly used MREM classes are described below. The abstraction hierarchy is shown in Figure 2.

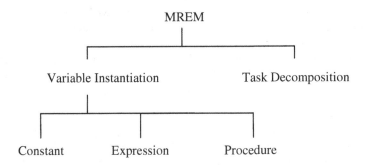

Fig. 2. A simple abstraction hierarchy of MREMs

VariableInstantiation: This is an abstract class that represents the instantiation of a single variable.

VariableInstantiationWithConstant: This class represents the instantiation of a variable with a numeric or symbolic constant. When no information is available as to why a particular value was chosen for a case variable, this MREM class is used.

VariableInstantiationWithExpression: This class represents the instantiation of a variable with an arithmetic expression. Expressions for computing values of variables are commonly found in design documents. However, reasons for choosing a particular expression and situations where it might be re-used are frequently not available.

VariableInstantiationWithProcedure: When simple expressions are not used to compute values of variables, it is better to code the method of computation as a procedure in a programming language. This class represents a method that uses an external procedure for computing the value of a variable.

TaskDecomposition: Most design tasks are decomposed hierarchically. This class represents a method that performs task decomposition. There is an MREM for carrying out each sub-task in the decomposition. These MREMs might belong to one of the classes described above.

The above classes correspond to programming language constructs available in standard languages. However, the object oriented representation of these constructs permit a higher level understanding of the processes used in a case. For example, when a case-method is represented as an instance of *TaskDecomposition*, the subtasks and their objectives are explicitly stored. This information is not readily available if the method is coded in a low level programming language.

The abstract classes described above can be used to represent both generic and case-specific processes. However, the derivational analogy system treats all instances of these classes which are found in cases, as case-specific. The retrieval engine determines the re-usability of case-specific methods in new situations.

2.2   Ways of Simplifying Representation

Potentially, any method can be represented using the MREM scheme. It is enough to derive a new class if existing classes are inadequate. However, this is not practical because end-users of CBR systems are non-programmers. It has been observed that many engineers are unable to input even simple expressions in syntactically correct forms. Even though, this can be overcome by training, a CBR system is unlikely to be used effectively if end-users have to write complex procedures.

Complementary Views

A solution to reducing the complexity of representation is proposed here using a concept called complementary views (CV) [7,8]. A view is a mapping between the attributes of an object and the attributes of another class. Many views might be defined for an object and these are analogous to observing the object from different perspectives. The views are complementary and help to improve the understanding of the object. For example, the attribute *topFlangeArea* of the class *BuiltupBeam* is mapped to the attribute *area* of the class *CompressionMember*. This mapping helps to view the top flange of a simply supported built-up beam as a compression member.

In the context of representing methods, a CV brings out the correspondence between a case-method and an existing MREM class. The case-method is not instantiated from the class. Instead, a CV is used to explain qualitatively what is done by the method. Consider an MREM called *adaptDepthOfWeb* that performs this activity: "The depth of the web is incremented by 10 mm until deflections are within limits".

Now, consider a generic MREM class *ConstraintSolver* that implements a rigorous method of solving constraints. The MREM *adaptDepthOfWeb* is not an instantiation of the class *ConstraintSolver*. Details of the two methods vary in many aspects. However, the MREM *adaptDepthOfWeb* may be viewed as a constraint solving procedure using the CV by mapping the attributes of *ConstraintSolver* to the attributes of *adaptDepthOfWeb* (Figure 3). This CV brings out the correspondence between the two classes. This information is enough for reusing the *ConstraintSolver* MREM instead of *adaptDepthOfWeb* in a new context.

How Do Complementary Views Simplify Representation?

The success of a derivational analogy system depends on the availability of methods that can be reused. However, rigorous representation of methods tends to be complex. Complementary views help avoid the problem. There is no need to represent the method in a syntactically correct form. It might be described as plain text. Even though such a form does not permit reuse of the method, the CV enables the CBR system to apply an alternative method for performing the same task. The advantage is that the person is responsible for inputting cases need not have the expertise to code all the methods in a formal language. When a method is not formally represented, the system ignores this method and uses the mappings provided in the CVs to create an instance of an equivalent class.

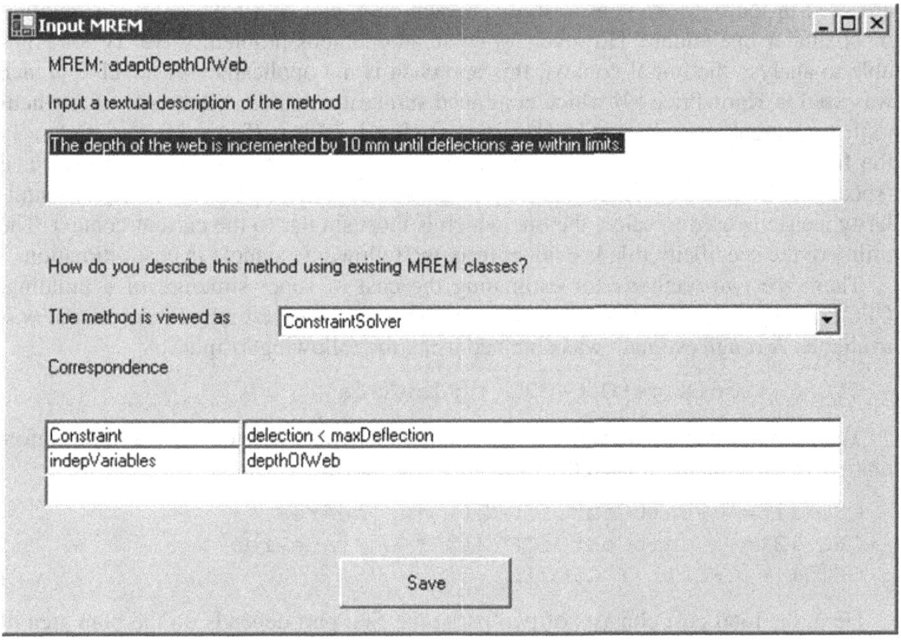

Fig. 3. A dialogue window for expressing views about an MREM

Graphical User Interfaces

Another way to simplify the task of inputting methods is to provide a rich set of classes and a graphical user interface for manipulating them. Commonly used programming language constructs should have equivalent classes in the MREM library. With this, users simply have to input values of class variables and there is no need to write code in the syntax of a programming language. For example, consider a class called *SimpleLoop*. It has three attributes, *numIterations*, *loopVariable*, *repeatedTask*. The first attribute denotes the number of times the loop is repeated, the second attribute the loop variable, and the third attribute the MREM class that is instantiated in the loop. Users input values of these variables through a GUI which is much simpler than expressing the method in a C like syntax such as given below:

```
For (loopVariable=0; loopVariable < numIterations;
loopVariable++) repeatedTask(loopVariable);
```

3 Difficulties Associated with Retrieval

When solutions are represented in the form of attribute-value pairs, simple nearest neighbour similarity metrics might be used for retrieving and ranking cases. More recently, fuzzy similarity functions have been used [10]. Retrieval of methods is not that straight forward. Carbonnell [2] suggested the following strategy: "Two problems are considered similar if their analysis results in equivalent reasoning process, at least in its initial stages". This strategy is more appropriate in an autonomous problem solver: Allow the problem solver to make initial inferences, compare the operators that are used in the new problem to those in each case, and select the case that involves most similar operations. However, when an autonomous problem solver is not available to analyse the initial context, this approach is not applicable. A novel approach was used in KnowPrice [9] which combined similarity metrics and the specificity heuristic (commonly used in rule based systems). Retrieval is performed in two stages. In the first stage, methods that use maximum number of context variables are selected (specificity heuristic). When many methods use the same number of variables, a similarity metric is used to select the one which is most similar to the current context. The utility of the specificity rule is evident from the following example in cost estimation.

There are two methods for estimating the cost of super structure of a building, CFC21. The first one (MREM1) was used in a case where limited information was available. A rough estimate was obtained using the following formula

```
CFC21 = unitCostOfCFC21 * planArea
```

The second method (MREM2) used a more elaborate procedure since more information was available. It used the following equations:

```
CFC211 =    unitCostOfCFC211 * planArea
CFC212 =    unitCostOfCFC212 * surfaceArea
CFC21 = CFC211 + CFC212
```

Here, the total cost consists of two parts; the first part depends on the plan area of the building and the second on the surface area. In order to apply this method, two variables are required; surface area and plan area. When both variables are known, this method gives a better estimate than MREM1. The specificity rule results in the selection of MREM2 because it uses more variables than MREM1.

However, there is no perfect algorithm for selecting the most relevant method in a given context. Determining the relevance and applicability of a method in a new situation requires considerable expertise. One way of capturing this knowledge is through the use of cases of retrieval. This was first proposed in CADREM [7,8]. CADREM learns to retrieve accurately through retrieval examples that are provided by human experts. With more and more retrieval examples, the system learns to select the right method in each situation. This takes considerable amount of time and effort.

4 System Validation Issues

Many CBR systems simply retrieve past cases and present them to the user. It is the responsibility of the user to adapt the solution to the new context. This considerably reduces system validation problems because users do not expect accurate answers. Even in the derivational analogy approach, users have to critically analyze proposed solutions and adapt them to take care of requirements that have not been considered by the case method. However, users expect a derivational analogy system to propose the correct answer all the time because it appears that solutions have already been adapted. This makes system validation all the more important. The system should be thoroughly tested for situations where wrong solutions are proposed.

There are many reasons why a derivational analogy system might propose solutions that are far from ideal. First of all, methods within cases usually contain only operational knowledge, that is, what was done in the past. Deeper knowledge related to reasons for choosing a method and its range of validity is missing. If methods are retrieved using only dependency relationships and similarity, wrong methods might be chosen. Better results are obtained using retrieval examples. However, this requires a developer to spend considerable amount of time examining the methods that are retrieved in a variety of situations and providing new retrieval examples whenever wrong methods are selected.

Another reason for getting incorrect answers is lack of case coverage. Adequate number of cases that can handle all possible situations may not be available. Such conditions should be recognized during system validation. If there are missing methods for accommodating special situations, fictitious cases might be added.

It is easy to overlook system validation issues. Many CBR systems have failed because developers did not anticipate the complexity of these knowledge engineering tasks. Sufficient time should be budgeted for creating and fine-tuning the case base and to ensure that its performance matches expectations.

5 Unique Opportunities

Rule based systems failed because it was difficult to find rules that are generic and to maintain the consistency of the rule base when more rules are added. Readers might suspect that similar problems exist with the derivational analogy approach. This is not true because cases contain methods which are case-specific and do not depend on methods in other cases. It fact, it is possible to find methods having varying levels of generality for generating the same case-solution. The more generic a method, the more reusable it is; but at the cost of increasing complexity. Even with less generic methods, solutions that satisfy vital integrity constraints can be generated in a new context. This is not possible with transformational analogy, unless the adaptation method simply regenerates a new solution.

A simple example is taken to demonstrate the idea that through the reuse of a method in a new situation, a solution that possesses essential qualities of the original case can be generated. Consider the truss configuration that is shown in Figure 3. The coordinates (x,y) of the nodes are generated using the method

```
Nodes = { {0,0}, {0,D}, {L/2,0}, {L/2,D/2}, {L,0} }
```

Reusing this method in a new situation where L and D are different, results in essentially the same truss configuration. Now, suppose that only the solution is represented as follows:

```
Nodes = {{0,0}, {0,1}, {3,0}, {3,0.5}, {6,0} }
```

Adapting this solution to a new situation is not easy without knowledge of the original method.

6 Concluding Remarks

The derivational analogy approach presents several challenges as well as opportunities. If careful attention is paid to issues such as representation and retrieval, this approach offers unique opportunities and permits going beyond what is possible with transformational analogy.

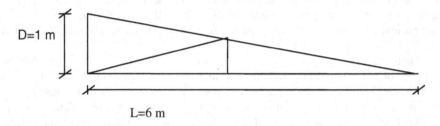

Fig. 4. A truss

Acknowledgements

The author wishes to thank Prof. Kalayanaraman and Mr. Rama Mohan for discussions on the use of derivational analogy for steel plate girder design. Collaborations with Dr. B. Domer, Mr. S. Saitta, Prof. B. Kumar and Prof. I.F.C. Smith on various CBR applications.are also gratefully acknowledged. This work would not have been possible without the financial support of School of Design and Environment, NUS.

References

1. Carbonell J.G.: Learning by analogy: formulating and generalizing plans from past experience, Machine Learning, An artificial intelligence approach, Michalski, Carbonell and Mitchell (Eds.), Morgan Kaufmann, Boston, (1983).
2. Carbonell, J.: Derivational analogy: A theory of reconstructive problem solving and expertise acquisition, Machine Learning, An artificial intelligence approach, Vol 2, Michalski, Carbonell and Mitchell (Eds.), Morgan Kaufmann, Boston, (1986).
3. Muñoz-Avila H., Ricci F. (ed.): Case-Based Reasoning Research and Development, 6th International Conference on Case-Based Reasoning, ICCBR 2005, Chicago, IL, USA, August 23-26, 2005. Proceedings, Lecture Notes in Computer Science, Vol. 3620. Springer-Verlag, Berlin Heidelberg New York (2005).
4. Funk P., Calero P.A.G. (ed.): Advances in Case-Based Reasoning: 7th European Conference, ECCBR 2004, Madrid, Spain, August 30 - September 2, Lecture Notes in Computer Science, Vol. 3155. Springer-Verlag, Berlin Heidelberg New York (2004).

5. Kolodner J.L., Case-based Reasoning, Morgan Kaufmann, San Mateo, CA, 1993.
6. Veloso M. and Carbonell J., Derivational analogy in PRODIGY: Automating case acquisition, storage, and utilization, Machine Learning, 10:249-278, 1993
7. Raphael B., Reconstructive Memory in Design Problem Solving, PhD thesis, University of Strathclyde, Glasgow, UK, 1995
8. Kumar B. and Raphael B., Derivational Analogy Based Structural Design, Saxe-Coburg Publications, UK, 2002
9. B. Raphael and S. Saitta, Knowprice: using derivational analogy to estimate project costs, Proceedings of the eighth international conference on the application of artificial intelligence to civil, structural and environmental engineering, B.H.V. Topping, (ed.), Civil-Comp Press, 2005.
10. Y. Cheng and H.G.Melhem, Monitoring bridge health using fuzzy case-based reasoning, Advanced Engineering Informatics, 19, 4, 2005.
11. E. Plaza, Cooperative Reuse for Compositional Cases in Multi-agent Systems, Muñoz-Avila H., Ricci F. (ed.): Case-Based Reasoning Research and Development, 6th International Conference on Case-Based Reasoning, ICCBR 2005, Chicago, IL, USA, August 23-26, 2005. Proceedings, Lecture Notes in Computer Science, Vol. 3620. Springer-Verlag, Berlin Heidelberg New York, 2005.
12. M.T. Cox and M. Veloso, Supporting Combined Human and Machine Planning: An Interface for Planning by Analogical Reasoning, In D. Leake & E. Plaza (Eds.), Case-Based Reasoning Research and Development: Second International Conference on Case-Based Reasoning (pp 531-540). Berlin: Springer-Verlag, 1997.

Civil Engineering Communication – Obstacles and Solutions

Danijel Rebolj

Faculty of Civil Engineering, University of Maribor
Smetanova 17, SI-2000 Maribor, Slovenia
danijel.rebolj@uni-mb.si

Abstract. Communication is what is giving value to information, and is thus crucial in any engineering decision making. But although Information and Communication technology (ICT) has improved enormously, civil engineering communication has by far not used its huge potentials. In the paper we are identifying some reasons for the slow introduction of emerging ICT in AEC sector; we emphasize some emerging technologies, their potentials, and implementation attempts, and then present a set of activities, which could improve the current situation.

1 Introduction

Communication plays a very important role when people have to solve a problem together. Collaboration and thus communication is a normal way of working in civil engineering. Before computers were introduced, all information in a construction project has been communicated either in printed form, as text and drawings on paper, or by voice communication between actors. The only way of transmitting information was by carrying paper from one person to another, therefore centralised hierarchical organisation has been a necessity to ensure effective decision making.

Telephones made communication possible on long distances and fax machines did the same for text and drawings on paper, but no significant improvement in communication patterns has been introduced. In civil engineering communication has not been systematically improved even by the introduction of computers, or by any other information or communication technology. It is still following the same traditional hierarchical patterns although it could become much more flexible and dynamic.

Various reasons are causing the delay in application of information technology and are preventing the quantum leap in ICT based communication. According to our experiences the main reasons are the lack of ICT innovation and standardisation, lack of R&D cooperation in AEC industry, and deficiencies in civil engineering education.

Some companies are trying to apply the ICT potentials in a higher extent, like in Japan, where the Daito Trust Construction Company developed a large-scale mobile computing system called the DK Network [1]. But problems occur when trying to use advanced technology in projects with other partners, who have implemented ICT on different levels. This can lead to even more complicated communication in joint projects, where different technologies and forms are used for information representation, then in projects where no sophisticated ICT is used at all. Experiencing such problems certainly discourages AEC companies from further investments in ICT related

I.F.C. Smith (Ed.): EG-ICE 2006, LNAI 4200, pp. 554–558, 2006.

innovation. Another discouraging fact is that research and development in Construction IT has not lead to any significant improvement in efficiency of construction projects.

Some important efforts were done in higher education with student projects that were related to the AEC industry ([2], e-site project described in [3]), but no significant break through could be noticed either. Our conclusion is that all these efforts would bring much more effect if they would become harmonized, and if innovation would become an indispensable process in every AEC company.

2 Civil Engineering Communication

Regarding innovation we believe a more open approach to new ways of performing business should be considered. Typically the AEC sector is bound to traditional ways of thinking and doing and has quite limited resources for ICT related innovation. Appearance of personal computers, fax machines, mobile phones, PDAs, IP telephony, instant messaging have all went by without being noticed by the management, and have typically been used by individuals first and systematically accepted much later, if at all.

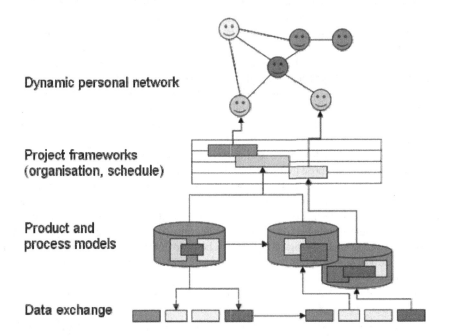

Dynamic personal network

Project frameworks
(organisation, schedule)

Product and
process models

Data exchange

Fig. 1. Dyce architecture is enabling user tailored context sensitive access to information as well as communication with contacts, which assures optimal decision making support.

Mobile and ubiquitous computing are proven to become a very important technology for the AEC sector, since they are extending the information systems to construction sites [3][4]. More sophisticated integrated communication systems with hidden complexity are already emerging [5], which shall finally free the humans from the limited modes of using computers, and assure creative collaboration and sharing of ideas.

With mobile computing location of the user can become vital information in an information system, delivering a new context parameter to the system [6]. Being overloaded with input, users need context sensitive systems to work effectively and to be able to communicate efficiently [7]. Development of DyCE, the dynamic communication system [8] (Fig. 1 is presenting the architecture of the system), has not only proven the effectiveness of context sensitivity, but has also shown, that new, more efficient organisational patterns are now possible, since information for decision making can be brought to any actor in the organisation. A network organisation (in opposite to hierarchical) is assuring better use of knowledge and expertise, and a higher level of innovation and co-operation.

3 A New Profile for Managing Information in Construction

In most of current organisations in AEC sector managers and engineers are not able to recognize and to turn to advantage the full potentials of ICT and are generally not aware of achievements in the area of construction informatics. Froese [9] is suggesting to introduce a new profile, a project information officer, who would focus on efficient flow of information in construction projects and assure the effective technology to support it. Of course such profile needs a solid background in civil engineering knowledge as well as the knowledge in the field of computer and information science.

Certainly civil engineers have some knowledge in the area of computer and information science, but by far not enough to be able to decide about applying ICT in a construction project. In our opinion there is some space for improvement in civil engineering curricula, if related subjects would emphasize more on information and communication aspects of construction objects and processes. The lack of such knowledge is causing neglecting of data structures and flows, whereby functional aspects of systems are overemphasized. This in turn is causing problems in integration attempts and leaves us with most of those automation islands from the previous century.

But this would still not be the knowledge a project information officer (or manager) would need. Therefore we have suggested a postgraduate program intended to educate a new engineering profile to focus on project information management (including information integration). In 2001 a consortium of seven universities has started to develop such a program, which is being offered to civil engineering graduates since 2004. The program and the teaching and learning environment are further described in [10], and it's location in [11].

The accreditation process of a joint study program proved to be a problem, since different rules are in power in such many different countries and universities. To overcome formal obstacles and to open the program to the global community we have decided to form an open pool of IT in Construction (ITC) related courses. The initial ITC course pool has started to accept courses developed in the ITC Euromaster project. However, any institution with knowledge in the ITC field is welcomed to offer a course to the pool. Once accepted by the steering committee, the new partner institution can include any number of existing courses in its own program, since the pool is based on reciprocity (Fig. 2). Any unbalances in students and courses will be regulated by the steering committee. One way of balancing is for example by requesting further supporting staff from a partner that has a significantly higher number of students in a specific course given by another partner.

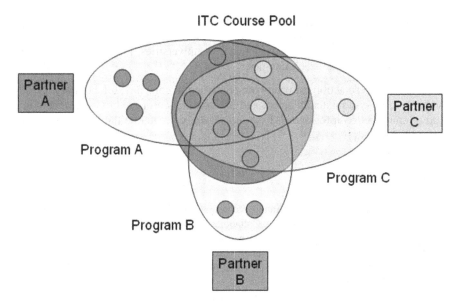

Fig. 2. The ITC Course pool concept.

From the technical viewpoint joining the ITC course pool is easy. The agreement is available at the ITC@EDU network web page (www.itcedu.net). The appendix of the agreement consists of the template of the Accession declaration to the ITC Course Pool, which includes the institution wanting to join, the offered courses, and nomination of the steering committee member. Once the document is accepted by the steering committee the courses are at disposal for all members. The acceptance procedure shall among others include the quality check of the offered course materials.

The ITC course pool will need a strong support from collaborating institutions. The experiences in the current ITC Euromaster program showed that much effort is necessary to prepare high quality e-learning materials, to become familiar with the on-line communication, to manage and further develop the e-learning system, and to coordinate the whole program. But even in the short term the investments give a high return. Having a whole pool of courses at hand certainly gives each partner a strong background to form a whole new program and to offer their students specialized knowledge and skills which they could possibly never be able to offer only by themselves.

4 Conclusion

Engineering communication is crucial for the efficiency of AEC industry. Not only to support decision making in a project, but also to support innovation. To improve civil engineering communication we suggest a concerted set of actions:

- introduction of a new engineering profile, project information officer, to focus on project communication support,
- higher education shall offer the relevant profile,

- organisational changes should support the possibility to make more effective decisions, on site and in time,
- AEC industry shall start, together with universities, more courageous experimenting with new technologies.

To support these actions, a prototype construction company, a laboratory, could be established, to systematically select and use solutions proposed by researchers in the field of construction informatics. It would certainly accelerate the innovation process in the AEC industry.

References

1. Daito Trust Construction Co.: Annual Report (2000)
2. Fruchter, R.: A/E/C Teamwork: A Collaborative Design and Learning Space. J. Comp. in Civ. Engrg., ASCE, New York, NY, Vol. 13 (1999) 261-269
3. Rebolj, D., Magdič, A., Čuš Babič, N.: Mobile computing in construction. In: Roy, R. (ed.), Prasad, B. (ed.). Advances in concurrent engineering. Tustin: CETEAM International (2001) 402-409
4. Satyanarayanan, M..: Pervasive Computing: Vision and Challenges. IEEE Personal Communications, Vol. 8. (Aug 2001) 10-15
5. Sousa, J. P., Garlan, D.: Aura: an Architectural Framework for User Mobility in Ubiquitous Computing Environments. Software Architecture: System Design, Development, and Maintenance. Jan Bosch, Morven Gentleman, Christine Hofmeister, Juha Kuusela (Eds), Kluwer Academic Publishers, August 25-31 (2002) 29-43
6. Judd, G, Steenkiste, P.: Providing Contextual Information to Pervasive Computing Applications. IEEE International Conference on Pervasive Computing (PERCOM), Dallas, March 23-25 (2003) 133-142
7. Čuš Babič, N., Rebolj, D., Magdič, A., Radosavljević, M.: MC as a means for supporting information flow in construction processes. Concurr. eng. res. appl. Vol. 11. (2003) 37-46
8. Magdič, A., Rebolj, D., Šuman, N.: Effective control of unanticipated on-site events: a pragmatic, human-oriented problem solving approach. Electron. j. inf. tech. constr., Vol. 9. (2004) 409-418 (available at http://www.itcon.org/cgi-bin/papers/Show?2004_29)
9. Froese, T.: Help wanted: project information officer. eWork and eBusiness in Architecture, Engineering and Construction A.A. Balkema, Rotterdam, The Netherlands. (2004).
10. Rebolj, D., Menzel, K.: Another step towards a virtual university in construction IT. Electron. j. inf. tech. constr. Vol. 9 (2004) 257-266 (available at http://www.itcon.org/cgi-bin/papers/Show?2004_17).
11. ITC EUROMASTER: The programme portal. (2006) (available at http://euromaster.itcedu.net).

Computer Assistance for Sustainable Building Design

Hugues Rivard

Canada Research Chair in Computer-Aided Engineering for Sustainable Building Design,
Department of Construction Engineering, École de technologie supérieure (ETS),
1100 Notre-Dame Street West, Montreal, H3C 1K3, Canada
hugues.rivard@etsmtl.ca

Abstract. The greatest opportunity for sustainable building design strategies occur in the early stages of design when the most important decisions are taken. Nevertheless, it is the stage with the least computer support. This paper presents recent research efforts at ETS toward the long term goal of developing the next generation of computer assistance to designers of sustainable buildings. Two research thrusts are presented: (1) to provide assistance to designers earlier in the design process, and (2) to provide better means to support collaboration among the various stakeholders. Two research projects are briefly presented that belong to the first thrust: an approach that allows structural engineers to propose feasible structural systems earlier directly from architectural sketches; and another that provides a means to optimize aspects of a building selected by the designer. A new laboratory is presented that will provide the infrastructure to carry out research in the second thrust.

1 Introduction

The earth's environment is undergoing alarming changes due to human activity [1], and buildings account for a large portion of the environmental impacts: at the global scale (ozone depletion, global warming, acid rain, and resource depletion); at the local scale (urban sprawl, solid waste, smog, and water run-offs); and at the indoor scale (indoor air pollution, hazardous materials, and workplace safety) [2]. A case in point, in 2001, buildings accounted for 30 percent of the total secondary energy use and of the CO_2 equivalent greenhouse gas emissions in Canada [3]. The urgent reduction of the environmental burdens of buildings can only be achieved by considering sustainable development for buildings. This concept, also known as green buildings, implies planning, designing, constructing, operating and discarding buildings in a manner to meet the needs of people today without compromising those of future generations.

The CIB Working Commission W82 investigated the needs for research in construction with respect to sustainable development and made the following two recommendations: 1) researchers need to develop adapted tools to assist designers in considering sustainability concepts; and 2) designers need to adopt an integrated approach to building design [4].

A new Canada Research Chair in computer-aided engineering for sustainable building design has recently been established at ETS. The objective of the paper is to present recent efforts toward the long term goal of this Chair for **developing the next generation of computer assistance to designers of sustainable buildings**. This

I.F.C. Smith (Ed.): EG-ICE 2006, LNAI 4200, pp. 559–575, 2006.
© Springer-Verlag Berlin Heidelberg 2006

long-term objective is addressed through two research thrusts each addressing one of the two recommendations of W82: 1) Conceptual building design support to provide assistance to designers in considering sustainability concepts early in the process; and 2) Design collaboration support to assist designers that do adopt an integrated building design approach. Each thrust is presented in the next two sections.

2   Conceptual Building Design Support

Designing green buildings is more complex than traditional buildings because of the many conflicting issues that need to be considered together. In order to achieve a successful sustainable building, particular attention needs to be paid to the conceptual design stage where potential design alternatives are generated and roughly evaluated in order to obtain the most promising solution. The decisions taken in the early stages of design have the greatest impact on the final form, constructibility, costs, and overall performance of the building over the whole life-cycle. Yet, the time, people, tools and resources allocated to this phase are very limited. During this stage, designers are faced with an overwhelming amount of requirements and data and must evaluate realistically the performance of different design alternatives. Consequently, designers design by intuition and experience rather than by exploring the unbounded space of possibilities in a more systematic manner.

Designers have a set of tools at their disposal to assess environmental outcomes of decisions taken: building energy simulation programs; envelope analysis programs; construction costs estimating programs; indoor air quality analysis programs; computational fluid dynamics programs; and new tools based on the Life Cycle Analysis (LCA) method to consider embodied energy and greenhouse gas emissions in building products [5]. Designing a building utilizing these heterogeneous tools is difficult because they typically cannot communicate among themselves, require time-consuming data input, and have steep learning curves. The available tools often address problems that are specific to one building specialty and do not support an integrated approach considering the analysis of many parameters together. Finally, most of these tools address more detailed design stages when important decisions have already been taken and do not support conceptual design appropriately.

Integration of some of these tools has been partially addressed in the literature. A case study from [6] has shown that the IFC (Industry Foundation Classes), a data interoperability standard, have allowed designers to share information between applications for the sustainable design of an auditorium building in Finland. In Australia, a system called LCADesign has been developed to also take advantage of the IFC by automating the take-off aspect of LCA from 3D object-oriented CAD files [2].

There is clearly a need for better decision support system to support the earlier stages of design. The objective of this thrust is to provide computer-support to assist building designers in exploring and formulating the designs of sustainable buildings at the conceptual stage. The envisioned environment will tap the designers' creativity, assist them in investigating a wide array of alternatives, foster collaboration among the various building designers involved, increase their expertise by volunteering additional knowledge, assess the environmental impacts of designs and provide easy access to knowledge stored within a growing library of past designs. Such a computer environment needs to be user-friendly and intuitive. Here, computer support is meant to assist human designers in solving problems and not to replace them.

Two recent research projects concentrate on different steps of the conceptual design stage: design synthesis when designers define potential solutions; and design exploration when many solutions are generated and evaluated.

Design synthesis: The creation of a design solution happens mostly in the mind of the designer. Sketches are often used as a dialogue between the designer and the unfolding design to help grasp the many different issues and formulate a working solution. By using current computer technology, efforts have been made over the last few years to provide support to designers earlier than what is currently available.

The Figure 1 below maps recent research efforts lead by the author toward the development of assistance for the early stages of architectural design and structural engineering. The efforts in gray are those that have achieved significant progress. The efforts shown with a white background are on-going projects. Research efforts from the Figure will be referred to in the text with a number between parentheses (e.g., (1) represents the first research effort).

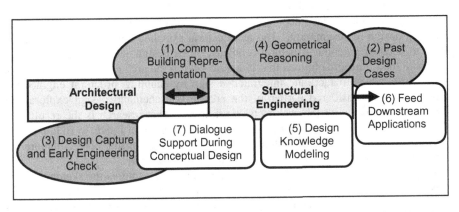

Fig. 1. Recent research efforts to assist the early stages of architectural and structural design

First, a building design representation was established as an expressive language to describe a wide variety of structures and to capture the more abstract concepts of structural systems as well as their refinements [7]. This representation has been expanded to include concerns from architecture (1). This common building representation allows architects and structural engineers to work together in defining volumes and structural systems [8]. Second, a case-based design tool (2) was developed to retrieve past cases similar to the problem being solved and adapt the most appropriate case to solve the problem [9]. Then, a protocol analysis focused upstream to study architects at work and resulted in a set of specifications for computer tools in early design [10]. A prototype software was implemented to address some of these specifications, to capture the design, and to bring engineering knowledge earlier in the design process to verify the proposed design (3) [11]. Downstream, a prototype called StAr was developed to provide geometric modeling and reasoning capabilities to support the elaboration of a structure from the overall geometry and functional requirements to specific components in 3D (4) [12]. Now, an on-going research effort (5) focuses on the modeling and capture of design knowledge based on the concept of technology nodes [13] or knowledge rules. This tool will assist the engineer in making design decisions in an interactive manner [14]. Knowledge modeling capabilities are being expanded beyond modeling heuristic knowledge to include look-up tables, curve-fitting, parameterized equations, interactive dialog, and others. Conceptual structural design knowledge has been elicited for steel and concrete structures. A mechanism is planned to translate the design synthesis results generated within the environment for current more specialized downstream applications (6). The Industry Foundation Classes or CIMSteel data standards will be used for this purpose since they are currently supported by several applications. All these complementary efforts provide the foundation to support a dialogue between structural engineers and architects during design synthesis. This is the purpose of a recent research effort elaborated further here (7).

A collaborative research effort is proposing an integrated approach to support earlier dialogues between architects and structural engineers who have complementary expertise but different working styles. Architects use sketches during conceptual design for exploration and development of their own ideas, and for communicating them [10]. Engineers must accommodate the architects' work pace as well as their evolving design representation. The architectural sketches can be used by the engineers to uncover potential structural problems in the design and devise and compare structural load transfer solutions that integrate well to the architecture. The objective of this research is to facilitate integrated design by enabling structural engineering concerns to be considered earlier in the architects' schematic design explorations without interfering with the architects' workflows. This research is the result of a collaboration between the LuciD group from the University of Liege, in Belgium, and the research team lead by the author at ETS in Montreal, Canada. The collaboration platform relies on two software prototypes: EsQUIsE and StAr.

EsQUIsE is a tool developed by the LuciD group that captures the sketches of the architect, analyzes and interprets them as they are drawn into a set of spaces, their topologies and wall components [15]. The user interface is designed to be as close as possible to the architect's traditional and natural way of work, thus relying on an electronic pen on a digital tablet, and eliminating mouse, keyboard and menus. Once the drawing elements are recognized, EsQUIsE constructs automatically a 3D model of the building and can provide rough evaluations such as virtual walk-throughs, yearly building energy needs; and construction costs. The interface of EsQUIsE is shown in Figure 2 below with a building example being a firehouse.

StAr is a tool developed at ETS that assists engineers in the inspection of a 3D architectural model while searching for continuous load paths to the ground and for structural opportunities. It also assists in the configuration of structural solutions from the system level to the element level. Assistance is based on geometrical reasoning algorithms and an integrated architecture-structure representation model (elements (1) and (4) in Figure 1) [12]. The output of StAr will be transferred to existing downstream advanced analysis packages.

The process of bringing free-hand sketches to the precision required in structural design involves difficulties that have not been tackled by existing tools. It is the premise of this research project that computers alone cannot and should not automatically transform imprecise architectural sketches into precise representations to be used for structural design. This process should be carried out by the computer with implicit guidance from both the architect and the engineer. This is done in two stages: (1) bottom-up automatic sketch interpretation in EsQUIsE, and (2) top-down interactive architectural-structural integration in StAr. Figure 3 below shows the previous building example in StAr where the engineer has recognized implicit project grid lines from the architecture and identified walls and columns that can be used as vertical supports. Figure 4 shows the 3D architectural model generated from EsQUIsE on the left and the structural system generated by StAr on the right [16]. This approach could be expanded in the future to include other stakeholders such as HVAC design, envelope design, etc.

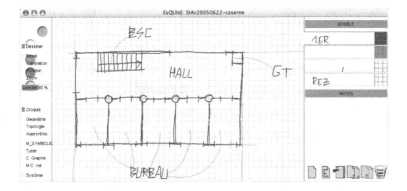

Fig. 2. The sketch of the second floor of a firehouse in EsQUIsE

There are many other relevant and promising research efforts that have focused on conceptual building design, a few of which are briefly mentioned here. M-RAM, a case-based design system, combines heuristic rules for case classification; and a genetic algorithm for case adaptation to recommend the type of structure for the building (e.g., braced frame, moment-resisting frame, or shear wall) [17]. CADREM, another case-based design system, uses derivational analogy; upon retrieval the procedures stored in the case are re-executed with the new system's parameters [18]. SHIQD- is a representative logic-based system that uses description logic and ordered planning to design structural components [19]. Genetic algorithms (GA) have been proposed for solving conceptual design problems. Grierson and Khajehpour used a multi-objective genetic algorithm to search for Pareto-optimal design solutions for office buildings [20]. Three cost-revenue objectives were considered: (1) minimize initial capital cost, (2) minimize annual operating cost, and (3) maximize annual income revenue. BGRID is another GA-based prototype for conceptual design of steel office buildings [21]. Its inputs are plan dimensions, number of floors, site location, dimensional constraints, and position of cores and atria. A structured genetic algorithm (SGA) has been developed for conceptual design of steel and concrete office buildings [22]. SGA is a variant of GA in which design solutions are encoded in a hierarchy of interrelated genes in order to better model the hierarchical aspect of building systems. Packham et al. have developed an interactive and visual mean to understand the results of complex design optimization [23]. Finally, INTEGRA is a prototype that once is given overall building information and constraints; a sketching system is able to provide immediate feedback on whether the design satisfies the imposed constraints [24].

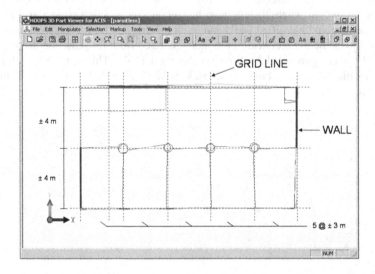

Fig. 3. The sketch of the firehouse being refined in StAr

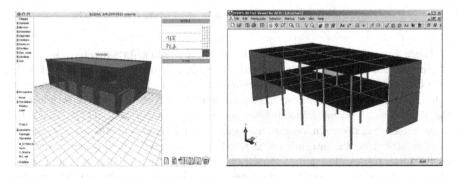

Fig. 4. The 3D models for the architecture on the left and the structural system on the right

Design exploration: Once the main aspects of the green building design are decided and the design space has been circumscribed, some of the unsolved design parameters could be refined through multi-objective optimization. There is always more than one possible solution to a given design problem. This almost infinite number of possible solutions defines a solution or design space. Hence, an important aspect of conceptual design is the generation of alternatives to explore this design space in order to find the "best" solution. The greater the number of alternatives that are generated, the greater is the odd to find a near optimal solution. Optimization can be used to provide assistance in exploring this design space in a systematic manner.

Optimization is applied here to find building shapes that perform better. Shape is an important consideration in building design due to its significant impacts on energy performance and construction costs. There are a number of previous studies on building shape optimization. A number of studies construct a building shape from its spatial constituents thus adopting a part-whole approach: one optimized whole building 2D plans from elements such as space units, rooms and zones [25]; another optimized 3D architectural forms that maximizes daylighting use and minimizes operating energy consumption based on basic blocks [26]; and finally another optimized 3D shapes of apartment buildings obtained from the aggregation of two fundamental shapes, a circulation unit and an apartment unit [27]. This part-whole approach is capable of defining a wide range of shapes, some of which may be innovative solutions for a design problem. In contrast, the whole-part approach defines a building shape by its external boundaries and represents its internal spatial elements implicitly. An advantage of the whole-part approach is that it can easily describe the building geometry for energy simulation programs. This approach is adopted in several shape optimization studies focusing on energy performance: two studies assumed a rectangular building plan and optimized its aspect ratio [28-29]; one optimized a building with a symmetrical octagonal plan by chamfering a rectangle [30]; and another considered both L-shape and rectangular shape [31]. These previous studies using the whole-part approach are limited to simple shapes thus precluding some possibly more promising shapes from the design space right from the start.

A multi-objective optimization software has been developed that searches the design space to find an optimal solution for the building envelope and the building shape defined as a n-sided polygon with the whole-part approach [31, 32, 33]. The software uses a multi-objective genetic algorithm and considers two objective functions: life-cycle costs and environmental impacts. It allows the visualization of the trade-off options between the two objectives. It lets the designer select which aspects of the building design to optimize (e.g., the shape, the wall section, or the window ratio) and which are fixed as well as their possible range of values. The optimization problem is formulated in terms of variables and objective functions, as presented below.

Variables
The variables are categorized into four groups: shape, structure, envelope configuration, and overhang. The shape being optimized is the building footprint or typical floor shape that can be defined as a simple n-sided polygon with no intersection of non-consecutive edges. The simple polygon is adopted because most energy simulation programs model walls as line segments, and a curve can be approximated with line segments. The n edges of a polygon are defined with a length-bearing representation [33] that includes the edge length (m) and the edge bearing (degree). The optimization is carried out here with n=5 for a pentagon.

A variable for the structure defines different available alternatives for the building structural system (e.g., steel frame vs. concrete frame). Its purpose is to ensure the compatibility between walls, roofs, floors and overhangs [31].

The following variables for the building envelope are considered: window types; window ratio for each façade; wall types (e.g., concrete block wall vs. steel stud wall), roof types, and floor types for each considered structural system; and each layer (e.g., type of insulation and its thickness) of the wall types, roof types, and floor types. A detailed description of these variables can be found in [32]. The type and layer variables are dealt as discrete variables.

An overhang is a passive solar architectural element installed to reduce the direct solar radiation through windows in the summer. The design of the overhang for each facade is defined by two variables: its type (e.g., aluminum overhang or no overhang); and its depth (i.e., the distance in meters between the wall and the outer edge of the overhang).

Objective functions

Life cycle analysis is employed to evaluate design alternatives for both economical and environmental criteria. Thus, the two objective functions to be minimized are life cycle cost (LCC) and life cycle environmental impact (LCEI). The life cycle cost includes the initial construction costs of the building envelope including exterior walls, windows, roof, and floor; and the life cycle operating cost including both demand and energy consumption costs. A period of 40 years is considered in the life cycle analysis. The energy consumption is obtained by coupling the optimization program with a building energy simulation program.

The life cycle environmental impacts are evaluated in terms of expanded cumulative exergy consumption. Integrating various environmental impact categories with different units and magnitudes into a unique objective function is not a trivial task. Some studies overcome this problem by considering energy consumption only, while others use weights or monetary values to aggregate different impacts into a normalized value [34]. The drawbacks are that there are no generally agreed weights or reference situation. A novel approach is to use the concept of exergy to evaluate the environmental impacts.

Exergy is "the maximum theoretical work that can be extracted from a combined system consisting of the system under study and the environment as the system passes from a given state to equilibrium with the environment---that is, passes to the dead state at which the combined system possesses energy but no exergy" [35]. Unlike energy, exergy is always destroyed because of the irreversible nature of the process. Therefore, the evaluation of exergy depends on both the state of a system under study and the conditions of the reference environment.

Exergy is adopted here as a unifying indicator to evaluate life cycle environmental impacts. Cumulative exergy consumption, proposed by Szargut et al. [36], sums the exergy of all natural resources consumed in all the steps of a production process. Unlike considering only energy consumption, it also takes into account the chemical exergy of the nonfuel raw materials extracted from the environment. Therefore, cumulative exergy consumption is a measure of natural resource depletion. This concept of cumulative exergy consumption is further expanded to include abatement exergy as a measure of waste emissions. Abatement exergy evaluates the required exergy to remove or isolate the emissions from the environment. As indicated by Cornelissen [37], it is feasible to determine an average abatement exergy for each emission based on current available technologies. Thus, the resulting expanded cumulative exergy consumption represents a single objective function that considers both resource inputs and waste emissions to the environment, across all life cycle phases. Some of its advantages are that it avoids the subjectivity of weights setting in the evaluation of environmental impacts and it combines fuel and nonfuel materials together to characterize the resource depletion [32].

In this study, the scope of the life-cycle analysis is limited to natural resource extraction, building material production, on-site construction, operation, and transportation associated with the above phases. Only global and long-lasting impact categories are considered and they include natural resources consumption, global warming, and acidification. The emissions considered are restricted to three major greenhouse gases (CO_2, CH_4, N_2O) and two major acidic gases (SO_x and NO_x).

Results
The results of the optimization are shown in Figure 5 in terms of both objective functions: life-cycle environment (LCEI) and life-cycle costs (LCC). The Pareto front is illustrated from non-dominated individual solutions for the 300th and last generation.

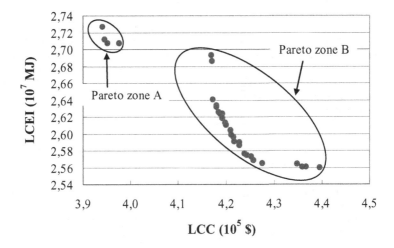

Fig. 5. The Pareto front obtained after 300 generations

It can be seen that the Pareto front is divided into two isolated zones. These two zones correspond to the two structural systems considered. In this study, the steel frame system (corresponding to zone A) is found to have lower costs but higher environmental impacts than the concrete frame system (zone B).

The building footprint corresponds to a typical floor with a fixed floor area of 1000 m^2 and a fixed height. The building footprint, represented by a five-sided polygon, takes different shapes along the Pareto front. The building shape corresponding to the lowest cost and highest environmental impact (the point on the upper left of the Pareto front in Figure 5) converges toward a regular polygon with a minimum perimeter. The building shape corresponding to the highest cost and lowest environmental impact (the point on the lower right of the Pareto front in Figure 5) is elongated along the East-West axis with a longer south-facing wall to benefit from the passive solar heating in the cold climate of Montreal. Figure 6 shows a three-dimensional rendering of the two extreme solutions. The two corresponding buildings of three typical floors are shown with the last floor visible. The renderings also help visualize the implications of the optimization in terms of windows and overhangs. The window ratios have converged to the lower bound of 20% for most facades except for the south-facing wall for the building with the lowest environmental impact. The last solution is also found to have an overhang on the south façade to shade the sun in the summer.

A multi-objective optimization tool can assist building designers by letting them explore more design alternatives and the resulting Pareto front can help them understand the trade-off relationship between the performance criteria. Additional information about the hypotheses, discussions and results can be found in [33].

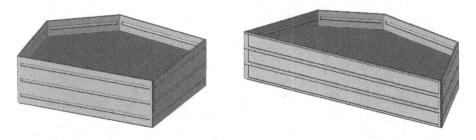

Fig. 6. Building footprints for minimum cost and for minimum environmental impact

3   Design Collaboration Support

It is generally agreed that in order to achieve a successful sustainable integrated building design, all important project participants (e.g., the client, the future occupants, the architect, the HVAC engineer, the building engineer, the facility manager, and the general contractor) must collaborate together from the beginning. Otherwise, specific subsystems may get locked into suboptimal solutions because of higher level decisions which cannot be changed in the later design stages (which occurs too often in traditional building design). A novel approach, the integrated design process, involves greater collaboration from the beginning to the end with several meetings, phone discussions, exchange of documents, and so on. Existing computer support for such multidisciplinary collaboration is still limited and a relatively recent research field aspires to address this hiatus. Although some collaboration support is used in the manufacturing industry, the construction industry could benefit more because of its unparalleled fragmentation. The objective of this research thrust is to conceive a computer-aided collaborative environment for designers and stakeholders working together on the conceptual designs of sustainable buildings.

Many recent research projects are focusing on collaboration support, some of which are briefly reviewed here. An example of synchronous collaboration for mechanical design is presented in [38] where control points to a single 3D model are distributed among several users who control them independently. Another important aspect of collaboration is how the interaction among the participants is managed. The system proposed by [39] offers three modes for participants in construction to interact: chairman; agenda and brainstorming. When conflicts or disagreements arise, the design environment should provide a mechanism to resolve them. A system proposed by [40] focuses on a mechanism for controlling when and how negotiation proceeds in an environment where humans and software agents collaborate. Another approach proposed by [41] helps settle conflicts that arise from unarticulated concerns from participants. A number of sources of conflicts were identified. Other studies have looked at the infrastructure to support collaboration. A collaborative design space with a flexible arrangement of networked computers and tables has been used for teaching and experimenting with collaboration along with specifically developed Internet support for distant collaboration [42]. An interactive information workspace has been demonstrated to be more intuitive and efficient in sharing information and establishing common focus in multi-disciplinary project team meetings than traditional meetings [43]. A telepresence environment has been proposed to search for available professionals and to offer a space for interactions [44]. Finally, a CAVE (a virtual reality visualisation space) was successfully used in the design and visualisation by stakeholders of an auditorium building in Finland [6].

Project collaboration can occur either synchronously (at the same time) or asynchronously (at different times) and either co-located (in the same room) or distributed (at a distance). IT offers still untapped potentials to support and facilitate all of these types of collaboration. This thrust focuses on the more challenging synchronous collaboration support. Still, much research is needed to understand the process of collaboration among designers, to formalize mechanisms to support such collaboration, to implement intuitive systems, and to assess the assistance provided by these new technologies. Synchronous collaboration may either occur in person in the same room or with people geographically distant through the Internet.

Synchronously co-located collaborations typically occur in traditional meeting rooms. In green building design, it is becoming common practice to have meetings at the beginning of a project, called "design charette", where all stakeholders get together to collaboratively plan the project and brainstorm possible solutions. Later, meetings are necessary to handle the complexity of green building design and to discuss, interact, and negotiate in order to arrive to an integrated solution. The traditional meeting room needs to be re-hauled and augmented with appropriate IT. The aim is to better support the collaboration of the project team particularly with respect to large quantity of disparate design information, complex 3D issues, automatic rapid analyses, and decision-making. Three important steps are necessary to further research in these areas. The first step is to study collaboration in situ or in a meeting room setting and develop an appropriate theoretical and practical foundation. The second step is to develop and test tools and methodologies to support and facilitate collaboration. And the third step is to validate the new environment in actual settings with practitioners.

Synchronously distributed collaborations can solve problems that could not be solved in a timely fashion otherwise between actual meetings. Current technologies support videoconferencing and whiteboards to allow discussions in real time as well as discussion boards and Web portals to keep track of asynchronous discussions and organize common documents. Beyond current support, improved collaboration could be supported by allowing designers to interact in real time with building sketches, 3D virtual models, analyses, and so on. This is an even more complicated problem than co-located collaboration.

A new laboratory is under construction at ETS that will provide the basic research infrastructure to understand, to develop, and to test synchronous collaborative design environment. The laboratory is called Computer-Aided Collaborative Conceptual Design Laboratory. The infrastructure envisaged consists in a collaborative environment built with low-cost and off-the-shelf hardware and software because the environment developed must be affordable for consulting engineering firms, manufacturers, and architectural firms. The laboratory will have three main components.

- **A collaborative design room** will be constructed in order to develop a supporting environment for and experiment with live collaboration among a group of designers. The room will be an "augmented meeting room" equipped with the latest technology to support collaboration among a group of designers for brainstorming, presenting designs, and decision-making. A large backlit 3D virtual reality screen and tracking devices will let participants see and interact with shapes and models, discuss spatial issues, and visualize results of complex analysis. A smart board will be used to capture handwritten notes, and ideas could be sketched. With a video and audio conference system tied to an Access Grid node, the facility will allow distant people to virtually join a group in the room. Thus, this room will be able to support both synchronously co-located and distributed collaboration. The room will be instrumented (video cameras and interaction capture) so that activities and interactions can be recorded for detailed post-analysis.
- **Collaboration software usability testing dual rooms** will consist of two separate and isolated cubicles equipped with networked computers and capturing devices to carry out protocol analysis of users testing an environment for synchronously distributed design collaboration.
- **A virtual design desk** will work like a drafting board without paper. The sketches will be captured and analyzed by a computer to provide quick feedback to the designer. This section of the laboratory will provide another testing facility to experiment distant collaboration over the Internet, but this time using a virtual desk to discuss over sketches. The virtual design desk could also be used to discuss and collaborate with 3D models or analysis results. This component of the laboratory was developed by the LuciD Research group at the University of Liege mentioned earlier. Figure 7 illustrates the virtual design desk.

This research facility will provide the infrastructure to experiment and develop a computer-aided collaborative environment that will support live collaboration of project teams either in person or with a distance. The proposed infrastructure, shown in Figure 8, will be located in a permanent 100 m^2 space in the main ETS building in downtown Montreal.

Fig. 7. The virtual design desk by the LUCID group at the University of Liege

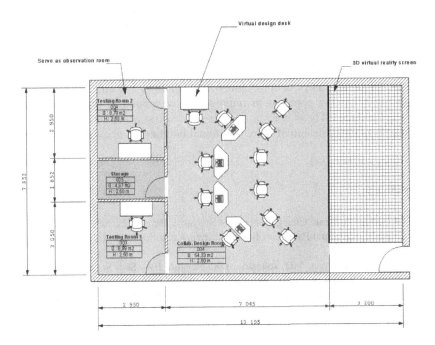

Fig. 8. Plan of the proposed collaborative design laboratory

4 Conclusions

The construction industry's output represents about 10% of the national gross domestic product. The building sector accounts for 60% of it. This industry has been plagued with plummeting productivity and lack of innovation. Improved design practices are direly needed. To compound the situation, the environment has become a priority and every nation must ensure that new buildings and building renovations are more sustainable.

The goal of a Canada Research Chair recently established at ETS, in Montreal, is to address this situation by developing the next generation of computer assistance for designers of sustainable buildings. This goal will be achieved by two research thrusts, which were presented here. The first thrust focuses on providing computer support in the early stages of design when the most important decisions are taken. The second thrust focuses on facilitating collaboration among designers involved in an integrated design process. The presented research projects can improve collaboration between architects and engineers by allowing earlier feedback and by relieving design conflicts as they appear. The projects introduce computer-supported design synthesis and exploration much earlier in the process while the presented laboratory will provide the basic infrastructure to research better collaboration support. The potential benefits are a more efficient and integrated design process resulting in better and more integrated buildings.

The construction industry is bound to benefit from collaborative support technology because of its unparalleled fragmentation. A new generation of IT collaborative support would facilitate collaborative meetings as well as distance collaboration. The approach would introduce computer-aided collaboration earlier in the process to a level that does not currently exist with current technologies. The end results will be a more productive and efficient design process with a more successful outcome in terms of energy efficiency, users comfort, integrated building, and environmental impacts. This approach has the potential to improve the way building designers and stakeholders collaborate. These innovations are geared toward sustainable buildings, but could also benefit the design of more traditional buildings, and even other engineering fields.

References

1. IPCC (2001) "Climate Change 2001: The Scientific Basis", Edited by Houghton JT et al., the Intergovernmental Panel on Climate Change, Cambridge University Press, Cambridge, UK.
2. Tucker, SN, Ambrose, MD, Johnston, DR, Newton, PW, Seo, S, and Jones, DG (2003) "LCADesign", Proceedings of the CIB W78 20th Int'l Conf. on IT for Construction, R Amor (Editor), Waiheke Island, New Zealand, CIB Report: Publication 284, pp. 403-412.
3. NRCAN (2003) "Energy Efficiency Trends in Canada, 1990 to 2001", Natural Resources Canada.
4. Bourdeau, L, Huovila, P, Lanting, R, and Gilham, A (1998) "Sustainable Development and the Future of Construction", CIB Working Commission W82, CIB Report, Publ. 225
5. Baouendi, R, Rivard, H, and Zmeureanu, R (2001) "Use of Life-Cycle Assessment Tools During the Design of Green Buildings", 4th Conference of the Canadian Society for Ecological Economics organized by F. Müller and T. Naylor at the McGill School of Environment, Montreal, p. 46.
6. Kam, C, Fischer, M, Hänninen, R, Karjalainen, A and Laitinen, J (2003) "The product model and Fourth Dimension project", ITcon, Vol. 8, pp. 137-166, http://www.itcon.org/2003/12

7. Rivard, H and Fenves, SJ (2000) "A Representation for Conceptual Design of Buildings", Journal of Computing in Civil Engineering, ASCE, Vol. 14, No. 3, pp. 151-159.

8. Mora, R, Rivard, H and Bédard, C (2006) "A Computer Representation to Support Conceptual Structural Design within a Building Architectural Context" Journal of Computing in Civil Engineering, Vol. 20, No. 2, pp. 76-87.

9. Fenves, SJ, Rivard, H and Gomez, N (2000) "SEED-Config: A Tool for Conceptual Structural Design in a Collaborative Building Design Environment", Artificial Intelligence in Engineering, Elsevier, Vol. 14, No. 3, pp. 233-247.

10. Meniru, K, Rivard, H, and Bédard, C (2003) "Specifications for Computer-Aided Conceptual Building Design", International Journal of Design Studies, Elsevier Science, Vol. 24, No. 1, pp 51-71.

11. Meniru, K, Rivard, H, and Bédard, C (2006) "Digitally Capturing Design Solutions", Proceedings of the Joint International Conference in Computing and Decision Making in Civil and Building Conference, Rivard, Miresco, and Melhem (Editors), Montreal, Canada, June 14-16.

12. Mora, R, Rivard, H, and Bédard, C (2004) "Computer-Aided Conceptual Design of Building Structures – Geometric Modeling for the Synthesis Process" First International Conference on Design Computing and Cognition, John Gero (Editor), Kluwer Academic Pub., held at MIT, Boston, on July 19-21, pp. 27-55.

13. Gomez, N (1998) "Conceptual Structural Design Through Knowledge Hierarchies." Ph.D. Thesis, Dept. of Civil Eng., Carnegie Mellon University.

14. Parent, S, Rivard, H, and Mora, R (2006) "Acquisition and Modeling of Conceptual Structural Design Knowledge", Proceedings of the Joint International Conference in Computing and Decision Making in Civil and Building Conference, Rivard, Miresco, and Melhem (Editors), Montreal, Canada, June 14-16.

15. Leclercq, P (1999) "Interpretative Tool for Architectural Sketches", Int'l Conference on Visual and Spatial Reasoning in Design, MIT, Cambridge, J S Gero and B Tversky (Editors), pp. 69-80.

16. Mora, R., Juchmes, R., Rivard, H., and Leclercq, P. (2006) "From Architectural Sketches to Feasible Structural Systems", 2nd International Conference on Design Computing and Cognition, Technical University of Eindhoven, J.S. Gero (Editor).

17. Soibelman L and Peña-Mora F (2000) "Distributed Multi-Reasoning Mechanism to Support Conceptual Structural Design", Journal of Structural Engineering, ASCE, Vol. 126, No. 6, pp.733-742.

18. Kumar B and Raphael B (1997) "CADREM: A Case-based System for Conceptual Structural Design", Engineering with Computers, Vol. 13, no. 3, pp. 153-164.

19. Eisfeld M, Scherer R (2003) "Assisting Conceptual Design of Building Structures by an Interactive Description Logic Based Planner" Advanced Engineering Informatics, Elsevier, Vol. 17, pp. 41-57.

20. Grierson DE, Khajehpour S (2002) "Method for conceptual design applied to office buildings" Journal of Computing in Civil Engineering, Vol. 16, No. 2, pp.83-103.

21. Sisk, GM, Miles, JC, Moore, CJ (2003) "Designer centered development of GA-based DSS for conceptual design of buildings" ASCE, Journal of Computing in Civil Engineering, Vol. 17, No. 3, pp. 159-166.

22. Rafiq, MY, Mathews, JD, Bullock, GN (2003) "Conceptual building design–Evolutionary approach" ASCE, Journal of Computing in Civil Engineering, Vol. 17, No. 3, pp. 150-158.

23. Packam, ISJ, Rafiq, MY, Borthwick, MF, and Denham, SL (2005) "Interactive visualization for decision support and evaluation of robustness in theory and practice", Advanced Engineering Informatics, Vol. 19, No. 4, pp. 263-280.

24. Miles, JC, Cen, M, Taylor, M, Bouchlaghem, NM, Anumba, CJ and Shang, H (2004) "Linking sketching and constraint checking in early conceptual design", in Beucke K. et al. (Editors) 10th Int. Conf. on Computing in Civil & Building Eng, 11pp.

25. Rosenman MA, Gero JS. (1999) "Evolving designs by generating useful complex gene structures." In: Bentley PJ, editor. Evolutionary Design by Computers, San Francisco: Morgan Kaufmann Publishers, 1999, p. 345-64.

26. Caldas, L (2002) "Evolving three-dimensional architecture form: An application to low-energy design." In: Gero JS, editor. Artificial Intelligence in Design'02, Dordrecht, The Netherlands: Kluwer Publishers, p. 351-70.

27. Chouchoulas, O (2003) "Shape evolution: An algorithmic method for conceptual architectural design combining shape grammars and genetic algorithms." Ph.D. Thesis, Department of Architecture and Civil Engineering, University of Bath, UK.

28. Bouchlaghem N (2000) "Optimizing the design of building envelopes for thermal performance." Automation in Construction Vol. 10, No.1, pp.101-112.

29. Peippo K, Lund PD, Vartiainen E. (1999) "Multivariate optimization of design trade-offs for solar low energy buildings." Energy and Building, Vol. 29, No.2, pp.189-205.

30. Jedrzejuk H, Marks W (2002) "Optimization of shape and functional structure of buildings as well as heat source utilization: Partial problems solution." Building and Environment, Vol. 37, No. 11, pp. 1037-43.

31. Wang, W, Rivard, H, and Zmeureanu, R (2005) "An Object-Oriented Framework for Simulation-Based Green Building Design Optimization with Genetic Algorithms" Journal of Advanced Engineering Informatics, Vol. 19, No. 1, pp. 5-23.

32. Wang, W, Zmeureanu, R, and Rivard, H (2005) "Applying Multi-Objective Genetic Algorithms in Green Building Design Optimization", Building and Environment, Vol. 40, No. 11, pp. 1512-1525.

33. Wang, W, Rivard, R, and Zmeureanu, R (2006) "Floor Shape Optimization for Green Building Design" In press in the Journal of Advanced Engineering Informatics.

34. Finnveden, G (1994) "Methods for describing and characterizing resource depletion in the context of life cycle assessment." Technical Report. Swedish Environmental Research Institute, Stockholm, Sweden.

35. Moran, MJ (1982) "Availability analysis: a guide to efficient energy use." Englewood Cliffs, NJ: Prentice-Hall.

36. Szargut, J, Morris, DR, Steward, FR (1988) "Exergy analysis of thermal, chemical, and metallurgical process." New York: Hemisphere Publishing.

37. Cornelissen, RL (1997) "Thermodynamics and sustainable development---the use of exergy analysis and the reduction of irreversibility." Ph.D. Thesis. Laboratory of Thermal Engineering, University of Twente, Enschede, The Netherlands.

38. Cera, CD, Regli, WC, Braude, I, Shapirstein, Y, Foster, CV (2002) "A Collaborative 3D Environment for Authoring Design Semantics", IEEE Computer Grapics and Applications, May, pp. 43-55.

39. Pena-Mora, F, Dwivedi, GH (2002) "Multiple Device Collaborative and Real Time Analysis System for Project Management in Civil Engineering", ASCE, Journal of Computing in Civil Engineering, Vol. 16, No. 1, pp. 23-38.

40. Cooper, S, Taleb-Bendiab, A (1998) "CONCENSUS: Multi-Party Negotiation Support for Conflict Resolution in Concurrent Engineering Design" Journal of Intelligent Manufacturing, Vol. 9, No. 2, pp. 155-159.

41. Adelson, B (1999) "Developing Strategic Alliances: A Framework for Collaborative Negotiation in Design" Research in Engineering Design, Springer Publ., Vol. 11, No. 3, pp. 133-144.

42. Fruchter R (1999) "AEC Teamwork: A Collaborative Design and Learning Space", Journal of Computing in Civil Engineering, ASCE, Vol. 13, No. 4, pp. 261-269.
43. Liston K, Fischer M and Winograd T (2001) "Focused Sharing of Information for Multidisciplinary Decision Making by Project Teams", ITcon Vol. 6, pg. 69-82, http://www.itcon.org/2001/6
44. Anumba CJ and AK Duke (2000) "Telepresence in Concurrent Lifecycle Design and Construction", Artificial Intelligence in Engineering, Elsevier, Vol. 14, pp. 221-232.

Interoperability in Building Construction Using Exchange Standards

W.M. Kim Roddis[1], Adolfo Matamoros[2], and Paul Graham[3]

[1] Professor and Chair, Civil & Environmental Engineering, The George Washington University, Washington, DC
roddis@gwu.edu
[2] Associate Professor, Civil, Environmental, and Architectural Engineering, The University of Kansas, Lawrence, KS
matamor@ku.edu
[3] Graduate Research Assistant, Civil, Environmental, and Architectural Engineering, The University of Kansas, Lawrence, KS

Abstract. Standard product and/or process models are a key enabling technology for the AEC industry to realize many of the benefits of more advanced computing approaches. The steel building industry has standardized on CIS/2. More broadly AEC has been striving to move IFCs into practice. STEP, the international product data standard ISO 10303, serves as the basic format for both CIS/2 and the IFCs. An overview of both formats accompanies example exchange files. An example one-way translation from CIS/2 to IFC2x3 illustrates some of the difficulties that must be overcome if the sought for harmonization of these standards is to be achieved.

1 Introduction

For decades, those involved in Architecture/Engineering/Construction (AEC) computing have bemoaned the lack of interoperability, classically phrased as the "islands of automation" problem. Software products used commercially typically address a part of the constructed facilities product or process, but there is no provision for systematic interaction and integration of the isolated individual implementations. This lack of a "common language" has proved a persistent barrier to realizing in practice the possible benefits of more advanced computing approaches.

The need for standard product and/or process models has been well known for many years. Despite many research demonstrations of feasibility and several major standardization efforts, progress was markedly slow during the 1980's and 1990's. In contrast the speed of transfer from theory to practice has picked up rapidly since 2000, particularly in the North American steel building industry. Interoperability and the exchange of product and/or process models between players and phases is a key enabling technology necessary to move more advanced computing approaches into practice. An understanding of the knowledge representation approaches used in existing standards is informative about what can be implemented now. Description of the development trajectory of the standards gives insight as to what may be implemented in the near future.

I.F.C. Smith (Ed.): EG-ICE 2006, LNAI 4200, pp. 576–596, 2006.

2 AEC EDI Standards: STEP, CIS/2, and IFCs

The benefits have long been recognized of the project team producing an integrated building model incorporating multiple aspects through the sharing of data. In the 1960's and 1970's the computing infrastructure was based on mainframe machines. The associated AEC software tools were proprietary integrated systems developed by large firms that both could afford them and benefit directly from them due to in-house inclusion of multiple disciplines and project phases. These tools were limited in distribution due to their proprietary nature as well as being cumbersome to maintain and expand, with software development practices and data structures that often proved inadequate for fully integrated solutions. The arrival of the desktop machine in the 1980's shifted the software mix toward individual tools addressing a particular AEC niche. Practitioners quickly recognized that time and money could be saved through electronic data exchange (EDI). A firm's interoperability system of choice was often to use commercial, or "third-party", software with converters allowing file translation from one program to the next (Gibson and Bell 1990).

2.1 STEP

Development of current AEC EDI standards began with STEP, an open set of standards for data exchange and sharing used to help engineering coordination. International adoption of the standard began in 1994 through the International Standards Organization as ISO-10303. The standard is now known as 'Industrial Automation Systems and Integration: Product Data Representation and Exchange' (Eastman 1999). The standard consists of a number of Parts, Resources, and Application Protocols (APs). APs are a set of exchange standards governed by a product model in the EXPRESS language. Examples of APs include: AP230 "Building Structural Frame: Steelwork" and AP228 "Heating, Ventilation and Air Conditioning" protocol. Parts can be considered specifications for STEP. Part 21 governs the format of the STEP File Structure. A STEP data-exchange file is divided into two sections: Header and Data. The Header contains exchange structure data, such as file conformance and file name. The Data contains the information to be transferred, including physical project data. The project data, such as members type, attributes, and locations, is represented using EXPRESS. Part 11 specifies the EXPRESS modeling language (Crowley and Watson 1997).

2.2 CIMsteel Integration Standard

The CIMsteel project (Crowley 1999), also known as the EUREKA Project EU 130, began in Europe with the collaboration of nine countries and 70 organizations. The project objectives were to help the growth of the steel industry, reduce design and construction times, and produce more economic steel structures. A result of the project is the CIMsteel Integration Standards (CIS), which allows for exchange of information throughout the steel design and construction process. In 1999, the American Institute of Steel Construction chose its second release, CIS/2, as the interoperability interface of choice for the AISC EDI initiative. To use CIS/2 data software companies must develop translators that map the model from the application to that of the common Logical Product Model. These mappings and the standards mentioned above are used to create CIS/2 files to import and export information between programs (Eastman, Sacks, and Lee 2002). The following discussion of CIS is intended to provide an overview while including details needed for comparisons below (Crowley and Watson 2003a).

CIS defines its supply chain as information contained within the design, detailing, scheduling, tendering, ordering, purchasing, and payment of structural steel buildings. CIS is similar to AP230 in that it relates information about the steelwork in structural frame buildings. However, it is a less formal version of the STEP protocol. This reduced the time necessary to establish the AP and made CIS more practical. CIS uses the STEP Part 21 exchange format as its file format.

2.2.1 Logical Product Model 6

The current version of CIS/2 uses edition 6 of the CIMsteel Logical Product Model (LPM/6) (Crowley and Watson 2003c). This is a computer product model defined in EXPRESS that uses all of the agreed upon requirements necessary in structural engineering. LPM/6 consists of four domains for exchange of information: analysis, member design, connection design, and detailing. This product model defines a wide range of information that may be used during the design of a steel structure, including sign conventions, loading, structural response, parts, joints, materials, and geometry.

Within LPM/6 the structure can be broken down into parts for design of frames, members, and connections. LPM/6 also divides the information into analysis model, design model, and manufacturing model. Structural analysis models consist of elements and nodes. Design models are comprised of design parts and design joint systems. Manufacturing models can be broken down into manufacturing assemblies containing local parts and local joint systems.

CIS/2 extensively describes joint systems. Joint systems may contain a set of bolts, welds, welded fasteners, sealants, or grout. To define these joints LPM/6 divides them into joint types (e.g. weld_alignment, weld_backing_type, weld_configuration) and entities (e.g. fastener_mechanism_with_position, joint_system_welded_with_shape, weld_mechanism_complex).

LPM/6 uses STEP Part 42 to define the geometry of a member explicitly. Explicit definition of geometry refers to defining the geometry of a member within the exchange file without referencing the shape to a specific identity. Originally, the LPM formally used implicit geometries, however because many CAD programs define shapes using explicit geometries the CIS had to be adapted to facilitate transfer of data between the two.

Within the CIS/2 file, LPM/6 is assigned the name 'Structural_Frame_Schema' which ensures unique data sharing of project information. This unique naming system is known as the Object Identifier. Developers adapted the format of STEP Part 41 to fit into the LPM/6. LPM/6 uses the STEP method for assigning units to measurements. LPM/6 can reference items by converting them into entities passed by the STEP Part 21 file. Items are divided into four classes: standard, proprietary, library, and non-standard. In order for these items to be read by both applications each application must have a reference to the item.

2.2.2 Conformance Requirements

There are varying degrees to which CIS/2 can be implemented (Crowley and Watson 2003d). First, software developers can produce Basic CIS Translators that exchange data via physical files. Second is the development of Data Management Conformant (DMC) Translators. Third are Incremental Data Import (IDI) Translators. Fourth are Product Model Repository (PMR) Translators that support CIS product model sharing and management. Finally, the last level of implementation is the development of the PMR, which consists of a database of project information. If implemented fully the standards offer continuous communication between physical files, databases, and direct procedure calls. Typically, commercial applications are limited to Basic Translators.

To implement a CIS/2 translator software engineers must follow the levels of STEP implementation. Basic CIS Translators fall under Level 1, the exchange of data files. Under STEP Level 1 guidelines the translator allows the exchange of data held in neutral file format specified by STEP Part 21 and structured in accordance with the LPM. At this level the export translators must be able to convert information from the native format of the specific application to neutral format, while import translators must convert data from neutral to native format.

At Level 1 implementation translators are broken into various components needed for file transfer (Crowley and Watson 2003b). Import components include a file reader and file parser. Together they read the STEP file and break up and move the data to the correct location within the application. Export components include a file formatter and file writer. These components put the engineering data into the neutral file format and create the file for exchange.

To develop a Basic CIS Translator, the data structure of an application must be mapped into the neutral structure of LPM/6. Developers must understand the original data structure of an application with possible creation of an EXPRESS model of the structure, define the scope of the translation, write an 'export-mapping' table to facilitate data exchange, and use the STEP Toolkit to gather the necessary LPM/6 schema.

The degree to which an application conforms to the CIS specifications depends on the level of implementation of the Conformance Classes (CCs) of the translator implemented (Crowley and Watson 2003d). A translator can conform to one or more CC. A CC is a simple 'short form' version of a schema. CCs check to see if instances of entities are created, imported correctly into the system, and that the result would be a valid STEP file. These classes are a departure from the data-exchange protocols (DEPs) used in previous versions of CIS. DEPs were considered too broad to enable strong testing of translators. The use of CCs is more like STEP, and has a major effect on the size and style of the data used for implementation.

The CIS/2 documentation lays out the conformance requirements for a number of features needed during exchange, such as the physical files and basic translators. Each feature must meet implementation, operational, and documentation requirements. The implementation requirements of the physical file are to create the file in accordance with Part 21, use .stp file extension, data structure must be a "…sequential file using clear text encoding", support CC1, and it must 'populate' one instance of File_Description, File_Name, and File_Schema with the Header implemented (Crowley and Watson 2003d). An operational requirement of this file is to create a 'log file' in ASCII text format. The physical file has no documentation requirements.

Conformance testing is the last requirement before a translator is considered adequate for commercial use. Testing of an import translator is more difficult than for an export translator since after a neutral file is imported the data can spread throughout the application or may not be held within the application. An additional requirement of a translator is to produce specific error messages when the application does not understand the data imported. The CIS categorize these as Intelligent Translators and developers spend a great deal of time to create this feature to allow better flow of information implemented (Crowley and Watson 2003d).

2.3 Industry Foundation Classes

In 1994, a group of construction industry representatives, interested in modernizing the information technology portion of the industry, formed the Industry Alliance for Interoperability. The organization soon became public and changed its name to the International Alliance for Interoperability (IAI) in 1996. The group developed the specifications known as the Industry Foundation Classes (IFCs), a building product model-based information sharing and exchange for AEC/FM industries. An IFC object is the instance of an IFC class. (Liebich 2004). As with the above CIS discussion, the following discussion of the IFCs is intended to provide an overview while including details needed for comparisons below. The goal of the IFCs is to share project information "...throughout the project life cycle, and across all disciplines and technical applications (Bazjanac 1998)." They include architecture, engineering, construction, and facilities-management. The IFCs include features beyond the structure, from windows and wall type to the heating, ventilation and air conditioning (HVAC) systems.

Much like CIS/2, the IFCs information exchange occurs through a file format that uses the standards set forth by Part 21 of STEP. Files are converted from one application into .ifc format, developed by the IAI, and then transferred to the next application. IFCs are organized in a similar way as STEP and use the EXPRESS modeling language, but are not considered compatible with those standards.

2.3.1 IFC Layers

IFC Release 2 Edition 3, designated IFC2x3, was released in February 2006 (IAI 2006). The architecture of the IFCs is defined by four layers that provide information about a project (Liebich and Wix 2000). The Resource layer includes information needed for upper layers and provides geometry, topology, dimensions, materials, and other generic information. The Core layer includes the Kernel as well as the Process, Product, and Control Extensions. The Kernel contains all non-specific AEC and Facilities Management (FM) information needed to produce the model. The other extensions provide all AEC/FM specific classes. The Interoperability layer contains a set of five Shared Elements grouping commonalities across multiple AEC applications. Dividing the classes into these groupings makes the information more specialized for each application. Examples are the Shared Building Elements section, including information shared by domain or application models, and the Shared Building Services Elements, including information needed for interoperability such as that for heating, ventilating, and air conditioning (HVAC). The Domain layer consists of extensions that represent application specific information.   These

"Application extensions facilitate exchange with application models that have a software architecture different from that of IFCs (Bazjanac 1998)." The domains include: Building Controls, Plumbing Fire Protection, Structural Element, Structural Analysis, HVAC, Electrical, Architecture, Construction Management, and Facilities Management. As development of the IFCs continues the number and complexity of the domain extensions grow to allow for enhanced interoperability. The original IFCs consisted of four domains. Edition 2 of the IFCs represented a significant improvement with respect to the structure of a building, adding the structural domains (Structural Elements Domain, Structural Analysis Domain) into the schema (Liebich 2004).

Each of the IFCs four layers: Resource, Core, Interoperability, and Domain, relies on the next layer to produce exchangeable data. A layer may only reference layers below it in the information architecture. For example, classes within the Interoperability layer may reference the Core and Resource layers but not the Domain layer (Liebich and Wix 2000).

2.3.2 Core Layer: The Kernel

The Kernel establishes the information that defines the information directly used in an exchange context. This set of information is known as the Leaf Node classes (Liebich 2004). These are the end-user relevant classes containing the basic object information, relationship information, type information, property information, and connection relationships needed for meaningful data exchange. With the use of this information the Kernel provides object class types, relationships between classes, by using data sets. Leaf Nodes can be considered the place where the IFC model ends (Liebich and Wix 2000). They are the last line of the schema before the data is exchanged and read by a different application.

All Leaf Node information begins at the Root entity level. The root level provides the first level of specialization for the IFC classes and is divided into three types: Object, Relationship, and Property definition (Liebich 2004). There are seven entity types, all under Object; Products, Processes, Controls, Resources, Actors, Project, and Group. Entity types can be broken up or associated with other types. There are five Relationship types: Assignment, Association, Decomposition, Definition, and Connectivity. These relationships define the way objects interact with other objects. There are two types of Property Definitions: Type Object and Property Set. These constitute a mechanism to allow definition, connection, and use of types of information missing in current IFCs.

The structures and contents of the Leaf Node classes thus constitute the IFC Object Model, giving a data exchange structure. The IFCs define how information should be broken down in order for data exchange through Part 21 or Part 22, or other types of encoding. The IFCs are defined in such a way that this information is expandable to allow users to define information that is not available in the current IFC model. The Project Definition allows new information to be defined by the developer (Liebich 2004). Project Definitions can be defined and shared among multiple instances, defined and shared among a specific instance, or extended. The Property Definition relates the object type to a set of properties, shares a set of values for multiple instances of a class, and defines different property values for each occurrence of a class.

2.3.3 Resource Layer: Geometric and Unit Transformations
To define the product geometry, the IFCs defines six Resources: Geometry, Topology, GeometricModel, Representation, GeometryConstraint, and Profile. Geometry, Topology, and GeometricModel are defined in ISO/IS 10303-42 "Integrated Generic Resources: Geometrical and topological representations" (Liebich 2004). Representation is defined in association with ISO/IS-43 "Integrated Generic Resources: Representation Structures". The last two geometric Resources are IFC defined.

The Measure resource define units and measure types based on ISO 10303-41 "Integrated Generic Resources: Fundamentals of Data Description and Support". Basic units are defined using ISO 1000. The default standard within the IFCs is the use of SI units, however, full unit conversion is available.

2.3.4 Unique Objection Identification
A key element within the IFCs is the unique definition of an object. An object's identification should remain unique, not only within exchange, but persistent across exchanges, in order for proper information exchange. This capability allows information to be exchanged and stored without errors to occur in identification along the project life. The IFCs use an algorithm to create a Globally Unique Identifier (GUID) or a Universal Unique Identifier (UUID) (Liebich 2004). The GUID is the stored, compressed value. Each time an instance is created a new value must be created.

2.3.5 Spatial Structure
The IFC2x Implementation Guide defines the spatial structure as the "breakdown of the project model into manageable subsets according to spatial arrangements (Liebich 2004)." IFCSpatialStructureElement divides the structure into a five part hierarchy: Project, Site, and Building, Building Story, and Space. The hierarchical spatial structure division of the IFC2x model allows building information from different disciplines to come together to produce an integrated model.

IFCProject, IFCBuilding, and IFCBuildingStorey are mandatory exchange attributes. IFCBuilding defines the elevation with reference to the height of the structure by its change from plus or minus zero. This schema also gives the elevation of terrain and building address. IFCBuildingStorey contains references to the spaces within a story and defines the elevation of the story based on slab elevation. All building elements are assigned to a building story.

Spaces contain all building services and interior design elements. Large distribution elements, however, are contained by stories. These elements may include air ducts or water pipes.

2.3.6 Building Elements
The IFC schema beginning with Root, uses the schemas for Object Definition, Object, Product, then Element. Element covers all components of an AEC product, so a collection of all Elements contains all building features making up the spatial hierarchy. The schema Building Element includes Beam, Column, Curtain Wall, Door, Member, Plate, Railing, Ramp, Ramp Flight, Roof, Slab, Stair, Stair Flight, Wall, and Window (Liebich 2004). These structural schemas are also included within the Shared Building Elements portion of the Interoperability layer. The Shared Building Elements also had associated Type schemas, such as Beam Type, Column Type, to define common section properties. The IFCs also have Profile Property Resource and Profile Resource are contained within

the Resource layer for various structural member profiles. Profile Property Resource defines properties for structural members, such as, section weight and cross-sectional area. Profile Resource includes schemas for various section profiles, including W-sections, L-sections, T-sections, and hollow tube sections. Thus, properties for structural members may be specified at different levels of the IFC hierarchy.

Likewise, properties of structural members may be specified using multiple geometric representations. The simplest representation is a Bounding Box defining a rectangular solid enclosing the member. Another representation is a Surface Model specified by surfaces or faces. Bounding Box and Surface Model are general representation types of the Building Element. In addition to these two representations, a Beam or Column may have Swept Solid (with or without Clipping), Boundary Representation (Brep), and Mapped Representation. Swept Solid uses an extruded solid approach. Clipping may be combined with a Swept Solid to cut off pieces sliced by specified planes. Brep uses formal boundary facets with or without voids. Mapped Representation allows an existing representation to be reused for more than one element.

3 Comparison of CIS/2 and IFCs

Table 1. Differences in the overall scope of CIS/2 and IFC2x3

Implementation Level	CIS/2	IFC2x3
Scope	Structural Steel	AEC/FM
Part 21 Exchange	●	●
Database Management	●	
Intelligent Translators	●	
Testing Requirements	CCs	Non-Standardized

(Note: ● denotes the file contains specifications for these elements).

Table 2. Differences in connection definitions

Joint Detail	CIS/2	IFC2x3
Connections	●	●
Boundary Conditions	●	●
Bolts	●	
Welds	●	
Fasteners	●	

(Note: ● denotes the file contains specifications for these elements).

3.1 Similarities and Differences

The data structures of CIS/2 and IFCs are quite similar. Analyzing the main elements that define both sets of standards gives a better understanding of these structures. By looking at their differences and similarities an approach can be formulated to begin merging the two standards to create a common data model for interoperability. Breaking down the structure of each exchange file also provides an objective way to study the standards.

3.1.1 The Use of STEP

Both CIS/2 and the IFCs rely heavily on STEP, incorporating numerous STEP Resources to define their respective product models. The Parts integrated into both standards include: 11, 21, 22, Integrated Generic Resources, and 225. Part 11 – EXPRESS Language describes how information should be structured using the EXPRESS modeling language. Using the direction provided by this Part allows project data structuring to remain consistent between standards. Part 21 – Clear Text Encoding of the Exchange Structure describes how project information should be laid out before it is transferred to another system. Part – 22 Standard Data Access Interface specifies how an interface operates and how information should be modeled. CIS/2 and the IFCs transfer information between applications in a comparable way because the structures of their exchange files use these Parts.

The Integrated Generic Resources are defined in Parts 41-49 (Eastman 1999). These Parts format how general project information should be formatted. The scope of the Integrated Generic Resources includes: Product Description and Support, Geometric and Topological Representation, Representation Structures, Product Structure Configuration, Materials, Visual Presentation, Shape Tolerances, Form Features, and Process Structure and Properties. A great deal of the IFCs comes from these Parts while half of LPM/6 comes from these Parts. CIS/2 identify material properties using a set of definitions, such as, Material_Isotropic and Material_Strength. IfcMechanicalSteelMaterialProperties defines the yield stress, ultimate stress, ultimate strain, hardening, plastic strain, and stress relaxation (IAI 2006).

3.1.2 Scope

CIS/2 and IFC2x have different scopes. CIS/2 only defines information related to structural steel while the plan for the IFCs is to define a Building Information Model (BIM) that relates all facets of AEC/FM industry. Currently, the IFCs are behind the CIS in terms of implementation of structural steel elements because IFC2x2 was the first release to aggressively address this issue. In addition to differences in structural definitions, CIS/2 also provides specifications for database management, partial exchanges through Intelligent Translators, and more rigorous testing standards, the CCs. Table 1 provides an overview of the differences in the overall scope of the standards.

3.1.3 Section Geometries

CIS/2 references a section name and catalog, while IFC's use explicit profiles. However, the IFCs made significant progress with IFC2x2, defining beam and column types to associate elements with specific sections. Both define the use of various sections. CIS/2 references the use of the standard shapes while the IFCs lack the definition of standard shapes. Sections in the CIS/2 file are referenced using Item_Reference_Standard. IFC2x define the Cartesian points of each element of the section. It then uses these points to define other properties of the section, for example, area. When defining these properties the IFCs omit the smaller features of a section. CIS/2 also allows for creation of non-standard sections. CIS/2 has specifications for even the smallest properties of a section, such as, flange edge radius. The specifications also base the geometry of a section of the geometric centroid of the section (Crowley and Watson 2003c). For these generic sections in CIS/2 a Bounding box defines the shape, just as in IFC2x2. The box is rectangle that defines the extreme dimensions of a shape.

3.1.4 Connections

The differences in connection entities are another issue. As mentioned, the CIS define even the smallest part of a joint system. The IFCs define connections in a very simple way, by defining the location along an element that another element intersects it. This enables structural analysis information to be exchanged but because IFC2x lacks the schemas for bolts and welds detailing information cannot be transferred. Table 2 diagrams the differences in connection definitions. A simple way to describe the difference between the IFCs and CIS/2 is to describe them in terms of analysis, design, and manufacturing models. CIS/2 incorporates all these models into its steel specifications. The IFCs incorporate only analysis and design models, which define elements, nodes, boundary conditions, and parts but do not define part features or connection assemblies.

3.1.5 Structural Analysis Features

The standards are at the same level of definition for structural analysis features. CIS/2 has a full set of analysis conditions to produce force information. The IFCs also have a set of structural analysis schemas located in the Domain layer, IfcStructuralAnalysisDomain. This allows the transfer of associations between members, boundary conditions, reactions, loads, and displacements.

3.2 Exchange File Comparison

Two identical frames were constructed to compare the exchange file structures of CIS/2 and IFC2x. The frames consisted of two 12-foot tall W10x33 columns located 20 feet apart with a W12x40 beam connected to the tops of columns. A simple frame was used to evaluate the differences in the exchange files.

One model was produced in RAM Structural System, Appendix A, and the other in Architectural Desktop 2004, Appendix B. RAM uses CIS/2 specifications to exchange project information while a plug-in can be installed to create IFC2x2 files in Architectural Desktop. Since IFC 2x3 was released February 2006, a third file was generated CIS/2 file in Appendix A to a IFC2x3 file (not included due to space) using the CIS/2 to VRML and IFC Translator based on research at NIST downloadable from http://ciks.cbt.nist.gov/cgi-bin/ctv/ctv_request.cgi. This demonstrated that the difference between IFC2x2 and IFC2x3 were not important for this particular file comparison. This also illustrates some of the results of translation between standards.

As discussed, both sets of standards define their exchange files using STEP Part 21. The similarities between these files are easy to see once the files are broken down into sections, as is done with annotations in the appendices. For example, both files begin with the Header section and then move to the Data section. Also, the first line of each file is 'ISO-10303-21' and the last line is 'END-ISO-10303-21', referencing the standard from which they get their structure. Table 3 compares the two Part 21 files.

Within the Header section both files contain File_Description, File_Name, and File_Schema. File_Description describes the file's level of conformance to the standards. In CIS/2 the file lists the CCs in which the file conforms. For example, the file references CC003, a generic CC for Cartesian_point, and CC305, a specific CC for Material_Isotropic. File_Name details file name, time, author, organization, preprocessor version, originating system, and authorization. In the case of CIS/2, the processor version is 'ST-Developer V10' which is used to keep the file in line with STEP. File_Schema describes the standard from which the frame gets its information. For example, the File_Schema for the IFC file is 'Ifc2x_Final'.

The rest of the CIS/2 file is organized in the following way. The Header ends with the file schema, Structural_Frame_Schema. The remained of the file is in the Data section, beginning with global geometry representations and the definition of units. The next portion includes general geometry, defining connectivity of element nodes by referencing schemas defined later in the file. Element schema follow, defining the geometry, section type, and material of each element. Next, the material properties and then the node points for each member are defined. The next set of schema contains the references to the beam and column sections. These include Item_Reference_Assigned, Section_Profile, Item_Reference_Standard, and Item_Ref_Source_Standard. Item_Ref_Source_Standard references the AISC EDI Standard Nomenclature, the standard for naming sections. Forces and specific units are defined. Finally, the specific assembly geometry for each member is called out. For example, the Cartesian points of each node are identified to give the unit length and orientation. The angles of each member are also included in this section.

The general layout of the IFC file is almost identical. The file begins with the Header and specifies general file information. The Data section then begins. The first portion of this section is the definition of units and conversion. The IFC file requires a conversion from SI units to English units when applicable. The next portion of the file is global axis geometry and file information, including the introduction of IfcProject. Much like the CIS/2 file, the specific member information is the next portion of the file. However, this portion includes additional section information unique to the IFC file. The IFCs do not include standard references to specific 'W' shapes. Therefore, the geometry of a section must be defined by using Cartesian points for each edge point of the section. An edge's distance from the centroid of the shape defines its point. For example, the point (3.98, 4.865), or (b$_f$/2, d/2), corresponds to the corner of a W10x33 section, highlighted in Appendix B. The points are located in the local axes of each element and then referenced to a global point later in the file. CIS/2 only defines global geometry. IfcPolyline then uses these points to construct the shape of the section by connecting each point with a bounding line. The cross-sectional area can then be defined using this information. The element lengths and directions are also defined in this portion of the file. The final part of the IFC file

Table 3. Graphical comparison of the Part 21 files

Item	CIS/2	IFCs
ISO - 10303 - 21	●	●
Header	●	●
File_Description	●	●
File_Name	●	●
File_Schema	Structural_Frame_Schema	IFC2x3
Data	●	●
Global Geometry	●	Defines global coordinate system
Units	References units defined later in file	●
Conversion		Converts SI to English using ratio
Project Definition		●
Connectivity	●	
Element Definition	Assigns properties to element	
Specific Element Geometry		●
Local Geometry		●
Cartesian Points		Defines points of W-section
Direction		Establishes member orientation
Area Definition		Defines W-section
Shape Extrusion		Defines length from area
Internal connectivity		Connects ends of each member
W - Shape Reference	References standard shape library	
Force and Unit Definition	●	
Assembly Geometry	●	
Model Definition	●	
Cartesian Points	Defines lengths and locations	
Direction	Establishes member orientation	
File Property Definition		Assigns file to IfcBuilding
END - ISO - 10303 – 21	●	●

(Note: ● denotes the file contain specifications for these elements)

assigns the frame a specific spatial structure, using IfcRelContainedInSpatialStructure, and then defines the properties of the structure. The frame was assigned the building spatial structure before export of the project data.

The third file was a translation of the CIS/2 file to IFC2x3 (Lipman 2006). As pointed out by the translator originator, Dr. Robert Lipman of NIST, the generated IFC2x3 file is just one of many different ways to write out the information contained in the CIS/2 file in IFC. This is a natural result of the provision in the IFCs of many ways to represent information such as units, and coordinate systems, and member shapes. This means a translation of a single item from CIS/2 is inherently ambiguous and the choice of which possible one-to-many mapping is correct is unclear.

Of interest in the third file was shuffling of the sections of the Data portion into significantly different order as well as scattering of information associated with a single element. Another result was the addition of multiple transformations of units and coordinates. Since the IFCs have many ways to represent unit and coordinate information, any translation can introduce nested transformations that collapse to match the input. If a file is translated multiple times, these arbitrary transformations will grow. This is like trying to deal with many manipulations of mathematical equations where no canonical form is defined. Simply determining equivalence becomes a daunting task. Another example of the possible results of the multiplicity of the IFCs representation options is the representation of the standard structural shapes. In this case, the translated file maps the CIS/2 wide flange information to a Swept Solid. This extruded representation is most natural, another possible translation is a Brep. These types of semantically invalid mappings are common in translation of natural language. Development of the standards must include agreement on translation choice preferences, as is done for mathematical operation precedences, and standard representation mappings, as is done for useful clichés in programming.

4   Common Data Model for Interoperability

Development of CIS began in part because industry practitioners wanted a standard for information exchange before all issues needed to produce a legitimate International Standard were resolved. When CIS/1 emerged it was seen as a stepping-stone to the future of industry data exchange. Crowley states that CIS/2 can be considered a short-term solution, the IFCs as a mid-range solution, and STEP as the ultimate long-term goal to achieve interoperability within the AEC/FM industries (Crowley 1999).

There are a variety of barriers that are preventing the creation of this model. For example, CIS/2 is currently available in a variety of commercial applications. Convincing software developers to spend more money on development of a new set of standards is not an easy task because these companies do not see the monetary benefits. Another obstacle is the cooperation between the two organizations that develop the standards, SCI and IAI.

There have been several efforts to create a single model for exchange of project data by merging CIS/2 and the IFCs. However, due to the barriers discussed above the research did not progress. Specifically, the development of entirely new translators to fit the IFCs was seen as a major obstacle, not only because of costs involved but also the time needed to create an IFC-compatible translator from a CIS/2 compatible translator (Crowley 1999). Research is currently underway in developing an intermediate translator at Georgia Tech (Eastman 2006). Under a plan developed by Eastman, the

IFCs would be used at the design level. The data would then be passed on to a CIS/2 file for detailing because CIS/2 provides better detailing guidelines than the IFCs. Finally, the files would be translated back to IFC format for checking. The development of this translator is very time consuming due to the differences between the standards and the complexity of the language. There remain substantial issues to resolve to include bi-directional translation and round-tripping of exchange files.

A simpler approach is the mapping approach taken in the NIST CIS/2 to VRML and IFC Translator. This permits a CIS/2 exchange file to be translated to IFC2x3 as a one-to-many mapping while avoiding the much more problematic many-to-one mapping entailed in an IFC2x3 file translation to CIS/2 (Lipman 2006).

Some believe the Extensible Markup Language (XML) will replace current data exchange file formats. In fact, the CIMsteel Integration Standards Release 2: Second Edition – Overview states XML is an accepted alternative to STEP Part 21 file format (IAI 2006).

Many researchers have their own opinions on the future of interoperability, but what remains is the need for a full scale study and actual implementation. AISC and IAI recently began working together to further develop exchange standards for the AEC/FM industry. The newly formed team will be mainly concerned with 'harmonizing' CIS/2 and the IFCs, allowing structural steel to be incorporated into a building information model (BIM) of the IFCs ("International" 2004). Integrated all portions of the industry into a BIM is the goal of the next generation of interoperability standards.

Acknowledgement

Dr. Robert Lipman of NIST provided both the translation of the CIS/2 file in Appendix A to IFC2x3 using the CIS/2 to VRML and IFC Translator based on research at NIST as well as a preprint of his 2006 paper cited in the references.

References

1. Gibson Jr., G. E. and Bell, L. C. "Electronic Data Interchange in Construction." *Journal of Construction Engineering and Management* 116. 4 (December 1990): 727-737.
2. Eastman, C. M. *Building Product Models: Computer Environments Supporting Design and Construction.* Boca Raton, FL: CRC Press, 1999.
3. Crowley, A. J. and Watson, A. S. "Representing Engineering Information for Constructional Steelwork." *Microcomputers in Civil Engineering* 12. 1 (January 1997): 69-81.
4. Crowley, A. J. "The Evolution of Data Exchange Standards: The Legacy of CIMsteel." (1999). 5 Nov. 2004. http://www.cis2.org/faq/crowley1999.
5. Eastman, C. M., Sacks, R., and Lee, G. (September 2002). "Strategies for Realizing the Benefits of 3D Integrated Modeling of Buildings for the AEC Industry." *19th International Association for Automation and Robotics in Construction.* Washington D.C. Sept. 2002.
6. Crowley, A. J., and Watson, A. S. "CIMsteel Integration Standards Release 2: Second Edition, P265: CIS/2.1:Volume 1 – Overview", The Steel Construction Institute, 2003a.

7. Crowley, A. J., and Watson, A. S. "CIMsteel Integration Standards Release 2: Second Edition, P268: CIS/2.1:Volume 3 – LPM/6", The Steel Construction Institute, 2003c.
8. Crowley, A. J., and Watson, A. S. "CIMsteel Integration Standards Release 2: Second Edition, P269: CIS/2.1:Volume 4 – Conformance Requirements", The Steel Construction Institute, 2003d.
9. Crowley, A. J., and Watson, A. S. "CIMsteel Integration Standards Release 2: Second Edition, P266: CIS/2.1:Volume 2 – Implementation Guide", The Steel Construction Institute, 2003b.
10. Liebich, T. (ed.), IFC 2x Edition 2 Model Implementation Guide Version 1.7,. IAI. http://www.iai-international.org/Model/files/20040318_Ifc2x_ModelImplGuide_V1-7.pdf (March 2004)
11. Bazjanac, V. "Industry Foundation Classes: Bringing Software Interoperability to the Building Industry." *The Construction Specifier* 15. 6 (June 1998): 47-54.
12. IAI, IFC 2x Edition 3. http://www.iai-international.org/Model/R2x3_final/index.htm (February 2006)
13. Liebich, T., and Wix, J. (eds.), IFC Technical Guide, Release 2x.. IAI. http://www.iai-international.org/Model/documentation/IFC_2x_Technical_Guide.pdf (October 2000)
14. Lipman, R. R., "Mapping Between the CIMsteel Integration Standards and Industry Foundation Classes Product Models for Structural Steel, ICCCBE-XI, 2006, preprint.
15. Eastman, C.M. "Harmonization for CIS/2 and IFC", http://www.coa.gatech.edu/~aisc/cisifc 2006
16. "International Model for EDI." *Structure Magazine*, CASE, NCSEA, and ASCE. (September 2004): 33.

Appendix A

CIS/2 File

```
ISO-10303-21;
HEADER;
/* Generated by software containing ST-Developer
 * from STEP Tools, Inc. (www.steptools.com) */
FILE_DESCRIPTION(
/* description */ ('','CC003, CC005, CC014, CC019, CC024, CC026,
CC029, CC030, CC031, CC032, CC034, CC035, CC110, (CC166, +CC167),
CC170, CC305, CC306, (CC177, +CC307), CC310, CC325, CC327, CC331'),
/* implementation_level */ '2;1');
FILE_NAME(/* name */ 'frame2',
/* time_stamp */ '2004-11-12T11:02:32-06:00',
/* author */ ('Paul R. Graham'),/* organization */ (''),
/* preprocessor_version */ 'ST-DEVELOPER v10',
/* originating_system */ 'RAM Structural System',
/* authorisation */ '');
FILE_SCHEMA (('STRUCTURAL_FRAME_SCHEMA'));
ENDSEC;
DATA;
#10=REPRESENTATION('representation for
all',(#71,#72,#73,#74,#75,#76,#77,#69,#70,#68),#11);
#11=(GEOMETRIC_REPRESENTATION_CONTEXT(3)
GLOBAL_UNIT_ASSIGNED_CONTEXT((#57))
REPRESENTATION_CONTEXT('linear_units_context','linear_units'));
#12=GLOBAL_UNIT_ASSIGNED_CONTEXT('force_units_context',
'force_units',(#49));
#13=DERIVED_UNIT((#14,#15));
#14=DERIVED_UNIT_ELEMENT(#16,1.);
#15=DERIVED_UNIT_ELEMENT(#57,-3.);
#16=(CONTEXT_DEPENDENT_UNIT('POUND')MASS_UNIT()NAMED_UNIT(#91));
#17=ASSEMBLY_MAP(#39,(#26));
#18=ASSEMBLY_MAP(#40,(#27));
#19=ASSEMBLY_MAP(#41,(#28));
#20=ELEMENT_NODE_CONNECTIVITY(1,'Start Node',#33,#26,$,#58);
#21=ELEMENT_NODE_CONNECTIVITY(2,'End Node',#34,#26,$,#58);
#22=ELEMENT_NODE_CONNECTIVITY(1,'Start Node',#35,#27,$,#58);
#23=ELEMENT_NODE_CONNECTIVITY(2,'End Node',#36,#27,$,#58);
#24=ELEMENT_NODE_CONNECTIVITY(1,'Start Node',#37,#28,$,#58);
#25=ELEMENT_NODE_CONNECTIVITY(2,'End Node',#38,#28,$,#58);
#26=(ELEMENT('Flr1Bm3',$,#66,1)ELEMENT_CURVE($)
ELEMENT_CURVE_SIMPLE(#44,#78)ELEMENT_WITH_MATERIAL(#29));
#27=(ELEMENT('Flr1Col1',$,#66,1)ELEMENT_CURVE($)
ELEMENT_CURVE_SIMPLE(#45,#78)ELEMENT_WITH_MATERIAL(#29));
#28=(ELEMENT('Flr1Col3',$,#66,1)ELEMENT_CURVE($)
ELEMENT_CURVE_SIMPLE(#45,#78)ELEMENT_WITH_MATERIAL(#29));
#29=MATERIAL_ISOTROPIC(0,'steel',$,#30);
#30=MATERIAL_REPRESENTATION('Fy 50.00',(#32),#53);
#31=MATERIAL_REPRESENTATION('material representation for all',
#32),#53);
#32=MATERIAL_STRENGTH('yield strength',50.);
#33=NODE('np0',#72,$,#66);
#34=NODE('np1',#73,$,#66);
#35=NODE('np2',#74,$,#66);
#36=NODE('np3',#75,$,#66);
#37=NODE('np4',#76,$,#66);
#38=NODE('np5',#77,$,#66);
```

Right-side bracket labels:
Conformance
File Information
Global Geometry and Units
Connectivity
Elements
Materials
Assembly

```
#39=ASSEMBLY_DESIGN_STRUCTURAL_MEMBER_LINEAR(0,'Flr1Bm3',$,$,$,$,
.T.,.F.,(),(),$,.COMBINED_MEMBER.,.UNDEFINED_CLASS.,.BEAM.);
#40=ASSEMBLY_DESIGN_STRUCTURAL_MEMBER_LINEAR(1,'Flr1Col1',$,$,$,$,
.T.,.F.,(),(),$,.COMBINED_MEMBER.,.UNDEFINED_CLASS.,.COLUMN.);
#41=ASSEMBLY_DESIGN_STRUCTURAL_MEMBER_LINEAR(2,'Flr1Col3',$,$,$,$,
.T.,.F.,(),(),$,.COMBINED_MEMBER.,.UNDEFINED_CLASS.,.COLUMN.);
#42=ITEM_REFERENCE_ASSIGNED(#46,#44);
#43=ITEM_REFERENCE_ASSIGNED(#47,#45);
#44=SECTION_PROFILE(0,'W12X40',$,$,8,.F.);
#45=SECTION_PROFILE(1,'W10X33',$,$,5,.F.);
#46=ITEM_REFERENCE_STANDARD('W12X40',#48);    <----Section Reference
#47=ITEM_REFERENCE_STANDARD('W10X33',#48);
#48=ITEM_REF_SOURCE_STANDARD('AISC','AISC EDI Standard Nomencla-
ture',2001,'1');
#49=(CONTEXT_DEPENDENT_UNIT('KIP')FORCE_UNIT()NAMED_UNIT(#89));
#50=FORCE_MEASURE_WITH_UNIT(FORCE_MEASURE(0.),#49);
#51=ANALYSIS_METHOD_STATIC('Static Analysis Method: Elastic 1st
Order ',$,.ELASTIC_1ST_ORDER.);
#52=(GLOBAL_UNIT_ASSIGNED_CONTEXT((#56))MATERIAL_PROPERTY_CONTEXT()
MATERIAL_PROPERTY_CONTEXT_DIMENSIONAL(0.,9999999.)
REPRESENTATION_CONTEXT('dimensional_context','material'));
#53=(GLOBAL_UNIT_ASSIGNED_CONTEXT((#55))MATERIAL_PROPERTY_CONTEXT()
MATERIAL_PROPERTY_CONTEXT_DIMENSIONAL(0.,9999999.)
REPRESENTATION_CONTEXT('pressure_units_context','pressure_units'));
#54=(GLOBAL_UNIT_ASSIGNED_CONTEXT((#13))MATERIAL_PROPERTY_CONTEXT()
MATERIAL_PROPERTY_CONTEXT_DIMENSIONAL(0.,9999999.)
REPRESENTATION_CONTEXT('density_units_context','density_units'));
#55=(CONTEXT_DEPENDENT_UNIT('KIPS_PER_SQUARE_INCH')NAMED_UNIT(#90)
PRESSURE_UNIT());
#56=(CONTEXT_DEPENDENT_UNIT('INCH')LENGTH_UNIT()NAMED_UNIT(#87));
#57=(CONTEXT_DEPENDENT_UNIT('INCH')LENGTH_UNIT()NAMED_UNIT(#87));
#58=RELEASE_LOGICAL('ffffff',$,.F.,.F.,.F.,.F.,.F.,.F.);
#59=RELEASE_LOGICAL('fffffp',$,.F.,.F.,.F.,.F.,.F.,.T.);
#60=RELEASE_LOGICAL('ffffpf',$,.F.,.F.,.F.,.F.,.T.,.F.);
#61=RELEASE_LOGICAL('ffffpp',$,.F.,.F.,.F.,.F.,.T.,.T.);
#62=RELEASE_LOGICAL('fffpff',$,.F.,.F.,.F.,.T.,.F.,.F.);
#63=RELEASE_LOGICAL('fffpfp',$,.F.,.F.,.F.,.T.,.F.,.T.);
#64=RELEASE_LOGICAL('fffppf',$,.F.,.F.,.F.,.T.,.T.,.F.);
#65=RELEASE_LOGICAL('fffppp',$,.F.,.F.,.F.,.T.,.T.,.T.);
#66=(ANALYSIS_MODEL('frame',$,.SPACE_FRAME.,$,3)
ANALYSIS_MODEL_3D()ANALYSIS_MODEL_LOCATED(#67));
#67=COORD_SYSTEM_CARTESIAN_3D('global coordinate system',
'coordinate system for all',$,3,#68);
#68=AXIS2_PLACEMENT_3D('axis for analysis_model',#71,#69,#70);
#69=DIRECTION('unit x vector',(1.,0.,0.));
#70=DIRECTION('unit y vector',(0.,1.,0.));
#71=CARTESIAN_POINT('cp1',(0.,0.,0.));
#72=CARTESIAN_POINT('cp1',(0.,0.,144.));
#73=CARTESIAN_POINT('cp2',(240.,0.,144.));
#74=CARTESIAN_POINT('cp3',(0.,0.,144.));
#75=CARTESIAN_POINT('cp4',(0.,0.,0.));
#76=CARTESIAN_POINT('cp5',(240.,0.,144.));
#77=CARTESIAN_POINT('cp6',(240.,0.,0.));
#78=PLANE_ANGLE_MEASURE_WITH_UNIT(PLANE_ANGLE_MEASURE(0.),#85);
#79=PLANE_ANGLE_MEASURE_WITH_UNIT(PLANE_ANGLE_MEASURE(90.),#85);
#80=PLANE_ANGLE_MEASURE_WITH_UNIT(PLANE_ANGLE_MEASURE(
1.5707963267949),#85);
#81=PLANE_ANGLE_MEASURE_WITH_UNIT(PLANE_ANGLE_MEASURE(180.),#85);
#82=PLANE_ANGLE_MEASURE_WITH_UNIT(PLANE_ANGLE_MEASURE(
3.14159265358979),#85);
```

Section Ref.

Forces and Units

Assembly Geometry

```
#83=PLANE_ANGLE_MEASURE_WITH_UNIT(PLANE_ANGLE_MEASURE(270.),#85);
#84=PLANE_ANGLE_MEASURE_WITH_UNIT(PLANE_ANGLE_MEASURE(
2.0943951023932),#85);
#85=(CONTEXT_DEPENDENT_UNIT('DEGREE')NAMED_UNIT(#88)
PLANE_ANGLE_UNIT());
#86=LENGTH_UNIT(#87);
#87=DIMENSIONAL_EXPONENTS(1.,0.,0.,0.,0.,0.,0.);
#88=DIMENSIONAL_EXPONENTS(0.,0.,0.,0.,0.,0.,0.);
#89=DIMENSIONAL_EXPONENTS(1.,1.,-2.,0.,0.,0.,0.);
#90=DIMENSIONAL_EXPONENTS(-1.,1.,-2.,0.,0.,0.,0.);
#91=DIMENSIONAL_EXPONENTS(0.,1.,0.,0.,0.,0.,0.);
ENDSEC;
END-ISO-10303-21;
```

Assembly Geometry

Appendix B

IFC2x2 File

```
ISO-10303-21;
HEADER;
FILE_DESCRIPTION(('IFC 2x'),'2;1');    ◄──── File Conformance
FILE_NAME('C:\\Documents and Settings\\student\\Desktop\\paul''s
frame\\frame5.dwg','2004-11-12T15:15:38',(''),
('University of Kansas'),'IFC-Utility 2x for ADT V. 2, 0, 2, 16
(www.inopso.com)
- IFC Toolbox Version 2.x (00/11/07)','Autodesk Architectural
Desktop','');
FILE_SCHEMA(('IFC2X_FINAL'));
ENDSEC;
DATA;
#1=IFCSIUNIT(*,.TIMEUNIT.,$,.SECOND.);
#2=IFCSIUNIT(*,.MASSUNIT.,$,.GRAM.);
#3=IFCDIMENSIONALEXPONENTS(1,0,0,0,0,0,0);
#4=IFCSIUNIT(*,.LENGTHUNIT.,$,.METRE.);
#5=IFCMEASUREWITHUNIT(IFCRATIOMEASURE(0.0254),#4);
#6=IFCCONVERSIONBASEDUNIT(#3,.LENGTHUNIT.,'Inch',#5);
#7=IFCSIUNIT(*,.AREAUNIT.,$,.SQUARE_METRE.);
#8=IFCSIUNIT(*,.VOLUMEUNIT.,$,.CUBIC_METRE.);
#9=IFCUNITASSIGNMENT((#6,#7,#8,#1,#2));
#10=IFCCARTESIANPOINT((0.,0.,0.));
#11=IFCDIRECTION((0.,0.,1.));
#12=IFCDIRECTION((1.,0.,0.));          ─── Global Axis Definition
#13=IFCAXIS2PLACEMENT3D(#10,#11,#12);
#14=IFCGEOMETRICREPRESENTATIONCONTEXT('TestGeometricContext',
'TestGeometry',3,0.,#13,$);
#15=IFCPERSON('','','',$,$,$,$,$);
#16=IFCORGANIZATION('','University of Kansas','',$,$);
#17=IFCPERSONANDORGANIZATION(#15,#16,$);
#18=IFCAPPLICATION(#16,'IFC-Utility 2x for ADT V. 2, 0, 2, 16
(www.inopso.com)','Autodesk Architectural Desktop','');
#19=IFCOWNERHISTORY(#17,#18,$,.ADDED.,0,$,$,1100294138);
#20=IFCPROJECT('3KSvRQcWT9p9vDPtVRdlEm',#19,'frame5','','',$,$,
(#14),#9);
#32=IFCCARTESIANPOINT((-3.98,-4.865));    ◄──── (bf/2,d/2)
#33=IFCCARTESIANPOINT((3.98,-4.865));
#34=IFCCARTESIANPOINT((3.98,-4.430000000000001));  ◄──── (bf/2,d/2-tf)
#35=IFCCARTESIANPOINT((0.145,-4.430000000000001));
```

File Information

Units

Global Geometry and Info.

1st Column Geometry

```
#36=IFCCARTESIANPOINT((0.145,4.430000000000001));     ◄──── (tw/2,d/2-tf)
#37=IFCCARTESIANPOINT((3.98,4.430000000000001));
#38=IFCCARTESIANPOINT((3.98,4.865));
#39=IFCCARTESIANPOINT((-3.98,4.865));
#40=IFCCARTESIANPOINT((-3.98,4.430000000000001));
#41=IFCCARTESIANPOINT((-0.145,4.430000000000001));
#42=IFCCARTESIANPOINT((-0.145,-4.430000000000001));
#43=IFCCARTESIANPOINT((-3.98,-4.430000000000001));
#44=IFCCARTESIANPOINT((-3.98,-4.865));
#45=IFCPOLYLINE((#32,#33,#34,#35,#36,#37,#38,#39,#40,#41,#42,#43,
#44));
#46=IFCARBITRARYCLOSEDPROFILEDEF(.AREA.,$,#45);
#47=IFCCARTESIANPOINT((0.,0.,0.));                   ┐
#48=IFCDIRECTION((1.,0.,0.));                         │
#49=IFCDIRECTION((0.,1.,0.));                         ├─Local Axis Definition
#50=IFCAXIS2PLACEMENT3D(#47,#48,#49);                 │
#51=IFCDIRECTION((0.,0.,1.));                        ┘
#52=IFCEXTRUDEDAREASOLID(#46,#50,#51,144.);
#54=IFCSHAPEREPRESENTATION(#14,'Body','SweptSolid',(#52));
#31=IFCLOCALPLACEMENT(#25,#30);
#30=IFCAXIS2PLACEMENT3D(#27,#28,#29);
#27=IFCCARTESIANPOINT((0.,0.,0.));
#28=IFCDIRECTION((-1.,2.220446049250313E-016,0.));
#29=IFCDIRECTION((0.,0.,1.));
#57=IFCCARTESIANPOINT((0.,-3.98,-4.865));
#58=IFCBOUNDINGBOX(#57,144.,7.96,9.73);
#59=IFCSHAPEREPRESENTATION(#14,'','BoundingBox',(#58));
#55=IFCPRODUCTDEFINITIONSHAPE($,$,(#54,#59));
#56=IFCCOLUMN('3625XMWnz9qwE2_9bETn3Z',#19,'','','',#31,#55,$);
#60=IFCPROPERTYSINGLEVALUE('Layername',$,IFCLABEL('S-Cols'),$);
#61=IFCPROPERTYSINGLEVALUE('Red',$,IFCINTEGER(204),$);
#62=IFCPROPERTYSINGLEVALUE('Green',$,IFCINTEGER(204),$);
#63=IFCPROPERTYSINGLEVALUE('Blue',$,IFCINTEGER(0),$);
#64=IFCCOMPLEXPROPERTY('Color',$,'Color',(#61,#62,#63));
#65=IFCPROPERTYSET('3_9JlNKUzE9gwuMCx_JSfx',#19,'PSet_Draughting',
$,(#60,#64));
#66=IFCRELDEFINESBYPROPERTIES('31d_2JRMXCJx5tEd8PFRsi',#19,$,$,
(#56),#65);
#72=IFCCARTESIANPOINT((-3.98,-4.865));
#73=IFCCARTESIANPOINT((3.98,-4.865));
#74=IFCCARTESIANPOINT((3.98,-4.430000000000001));
#75=IFCCARTESIANPOINT((0.145,-4.430000000000001));
#76=IFCCARTESIANPOINT((0.145,4.430000000000001));
#77=IFCCARTESIANPOINT((3.98,4.430000000000001));
#78=IFCCARTESIANPOINT((3.98,4.865));
#79=IFCCARTESIANPOINT((-3.98,4.865));
#80=IFCCARTESIANPOINT((-3.98,4.430000000000001));
#81=IFCCARTESIANPOINT((-0.145,4.430000000000001));
#82=IFCCARTESIANPOINT((-0.145,-4.430000000000001));
#83=IFCCARTESIANPOINT((-3.98,-4.430000000000001));
#84=IFCCARTESIANPOINT((-3.98,-4.865));
#85=IFCPOLYLINE((#72,#73,#74,#75,#76,#77,#78,#79,#80,#81,#82,#83,
#84));
#86=IFCARBITRARYCLOSEDPROFILEDEF(.AREA.,$,#85);
#87=IFCCARTESIANPOINT((0.,0.,0.));
#88=IFCDIRECTION((1.,0.,0.));
#89=IFCDIRECTION((0.,1.,0.));
#90=IFCAXIS2PLACEMENT3D(#87,#88,#89);
#91=IFCDIRECTION((0.,0.,1.));
#92=IFCEXTRUDEDAREASOLID(#86,#90,#91,144.);
```

1st Column (Continued)

2nd Column Geometry

```
#94=IFCSHAPEREPRESENTATION(#14,'Body','SweptSolid',(#92));
#71=IFCLOCALPLACEMENT(#25,#70);
#70=IFCAXIS2PLACEMENT3D(#67,#68,#69);
#67=IFCCARTESIANPOINT((240.,0.,0.));
#68=IFCDIRECTION((-1.,2.220446049250313E-016,0.));
#69=IFCDIRECTION((0.,0.,1.));
#97=IFCCARTESIANPOINT((0.,-3.98,-4.865));
#98=IFCBOUNDINGBOX(#97,144.,7.96,9.73);
#99=IFCSHAPEREPRESENTATION(#14,'','BoundingBox',(#98));
#95=IFCPRODUCTDEFINITIONSHAPE($,$,(#94,#99));
#96=IFCCOLUMN('3PDpeaQxX2oux8Nk96okU7',#19,'','','',#71,#95,$);
#100=IFCPROPERTYSINGLEVALUE('Layername',$,IFCLABEL('S-Cols'),$);
#101=IFCPROPERTYSINGLEVALUE('Red',$,IFCINTEGER(204),$);
#102=IFCPROPERTYSINGLEVALUE('Green',$,IFCINTEGER(204),$);
#103=IFCPROPERTYSINGLEVALUE('Blue',$,IFCINTEGER(0),$);
#104=IFCCOMPLEXPROPERTY('Color',$,'Color',(#101,#102,#103));
#105=IFCPROPERTYSET('0kUOAwTwT6yxDhnOJtId11',#19,
'PSet_Draughting',$,(#100,#104));
#106=IFCRELDEFINESBYPROPERTIES('0Bg_1AaynC88cgG6cJip16',#19,$,$,
(#96),#105);
#26=IFCBUILDING('20H_yOw7P0PA$aA_vbhgZT',#19,'frame5','','',#25,
$,'',.ELEMENT.,$,$,$);
#112=IFCCARTESIANPOINT((-4.0025,-5.97));
#113=IFCCARTESIANPOINT((4.0025,-5.97));
#114=IFCCARTESIANPOINT((4.0025,-5.455));
#115=IFCCARTESIANPOINT((0.1475,-5.455));
#116=IFCCARTESIANPOINT((0.1475,5.455));
#117=IFCCARTESIANPOINT((4.0025,5.455));
#118=IFCCARTESIANPOINT((4.0025,5.97));
#119=IFCCARTESIANPOINT((-4.0025,5.97));
#120=IFCCARTESIANPOINT((-4.0025,5.455));
#121=IFCCARTESIANPOINT((-0.1475,5.455));
#122=IFCCARTESIANPOINT((-0.1475,-5.455));
#123=IFCCARTESIANPOINT((-4.0025,-5.455));
#124=IFCCARTESIANPOINT((-4.0025,-5.97));
#125=IFCPOLYLINE((#112,#113,#114,#115,#116,#117,#118,#119,#120,
#121,#122,#123,#124));
#126=IFCARBITRARYCLOSEDPROFILEDEF(.AREA.,$,#125);
#127=IFCCARTESIANPOINT((0.,0.,-5.97));
#128=IFCDIRECTION((1.,0.,0.));
#129=IFCDIRECTION((0.,1.,0.));
#130=IFCAXIS2PLACEMENT3D(#127,#128,#129);
#131=IFCDIRECTION((1.729958125484675E-033,0.,1.));
#132=IFCEXTRUDEDAREASOLID(#126,#130,#131,228.);
#134=IFCSHAPEREPRESENTATION(#14,'Body','SweptSolid',(#132));
#111=IFCLOCALPLACEMENT(#25,#110);
#110=IFCAXIS2PLACEMENT3D(#107,#108,#109);
#107=IFCCARTESIANPOINT((6.,-1.776356839400251E-015,144.));
#108=IFCDIRECTION((0.,0.,1.));
#109=IFCDIRECTION((1.,1.558207753859869E-017,0.));
#25=IFCLOCALPLACEMENT($,#24);
#24=IFCAXIS2PLACEMENT3D(#21,#22,#23);
#21=IFCCARTESIANPOINT((0.,0.,0.));
#22=IFCDIRECTION((0.,0.,1.));
#23=IFCDIRECTION((1.,0.,0.));
#137=IFCCARTESIANPOINT((0.,-4.0025,-11.94));
#138=IFCBOUNDINGBOX(#137,228.,8.005000000000001,11.94);
#139=IFCSHAPEREPRESENTATION(#14,'','BoundingBox',(#138));
#135=IFCPRODUCTDEFINITIONSHAPE($,$,(#134,#139));
#136=IFCBEAM('3YzDSDki9E4eSGD1VTj8Ny',#19,'','','',#111,#135,$);
```

2nd Column Geometry (Continued)

Beam Geometry

```
#140=IFCPROPERTYSINGLEVALUE('Layername',$,IFCLABEL('S-Beam'),$);
#141=IFCPROPERTYSINGLEVALUE('Red',$,IFCINTEGER(204),$);
#142=IFCPROPERTYSINGLEVALUE('Green',$,IFCINTEGER(0),$);
#143=IFCPROPERTYSINGLEVALUE('Blue',$,IFCINTEGER(0),$);
#144=IFCCOMPLEXPROPERTY('Color',$,'Color',(#141,#142,#143));
#145=IFCPROPERTYSET('0qQ8E9gjv8dh94v2MLCO$K',#19,
'PSet_Draughting',$,(#140,#144));
#146=IFCRELDEFINESBYPROPERTIES('3kVlLf6mn1Lv2_95wWeM4I',#19,$,$,
(#136),#145);
#147=IFCRELCONTAINEDINSPATIALSTRUCTURE('37ElBQvVLCqBmnelJ1Huak',
#19,$,$,(#56,#96,#136),#26);
#148=IFCPROPERTYSINGLEVALUE('Layername',$,IFCLABEL('IfcBuilding'),
$);
#149=IFCPROPERTYSINGLEVALUE('Red',$,IFCINTEGER(255),$);
#150=IFCPROPERTYSINGLEVALUE('Green',$,IFCINTEGER(255),$);
#151=IFCPROPERTYSINGLEVALUE('Blue',$,IFCINTEGER(255),$);
#152=IFCCOMPLEXPROPERTY('Color',$,'Color',(#149,#150,#151));
#153=IFCPROPERTYSET('2zzeiNzYbApe3MQP11rjhM',#19,
'PSet_Draughting',$,(#148,#152));
#154=IFCRELDEFINESBYPROPERTIES('3gW2BgDbv8_9n1S63hMhYY',#19,$,
$,(#26),#153);
#155=IFCRELAGGREGATES('1AGdlyyYTFyAT27fv$zm2I',#19,$,$,#20,
(#26));
#156=IFCPROPERTYSINGLEVALUE('Layername',$,
IFCLABEL('IfcProject'),$);
#157=IFCPROPERTYSINGLEVALUE('Red',$,IFCINTEGER(255),$);
#158=IFCPROPERTYSINGLEVALUE('Green',$,IFCINTEGER(255),$);
#159=IFCPROPERTYSINGLEVALUE('Blue',$,IFCINTEGER(255),$);
#160=IFCCOMPLEXPROPERTY('Color',$,'Color',(#157,#158,#159));
#161=IFCPROPERTYSET('0_gA3JK8f0uhlRqvGVIgRl',#19,
'PSet_Draughting',$,(#156,#160));
#162=IFCRELDEFINESBYPROPERTIES('0dTWBHEPDDOulLk0o$MseX',#19,$,
$,(#20),#161);
ENDSEC;
END-ISO-10303-21;
```

File Property Definition

A Conceptual Model of Web Service-Based Construction Information System

Seungjun Roh[1], Moonseo Park[2], Hyunsoo Lee[3], and Eunbae Kim[4]

[1] M.Sc. student, Dept. of Architecture, Seoul National Univ., Seoul, 151-742, Korea
[2] Assistant professor, Dept. of Architecture, Seoul National Univ., Seoul, 151-742, Korea
[3] Associate professor, Dept. of Architecture, Seoul National Univ., Seoul, 151-742, Korea
[4] Ph.D. student, Dept. of Architecture, Seoul National Univ., Seoul, 151-742, Korea
rohsj97@snu.ac.kr, mspark@snu.ac.kr, hyunslee@snu.ac.kr,
unbai92@snu.ac.kr

Abstract. Business patterns of construction require diverse collaborations among participants and systems. To improve communication and the use of information, construction companies have invested on building their own information system. However, the past construction information system has focused on a point-to-point system which has limitation in data exchange with distributed systems. As a result, construction information systems lack of business agility and bring low Return on Investment (ROI). To address this challenging issue, Web Services is used as an independent platform to integrate different system components. With Web Services standards, a conceptual model of Web Service-based construction information system is suggested to improve transfer and search of construction information, which has the purpose of providing a pertinent collaborative interface for communication with partners. This paper proposes a new possibility of adapting Web Services in the construction information systems.

1 Introduction

The Architecture, Engineering, and Construction (AEC) industry is fragmented geographically and functionally [1], [2]. Furthermore, for the characteristics of construction industries; uniqueness of projects, difficulty of data collection on site and varying processes [3], a construction information system which has the potential to facilitate communication among participants is considered as one of the key factors for the success of construction projects. Managing knowledge is important to construction industry, because of unique characteristic of its projects; the temporary teams and heavy reliance on experiences [4]. For this potential benefit, construction enterprises have been forced to integrate their information systems with an aim to increase making great use of construction information and knowledge. For example, Enterprise Resource Planning (ERP) and Project Management Information System (PMIS) solutions have been attempted to provide various applications with an integrated platform. Such information systems are developed to manage system integration and share different kinds of functions.

However, most applications rarely consider interoperability with other applications. Construction companies have their own legacy systems and point-to-point

I.F.C. Smith (Ed.): EG-ICE 2006, LNAI 4200, pp. 597–605, 2006.
© Springer-Verlag Berlin Heidelberg 2006

integrations are increasing IT investment volume more and more [5]. Accordingly, compared to other industries, the construction information system brings lower Return on Investment (ROI) and high total cost of ownership. Moreover, users are not familiar with information systems as a result of the complexity, inefficiency, and inconvenience of the systems. Particularly, financially less capable construction companies do not have their own information systems, although they have the same needs for IT systems.

To address these problems of the current construction information system, a Web Services technology is suggested in this paper. To deal with diverse project management functions and facilitate to use construction information, three main functional features of the proposed Web Service-based construction information system are discussed. A conceptual model of construction information system is introduced to enhance interoperability and internetworking for AEC industry participants. Using Web Services standards, the proposed model is designed to intelligently search for construction information and timely transfer the acquired information according to workflow and users. The research result would provide a platform for Web Service-based construction information system by exploring a new direction for integration, transfer, and search of construction information.

2 IT Paradox

Construction enterprises also firmly believe that the invested construction information system gives them lots of benefits [6]. However, most of enterprises are unsatisfied with the performance of their systems. Their dissatisfaction is attributed both to ineffective operation and system-itself problems. Ineffective operation is mostly caused by the lack of employee education, slow adaptation speed by employee's psychological resistance to new work systems. The other problem is due to the limitation of IT technology and difficulties in describing the construction business process, which is focused in my research.

As an effort to meet the need for improving productivity, IT investment is done mainly for hardware and software upgrade, collaboration information system, on-line Decision-making Support System (DSS) and integration effort. However, each investment has rarely reached the invest goal. While investment on hardware and software upgrading increases data processing capacity and productivity, such upgrading also increases communication incompatibility with other work divisions or companies. In consequence, incompatibility reduces productivity by requiring additional effort to convert data and integrate heterogonous systems. As business size grows, collaboration information systems are required as they can improve productivity with data transferring automation. However, the problems inherited from data transferring and integration cause the low usage of the collaboration system. Moreover, the current on-line DSS is not widely used due to insufficient knowledge, information and links. Long search time is another reason for the low level of usage. As a result, IT investment in construction companies has not been associated with the expected productivity improvement. The main reason of IT paradox is unreasonable system integration with no consideration of information usefulness. The one way to overcome the IT paradox is building an information system with more interoperable components.

3 Web Services

Web Services approach can be an alternative to deal with the IT paradox. Web Services provides a much more efficient way to manage current various construction information systems and supports more flexible collaboration, both among a company's own units and partners. Therefore, construction information systems that are linked with Web Services standards would be enabling construction participants to use a higher level of integrated information.

3.1 Definitions

Abrams [7] defines Web Services as software components that perform distributed computing using standard internet protocols. W3C [8] defines a software system designed to support interoperable machine-to-machine interaction over a network. To summarize, Web Services is defined by a collection of web and object-oriented technologies for linking Web-based applications running on different hardware, software, database, or network platforms.

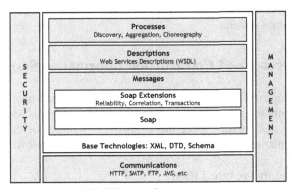

• WSDL : Language to describe Web service interface
• UDDI : Public (or private) repository to publish and locate Web services
• SOAP : Protocol to call a Web service

Fig. 1. Web Services Architecture Stack (W3C, 2004)

A Web Services model uses Web Service Definition Language (WSDL), Universal Description Discovery Integration (UDDI), and Simple Object Access Protocol (SOAP)/eXtensible Markup Language (XML) that allow information to be exchanged easily among different applications [9]. These tools provide the common languages for Web services, enabling applications to connect freely to other applications and to read electronic messages from them. As shown in Fig. 1, Web Services architecture involves many layered and interrelated technologies. A WSDL description is retrieved from the UDDI directory. WSDL descriptions allow the software to extend to use those of the other directly. The services are invoked over the World Wide Web using the SOAP/XML protocol. Where two companies know about each other's web services they can link their SOAP/XML protocol interfaces.

3.2 Benefits

Web Services can be used to enable existing systems to inter-operate without replacement and evolve along other applications [10]. The potential benefits include [11]:

- **Interoperability in a heterogeneous environment:** the key benefit of the Web Service model is that it permits different systems to operate on a variety of software platforms. Construction enterprises add systems that require different platforms and frequently don't communicate with each other. Later, due to a consolidation or the addition of another application, it becomes necessary to tie together.
- **Integration with existing systems:** Basing on interoperability with other systems, Web Services let enterprise application developers reuse and commoditize an enormous amount of data stored in existing information systems.
- **Support more client types:** Since a main objective of Web Services is improving interoperability, existing applications as Web Services increases their reach to different client types.

The development of information system has been complicated by the requirement that a particular application support a specific type of project or business process [12]. Web services, because they promote interoperability, is a solution of developed construction information system.

4 Web Services-Based Construction Information System

Construction companies are demanding an information system that can support the integration and communication with participants. Web Services is adopted in terms of its capability. In this context, three main functional features of the proposed Web Service-based construction information system are mentioned below.

4.1 Interoperability for Integration

Web Services standardize data exchanges by replacing the communication protocols with the UDDI, WSDL and SOAP. Benefits for construction enterprises adopting Web Services include enhanced capabilities for system integration among key project partners, and economic benefits due to reduced middle-ware and infrastructure costs. Interoperability of construction information can have the following benefits:

- **Interacting with partners:** The current information system requires a significant up-front investment by all parties. However, it's hard to expect many partners to invest on implementing communication infrastructure. Web Services-based interaction can be a good solution for small business partners that have no investment, and more cost effective for larger enterprises having the existing information systems.
- **Integration with existing enterprise systems:** So far, construction information system have been developed and fine-tuned to company's unique needs. Since Web Services are based on universally accepted, platform-independent standards, it makes a natural integration layer.

- **Aggregation of partner data:** To provide users with construction information, construction enterprise's system should gather information from its partners. Web Services enables a truly dynamic way to build and maintain a catalog of information.

4.2 Advanced Search for Construction Information

Construction information are scattered, unformailzed, and temporized. Thus, it's difficult to search information without efforts to adjust its formats. Advanced search for construction information through Web Services facilitates its usefulness. Web Service-based search makes it easy to create a unified search index across multiple data sources, programming platforms and software applications. Because it allows local customization of search results and search options no matter where content is stored, it is valuable for departmental or user-specific adjustments of search results.

Additionally, advanced search supports to facilitate knowledge management system. Construction knowledge which is consisted with explicit and tacit knowledge can be easily obtained and transferred from a current project management system. Facilitating to share and search knowledge is more useful and effective for managing construction information. The search result is finally stored at knowledge management systems for reusing and categorizing. As information and knowledge accumulate and categorize at the knowledge system, advanced search activates to share knowledge by providing the opened communities and dashboards among participants.

4.3 Real-Time Information Transfer

The value of reliable information is increased by instantly detecting the main events and removing delay time of information. Real-time information transfer is fundamental element considering that well-timed construction information is the key to the success of construction projects. Connecting with project process and knowledge map, real-time information transfer system automatically brings fitted references and history data in an integrated system where events are responded to as they happen. The main construction information which is essential in decision making and project carrying out is sent to managers before the planned work is performed through any types of web devices. Therefore, Web Services provide high levels of automation to the solution of information delivery process.

5 System Architecture of Web Service-Based Construction Information System

The structure of Web Service-based construction information system is mapping as Fig. 2. It has the three main components: integration system, real-time information transfer and construction information search system.

The enterprise integration system supports to integrate existing information systems using Web Services standards. Diverse management systems that are developed to support specific work processes are linked up to the internet using the SOAP/XML protocol. Any types of data and information which have stored at each system are available in the requirement of data integration and exchange.

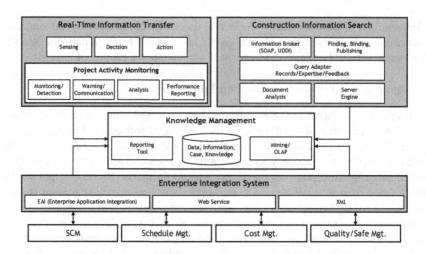

Fig. 2. The structure of Web Service-based Construction Information System

The basic function of real-time information transfer is managing the project process with work references and knowledge that are automatically sent to users. The status of the project process is mapped as a stream of construction information in order to monitor and analyze project site activities. Real-time construction information is transferred to diverse information and enforces to stack on database and knowledge management storage. Knowledge management system is naturally growing to store and retrieve from transferred information. Specific work supporting systems synchronize with knowledge management system to serve real-time exchanging data. Construction information search allows the information broker which is maintaining a UDDI repository to gather and structure construction information. Categorized construction information is also linked with knowledge management system in common.

Fig.3 illustrates the flow of information in Web Services adopting real-time information transfer and construction information search system. On the web browser, users can access to Web Service-based construction information system through Single Sign-On (SSO) and Extranet Access Management (EAM). ① Real-time information transfer system detects the work processes from project management systems. ② To instantly serve information to user before performance, this system request for existing systems and knowledge management system to transfer data, manual and know-how. ③ Gathered information is analyzed and matched with project processes. ④ Automatically, work information and process monitoring data, e.g. schedule, specifications, site performance and web cam, is transferred on the user interface. ⑤ After a work is finished, the results and data are retrieved to the knowledge management system to reuse. For example of submitting material purchasing requisitions, past application allows managers to fill in a purchase items using a browser, but it does not automate the ordering process by interoperability. Using the Web Services, the rules for purchase ordering process are described in a WSDL document. A partner of construction enterprise would request the WSDL document over the internet as an

XML/SOAP. The partner would use this WSDL document to create a bid and RFQ application. This application would be linked with real-time information transfer system to automate the ordering process. Real-time information transfer system gives a manager the purchasing progress to check out and the information of purchasing items.

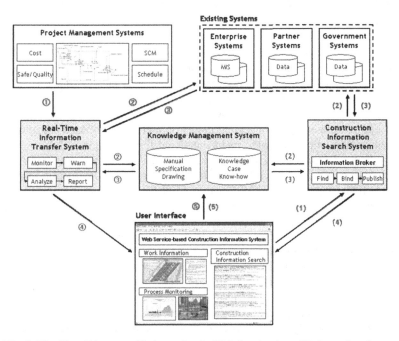

Fig. 3. The Flow Diagram of Information in Web Service-based Information System

The other system, construction information search, gives user more accurate information and knowledge at once. Using one of system, it's difficult to directly search construction information. Users wander usually related systems without any rules. Contrary to a common research, the information broker will support the reliable search of information that is held in heterogeneous forms. (1) When users input a key word, (2) information broker requests related information to existing systems and knowledge management system. (3) It allows each system to response via common data format and protocols, as the SOAP. (4) Before output is sent, the information broker categorizes and melds it on demand. (5) Served information is stored at knowledge management system to increase information searching time. To continue using above example, a project manager additionally wants quantity surveyor to request original quotation to five databases stored in separate partners and access to additional items. Using Web Services, a single interface can run a variety of disparate systems, and the information broker instantly pulls information from many databases and the knowledge management system. With categorized results, quantity surveyor can compare and evaluate items cost information to support a project manager's decision-making.

In the Web Service-based construction information system, the process is simplified through automatic notification of relevant participants and partners, and the performance of work process is simultaneously monitoring and evaluating. Organic integration of knowledge stored in legacy system and project work process is realized to improve project management.

6 Conclusions

IT investment in construction industry is focused to integrate and upgrade systems. Nevertheless, current construction information systems have difficulty to deal with the complexity and distribution of information. The result of investment has not been associated with the expected productivity improvement. As a solution of these problems, Web Services is suggested to provide a standard hub for interoperability.

This research presented a conceptual model of Web Service-based construction information system which has three main functional features, interoperability for integration, advanced construction information system and real-time information transfer. Proposed system has components to intelligently search for construction information and timely transfer the acquired information. Real-time information transfer and construction information search system share a knowledge management system so that it enables companies seeking developed information usage.

The Web Service-based construction information system provides a new platform of interoperability with partners and system. Moreover, transfer and search of construction information would enable to facilitate the development of knowledge management system. Due to above possibilities of improvement, Construction information system can still benefit from moving to a web services model for certain functions. Therefore, the suggested model would facilitate efficient interoperable environments to integrate diverse projects and systems that promote interaction of many stakeholders with different cognitive experiences and knowledge. For the future study, a conceptual model needs to be developed for the adaptation in the case project process and to meet the specific requirements of construction enterprises and partners.

Acknowledgement

The authors would like to acknowledge the support for this research from the Korean Ministry of Construction and Transportation, Research Project 05 CIT D05-01.

References

[1] Howard, H.C., Levitt, R.E., Paulson, B.C., Tatum, C.B.: Computer-Integration: Reducing Fragmentation in AEC Industry, Journal of Computing in Civil Engineering, Vol. 3, No. 1, January (1989) 18-32
[2] Nitithamyong, P., Skibniewski, M. J.: Web-based construction project management systems: how to make them successful?, Automation in Construction, Vol. 13, No. 4, July (2004)

[3] Koskela, L.: Application of the new production philosophy to construction, CIFE Technical Reprot #72, Stanford University (1992)

[4] Khalfan, M. M. A., Bouchlaghem, N. M., Anumba, C. J., Carrillo, P. M.: Knowledge Management for Sustainable Construction: The C-CanD Project, Journal of Management in Engineering, Vol. 22, No. 1, January (2006), pp. 2-10

[5] Zhu, Y.: Web-based construction document processing through a malleable Frame, PhD thesis, University of Florida (1999)

[6] Deng, Z. M., Li, H., Tam, C. M., Shen, Q. P., Love, P. E. D.: An application of the Internet-based project management system, Automation in Construction, Vol. 10, No. 2 (2004), pp. 239-246

[7] Abrams, C.: Web Services Scenario: Setting and Resetting Expectations, Gartner Symposium ITXPO, Cannes, France, 4-7 November (2002)

[8] W3C: Web Services Architecture, W3C Working Group Note, 11 February (2004)

[9] Kreger, H.: Web Services Conceptual Architecture, IBM Software Group (2001)

[10] Hagel, J., Brown, J.S.: Your Next IT Strategy, Harvard Business Review, Vol. 79, No. 9 (2001) 105-115

[11] Singh, I., Strearns, B., Brydon, S., Murray, G., Ramachandran, V.: Designing Web Services with the J2EE 1.4 Platform (The Java Series): JAX-RPC, SOAP, and XML Technologies, Addison-Wesley (2004)

[12] Wilkes, L.: Web Services-Right Here, Right Now, CBDI Forum, 16 December (2001)

Combining Two Data Mining Methods for System Identification

Sandro Saitta[1], Benny Raphael[2], and Ian F.C. Smith[1]

[1] Ecole Polytechnique Fédérale de Lausanne, IMAC, Station 18, CH-1015 Lausanne,
Switzerland
{sandro.saitta, ian.smith}@epfl.ch
[2] National University of Singapore, Department of Building, Singapore, 117566
bdgbr@nus.edu.sg

Abstract. System identification is an abductive task which is affected by several kinds of modeling assumptions and measurement errors. Therefore, instead of optimizing values of parameters within one behavior model, system identification is supported by multi-model reasoning strategies. The objective of this work is to develop a data mining algorithm that combines principal component analysis and k-means to obtain better understandings of spaces of candidate models. One goal is to improve views of model-space topologies. The presence of clusters of models having the same characteristics, thereby defining model classes, is an example of useful topological information. Distance metrics add knowledge related to cluster dissimilarity. Engineers are thus better able to improve decision making for system identification and downstream tasks such as further measurement, preventative maintenance and structural replacement.

1 Introduction

The goal of system identification [6] is to determine the properties of a system including values of system parameters through comparisons of predicted behavior with measurements. By definition, system identification is an abductive task and therefore, computational complexity may hinder the success of full-scale engineering applications. Abduction can be supported by multi-model reasoning since many causes (model results) usually lead to the same consequences (sensor data). In addition, several kinds of assumptions and measurement errors influence the reliability of system identification tasks. The effect of incorrect modeling assumptions may be compensated by measurement error and this may lead to inaccurate system identification when single-model reasoning is carried out. Therefore, instead of optimizing one model, a set of candidate models is identified in our approach. These candidate models lie below a threshold which is defined by an estimate of the upper bound of errors due to modeling assumptions as well as measurements. When data mining techniques [4] [17] [18] are applied to model parameters, engineers may obtain useful knowledge for system identification tasks. In general, the objective is to inform engineers of the accuracy of their diagnoses given the information that is available. For example, if it

I.F.C. Smith (Ed.): EG-ICE 2006, LNAI 4200, pp. 606–614, 2006.

is not possible to obtain a unique identification, classes of possible solutions are proposed.

Using data mining in engineering is not new [1] [7]. Examples of application include oil production prediction [8], traffic pattern recognition [20], composite joint behavior [16] and joint damage assessment [21]. However, all of these contributions use data mining as a predictive tool. In previous work at EPFL [15], data mining techniques have been applied to models to extract knowledge. Although the information obtained has been useful, it is limited to confirming relationships between parameters and their importance. For example, correlation and principal component analysis help indicate independent parameters. Decision tree analysis is capable of finding parameters that are able to separate models. However, no information related to the space of possible models is provided. Furthermore, no guidelines are provided to use the results and the limitations of such a methodology have not been clearly evaluated. For example, it may be useful for the engineer to know how many clusters there are in the space of possible models. For these assessments, clustering techniques such as k-means are useful. Even though clustering is often proposed for various applications by the data mining community, it is not straightforward; there are many open research issues. K-means clustering has been successfully applied in domains such as speech recognition [11], relational databases [9] and gene expression data [2].

Hybrid data mining methods are proposed in the literature, for example, [10] and [19]. Most work combines data mining methods for better prediction. For example, [3] propose a combination of PCA and k-means to improve the prediction accuracy. However, the visualization improvement is not taken into account. Work that aims to generate better descriptions of spaces of models is rare. This work combines principal component analysis (PCA) [5] with k-means clustering in order to obtain better understandings of the space of possible models. An important objective is to obtain a view of model-space topologies. No work known has been found that combines PCA and k-means to increase understanding of model spaces.

While there are well accepted methods for evaluating predictive models such as cross-validation [18], clustering of possible models has not been investigated and quantitative methods are not available for evaluating this task. The criterion for assessing the capability of algorithms is subjective and dependent on the final goal of the knowledge discovery task. In this paper, three issues are addressed. Firstly, an evaluation of the quality of a set of clusters is performed in terms of the task of system identification. Secondly, the choice of the number of clusters is discussed. Finally, issues of limitations of knowledge discovery are presented.

The methodology is illustrated through a case study of a two-span beam. The structure is two meters long and its middle support is a spring. Using the methodology described in [15], several models are generated. There are seven parameters. They consist of three loads (position on the beam and magnitude) as well as the stiffness of the central spring. According to the error threshold discussed in [15], 300 models are identified. These models are used in this paper and their parameter values are considered as input points for two data mining techniques.

The paper is structured as follows. In Section 2, a clustering procedure using PCA and k-means is presented. Section 3 describes the evaluation of the methodology and the interpretation of the results obtained from an engineering perspective. Section 4 contains the results of an application of the methodology as well as a discussion of limitations. The final section contains conclusions and a description of future work.

2   Combining Data Mining Techniques

2.1   Principal Component Analysis (PCA)

PCA is a linear method for dimensionality reduction [5]. The goal of PCA is to generate a new set of variables - called Principal Components (PC) - that are linear combinations of original variables. Ultimately, PCA finds a set of principal components that are sorted so that the first components explain most of the variability of the data. A property of principal components is that each component is mutually uncorrelated.

PCA begins with determining S, the covariance matrix of the normalized data (for more details see [5]). The term *parameter* refers to the parameter in its normalized form. To obtain the principal components, the covariance matrix S is decomposed such that $S = VLV^T$ where L is a diagonal matrix containing the eigenvalues of S and the columns of V contains the eigenvectors of S. Finally, the transformation is carried out as follows:

$$x^{new} = \sum_{i=1}^{PC} \alpha_i x \qquad (1)$$

where x is a point in the normal space, α_i is the ith eigenvector, PC is the number of Principal Components and x^{new} is the point in the feature space. A point used by the data mining algorithm represents set of parameter values (models) in the system identification perspective. Principal components (PCs), which are linear combination of the original variables, are represented as an orthogonal basis for a new representational space of the data. PCs are sorted in decreasing order of their ability to represent the variability of the data. Finally, each sample is transformed into a point in the feature space using Equation 1.

2.2   K-Means Using Principal Components

K-means is a widely applied clustering algorithm. Although it is simple to understand and implement, it is effective only if applied and interpreted correctly. The k-means algorithm divides the data into K clusters according to a given distance measure. Although the Euclidean distance is usually chosen, other metrics may be more appropriate. The algorithm works as follows. First, K starting points (named centroids) are chosen randomly. The number of clusters, K, is chosen a-priori by the user. Then each point is assigned to its closest centroid. Collections of points form the initial clusters. The centroid of each cluster is then updated using the positions of the points assigned to it. This process is repeated until there is no change of centroid or until point assignments remain the same.

The complete methodology - combining PCA and k-means - is described next. First, the PCA procedure outlined in Section 2.1 is applied to the models. Using the principal components the complete set of model predictions is mapped into the new feature space according to Equation 1. Then, the k-means algorithm is applied to the data in the feature space. The final objective is to see if it is possible to separate models into clusters. Table 1 present the pseudo-code of the methodology used.

Table 1. Pseudo-code algorithm to separate models into classes

Clustering procedure
1. Transform the data using principal component analysis.
2. Choose the number K of clusters (see Section 3).
3. Randomly select K initial centroids.
4. **Do**
5.   Assign each point to the closest cluster.
6.   Recompute centroid of the K clusters.
7. **While** centroid position change

In addition to the limitations mentioned in [17], this methodology has two drawbacks. Firstly, the number of clusters has to be specified by the user a-priori. Strategies for estimating the number of clusters have been proposed in [17] and [18]. One of these method is chosen here and adapted to the system identification context in Section 3. Secondly, as stated above, the K initial centroids are chosen randomly. Therefore, running P times will result in P different clustering of the same data. A common technique for avoiding such a problem is described in Section 3.

3   Evaluation and Significance of the Methodology

The number of clusters of models is useful information for engineers performing system identification. When the methodology defined in [14] outputs M possible models, it does not mean that there are M different models of the structure. These M models might only differ slightly in a few values of parameters while representing the same model. In other situations, models might have important differences representing distinct classes which are referred to as clusters.

When predictive performances are evaluated, the classification error rate is usually used. If the aim is to make predictions on unseen data sets, the most common way to judge the results is through cross-validation [17]. In this work, the evaluation process is different since the goal is not prediction. Results are evaluated in two ways. Firstly, a criterion is used to evaluate the performance of the clustering procedure. Secondly, from a decision-support point-of-view, the performance is evaluated by users.

The main theme of this section is to develop a metric in order to evaluate results obtained with the methodology described in Section 2. Without a metric, the way clusters are seen and evaluated is subjective. Furthermore, it is not possible to know the real number of cluster in the data since the task is unsupervised learning and this means that the answer - the number of clusters - is unknown. In this paper, the results obtained by the clustering technique are evaluated using a score function (SF). The score function combines two aspects: the compactness of clusters and the distance between clusters. The first notion is referred to as within class distance (WCD) whereas the second is the between class distance (BCD). Since objectives are to minimize the first aspect and to maximize the second aspect, combining the two is possible through maximizing $SF = BCD/WCD$. This idea is related to the Fisher criterion [4].

In this research the WCD and the BCD are defined in Equation 2. It is important that an engineering meaning in terms of model-based diagnosis can be given to these two distances. They are both directly related to the space of models for the task of system identification using multiple models. The WCD represents the spread of model predictions within one cluster. Since it gives information on the size of the cluster, a high WCD means that models inside the class are widely spread and that the cluster may not reflect physical similarity. The BCD is an estimate of the mean distance between the centers of all clusters and therefore, it provides information related to the spread of clusters. For example, a high BCD value means that classes are far from each other and that the system identification is not currently reliable. The detailed score function is given by Equation 2.

$$SF = \frac{\sum_{i=1}^{K} dist(c_i, c_{tot})^2 \cdot size(c_i)}{\sum_{i=1}^{K} \frac{\sum_{x \in C_i} dist(c_i, x)^2}{size(c_i)}} \tag{2}$$

where K is the number of clusters, C_i the cluster i, c_i its centroid and c_{tot} the centroid of all the points. The function $dist$ and $size$ define respectively the Euclidean distance between two points (each point is a model which is represented by parameter values) and the number of points in a cluster. From a system identification point of view, BCD values indicate how different the K situations are. Values of WCD give overviews of sizes of groups of models. As explained in Section 2, the number of clusters - in our application, this is the number of classes of models - for a data set is unknown. The idea to determine the most reliable number of clusters is to run the procedure for P different predefined number of clusters. The criterion used to see if the number of cluster is appropriate is the same as the score function described above. The higher the value of the criterion, the more suitable the number of clusters.

The second weakness of the procedure is the random choice of the K first centroids. One solution is to run the algorithm N times and to average the value

Table 2. Procedure to limit the effect of the random choice of the starting centroids

Controlling Randomness
1. **Loop** i from 1 to N
2.   **Loop** j from 2 to P
3.   Run clustering procedure described in Table 1 with j centroids.
4.   Calculate score function.
5.   **End**
6. **End**
7. Average score function.

of the score function. Therefore, randomness is controlled by N. The pseudo-code of the mentioned procedure is given in Table 2.

To conclude this section, the score function defined above serves two purposes. First, it gives an idea of the performance of the clustering procedure. Second, it allows choice of a realistic value for the number of clusters. Although values could have physical significance, this number must be interpreted with care as explained in Section 4. Reducing the random effect of the procedure is done through several run of the algorithm to compute the score function value. Finally, the number of clusters could be fixed by the expert and therefore this may be considered to be domain knowledge.

4   Results and Limitations

The case study described in this section is a beam structure that was presented in [15]. It is used to illustrate the methodology described in Section 2.2. Although this study focuses on bridge structures, it can be applied to other structures and in other domains. The procedure for generating models from modeling assumptions is given in [15]. In this particular example, six parameters consisting of position and magnitude of three loads have been chosen. After running the procedure described in Section 3, the number of clusters is chosen to be three in this case. The results are shown in Table 3.

Table 3. Comparison of values for between class distance (BCD), within class distance (WCD) and score function (SF) for various numbers of clusters

Clusters	BCD	WCD	SF
2	78.43	1.65	47.67
3	127.70	1.96	**65.22**
4	147.59	2.34	63.09
5	166.35	2.66	62.68
6	181.26	2.89	62.80
7	192.11	3.12	61.60
8	202.83	3.31	61.39

It can be seen that the maximum value for the score function is reached with three clusters and a local maximum can be observed with six clusters. This effect can also be seen on the right part of Figure 1 where the three clusters could be divided into two to obtain six clusters. Furthermore, BCD and WCD are always increasing. This is due to the fact that at maximum there could be one cluster for each point.

Once the number of clusters is fixed, the procedure outlined in Table 1 is followed. To judge the improvement of the methodology with respect to the standard k-means algorithm, the two techniques are compared. Figure 1 shows the improvement in a visualization point of view. The left part of Figure 1 corresponds to standard k-means whereas the right part is the result of the methodology described in this paper. It is easy to see that our methodology is better able to present results visually.

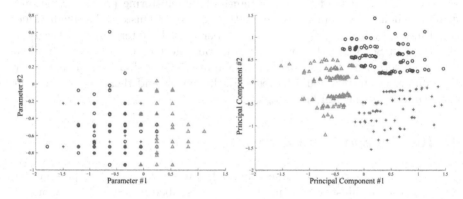

Fig. 1. Visual comparison of standard k-means (left) with respect to the proposed methodology (right). Every point represents a model and belongs to one of the three possible clusters (+, O or △).

This methodology has a number of limitations. Firstly, results of data mining have to be interpreted carefully. The user thus has an important role in ensuring that the methodology is successful. Secondly, even if the methodology is well applied, results are not necessary the most appropriate. For example, data might be noisy (poor sensor precision), or may have missing values (low sensor quality) or may be missing useful information (bad sensor configuration) and this may preclude obtaining useful results.

An example of challenges associated with applying data mining to system identification is given below. Assume that, after applying data-mining methodology, three clusters of models are obtained. The methodology alone is not able to *interpret* these clusters. Suppose that two clusters group similar information. Although the clustering algorithm has generated three clusters, only the user is able to identify that there are only two clusters that have physical meaning. Therefore, data mining is only able to suggest possible additional knowledge.

The process of acquiring knowledge that is of practical use for decision-support is left for the engineer.

5 Conclusions

A methodology that combines PCA and k-means for studying the solution space of models obtained during system identification is presented in this paper. The conclusions are as follows:

- Combining the data mining techniques, PCA and k-means, helps improve visualization of data
- Evaluation of results obtained through clustering is difficult The metric that has been developed in this work helps in the evaluation
- Application of data mining to complex tasks such as system identification requires an expert user

Future work involves the use of more complex data mining methods to obtain other ways for separating and clustering models. Furthermore, better visualization of solution spaces needs to be addressed in order to improve engineer/computer interaction. Finally, strategies for models containing a varying number of parameters are under development.

Acknowledgments

This research is funded by the Swiss National Science Foundation through grant no 200020-109257. The authors recognize Dr. Fleuret for several fruitful discussions on data mining techniques. The two anonymous reviewers are acknowledged for their propositions which have modified the direction of the paper.

References

1. Alonso C., Rodriguez J.J. and Pulido B. *Enhancing Consistency based Diagnosis with Machine Learning Techniques*. LNCS, Vol. 3040, 2004, pp. 312-321.
2. Chan Z.S.H., Collins L. and Kasabov N. *An efficient greedy k-means algorithm for global gene trajectory clustering*. Exp. Sys. with Appl., 30 (1), 2006, pp. 137-141.
3. Ding C. and He X. *K-means clustering via principal component analysis*. Proceedings of the 21st International Conference on Machine Learning, 2004.
4. Hand D., Mannila H. and Smyth P. *Principles of Data Mining*. MIT Press, 2001, 546p.
5. Jolliffe I.T. *Principal Component Analysis*. Statistics Series, Springer-Verlag, 1986, 271p.
6. Ljung L. *System Identification - Theory For the User*. Prentice Hall, 1999, 609p.
7. Melhem H.G. and Cheng Y. *Prediction of Remaining Service Life of Bridge Decks Using Machine Learning*. J. Comp. in Civ. Eng., 17 (1), 2003, pp. 1-9.
8. Nguyen H.H. and Chan C.W. *Applications of data analysis techniques for oil production prediction*. Art. Int. in Eng., Vol 13, 1999, pp. 257-272.

9. Ordonez C. *Integrating k-means clustering with a relational DBMS using SQL.* IEEE Trans. on Know. and Data Eng., 18 (2), 2006, pp. 188-201.
10. Pan X., Ye X. and Zhang S. *A hybrid method for robust car plate character recognition.* Eng. Appl. of Art. Int., 18 (8), 2005, pp. 963-972.
11. Picone J. *Duration in context clustering for speech recognition.* Speech Com., 9 (2), 1990, pp. 119-128.
12. Raphael B. and Smith I.F.C. *Fundamentals of Computer-Aided Engineering.* John Wiley, 2003, 306p.
13. Reich Y. and Barai S.V. *Evaluating machine learning models for engineering problems.* Art. Int. in Eng., Vol 13, 1999, pp. 257-272.
14. Robert-Nicoud Y., Raphael B. and Smith I.F.C. *Improving the reliability of system identification.* Next Gen. Int. Sys. in Eng., No 199, VDI Verlag, 2004, pp. 100-109.
15. Saitta S., Raphael B. and Smith I.F.C. *Data mining techniques for improving the reliability of system identification.* Adv. Eng. Inf., 19 (4), 2005, pp. 289-298.
16. Shirazi Kia S., Noroozi S., Carse B. and Vinney J. *Application of Data Mining Techniques in Predicting the Behaviour of Composite Joints.* Eighth AICC, 2005, Paper 18, CD-ROM.
17. Tan P.-N., Steinbach M. and Kumar V. *Introduction to Data Mining.* Addison Wesley, 2006, 769p.
18. Webb A. *Statistical Pattern Recognition.* Wiley, 2002, 496p.
19. Xu L.J., Yan Y., Cornwell S. and Riley G. *Online fuel tracking by combining principal component analysis and neural network techniques.* IEEE Trans. on Inst. and Meas., 54 (4), 2005, pp. 1640-1645.
20. Yan L., Fraser M., Oliver K., Elgamal A., Conte J.P. and Fountain T. *Traffic Pattern Recognition using an Active Learning Neural Network and Principal Components Analysis.* Eighth AICC, 2005, Paper 48, CD-ROM.
21. Yun C.-B., Yi J.-H. and Bahng E.Y. *Joint damage assessment of framed structures using a neural networks technique.* Eng. Struct., Vol 23, 2001, pp. 425-435.

From Data to Model Consistency in Shared Engineering Environments

Raimar J. Scherer and Peter Katranuschkov

Institute of Construction Informatics, TU Dresden,
01062 Dresden, Germany
{Raimar.Scherer, Peter.Katranuschkov}@cib.bau.tu-dresden.de

Abstract. Collaborative and concurrent engineering can be greatly facilitated by enhanced interoperability and consistency of data and tools based on the use of common shared models. In the paper, we examine the situation in model-based collaborative work and discuss the different types of models that are needed and have to be dealt with in a shared engineering environment. On the basis of a generalized cooperation scenario we then address consistency problems related to the model data and present a set of methods for their solution, including model view extraction, model mapping, knowledge-based consistency checking, model matching, model reintegration and model merging.

1 Introduction

Efficient collaborative and concurrent engineering are vital prerequisites for competitive advantage in today's global AEC business. They require achievement of adequate asynchronous teamwork in the highly heterogeneous project environments of one-off virtual organizations. This requirement can be met if interoperability and consistency of data and tools based on the use of common shared models are successfully realized.

Indeed, the main criterion for the efficiency of collaboration is the degree of common understanding of the communicated information. Connecting computers by fast communication channels is only a preliminary first step. Interoperability and consistency of the data require to go much further [1].

According to the ICH Glossary[1] interoperability is defined as "*the ability of information systems to operate in conjunction with each other encompassing hardware, communication protocols, applications and data compatibility layers*". On software level this includes syntactic, structural and semantic aspects that must be taken into account.

Achieving interoperability has been a major research and development issue in the past decades, giving rise to various new methods of model-based IT-supported work. It has led to the idea of a common standard Building Information Model (BIM) that can serve as reference for business process aware information delivery (Fig. 1). With the advance of the IFCs, this idea is now approaching fruition [2, 3].

---

[1] (c) ICH Architecture Resource Center, http://www.ichnet.org.glossary.htm

I.F.C. Smith (Ed.): EG-ICE 2006, LNAI 4200, pp. 615–626, 2006.

616   R.J. Scherer and P. Katranuschkov

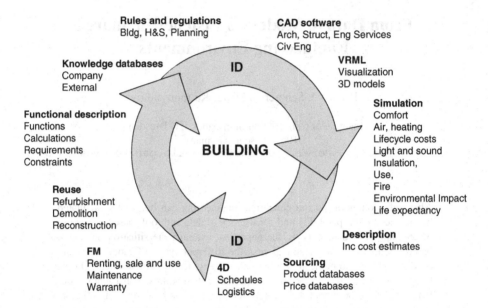

Fig. 1. Vision of interoperability through continuous process aware information delivery (adopted from the BuildingSMART initiative - http://www.buildingsmart.de)

However, successful collaboration requires more. It is an objective that encompasses the idea of interoperability and extends it further to enable cooperative working (1) between the organizations engaged in the realization of a construction project, (2) between the divisions of each organization, (3) in the supply chain, (4) in the knowledge supply chain, and (5) throughout the processes of the project life cycle [1, 4]. To facilitate such a comprehensive level of collaboration a series of components can be perceived. These are:

- a standard global schema, such as IFC in the AEC domain
- ontologies providing common agreements for concepts that exist within and beyond the global schema
- standardized constructs extending the global schema to support domain and discipline specific needs
- model views defining reusable units of information from the global schema or its domain extensions
- rules and methods to ensure consistency of the data throughout the life cycle.

The focus of this paper is particularly on the last component. Indeed, sophisticated interoperability and data processing methods would not fully achieve their goal if consistency is not guaranteed. In fact, the lack of efficient methods enabling data and model consistency are among the main causes of waste of resources and time, often amounting to nearly 20% of the total costs of the overall construction process.

In the following we suggest an approach based on a generalized collaboration scenario, which encompasses a set of methods targeting achievement of *model consistency*. Collaborative work subsumes some kind of distributed processing environment (client-server, peer-to-peer, grid) in which information is exchanged or shared. In the model-based paradigm the basis for successful exchange/sharing of information is provided by Building Information Models (BIM). Therefore, we first provide a brief overview of the different types of models that are needed and have to be dealt with and then present the suggested interrelated methods for tackling the model consistency problems.

2 Data Models and Engineering Models

2.1 Initial Definitions

Regarding the problems of interoperability and consistency in shared engineering environments we first have to distinguish between the *different kinds of models* that need to be considered, namely (1) the *meta data model* which describes the conceptual constructs like object, relationship, attribute, rules and inheritance, (2) the *data model* which describes the conceptual schema, and (3) the *product model* which describes the particular product, such as "The House of John Doe". Ideally the product model should be the complete formal description of the product. Consequently, it would have to subsume all different discipline-specific (engineering, architectural ...) models in one big *multi-model*. In reality, this does not happen for at least two reasons: (1) harmonization of the information structures of such enormous complexity is hardly achievable on large sociotechnical scale, and (2) it is not acceptable to overburden specific engineering applications with all the concepts of the full industry domain. Therefore, beside a global reference data model such as IFC, there will continue to exist a number of more specific data models. Achieving interoperability and consistency of the full modeling framework needed in a shared engineering environment will continue to be one of the major challenges of concurrent model-based work. In the following, we briefly outline the principal types of models that have to be taken into account and indicate the main consistency problems. A more detailed analysis is provided e.g. in [5].

2.2 Standardized Global Models

Standardized data models are based on a normative meta model and are typically structured into several inter-connected sub-schemas to allow shared development and maintenance of the data model. The main objectives of such models are (1) to serve a broad range of applications, (2) to be lean, (3) to be flexible and (4) to be extensible. The largest and most widely accepted model in AEC/FM is currently IFC [2], based on the ISO STEP methodology [6]. Standardized global data models should neither be too expressive, nor too sophisticated or too much focused on a specific use. However, several models that do not fulfill all these criteria have also been standardized to answer particular industry needs. Examples are CIS/2 (in the domain of steel construction) and OKSTRA (for road works). In the context of general standardized data models like IFC, such models can be considered as special kinds of domain models that are not harmonized with the global schema and hence require mapping, if their data are to be shared.

2.3 Proprietary Models

Proprietary data models are developed and used for a specific software tool or for several tools of one software provider. Such data models are tightly focused and strongly optimized to the functionality of the respective tool(s). Typically, they are formalized only to an extent needed for in-house software development and often do not expose a clear meta model. They are also very detailed and granular but not necessarily very explicit. Consequently, non-trivial mappings between proprietary and standardized data models are necessary leading to sophisticated and often ambiguous, potentially error-prone procedures [7, 8].

As remedy to that problem, both proprietary models and standardized global models are providing constructs that are intended to facilitate mapping and harmonization. Such constructs include object containers (proxies) that can hold instances of arbitrary classes, or – as in IFC – generalized property objects (property sets) that enable extended, non standardized agreements. However, whilst this helps to simplify the mapping process and to agree upon data exchange requirements not foreseen in advance, recognition of changed data and consistency management become even more difficult as in many cases it cannot be clearly determined what exactly has been changed and by whom. Therefore, beside the efforts within the models, adequate server-side mapping and matching methods are required to ensure consistency of the model data.

2.4 Domain Models

A domain model is usually understood as the data model established for a certain specific domain of interest (architecture, structural engineering, building services engineering …) and explicitly linked to other domain models. Generally, in the development of such models the core modeling approach is used where a lean core model is the overarching bridge for various domain modeling extensions. Like global core models, domain models should also be as lean as possible, extensible and broadly applicable data models. Therefore concepts including specific engineering knowledge should generally be avoided. For example, in a structural domain model the definition of a *frame*, which depends on the domain theory and the specifically used approximation method, should be avoided, whereas linear and planar elements are to be included as basic entities of the targeted engineering models of the domain [9].

The modern modeling approach is to clearly separate data and knowledge models. Knowledge models use data models but not vice versa. Data models should be timeless, whereas knowledge models are evolutionary, described by ontologies, and there may exist different ontology models for one and the same domain, depending upon the particular objectives and focus. Therefore, conceptually, domain models merely provide for further structuring of a global schema, attempting larger degree of harmonization and enabling easier definition of useful model views. They extend and partition the agreements in a shared environment but they do not resolve the consistency problems as such.

2.5 Model Views

A model view is a specific projection of a global model and/or one or more domain models. To help understand the meaning and use of model views better, below we take an IFC-based environment as example.

The IFC schema is a standardized global model that can capture project data about AEC projects over the complete life cycle. Potentially, it can support a broad spectrum of business requirements, but it does not particularize on any of these. This is illustrated on the left side of Fig. 2 where the 'periscope' allows to see everything in general but nothing in particular. However, from the perspective of an end user, the real need is to support *specific* business requirements in *specific* processes over one or more project stages. This produces a different periscope view in which support for particular business requirements can be seen (Fig. 2, right).

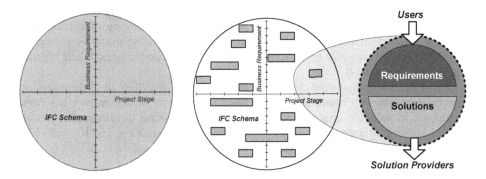

Fig. 2. Support of business requirements on the example of the IFC model (*left*: IFC supports all business requirements in all life cycle phases without particular differentiation; *right*: Specific business requirements at a specific project stage can be supported via dedicated model views).

The technical content required to support a business requirement is a set of units of information and this is what is generally termed *functional part* or *model view* [3, 4].

A model view describes the information in terms of the schema from which it is derived. Thus, a model view derived from the IFC schema will be a legal IFC subset schema. However, it could equally be derived from other schemas. Examples of model views are "a wall with associated basic geometry and material information", "associating classification codes to a set of heating elements", "grouping of structural elements to a frame system" etc.

On class level, model views are generally reusable, whereas their actual instantiation is specific for the targeted use case and process context. Model views can be referenced on high level by assigning specific services to perform certain actions on them, such as *createView*, *checkIn*, *checkOut*, *export*, *import*, *checkConsistency* and so on. Hence, as will be shown in the next chapter, they play an essential role in almost every collaborative work scenario in AEC/FM. In fact, it is believed that model views will be the feature that makes the IFCs work [3].

3   Achieving Model Consistency

The model diversity outlined above and the requirements of asynchronous concurrent work impose heavy constraints to model interoperability and consistency, especially due to the inherent long transactions in engineering design. During the last decade there have been various efforts to tackle the related problems on language level, meta model level or model level, by server-side methods, knowledge-based systems etc., but no final solutions are yet available – cf. [5, 8].

We propose an approach based on a set of inter-related methods to support model interoperability and consistency in cooperative work environments. It does not prom- ise a perfect environment providing faultless data integrity. The strategy is to mitigate the requirements to the involved engineering applications, reduce data loss caused by data mapping and other data conflicts, and at the same time take into account practical deficiencies in current data models and their software implementations. The suggested generalized process is based on the use of *long transactions* allowing off-line modifi- cations and parallel work, which involves versioning and merging of concurrently changed data. Specifically considered are problems related to the use of *legacy appli- cations* which are typically constrained by file based data exchange and their internal proprietary data models.

The suggested methods are outlined in the next subsections. For all of them *version management* plays an important role. Substantial research in that respect has been conducted in software engineering in the areas of database development, version control and configuration management. Details can be found e.g. in [10] and [11], where comprehensive studies of related efforts are also provided.

3.1   Generalized Collaboration Scenario

Whilst there are many different use cases where the suggested methods can be ap- plied, they can all be derived from the *generalized collaboration scenario* presented on Fig. 3 below. It starts at time point t_i with a consistent shared model version M_i, based on the product data model M defining the data that have to be shared, and pro- ceeds until the next coordination point t_{i+c} is reached. The data processing sequence for a single actor is comprised of the following seven steps:

1. *Model view extraction*: $Ms_i = extractView (M_i , viewDef (M_i))$
2. *Mapping* of the model view Ms_i to the (proprietary) discipline-specific model S_i, an instantiation of the data model S: $S_i = map (Ms_i , mappingDef (M, S))$
3. *Modification by the user* of S_i to S_{i+1} via some legacy application, which can be expressed abstractly as: $S_{i+1} = userModify (S_i , useApplication (A , S_i))$
4. *Backward mapping* of S_{i+1} to Ms_{i+1}, i.e.: $Ms_{i+1} = map (S_{i+1}, mappingDef (S , M))$
5. *Matching* of Ms_i and Ms_{i+1} to find the differences: $\Delta Ms_{i+1,i} = match (Ms_i , Ms_{i+1})$
6. *Reintegration* of $\Delta Ms_{i+1,i}$ into the model: $M_{i+1} = reintegrate (M_i , \Delta Ms_{i+1,i})$
7. *Merging* of the final consistent model M_{i+1} with the data of other users, that may concurrently have changed the model, to obtain a new stable model state, i.e.: $M_{i+c} = merge (M_{i+1} , M_{i+2} , ... , M_{i+k})$, with k = the number of concurrently changed checked out models.

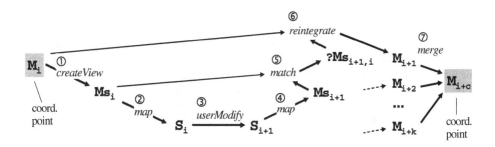

Fig. 3. Generalized collaboration scenario

3.2 Model View Extraction

Model View Extraction is the first step in the presented generalized collaboration scenario. To be usefully applied, a model view should (1) be easily definable with as few as possible formal statements, (2) allow for adequate (run-time) flexibility on instance and attribute level and (3) provide adequate constructs for subsequent reintegration of the data into the originating model. To meet these requirements, a *Generalised Model Subset Definition* schema (GMSD) has been developed [12]. It is a neutral definition format for EXPRESS-based models comprised of two subparts which are almost independent of each other with regard to the data but are strongly inter-related in the overall process. These two parts are: (1) *object selection*, and (2) *view definition*. The first is purely focused on the selection of object instances using set theory as baseline. The second is intended for post-processing (filtering, projection, folding ...) of the selected data in accordance with the specific partial model view. Fig. 4 shows the top level entities of the GMSD schema and illustrates the envisaged method of its use in run-time model server environments. More details on GMSD are provided in [11] and [12], along with references to other related efforts such as the PMQL language developed by Adachi.

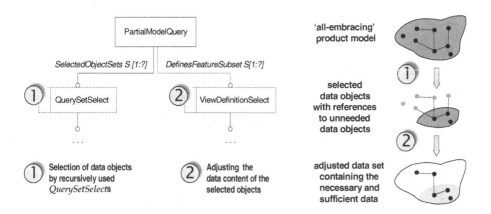

Fig. 4. Top level structure of the GMSD schema and the related instance level operations

3.3 Model Mapping

Model Mapping is needed by the transformation of the data from one model schema to another. Typically, this would happen in the transition from/to an agreed shared model or model view to/from the proprietary model of the application the user works with (see Fig. 3). The overall mapping process consists of four steps: (1) Detection of schema overlaps, (2) Detection of inter-schema conflicts, (3) Definition of the inter-schema correspondences with the help of formal mapping specifications, and (4) Use of appropriate mapping methods for the actual transformations on entity instance level at run-time.

Mapping patterns allow to understand better the mapping task and to formalize what and how has to be mapped in each particular case. By examining the theoretical background of object-oriented modeling the following types of mapping patterns can be identified: (1) *Unconditional class level mapping patterns,* depicting the most general high-level mappings, (2) *Conditional instance level mapping patterns,* including logical conditions to select the set of instances to map from the full set of instances in the source model, and (3) *Attribute level mapping patterns,* specifying how an attribute with a given data type should be mapped. For each of these categories, several sub-cases have been identified in [8]. Examples on attribute level include *simple equivalence, set equivalence, functional equivalence, homomorphic mapping, transitive mapping, inverse transitive mapping, functional generative mapping* and so on.

All mapping patterns can be defined by means of the developed formal mapping language CSML. Fig. 5 below shows the formalism to present these patterns *graphically,* along with respective typical examples. More details and examples are provided in [8].

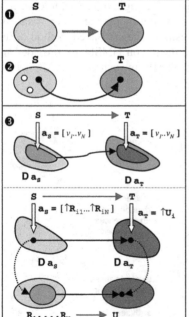

❶ *Class level mappings*
S and T denote the source and target class respectively. The ovals show the domains of the classes, the grey arrow the direction of the mapping. The given example depicts a 1:1 mapping S → T.

❷ *Conditional entity instance level mappings*
Black dots represent the instances being mapped, and white dots the instances discarded from the mapping as a result of an applied *condition.* The particular mapping equivalences are shown with black arrows. The given example refines the above 1:1 class mapping.

❸ *Attribute level mapping patterns*
Here class mapping (S → T) is indicated merely with an arrow. The domains of the attributes (Da) are shown as free-form areas, and the attribute values with dots (for single values) or with darker areas (for lists or sets). The mapping equivalences are indicated with black arrows. Thick white arrows denote the attribute's designation to a given class (S, T); pointer references to other objects are shown with dotted arrows. The first example depicts a 1:1 set equivalence. The second shows a homomorphic mapping, involving two pairs of mappings due to the references from a_S and a_T to the resp. object instances $[R_1,...,R_N]$ and U.

Fig. 5. Symbolism for graphical presentation of mapping patterns with illustrative examples

3.4 Knowledge-Based Consistency Checking

Knowledge-Based Consistency Checking plays a vital role in the overall collaboration scenario shown in Fig. 3. It is related to step 3 of the scenario where the user, while modifying the model data, needs to make sure that his changes are consistent with the engineering knowledge encoded in codes of practice and design/construction rules. Unlike the other methods described in this chapter, it is typically performed *locally*: (1) as a dedicated application, or (2) as part of a design assistant tool.

There are several known realizations of the first approach as e.g. Solibri's Model Checker, EPM's EDMmodelChecker, ISPRAS's STEP Semantic Checker etc. (see http://www.iai-international.org). However, such tools are not always efficient as they are decoupled from the actual user process. In our opinion, the second approach of embedding consistency checking within a knowledge-based planning system is more promising because it enables use of ontological concepts and rules both for planning and for checking the correctness and consistency of each planning action. This has been demonstrated in [13] where the output of each conceptual design step suggested by a knowledge-based planning algorithm is checked to be a *correct refinement* of the initial concept, fulfilling requirements for completeness and consistency. Such consistency checking methods cannot be developed generically but their inter-relationship with the other methods in the collaboration scenario can be well-defined and formalized.

3.5 Model Matching

Model matching has to deal with the identification of the changes made by one or more applications to the used model data (step 4 of the generalized collaboration scenario on Fig. 3). This may be done by a dedicated client application but a more natural implementation is a server-side procedure within a product model server.

In our approach, matching exploits the object structure without considering its semantic meaning. Hence, it can be applied to different data models and different engineering tasks. It does not require nor involve specific engineering knowledge.

Comparison of the model data of the old and new model versions begins with the identification of pairs of potentially matching objects, established by using their unique identifiers or some other key value. However, such identifiers are not always available for all objects. Moreover, unidentifiable objects may also be shared via references from different identifiable objects. The general complexity of this problem is shown in [14] where a fully generic tree-matching algorithm is shown to be NP-complete. Therefore, we have developed a *pragmatic algorithm* that provides a simple scalable way for finding corresponding data objects. Its essence is in the iterative generation of object pairs by evaluating equivalent references of already validated object pairs. The first set of valid object pairs is built by unambiguously definable object pairs. Any newly found object pair is then validated in a following iteration cycle, depending on its weighting factor derived from the *type* of the reference responsible for its creation. Attribute values are only used if ambiguities of aggregated references do not allow the generation of new object pairs. To avoid costly evaluation of attribute values a hashcode is used indicating identical references. In this way, the pair-wise comparison of objects can be significantly reduced with regard to other known approaches.

Fig. 6 illustrates the outlined procedure. Before starting any comparison of objects the set VP of validated object pairs and the set UO of unidentifiable objects are initially created using available unique identifiers. After that the object pairs of VP are compared as shown on the right side of the figure for $\{A_1, A_2\}$. Using their equivalent references "*has_material*" a new object pair can be assumed for the objects E_1 and G_2 of UO. A weighting factor indicating the validity of this assumption is then derived from the *reference type*[2] of "*has_material*" and added to the newly created object pair, which is then placed in the set AP containing all such derived matching pairs. After comparison of all object pairs of VP, the highest weighted object pairs of AP are moved from AP (and UO) to VP and a new iteration cycle is started. However, now the weighting factors of newly created object pairs are combined with the weighting factors of already validated object pairs. More details on the developed algorithm along with an overview of related efforts are provided in [11] and [15].

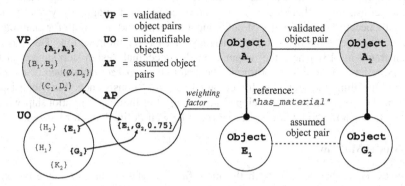

Fig. 6. Iterative derivation of object pairs

3.6 Model Reintegration

Reintegration of the changed data is always necessary when model views are used, as shown in the generalized collaboration scenario in Fig. 3. Model view extraction creates a model view by removing data objects, cutting off or reducing references, filtering attributes etc. Reintegration means to invert the process, i.e. to add removed data objects, restore cut-off or reduced references and re-create filtered out attributes. In our approach, this is strongly dependent on the model view definition achieved via GMSD, the results of the model comparison and the adopted version management. However, any other approach would rely similarly on such interdependencies.

Fig. 7 below illustrates the reintegration process on an example, assuming that by applying a GMSD-based *createView( )* operation to a given model some objects and attributes will be removed. For object O_1 this results in a new version O_{S1} in which the simple reference '*a*' is removed and the aggregated reference '*b*' is downsized by one element. This object is then modified externally by some user application to O_{S2} which differs from O_{S1} in the aggregated reference '*b*', downsized by another element, and the simple references '*c*' and '*d*'. The reintegration process adds all objects and attributes from O_1 that have been removed according to the model view definition. In this particular case, this will recreate the cut/downsized references '*a*' and '*b*'. However, it will take care not to add objects/attributes that have been modified by the used application.

---

[2] Reference types are differentiated in *simple references, ordered aggregated references* and *unordered aggregated references*. In this example, only a simple reference is shown.

Whilst this procedure is pretty clear and more or less the same for various different scenarios, there are several detailed problems that have to be dealt with. These can be subdivided into (1) *structural problems* (1:m version relationships for reference attributes, n:1 version relationships, change of the object type in a version relationship), and (2) *semantic problems*. The latter cannot be resolved solely by generic server-side procedures but require domain knowledge and respective user interaction. Such problems typically occur when a change to a model view requires propagation of changes to another part of the overall model in order to restore consistency. Therefore data consistency must be evaluated by all involved actors during a final merging process [11, 15].

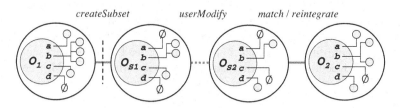

Fig. 7. Reintegration of model views

3.7 Model Merging

Model Merging has to deal with the consistency of *concurrently changed data* that exist in two or more divergent versions. It should be performed at a (pre)defined coordination point in cooperation of all involved users. The aim is to provide a procedure by which modifications can be reconciled and appropriately adjusted to a consistent new model state, marking the begin of a new collaboration cycle.

Fig. 8 schematically shows the suggested approach of using available prior knowledge to enable efficient management of the iterative agreement process. However, this is a highly complex process where world-wide research is still at an early phase.

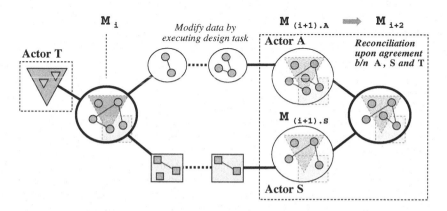

Fig. 8. Principal schema for the merging of concurrently changed design data

4 Conclusions

We described a set of inter-related methods to support consistency of the distributed project data in a shared collaborative work environment. These data are represented in various data models and are continuously transformed and moved in the process of their evolution from initial estimates to detailed specifications of the designed/built facility.

The developed generic server-side methods which rely mainly on the model structures can help a lot to achieve correct data management but they are not solely sufficient to warrant consistency. Whilst at many places theoretical complexity can be overcome by taking into account the actual features of BIM, realistic use cases and common sense, it is nevertheless not possible to solve all consistency problems without deeper engineering knowledge. The transitions between global and local models, man-machine interaction in the decision-making process and the combination of structure and semantics are issues that require further research. Enabling smooth and correct data handling from global shared models down to engineering ontologies where knowledge encoded in rules can be applied will be a huge step towards achievement of model consistency.

References

1. Ducq, Y, Chen, D., Vallespir, B.: Interoperability in Enterprise Modeling: Requirements and Roadmap. J. Advanced Engineering Informatics, Vol. 18, No. 4, pp 193-204 (2004).
2. IAI: IFC2x Edition 3, Online documentation. © International Alliance for Interoperability (1996-2006). Available at: http://www.iai-international.org/Model/R2x3_final/index.htm
3. Wix J. /ed./: Information Delivery Manual: Using IFC to Build Smart (2005). Available at: http://www.iai.no/idm/learningpackage/idm_home.htm
4. Gehre, A., Katranuschkov, P., Wix, J., Beetz, J.: InteliGrid Deliverable D31: Ontology Specification. The InteliGrid Consortium, c/o University of Ljubljana, Slovenia (2006).
5. Eastman, C. M.: Building Product Models: Computer Environments Supporting Design and Construction. CRC Press, Boca Raton, Florida, USA (1999).
6. Fowler, J.: STEP for Data Management, Exchange and Sharing. Technology Appraisals Ltd., Twickenham, UK (1995).
7. Amor, R., Faraj, I.: Misconceptions about Integrated Project Databases. ITcon, Vol. 6 (2001).
8. Katranuschkov, P.: A Mapping Language for Concurrent Engineering Processes. Diss. Report, Institute of Construction Informatics, TU Dresden, Germany (2001).
9. Weise, M., Katranuschkov, P., Liebich, T., Scherer, R. J.: Structural Analysis Extension of the IFC Modelling Framework. ITcon, Vol 8 (2003). Available at: http://www.itcon.org
10. Zeller, A. Configuration Management with Version Sets. Ph.D. Thesis, TU Braunschweig, Germany (1997).
11. Weise, M. An Approach for the Representation of Design Changes and its Use in Building Design (in German). PhD Thesis (submitted 05/2006), TU Dresden, Germany (2006).
12. Weise, M., Katranuschkov, P., Scherer R. J.: Generalised Model Subset Definition Schema. Proc. CIB-W78 Conference 2003, Auckland, NZ (2003).
13. Eisfeld, M.: Assistance in Conceptual Design of Concrete Structures by a Description Logic Planner. PhD Thesis, Institute of Construction Informatics, TU Dresden, Germany (2005).
14. Spinner, A.: A Learning System for the Creation of Complex Commands in Programming Environments (in German). PhD Thesis, TH Darmstadt, Germany (1989).
15. Weise, M., Katranuschkov, P., Scherer, R. J.: Generic Services for the Support of Evolving Building Model Data. Proc. X[th] ICCCBE, Weimar, Germany (2004).

Multicriteria Optimization of Paneled Building Envelopes Using Ant Colony Optimization

Kristina Shea[1], Andrew Sedgwick[2], and Giulio Antonuntto[2]

[1]Product Development, Technical University of Munich
Boltzmannstraße 15, 85748 Garching, Germany
kristina.shea@pe.mw.tum.de
[2]Arup, 13 Fitzroy St., London W1T 4BQ, UK
{andrew.sedgwick, giulio.antonuntto}@arup.com

Abstract. Definition of building envelopes is guided by a large number of influences including structural, aesthetic, lighting, energy and acoustic considerations. There is a need to increase design understanding of the tradeoffs involved to create optimized building envelope designs considering multiple viewpoints. This paper presents a proof-of-concept computational design and optimization tool aimed at facilitating the design of optimized panelized building envelopes for lighting performance and cost criteria. A multicriteria ant colony optimization (MACO) method using Pareto filtering is applied. The software Radiance is used to calculate lighting performance. Initial results are presented for a benchmark and project-motivated scenario, a media center in Paris, and show that the method is capable of generating Pareto optimal design archives for up to 11 independent performance criteria. A preliminary GUI for visualizing the Pareto design archives and selecting designs is shown. The results illustrate that for desired values of lighting performance in different internal spaces, there is often a range of possible panel configurations and costs.

1 Introduction

Building envelopes are molded by a large number of design influences including structural, aesthetic, lighting, energy and acoustic considerations. While there have been significant amounts of research in structural optimization of buildings, there has been significantly less research on aspects of building physics. Further, compared to structures, less expert knowledge exists about optimizing building physics, especially for complex projects, thus providing an opportunity to use computational optimization. A need exists to increase design understanding of the tradeoffs involved to create optimized building envelope designs considering multiple viewpoints. This paper presents work carried out at Arup, London towards developing a computational design and optimization tool that facilitates the design of optimized building envelopes for lighting, energy, cost and architectural criteria. The initial focus presented in this paper is the optimization of paneled building envelopes mainly considering lighting performance and cost.

Previous approaches to building envelope optimization include both evolutionary and numerical optimization approaches. Caldas and Norford [1] describe a method that combines genetic algorithms with detailed building energy simulation, via DOE-2, to optimize the placement and sizing of windows in an office building. The

I.F.C. Smith (Ed.): EG-ICE 2006, LNAI 4200, pp. 627–636, 2006.

work is extended to a multiobjective approach to generate Pareto archives that illustrate the tradeoff between energy consumption and initial cost considering additional design variables of wall materials [2]. Further studies consider building form variables, including roof tilt, to generate 3D building envelopes that tradeoff lighting energy and heating energy [2]. A similar approach focusing on developing a flexible framework for building optimization studies involving life-cycle cost and environmental impact is presented in [3]. Other research has considered the optimization of thermal performance of building envelopes as a parametric design optimization task employing the simplex method to optimize variables that define building geometry, material properties, as well as glazing and shading parameters [4]. Finally, a multi-level mathematical approach has been taken to optimize energy use and thermal comfort in a three zone office space [5].

The task considered in this paper is to optimize the configuration of paneled building envelopes, where each panel type has different lighting properties, e.g. opaque or clear glass, and is chosen from a defined discrete set. The main differences between previous work and the work presented here are the optimization of pixilated panel configurations on building roofs and walls and the focus on optimizing individual lighting performance at defined response points in building spaces separately. This model enables designers to tune a panel configuration to meet and balance lighting performance criteria within a number of internal spaces.

2 Method

The task of optimizing panel configurations on building walls and roofs can be classified as a discrete, combinatorial optimization task. A multicriteria ant colony optimization (MACO) method using Pareto filtering is applied. The software Radiance is used to calculate lighting performance.

2.1 Multicriteria Ant Colony Optimization (MACO)

Ant colony optimization (ACO) algorithms are discrete optimization methods inspired by the behavior of natural ant colonies [6,7]. They solve tasks by multi-agent cooperation using indirect communication through modifications in the environment, just like other social insects. Natural, or real, ants release a certain amount of pheromone while walking, e.g. from their nest to food. Subsequent ants preferentially follow directions and paths with higher pheromone concentration. The main idea is to use repeated and often recurrent simulations of artificial ants to dynamically generate new solutions.

An overview of the method, which is implemented in Matlab, is shown in Fig. 1. As this paper focuses on its application, only basic information will be given. Artificial ants modify some aspects of their environment in the same way as real ants do. By analogy, this numeric information is called an artificial pheromone trail. A search space can be formulated by a finite number of states where the connectivity of each state to all others is fully defined to completely model the movement of ants in the search space. In this application, a state is a unique design defined by its configuration of panels. The search space is then the combinatorial expansion of all possible panel configurations. A "nest" in the search space defines the position in the search space from which a group of ants start their exploration. Given a set of initial states, or "nests", ants are "let out" from these positions to search for improved states. Moving from one state to another is controlled by a set of "moves", or design modifications, and in this application a single panel material in a configuration is changed to a different panel material.

The pheromone table at each state contains probabilities representing the strength of pheromone, which are updated as soon as an ant reaches a new state. Updating the probabilities thus mimics pheromone laying. An ant-decision table is obtained by the composition of the local pheromone trail values with local heuristic values. The probability with which an ant chooses to go from one state to another is then calculated from this ant-decision table and a move is selected using a weighted roulette wheel technique. After each ant makes its move, pheromones are exponentially decreased to mimic evaporation in the natural system.

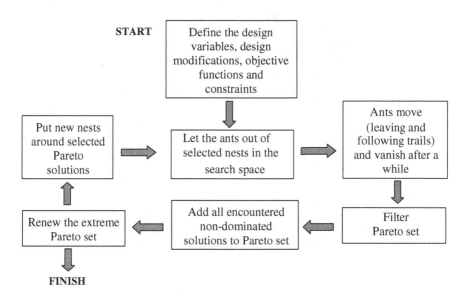

Fig. 1. Multicriteria Ant Colony Optimization (MACO) algorithm

To create a multicriteria ACO method, Pareto tests and a Pareto archive are used. Here, the design criteria, or objectives, are treated independently in the optimization and are not weighted, although they are scaled. The goal of the multicriteria optimization is then to generate a well spread archive of Pareto optimal designs that represents the tradeoffs among the design performance criteria.

Given all feasible solutions to a multicriteria optimization problem, a Pareto optimal, or non-dominated, solution is defined as one where no other feasible solution exists that is at least as good in all criteria and for at least one criterion the solution is better than all other designs. The complete set of Pareto optimal solutions is called the Pareto front. In theory there is no limit to the number of criteria that can be included. However, with the addition of each criterion the computational time for testing non-dominated status increases and the archives can become extremely large. A method for filtering evolving Pareto archives is implemented based on the smart Pareto filtering method described in [8] and extended to consider 11 independent criteria. The method works by constructing regions of insignificant tradeoff and maintaining only one Pareto solution in that region while discarding the rest. The regions are constructed using two control parameters set for each criterion that define the size of the regions of insignificant tradeoff. Filtering can be used both during optimization and after to reduce the size of the archives for easier visualization and interaction. Further details can be found in [8].

2.2 Lighting Analysis

Daylight factor is one of the key quantitative metrics for analyzing day lighting in building spaces. It is a measure of the proportion of available exterior illumination that reaches an internal surface under overcast conditions. Therefore, if a point has a daylight factor of 1%, then, when the external available horizontal illuminance is 10,000 lux, that point is illuminated by daylight to 100 lux. *Sun hours* is defined as the number of hours that a point in a space receives direct sunlight. This is expressed on a per annum basis or for a given time period, e.g. summer months. As a reference, at the latitude of London the probable annual sun hours are 1486.

Radiance is a software program for the analysis and visualization of lighting and is used by architects and engineers to predict illumination, visual quality and appearance of design spaces. Lighting performance is calculated using precompiled influence matrices constructed from lighting simulation output from Radiance. Radiance has been used to calculate the single contribution of each building envelope panel, for each different configuration, and for each reference point. This has been done by means of a script. In particular the script used allows parametric subdivision of any surface type, since it is based on mixtures of patterns and not on exact geometry. Results for both daylight factor and sun hours are collected into matrices and used as input to the optimization algorithm. Daylight factors and sun hours metrics are calculated within the optimization for each response point by summing the contribution of daylight factor and sun hours from each panel in the configuration. Since lighting analysis can be computationally expensive, the advantage of calculating lighting performance "off-line" is that it only needs to be carried out once for a given envelope definition and can be used for multiple optimization studies as long as design parameters that influence the calculations, e.g. interior wall height, do not change.

3 Results

The MACO method is now applied to two examples: one benchmark and one project-motivated scenario. In both examples the internal walls separating the internal spaces do not reach the ceiling, allowing light to interact by passing between the spaces.

3.1 Benchmark

The benchmark explores the known tradeoff between maximizing daylight factor and minimizing the number of direct sun hours for a small, 48 panel roof. The scenario is shown in Fig. 2 and consists of a 6m x 8m x 3m parallelepiped shaped room, divided into two equal spaces by a partition 1.5m high. The 6m x 8m roof is divided into square cells of 1m x 1m, or 48 panels. Each panel can take one of four defined panel materials (Table 1), yielding 8×10^{28} possible designs. Two response points, one in each space, are defined as reference points for calculating daylight factor and number of direct sun hours. The model has been geographically set to a site in London at latitude 51.5°N and oriented with the long side in direction North – South.

The optimization model is formulated as follows:

$$\text{minimize: } \{(1/((DF\_P1 + DF\_P2)*10000)),$$
$$((SH\_P1 + SH\_P2)/100) + num\_opaque\_panels)\} \quad (1)$$

where DF_P1 and DF_P2 are the daylight factor at point one and point two respectively and SH_P1 and SH_P2 are the number of direct sun hours at point one and point two respectively. Minimizing the number of opaque panels, in addition to sun hours, favors solutions where glass panels do not contribute to the number of direct sun hours at the response points due to their location.

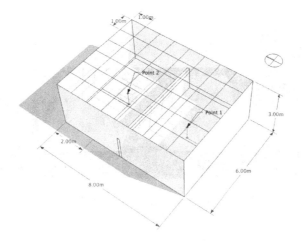

Fig. 2. Benchmark space specification

Table 1. Panel materials for benchmark

Panel material	Light transmission	Direct Sun?
00. Opaque	0%	no
01. Clear glass	75%	yes
02. Diffusing glass	75%	yes
03. Low transmission glass	25%	yes

The best archive of designs generated for the benchmark task is shown in Fig. 3. The extreme points (top left and lower right) of the archive indicate the best design found for each single objective.   While it is often difficult for multicritieria optimization to find the exact extreme solutions, in this case the extremes, which are well known to designers, occur when all panels are either opaque or clear glass. Of more interest to designers are the solutions in between the extremes that tradeoff the two criteria. A designer can select designs based on the compromise in design criteria that they are willing to make.

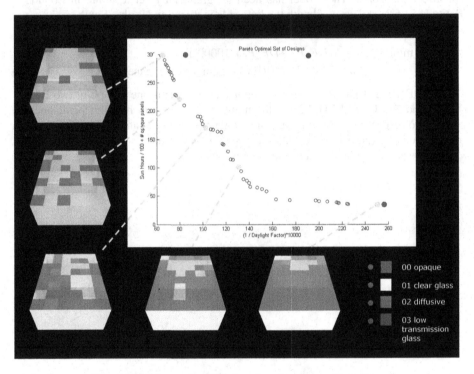

Fig. 3. Resulting trade-off for benchmark and sample designs

3.2 Paris Media Centre Design Scenario

The method is now applied to a larger, project-driven scenario for a media centre in Paris. The scenario is shown in Fig. 4 and consists of a 12m x 20m x 8m parallelepiped shaped space, which is located at the corner of a larger building. The space defined is split into five internal spaces (gallery wall 1, gallery wall 2, meeting area, reception, office) all with different lighting performance requirements, e.g. gallery walls must have no direct sun and low daylight factors. The roof and two walls are divided into 1m x 1m panels, yielding a total of 496 panels. The roof and wall panels can be made of four different materials (Table 2), hence 4.2×10^{298} possible designs exist. The other two walls are internal and are not considered. The building is located in Paris at longitude 2°21'E and latitude 48°51'N.

The optimization model is defined as:

$$\text{Maximize \{daylight factor (P1, P2, P3, P4, P5), view (P4, P5)\}} \tag{2}$$

Minimize {afternoon direct sun hours in summer (P3, P4, P5),
    average U-value, cost}   (3)
Subject to: all daylight factors ≤ 15 %, direct sun hours (P1, P2) = 0 hours,
    afternoon sun hours in summer (P3, P4, P5) ≤ 100 hours.   (4)

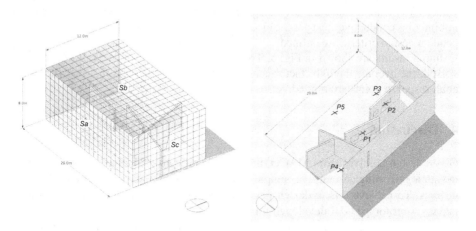

Fig. 4. Paris media centre space and response point (P1-P5) specification

Table 2. Panel materials for media centre

Panel material	Light transmission	Direct Sun?	U-value (W/m²k)	Cost (Euros)
00. opaque	0%	no	0.3	300
01. clear glass	75%	yes	1.8	450
02. diffusing glass	60%	no	1.8	550
03. shaded glass	25%	no	1.6	600

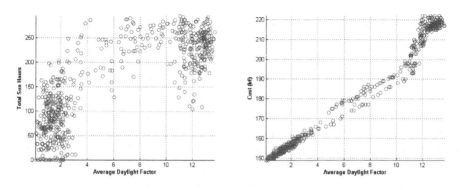

Fig. 5. Resulting performance trade-offs for Media Centre. (3D view given in Fig. 6)

View is calculated as a single criterion, using predefined matrices, that scores the view from response points four (P4) and five (P5) through the smallest wall, *Sc*, to a defined location. This indicates where clear glass panels are required on wall *Sc* to view a desired object in the distance. To illustrate the possibility of including a thermal performance criterion within the optimization studies, the average insulation value, or U-value (Table 2), over all panels is minimized as a design indicator for minimizing conductive heat loss. Since view is a single criterion, a total of 11 independent design criteria are defined.

Initial results are shown in Fig. 5. To aid understanding of the overall tradeoffs involved, the average daylight factor across all response points and the total sun hours at all response points are plotted versus cost.

4 Discussion

Given design archives from MACO, the most significant challenge for using them in design is providing an intuitive, simple GUI for exploring the trade-offs and selecting designs. To achieve this, a design perspective rather than data perspective, must be taken. A prototype GUI developed for this project and written in Matlab is shown in Fig. 6. The GUI allows designers to select designs by activating individual response point criteria (1-5) as well as adjusting sliders to the desired value for each design criterion within the range provided in the archive. A design is then selected by finding the nearest solution to the set values, calculated as the average of normalized distances, and displayed as a 3D model with all criteria values listed. If multiple, equivalent designs exist the user is notified. The corresponding point in the Pareto archive is highlighted in the performance view (upper right) using a yellow marker (Fig. 6). Ranges of individual performance criteria and their corresponding panel configurations can be studied by deselecting all other criteria. Future improvements to visualization and interaction include enabling designers to iteratively select performance criteria and ranges to reduce the solution space and focus in on a desired performance region. Further, means to create customized graphs of the performances, e.g. daylight factor at response point 2 vs. sun hours at response point 3 vs. cost, would help to visualize design tradeoffs.

In the results presented, the spread of solutions in design archives is better for daylight factor as the range is smaller (0-15) than for sun hours (0-100): see Fig. 5 and Fig. 6. Since the space of potential solutions in the second scenario described is so vast, improvements to the MACO are necessary to create a robust design tool that can be used on live building projects. One approach is to improve the Pareto archive filtering technique and how Pareto solutions are selected as new nests from which the ants start their search (Fig. 1). Convergence tests and MACO parameter studies also need to be carried out. Further extensions to the calculation of views as well as adding aesthetic models will extend the applicability of the method to architectural design criteria. Accurate consideration of energy criteria would require full building energy performance simulation, which can be very computationally intensive and pose difficulties for integration within design processes.

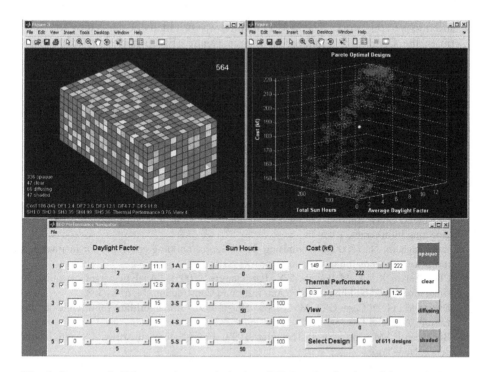

Fig. 6. Prototype building envelope optimization GUI for visualization of Pareto design archives and design selection

5   Conclusions

A multiobjective ant colony optimization (MACO) method was presented and applied to optimize paneled building envelopes for individual lighting performance and cost criteria. Examples of a thermal performance criterion, average U-value, and a view criterion were also explored. Initial results confirmed a known tradeoff for a benchmark example and produced promising results for a project-motivated scenario, which included 11 independent design criteria defined over five individual spaces and response points. A prototype GUI developed to visualize Pareto archives and select designs enabled designers to explore the issues involved in using such an approach for live building projects. Further work is required to enable the MACO method to better handle the large design spaces and large number of independent design criteria involved in the application explored. However, visualization of and interaction with the high dimensional performance spaces produced by MACO are the most important factor for creating a beneficial design tool.

Acknowledgements

The authors would like to thank Arfon Davies and Jeff Shaw in the Arup Lighting Group and Gianni Botsford, Gianni Botsford Architects for their collaboration and contributions to this work. The MACO method was originally developed by Dr. Onur Cetin in the Cambridge Engineering Design Centre (EDC) under an IMRC grant funded by the UK EPSRC.

References

1. Caldas, L. G., L. K. Norford (2002), "A design optimization tool based on a genetic algorithm", Automation in Construction, 11: 173–184.
2. Caldas, L. G., L. K. Norford (2003), "Genetic Algorithms for Optimization of Building Envelopes and the Design and Control of HVAC Systems", ASME Journal of Solar Energy Engineering, 125: 343-351.
3. Wang, W., H. Rivard, R. Zmeureanu (2005), "An Object-Oriented Framework for Simulation-Based Green Building Design Optimization with Genetic Algorithms", Advanced Engineering Informatics, 19: 5-23.
4. Bouchlaghem, N., (2000), "Optimising the design of building envelopes for thermal performance", Automation in Construction, 10(1): 101-112.
5. Choudharya, R., A. Malkawib, P.Y. Papalambros (2005), "Analytic target cascading in simulation-based building design", Automation in Construction, 14: 551– 568.
6. Dorigo M., G. Di Caro, G., L.M. Gambardella, (1999), "Ant algorithms for discrete optimization", Artificial Life, 5(2):137-172.
7. Bonabeau E., M. Dorigo & T. Theraulaz (1999), From Natural to Artificial Swarm Intelligence, New York.
8. Mattson, C. A., Mullur, A. A., and Messac, A., (2004), "Smart Pareto Filter: Obtaining a Minimal Representation of Multiobjective Design Space," Engineering Optimization, 36(6): 721–740.

Data Analysis on Complicated Construction Data Sources: Vision, Research, and Recent Developments

Lucio Soibelman[1], Jianfeng Wu[1], Carlos Caldas[2], Ioannis Brilakis[3], and Ken-Yu Lin[4]

[1] Department of Civil and Environmental Engineering, Carnegie Mellon University,
Pittsburgh, PA 15213, USA
{lucio, jianfen1}@andrew.cmu.edu
[2] Department of Civil, Architecture and Environmental Engineering,
University of Texas at Austin, Austin, TX 78712, USA
caldas@mail.utexas.edu
[3] Department of Civil and Environmental Engineering, University of Michigan
Ann Arbor, MI 48109, USA
brilakis@engin.umich.edu
[4] Ming-Jian Power Corporation, Taipei, Taiwan
kengyuli@hotmail.com

Abstract. Compared with construction data sources that are usually stored and analyzed in spreadsheets and single data tables, data sources with more complicated structures, such as text documents, site images, web pages, and project schedules have been less intensively studied due to additional challenges in data preparation, representation, and analysis. In this paper, our definition and vision for advanced data analysis addressing such challenges are presented, together with related research results from previous work, as well as our recent developments of data analysis on text-based, image-based, web-based, and network-based construction sources. It is shown in this paper that particular data preparation, representation, and analysis operations should be identified, and integrated with careful problem investigations and scientific validation measures in order to provide general frameworks in support of information search and knowledge discovery from such information-abundant data sources.

1 Introduction

With the ever-increasing application of new information technologies in the construction industry, we have seen an explosive increment of data in both volume and complexity. Some research efforts have been focused on applying data mining tools on construction data to discover novel and useful knowledge, in support of decision making by project managers. However, a majority of such efforts were focused on analyzing electronic databases, such as productivity records and cost estimates available in spreadsheets or single data tables. Other construction data sources, such as text documents, digital images, web pages, and project schedules were less intensively studied due to their complicated data structures, even though they also carry important and abundant information from current and previous projects.

Such a situation may thwart further efforts in improving information search or lessons learning in support of construction decision making. For example, a productivity

I.F.C. Smith (Ed.): EG-ICE 2006, LNAI 4200, pp. 637–652, 2006.

database could help project planners obtain an approximate range of productivity values for a specific activity, say, building a concrete wall based on historical records. However, other contextual information influencing the individual values of previous activities may not be available in the database for various reasons: specifications for different walls might be available only in text formats; correlations between wall-building activities and other preceding/succeeding/parallel jobs are stored in schedules; and pictures verifying such correlations, e.g., space conflicts, could be buried in hundreds of, even thousands of digital images taken in previous projects. This demonstrates the need for the development of advanced tools for effective and efficient information search and knowledge discovery from these texts, schedules, images, so that construction practitioners could make better informed decisions on more comprehensive data sources.

In this paper, our vision for data analysis on complicated construction data sources is first introduced, including its definition, importance, and related issues. Previous research efforts relevant to data analysis on various construction data sources are reviewed. Our recent work in graphical analysis on network-based project schedules is also shown, and initial results from this ongoing research are demonstrated for further discussions.

2 Definition and Vision

Traditionally, research in construction data analysis has been focused on transactional databases, in which each object/instance is represented by a list of values for a given set of features in a data table or a spreadsheet. Comprehensive Knowledge Discovery in Databases (KDD) processes, as well as many statistical and machine learning algorithms, have been introduced and applied in previous research [1,2].

In reality, there are a variety of construction data sources besides those available in spreadsheets or data tables, including semi-structured or unstructured text documents such as contracts, specifications, change orders, requests for information, and meeting minutes; unstructured multimedia data such as 2D or 3D drawings, binary pictures, as well as audio and video files; and structured data for specific applications that need more complicated data structures for storage and analysis than single data tables, e.g., network-based schedules. Currently a large percentage of project information is still kept in these sources for two major reasons. First, the large numbers of features, or high dimensionalities of such information make it inefficient to store and manage related data, such as text files with hundreds of words or images with thousand of pixels, in transactional databases. Second, information contained in the original data formats, such as graphical architectures of schedules and texture information of images could be lost if they are transferred into a single data table or spreadsheet. This situation requires that KDD processes, originally developed for transactional databases, be adapted to be applicable for more complicated construction data sources.

Compared with transactional construction databases that have been analyzed in a number of existing research efforts, construction data sources with complicated data structures are relatively less studied for information retrieval or lessons learning, even though they contain important and abundant historical information from ongoing and previous projects. As a result, useful data could be buried in large volumes of records due to difficulties in extracting relevant files for various purposes in a timely manner, while many information-intensive historical data could be left intact after projects are finished, without being further analyzed to improve practices in future projects.

Three major challenges could be attributed to such a relative scarce of related research. First, in addition to data preparation operations for general KDD processes [3], special attention should be taken to filtering out incorrect or irrelevant data from available construction data sources, while ensuring integrity and completeness of original information and further integrating it with other contextual information from other data sources if necessary. Second, extra efforts are needed to reorganize project data, originally recorded in application-oriented formats, into analysis-friendly representations that support efficient and effective KDD implementation. Third, data analysis techniques, which were initially developed to identify patterns and trends from single tables or spreadsheets, should be adapted in order to learn useful knowledge from text-based, image-based, web-based, network-based, and other construction data sources.

In our vision, a complete framework should be designed, developed, and validated for data analysis on particular types of complicated construction data sources. Three essential parts should be included in such a framework. First, careful problem investigations should be done in order to understand data analysis issues and related work in previous research. Second, detailed guidelines for special data preparation, representation, and analysis operations are also needed to ensure generality and applicability of the developed framework. And eventually, scientific criteria and techniques should be employed to evaluate validity of data analysis results and feasibility of extending such a framework to same or similar types of construction data sources.

3 Related Research Areas and Previous Work

In this section, previous research for analysis of text-based, web-based, and image-based construction data are presented, reviewing work done by other researchers, together with frameworks developed in our research group to support advanced data analysis on complicated construction data sources.

3.1 Text Mining for Management of Project Documents

In construction projects, a high percentage of project information is exchanged using text documents such as contracts, change orders, field reports, requests for information, and meeting minutes, among many others [4]. Management of these documents in model-based project management information systems is a challenging task due to difficulties in establishing relations between such documents and objects. Manually building the desired connections is impractical since these information systems typically store thousands of text documents and hundreds of model objects. Current technologies used for project document management, such as project websites, document management systems, and project contract management systems do not provide direct support for this integration. There are some critical issues involved, most of them due to the large number of documents and project model objects and differences in vocabulary. Search engines based on term match are available in many information systems used in construction. However, the use of these tools also has some limitations in cases where multiple words share the same meaning, where words have multiple meanings, and where relevant documents don't contain the user-defined search terms [4].

Recent research addressed some issues related to text data management. Information systems and algorithms were designed to improve document management [5,6,7]; controlled vocabularies (thesauri) were used to integrate heterogeneous data representations including text documents [8]; various data analysis tools were also applied on text data to create thesauri, extract hierarchical concepts, and group similar files for reusing past design information and construction knowledge [9,10,11]. In one of the previous studies conducted by our research group, a framework was devised to explore the linguistic features of text documents in order to automatically classify, rank, and associate them with objects in project models [4,12]. This framework involved several essential steps as discussed in our vision for data analysis on complicated construction data sources:

- Special **Data Preparation** operations for text documents were identified, such as transferring text-based information into flat text files from their original formats, including word processors, spreadsheets, emails, and PDF files; removing irrelevant tags and punctuations in original documents; removing stop words that are too frequently used to carry useful information for text analysis like articles, conjunctions, pronouns, and prepositions; and finally, performing word stemming to get rid of prefixes and/or suffixes and group words that have the same conceptual meanings.

- The preprocessed text data were transformed into a specific **Data Representation** using a weighted frequency matrix $A=\{a_{ij}\}$, where a_{ij} was defined as the weight of a word i in document j. Various weighting functions were investigated based on two empirical observations regarding text documents: 1) the more frequent a word is in a document, the more relevant it is to topic of the document; and 2) the more frequent a word is throughout all documents in the collection, the more poorly it differentiates between documents. By selecting and applying appropriate weighting functions, project documents were represented as vectors in a multi-dimensional space. Query vectors could then be constructed to identify similar documents based on similarity measures such as Euclidian distance and the cosine between vectors.

- **Data Analysis** tools were applied to build document classifiers, as well as document retrieval, ranking, and association mechanisms. The proposed methods require the definition of classes in the project model. These classes are usually represented as items of a construction information classification system (CICS) or components of a work breakdown structure (WBS). Classification models are created for each of these classes and project documents are automatically classified. Project document classification creates a higher semantic level that facilitates the identification of documents that are associated to selected objects. This is one of the major characteristics of the proposed methodology and represents a major distinction from existing project document management approaches. In order to retrieve documents that are related to selected objects, data that characterize the object is extracted from the model and used by the retrieval and ranking mechanisms. In the retrieval and ranking phase, all project documents are analyzed, but a higher weight is given to documents belonging to the object's class. In the Association step, documents are linked to the model objects.

The proposed methodology uses both the object's description terms and the object's class as input for the data analysis process. In methods based just on text search, only documents that contain the search terms can be located. Documents that use synonyms would not be found. If only the object's class is considered, some documents that belong to the object's class but are not related to the object would be mistakenly retrieved. By analyzing all model classes, but giving a higher weight to the object's class, related documents in other classes can also be identified.

A prototype system, Unstructured Data Integration System (UDIS), was developed to classify, retrieve, rank, and associate documents according to their relevance to project model objects [12]. Figure 1 in the next page illustrates UDIS.

The validation of the proposed methodology was based on the analysis of more than 25 construction databases and 30,000 electronic documents. The following two measures were used for comparison: recall and precision, which represent the percentage of the retrieved documents that were related to the selected model object, and the percentage of the related documents that were actually retrieved, respectively. Figure 2 in the following page shows how recall and precision are defined in information retrieval problems. The UDIS prototype achieved better performance in both recall and precision than typical project contract management systems, project websites, and commercial information retrieval systems [12].

Some of the major advantages of this framework include: 1) it did not involve manual assignments of metadata to any text documents; 2) it did not require a controlled vocabulary that would only be effective if it is adopted by all users of an information system; and 3) it could provide automated mechanisms to map text documents to project objects using their internal characteristics instead of user-defined search terms.

3.2 Text Mining for Ontology-Based Online A/E/C Product Information Search

As apposed to managing a set of in-house document collection in Section 3.1, online documents, especially those containing Architecture/Engineering/Construction (AEC) product information, have also received increasing attention while more and more manufacturers/suppliers host such information on the Internet. This is because the web provides boundary-less information sources, which could be used by industry practitioners to survey the A/E/C product market for product specification, selection, and procurement applications. Challenges similar to those encountered when dealing with in-house project document collection include the ambiguity of natural languages (e.g., polysemy, synonym, and structural ambiguity), and the lack of consistent product descriptions (e.g., definitions of product properties, domain lexicons for describing the products, and product information formatting) within A/E/C. Besides those, the enormous search space of the Internet, together with industry practitioners' varied ability to conduct effective queries, presents additional difficulties to the search of online product information. Concerns about the effectiveness of general search engines for retrieving domain specific information were also raised [13].

Domain knowledge represented in reusable formats can be of great help to tackle the above mentioned challenges. In knowledge engineering, ontology is used to represent the knowledge-level characterization of a domain. *Taxonomy* serves as the simplest ontology, which organizes a set of control vocabularies into a hierarchical structure. The employment of machine learning methods on training data sets to generate a classification system represented by hierarchical clusters of grouped project documents [4] is one noble transformed example of domain knowledge utilization. Other applications of domain knowledge represented in the form of taxonomy have also been reported [14,15,16].

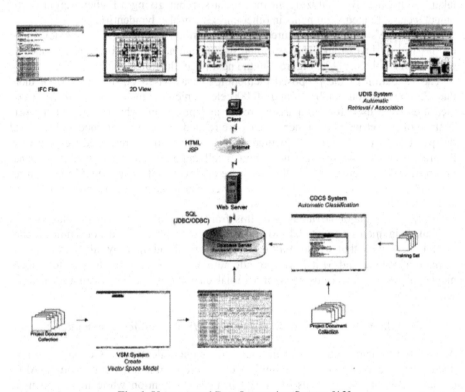

Fig. 1. Unstructured Data Integration System [12]

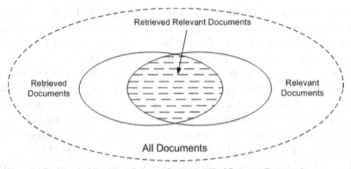

Recall = # of Retrieved Relevant Documents/# of Relevant Documents

Precision = # of Retrieved Relevant Documents/# of Retrieved Documents

Fig. 2. Definitions of recall and precision in Information Retrieval (IR)

Our research in [17] used *thesaurus* as a more elaborate form of ontology to help organize domain concepts related to a given A/E/C product. This application demonstrated how domain knowledge in the form of a thesaurus was utilized to support data preparation, representation, and analysis to identify highly relevant online product documents for a given product domain. Main steps as defined by our vision for analyzing complicated construction data sources in [17] are described below.

- **Data Preparation** operations were conducted automatically at two levels. Level one leveraged the represented thesaurus to identify a high quality pool of possibly relevant online documents. This was achieved by first expanding a single query into multiple queries (i.e., the growing strategy) and then pooling (i.e., the trimming strategy) the top-k documents returned by a general search engine after the expanded queries were entered as inputs. Level two performed word operations using the linguistic techniques such as removing stop words and stemming, similar to that introduced in Section 3.1.
- The prepared textual data were converted into document vectors for **Data Representation** purposes. The document vectors were assumed to exist in a multi dimensional space whose dimensions were defined by the concepts derived from the product thesaurus. The values of the vectors were defined by a weighted term-frequency matrix $A=\{a_{ij}\}$ where a_{ij} was the weight of a thesaurus term i in document j.
- The **Data Analysis** was achieved by two stages. In the first stage, each document vector was evaluated based on each expanded query, considering the thesaurus terms as well as the Boolean connectors resided within the query. In the second stage, a synthesis evaluation was calculated for a given document vector, considering all the possibly expanded queries with the use of domain thesaurus.

A semi-automated thesaurus generation method was developed in this research in order to supply domain knowledge for retrieving relevant A/E/C online product information. With the aid of domain knowledge in the generated thesaurus, the enormous Web space was successfully reduced into a manageable set of candidate online documents during data preparation. The use of thesauri in data representation also helped solve the language ambiguity problems so that synonyms like "lift" and "elevator" were treated indifferently. In this research for the purpose of validation, five data sets originating from different A/E/C product domains were processed by a developed prototype to return a list of search results that contained highly relevant online product information. To assess the true relevance of the returned search results, each web document was examined and labeled manually as 'relevant' or 'nonrelevant'. Then retrieval performance was evaluated accordingly. The performance evaluation focused particularly on the number of distinct product manufacturers identified because the goal of this research is to help A/E/C industry practitioners obtain more product information from the available marketplace. Hence for each data set, the number of distinct product manufacturers derived from 1) the search results generated from the established prototype; 2) the searching results returned by Google; and 3) an existing catalog compiled by SWEETS were compared as displayed in Table 1. It was observed that in most cases, the developed research prototype consistently retrieved more distinct product manufactures than the other two approaches.

Table. 1. Prototype performance comparisons for each of the five data sets [17]

Num of Distinctive Product Manufacturers Discovered	1	2	3	4	5
This research	8	12	28	26	30
Searching via Google	3	7	19	17	36
SWEETS	5	1	2	6	1

Compared with approaches that enforce standardized data representation, the developed framework in [17] has more flexibility in dealing with text-based and heterogeneous A/E/C product specifications on the Internet. The automatic query expansion process in the framework also relieved searchers' burden to structure complex queries using highly related terms. Therefore, an industry practitioner can take advantage of the developed A/E/C domain-specific search engine to survey the virtual product market for making more informed decisions.

3.3   Image Reasoning for Search of Jobsite Pictures

Pictures are valuable sources of accurate and compact project information [18]. It is becoming a common practice for jobsite images to be gathered periodically, stored in central databases, and utilized in project management tasks. However, the volume of images stored in an average project is increasing rapidly, making it difficult to manually browse such images for their utilization. Moreover, the labeling systems used by digital equipment are tweaked to provide distinct labels for all images acquired to avoid overlaps, but they do not record any information regarding the visual content. Consequently, images stored in central databases cannot be queried using conventional data and text search methods and thus, retrievals of jobsite images have so far been dependent on manual labeling and indexing by engineers.

Research efforts have been developed to address the image retrieval concerns described above. A prototype system, for example, was developed by Abudayyeh [19] based on a MS Access database that allowed engineers to manually link construction multimedia data including images with other items in an integrated database. The use of thesauri was also proposed by Kosovak et al. [8] to assist image indexing by enabling users to label images with specific standards. However, both solutions are difficult to use in practice due to the large number of manual operations that is needed to classify images in meaningful ways. Also, besides manual operations that are still involved in both solutions, the issue of how to index images in different ways was not addressed. This issue is important since it is not unusual that an image could be used for multiple purposes, even if it had been taken for just one use. For example, an engineer may have taken a picture of protective measures to prove their existence, but the same picture might contain other useful content needed in the future, such as materials or structural components included in its background [18].

A novel construction site image retrieval methodology based on material recognition was developed to address these issues [18,20]. A similar framework was developed based on our vision for data analysis on complicated construction data sources:

- **Data Preparation:** images from construction site were first preprocessed by extracting basic graphical features, such as intensity, color histograms, and texture presentations using various filtering methods. Features unrelated to image content such as intensity of lights and shades were then removed automatically.
- **Data Representation:** preprocessed images were first divided into separate regions representing different construction objects using appropriate clustering methods; graphical features of each region were further compressed into quantifiable, compact, and accurate descriptors, or signatures. As a result, a binary image was transformed into several clusters, each with a list of values for a given set of feature signatures.
- **Data Analysis:** the feature signatures of each image region were compared with signatures in an image collection of material samples. Using proper threshold and/or distance functions, the actual material of each region was recognized. An example is given in the next page (Figure 3) to show how a site image is decomposed into clusters, represented with signatures, and identified with actual materials in each separated region.

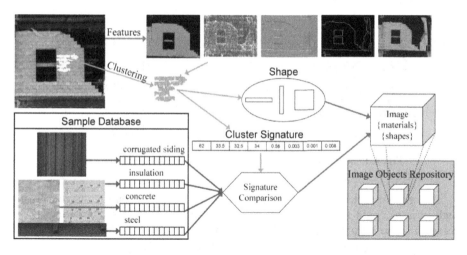

Fig. 3. An example for image data preparation, representation, and analysis [20]

The material information was then combined with other spatial, temporal, and design data to enable automated, fast, and accurate retrieval of relevant site images for various management tasks in a developed system as illustrated below (Figure 4).

As shown in Figure 4, in addition to image attributes (e.g., materials) that can be automatically extracted using the above method, time data can be obtained from digital cameras when image are taken on construction sites; 2D/3D locations can be obtained from positioning technologies; and as-planned and as-built information for construction products can be retrieved from a model-based system or a construction database. By combining all such information together, images in the database can be compared with the query's criteria in order to filter out irrelevant images and rank the selected images according to their relevance. The results are then displayed on the screen, and the user could easily browse through a very limited amount of sorted images to identify and choose the desired ones for specific tasks. Figure 4 provides a detailed example of retrieving images for a brick wall using the developed prototype. This method was tested on a collection of more than a thousand images from several projects and evaluated with similar criteria like recall and precision in Section 3.1. The results showed that images can be successfully classified according to the construction materials visible within the image content [20].

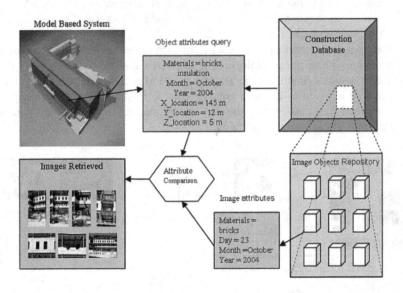

Fig. 4. Image retrieval based on image contents and other attributes [20]

The original contribution of this research work on construction site image indexing and retrieval could be extended to solve other related problems. For example, our research group is currently investigating a new application of image processing, pattern recognition, and computer vision technologies for fully automated detection and classification of defects found in acquired images and video clips of underground pipelines.

A major motivation of our new research work resides on the rapid aging of pipeline infrastructure systems. It is desirable to develop and use more effective methods for assessing the condition of pipelines, evaluating the level of deterioration, and determining the type and probability of defects to facilitate decisions making for maintenance/repair/re-habilitation strategies that will be least disruptive, most cost-effective and safest [21]. Currently, trained human operators are required to examine acquired visual data and identify the detailed defect class/subclass according to the standard pipeline condition grading system, such as Pipeline Assessment and Certification Program (PACP) [22], which makes the process error-prone and labor-intensive. Building on our previous research efforts in image analysis [18] and former research efforts in defect detection and classification of pipe defects [23,24,25], we are studying more advanced data preparation, representation, analysis solutions to boost automated detection and classification of pipe defects based on improved problem investigation, feature selection, data representation, and multi-level classifications of pipeline images.

4 A Recent Development in Scheduling Data Analysis

This section presents our recent work applying the same vision, which has been successfully applied for analysis of unstructured text and image data, for analysis of network-based project schedules. A preliminary study including problem investigation, data preparation, data representation, data analysis, and pattern evaluations are detailed below.

4.1 Problem Investigation

Since computer software for Critical Path Method, or CPM-based scheduling started to be used in the construction industry decades ago, planning and schedule control history of previous projects has been increasingly available in computerized systems in many companies. However, there is a current lack of appropriate data analysis tools for scheduling networks with thousands of activities and complicated dependencies among them. As a result, schedule historical data are left intact without being further analyzed to improve future practices. At the same time, companies are still relying on human planners to make critical decisions in various scheduling tasks manually and in a case-by-case manner. Such practices, however, are human-dependent and error-prone in many cases due to subjective, implicit, and incomplete scheduling knowledge that planners have to use for decision making within a limited timeframe.

Many research efforts have been focused on improving current scheduling practices. Delay analysis tools were applied to identify fluctuations in project implementation, responsible parties for the delays, and predict possible consequences [26,27]; machine learning tools like genetic algorithms (GA) were applied to explore optimal or near-optimal solutions for resource leveling and/or time-cost tradeoff problems, by effectively searching only a small part of large sample space for construction alternatives [28,29,30]; simulation and visualization methods were developed on domain-specific process models to identify and prevent potential problems during implementation [31,32]; and knowledge-based systems were also introduced to collect, process, and represent knowledge from experts and other sources for automated generation and reviewing of project schedules [33,34]. Such research efforts helped improve efficiency and quality of scheduling work in many aspects, but none of them addressed the above "data rich but knowledge scarce" issue, i.e., no research described here has worked on learning explicit and objective knowledge from abundant planning and schedule control history of previous projects.

Our research is intended to address this issue related to scheduling knowledge discovery from historical network-based schedules, so that more explicit and objective patterns could be identified from previous schedules in support of project planners' decision making. A case study was implemented on a project control database to identify special problems in preparing, representing, and analyzing project schedules for this purpose.

4.2 Data Preparation

The project control database for this case study was collected from a large capital facility project. Three data tables were used in the scheduling analysis: one data table with detailed information about individual activities; one with a complete list of predecessors and successors of dependencies among the activities; and a third table with historical records about activities that were not completed as planned during implementation, and reasons for their non-completions.

In this step, we first separated nearly 21,000 activities into about 2,600 networks based on their connectivity. Another special data preparation task identified was the removal of redundant dependencies in networks [35]. As shown in Figure 5, a dependency from activity X to Y is redundant if there is another activity Z, such that Z is succeeding to X, and Y is succeeding to Z, directly or indirectly. Obviously, the dependency from X to Y is unnecessary because Y can not be started immediately after X anyway.

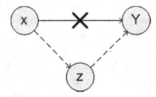

Fig. 5. An example for redundant dependencies in scheduling networks

4.3 Data Representation

Architectures of scheduling networks may influence their risks and performance in implementation to a great extent. A challenge here was to describe such networks in a concise and precise way, so that networks having similar architectures and performance could be abstracted into same or close descriptions. On the other hand, it would also be difficult, if not impossible, to identify common patterns among all 2,600 scheduling networks with varied size and complicated graphical architectures by just reviewing them visually or in their original formats. Like the developed solutions in data analysis for text-based and image-based construction data sources, a general data representation is necessary to transform scheduling information into appropriate formats for desirable data analysis.

Intuitively, how a network is split into branches or converged from branches could be a good indicator of its reliability during implement. A possible explanation for such an intuition is that such splits or converges generate or eliminate parallelisms, if we assume that parallel implementation is a major cause for resource conflicts, and thus variability during actual implementation. We used such an intuition as our major hypothesis in this case study, and developed a data representation to describe scheduling networks in an abstract way focusing on how a work flow goes through it by splitting/converging, and finally arriving at its end. In this study, we developed a novel abstraction process comprising of two major stages:

– Recursive Divisions in a top-down manner: the network in Figure 6 could be divided into two sequential sub-networks, N1/N2, so that N1 is preceding to N2 only through activity B, and N2 could be further divided into two parallel sub-networks, N21/N22, which only have common starting task B and ending tasks C; such divisions could continue until a sub-network eventually consists of only 'atomic' sequences of tasks without any branches (e.g., all the three sequences from A to B).

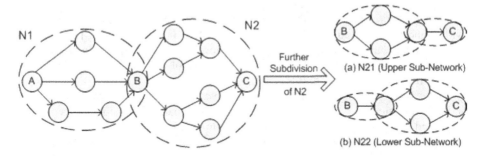

Fig. 6. An example for recursive division of a scheduling network

– Type description in a bottom-up manner: in this stage, task identifications were removed since we only concerned the abstract description of the network architecture. In a reverse direction to recursive divisions, the type description for a network/sub-network could be obtained by following the rules illustrated in Figure 7.

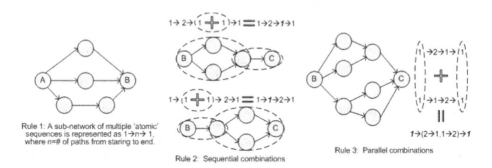

Fig. 7. Examples for type descriptions rules

4.4 Data Analysis and Primary Pattern Evaluations

Type descriptions for all 2,600 networks in this study were generated using a computer program developed on a Matlab® 7.0 platform. By comparing type descriptions of networks with their probabilities of non-conformances, i.e., having non-completion tasks during implementation, some interesting patterns were found. For example, in this project, the non-conformance rate of networks that started with a single task (type description: "1→n→…") was significantly lower than that of networks started with multiple tasks (type description: "n→1→..."). This pattern is simple in its statement, but it could be hard to discover if the complicated network structures had not been abstracted into precise and concise descriptions. And it makes sense considering that a schedule that starts with multiple parallel tasks at the beginning could be more risky due to lack of sufficient preparations by managerial team at an early stage. Also, comparisons between non-completion rates of activities at different positions turned out some meaningful results. The non-completion rate of tasks connecting sequential or parallel sub-networks, i.e., those with more than one preceding and/or succeeding tasks, was significantly larger than that of tasks among 'atomic' sequences, i.e., those with at most one preceding and succeeding activities. It also verify our previous hypothesis made during data representation that how a network is split into branches or converged from branches could be a good indicator of reliability during implement for specific networks and activities.

These initial results also demonstrated the potentials of extending our current work in many directions: 1) the recursive division method makes it possible to decompose large and complicated scheduling networks into multiple levels of components for further representation of project planning and schedule control history; 2) the abstract descriptions could be viewed as extracted graphical features of scheduling networks, which could be integrated with other features of activities and dependencies for identifying other general and useful patterns in scheduling data; and 3) the complete process of scheduling data preparation, representation, and analysis provided primary insights and experiences for the following development of this research on a larger scale.

5   Conclusion and Future Work

Complicated construction data sources are different from transactional databases in their information contents, data structures, and storage sizes. In this paper, text-based, image-based, web based, and network-based construction data were used as examples to present our vision for possible advanced data analysis on these types of data sources. Several required steps essential for achieving this vision were presented, including problem investigation to define relevant issues; data preparation to remove features not carrying useful information; data representation to transform data sources into new precise and concise descriptions; and proper data analysis for information search or for extraction of lessons learned. The success obtained by the implementation of the prototypes developed to test our vision suggests that further research should be developed to extend the results obtained and to allow the development of the required tools to deal with other types of construction data sources, such as audio records of construction meetings (e.g., for automatically extracting and classifying useful information based on voice recognition) or video data from construction sites (e.g., automated estimations of overall and/or specific productivity values of construction workers).

Further research efforts on advanced data analysis on complicated construction data sources could be integrated with a broad range of related work. For example, retrieved data or identified patterns from such data sources could be fed into existing research in construction decision support systems (DSS) to enhance current solutions with improved access to historical data and/or knowledge; such research efforts may also be combined with current data modeling and information management research to integrate construction data sources with various formats and broad contexts for data retrieval, queries, and analysis; and it is expected that new challenges and pattern recognitions tasks could be identified and inspire innovative research methods in data representation and data mining communities as well.

Acknowledgements

The authors would like to thank US National Science Foundation for its support under Grants No. 0201299 (CAREER) and No. 0093841, and to all industrial collaborators who provided the data sources for our previous and current studies.

References

1. Buchheit, R.B., Garret J.H. Jr., Lee, S.R. and Brahme, R.: A Knowledge Discovery Case Study for the Intelligent Workplace. *Proc. of the 8th Int. Conf. on Comp. in Civil and Building Eng.*, Stanford, CA (2000), 914-921
2. Soibelman, L. and Kim, H.: Data Preparation Process for Construction Knowledge Generation through Knowledge Discovery in Databases. *J. of Comp. in Civil Eng.*, ASCE, 1(2002), 39-48.
3. Soibelman, L, Kim, H., and Wu, J.: Knowledge Discovery for Project Delay Analysis. *Bauingenieur*, Springer VDI Verlag, Feburary 2005
4. Caldas, H.C., Soibelman, L., and Han, J.: Automated Classification of Construction Project Documents. *J. of Comp. in Civil Eng.*, ASCE, 4(2002), 234-243
5. Hajjar, D. and AbouRizk, M.S.: Integrating Document Management with Project and Company Data. *J. of Comp. in Civil Eng.*, ASCE, 1(2000), 70-77.
6. Zhu, Y., Issa, R.R.A., and Cox, R.F.: Web-Based Construction Document Processing via Malleable Frame. *J. of Comp. in Civil Eng.*, ASCE, 3(2001), 157-169.
7. Fruchter, R., and Reiner, K.: Model-Centered World Wide Web Coach. *Proc. of 3rd Congress on Comp. in Civil Engineering*, ASCE, Anaheim, CA (1996), 1-7.
8. Kosovac, B., Froese, T.M. and Vanier, D.J.: Integration Heterogeneous Data Representations in Model-Based AEC/FM Systems. *Proc. of CIT 2000*, Reykjavik, Iceland (2000), 556-567
9. Yang, M.C., Wood, W.H., and Cutkosky, M.R.: Data Mining for Thesaurus Generation in Informal Design Information Retrieval. *Proc. of Int. Congress on Comp. in Civil Eng.*, ASCE, Boston, MA (1998), 189-200
10. Scherer, R.J. and Reul, S.: Retrieval of Project Knowledge from Heterogeneous AEC Documents. *Proc. of the 8th Int. Conf. on Comp. in Civil and Building Eng.*, Stanford, CA (2000), 812-819
11. Wood, W.H.: The Development of Modes in Textual Design Data. *Proc. of the 8th Int. Conf. on Comp. in Civil and Building Eng.*, Stanford, CA (2000), 882-889
12. Caldas, H.C., Soibelman, L., and Gasser, L.: Methodology for the Integration of Project Documents in Model-Based Information Systems. *J. of Comp. in Civil Eng.*, ASCE, 1(2005), 25-33
13. Dewan, R., Freimer, M., and Seimann, A.: Portal Kombat: The Battle between Web Pages to become the Point of Entry to the World Wide Web. *Proc. of the 32nd Hawaii Intl. Conf. System Science*, IEEE, Los Alamitos, CA, USA (1999)

14. El-Diraby, T.A., Lima, C., and Feis, S.: Domain Taxonomy for Construction Concepts: Toward a Formal Ontology for Construction Knowledge. *J. of Comp. in Civil Eng.*, ASCE, 4 (2005), 394–406.
15. Lima, C., Stephens, J., and Böhms, M.: The bcXML: Supporting Ecommerce and Knowledge Management in the Construction Industry. *ITCON*, (2003), 293–308.
16. Staub-French, S., Fischer, M., Kunz, J., Paulson, B., and Ishii, K.: An Ontology for Relating Features of Building Product Models with Construction Activities to Support Cost Estimating", CIFE Working Paper #70, July 2002, Stanford University.
17. Lin, K. and Soibelman, L.: Promoting Transactions for A/E/C Product Information. *Automation in Construction*, Elsevier Science, in print
18. Brilakis, I. and Soibelman, L.: Material-Based Construction Site Image Retrieval. *J. of Comp. in Civil Eng.*, ASCE, 4(2005), 341-355
19. Abudayyeh, O.: Audio/Visual Information in Construction Project Control. *J. of Adv. Eng. Software*, Elsevier Science, 2(1997), 97-101
20. Brilakis, I. and Soibelman, L.: Multi-Modal Image Retrieval from Construction Databases and Model-Based Systems, *J. of Constr. Eng. and Mgmt.*, ASCE, July 2006, in print
21. Water Environment Federation (WEF): Existing Sewer Evaluation and Rehabilitation. *ASCE Manuals and References on Eng. Practices, Ref. No. 62*, Alexandria, VA, 1994.
22. Pipeline Assessment and Certification Program (PACP) Manual, Second Edition Reference Manual, NASSCO, 2001
23. Chae, M.J. and Abraham, D.M.: Neuro-Fuzzy Approaches for Sanitary Sewer Pipeline Condition Assessment. *J. Comp. in Civil. Eng,* ASCE, 1(2001), 4–14
24. Sinha, S.K. and Fieguth, P.W.: Automated Detection of Cracks in Buried Concrete Pipe Images. *Automation in Construction*, Elsevier Science, 1(2005), 58-72
25. Sinha, S.K. & Fieguth, P.W.: Neuro-Fuzzy Network for the Classification of Buried Pipe Defects. *Automation in Construction*, Elsevier Science, 1(2005), 73-83
26. Hegazy, T. and Zhang, K.: Daily Windows Delay Analysis. *J. of Constr. Eng. and Mgmt.*, ASCE, 5(2005), 505-512
27. Shi, J.J., Cheung, S.O. and Arditi, D.: Construction Delay Computation Method. *J. of Comp. in Civil Eng.*, ASCE, 1(2001), 60-65
28. Hegazy, T. and Kassab, M.: Resource Optimization Using Combined Simulation and Genetic Algorithms. *J. of Constr. Eng. and Mgmt.*, ASCE, 6(2003), 698-705
29. Feng, C., Liu, L. and Burns, S.A.: Using Genetic Algorithms to Solve Construction Time-Cost Trade-Off Problems. *J. of Comp. in Civil Eng.*, ASCE, 3(1997), 184-189
30. El-Rayes, K. and Kandil, A.: Time-Cost-Quality Trade-off Analysis for Highway Construction. *J. of Constr. Eng. and Mgmt.*, ASCE, 4(2005), 477-486
31. Akinci, B., Fischer, M. and Kunz, J.: Automated Generation of Work Spaces Required by Construction Activities. *J. of Constr. Eng. and Mgmt.*, ASCE, 4(2002), 306-315
32. Kamat, V.R. and Martinez, J.C.: Large-Scale Dynamic Terrain in Three-Dimensional Construction Process Visualizations. *J. of Comp. in Civil Eng.*, ASCE, 2(2005), 160-171
33. Dzeng, R. and Tommelain, I.D.: Boiler Erection Scheduling Using Product Models and Case-Based Reasoning. *J. of Constr. Eng. and Mgmt.*, ASCE, 3(1997), 338-347
34. Dzeng, R. and Lee, H.: Critiquing Contractors' Scheduling by Integrating Rule-Based and Case-Based Reasoning. *Automation in Construction*, Elsevier Science, 5(2004), 665-678
35. Kolisch R., Sprecher A. and Drexl A.: Characterization and Generation of a General Class of Resource-Constrained Project Scheduling Problems. *Mgmt. Science*, (1995), 1693-1703

Constructing Design Representations

Rudi Stouffs and Albert ter Haar

Design Informatics Chair, Dept. of Building Technology, Faculty of Architecture, Delft
University of Technology, Berlageweg 1, 2628 CR Delft, The Netherlands
{r.m.f.stouffs, a.terhaar}@tudelft.nl

Abstract. Supporting the early phases of design requires, among others, support
for the specification and use of multiple and evolving representations, and for
the exchange of information between these representations. We consider a
complex adaptive system as a model for the development of design representations, and present a semi-constructive algebraic formalism for design representations, termed *sorts*, as a candidate for supporting this approach. We analyze
sorts with respect to the requirements of a complex adaptive system and compare it to other representational formalisms that consider a constructive approach to representations.

1 Introduction

Design is a multi-disciplinary process, involving participants, knowledge and information from various domains. As such, design problems require a multiplicity of
viewpoints each distinguished by particular interests and emphases, and each of these
views, in turn, requires a different representation of the design entity. Even within the
same task and for the same person, various representations may serve different purposes defined within the problem context and the selected approach. Especially in the
early phases of design, the exploratory and dynamic nature of the design process
invites a variety of approaches and representations, and any particular representation
may be as much an outcome of as a means to the design process. Therefore, supporting the designer in these early phases requires, among others, support for the specification and use of multiple and evolving representations, and for the exchange of information between these representations.

Various modeling schemes for defining product models and ontologies exist (e.g.,
[1], [8]). These allow for the development of representations in support of different
disciplines or methodologies, and enable information exchange between representations and collaboration across disciplines. However, they still require an a priori effort
at establishing an agreement on concepts and relationships, which offer a complete
and uniform description of the project data, mainly independent of any project specif-
ics. We are particularly interested in providing the user access to the specification of a
design representation, and the means to create and adapt design representations according to the designer's intentions in the task at hand, in support of creativity. Creativity, as an activity in the design process, relies on a restructuring of information that
is not yet captured in a current information structure — that is, emergent information
— for example, when the design provides new insights that lead to a new interpretation of constituent design entities.

I.F.C. Smith (Ed.): EG-ICE 2006, LNAI 4200, pp. 653–662, 2006.

This flexibility in using design representations necessitates a solution for dealing with the complexity of representations. For this purpose, we consider a complex adaptive system as a model for the development of design representations, in particular, its three key principles [5]. First, the outcome of the design process is generally *unpredictable*, as it is indeterminately related to the design requirements and the design process. Under the assumption that the design representation is an intricate part of the design outcome, this representation is necessarily also unpredictable. Secondly, the state or history of a design representation is in principle *irreversible* as changes to the representational structure can result in data loss. Finally, with respect to the *emergence of order* in the development of a design representation, we refer to Prigogine and Stengers [11]: "Order arises from complexity through self-organization." In the context of constructing a design representation, the process of self-organization can take on the form of human communication or correspondence, leading to an agreement on the representation that prevails in the system (see also [6]). This communication may be considered among different users or between the user and the design application (correspondence between the user's mental model and the application's design representation).

Typically, a representation is a complex structure of properties and constructors, and a representation may be a construction of another [16]. As such, an approach to constructing representations in terms of other representations can be considered, in which correspondence can be achieved through the adaptation of the representational structure of properties and constructors and by agreement on the naming of representations, or parts thereof. Such an approach can benefit from a formal framework that allows for alternative representations of the same entity to be compared and related with respect to scope and coverage, in order to support translation and identify where exact translation is possible. Considering a representation as a complex structure of properties and constructors, comparing alternative representations requires a comparison of their respective properties and their mutual relationships, and of the overall construction. Such a comparison will not only yield a possible mapping in support of information exchange, but also uncover potential data loss when moving data from less-restrictive to more-restrictive representations. At the same time, the vocabulary of available properties and constructors defines the expressive power of the representational framework.

In this paper, we consider a semi-constructive algebraic formalism for design representations, termed *sorts* [15], as a candidate for supporting a constructive approach to design representations, and analyze *sorts* with respect to the requirements of a complex adaptive system. In particular, we consider the ability of *sorts* to support correspondence on design representations, to compare representations with respect to scope and coverage and detect data loss, and consider its potential to support the design process and the design outcome. Critical in this respect is the formal specification of *sortal* representations in terms of other *sortal* representations under formal compositional operations, the behavioral specification of *sortal* representations enabling the comparison of alternative representations and the detection of data loss, and the ability to integrate data functions into *sortal* representations. *Sorts* are also compared to other representational formalisms that consider a constructive approach to representations.

2 Incrementally Constructing Design Representations

Van Leeuwen et al. [17] describe a property-oriented data modeling approach, in which design concepts are represented as flexible networks of objects and properties. In contrast to traditional modeling approaches, an object has no predefined set of properties and the composition of properties defining an object can be changed at any time. Under the property-modeling approach, correspondence can be achieved through the development (over time) of the network of objects and properties and by agreement on the naming of objects. Such an approach can greatly benefit from a formal framework that allows for representations to be compared and related, formally, in order to support translation and identify where exact translation is possible. For example, Stouffs et al. [16] were able to show, using a subsumption relation defined on well-known solid models, that information loss between some of these solid models is inevitable. Subsumption is a powerful mechanism for comparing alternative representations of the same entity. When a representation subsumes another, the entities represented by the latter can also be represented by the former representation, without any data loss.

There are many representational formalisms that consider the subsumption relationship in order to achieve partially ordered representational structures; most are based on first-order logic. Applied to building design, a good example is Woodbury et al. [18], who adopt typed feature structures as a model for design space exploration. Like many other formalisms, typed feature structures consider a record-like data structure for representing data types. Record-like data structures facilitate the encapsulation of property information in (a variation of) attribute/value pairs [2]. Furthermore, the properties may themselves be typed feature structures, i.e., expressed in terms of record-like data structures, containing (sub)properties. Then, the subsumption relationship defines a partial ordering on feature structures. Furthermore, the algebraic operations of intersection and union (or others similar) may be defined on feature structures so that the intersection of two feature structures is subsumed by either structure, and the union of two feature structures subsumes either structure.

Key to typed feature structures is the notion of partial information structures and the existence of a unification procedure that determines if two partial information structures are consistent and, if so, combines them into a single, new (partial) information structure. Typed feature structures further consider a type hierarchy and a description language, where each type defines a corresponding description. The subsumption relation between feature structures extends the subsumption ordering on types inherent to the type hierarchy. Woodbury et al. [18] also specify a generating procedure that relates feature structures with a description (or type) that they satisfy, and that incrementally generates more complete design structures. This fact — that the generating procedure monotonically generates more complete information structures — could be interpreted as excluding the possibility for information loss and thus making design states reversible. However, the inclusion of an information removal operator is possible providing more flexibility at the cost of limiting search strategies [18]. Datta [4] also presents a visual notation for representing design correspondence between designer and typed feature structures, using the concepts of mixed initiative and rational conversation.

3   A Subsumption Relationship over *sorts*

Sorts [15] provide a semi-constructive algebraic formulation of design representations that enables these to be compared with respect to scope and coverage and that presents a uniform approach to dealing with and manipulating data constructs. *Sorts* are class structures identified by compositions of properties [16], where properties are named entities identified by a type specifying the set of possible values. Exemplar types are labels and numeric values, and spatial types such as points, line segments, plane segments and volumes. In the construction of *sorts*, every composition of properties is considered a *sort*. Even a single property defines a *sort* — thus, a *sort* is typically a composition of other *sorts*. We denote a *sort* identified by a single property as primitive and all other *sorts* as composite. A primitive *sort* necessarily has a name; a composite *sort* can also have a name assigned. Named *sorts* can be conceived to define object classes. Similarly to the property-oriented modeling approach [17], the collection of properties of a class is not predefined. This allows class structures easily to be modified, both by adding and removing properties, and by altering the constructive relationships. For this purpose, we consider even property relationships and data functions (see Section 5) as properties, such that these can be dealt with in the same way.

Properties are composed using one or more constructors. We consider an *attribute* operator (denoted '$\wedge$'), resulting in a conjunctively subordinate composition of properties, and an operation of *sum* (denoted '+'), resulting in a disjunctively co-ordinate composition. For example, a *sort* of colored labels may be defined as a composition of labels and colors under the attribute operator, such that each label has one (or more) colors assigned as attribute. On the other hand, a *sort* of points and line segments may be defined as a composition of points and line segments under the operation of sum; a resulting data entity can be either a point or a line segment. The operation of sum defines a subsumption relationship (denoted '$\leq$') over *sorts*, as follows:

$$a \leq b \Leftrightarrow a + b = b. \tag{1}$$

The typed feature structures formalism, like most logic-based formalisms, links subsumption directly to information specificity, that is, a structure is subsumed by another, if this structure contains strictly more information than the other. One consequence of subsumption is that the absence of information in a design representation does not necessarily imply the absence of this information in the design, that is, representations are automatically considered to be incomplete. As a result, when searching for a design (representation) that satisfies certain information, less specific representations cannot automatically be excluded (e.g., [3]).

The subsumption relationship over *sorts* does not formally apply over the attribute operator. Though $a \wedge b$ is more information specific than either a or b, $a \wedge b$ is not subsumed by a or b (nor $a + b$), i.e., $(a \wedge b) + a + b \neq a + b$ — algebraically, the attribute operator corresponds to the Cartesian product operator; $a \wedge b$ is the *sort* of all proper 2-tuples of which the first member belongs to a and the second to b. In logic formalisms, a relational construct is used to represent such tuples. For example, in description logic [3], roles are defined as binary relationships between individuals. Consider a concept Label and a concept Color; the concept of colored labels can then be represented as Label $\cap$ $\exists$hasAttribute.Color[1], denoting those labels that have an attribute that is a color. Here, $\cap$ denotes intersection and $\exists$R.C denotes full existential quantification with respect to role R and concept C. It follows that Label $\cap$ $\exists$hasAttribute.Color $\subseteq$ Label; the concept of labels subsumes the concept of colored labels. A similar construct is not considered with respect to *sorts* — e.g., when looking for a yellow square, any square will not do, unless it has the yellow color assigned. In other words, logic-based models adhere to an open world — that is, nothing can be excluded unless it is explicitly excluded — *sorts*, on the other hand, adhere to a closed world, any reasoning is based purely on present or emergent (under a part relationship, see Section 4) information. *Sorts* only represent data; logic-based models essentially represent knowledge.

4 A Behavioral Specification for *sorts*

An important ingredient of *sorts* is behavioral specification. Behavioral specification is a prerequisite for the effective exchange of data between various representations. When an application receives data along with its behavioral specification, the application can correctly interpret, manipulate, and represent this information without unexpected data loss. For instance, at the representational level, operations that may otherwise seem trivial, such as adding or removing data elements, become resolutely non-trivial — for instance, the addition of two numbers when these represent cardinal values (e.g., a number of columns that is increased) and when these represent ordinal values (e.g., for a given space, determining the minimum distance to a fire exit or the (maximum) amount of ventilation required given a variety of activities), and similarly, additive versus subtractive colors, depending on whether these refer to the mixing of surface paints or colors of light, respectively. Fortunately, behavioral specification is reasonably limited to the common arithmetic operations of addition, subtraction, and product. It turns out that the more common CAD operations of creation and deletion, and selection and deselection, can all be expressed as some combination of addition and subtraction from one design space (*sort*) to another. The complex operations of grouping and layering can be treated likewise [14].

The simplest specification of a part relationship corresponds to the subset relationship on mathematical sets. This part relationship particularly applies to points and labels, e.g., a point is part of another point only if the two are identical, and a label is a part of a collection of labels only if it is identical to one of the labels in the collection. Then, operations of addition (combining elements), subtraction, and product (intersecting elements) correspond to set union, difference, and intersection, respectively.

Another kind of behavior arises when we consider the part relationship on line segments. A line segment is an interval on an infinite line carrier; in general, one-dimensional quantities such as time may be considered as intervals. An interval is a part of another interval if it is embedded in this interval; intervals on the same carrier that are adjacent or overlap combine into a single interval. Specifically, interval behavior can be expressed in terms of the behavior of the boundaries of intervals [7].

---

[1] Note that this syntax differs slightly from the syntax adopted by Baader et al. [3], which, for example, differentiates the intersection constructor on concepts from the operation of intersection on interpretations. Interpretations do not play a role in this example.

Behaviors also apply to composite *sorts*, that is, a part relationship can be defined for its component data elements belonging to a composite *sort* defined under a conjunction (attribute operator) or disjunction. The composite *sort* inherits its behavior from its components in a manner that depends on the compositional relationship.

The disjunctive operator distinguishes all operand *sorts* such that each data element belongs explicitly to one of these *sorts*. Consequently, a data element is part of a disjunctive data collection if it is a part of the partial data collection of elements from the same component *sort*. In other words, data collections from different component *sorts*, under the disjunctive operator, never interact; the resulting data collection is the set of collections from all component *sorts*. When the operation of addition, subtraction or product is applied to two data collections of the same disjunctive *sort*, the operation instead applies to the respective component collections.

Under the attribute operator a data element is part of a data collection if it is a part of the data elements of the first component *sort*, and if it has an attribute collection that is a part of the respective attribute collection(s) of the data element(s) of the first component *sort* it is a part of. When data collections of the same composite *sort* (under the attribute operator) are pairwise summed (differenced or intersected), identical data elements merge, and their attribute collections combine, under this operation. Elements with empty attributes are removed.

When reorganizing the composition of *sorts* under the attribute operator, the corresponding behavior may be altered in such a way as to trigger data loss. Consider a behavior for weights [12] (e.g., line thickness or surface tones) as becomes apparent from drawings on paper — a single line drawn multiple times, each time with a different thickness, appears as if it were drawn once with the largest thickness, even though it assumes the same line with other thickness. When using numeric values to represent weights, the part relation on weights corresponds to the less-than-or-equal relation on numeric values. Thus, weights can combine into a single weight, which has as its value the least upper bound of all the respective weight values, i.e., their maximum value. Similarly, the common value (intersection) of a collection of weights is the greatest lower bound of all the individual weights, i.e., their minimum value. The result of subtracting one weight from another is either a weight that equals the numeric difference of their values or zero (i.e., no weight), and this depends on their relative values.

Now consider a *sort* of weighted entities, say points, i.e., a *sort* of points with attribute weights, and a *sort* of pointed weights, i.e., a *sort* of weights with attribute points. A collection of weighted points defines a set of non-identical points, each having a single weight assigned (possibly the maximum value of various weights assigned to the same point). These weights may be different for different points. On the other hand, a collection of pointed weights is defined as a single weight (which is the maximum of all weights considered) with an attribute collection of points. In both cases, points are associated with weights. However, in the first case, different points may be associated with different weights, whereas, in the second case, all points are associated with the same weight. In a conversion from the first to the second *sort*, data loss is inevitable.

An understanding of when and where exact translation of data between different *sorts*, or representations, is possible, becomes important for assessing data integrity and controlling data flow [16]. Data loss can easily be assessed under the subsumption relationship. If one *sort* subsumes another, exact translation is trivial from the part to the whole. If two *sorts* subsume a third, exact translation only applies to the data that can be said to belong to the third *sort*. When the subsumption relationship doesn't apply, such as under the attribute operator — as is the case in the examples above — *sorts* can still be compared, roughly, as equivalent, similar, convertible and incongruent [15]. Two *sorts* are said to be equivalent if these are semantically derived from the same *sort* — through renaming. Equivalent *sorts* are syntactically identical; this guarantees exact translation of data, except for a loss of semantic identity. Two *sorts* are denoted similar if these are similarly constructed from equivalent *sorts*. The similarity of *sorts* relies on the existence of a semi-canonical form of a composite *sort* as a disjunctive composition over *sorts*, each of which is either a primitive *sort* or composed of primitive *sorts* under the attribute operator [15]. Associative and distributive rules with respect to both compositional operators allow for a syntactical reduction of *sorts* to this semi-canonical form. If two *sorts* reduce to the same semi-canonical form, then these *sorts* are considered similar, and exact translation, except for a loss of semantic identity, applies. Otherwise, two *sorts* are either convertible or incongruent. If two *sorts* are convertible, data loss depends also on their behavioral specification, as in the examples above.

5 Functional Descriptions

The part relationship that underlies the behavioral specification for a *sort* enables data recognition to be implemented for this *sort*; since composite *sorts* inherit their behavior and part relationship from their component *sorts*, any technical difficulties in implementing data recognition apply just once, for each primitive *sort*. Data recognition plays an important role in the specification of design queries. So does counting. Stouffs and Krishnamurti [13] indicate how a query language for querying graphical design information can be built from basic operations and geometric relations that are defined as part of a maximal element representation for weighted geometries, augmented with operations that are derived from techniques of counting and data recognition. For example, by augmenting networks of utility pipes, represented as volumes (or plane segments) with appropriate behavioral specification, with labels as attributes, and by combining these augmented geometries under the operation of sum, colliding pipes specifically result in geometries that have more than one label as attribute. These collisions can easily be counted, while the labels on each geometry identify the colliding pipes, and each geometry itself specifies the location of the collision [13].

In order to consider counting and other functional behavior as part of the representational approach, *sorts* consider data functions as a data kind, offering functional behavior integrated into data constructs. Data functions are assigned to apply to one or more selected *sorts* — specifically, they apply over tuples of data entities, one from each selected *sort*, where these data entities relate to the function under a sequence of one or more compositional relationships. Then, the result value of the data function is computed (iteratively) from the values of these tuples of data entities. The value of a data entity used in the computation is the actual value of the entity, such as its numeric value, or the position vector for a point, but may also be a derived value, such as the length of a line segment, or its direction vector. The data function's result value is automatically recomputed each time the data structure is traversed, e.g., when visualizing the structure. As a data kind, data functions specify both a functional description, a result value, and one or more *sorts* and their respective value methods.

Data functions can introduce specific behaviors and functionalities into representational structures, for the purpose of counting or other numerical or geometric operations. Consider, for example, a *sort* of linear building elements, represented as line segments, with an attribute *sort* specifying the cost of each element per unit length. Then, by augmenting the corresponding data construct with a sum-over-product function applied to the numeric value of the cost entities and the length value of the linear elements, the value of this function is automatically computed as the total cost of all the building elements. As another example, consider a composite *sort* specifying both a reference point and a number of emergency exits represented as line segments. Then, a minimum-value function in combination with a function that computes the distance between a position vector and a line segment, specified by two end vectors, will yield the minimum distance from the reference point to any emergency exit.

Moving data functions in the data construct, by altering the compositional structure of the representation, alters the scope of the function — that is, the sorts' data entities that relate to the function under a sequence of one or more compositional relationships — and thereby its result. In this way, data functions can be used as a technique for querying design information, where moving the data function alters the query.

6 Discussion

Sorts present an algebraic formalism for constructing design representations. A *sortal* representation is a composition of *sorts* that can easily be modified by adding and removing *sorts* or by altering the constructive relationships, and that can be given a name. A subsumption relationship over *sorts*, in combination with a behavioral specification of *sorts*, allows *sortal* representations to be compared and related with respect to scope and coverage, and data loss to be assessed when converting data from one representation to another. Data functions can be integrated into data constructs in order to query design information. In this way, the design representation is an intricate part of the design outcome, and the construction of a design representation can be the result of correspondence that forms part of the design process. This naturally raises the question how such correspondence can be facilitated through an application interface.

In developing such an application interface, we consider three aspects in particular; these are the ability to conceptualize representational structures, the need for effective visualizations of these structures and the embedding of the application in a practical context. First, we're considering the definition of *sorts* as the specification of a concept hierarchy that, subsequently, can be detailed into a representational structure consisting of primitive *sorts* and constructive relationships. By separating the specification of the representational semantics (the names of the structures and their hierarchical relationships) from the specification of the nuts and bolts (the data types and their behaviors, and the distinction between disjunctively compositional and attribute relationships) we aim to ease a conceptualization of the intended representational structures that facilitates their development. Secondly, we're exploring effective (graphical) visualizations of (parts of) the representational and data structures that can offer the user insight into these structures. In particular, we're implementing a dynamic visualization of these structures with variable focus and level of detail.

Thirdly, we're investigating practical applications of *sorts* in order to illustrate their strengths in practical contexts. Specifically, we're looking into the context of collaborative building design projects where a CAD program is used to express the design but where the design process also involves other information that is collected and stored in the form of documents. We're investigating the use of *sorts* to specify relationships between these documents and elements within the CAD model that help to organize this information. Given a *sort element_ids ^ element_descriptions* that reflects on (part of) the CAD model, the data for this *sort* can be automatically generated from the CAD data. This representation can then be extended using the *sort element_ids ^ (element_descriptions + document_references)* to represent CAD elements with associated document references. Using a graphical interface, the user can specify both the references and their associations to CAD elements. When the CAD model is changed, the data for *element_ids ^ element_descriptions* can be regenerated, while the data for *element_ids ^ document_references* can be retrieved from *element_ids ^ (element_descriptions + document_references)* using an automatic conversion based on the matching of both *sorts*. Since the first *sort* is subsumed by the second ('^' distributes over '+'), exact translation applies. Merging both data forms re-associates the document references to the CAD elements, on condition that the respective element IDs have not changed.

Park and Krishnamurti investigate the use of sorts in the context of building construction, within a larger project that investigates ways of integrating "suites of emerging evaluation technologies to help find, record, manage, and limit the impact of construction defects" [10]. The project considers an Integrated Project Model (IPM) that is continuously updated to reflect on both the as-designed and as-built building models. "The as-designed model is an IFC file obtained from a commercial parametric design software. Laser scanning provides accurate 3D geometric as-built information (e.g., component identity); similarly, embedded sensors provide frequent quality related information (e.g., thermal expansion)" [10]. *Sorts* are adopted to provide the flexibility to generate both pre-defined and user-defined views. For example, Park and Krishnamurti consider the use of sorts to generate different information views of a target object. "The embedded sensor planner needs the geometric information, location, material type, and construction method of the target object. On the other hand, the laser scanning technician needs two-dimensional geometric information of the target region, geometric information, and location of the target object." [9]

Sortal representations can complement a Building Information Model (BIM). Kuhn and Krishnamurti argue that "current methods of obtaining precise information from a BIM are cumbersome. Furthermore, it is computationally expensive to produce a representation from a BIM" [10]. We aim to put forward the concept of a *sortal* building model as an extension to a BIM, offering the user the means to build up design representations, in support of common or interdisciplinary views, and to use such representations for querying building information.

Acknowledgments

The first author wishes to thank Ramesh Krishnamurti for his collaboration on this research.

References

1. AIA Model Support Group: IFC2x Edition 3. International Alliance for Interoperability (2006). http://www.iai-international.org/Model/R2x3_final/index.htm (1 May 2006)
2. Aït-Kaci, H.: A lattice theoretic approach to computation based on a calculus of partially ordered type structures (property inheritance, semantic nets, graph unification). Ph.D. Diss. University of Pennsylvania, Philadelphia, PA (1984)
3. Baader, F., Calvanese, D., McGuinness, D., Nardi, D., Patel-Schneider, P.: The Description Logic Handbook: Theory, Implementation and Applications. Cambridge University Press, Cambridge (2003)
4. Datta, S.: Modeling dialogue with mixed initiative in design space exploration. Artificial Intelligence for Engineering Design, Analysis and Manufacturing 20 (2006) 129-142
5. Dooley, K.J.: A complex adaptive systems model of organization change. Nonlinear Dynamics, Psychology, and Life Sciences 1 (1997) 69-97
6. Kooistra, J.: Flowing. Systems Research and Behavioral Science 19 (2002) 123-127
7. Krishnamurti, R., Stouffs, R.: The boundary of a shape and its classification. The Journal of Design Research 4(1) (2004)
8. Manola, F., Miller, E. (eds.): RDF Primer. W3C World Wide Web Consortium (2004). http://www.w3.org/TR/rdf-primer/ (1 May 2006)
9. Park, K., Krishnamurti, R.: Flexible design representation for construction. In: Lee, H.S., Choi, J.W. (eds.): CAADRIA 2004. Yonsei University Press, Seoul, South Korea (2004) 671-680
10. Park, K., Krishnamurti, R.: The digital diary of a building. In: Bhatt, A. (ed.): CAADRIA'05, Vol 2. TVB School of Habitat Studies, New Delhi (2005) 15-25
11. Prigogine, I., Stengers, I.: Order out of Chaos. Bantam Books, New York (1984)
12. Stiny, G.: Weights. Environment and Planning B: Planning and Design 19 (1992) 413-430
13. Stouffs, R., Krishnamurti, R.: On a query language for weighted geometries. In: Moselhi, O., Bedard, C., Alkass, S. (eds.): Third Canadian Conference on Computing in Civil and Building Engineering. Canadian Society for Civil Engineering, Montreal (1996) 783-793
14. Stouffs, R., Krishnamurti, R.: The extensibility and applicability of geometric representations. In: Architecture proceedings of 3rd Design and Decision Support Systems in Architecture and Urban Planning Conference. Eindhoven University of Technology, Eindhoven, The Netherlands (1996) 436-452
15. Stouffs, R. Krishnamurti, R., Cumming, M.: Mapping design information by manipulating representational structures. In: Akın, Ö., Krishnamurti, R., Lam, K.P. (eds.): Generative CAD Systems. School of Architecture, Carnegie Mellon University, Pittsburgh, PA (2004) 387-400
16. Stouffs, R., Krishnamurti, R., Eastman, C.M.: A Formal Structure for Nonequivalent Solid Representations. In: Finger, S., Mäntylä, M., Tomiyama, T. (eds.): Proc. IFIP WG 5.2 Workshop on Knowledge Intensive CAD II. IFIP WG 5.2, Pittsburgh, PA (1996) 269-289
17. van Leeuwen, J.P., Hendrickx A., Fridqvist, S.: Towards dynamic information modelling in architectural design. Proc. CIB-W78 International Conference IT in Construction in Africa. CSIR, Pretoria (2001) 19.1-19.14
18. Woodbury, R., Burrow, A., Datta, S., Chang, T.: Typed feature structures and design space exploration. Artificial Intelligence in Design, Engineering and Manufacturing 13 (1999) 287-302

Methodologies for Construction Informatics Research

Žiga Turk

University of Ljubljana, FGG, Construction Informatics
Jamova 2, 1000 Ljubljana, Slovenia
www.zturk.com

Abstract. Construction informatics is a rather new topic in civil engineering and as such its rules of scientific investigation are not as mature as with topics with a longer tradition. The paper argues that construction informatics has elements in common with both natural sciences, mathematics, technology as well as humanities and therefore a broad scope of methodological apparatus is available, including that of humanities. Some current methods are criticized and action research and Socratic methods suggested as an alternative.

1 Introduction

Construction informatics[1] research is pursuing a direction much along the demarcation line between scientific research and engineering problem solving. The goal of the first is to help us understand the fundamental phenomena in nature, man and society while the goal of engineering is practical and related to useful products. The research work in construction informatics is winding across this vaguely understood line without realizing the differences in the methodological apparatus applicable for the two different missions. As the topic of construction informatics is maturing, a critical overview of the appropriate methodologies presented. This paper - a position paper by nature - should contribute to the improvement of research methods that are used in the community.

1.1 Classification of Science

After making a distinction between science and engineering we are further classifying kinds of scientific research.

Good source of definitions and classification is the OECD's Frascatti Manual [1]. It classifies science and technology into natural sciences, engineering and technology, medical sciences, agricultural sciences, social sciences and humanities. It identifies peculiarities of R&D in the software and informatics from research policy management perspective: "*Software development has since become a major intangible innovation activity with a high R&D content. In addition, an increasing share of relevant activities draws on the social sciences and humanities, and, together with advances in computing, leads to intangible innovations in service activities and products, with a*

---

[1] Also referred to as computing in civil engineering, construction information technology or information and communication technology in construction.

I.F.C. Smith (Ed.): EG-ICE 2006, LNAI 4200, pp. 663–669, 2006.
© Springer-Verlag Berlin Heidelberg 2006

growing contributions from service industries in the business enterprise sector. The tools developed for identifying R&D in traditional fields and industries are not always easy to apply to these new areas."

By method the main distinction is between the theoretical and empirical research. By result type research can be analytic or synthetic (constructive). By what it studies we can distinguish between the study of objective real world and the study of subjective interpreted world. Other taxonomies [2] refer to philosophical grounding (positivism, negativism, historicism, critical rationalism, hermeneutics and interpretivism) and research strategies (induction, deduction, retroduction and abduction).

This paper argues that the main methodological problems in construction informatics research can be traced back to the wrong categorization of research according to these three criteria, in particular in the wrong assumption that it studies the objective real world.

1.2   Item of Concern in Construction Informatics

Schutz [3] distinguishes between first and second order constructs - the first corresponding to what some call "real world" and the second by what some would call "subjective, interpreted, constructed" world. Heidegger opened the debate weather not everything is in fact constructed, but to argue appropriate research methods in construction informatics we do not need to share this concern. We can safely assume that natural sciences (the study of nature including man, including physics, chemistry, biology, medicine) concern itself with first order constructs. They exist a-priori the observer, independently of him. They are not changed or influenced by observation or intellectual manipulation. For example, in a limited context of a structural mechanics theory, a structural beam is a first order construct. No matter what kind of theory we set about its behavior, in a lab, under weight, it would bend regardless of the favorite theory of the structural mechanics researcher.

The second order constructs are "constructs of the constructs made by actors" and are therefore influenced and changed by the observer(s). They are constructed within a personal, organizational, social or historical context. The concepts studied by construction IT (organizations, processes, works, information, representations, engineers, technologies) are such second order constructs. A model of how an architect and engineer collaborate, a new collaboration method or tool would change what was initially observed. A data structure describing a generic "beam" would influence the way this concepts is understood by the engineers.

2   1st Order Construction Informatics

Some fields of construction informatics are concerned with first order constructs. For example simulation is predicting some real world behavior and sensing is measuring the real world. Information technology is in this case supporting a natural science, for example mechanics, building physics etc. The "scientific method", in its best positivist tradition, can and should be used.

2.1   Scientific Method for the 1st Order Construction Informatics

The study of first order constructs fits well the traditional scientific method also referred to as "positivist science". For the illustration purposes we will use the famous Galileo's foundations of structural mechanics and materials science in 1638 [4]. The steps in the scientific method [5] are as follows:

1. Wonder. Pose a question. How is the load that a cantilever can sustain (Figure 1, middle), related to the tensile strength of that same beam (Figure 1, left).

2. Hypothesis. Suggest a plausible answer (a theory) from which some empirically testable hypothetical propositions can be deduced. Gallileo's proposed answer was

$$E = Sbh^2 / 2l \tag{1}$$

where S is the tensile strength so that

$$E = Sbh \tag{2}$$

3. Testing. Construct and perform an experiment which makes it possible to observe whether the consequences specified in one or more of those hypothetical propositions actually follow when the conditions specified in the same proposition(s) pertain. If the experiment fails, return to step 2, otherwise go to step 4. Galileo confirmed the formula by an experiment shown in Figure 1, right and which gave the appropriate relative strengths of both cantilevers.

4. Accept the hypothesis as provisionally true. Return to step 3 if there other predictable consequences of the theory which have not been experimentally confirmed. In 1729 the strength was reduced to $Sbh^2 / 3l$ and only around 1800 a correct formula $Sbh^2 / 6l$ came into use.

5. Act accordingly. Until the end of 18th century, structures were built using formulae that ware two or three fold on the dangerous side. Many of those structures still stand, including the pipelines for the water of Versailles that were built according to the 1729 equation.

Fig. 1. Gallileo's cantilever

Informatics research, to support a trivial case like this, would include:

1. A question. How can we quickly calculate cantilevers.
2. A hypothesis. We can write a computer program. A more specific hypothesis may include a particular technology that would be used - a spreadsheet or object oriented programming or grid computing or a neural network.
3. Testing. We write the program, make sure it runs and check, that the results are the same as if we were doing the calculation by hand.
4. The hypothesis would be confirmed.
5. And we would keep using the program and using new ones.

Another and equally trivial construction informatics research scenario could be:

1. A question. What is a cantilever? How do we model it?
2. A hypothesis. Cantilever is a such and such information structure. We draw (in the order as they came into fashion) a NIAM diagram, and E-R diagram, an EXPRESS-G diagram, a UML class diagram, write an XML schema or show a graphic representation of OWL.
3. Testing. We write a program that uses or reads such definition of a cantilever. The program is tested. Weather the results match the theoretical ones is of secondary importance since the hypothesis was about being able to model the cantilever structure and include all the information that a particular algorithm requires.
4. The hypothesis would be confirmed.
5. And we would keep using the program and using new ones using that particular modeling language or model.

In this last example, the modelers had the luxury of not having to model the cantilever as it is, but a model of a cantilever as defined in a simple, 17th century structural mechanics theory. A cantilever as such is a second order construct discussed in the next section.

3   2nd Order Construction Informatics

Understanding cantilevers as second order constructs would require to model a convention of what a cantilever is and this is essentially a philosophical question. Our thinking of it would probably affected by the Aristotle's four causes - the *causa materialis*, the *causa formalis*, the *causa finalis* and the *causa efficiens*. What is a cantiliver made of, what is the form of a cantilever, who makes cantilevers and what are cantilevers used for. These causes were accepted throughout the history of philosophy and are, last but not least, echoed in Geilling's [6] functional unites (*causa efficiens*) and technical solutions (*causa materialis, causa formalis*) or form-function-behavior models in the design sciences [7,8,9]. The point, however, is that one does not verify Aristotelian four causes by building software prototypes and, likewise, prototypes would have a limited credibility in verifying conceptual models.

To make matters worse, important areas of construction informatics are not concerned with what is objectively in the real world, but with socio technical constellations (to avoid a prejudice towards systems thinking). For example making models of organizations and implementing those models in software. In this case information technology is supporting what is essentially studied by social sciences (including organization sciences, economics, law), humanities (including philosophy, language studies) and art (including aesthetic, designing etc.).

Research methodology should be based on the research methodologies invented for those areas. Possible research approaches include: conceptual studies, prototyping (laboratory and field), surveys, case studies, phenomenological/hermeneutic research, action research and soft systems methodology. Underlying to all these approaches is the Socratic research method. Its main difference to the scientific method is that it is not looking outside of the system (e.g. to nature) for verification, but does so through argument and dialogue.

3.1 The Socratic Method

The Socratic process is as follows:

1. Wonder. Pose a question. I.e. what is a cantilever?

2. Hypothesis. Suggest a plausible answer (a definition or definiens) from which some conceptually testable hypothetical propositions can be deduced. I.e. we structure cantilever's information into form, function and behaviour properties.

3. Elenchus ; "testing," "refutation," or "cross-examination." Perform a thought experiment by imagining a case which conforms to the definiens but clearly fails to exemplify the definiendum, or vice versa. Such cases, if successful, are called counterexamples. If a counterexample is generated, return to step 2, otherwise go to step 4. I.e. we are trying to find examples where the accepted definition would fail. Border cases that, of example, exhibit some functional properties of a cantilever but have different structure. A SWOT analysis of the proposed model is performed.

4. Accept the hypothesis as provisionally true. Return to step 3 if you can conceive any other case which may show the answer to be defective.

5. Act accordingly. Write software according to that definition, use the form-function-behavior reference model for other cases.

3.2 Action Research

Action research is an increasingly popular method in informatics. It is a kind of research where researchers try out their theories with practitioners in real situations and in real organizations [10]. The process in action research is:

1) Enter the problem situation (also called area of concern) with personal interest. E.g. calculating cantelivers is my bread and butter.

2) Establish roles. E.g. I am the structural engineer.

3) Declare, explicitly, the framework of ideas and the methodology.

4) Take part in the change process and

5) repeat steps 2,3 and 4.

6) Exit the problem situation.

7) Reflect on experience and record learning that occurred in the process.

Action research combines the best of both epistemological approaches. The positivistic perspective of an objective, detached observer in step 7 and a phenomenological perspective of an involved, concerned, being-in-the-world practitioner in steps 1-5. A problem of the approach is that in many cases the researchers do not genuinely enter the problem situation.

3.3 Other Methods

Prototyping: We have raised some questions on the relevance of the prototypes in Section 2.1. There are few reports of failed prototypes or architectures that were later found impossible to build. There are, however, numerous research results with successful prototypes and demonstrators that to date did not materialize in a commercial solution. Verification of any hypothesis through prototype development generally also fails Popper's criteria for some research being scientific or not. He claimed a theory, a hypothesis, needs to be refutable. It should be possible to fail it [11]. Prototypes reported in construction informatics papers are generally succeeding. Few researchers state criteria at which the prototype would be considered a failure. Another difficulty is that if some process ran in the past without IT support, how relevant is a finding, that it can also run with IT support.

Conceptual modeling: One of the problems of research in construction informatics is that while in physics or medicine the general conceptualization of the studied world changes very very seldom (called paradigm shift by Kuhn) in informatics where novel information systems are first constructed and then, perhaps, analyzed, we seem to be working in a constant paradigm shift. Almost every new synthesized world presents itself for analysis in a way where another paradigm is invented, just for that one case or a generation of cases. This, however, is only true in the context of a positivist method. Action research frames the research with another intellectual structure where the model and the paradigm are systematically changeable.

4 Conclusions

In one of the key action research papers Susman and Evered [12] wrote: "...we suggest that the researcher ought to be skeptical of positivist science when (a) the unit of analysis is, like the researcher, a self-reflecting subject, (b) when relationships between subjects are influenced by definitions of the situation, or (c) when the reason for undertaking the research is to solve a problem which the actors have helped to define".(a,b,c added by author for clarity). Many problems in construction informatics research fit this definition perfectly, particularly they fit into (b) and (c).

Most researchers in construction informatics have backgrounds in architecture, engineering and computer science and have been brought up in the positivistic traditions of physics, mechanics and mathematics. We are trying hard to map the methods of those traditions into research areas that are of different nature. Most of the philosophical discussion on the scientific methods (e.g. Popper, Feyerabend, Kuhn) has been dealing with sciences that observe nature and whose hypotheses and theories are confirmed or refuted by observing some real world phenomena.

The experiences with construction informatics research paradigm evolution, particularly the one that takes information modeling as its underlying methodology, are inviting a conclusion concurring with Feyerabend. He claimed [13] that a theory of science in general is too crude for application to specific problems; specific in this case means the different paradigms that govern in astronomy, biology, mathematics, physics or, of course, construction informatics. It needs its own theory and methodology and should borrow not only from natural sciences but from humanities and social sciences as well. It is not only applied computer science, it has elements of applied humanity as well. Such understanding would most likely bring some stability into the paradigms and methods being used.

4.1 Further Work

Further work should review the research work in construction informatics and statistically confirm some of the generalized statements about construction informatics, such as the one in the first paragraph of this paper as well the statements about the success/failure of research work. A survey of a bibliographic database of construction informatics research and classifying papers by the used research method and success/failure or research is a possible method. Another possible method would a be a survey or a Delphi study among the statistically relevant sample of the researchers in the field.

References

1. OECD (2002) Frascati Manual, OECD Publication Services.
2. Blaikie, N. (1993). Approaches to Social Enquiry, Polity Press, Cambridge
3. Schutz, A. (1962). Common-Sense and Scientific Interpretation of Human Action, Collected Papers, Vol. 1: The Problem of Social Reality, Martinus Hijhoff, The Hague, Netherlands.
4. Petroski H. (1994). Design Paradigms, Cambridge University Press, UK.
5. Dye, J. (1996). Socratic Method and Scientific Method, http://www.soci.niu.edu/~phildept/Dye/method.html
6. Gielingh W (1988) General AEC reference model (GARM), Christiansson P, Karlsson H (ed.); Conceptual modelling of buildings. CIB W74+W78 seminar, October, 1988. Lund university and the Swedish building centre. CIB proceedings 126 http://itc.scix.net/cgi-bin/works/Show?w78-1988-165
7. de Kleer, J. and J. S. Brown (1983), "Assumptions and Ambiguities in Mechanistic Mental Models," Mental Models, D. Gentner and A. L. Stevens (Eds.), Lawrence Erlbaum Associates, New Jersey, pp. 155-190.
8. Henson, B., N. Juster and A. de Pennington (1994), "Towards an Integrated Representation of Function, Behavior and Form," Computer Aided Conceptual Design, Proceedings of the 1994 Lancaster International Workshop on Engineering Design, Sharpe J. and V. Oh (eds.), Lancaster University EDC, pp. 95-111.
9. Qian L. and J. S. Gero (1996), "Function-Behavior-Structure Paths and Their Role in Analogy-Based Design," Artificial Intelligence for Engineering Design, Analysis and Manufacturing, Vol. 10, No. 4, pp. 289-312.
10. Avison, D., Lau, F., Myers, M. and Nielsen, P. A. Action research, Communications of the ACM, vol. 42, no. 1, 1999.
11. Popper, K. (1935). The logic of scientific discovery, Routledge Classics, 2002 (first published in German in 1935).
12. Susman, G.I. and Evered, R.D. (1978). An Assessment of the Scientific Merits of Action Research, Administrative Science Quarterly, Vol. 23, No. 4 (Dec., 1978) , pp. 582-603.
13. Feyerabend, P. (1978). Against Method Verso, London, 1978.

Wireless Sensing, Actuation and Control – With Applications to Civil Structures

Yang Wang[1], Jerome P. Lynch[2], and Kincho H. Law[1]

[1] Dept. of Civil and Environmental Engineering, Stanford Univ., Stanford, CA 94305, USA
[2] Dept. of Civil and Environmental Engineering and Dept. of Electrical Engineering and Computer Science, Univ. of Michigan, Ann Arbor, MI 48109, USA
wyang98@stanford.edu, jerlynch@umich.edu, law@stanford.edu

Abstract. Structural monitoring and control have been subjects of interests in structural engineering for quite some time. Structural sensing and control technologies can benefit in terms of installation cost and time from wireless communication and embedded computing. The hardware and software requirements pose an interesting, interdisciplinary research challenge. This paper describes a low-cost wireless sensing system that is judiciously designed for large-scale applications in civil structures. Laboratory and field tests have been conducted to validate the performance of the prototype system for measuring vibration responses. By incorporating an actuation signal generation interface, the wireless sensing system has the capabilities to perform structural actuation and support structural control applications. Structural control tests have been performed to validate the wireless sensing and actuation system.

1 Introduction

Ensuring the safety of civil structures, including buildings, bridges, dams, tunnels, and others, is of utmost importance to society. Developments in many engineering fields, notably electrical engineering, mechanical engineering, material science, and information technology are now being explored and incorporated in today's structural engineering research and practice. For example, in the last couple of decades, structural sensors, such as micro-electro-mechanical system (MEMS) accelerometers, metal foil strain gages, fiber optic strain sensors, linear variable displacement transducers (LVDT), etc., have been employed to collect important information that could be used to infer the safety conditions or monitor the health of structures [1-3].

To limit the response of structures subjected to strong dynamic loads, such as earthquake or wind, structural control systems can be used. There are three basic types of structural control systems: passive, active and semi-active [4-6]. Passive control systems, e.g. base isolators, entail the use of passive energy dissipation devices to control the response of a structure without the use of sensors and controllers. Active control systems use a small number of large mass dampers or hydraulic actuators for the direct application of control forces. In a semi-active control system, semi-active control devices are used for indirect application of control forces. Examples of semi-active structural actuators include active variable stiffness (AVS) systems, semi-active hydraulic dampers (SHD), electrorheological (ER) and magnetorheological (MR) dampers. Semi-active control is currently preferred by many researchers, because of its reliability, low

I.F.C. Smith (Ed.): EG-ICE 2006, LNAI 4200, pp. 670–689, 2006.

power consumption, and adequate performance during large seismic events. In active or semi-active control systems, sensing devices are installed to record real-time structural response data for the calculation of control decisions.

In order to transfer real-time data in a structural monitoring or control system, coaxial cables are normally deployed as the primary communication link. However, cable installation is time consuming and can cost as much as $5,000 US dollar per communication channel [7]. Large-scale structures, such as long-span cable-stayed bridges, could easily require over thousands of sensors and miles of cables [8]. To eradicate the high cost incurred in the use of cables, wireless systems could serve as a viable alternative [9]. Wireless communication standards, such as Bluetooth (IEEE 802.15.1), Zigbee (IEEE 802.15.4), Wi-Fi (IEEE 802.11b), etc. [10], are now mature and reliable technologies widely adopted in many industrial applications. Potential applications of wireless technologies in structural health monitoring have been explored by a number of researchers [11-19]. A comprehensive review of wireless sensors and their adoption in structural health monitoring can be found in reference [20].

As opposed to structural monitoring, where sensors are used in a passive manner to measure structural responses, researchers have now begun to incorporate actuation interface in wireless sensors for damage detection applications [21-23]. For example, actuation interfaces can be used to induce stress waves in structural elements by wireless "active" sensors. Corresponding strain responses to propagating stress waves can be used to infer the health of the component. An integrated actuation interface can also be used to potentially operate actuators for structural control [24-26].

Compared to traditional cable-based systems, wireless structural sensing and control systems have a unique set of advantages and technical challenges. Portable energy sources, such as batteries, are a convenient, albeit limited, supply of power for wireless sensing units. Nevertheless, the need for reliable and low-cost energy sources remains a key challenge for wireless sensors [27-29]. Furthermore, data transmission in a wireless network is inherently less reliable than that in cable-based systems, particularly when node-to-node communication ranges lengthen. The limited wireless bandwidth can also impede real-time data transmission as required by feedback structural control systems. Last but not least, the time delay issues due to transmission and sensor blockage need to be considered [25,26,30]. These issues need to be resolved with a system approach involving the selection of hardware technologies and the design of software/algorithmic strategies.

A "smart" sensor combines both hardware and software technologies to provide the capabilities that can acquire environmental data, process the measured data and make "intelligent" decisions [18]. The development of autonomous, self-sensing and actuating devices for structural monitoring and control applications poses an intriguing, interdisciplinary research challenge in structural and electrical engineering. The purpose of this paper is to describe the design and implementation of a modular system consisting of autonomous wireless sensor units for civil structures applications. Designed for structural monitoring applications, the wireless sensor consists of a sensing interface to which analog sensors can be attached, an embedded microcontroller for data processing, and a spread spectrum wireless radio for communication. Optionally, for field applications where signals subject to environmental effects and ambient vibrations are relatively noisy, a signal conditioning board is designed to interface with the wireless sensing unit for signal amplification and filtering. To support active sensing and control applications, a signal generation module is designed to interface with the wireless sensing unit. This wireless actuation unit combining sensing, data processing, and signal generation, can be used to issue desired actuation commands for real-time feedback structural control. Laboratory and field validation tests are presented to assess the performance of the wireless sensing and actuation unit for structural monitoring and structural control applications.

2   Hardware Design of Wireless Sensing and Actuation Units

The building block of a wireless monitoring system is the wireless sensing unit. Fig.1 shows the overall hardware of the wireless sensing unit, and the two optional off-board auxiliary modules for conditioning analog sensor outputs and actuation signal generation. This section first describes in detail the key characteristics and components of the wireless sensing unit design. Off-board modules for signal conditioning and actuation command generation are then presented.

2.1   Wireless Sensing Unit

A simple star-topology network, which is adopted for the prototype wireless sensing system, includes a server and multiple wireless sensing units (Fig. 2). The functional diagram of the proposed wireless sensing unit is shown in the top part of Fig. 1. The wireless sensing unit consists of three functional modules: sensor signal digitization, computational core, and wireless communication. The sensing interface converts analog sensor signals into digital data which is then transferred to the computational core through a high-speed Serial Peripheral Interface (SPI) port. Besides a low-power 8-bit Atmel ATmega128 microcontroller, external Static Random Access Memory (SRAM) is integrated with the computational core to accommodate local data storage and analysis. The computational core communicates with a wireless transceiver through a Universal Asynchronous Receiver and Transmitter (UART) interface.

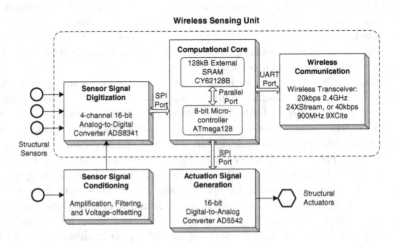

Fig. 1. Functional diagram detailing the hardware design of the wireless sensing unit. Additional off-board modules can be interfaced to the wireless sensing unit to condition sensor signals and issue actuation commands.

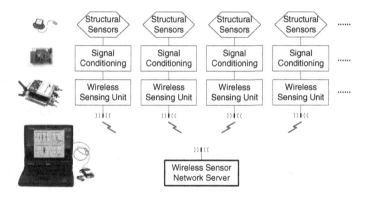

Fig. 2. An overview of the prototype wireless structural sensing system

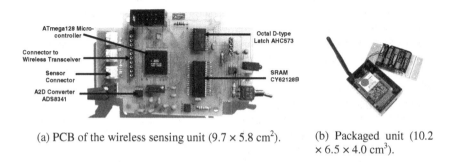

(a) PCB of the wireless sensing unit (9.7 × 5.8 cm²).

(b) Packaged unit (10.2 × 6.5 × 4.0 cm³).

Fig. 3. Pictures of the wireless sensing unit

A simple two-layer printed circuit board (PCB) is designed and fabricated. As shown in Fig. 3, the PCB, wireless transceiver, and batteries are stored within an off-the-shelf weatherproof plastic container, which has a dimension of 10.2 by 6.5 by 4.0 cm³. Each sensing unit acts as an autonomous node capable of collecting, processing, and wirelessly transmitting data to other sensing units and the central server.

Sensing Signal Digitization Module

The main component of the sensor signal digitization module is a 4 channel 16 bit analog-to-digital (A/D) converter, Texas Instruments ADS8341. Each wireless sensing unit can accommodate signals from a heterogeneous set of up to 4 analog structural sensors (e.g. accelerometers, strain gages, among others). The 16 bit A/D resolution is sufficient for most applications in structural sensing. One requirement from the ADS8341 A/D converter is that the sensor signal should be between 0 and 5V. The highest sampling rate supported by this A/D converter is 100kHz, which is much higher than the sampling frequency typically needed for monitoring civil structures. This rate determines that each sampling consumes only 10μs. Therefore, the A/D conversion can be finished swiftly through the timer interrupt service of the micro-controller (ATmega128), without interrupting the execution of wireless communication or data processing programs.

Computational Core

The computational core of a wireless unit is responsible for executing embedded software instructions as required by the application end-user. A low-cost 8-bit microcontroller, Atmel ATmega128, is selected for this purpose. The key objective for this selection is to balance the power consumption and hardware cost versus the computation power needed by software applications. Running at 8MHz, the ATmega128 consumes about 15mA when it is active. Considering the energy capacity of normal batteries in the market, which is usually a few thousand milliamp-hours (mAh), normal AA batteries can easily support the ATmega128 active for hundreds of hours. Running in a duty cycle manner, with active and sleep modes interleaved, the ATmega128 microcontroller may sustain even longer before battery replacement is needed.

The ATmega128 microcontroller contains 128kB of reprogrammable flash memory for the storage of embedded software, which, based on our laboratory and field experiments, is sufficient to incorporate a wide variety of structural monitoring and control algorithms. One serial peripheral interface (SPI) and two universal asynchronous receiver and transmitter (UART) interfaces are provided by the ATmega128 to facilitate communication with other hardware components. The timer and interrupt modules of the ATmega128 are employed for executing routines that need to be precisely timed, e.g. sampling sensor data or applying actuation signal at specified frequencies.

The microcontroller also contains 4kB static random access memory (SRAM) for storing stack and heap variables, which as it turns out, is often insufficient for the execution of embedded data interrogation algorithms. To address this issue, an external 128kB memory chip, Cypress CY62128B, is incorporated within the wireless sensing unit design. Furthermore, hardware and software procedures are implemented to bypass the 64kB memory address space limitation of the ATmega128, to ensure that the full 128kB address space of the CY62128B can be utilized.

Wireless Communication Module

The wireless communication module provides the interface for the unit to exchange data with other wireless units, or a data server with a wireless transceiver attached. Sufficient communication reliability, range, and data transfer rate are needed to employ the wireless units in civil structures. On the other hand, due to stringent battery power constraints, the wireless module, which is the most power-consuming component of a typical wireless sensing unit, should not consume too much battery power while active. Certain trade-offs have to be achieved to delicately balance performance and low power requirements. Wireless frequency allocation regulated by the government is another factor that should be considered while selecting wireless transceivers.

The wireless sensing unit is designed to be operable with two different wireless transceivers: 900MHz MaxStream 9XCite and 2.4GHz MaxStream 24XStream. Pin-to-pin compatibility between these two wireless transceivers makes it possible for the two modules to share the same hardware connections in the wireless unit. Because of the different data rates, embedded software for using the two transceivers is slightly different. This dual-transceiver support affords the wireless sensing/actuation unit to have more flexibility in terms of not only geographical area, but also data transfer rate, communication range, and power consumption. Table 1 summarizes the key performance parameters of the two wireless transceivers. As shown from the table, the data transfer rate of the 9XCite is double that of the 24XStream, while 24XStream provides a longer communication range but consumes much more battery power. Both transceivers support peer-to-peer and broadcasting communication modes, rendering information flow in the wireless sensor network more flexible.

Table 1. Key performance parameters of the wireless transceivers

Specification	9XCite	24XStream
Operating Frequency	ISM 902-928 MHz	ISM 2.4000 – 2.4835 GHz
Channel Mode	7 frequency hopping channels, or 25 single frequency channels	7 frequency hopping channels
Data Transfer Rate	38.4 kbps	19.2 kbps
Communication Range	Up to 300' (90m) indoor, 1000' (300m) at line-of-sight	Up to 600' (180m) indoor, 3 miles (5km) at line-of-sight
Supply Voltage	2.85VDC to 5.50VDC	5VDC (±0.25V)
Power Consumption	55mA transmitting, 35mA receiving, 20µA standby	150mA transmitting, 80mA receiving, 26µA standby
Module Size	1.6" × 2.825" × 0.35" (4.06 × 7.17 × 0.89 cm^3)	1.6" × 2.825" × 0.35" (4.06 × 7.17 × 0.89 cm^3)
Network Topology	Peer-to-peer, broadcasting	Peer-to-peer, broadcasting

* For details about the transceivers, see http://www.maxstream.net.

2.2 Sensor Signal Conditioning Module

For field applications, a wireless monitoring system must be capable of recording both ambient and forced structural vibrations. With ambient vibrations typically defined by small amplitudes, a high-resolution (16-bit or higher) A/D converter is normally needed by a structural monitoring system. The placement of the low-cost 16-bit ADS8341 A/D converter leaves the A/D vulnerable to electrical noise present in the circuit. From experimental tests, the effective resolution for the A/D channels is found to be approximately 13-bit, which is likely insufficient for sampling low-amplitude vibration data. Additionally, for the ADS8341 A/D converter, the sensor signals must be within 0 to 5V. A signal conditioning module is thus needed to amplify signals, filter out noise, and shift sensor signals within range.

Sensor signals are fed through the signal conditioning module prior to the A/D conversion, as shown in the lower left part of Fig. 1. As shown in Fig. 4(a), the filtering circuits consist of a high-pass resistor-capacitor (RC) filter with a cutoff frequency of 0.02Hz and a low-pass fourth-order Bessel filter with a cutoff frequency of 25Hz. The linear-phase shift property of the Bessel filter ensures a constant time delay for signals in the pass band, thus maintaining the signal waveform in the time domain. Fig. 4(b) shows the complete signal conditioning board that includes circuit modules that support the filtering, offsetting, and amplification functions.

To illustrate the performance of the signal conditioning module, Fig. 5 shows two acceleration time histories, where the signal outputs are fed into the A/D converter with and without the signal conditioning (S.C.) module. As shown in Fig. 5(a), when the vibration amplitude is low, in which case the Signal-to-Noise-Ratio (SNR) is low, the sensor data with signal conditioning becomes much smoother than the data without signal conditioning. When the vibration amplitude is higher, i.e. when the SNR is high, the difference between the data collected with and without signal conditioning is almost negligible with respect to the signal amplitude, as shown in Fig. 5(b).

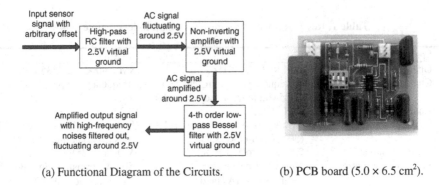

(a) Functional Diagram of the Circuits.   (b) PCB board (5.0×6.5 cm^2).

Fig. 4. Sensor signal conditioning module

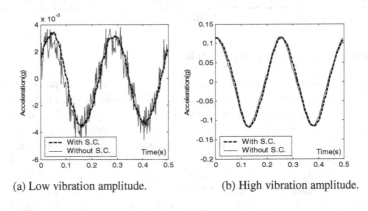

(a) Low vibration amplitude.   (b) High vibration amplitude.

Fig. 5. Wireless accelerometer data with and without signal conditioning

2.3 Actuation Signal Generation Module

The functionality of the wireless sensing unit can be extended to support structural actuation and control applications. The key component of the actuation signal generation module is the Analog Device AD5542 digital-to-analog (D/A) converter which converts unsigned 16-bit integer numbers issued by the microcontroller into a zero-order hold analog output spanning from -5 to 5V. It should be noted that the wireless sensor is based upon 5V electronics; this requires an auxiliary -5V power supply to be included in the actuation signal generation module. The switching regulator, Texas Instruments PT5022, is employed to convert the 5V voltage source from the wireless sensing unit into a regulated -5V signal. Another component included in the actuation signal generation module is an operational amplifier (National Semiconductor

LMC6484), to shift the output signal to have a mean of 0V. The actuation signal generation module is capable of outputting -5 to 5V analog signals within a few microseconds after the module receives the digital command from the microcontroller.

The actuation signal generation module is connected with the wireless sensing unit through two multi-line cables: an analog signal cable and a digital signal cable. The digital signal cable connects between the D/A converter of the signal generation module to the microcontroller of the wireless sensing unit via the SPI interface. The analog cable is used to transfer an accurate +5V voltage reference, from the wireless sensing unit to the actuation board. The generated actuation signal is transmitted to the structural actuator through a third output cable in the module. Fig. 6 shows the signal conditioning module, which is designed as a separate board readily interfaced with the wireless sensing unit for actuation and control applications.

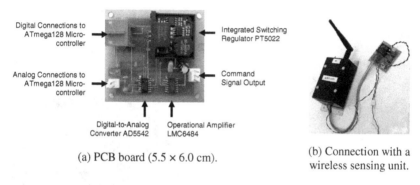

(a) PCB board (5.5 × 6.0 cm).

(b) Connection with a wireless sensing unit.

Fig. 6. Pictures of the control signal module

3 Wireless Sensing for Structural Testing and Monitoring

The wireless sensing unit prototypes (with and without the auxiliary modules), have undergone a number of large-scale validation tests to assess performance and verify the prototype design and implementation [31-33]. Field validation tests are particularly important, since they subject the wireless units under the complexities of real structural environments. For example, the long range communication of the wireless sensors is quantified. Furthermore, structural obstructions can pose significant challenges for propagation of wireless communication signals. Last but not least, for typical field tests, ambient responses from normal daily operations give very low signal amplitude, which could be difficult to measure with high precision. This section describes two validation results; first is a large shake table test conducted at the National Center for Research on Earthquake Engineering (NCREE) in Taipei, Taiwan, and the second is a field test conducted at Geumdang Bridge, Icheon, South Korea.

3.1 Laboratory Tests on a 3-Story Steel Frame at NCREE, Taiwan

In collaboration with researchers at NCREE, Taiwan, the wireless monitoring system is installed within a three-story steel frame structure mounted on a shake table. As shown in Fig. 7(a) and (b), the three-story single-bay steel frame structure has a 3 by $2m^2$ floor area and a 3m inter-story height. H150x150 x7x10 I-section elements are used for all columns and beams with each beam-column joint designed as bolted connections. Each floor is loaded with concrete blocks and has a total mass of 6,000kg. The test structure is mounted on a 5 by $5m^2$ shake table capable of applying base motion in 6 independent degrees-of-freedom.

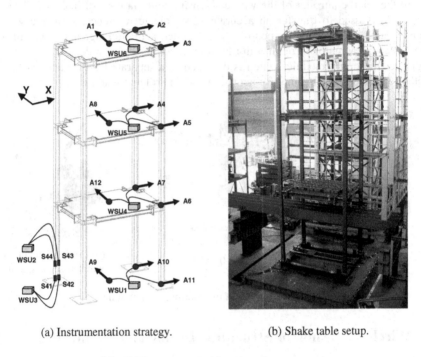

(a) Instrumentation strategy.   (b) Shake table setup.

Fig. 7. Three-story steel benchmark structure

As presented in Fig. 7(a), the test structure is instrumented with a wireless monitoring system consisting of 6 wireless sensing units. Because of local frequency band requirements, the MaxStream 24XStream wireless transceiver operating at 2.4GHz spectrum is employed for the wireless sensing unit. The instrumentation strategy of the wireless monitoring system is governed by an interest in both the acceleration response of the structure as well as the strain behavior at the base column. As shown in Fig. 7(a), one wireless sensing unit is responsible for the three accelerometers instrumented on a floor. For example, wireless sensing unit WSU6 is used to record the acceleration of the structure at locations A1, A2 and A3. This configuration of accelerometers is intended to capture both the longitudinal and lateral response of each floor, as well as any torsion behavior.

The accelerometers employed with the wireless sensing units are the Crossbow CXL01 and CXL02 (MEMS) accelerometers, which have acceleration ranges of ±1g and ±2g, respectively. The CXL01 accelerometer has a noise floor of 0.5mg and a sensitivity of 2V/g, while the CXL02 accelerometer has a noise floor of 1mg and a sensitivity of 1V/g. Additionally, 4 metal foil strain gages with nominal resistances of 120Ω and a gage factor of 2, are mounted on the base column to measure the column flexural response during base excitation. To record the strain response, a Wheatstone bridge amplification circuit is used to convert the changes in gage resistance into voltage signals. Two wireless sensing units (WSU2 and WSU3) are dedicated to recording the strain response with each unit connecting to two gages. As for comparison, Setra141-A accelerometers (with acceleration range of ±4g and a noise floor of 0.4mg) and 120Ω metal foil strain gages connecting to a traditional cable-based data acquisition system are installed side-by-side.

Various ambient white noise and seismic excitations, including El Centro (1940), Kobe (1995), and Chi-Chi (1999) earthquake records, were applied to excite the test structure [33]. The results shown in Fig. 8 are based on a 90 sec bi-directional white noise excitation of 1m/s and 0.5m/s standard deviation velocities in the X and Y directions respectively. The time history responses for both acceleration and strain measurements recorded (at locations A1, A2, and S44) by the wireless monitoring system are identical to those measured independently by the cable-based monitoring system.

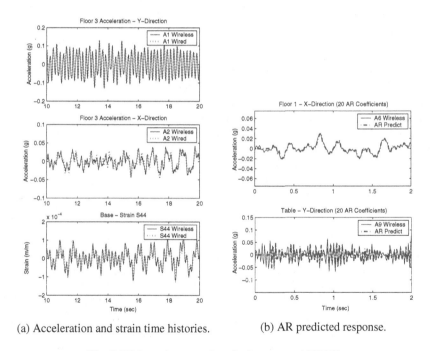

(a) Acceleration and strain time histories.   (b) AR predicted response.

Fig. 8. White noise ground excitation tests at NCREE

To illustrate the utilization of the on-board microcontroller for data interrogation, an auto-regressive (AR) time series model which is often used for damage detection applications [34] is implemented. During the test, the wireless sensing units determine the optimal AR model fit to the acceleration and strain data. Once the AR model is calculated, the model coefficients are then transmitted to the central server. As shown in Fig. 8(b), the acceleration time history responses reconstructed using 20 AR model coefficients are compared with the directly recorded raw time history data at sensor locations A6 and A9 (located on the first floor and table, respectively). The reconstructed time history using the AR model coefficients can accurately predict the response of the structure. That is, with the microcontroller, useful computations can be performed on the wireless sensing unit, and the amount of data (in this case, the AR coefficients) that need to be transmitted in real time can be significantly reduced.

3.2 Field Validation Tests at the Geumdang Bridge, South Korea

In collaboration with researchers at the Korea Advanced Institute of Science and Technology (KAIST), a field validation test has been conducted on the Geumdang Bridge in Icheon, South Korea. Designed and managed by the Korea Highway Corporation (KHC), the two-lane test road is heavily instrumented to measure the performance of the pavement systems [35]. One convenient feature of this testing venue is that KHC has the ability to open or close the test road at will and to reroute traffic.

The Geumdang Bridge has a total span of 273m, and is designed using two different section types. The northern portion is constructed with four independent spans (with span lengths of 31, 40, 40 and 40m respectively), each of which is designed using a 27cm concrete deck supported by four pre-cast concrete girders. The validation test is performed on the southern portion of the Geumdang Bridge. As shown in Fig. 9, the southern portion of the bridge is constructed with a continuous 122m long post-tensioned box girder. The depth of the box girder is 2.6m, and the width at the bridge deck is about 12.6m. The southern portion is subdivided into three sections (38, 46 and 38m respectively) which are supported by the abutment and the three piers.

As shown in Fig. 9(d), a total of 14 accelerometers are deployed for the wireless system. Furthermore, 13 cabled accelerometers are also instrumented with the tethered monitoring system. The accelerometers and their characteristics for the two systems are tabulated in Table 2. The accelerometers used in the cabled system are PCB piezoelectric accelerometers Piezotronics 393B12, which has a very low noise floor of only 50μg, and a high sensitivity of 10V/g and is well suited for use in ambient vibration applications, because of its low noise to signal level. Additionally, the cabled system employs a 16 channel PCB Piezotronics 481A03 signal conditioner which can simultaneously amplify (up to a gain of 200) and filter the sensor signal before digitization. A National Instruments 12-bit data acquisition card, NI DAQCard-6062E, is used to sample and collect the conditioned signal.

For the wireless system, an inexpensive PCB Piezotronics 3801D1FB3G MEMS accelerometer is selected. Operating at 5 VDC, this capacitive accelerometer can be conveniently powered by the wireless sensing unit. The noise floor of the PCB3801 accelerometers is 500μg, which is ten times that of the PCB383 accelerometers used in the cabled system. Meanwhile, the sensitivity of the PCB3801 is lower than the PCB383. Therefore, with a lower Signal-to-Noise-Ratio (SNR) ratio, signals from the PCB3801 are expected to be noisier than those from the PCB383. The wireless sensing units are installed with MaxStream 9XCite radios operating on the 900MHz frequency spectrum. To improve the SNR ratio, the optional signal conditioning module as described in section 2.2 is employed for the wireless sensing unit. A laptop

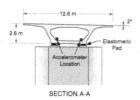

SECTION A-A

(a) Section view of the girder.

(b) Side view picture of the bridge.

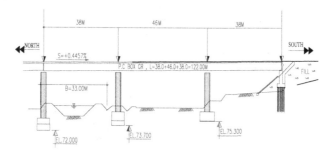

(c) Elevation view on the southern portion of the bridge.

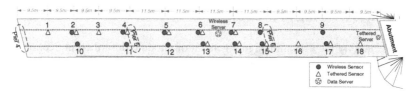

(d) Plan view of the accelerometer locations on the Geumdang Bridge

Fig. 9. Illustration of the Geumdang Bridge and wireless sensor deployment

Table 2. Parameters for accelerometers used in the cabled and wireless sensing systems

	PCB393 (Cabled System)	PCB3801 (Wireless System)
Sensor Type	Piezoelectric	Capacitive
Maximum Range	±0.5 g	±3 g
Sensitivity	10 V/g	0.7 V/g
Bandwidth	2000 Hz	80 Hz
RMS Resolution (Noise Floor)	50 μg	500 μg
Minimal Excitation Voltage	18 VDC	5 VDC

connected with a MaxStream 9XCite transceiver, located at around the middle of the bridge, is employed to collect sensor data from all the 14 wireless sensing units.

Vibration tests are conducted by driving a 40-ton truck at set speeds to induce structural vibrations into the system. For all the tests conducted, no data losses have been observed and the wireless sensing system proves to be highly reliable using the designed communication protocol for synchronized and continuous data acquisition. Fig. 10(a) shows the acceleration data recorded with a sampling rate of 200Hz at sensor location #17 when the truck was crossing the bridge at 60km/h. Despite the difference in the accelerometer and signal conditioning devices, the recorded output by the wireless system has the precision identical to that offered by a commercial cabled system. With the microcontroller, an embedded 4096 point FFT algorithm is used to determine the Fourier transform to the acceleration data. As shown in Fig. 10(b), the first three dominant frequencies can easily be identified as 3.0, 4.3 and 5Hz, which are very close to the bridge natural frequencies previously published [35].

Once the dominant frequencies are determined, the complex numbers of the Fourier transform in the interested frequency range are wirelessly transmitted to the central server, so that the operational deflection shapes (ODS) of the bridge under the truck loading can be computed. Fig. 10(c) illustrates the ODS for the first three dominant frequencies computed from the wireless sensor data. The ODS shapes are not the bridge mode shapes, since the external excitation by driving the truck along the bridge is difficult to accurately quantify. Nevertheless, the ODS shapes are dominated by the corresponding modes and are typically good approximations to the mode shapes.

4 Wireless Sensing and Control

Supplemented by the actuation signal generation module described in Section 2.3, the functionality of the wireless sensing unit can be extended to command structural actuators for structural control applications. With a wireless sensor network capable of exchanging real-time sensor data among the sensing units, feedback control decisions can be determined in real time to limit structural responses. The current prototype implementation focuses on the use of a wireless system for semi-active control using MR dampers as control actuators. Fig. 11 illustrates the operations of a wireless structural sensing and control system, termed herein as WiSSCon.

As shown in the figure, the system consists of multiple wireless sensing and actuation units. The wireless sensing units (S_1, S_2, and S_3) collect structural response data. At each time step, the wireless control unit (C_1), which consists of a wireless sensing unit and an actuation signal generation module, broadcasts a beacon signal to all the sensing units via the wireless communication channel. Upon receiving the beacon signal, the sensing units immediately send the sensor data to the control unit. The control unit processes the sensor data with the embedded microcontroller, and computes the control signals to be sent to actuate the MR damper.

To validate the concept of the WiSSCon system, experimental tests were conducted at NCREE, Taiwan, using the same three-story steel frame described in section 3.1. In the feedback control experiments, the test structure is implemented with accelerometers, velocity transducers, and LVDTs at all floors. Both accelerometers and velocity transducers are connected to the wireless sensing units. A cabled data acquisition system is used to collect the test data from the sensors for later analysis. Furthermore, a cabled control system is also available and is used to serve as the baseline reference system to which the WiSSCon system can be compared.

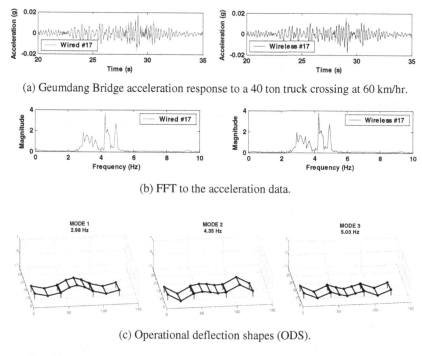

(a) Geumdang Bridge acceleration response to a 40 ton truck crossing at 60 km/hr.

(b) FFT to the acceleration data.

(c) Operational deflection shapes (ODS).

Fig. 10. Geumdang Bridge forced vibration data and frequency-domain analysis

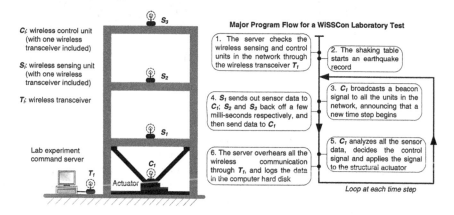

Fig. 11. Example illustration of the WiSSCon system instrumented on a 3-story test structure with one actuator

The laboratory setup for the structural control experiment is shown in Fig. 12. An MR damper with a maximum force capacity of 20kN and a piston stroke of ±0.054m is installed at the base floor. During a dynamic test, the damping coefficient of the MR damper can be changed in real time by issuing an analog command signal between 0 to 1V. This command signal controls the electric current of the electromagnetic coil in the MR damper, which in turn, generates the magnetic field that sets the viscous damping properties of the MR fluid inside the damper. The damper hysteresis behavior is determined using a modified Bouc-Wen model [36]. During dynamic excitation, the control unit, either cabled or wireless, has to maintain the time history of the damper model so that the damper hysteresis is known at all times and a suitable voltage can be determined for the MR damper.

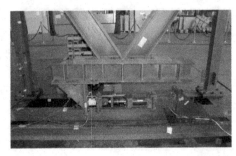

(b) The MR damper and its supporting brace.

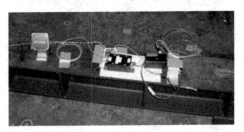

(a) Shake table setup.   (c) The wireless control unit.

Fig. 12. Wireless structural sensing and control test with one MR damper installed between the 1st floor and the base floor of the structure

In this structural control experiment, a discrete Linear Quadratic Regulator (LQR) control algorithm is employed to compute control forces applied by the MR damper. Detailed descriptions of the discrete LQR algorithm can be found in many control textbooks [37]. In essence, weighting matrices are selected for a scalar cost function that considers the state response of the structure and the energy required by the system actuators. The LQR algorithm determines a constant feedback "gain" matrix, which is used to optimally compute the desired control force based on the sensor output measurements. Prior to the dynamic tests, the constant "gain" matrix can be embedded into the microcontroller of the wireless control unit for real-time execution. Specifically, the experimental study is designed to execute velocity feedback control in real time. Each wireless sensing or control unit collects the floor velocity using a Tokyo Soku-shin VSE15-D velocity meter, which has a measurement range of ±1m/s, sensitivity of 10V/(m/s) and a dynamic frequency range of 0.1 to 70Hz. The transducer is well suited for the test structure whose primary modal frequencies fall well below 10Hz.

Time delay is an important issue in real-time feedback control. There are three major components that constitute the time delay: sensor data acquisition, control decision calculation, and actuator latency in applying the desired control force. Normally, the control decision calculation time is the minimum among the three, while the actuator latency being the maximum. In the LQR formulation, time delay from sensor data collection to control force application is assumed to be zero, even though a non-zero time delay always exists in practice. In active structural control, time delay may cause system instability, in which the control force could actually excite the structure. In semi-active control, the actuators normally dissipate vibration energy, without the capability to excite the structure. Nevertheless, large time delay remains an important issue in semi-active control, since it can degrade the performance of the system.

The difference between a wired control system and a wireless control system is mostly in the sensor data acquisition time. For the cabled control system, it is estimated that the time delay due to data acquisition is approximately 5ms. For the Max-Stream 24XStream wireless transceivers, a single wireless transmission time delay is about 20ms. For this experimental study, at each time step, four wireless transmissions are performed: a beacon signal sent by the control unit and 3 data packets transmitted one at a time by the three wireless sensors. Therefore, the WiSSCon system implemented upon the 24XStream wireless transceiver provides a control time step of about 80ms resulting in an achievable sampling frequency of 12.5Hz. Besides validating the concept of feedback wireless structural sensing and control, the experimental study attempts to investigate the influence of this time delay difference.

Fig. 13 shows the maximum absolute inter-story drift of each floor obtained from the tests conducted using the El Centro earthquake record with its peak acceleration scaled to 1m/s^2. Detailed description of the structural control tests can be found in reference [26]. As for comparison, two passive control tests when the MR damper voltage is fixed at 0 and 1V respectively, are also shown in the figure. The results illustrate that the LQR control with the cabled and wireless systems give more uniform maximum inter-story drifts for all three floors than the passive control tests. These preliminary results also show the wireless control system, even though with significantly larger time delay, suffers only minor performance degradation. Current investigation to improve wireless communication and minimize time delay is underway.

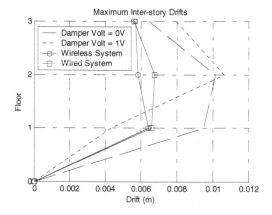

Fig. 13. Maximum inter-story drifts for tests with a scaled El Centro record as ground excitation

5 Summary and Discussion

Smart sensing devices must include capabilities that can interface with sensors, process acquired data and make decisions for a particular application of interest. Research in this area requires all facets of hardware technologies and software strategies to be selected, designed and implemented for the application. This paper describes the basic modules that compose a wireless sensing device. Building upon off-the-self components, the prototype wireless sensing unit described in this paper is capable of monitoring civil structures subjected to ambient and forced vibrations. The cost of wireless monitoring systems, including labor as well as installation efforts, is significantly lower than that of tethered systems that require installation of extensive lengths of coaxial cables. In addition, the performance features of a wireless sensing system differ greatly from the tethered counterparts. Wireless systems are highly decentralized with A/D conversion and data processing performed locally at the wireless sensing units, as opposed to at the central server. Embedded computations allow parallel processing of measurement data and lower energy consumption. Structural monitoring algorithms can be implemented on the sensor units for data processing and decision making. However, precise synchronization of raw time history data in a large scale wireless monitoring system remains a challenging task.

By including an actuation signal generation module in the wireless sensing unit, the potential application of a WiSSCon system for feedback structural control has been illustrated. The WiSSCon system could not only lead to significant reduction in system cost by eradicating cables in the control system but also, by using low cost microcontrollers, provide a highly flexible and adaptable system configuration because of wireless communication. With the embedded microcontroller, sophisticated "intelligent" computational strategies can also be incorporated [38,39]. For real-time feedback applications, the adverse effects of communication and computation time delay using the WiSSCon system could be mitigated by using algorithms that can specifically address the time delay issue. One possibility is to explore decentralized (or partially decentralized) structural control algorithms [40,41] that optimal control decisions are made using measurement output data from its own sensor or from only their neighboring units. Feasibility study and laboratory investigation of wireless decentralized controls are currently underway [42]. Further research includes investigation of "intelligent" strategies that can be implemented in the WiSSCon system for civil structure monitoring and control applications.

Acknowledgement

This research is partially funded by the National Science Foundation under grants CMS-9988909 (Stanford University) and CMS-0421180 (University of Michigan), and the Office of Naval Research Young Investigator Program awarded to Prof. Lynch at University of Michigan. The first author is supported by an Office of Technology Licensing Stanford Graduate Fellowship. Additional support was provided by the Rackham Grant and Fellowship Program at the University of Michigan. Prof. Chin-Hsiung Loh, Dr. Pei-Yang Lin, and Mr. Kung-Chun Lu at National Taiwan University provided generous support for conducting the shake table experiments at

NCREE, Taiwan. The authors would also like to express their gratitude to Professors Chung Bang Yun and Jin Hak Yi, as well as Mr. Chang Geun Lee, from the Korea Advanced Institute of Science and Technology (KAIST), for the access to Geumdang Bridge. During this study, the authors have received many valuable advices on the PCB layout from Prof. Ed Carryer at Stanford University. The authors appreciate the generous assistance from the individuals acknowledged above.

References

1. Chang, P.C., Flatau, A., Liu, S.C.: Review Paper: Health Monitoring of Civil Infrastructure. Struct. Health Monit. 2 (2003) 257-267
2. Farrar, C.R., Sohn, H., Hemez, F.M., Anderson, M.C., Bement, M.T., Cornwell, P.J., Doebling, S.W., Schultze, J.F., Lieven, N., Robertson, A.N.: Damage Prognosis: Current Status and Future Needs. Report LA-14051-MS, Los Alamos National Laboratory, NM (2003)
3. Elgamal, A., Conte, J.P., Masri, S., Fraser, M., Fountain, T., Gupta, A., Trivedi, M., El Zarki, M.: Health Monitoring Framework for Bridges and Civil Infrastructure. Proc. of the 4th Int. Workshop on Structural Health Monitoring, ed. F.-K. Chang. Stanford, CA (2003) 123-130
4. Yao, J.T.: Concept of Structural Control. ASCE J. of Struct. Div.. 98 (1972) 1567-1574
5. Soong, T.T., Spencer, B.F.: Supplemental Energy Dissipation: State-of-the-art and State-of-the-practice. Engng. Struct. 24 (2002) 243-259
6. Chu, S.Y., Soong, T.T., Reinhorn, A.M.: Active, Hybrid, and Semi-active Structural Control: a Design and Implementation Handbook. John Wiley & Sons, NJ (2005)
7. Celebi, M., Seismic Instrumentation of Buildings (with Emphasis on Federal Buildings). Report No. 0-7460-68170 United States Geological Survey (USGS). Menlo Park, CA (2002)
8. Solomon, I., Cunnane, J., Stevenson, P.: Large-scale Structural Monitoring Systems. Proc. of SPIE Nondestructive Evaluation of Highways, Utilities, and Pipelines IV. SPIE Vol. 3995, ed. A.E. Aktan, S.R. Gosselin (2000) 276-287
9. Straser, E.G., Kiremidjian, A.S.: A Modular, Wireless Damage Monitoring System for Structures. Report No. 128, John A. Blume Earthquake Eng. Ctr., Stanford Univ., (1998)
10. Cooklev, T.: IEEE Wireless Communication Standards: A Study of 802.11, 802.15, and 802.16. IEEE Press, NY (2004)
11. Kling, R.M.: Intel Mote: an Enhanced Sensor Network Node. Proc. of Int. Workshop on Advanced Sensors, Struct. Health Monitoring, and Smart Struct.. Keio Univ., Japan (2003)
12. Lynch, J.P., Sundararajan, A., Law, K.H., Kiremidjian, A.S., Kenny, T., Carryer, E.: Embedment of Structural Monitoring Algorithms in a Wireless Sensing Unit. Struct. Eng. Mech. 15 (2003) 285-297
13. Arms, S.W., Townsend, C.P., Galbreath, J.H. Newhard, A.T.: Wireless Strain Sensing Networks. Proc. of 2nd Euro. Work. on Struct. Health Monitoring. Munich, Germany (2004)
14. Glaser, S.D.: Some Real-world Applications of Wireless Sensor Nodes. Proc. of SPIE 11th Annual Inter. Symp. on Smart Structures and Materials. SPIE Vol. 5391, ed. S.C. Liu. San Diego, CA (2004) 344-355
15. Mastroleon, L., Kiremidjian, A.S., Carryer, E., Law, K.H.: Design of a New Power-efficient Wireless Sensor System for Structural Health Monitoring. Proc. of SPIE 9th Annual Int. Symp. on NDE for Health Monitoring and Diagnostics. SPIE Vol. 5395, ed. S.R. Doctor, Y. Bar-Cohen, A.E. Aktan, H.F. Wu. San Diego, CA (2004) 51-60

16. Ou, J.P., Li, H., Yu, Y.: Development and Performance of Wireless Sensor Network for Structural Health Monitoring. Proc. of SPIE 11th Annual Inter. Symp. on Smart Structures and Materials. SPIE Vol. 5391, ed. S.C. Liu. San Diego, CA (2004) 765-773
17. Shinozuka, M., Feng, M.Q., Chou, P., Chen, Y., Park, C.: MEMS-based Wireless Real-time Health Monitoring of Bridges. Proc. of 3rd Inter. Conf. on Earthquake Engng. Nanjing, China (2004)
18. Spencer, B.F. Jr., Ruiz-Sandoval, M.E., Kurata, N.: Smart Sensing Technology: Opportunities and Challenges. Struct. Control Health Monit. 11 (2004) 349-368
19. Wang, Y., Lynch, J.P., Law, K.H.: Wireless Structural Sensors using Reliable Communication Protocols for Data Acquisition and Interrogation. Proc. of 23rd Inter. Modal Anal. Conf. (IMAC XXIII). Orlando, FL (2005)
20. Lynch, J.P., Loh, K.: A Summary Review of Wireless Sensors and Sensor Networks for Structural Health Monitoring. Shock and Vibration Dig.. 38 (2005) 91-128
21. Lynch, J. P., Sundararajan, A., Sohn, H., Park, G., Farrar, C., Law, K.: Embedding Actuation Functionalities in a Wireless Structural Health Monitoring System. Proc. of 1st Inter.Workshop on Adv. Smart Materials and Smart Struct. Technology. Honolulu, HI (2004)
22. Grisso, B. L., Martin, L.A., Inman, D.J.: A Wireless Active Sensing System for Impedance-Based Structural Health Monitoring. Proc. of 23rd Inter. Modal Anal. Conf. (IMAC XXIII). Orlando, FL (2005)
23. Liu, L., Yuan, F.G., Zhang, F.: Development of Wireless Smart Sensor for Structural Health Monitoring. Proc. of SPIE Smart Struct. and Mat.. SPIE Vol. 5765. San Diego, CA (2005) 176-186
24. Casciati, F., Rossi, R.: Fuzzy Chip Controllers and Wireless Links in Smart Structures. Proc. of AMAS/ECCOMAS/STC Workshop on Smart Mat. and Struct. (SMART'03). Warsaw, Poland (2003)
25. Seth, S., Lynch, J. P., Tilbury, D.: Feasibility of Real-Time Distributed Structural Control upon a Wireless Sensor Network. Proc. of 42nd Annual Allerton Conf. on Comm., Control and Computing. Allerton, IL (2004)
26. Wang, Y., Swartz, A., Lynch, J.P., Law, K.H., Lu, K.-C., Loh, C.-H.: Wireless Feedback Structural Control with Embedded Computing, Proc. of SPIE 11th Inter. Symp. on Nondestructive Evaluation for Health Monitoring and Diagnostics. San Diego, CA (2006)
27. Churchill, D.L., Hamel, M.J., Townsend, C.P., Arms, S.W.: Strain Energy Harvesting for Wireless Sensor Networks. Proc. of SPIE 10th Annual Int. Symp. on Smart Struct. and Mat. SPIE Vol. 5055, ed. by Varadan, V.K. and Kish, L.B. San Diego, CA (2003) 319-327
28. Roundy, S.J.: Energy Scavenging for Wireless Sensor Nodes with a Focus on Vibration to Electricity Conversion. Ph.D. Thesis, Mech. Engng.. Univ. of California, Berkeley (2003)
29. Sodano, H.A., Inman, D.J., Park, G.: A Review of Power Harvesting from Vibration using Piezoelectric Materials. Shock and Vibration Dig.. 36 (2004) 197-205
30. Lei, Y., Kiremidjian, A.S., Nair, K.K., Lynch, J.P., Law, K.H.: Time Synchronization Algorithms for Wireless Monitoring System. Proc. of SPIE 10th Annual Int. Symp. on Smart Struct. and Mat.. SPIE Vol. 5057, ed. S.C. Liu. San Diego, CA (2003) 308-317
31. Lynch, J.P., Wang, Y., Law, K.H., Yi, J.H., Lee, C.G., Yun, C.B.: Validation of Large-Scale Wireless Structural Monitoring System on the Geumdang Bridge. Proc. of 9th Int. Conf. on Struct. Safety and Reliability. Rome, Italy (2005)
32. Lu, K.-C., Wang, Y., Lynch, J.P., Loh, C.-H., Chen, Y.-J., Lin, P.-Y., Lee, Z.-K.: Ambient Vibration Study of the Gi-Lu Cable-Stay Bridge: Application of Wireless Sensing Units. Procc of SPIE 13th Annual Symp. on Smart Struct. and Mat.. San Diego, CA (2006)

33. Lynch, J.P., Wang, Y., Lu, K.-C., Hou, T.-C., Loh, C.-H.: Post-seismic Damage Assessment of Steel Structures Instrumented with Self-interrogating Wireless Sensors. Proc. of 8th National Conf. on Earthquake Engng. San Francisco, CA (2006)
34. Sohn, H., Farrar, C.: Damage Diagnosis using Time-series Analysis of Vibrating Signals. J. of Smart Mat. and Struct.. 10 (2001) 446-451
35. Lee, C.-G., Lee, W.-T., Yun, C.-B., Choi, J.-S.: Summary Report - Development of Integrated System for Smart Evaluation of Load Carrying Capacity of Bridges. Korea Highway Corporation, Seoul, South Korea (2004)
36. Lin, P.-Y., Roschke, P.N., Loh, C.-H.: System Identification and Real Application of a Smart Magneto-Rheological Damper. Proc. of 2005 Int. Symp. on Intelligent Control, 13th Mediterranean Conf. on Control and Automation. Limassol, Cyprus (2005)
37. Franklin, G.F., Powell, J.D., Workman, M.: Digital Control of Dynamic Systems. Pearson Education (2003)
38. Domer, B. and Smith, I.F.C.: An Active Structure that Learns. J. Comput. in Civil Engng., 19 (2005) 16-24
39. Fest, E., Shea, K. and Smith, I.F.C.: Active Tensegrity Structure. J. Struct. Engng. 130 (2004) 1454-1465
40. Lynch, J.P., Law, K.H.: Decentralized Control Techniques for Large Scale Civil Structural Systems. Proc. of the 20th Int. Modal Analysis Conference (IMAC XX), Los Angeles, CA (2002)
41. Lynch, J.P., Law, K.H.: Decentralized Energy Market-Based Structural Control. Struct. Engng. and Mech., 17 (2004) 557-572
42. Wang, Y., Swartz, R.A., Lynch, J.P., Law, K.H., Lu, K.-C., and Loh, C.-H.: Decentralized Civil Structural Control using a Real-time Wireless Sensing and Control System. Proc. of the 4th World Conf. on Struct. Control and Monitoring, San Diego, CA (2006)

Author Index

Lecture Notes in Artificial Intelligence (LNAI)

Vol. 4087: F. Schwenker, S. Marinai (Eds.), Artificial Neural Networks in Pattern Recognition. IX, 299 pages. 2006.

Vol. 4068: H. Schärfe, P. Hitzler, P. Øhrstrøm (Eds.), Conceptual Structures: Inspiration and Application. XI, 455 pages. 2006.

Vol. 4065: P. Perner (Ed.), Advances in Data Mining. XI, 592 pages. 2006.

Vol. 4062: G. Wang, J.F. Peters, A. Skowron, Y. Yao (Eds.), Rough Sets and Knowledge Technology. XX, 810 pages. 2006.

Vol. 4049: S. Parsons, N. Maudet, P. Moraitis, I. Rahwan (Eds.), Argumentation in Multi-Agent Systems. XIV, 313 pages. 2006.

Vol. 4048: L. Goble, J.-J.C.. Meyer (Eds.), Deontic Logic and Artificial Normative Systems. X, 273 pages. 2006.

Vol. 4045: D. Barker-Plummer, R. Cox, N. Swoboda (Eds.), Diagrammatic Representation and Inference. XII, 301 pages. 2006.

Vol. 4031: M. Ali, R. Dapoigny (Eds.), Advances in Applied Artificial Intelligence. XXIII, 1353 pages. 2006.

Vol. 4029: L. Rutkowski, R. Tadeusiewicz, L.A. Zadeh, J.M. Zurada (Eds.), Artificial Intelligence and Soft Computing – ICAISC 2006. XXI, 1235 pages. 2006.

Vol. 4027: H.L. Larsen, G. Pasi, D. Ortiz-Arroyo, T. Andreasen, H. Christiansen (Eds.), Flexible Query Answering Systems. XVIII, 714 pages. 2006.

Vol. 4021: E. André, L. Dybkjær, W. Minker, H. Neumann, M. Weber (Eds.), Perception and Interactive Technologies. XI, 217 pages. 2006.

Vol. 4020: A. Bredenfeld, A. Jacoff, I. Noda, Y. Takahashi (Eds.), RoboCup 2005: Robot Soccer World Cup IX. XVII, 727 pages. 2006.

Vol. 4013: L. Lamontagne, M. Marchand (Eds.), Advances in Artificial Intelligence. XIII, 564 pages. 2006.

Vol. 4012: T. Washio, A. Sakurai, K. Nakajima, H. Takeda, S. Tojo, M. Yokoo (Eds.), New Frontiers in Artificial Intelligence. XIII, 484 pages. 2006.

Vol. 4008: J.C. Augusto, C.D. Nugent (Eds.), Designing Smart Homes. XI, 183 pages. 2006.

Vol. 4005: G. Lugosi, H.U. Simon (Eds.), Learning Theory. XI, 656 pages. 2006.

Vol. 3978: B. Hnich, M. Carlsson, F. Fages, F. Rossi (Eds.), Recent Advances in Constraints. VIII, 179 pages. 2006.

Vol. 3963: O. Dikenelli, M.-P. Gleizes, A. Ricci (Eds.), Engineering Societies in the Agents World VI. XII, 303 pages. 2006.

Vol. 3960: R. Vieira, P. Quaresma, M.d.G.V. Nunes, N.J. Mamede, C. Oliveira, M.C. Dias (Eds.), Computational Processing of the Portuguese Language. XII, 274 pages. 2006.

Vol. 3955: G. Antoniou, G. Potamias, C. Spyropoulos, D. Plexousakis (Eds.), Advances in Artificial Intelligence. XVII, 611 pages. 2006.

Vol. 3949: F. A. Savacı (Ed.), Artificial Intelligence and Neural Networks. IX, 227 pages. 2006.

Vol. 3946: T.R. Roth-Berghofer, S. Schulz, D.B. Leake (Eds.), Modeling and Retrieval of Context. XI, 149 pages. 2006.

Vol. 3944: J. Quiñonero-Candela, I. Dagan, B. Magnini, F. d'Alché-Buc (Eds.), Machine Learning Challenges. XIII, 462 pages. 2006.

Vol. 3937: H. La Poutré, N.M. Sadeh, S. Janson (Eds.), Agent-Mediated Electronic Commerce. X, 227 pages. 2006.

Vol. 3930: D.S. Yeung, Z.-Q. Liu, X.-Z. Wang, H. Yan (Eds.), Advances in Machine Learning and Cybernetics. XXI, 1110 pages. 2006.

Vol. 3918: W.K. Ng, M. Kitsuregawa, J. Li, K. Chang (Eds.), Advances in Knowledge Discovery and Data Mining. XXIV, 879 pages. 2006.

Vol. 3913: O. Boissier, J. Padget, V. Dignum, G. Lindemann, E. Matson, S. Ossowski, J.S. Sichman, J. Vázquez-Salceda (Eds.), Coordination, Organizations, Institutions, and Norms in Multi-Agent Systems. XII, 259 pages. 2006.

Vol. 3910: S.A. Brueckner, G.D.M. Serugendo, D. Hales, F. Zambonelli (Eds.), Engineering Self-Organising Systems. XII, 245 pages. 2006.

Vol. 3904: M. Baldoni, U. Endriss, A. Omicini, P. Torroni (Eds.), Declarative Agent Languages and Technologies III. XII, 245 pages. 2006.

Vol. 3900: F. Toni, P. Torroni (Eds.), Computational Logic in Multi-Agent Systems. XVII, 427 pages. 2006.

Vol. 3899: S. Frintrop, VOCUS: A Visual Attention System for Object Detection and Goal-Directed Search. XIV, 216 pages. 2006.

Vol. 3898: K. Tuyls, P.J. 't Hoen, K. Verbeeck, S. Sen (Eds.), Learning and Adaption in Multi-Agent Systems. X, 217 pages. 2006.

Vol. 3891: J.S. Sichman, L. Antunes (Eds.), Multi-Agent-Based Simulation VI. X, 191 pages. 2006.

Vol. 3890: S.G. Thompson, R. Ghanea-Hercock (Eds.), Defence Applications of Multi-Agent Systems. XII, 141 pages. 2006.

Vol. 3885: V. Torra, Y. Narukawa, A. Valls, J. Domingo-Ferrer (Eds.), Modeling Decisions for Artificial Intelligence. XII, 374 pages. 2006.

Vol. 3881: S. Gibet, N. Courty, J.-F. Kamp (Eds.), Gesture in Human-Computer Interaction and Simulation. XIII, 344 pages. 2006.

Vol. 3874: R. Missaoui, J. Schmidt (Eds.), Formal Concept Analysis. X, 309 pages. 2006.

Vol. 3873: L. Maicher, J. Park (Eds.), Charting the Topic Maps Research and Applications Landscape. VIII, 281 pages. 2006.

Vol. 3864: Y. Cai, J. Abascal (Eds.), Ambient Intelligence in Everyday Life. XII, 323 pages. 2006.

Vol. 3863: M. Kohlhase (Ed.), Mathematical Knowledge Management. XI, 405 pages. 2006.

Vol. 3862: R.H. Bordini, M. Dastani, J. Dix, A.E.F. Seghrouchni (Eds.), Programming Multi-Agent Systems. XIV, 267 pages. 2006.